nderstanding rystallisation

ıiversity of York, UK and online

3 – 30 March 2022

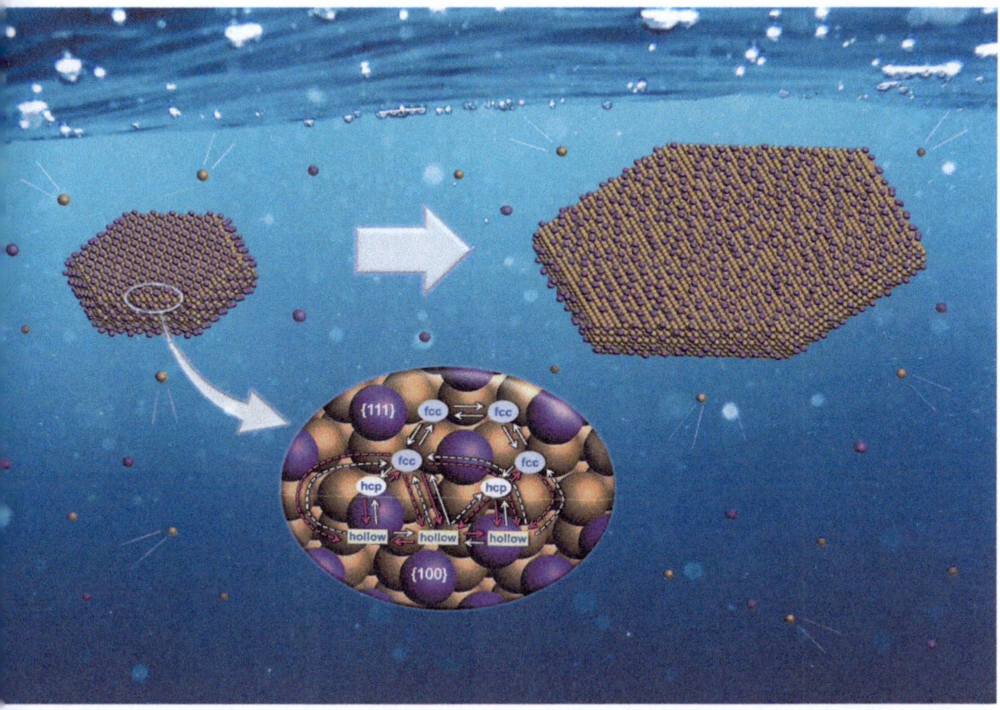

ARADAY DISCUSSIONS

olume 235, 2022

ROYAL SOCIETY
OF **CHEMISTRY**

The **Faraday Division** of the Royal Society of Chemistry, previously the Faraday Society, was founded in 1903 to promote the study of sciences lying between chemistry, physics and biology.

Editorial Staff

Faraday Discussions (Print ISSN 1359-6640, Electronic ISSN 1364-5498) is published 8 times a year by the Royal Society of Chemistry, Thomas Graham House, Science Park, Milton Road, Cambridge, UK CB4 0WF.

Volume 235 ISBN 978-1-83916-426-2

2022 annual subscription price: print+electronic £1133 US $1994; electronic only £1079, US $1899.
Customers in Canada will be subject to a surcharge to cover GST. Customers in the EU subscribing to the electronic version only will be charged VAT.

All orders, with cheques made payable to the Royal Society of Chemistry, should be sent to the Royal Society of Chemistry Order Department, Royal Society of Chemistry, Thomas Graham House, Science Park, Milton Road, Cambridge, CB4 0WF, UK
Tel +44 (0)1223 432398; E-mail **orders@rsc.org**

If you take an institutional subscription to any Royal Society of Chemistry journal you are entitled to free, site-wide web access to that journal. You can arrange access via Internet Protocol (IP) address at **www.rsc.org/ip**

Customers should make payments by cheque in sterling payable on a UK clearing bank or in US dollars payable on a US clearing bank.

Printed in the UK

MIX
Paper from responsible sources
FSC® C013604

ROYAL SOCIETY OF CHEMISTRY

Faraday Discussions

Faraday Discussions are unique internation discussion meetings that focus on rapidly developing areas of chemistry and its interfaces with other scientific disciplines.

Scientific Committee volume 235

Faraday Standing Committee on Conferences

Advisory Board

Information for Authors

nderstanding Crystallisation

'aday Discussions

/w.rsc.org/faraday_d

neral Discussion on Understanding Crystallisation was held in York, UK on the 28th, 29th and 30th arch 2022.

Royal Society of Chemistry is the world's leading chemistry community. Through our high ict journals and publications we connect the world with the chemical sciences and invest the 'ts back into the chemistry community.

NTENTS

1359-6640; ISBN 978-1-83916-426-2

Cover

See J. Kim & K. A. Fichthorn, *Faraday Discuss.*, 2022, **235**, 273–288.

Cu nanoplates grow with the adsorption of iodine atoms on the surface in aqueous solution. Iodine helps Cu atoms preferentially diffuse and deposit to the side facets of the plate so that the plate grows along the width direction.

Image reproduced by permission of Junseok Kim and Kristen Fichthorn from *Faraday Discuss.*, 2022, **235**, 273–288.

'RODUCTORY LECTURE

'ERS AND DISCUSSIONS

EXHIBITORS

Surface Measurement Systems
World Leader in Sorption Science

Technobis group

POSTER SPONSORSHIP

CrystEngComm

CONCLUDING REMARKS

ADDITIONAL INFORMATION

Faraday Discussions

ROYAL SOCIETY OF CHEMISTRY

PAPER

Spiers Memorial Lecture: Assembly-based pathways of crystallization

James J. De Yoreo, *[ab] Elias Nakouzi, [a] Biao Jin, [a] Jaehun Chun [ac] and Christopher J. Mundy [ad]

Received 14th March 2022, Accepted 21st March 2022
DOI: 10.1039/d2fd00061j

Solution crystallization of materials ranging from simple salts to complex supramolecular assemblies has long been viewed through the lens of classical nucleation and growth theories in which monomeric building blocks assemble into ordered structures through inherent thermal fluctuations that overcome a free energy barrier and continue to grow by the addition of such units to atomic steps. However, recent observations have revealed a rich set of hierarchical pathways during both nucleation and growth involving species of a higher order than monomers. While many studies have investigated and deduced the mechanisms underlying hierarchical nucleation pathways, much less research has been directed towards the development of a mechanistic picture of growth by the assembly of more complex units. Here, we review recent investigations into crystal growth by particle attachment, with an emphasis on oriented attachment. We discuss the relationship between interfacial structure, interparticle forces, and attachment dynamics, discuss the consequences of size dependent phase stability, and examine the impact of the ligand-functionalization of primary particles.

1. Introduction

The mechanisms by which crystalline structures form from solutions often diverge from the classical picture during both nucleation and growth. In that picture, nuclei of the final structure form through monomer-by-monomer addition to unstable clusters that overcome a free energy barrier *via* inherent thermal fluctuations,[1–4] then continue to grow by monomer addition, generally at atomic step edges.[5–7] However, research over the past two decades has revealed poorly understood, diverse crystallization pathways involving non-classical, hierarchical processes.[9–14] That is, formation occurs *via* the assembly of "higher-order" species,[9,11] including multi-ion complexes,[17,18] oligomeric units,[19–23] dense liquid

[a]*Physical Sciences Division, Pacific Northwest National Laboratory, Richland, WA, 99352, USA. E-mail: james.deyoreo@pnnl.gov*
[b]*Department of Materials Science and Engineering, University of Washington, Seattle, Washington, 98195, USA*
[c]*Benjamin Levich Institute, CUNY City College of New York, New York, 10031, USA*
[d]*Department of Chemical Engineering, University of Washington, Seattle, Washington, 98195, USA*

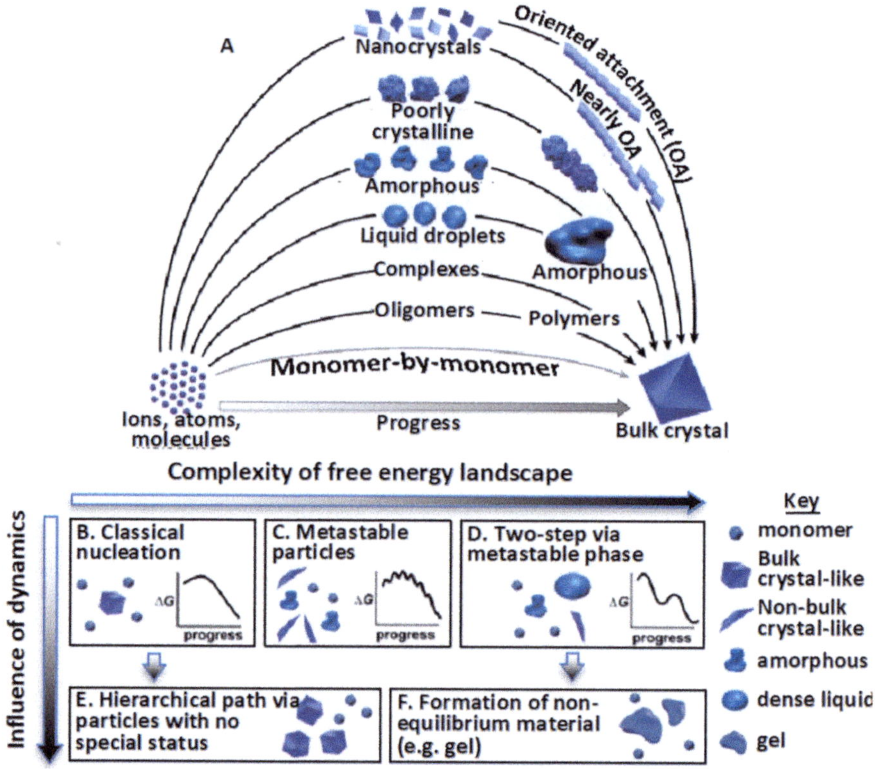

Fig. 1 Stable bulk crystals often form from species of a higher order than monomers.[11] (A) Pathways can be thermodynamic (B–D) or kinetic (E and F); (B) classical monomer-by-monomer addition; (C) metastable microscopic precursors; (D) metastable bulk precursors; (E) cluster/particle assembly; (F) kinetically trapped metastable or unstable products. In the previous project period, we focused primarily on complexes, oligomers, and nanocrystals. In the next, we will address the circled pathways in (A) corresponding to (D), (E), and (F). From ref. 11. Reprinted with permission from AAAS.

phase (DLP) droplets,[24–31] or amorphous[17,32–41] or ordered[42–62] particles (Fig. 1A). Reflecting the distinction between the addition of monomeric units and these higher order species, these hierarchical styles of nucleation and growth have been termed "crystallization by particle attachment" (CPA). The impact of CPA on the material's structure and properties is extensive because such pathways often create intermediate states which are not otherwise accessible and final crystals that exhibit complex, hierarchical morphologies that lead to novel properties.[36,63–65]

Deviations from the classical crystallization pathway (Fig. 1B) can be expected for two reasons.[11] First, complexities in the free energy landscape create metastable structures offering lower-barrier pathways to the final state (Fig. 1C and D). These can either be bulk phases on a phase diagram that are more stable than the solution but less stable than the final solid (Fig. 1C), or configurations that represent local minima that are less stable than the solution itself (Fig. 1D)—and thus can only exist transiently and microscopically—but that provide an environment in which the stable configuration appears more readily than in the bulk.

The anatase and brookite phases of TiO_2 are examples of the first type; they form more rapidly than the stable rutile phase because the lower surface energies of the *metastable* phases render the penalty associated with creating the new solid–solution interface of the nuclei smaller than that for rutile.[70] Examples of the second case include proteins that form *unstable* protein-rich liquid droplets within which the first crystals nucleate because of the higher concentration.[27–29]

The second reason deviations arise stems from dynamical effects of two kinds (Fig. 1E and F):[11] (1) high driving forces generate many pre- and post-critical particles that interact and assemble to form highly branched structures or coarsen to compact single crystals (Fig. 1E), and (2) low atomic mobility leaves the system stuck with particles of a metastable bulk phase, particles of an unstable (but locally metastable) microscopic phase, or in a globally unstable state (Fig. 1F), such as a gel. Though this conceptual framework of CPA is clear, the challenge we face in designing or interpreting crystallization processes is that predicting which pathway a system follows and why is generally beyond current understanding.

As Fig. 1A indicates, CPA encompasses hierarchical processes that occur during both the nucleation and growth phase. That is, the assembly of higher order species can occur prior to the establishment of a stable nucleus when the free energy is still increasing with cluster size, or it can occur in the post-nucleation stage, where it essentially constitutes colloidal aggregation. Numerous investigations have addressed the mechanisms through which hierarchical nucleation pathways arise. Here, we focus on the post-nucleation stage, with an emphasis on growth by oriented attachment (OA), a process first reported by Penn and Banfield in 1998 (ref. 53 and 71) in which nanocrystals aggregate into larger single crystals by attaching on identical facets with either crystallographic coalignment or, alternatively, between the two initial—or "primary"—particles.

OA leads to remarkable morphological outcomes, including the formation of tetrapods, chains and sheets, highly branched nanowires, and self-similar 3D mesocrystals. One study showed that OA coupled with strain relaxation is even responsible for the growth of commonly observed five-fold twin structures of noble metal nanoparticles.[72] OA and, more generally, the particle-mediated growth of single crystals, has now been widely observed in semiconductors,[73–76] metals,[58,77] silicates,[37,78] oxides,[9,43,44,50,52–54,59,62,79,80] fluorides,[81] carbonates,[50] organic compounds,[82] peptides,[83] and proteins.[84]

Many styles of OA have been observed since the first report, extending the concept beyond attachment with crystallographic coalignment to include (1) attachment with some degree of misalignment, followed by relaxation through atomistic processes into a coaligned state,[85] (2) attachment of metastable particles of a nanoscopic phase onto crystals of the bulk phase, followed by transformation to the bulk phase with the creation of a coherent interface,[43] and (3) a special class of #2 in which the nanoscopic phase is amorphous.[50] In all cases, the progression of OA reflects an interplay of the solution structure, forces, and motion (Fig. 2), in which crystal surfaces impose structure on the near-surface solution, leading to a set of interparticle forces, including hydration barriers, and van der Waals (vdW) and electrostatic forces. These interactions and forces are further modified by surface-bound ligands, which are a common feature of nanoparticle systems introduced during colloidal synthesis to stabilize the particles at the nanoscopic size. Moreover, these ligand functionalized interfaces create chemical gradients

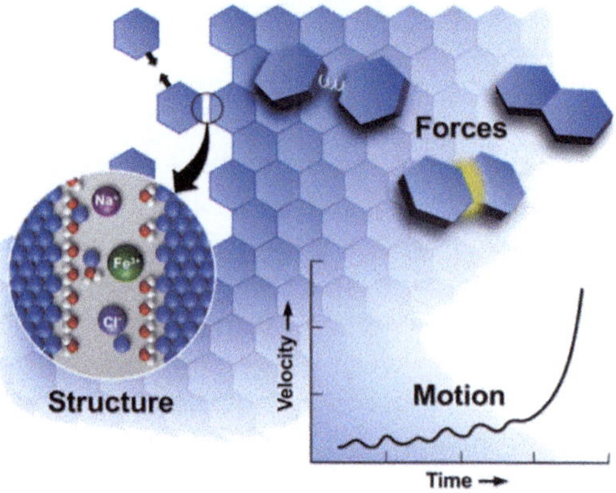

Fig. 2 OA is a mesoscale response to forces defined by the molecular details of the interfacial region.[11] These details create forces that drive particle motion. As particles move, the structure and forces evolve, transitioning from a regime of long-range to short-range interactions, leading to alignment and attachment. From ref. 11. Reprinted with permission from AAAS.

that can impact the formation of new primary particles and their subsequent attachment. In what follows, we discuss the relationship between interfacial structure, interparticle forces and particle dynamics, discuss the consequences of size dependent phase stability, and examine the impact of the ligand-functionalization of primary particles.

2. Structure, forces, and motion

While the development of a quantitative framework for predicting particle interactions and assembly has its roots in colloidal physics and continuum-based Derjaguin–Landau–Verwey–Overbeek (DLVO) theory, nanocrystals present additional complexities: the face-specificity of dielectric properties, the presence of structured nanoscale interfaces, and the consequent solvent responses that should be of a comparable length scale. In essence, crystal structure is imprinted upon solvent structure, creating forces between particles whose consequent motion alters the intervening solvent structure, changes the forces, further drives motion, and eventually leads to attachment. Thus, the energetics are distinct from those of conventional colloidal phenomena and lead to the non-trivial coupling of scales, giving rise to hydrodynamic phenomena, such as strong coupling between translational and angular velocities. As such, theoretical progress at the nanoscale requires treatment of hydrodynamics consistent with the molecular level length scale of solvent-response and the concomitant fluctuations.

2.1 Interfacial structure

As two approaching nanocrystals reach a separation of a few nanometers, the molecular details of the crystal–solution interface become increasingly important. Indeed, ions and water molecules at the interface behave differently from those in the bulk solution due to interactions with surface chemical groups, patterning against the crystal lattice, and—crucially in the context of CPA—confinement

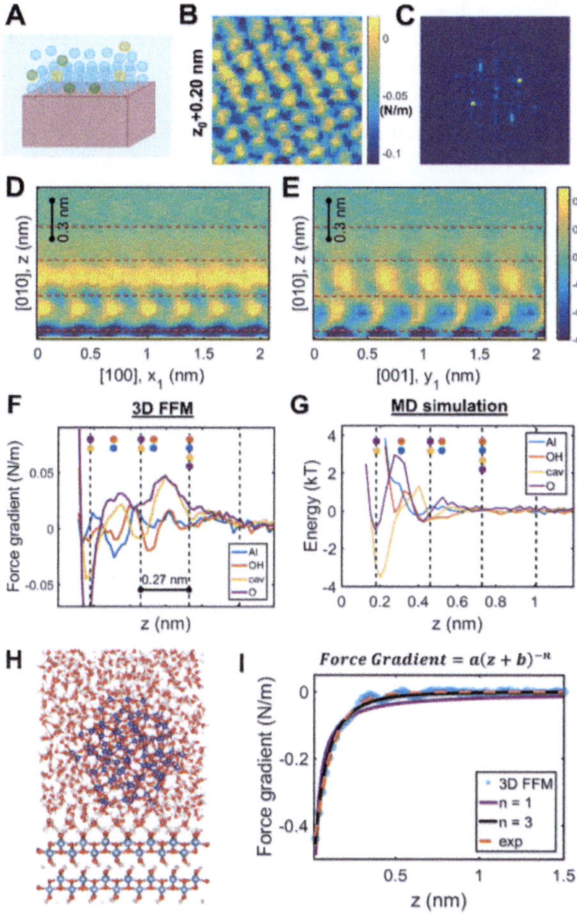

Fig. 3 Solution structure at crystal–solution interfaces. (A) Scheme representing ion and water distribution; (B) lateral slice from the boehmite (010)–water 3D FFM measurement[16] and (C) corresponding Fourier transform showing that the solution structure is templated by the underlying crystal lattice; (D and E) vertical slices of 3D FFM data along different crystallographic directions; (F) site-specific 3D FFM force gradient and (G) MD water potential of mean force profiles; (H) snapshot of a full-scale simulation of the boehmite–water–silica system; (I) fitting long-range interaction that is ascribed to hydrodynamic force due to confinement. Reprinted with permission from ref. 16. Copyright 2021 American Chemical Society.

between the two approaching surfaces (Fig. 3A). Accordingly, resolving the interfacial structure and dynamics at the molecular scale will be necessary towards building a predictive understanding of CPA. To date, such a rigorous theoretical framework does not exist. While electrical double layer theories predict ion concentrations as a function of distance from a charged surface, these formulations do not resolve the spatial distribution of solvent molecules[86] (despite studies involving a modified Poisson–Boltzmann formulation – see Section 2.2 for more details) beyond continuum approximations. Similarly, empirical models of ion adsorption and speciation remain largely informed at the bulk scale by experimental data and chemical intuition rather than robust local theoretical predictions.[87] These models thus fail to capture an important and ubiquitous feature of nanoparticles in solution, namely a molecular based description of the double layers in the vicinity of the solid–solution interface.

The "molecular layering of water at surfaces" was first reported by Israelachvili and Pashley, who used a surface force apparatus to measure short-range oscillatory forces between two mica surfaces immersed in aqueous solution.[88] Since then, a range of experimental and simulation techniques have been employed to investigate the interfacial solution structure. Most notably, studies using X-ray reflectivity and diffraction have resolved a multitude of mineral–aqueous interfaces by measuring electron density profiles as a function of distance from the surface and fitting the data to candidate solution structure profiles.[89–93] Spectroscopic methods, such as sum frequency generation spectroscopy, provide information on the water orientation at the interface, demonstrating the more limited degrees of freedom of surface-bound water molecules compared to those in the bulk.[94,95] More recently, an emerging technique known as 3D fast force mapping (3D FFM) has enabled the visualization of solution structure by measuring the forces experienced by a nanosized probe as it navigates the interfacial region.[96–103] These experimental methods are typically coupled to molecular dynamics (MD) simulations, which have evolved significantly in recent years to incorporate advanced force fields, reactivity parameters, and more accurate descriptions of surface chemistry.[104–109] While each of these techniques is beset by technical limitations and interpretation challenges, they collectively provide complementary insights that have vastly improved our understanding of crystal–solution interfaces.

One recent case study investigated the boehmite (010)–water interface, which is the preferred crystallographic plane for boehmite oriented attachment in alkaline solution conditions.[16] Using 3D FFM, the authors determined that the solution phase close to the surface showed the orthorhombic symmetry that is characteristic of boehmite, hence confirming that the underlying crystal lattice templates the distribution of interfacial water molecules (Fig. 3B and C). These patterns dissipate beyond one nanometer from the surface (Fig. 3D and E). Moreover, four laterally structured hydration layers were resolved with oscillatory features in the order of 0.27 nm, consistent with the size of a water molecule. The data showed distinct responses in the measured force gradients that were ascribed to water density distributions above each crystallographic lattice site (Fig. 3F). Comparison to MD simulations confirmed that the highest water densities occur at the sites adjacent to the boehmite hydroxyl groups, thus facilitating hydrogen bonding (Fig. 3G).

Beyond delineating the solution structure, these results provide important insights that are directly relevant to CPA.[16] For example, the 3D FFM data showed additional hydration layers compared to the MD simulation of the free boehmite–water interface (Fig. 3F and G). After creating a suite of simulations with increasing degrees of complexity, the authors were able to reproduce the experimental results by applying a full-scale simulation of the boehmite–water–silica system (Fig. 3H). The conclusion was that confinement between the two surfaces, namely the silica probe and the boehmite substrate, causes an enhanced structuring of the intervening solution, which cannot be probed by "non-intrusive" tools such as X-ray reflectivity and sum frequency generation spectroscopy. Indeed, earlier studies have demonstrated extensive "water packing" in simple

Monte Carlo simulations of rigid spheres confined between two infinite planes at close separations, even without incorporating the rigorous molecular and crystallographic details. Accordingly, the interface between two nanosized particles undergoing CPA is expected to follow a non-trivial behavior limited by two extreme cases: the intrinsic ordering of water molecules at a free-standing crystal surface and the extensive ordering due to close proximity between two finite-sized planes.

Another key observation pertains to a qualitative connection between the interfacial structure and other emergent interfacial properties.[16] Specifically, the study hypothesized that the interfacial solution structure can cause unusual transport properties due to confinement, leading to "conservative" hydrodynamic forces within $\sim$1 nm separation between the probe and the surface (Fig. 3I). Interestingly, this result is, in fact, consistent with previous experimental studies that reported chemistry-dependent transport properties at confined liquids, such as the viscoelasticity of water[110–113] as well as a reported unusual dielectric property of confined water ($\sim$2 instead of the well-known bulk value, $\sim$80).[114] Considering that these fundamental material/transport properties directly influence particle forces and dynamics, one can convincingly argue that the interfacial structure plays a determining role in the pathways and outcomes of CPA.

Perhaps the most elusive mystery is related to the dynamic nature of the crystal–solution interface, beyond simply an equilibrium distribution of water and ions. During a CPA event, the final "jump-to-contact" between the two nanoparticles predicates that the surface-bound water, ions, and ligands be released from the surface, or otherwise become included in the permanent crystal structure. The corresponding timescales and kinetic barriers for these processes remain difficult to resolve, and are perhaps best interrogated by molecular simulations that can provide valuable information on the desorption/exchange rates for the various interfacial species, as well as changes in the surface chemistry. On this note, recent simulations on anatase in vacuum and humid environments showed that the presence of adsorbed water in fact facilitates oriented attachment by delaying the particle collision and, thus, preventing random aggregation.[115]

2.2 Forces between crystalline particles

The propensity of nanoparticles to undergo CPA is dictated by a complex interplay between various short- and long-range forces, including electrostatics, dispersion, and hydration forces. These forces depend on fundamental material properties, such as the dielectric permittivity and dipole moments, as well as the physicochemical properties of the crystal surfaces (*e.g.*, the presence of foreign molecules, such as ligands, and surface roughness). Competing with dynamic contributions, such as hydrodynamic forces (described in Section 2.3), the interplay is pivotal to the local/transient responses of particles and the resultant assemblies/microstructures.

Long-range forces are mainly described by continuum models, such as the DLVO theory, which typically involve two components: the electrostatic interaction force and the dispersive van der Waals force (Fig. 4A). According to the DLVO theory, the electrostatic interaction force is described by the Poisson–Boltzmann (PB) model, which considers the electrostatic potential of the particle surface and

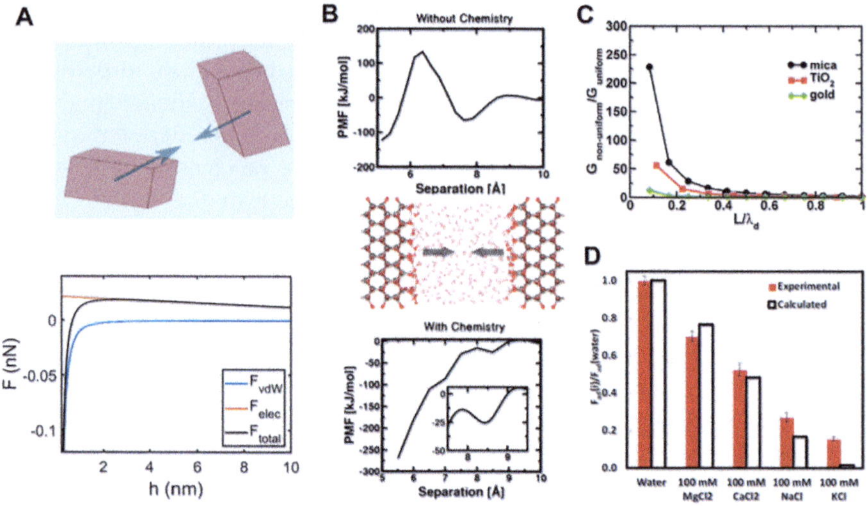

Fig. 4 Long- and short-range forces between crystalline particles. (A) Typical force–separation curves from DLVO theory along with a schematic of crystalline particles; (B) surface chemistry-dependent hydration forces between ZnO(10$\bar{1}$0) surfaces from repulsive (without hydroxylation) to attractive nature (with hydroxylation);[15]. (C) An increase in van der Waals interactions due to the interfacial structure as a function of the particle separation normalized with respect to the thickness of the interfacial layer (*i.e.*, L/λ_d) over different mineral surfaces;[68]. (D) Normalized adhesion forces (with respect to an adhesion force for pure water) over different ions as to ion specificity and valence.[69] (A and B) Reprinted with permission from ref. 15. Copyright 2018 American Chemical Society. (C) Reprinted with permission from ref. 68. Copyright 2015 American Chemical Society. (D) Reprinted with permission from ref. 69, copyright (2017) National Academy of Sciences.

the equilibrium distribution of mobile ions around the particle and relates the local average charge densities to the corresponding electrical fields and potentials. Key approximations implemented in the continuum PB model[116] are: (i) describing the electrolyte solution as a continuous dielectric medium characterized by an average density of electrolytes, effectively averaging out the degrees of freedom of solvent molecules and mobile ions, (ii) considering an equilibrium distribution for mobile ions, implying that the gradient of the electrochemical potential needs to be vanished, and (iii) neglecting considerations of ion size and interactions. As such, the electrostatic interaction force between two surfaces is described as an osmotic pressure difference associated with the inhomogeneous distributions of ions.

The second component of DLVO, namely the van der Waals force, is described by Lifshitz theory, which assumes that each particle (macroscopic entity) can be fully represented by its corresponding dielectric spectra (*i.e.*, as a function of frequency up to X-ray), inherently including many-body characteristics.[117,118] Here, the interaction between fluctuating electromagnetic fields of the interacting particles in the presence of "screening" due to an intervening medium results in the van der Waals force. It is important to note that the van der Waals force is dominated by dielectric spectra differences between interacting bodies in the UV range, whereas frequency differences between bodies in the IR range insignificantly contribute to the overall force. This continuum-based approach for long-range forces has been successfully implemented for characterizing the stability and aggregation of colloids in various industrial and geological settings.

Nevertheless, by ignoring molecular details, the predictive capability of DLVO theory is significantly limited. As discussed in the previous section, "molecular granularity" becomes critical at smaller particle separations where the intervening solution cannot be treated as bulk. For example, one manifestation of the interfacial structure is the emergence of a short-range hydration force at about $O(1)$ nm separation due to a competition between surface hydration and ion hydration.[119–121] Note that ions differ appreciably in their relative preference for water–ion interactions over water–surface interactions (kosmotropic or structure-making *vs.* chaotropic or structure-breaking ions[122]). Therefore, the hydration force is intrinsically expected to show ion specificity (*e.g.*, Hoffmeister series), in addition to its dependence on the ion concentration and valence; this nature has been confirmed using surface force apparatus (SFA) and atomic force microscopy (AFM) measurements.[103,120,121,123,124] Interestingly, a recent study based on quantum density functional theory (qDFT)-based potential of mean forces calculations by Shen *et al.*[15] clearly demonstrated that a hydration force becomes qualitatively different, depending on details of the surface chemistry of the ZnO $(10\bar{1}0)$ surface (*i.e.*, hydroxylation) – see Fig. 4B.

At sufficiently high ion concentrations, the competition between ion-surface and ion–ion interactions becomes more relevant. The average separation between ions scales as ${n_b}^{-1/3}$ (n_b is the number density of the ion at bulk), whereas the length scale for the electrostatic interaction between ions against thermal energy is the Bjerrum length, λ_B ($=q^2/\varepsilon kT$), where q is the ionic charge, ε is the dielectric permittivity, k is the Boltzmann constant, and T is the temperature. Therefore, when $\lambda_B \gg {n_b}^{-1/3}$, a strong ion correlation is expected, suggesting that ion–ion interactions must be considered. To this end, the incorporation of the molecular level detail of specific ion effects and loss of electrostatic screening due to the presence of an interface have been employed into modified PB (mPB) theory. This modification affords a more predictive tool for specific ion-surface segregation that gives rise to novel electrostatic phenomena.[125,126] These theories incorporate ion concentration, size, and valence consistent with the aforementioned simple scaling argument.

While these descriptions for both long-range and short-range forces are generally reasonable, some unique features are specifically associated with CPA. First, the distinction between short-range and long-range forces becomes blurred for particles whose size is comparable to the length scale where molecular granularity becomes important. This implies that the molecular level description needs to be implemented into the continuum-based approach, leading to coupling between the forces coming from different scales. One example is a modified Lifshitz theory by Chun *et al.*,[68] which implemented the inhomogeneity of dielectric properties near mineral–fluid interfaces into the conventional Lifshitz theory. The main conclusion was that the coupling of the solution response at different scales results in a much stronger van der Waals force, as demonstrated in Fig. 4C. Similarly, in the study of Li *et al.*,[69] it was revealed that the adhesion forces between two mica surfaces is strongly dependent on ion concentration and specificity, which is attributed to molecular scale phenomena such as ion hydration (see Fig. 4D).

Perhaps one of the key results in the context of CPA is the observation of orientation-dependent adhesion forces between crystalline surfaces, ascribed to the anisotropy in dielectric permittivity[127] and surface chemistry, giving rise to torque. Considering the orientation-dependent interaction potential (W), a torque balance between the interaction potential ($\partial W/\partial \boldsymbol{p}$) and Brownian ($kT$) contributions can be described by an orientational probability distribution function (ψ), combined with:

$$\frac{\partial \ln \psi}{\partial \boldsymbol{p}} = -\frac{1}{kT}\frac{\partial W}{\partial \boldsymbol{p}}$$

where ψ and W should depend on $\boldsymbol{p}$ (a director vector representing a crystallographic axis of interest) that is a key descriptor for the alignment. This formulation had been introduced to understand the 3-fold symmetric orientational adsorption of macromolecules on graphite surfaces,[128,129] and was then utilized to interpret the orientational dependence of adhesion forces between rutile TiO_2 crystals.[130]

We also note that unique material properties, such as a strong dipole moment or the existence of ligands on the nanocrystal surfaces, contribute additional forces. The former exhibited an attractive dipole–dipole force/torque at large separation that is responsible for ZnO oriented attachment (see Section 2.3),[8] whereas the latter triggered a repulsive "steric" hindrance force, analogous to that due to adsorbed polymers, preventing particle attachment[131,132] (see Section 4 for details).

2.3 Particle dynamics

Beyond simply characterizing the short-range and long-range forces, the coupling of these forces to particle dynamics is pivotal for obtaining a mechanistic understanding of CPA, a fact that has been largely underestimated or ignored (Fig. 5A). Such coupling can be represented *via* hydrodynamic mobility, which originates from a linear relationship between a translational/rotational velocity and force/torque.[133,134] The hydrodynamic mobility reflects the effects of the particle shape and dimensions, as well as the solution viscosity. This component also includes a separation dependence where it would be applied for "relative" velocities and "relative" force/torque between a pair of particles.

Assuming that particle inertia is negligible due to the size of nanocrystals, the linear relationship between a translational velocity ($\boldsymbol{U}$) and force ($\boldsymbol{F}$) for the translational motion can be described as $\boldsymbol{U} = \mathbf{M}^{\mathrm{t}}\boldsymbol{F}$, where $\mathbf{M}^{\mathrm{t}}$ is a hydrodynamic translational mobility tensor. An analogous relation exists for a rotational velocity ($\boldsymbol{\Omega}$) and torque ($\boldsymbol{T}$), $\boldsymbol{\Omega} = \mathbf{M}^{\mathrm{r}}\boldsymbol{T}$, where $\mathbf{M}^{\mathrm{r}}$ is a hydrodynamic rotational mobility tensor.[135] The hydrodynamic mobility becomes critical when two particles are close each other, within a lubrication regime (typically about one tenth of the particle size), where a large hydrodynamic pressure builds up at an intervening space between particles due to their relative motion. Typically, this nature exerts an additional "repulsive" force of a purely dynamic origin. Note that such a hydrodynamic "lubrication" force is expected to be largest for a relative approaching motion along a line between centers of particles, which incidentally is the same direction as the forces described in Section 2.2 (radially symmetric for electrostatic, dispersion, and hydration for spherical particles). Consequently, the resultant translational motions by such forces can be attenuated in the presence

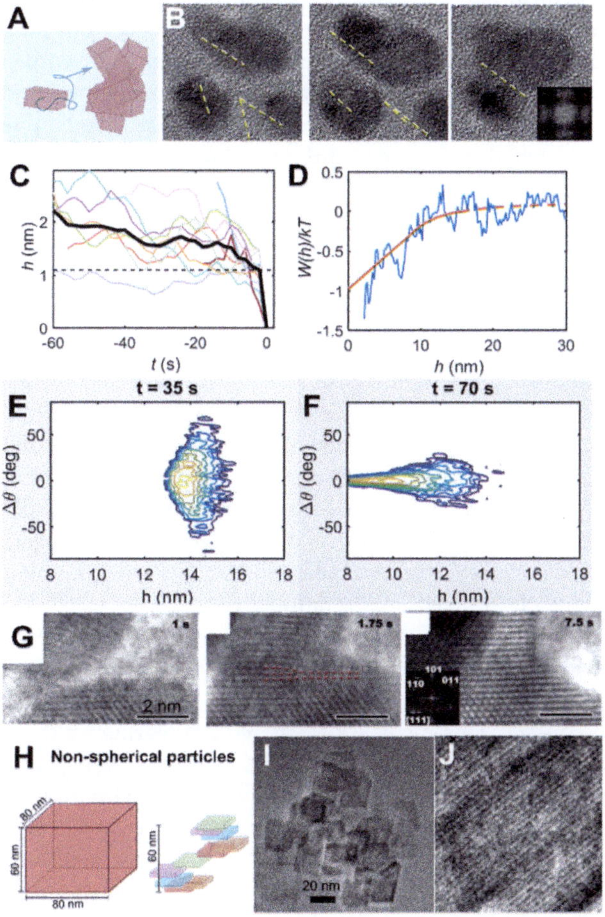

Fig. 5 Coupling particle energetics to the dynamics of CPA. (A) Scheme representing particle diffusion and collisions; (B) oriented attachment of ZnO nanocrystals;[8] (C) Evolution of particle separation during multiple attachment events; (D) interaction potential of ZnO attachment extracted from the ensemble assembly dynamics; (E and F) Langevin dynamics simulation starting from randomly aligned, far particles show convergence towards perfect alignment at long-range; (G) oriented attachment of iron oxyhydroxide;[66] (H) Scheme of non-spherical particle shapes with different diffusion properties and (I and J) examples of oriented attachment outcomes.[67] (A–E) Reproduced from ref. 8 with permission from Springer Nature. Copyright 2020. (G) From ref. 66. Reprinted with permission from AAAS. (H) Reprinted with permission from ref. 67. Copyright 2013 American Chemical Society.

of hydrodynamic mobility, leading to a decisive factor for the collision of two approaching particles. A similar nature is expected for the resultant rotational motion by applied torques.

One interesting point is that the hydrodynamic mobility for rotational motion is typically much higher than that for translational motion, especially at close separations. In the case of two interacting spheres, the hydrodynamic mobility remains invariant with respect to separation for rotational motion, in comparison to a decreasing linear dependence on separation for translational motion.[135] The observation by Liu *et al.*[8] illustrated that this rotational motion can be conducive for oriented attachment, forcing the nanocrystals to co-align, while a significant hydrodynamic attenuation for the approaching translational motion of the nanocrystals prevents the collision. For spherical nanocrystals, one can understand such coupling between forces/torques and translational/rotational motions based on Jeffrey and Onishi's[135] full mathematical descriptions and corresponding formulations for $\mathbf{M}^t$ and $\mathbf{M}^r$. Recent *in situ* LP-TEM studies for Ag and Au nanoparticles by Lee *et al.*[131,132] implemented hydrodynamic resistivity to rigorously understand the kinetics of the growth of particle clusters and the transient behavior of particles/clusters,[132] leading to the superlattice formation based on the coupling between energetics and dynamics.

Recently, Liu *et al.* presented a case study that connected the energetics and dynamics of ZnO oriented attachment (Fig. 5B and C).[8] Specifically, *in situ* TEM observations documented multiple attachment events and were used to calculate the interaction potential between ZnO crystals (Fig. 5D). The authors determined that dipole–dipole interactions, active at particle separations of >5 nm, are responsible for long-range alignment and subsequent aggregation with no appreciable barrier. This result was validated by Langevin dynamics simulations that coupled particle translational/rotational diffusivities to the experimentally measured potential (Fig. 5E and F). The main conclusion was that strongly polar nanocrystals are conducive for OA due to dipole–dipole forces/torques inducing their coalignment before the point where strong attractive potentials drive the final jump to contact. This result contrasts with an earlier study on dipole-free iron oxyhydroxide nanocrystals, wherein alignment was attributed to short-range interactions. In the latter case, the two approaching crystals linger at a separation of a couple nanometers, sample multiple relative orientations, and jump to contact once a crystallographic alignment has been achieved (Fig. 5G).[66]

An additional yet important point for particle dynamics comes from the shape of nanocrystals (Fig. 5H–J). Even without considering the crystallography-dependent forces described in Section 2.2, anisotropic particles exhibit cross-coupling between force/torque and translation/rotational velocities that is typically negligible for spherical particles. That is, translational motions can be induced by torques and *vice versa*. Furthermore, for nanocrystals, the characteristic time scales for translational and rotational motions, obtained from the respective diffusivities, are likely to be comparable,[67] indicating appreciable coupling between rotational and translational diffusive motions. Considering the tensorial nature of the hydrodynamic mobilities (*i.e.*, directionality), this suggests that the assembly kinetics of nanocrystals are heavily influenced by the orientation-dependent forces and hydrodynamic mobilities. A recent study[67] convincingly demonstrated that such orientational dependence in both particle forces and dynamics is responsible for the qualitative nature of the assembly from a diffusion-limited to a reaction-limited case by a combination of dynamic light scattering and scaling analysis based on irregular-shaped boehmite nanocrystals.

Despite the importance of particle dynamics, hydrodynamic mobilities for non-spherical particles are relatively unknown, while those of certain representative cases, such as cylinders and ellipsoids[133] and limited cases for sharp-edged particles (*e.g.*, the face–face configuration between two cubic particles[136]) are known. The effect of particle shape on hydrodynamic mobility can be significant in magnitude, in comparison to a spherical case; a recent study based on the boundary element method, along with an integral representation of the Stokes flow *via* a hydrodynamic Green's function (called the Oseen tensor), showed that the translational hydrodynamic mobility between two cubic particles in parallel configuration is much smaller than that between two spherical particles at the lubrication regime (*i.e.*, $O(\varepsilon^3)$ *vs.* $O(\varepsilon)$ scale, where ε is a normalized separation between two particles). It is worth noting that all descriptions above are based on continuum hydrodynamics where the infinitesimal volume of fluid is much larger than a molecular scale. Considering the unusual transport properties at confinement associated with interfacial structures from molecular nature, mentioned in Section 2.1, the fundamental concept for hydrodynamic mobility would need to be revisited and possibly reformulated, requiring further theoretical studies. This would be especially important because the last stage of particle assembly, the physical collision of nanocrystals, occurs at the lubrication regime due to the size of nanocrystals.

3. Nanoscopic phase primary particles

As noted in the Introduction, an important complexity of OA is that many systems start from primary particles of a phase that is stable only at the nanoscale,[70,137,138] but produce final crystals of the stable bulk phase. Some of the systems in which this pathway has been studied in detail include iron oxides, for which ferrihydrite (FeOOH; Fh) precedes goethite,[139] hematite (Fe_2O_3; Hm)[80] or magnetite[43] and akageneite precedes Hm,[140] titanium oxide, for which anatase precedes rutile,[48] and calcium sulfate, for which bassanite precedes gypsum.[57] Here again, the style of OA varies and includes the oriented assembly of bassanite ($CaSO_4 \cdot 0.5H_2O$) rods that then convert to gypsum ($CaSO_4 \cdot 2H_2O$), the misoriented assembly of akageneite (β-FeOOH) rods followed by coarsening to Hm, the growth of branched TiO_2 *via* the lattice-matched attachment of anatase onto rutile, followed by transformation (Fig. 6), and a coupled dissolution–reprecipitation reaction in

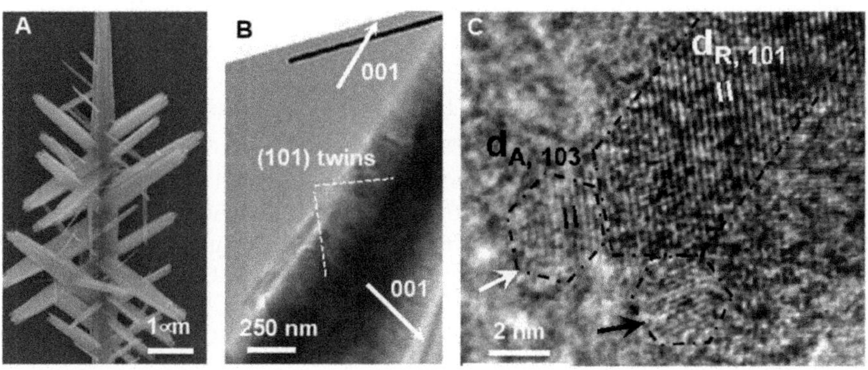

Fig. 6 *Ex situ* TEM data showing (A) a branched rutile nanowire, (B) the crystallographic relationship between the branches and trunk, and (C) attachment of the (103) faces of anatase nanoparticles onto the (101) faces of rutile nanowires, followed by transformation to rutile.[48] Reprinted with permission from ref. 48. Copyright 2013 American Chemical Society.

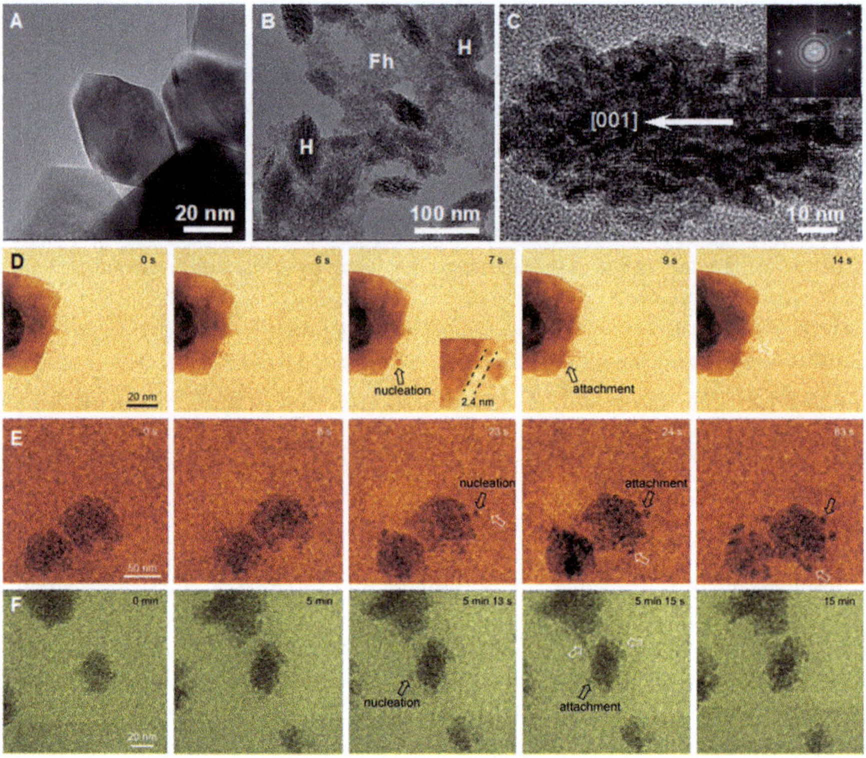

Fig. 7 (A–C) *Ex situ* TEM data showing (A) rhombohedral Hm formed during dissolution–reprecipitation from Fh in oxalate free solution; (B) Hm (H) spindles forming in a Fh (F) suspension containing sodium oxalate; (D–F) *in situ* TEM data showing the near-surface nucleation and attachment of Hm nanoparticles (D and E) during the initial period of Hm particle formation and (F) over an extended time period to show the development of Hm spindles.[80] Reproduced from ref. 80 with permission from Springer Nature. Copyright 2021.

which Fh is replaced by nucleating Hm particles, which undergo OA as the new particles form (Fig. 7).

While the appearance of precursor phases can occur for both kinetic and thermodynamic reasons, the latter is expected to be pervasive throughout systems with multiple polymorphs or closely related phases on the phase diagram. The underlying thermodynamic driver lies in the impact of surface energy α on the free energy of formation $\Delta G(R)$ of a particle with radius R. Sohnel *et al.*[141] showed that there is a rough inverse scaling between the equilibrium solubility C_e of a phase and α. Thus, less soluble phases are more stable, but have higher surface free energy, implying that the crystal is more poorly matched energetically to the surrounding solution. As the particle radius R is reduced, the proportion of growth units that lie within the bulk of the particle relative to the surface decreases, with the ratio $\propto R$. Consequently, as shown for a wide range of materials, at a sufficiently small particle size, the free energy per growth unit, which is the sum of the surface and bulk contributions, becomes smaller for metastable phase than for the bulk phase.[137,138] In other words, *the relative* *stabilities are reversed*. Compounding this phenomenon is the impact of the inverse relationship between C_e and α on the relative probabilities of nucleating particles of the two phases. This effect is most easily seen by considering nucleation within the framework of classical nucleation theory (CNT). Within CNT, the rate of nucleation J_n scales with the exponential of the free energy barrier ΔG^*, which in turn scales with the cube of the interfacial free energy α. That is, $J_n \propto e^{-\Delta G^*/kT}$ and $\Delta G^* \propto \alpha^3$, where k is Boltzmann's constant and T is the temperature.[3,4] Thus, as a particle first begins to nucleate and R is below the value at which the free energies of formation for the two phases cross over, the metastable bulk phase is both the most stable phase and faces a lower barrier to nucleation. Consequently, the first particles to appear are of the bulk metastable phase rather than the stable phase.

3.1 Nanoscopic phase as a particle source

The growth of branched rutile nanowires from anatase primary particles (Fig. 6A) presents a particularly interesting example of OA *via* a nanoscopic phase.[48] After the anatase particles form, the solution remains supersaturated with respect to rutile, because it is the more stable bulk phase with the lower solubility. Thus, rutile wires, once they begin to form, can grow by classical ion-by-ion processes. However, the anatase (103) face is a near-perfect match for the rutile (101) face. Consequently, concomitant with the ion-by-ion growth of the wires, anatase particles attach to the rutile nanowires with perfect lattice matching and, after doing so, undergo a pseudomorphic transformation to rutile. Depending on which (103) face attaches to the rutile trunk (Fig. 6B), wires either extend in length or form branches meeting the trunk on (101) twin planes (Fig. 6C). This process occurs repeatedly to produce multiple generations of branches, all under crystallographic control.

While the direct involvement of nanoscopic phase primary particles would seem to be a requirement for such phases to influence CPA, the growth of Hm mesocrystals starting from a Fh precursor illustrates a strikingly different process. When ferric nitrate and sodium hydroxide are mixed, a suspension of Fh rapidly forms. Eventually, this precursor phase dissolves as Hm nucleates and grows into well faceted rhombohedral Hm through classical growth processes (Fig. 7A). However, when sodium oxalate (Ox) is added to the initial Fh suspension, Hm forms a nanoporous spindle-shaped single crystal elongated along [001] and having a nearly constant aspect ratio, regardless of the absolute size or time after formation begins (Fig. 7B).[80] The spindles are hierarchically organized across two length scales. At the shortest length scale, they consist of ~5 nm crystallographically coaligned particles (Fig. 7C). These then form a second order structure consisting of chains of the Hm particles (Fig. 7C).

The first Hm particles always form either by the direct transformation of Fh to Hm or by the heterogeneous nucleation of Hm on Fh. However, as soon as a Hm particle forms, new particles begin to nucleate within a couple of nm of the first particle (Fig. 7D–F). Immediately after they nucleate, the daughter Hm particles attach to the parent particle next to which nucleation occurred (Fig. 7D–F). Repeated occurrences of these events lead to the formation of the self-similar Hm mesocrystals, which are elongated along the [001] direction because the rate of nucleation near the Hm (001) face of the particles exceeds that along other directions.

In this process of Hm mesocrystal formation, the Fh plays a number of key roles. First, it serves as a source of solute for the growing Hm particles. Second, it provides a buffer that fixes the Fe^{3+} concentration at the solubility limit of Fh. Due to the moderate difference in solubility between the two phases and the high Hm surface energy, the barrier to Hm nucleation is large ($\sim$80 kT) and, thus, nucleation in the bulk solution is rare, consistent with the observation that the nucleation of new Hm particles in the solution far from a Hm particle is never observed in our experiments. Finally, the Fh provides a favourable environment for the sporadic nucleation of Hm, either through direct transformation or heterogeneous nucleation. Why Ox then enables rapid nucleation to occur in the vicinity of these initially formed Hm particles, limits the Hm particle size, and drives the subsequent attachment events to create the mesocrystals is discussed below.

4. Control by ligands and interfaces

4.1 Ligands as a source of bias

CPA is always accompanied by relaxation processes, leading to an evolution in the final particle morphology. For example, rapid particle attachment and slow relaxation processes lead to the formation of dendric structures, while slow particle attachment and rapid relaxation may drive spherical particle formation (Fig. 8A and B). To balance particle attachment and relaxation rates, various ligands can be introduced to tune the ligand–ligand and ligand–particle interaction strengths, as well as the particle–particle interactions.[142–144] Taking Au nanocrystals as an example, the involvement of organic small molecules, such as CTAB, citrate, or ionic liquids, can drive the attachment of Au particles along the [111] direction and drive two-dimensional particle assembly.[145–150] Sequence-defined biomolecules have also been shown to be effective ligands in nanocrystal synthesis and bring the added benefit of tunability, as illustrated by the regulation of Au nanoparticle shapes by peptoids (Fig. 8B–H).[151,152] To develop a controllable and predictable synthesis strategy, an in-depth understanding of ligand control over particle attachment and relaxation is required.

One approach using sequence-defined peptoids is to tune the design by systematically varying the hydrophobicity, number of carboxylate and amino groups, and side-chain positions to control the nanoparticle–ligand interactions. When used for Au nanoparticle synthesis, this approach enabled the controlled and predictable synthesis of spherical to coral-shaped nanoparticles (Fig. 8A and B).[151] In all cases, the process starts with the random attachment of initially formed Au nanoparticles, resulting in the formation of distorted nanorods. These nanorods then randomly attach to one another to form clusters that are intermediates of the coral-shaped particles (Fig. 8C and D).

During the particle attachment process, the strength of the binding free energy of the peptoid onto the Au particle is directly correlated with the degree to which the interfacial free energy is lowered and, therefore, how low the driving force is for relaxation back to spherical particles. Moreover, the strong peptoid–Au binding affinity reduces the mobility of surface atoms, thus further slowing the kinetics of relaxation. Depending on how strong these effects are, the outcome can be coral-shaped particle formation or relaxation towards spherical particles. These results also highlight the important role of the peptoids in stabilizing small

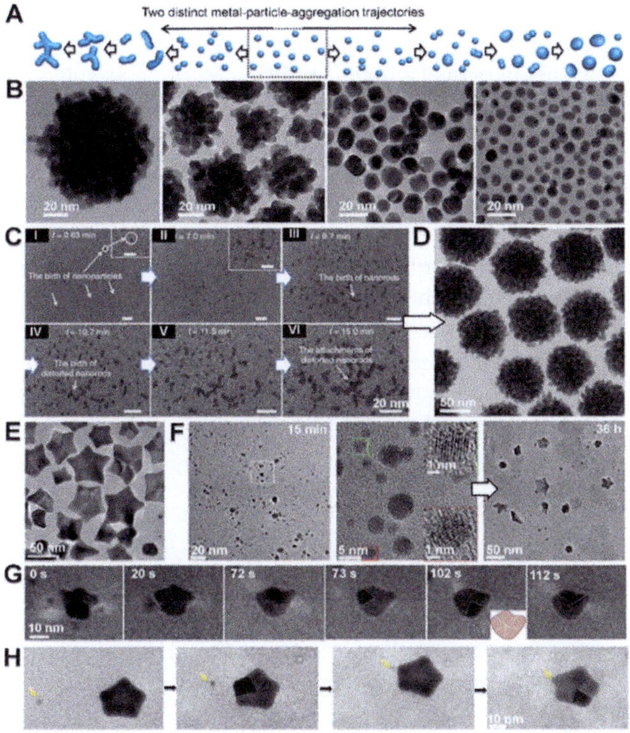

Fig. 8 The regulation roles of peptoids in Au nanocrystal formation. (A) Schemes showing two distinct, well-known metal particle aggregation trajectories. (B) TEM images of Au nanoparticle morphologies induced by different peptoid sequences at pH 5.5. The morphology evolves from coral-shaped to spherical particles as peptoids move from high to low binding affinities. (C) Time dependent LP-TEM images showing the birth of Au nanoparticles (I, II), nanorods (III), distorted nanorods (IV), and clusters of distorted nanorods (V, VI) during the early stages of coral-shaped particle formation. (D) TEM image showing typical monodispersed spherical coral-shaped nanoparticles.[151] (E) Representative TEM image showing five-fold twinned Au nanostars. (F) Time-dependent *ex situ* TEM images revealed some intermediate nanostructures during the Pep-1 induced formation of Au nanostars: Au nanoparticles with single crystalline characteristics formed at 15 min and Au stars emerged at 36 h. (G and H) Time dependent TEM images showing the formation and growth of one five-fold twinned nanocrystal through particle attachment (G) and the further growth of the regular five-fold twinned nanocrystal into a concave Au star by the attachment of small nanocrystals (marked by yellow arrow) on the corners (H).[152] (A–D) Reproduced from ref. 151 with permission from Springer Nature. Copyright 2018. (E–H) Reproduced from ref. 152 with permission from Wiley-VCH GmbH. Copyright 2022.

Au nanoparticles (*i.e.*, inhibiting their direct growth) and enhancing CPA. Moreover, this work also showed that the interaction strengths between the Au surface and the peptoids are tuneable through the choice of sequence, which provides facile control over the Au nanocrystal morphology.

Inspired by the ability to control Au–peptoid interaction strengths, a different set of peptoid-based ligands was discovered that produce star-shaped Au nanoparticles (Fig. 8E),[152] which consist of concave five-fold twinned particles with arms pointing along the [100] directions and bound by (111) side facets. Time dependent *ex situ* TEM investigations and direct observations of LP-TEM revealed a stepwise formation mechanism for the nanostars (Fig. 8F–H). The initial formation of small spherical single crystals (~5 nm) was followed by their attachment and relaxation to form spherical five-fold twinned particles. Further attachment and relaxation events, in which attachment occurred preferentially near the ends of the growing arms, led to the well-developed five-fold twinned stars. Experimental analysis and simulation of peptoid–Au binding confirmed that the formation of the Au star is caused by a strong energetic preference of peptoid binding to Au(111) over Au(100) facets. Variations in growth parameters and peptoid design showed that the outcome is strongly affected by the ratio of peptoid to Au and the absolute and relative strengths of peptoid binding to the Au faces. These factors work together to define a balance between the rate and location of particle attachment events, the kinetics of relaxation towards equilibrium, and (111) facet stabilization.

4.2 Interface-driven nucleation and assembly

Interface-driven nucleation due to local chemical gradients has been observed in a number of systems through a combination of *in situ* and *ex situ* characterization.[80,153,154] Those studies showed that, when ligands associate with a particle surface, they can bias nucleation to occur in the interfacial region, where attractive interactions then drive particle attachment. For example, pH-sensitive chemical gradients can drive near-surface nucleation and particle attachment of Au nanoparticles in the presence of citric anions.[153] Using *in situ* AFM and LP-TEM to follow transient structural changes during the nucleation and growth of Au nanoparticles, the formation of small particles through secondary nucleation (Fig. 9A, marked with blue arrows) on or near the surface of existing particles with a typical distance of ~1–2 nm (Fig. 9B) was documented. The new particles formed in the interfacial region, then attached to the existing particles, after which the aggregate either relaxed back towards a faceted morphology (Fig. 9A) or continued to expand outward through new aggregation events (Fig. 9C). The attachment of secondary particles to pre-existing particles proceeded through either a jump-to-contact at sufficiently small separation or the formation of a neck (Fig. 9C).

Both interfacial nucleation and neck formation imply a difference in the thermodynamic properties in the interfacial region. The molecular origin of this thermodynamic driving force lies in an interface-induced chemical potential gradient for citrate and $[AuCl_3(OH)]^-$ anions that promotes the nucleation of secondary particles either through an increase in the local supersaturation or a decrease in the interfacial free energy in the interfacial region. Furthermore, the gradient in the chemical potential drives $[AuCl_3(OH)]^-$ anions into the interparticle gap, which compensates for the energy cost of dehydration and attachment and/or promotes neck formation by promoting the adsorption and accumulation of citric ions on the Au surface.

Interface-driven nucleation has also been demonstrated in the precipitation process of barite at surfaces functionalized with organic monolayers, even in solutions undersaturated with respect to barite.[154] In supersaturated solutions, this phenomenon leads to Sr incorporation levels not possible during barite formation in bulk solution. This mechanism has been hypothesized to provide an

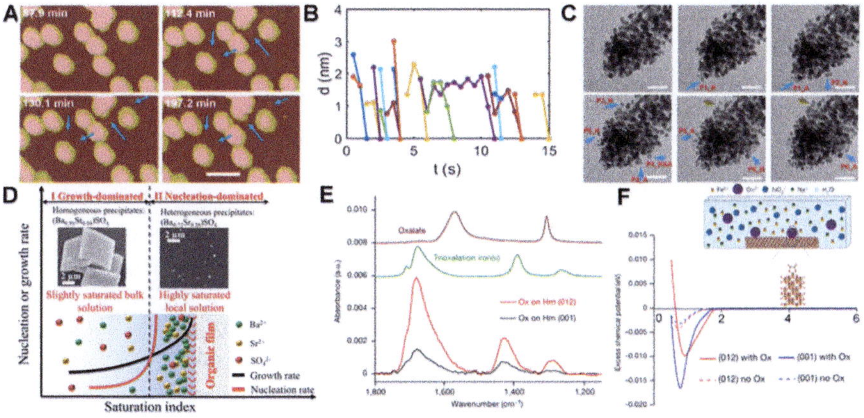

Fig. 9 Interface-driven nucleation and assembly process. (A) *In situ* AFM observation of the particle mediated growth of Au seed crystals; blue arrows highlight newly formed particles. Scale bar is 100 nm; (B) the change in separation with time of the newly nucleated particles from the nearest primary particle; (C) Time dependent LP-TEM images showing the attachment of newly formed secondary particles. Time interval is ~1 second between images. Secondary particles attaching to primary particles *via* diffusive jumps (particles 1, 2, 4–6, blue arrows) and *via* growing a neck (particle 3, brown arrows) were both evident. The suffixes _N and _A refer to particle nucleation and attachment, respectively. The scale bars are 20 nm.[153] (D) Schematic showing cation enrichment at organic–water interfaces. Under slightly supersaturated bulk solutions ($Ba_{50}Sr_{50}$, bulk SI = 1.53), barite precipitates in bulk solution (SEM image, left inset) and grows quickly to micrometer-sized crystals, while a nucleation-dominated precipitation process at a highly supersaturated solution can be achieved near organic–water interfaces due to significant cation enrichment. The STEM image (right inset) shows nanometer-sized, Sr-rich heterogeneous precipitates [$(Ba_{0.72}Sr_{0.28})SO_4$] formed at organic films from $Ba_{50}Sr_{50}$ solution (local SI = 8.10).[154] (E) Attenuated total reflection–Fourier transform infrared spectra of Ox adsorbed on Hm (001) *versus* Hm (012). Comparison of the spectra of adsorbed Ox to these reference spectra indicates that Ox is predominantly bound to Hm through direct surface Fe(III) coordination in a bidentate mononuclear fashion; (F) average potential of mean force (PMF) of Fe ions *versus* distance away from the surface, with (solid lines) and without (dashed lines) Ox. The inset is a schematic of the solution speciation and simulation set-up.[80] (A–C) Reproduced from ref. 153 with permission from RSC. Copyright 2018. D Reprinted with permission from ref. 154, Copyright (2019) National Academy of Sciences. (E–F) Reproduced from ref. 80 with permission from Springer Nature. Copyright 2021.

explanation for the wide occurrence of Sr-rich marine barites in seawater, which is globally undersaturated in barite. Investigations into the specific organic–mineral interaction at work during this process showed that the organic film induces significant cation enrichment near the film, implying, once again, that a change in interfacial chemistry associated with ion enrichment alters the thermodynamic properties of the interfacial region.

One of the most important findings of this study is that analysis of the phase diagram for the Sr-barite system implies that the formation of nanometer-sized, Sr-rich particles formed at the organic film (right inset in Fig. 9D) requires an increase in the local supersaturation to a saturation index (SI) of 8.10, when the SI in the bulk, where Sr-poor barite crystals form, is only 1.53 (left inset in Fig. 9D). Given that the chemical potential must be constant throughout the system and the increase in ion concentration near the film is a consequence of ion distribution to keep the chemical potential constant, these findings imply a decrease in the solubility of barite itself near the interface. As in the case of the Au-citrate system, these findings demonstrate that organic adsorbates can alter the local thermodynamic properties by concentrating ions, which affects the balance between nucleation and growth rates, causing the formation of crystals under conditions in which they do not form in the bulk or with sizes and compositions distinct from those that form in the bulk.

The understanding of interface-driven nucleation rate and growth rate differences suggests new approaches to achieving control over particle size and morphology during material synthesis. As discussed above (Fig. 7),[80] the interface-driven nucleation and assembly of Hm through the dissolution of Fh in the presence of Ox can lead to the synthesis of self-similar Hm mesocrystals (Fig. 7C). The appearance of spheroidal Hm particles only ~5 nm in size in the Fh aggregates or near Hm seeds, rather than far from any Fh or pre-existing Hm seeds, and the absence of large, faceted Hm crystals, shows that Ox plays an important role in stabilizing Hm nanoparticles and inhibiting their direct growth through classical monomer attachment. The *in situ* LP-TEM experiments demonstrate that new 'daughter' Hm particles repeatedly nucleate adjacent to—but not on—the Ox-covered seed surfaces, exhibiting an intervening edge-to-edge gap of ~2 nm. Following formation, the daughter particles quickly attach to the seed surface (Fig. 7D–F), thereby forming the mesocrystals.

Measurement of the Ox binding degree to the Hm faces confirms the existence of Ox at the Hm particle surface due to strong binding affinity (Fig. 9E). The surface-bound Ox drives Fe^{3+} accumulation approximately 1 nm from the surface by creating a layer of negatively charged sites and stabilizing the interfacial diffuse layer of solvated Fe^{3+} (Fig. 9F). In addition, the interfacial gradients at the Ox-covered surfaces create an attractive interparticle potential that assists in driving particle attachment. Whether the new Hm particles are coaligned upon nucleation, or align prior to, during or following attachment is unclear from the experiments and remains an open question for further exploration.

5. Conclusions

Developing a quantitative framework for CPA that correctly captures the underlying physics by linking atomistic details to ensemble outcomes is crucial both for interpreting mineralization processes in geochemical and biological systems and advancing material synthesis. As discussed in this review, that framework must, first and foremost, be rooted in an ability to define the relationship between interfacial structure, interparticle forces and the dynamics of aggregation. However, the impact of interfacial chemistry, especially in the presence of surface bound ligands, introduces a complicating factor that alters the thermodynamic properties of the solution and/or the crystal in ways that are currently unclear. Moreover, the dependence of phase stability on particle size and the interplay of free energetics and kinetics in determining nucleation rates brings about a convolution of phase evolution and aggregative growth that further complicates predictions, enriching the plethora of potential outcomes. Associated with the relationship between structure, forces and motion, the impact of ligands and interfaces on both structure and thermodynamic properties near the interface, and the evolution of phases with particle size must be considered. The challenges these considerations define can be encapsulated by three major knowledge gaps: (1) how do the complexities of a particle system, including (a) interfacial solution structure and the resulting unique interfacial solution properties, (b) anisotropies in particle shape, and (c) solvent dielectric properties, affect the forces that drive CPA? (2) How do surface-bound organic ligands and solution electrolytes affect interfacial solution structure and the resultant dielectric and transport properties and interparticle forces to enhance or hinder OA? (3) How do the solution structure, size dependent solubilities and free energies of formation determine the progression of phases and probabilities of nucleation, particularly in the interfacial region? Future progress in developing a quantitative framework to describe CPA will depend on our ability to fill these gaps.

Author contributions

All authors contributed to the writing of this review.

Conflicts of interest

There are no conflicts to declare.

Acknowledgements

This work was supported by the US Department of Energy (DOE), Office of Basic Energy Sciences, Division of Materials Science and Engineering, Synthesis and Processing Science Program at Pacific Northwest National Laboratory (PNNL). PNNL is a Multiprogram National Laboratory operated for the DOE by Battelle under contract no. DE-AC05–76RL01830.

Notes and references

1 J. W. Gibbs, *The Collected Works of J. Willard Gibbs*, Yale Univ. Press, New Haven, 1948.
2 J. W. Gibbs and A. W. Smith, in *Transactions of the Connecticut Academy of Arts and Sciences*, Volume III, Chapter V. and Chapter IX., Tuttle, Morehouse & Taylor, New Haven, New Haven, 1874, ch. V. & IX., pp. 108–248, 343–524.
3 D. Kashchiev, *Nucleation: Basic Theory with Applications*, Butterworth Heinemann, Oxford; Boston, 2000.
4 D. Kashchiev, *J. Chem. Phys.*, 2003, **118**, 1837–1851.
5 W. K. Burton, N. Cabrera and F. C. Frank, *Nature*, 1949, **163**, 398–399.
6 W. K. Burton, N. Cabrera and F. C. Frank, *Philos. Trans. R. Soc., A*, 1951, **243**, 299–358.
7 A. A. Chernov and E. I. Givargizov, *Modern Crystallography III: Crystal Growth*, Springer, Berlin, 1984.
8 L. L. Liu, E. Nakouzi, M. L. Sushko, G. K. Schenter, C. J. Mundy, J. Chun and J. J. De Yoreo, *Nat. Commun.*, 2020, **11**, 1045.
9 H. Cölfen, in *Handbook of Biomineralization: Biological Aspects and Structure Formation*, ed. E. Bauerlein, Weinheim, Germany, 2008, pp. 39–64.
10 H. Cölfen and M. Antonietti, *Mesocrystals and Nonclassical Crystallization*, Wiley, Chichester, England; Hoboken, NJ, 2008.

11 J. J. De Yoreo, P. U. P. A. Gilbert, N. A. J. M. Sommerdijk, R. L. Penn, S. Whitelam, D. Joester, H. Z. Zhang, J. D. Rimer, A. Navrotsky, J. F. Banfield, A. F. Wallace, F. M. Michel, F. C. Meldrum, H. Cölfen and P. M. Dove, *Science*, 2015, **349**, aaa6760.
12 M. A. Duran-Olivencia, P. Yatsyshin, S. Kalliadasis and J. F. Lutsko, *New J. Phys.*, 2018, **20**, 083019.
13 D. Erdemir, A. Y. Lee and A. S. Myerson, *Acc. Chem. Res.*, 2009, **42**, 621–629.
14 R. P. Sear, *Int. Mater. Rev.*, 2012, **57**, 328–356.
15 Z. Z. Shen, J. Chun, K. M. Rosso and C. J. Mundy, *J. Phys. Chem. C*, 2018, **122**, 12259–12266.
16 E. Nakouzi, A. G. Stack, S. Kerisit, B. A. Legg, C. J. Mundy, G. K. Schenter, J. Chun and J. J. De Yoreo, *J. Phys. Chem. C*, 2021, **125**, 1282–1291.
17 W. J. E. M. Habraken, J. H. Tao, L. J. Brylka, H. Friedrich, L. Bertinetti, A. S. Schenk, A. Verch, V. Dmitrovic, P. H. H. Bomans, P. M. Frederik, J. Laven, P. van der Schoot, B. Aichmayer, G. de With, J. J. DeYoreo and N. Sommerdijk, *Nat. Commun.*, 2013, **4**, 1507.
18 A. Dey, P. H. H. Bomans, F. A. Mueller, J. Will, P. M. Frederik, G. de With and N. A. J. M. Sommerdijk, *Nat. Mater.*, 2010, **9**, 1010–1014.
19 R. Demichelis, P. Raiteri, J. D. Gale, D. Quigley and D. Gebauer, *Nat. Commun.*, 2011, **2**, 590.
20 D. Gebauer and H. Cölfen, *Nano Today*, 2011, **6**, 564–584.
21 D. Gebauer, M. Kellermeier, J. D. Gale, L. Bergstrom and H. Cölfen, *Chem. Soc. Rev.*, 2014, **43**, 2348–2371.
22 D. Gebauer, A. Volkel and H. Cölfen, *Science*, 2008, **322**, 1819–1822.
23 X. Ma, S. Zhang, F. Jiao, C. J. Newcomb, Y. L. Zhang, A. Prakash, Z. H. Liao, M. D. Baer, C. J. Mundy, J. Pfaendtner, A. Noy, C. L. Chen and J. J. De Yoreo, *Nat. Mater.*, 2017, **16**, 767–774.
24 S. Chung, S. H. Shin, C. R. Bertozzi and J. J. De Yoreo, *Proc. Natl. Acad. Sci. U. S. A.*, 2010, **107**, 16536–16541.
25 L. B. Gower and D. J. Odom, *J. Cryst. Growth*, 2000, **210**, 719–734.
26 P. J. M. Smeets, A. R. Finney, W. J. E. M. Habraken, F. Nudelman, H. Friedrich, J. Laven, J. J. De Yoreo, P. M. Rodger and N. A. J. M. Sommerdijk, *Proc. Natl. Acad. Sci. U. S. A.*, 2017, **114**, E7882–E7890.
27 P. G. Vekilov, *J. Cryst. Growth*, 2005, **275**, 65–76.
28 P. G. Vekilov, *Nanoscale*, 2010, **2**, 2346–2357.
29 O. Galkin, K. Chen, R. L. Nagel, R. E. Hirsch and P. G. Vekilov, *Proc. Natl. Acad. Sci. U. S. A.*, 2002, **99**, 8479–8483.
30 A. F. Wallace, L. O. Hedges, A. Fernandez-Martinez, P. Raiteri, J. D. Gale, G. A. Waychunas, S. Whitelam, J. F. Banfield and J. J. De Yoreo, *Science*, 2013, **341**, 885–889.
31 X. L. Wang, I. M. Chou, W. X. Hu and R. C. Burruss, *Geochim. Cosmochim. Acta*, 2013, **103**, 1–10.
32 L. Addadi, S. Raz and S. Weiner, *Adv. Mater.*, 2003, **15**, 959–970.
33 E. Beniash, J. Aizenberg, L. Addadi and S. Weiner, *Proc. R. Soc. London, Ser. B*, 1997, **264**, 461–465.
34 E. Beniash, R. A. Metzler, R. S. K. Lam and P. U. P. A. Gilbert, *J. Struct. Biol.*, 2009, **166**, 133–143.

35 K. C. Cao, J. Biskupek, C. T. Stoppiello, R. L. McSweeney, T. W. Chamberlain, Z. Liu, K. Suenaga, S. T. Skowron, E. Besley, A. N. Khlobystov and U. Kaiser, *Nat. Chem.*, 2020, **12**, 921–928.
36 M. E. Davis, *Nature*, 2002, **417**, 813–821.
37 T. M. Davis, T. O. Drews, H. Ramanan, C. He, J. S. Dong, H. Schnablegger, M. A. Katsoulakis, E. Kokkoli, A. V. McCormick, R. L. Penn and M. Tsapatsis, *Nat. Mater.*, 2006, **5**, 400–408.
38 A. Navrotsky, *Proc. Natl. Acad. Sci. U. S. A.*, 2004, **101**, 12096–12101.
39 M. H. Nielsen, S. Aloni and J. J. De Yoreo, *Science*, 2014, **345**, 1158–1162.
40 Y. Politi, T. Arad, E. Klein, S. Weiner and L. Addadi, *Science*, 2004, **306**, 1161–1164.
41 P. J. M. Smeets, K. R. Cho, R. G. E. Kempen, N. A. J. M. Sommerdijk and J. J. De Yoreo, *Nat. Mater.*, 2015, **14**, 394–399.
42 J. F. Banfield, S. A. Welch, H. Z. Zhang, T. T. Ebert and R. L. Penn, *Science*, 2000, **289**, 751–754.
43 J. Baumgartner, A. Dey, P. H. H. Bomans, C. Le Coadou, P. Fratzl, N. Sommerdijk and D. Faivre, *Nat. Mater.*, 2013, **12**, 310–314.
44 N. D. Burrows, C. R. H. Hale and R. L. Penn, *Cryst. Growth Des.*, 2013, **13**, 3396–3403.
45 H. Cölfen and M. Antonietti, *Angew. Chem., Int. Ed.*, 2005, **44**, 5576–5591.
46 V. K. Ivanov, P. P. Fedorov, A. Y. Baranchikov and V. V. Osiko, *Russ. Chem. Rev.*, 2014, **83**, 1204–1222.
47 S. Kumar, Z. Wang, R. L. Penn and M. Tsapatsis, *J. Am. Chem. Soc.*, 2008, **130**, 17284–17286.
48 D. S. Li, F. Soberanis, J. Fu, W. T. Hou, J. Z. Wu and D. Kisailus, *Cryst. Growth Des.*, 2013, **13**, 422–428.
49 H. G. Liao, L. K. Cui, S. Whitelam and H. M. Zheng, *Science*, 2012, **336**, 1011–1014.
50 M. H. Nielsen, D. S. Li, H. Z. Zhang, S. Aloni, T. Y. J. Han, C. Frandsen, J. Seto, J. F. Banfield, H. Cölfen and J. J. De Yoreo, *Microsc. Microanal.*, 2014, **20**, 425–436.
51 J. Park, H. M. Zheng, W. C. Lee, P. L. Geissler, E. Rabani and A. P. Alivisatos, *ACS Nano*, 2012, **6**, 2078–2085.
52 R. L. Penn, *J. Phys. Chem. B*, 2004, **108**, 12707–12712.
53 R. L. Penn and J. F. Banfield, *Science*, 1998, **281**, 969–971.
54 M. Raju, A. C. T. van Duin and K. A. Fichthorn, *Nano Lett.*, 2014, **14**, 1836–1842.
55 L. Y. Ruan, H. Ramezani-Dakhel, C. Lee, Y. J. Li, X. F. Duan, H. Heinz and Y. Huang, *ACS Nano*, 2014, **8**, 6934–6944.
56 R. Q. Song and H. Cölfen, *Adv. Mater.*, 2010, **22**, 1301–1330.
57 A. E. S. Van Driessche, L. G. Benning, J. D. Rodriguez-Blanco, M. Ossorio, P. Bots and J. M. Garcia-Ruiz, *Science*, 2012, **336**, 69–72.
58 J. M. Yuk, J. Park, P. Ercius, K. Kim, D. J. Hellebusch, M. F. Crommie, J. Y. Lee, A. Zettl and A. P. Alivisatos, *Science*, 2012, **336**, 61–64.
59 V. M. Yuwono, N. D. Burrows, J. A. Soltis and R. L. Penn, *J. Am. Chem. Soc.*, 2010, **132**, 2163–2165.
60 H. Z. Zhang and J. F. Banfield, *Chem. Mater.*, 2002, **14**, 4145–4154.
61 H. Z. Zhang and J. F. Banfield, *J. Phys. Chem. Lett.*, 2012, **3**, 2882–2886.
62 H. Z. Zhang, J. J. De Yoreo and J. F. Banfield, *ACS Nano*, 2014, **8**, 6526–6530.

63 M. J. Bierman and S. Jin, *Energy Environ. Sci.*, 2009, **2**, 1050–1059.
64 R. R. Nair, H. A. Wu, P. N. Jayaram, I. V. Grigorieva and A. K. Geim, *Science*, 2012, **335**, 442–444.
65 D. B. Suyatin, J. Sun, A. Fuhrer, D. Wallin, L. E. Froberg, L. S. Karlsson, I. Maximov, L. R. Wallenberg, L. Samuelson and H. Q. Xu, *Nano Lett.*, 2008, **8**, 1100–1104.
66 D. Li, M. H. Nielsen, J. R. I. Lee, C. Frandsen, J. F. Banfield and J. J. De Yoreo, *Science*, 2012, **336**, 1014–1018.
67 E. Nakouzi, J. A. Soltis, B. A. Legg, G. K. Schenter, X. Zhang, T. R. Graham, K. M. Rosso, L. M. Anovitz, J. J. De Yoreo and J. Chun, *ACS Nano*, 2018, **12**, 10114–10122.
68 J. Chun, C. J. Mundy and G. K. Schenter, *J. Phys. Chem. B*, 2015, **119**, 5873–5881.
69 D. S. Li, J. Chun, D. D. Xiao, W. J. Zhou, H. C. Cai, L. Zhang, K. M. Rosso, C. J. Mundy, G. K. Schenter and J. J. De Yoreo, *Proc. Natl. Acad. Sci. U. S. A.*, 2017, **114**, 7537–7542.
70 A. Navrotsky, *ChemPhysChem*, 2011, **12**, 2207–2215.
71 R. L. Penn and J. F. Banfield, *Am. Mineral.*, 1998, **83**, 1077–1082.
72 M. Song, G. Zhou, N. Lu, J. Lee, E. Nakouzi, H. Wang and D. Li, *Science*, 2020, **367**, 40–45.
73 M. P. Boneschanscher, W. H. Evers, J. J. Geuchies, T. Altantzis, B. Goris, F. T. Rabouw, S. A. P. van Rossum, H. S. J. van der Zant, L. D. A. Siebbeles, G. Van Tendeloo, I. Swart, J. Hilhorst, A. V. Petukhov, S. Bals and D. Vanmaekelbergh, *Science*, 2014, **344**, 1377–1380.
74 K. S. Cho, D. V. Talapin, W. Gaschler and C. B. Murray, *J. Am. Chem. Soc.*, 2005, **127**, 7140–7147.
75 M. Kuno, *Phys. Chem. Chem. Phys.*, 2008, **10**, 620–639.
76 Z. G. Li, J. H. Sui, X. L. Li and W. Cai, *Langmuir*, 2011, **27**, 2258–2264.
77 H. G. Liao and H. M. Zheng, *J. Am. Chem. Soc.*, 2013, **135**, 5038–5043.
78 A. I. Lupulescu and J. D. Rimer, *Science*, 2014, **344**, 729–732.
79 D. Li, F. Soberanis, J. Fu, W. Hou, J. Wu and D. Kisailus, *Cryst. Growth Des.*, 2013, **13**, 422–428.
80 G. Zhu, M. L. Sushko, J. S. Loring, B. A. Legg, M. Song, J. A. Soltis, X. Huang, K. M. Rosso and J. J. De Yoreo, *Nature*, 2021, **590**, 416–422.
81 A. B. Bard, X. Z. Zhou, X. J. Xia, G. M. Zhu, M. B. Lim, S. M. Kim, M. C. Johnson, J. M. Kollman, M. A. Marcus, S. R. Spurgeon, D. E. Perea, A. Devaraj, J. Chun, J. J. De Yoreo and P. J. Pauzauskie, *Chem. Mater.*, 2020, **32**, 2753–2763.
82 S. L. Burkett and M. E. Davis, *J. Phys. Chem.*, 1994, **98**, 4647–4653.
83 J. J. Chen, E. B. Zhu, J. Liu, S. Zhang, Z. Y. Lin, X. F. Duan, H. Heinz, Y. Huang and J. J. De Yoreo, *Science*, 2018, **362**, 1135–1139.
84 A. E. S. Van Driessche, N. Van Gerven, P. H. H. Bomans, R. R. M. Joosten, H. Friedrich, D. Gil-Carton, N. A. J. M. Sommerdijk and M. Sleutel, *Nature*, 2018, **556**, 89–94.
85 D. S. Li, M. H. Nielsen, J. R. I. Lee, C. Frandsen, J. F. Banfield and J. J. De Yoreo, *Science*, 2012, **336**, 1014–1018.
86 O. M. Magnussen and A. Gross, *J. Am. Chem. Soc.*, 2019, **141**, 4777–4790.

87 W. Stumm, *Chemistry of the Solid-Water Interface: Processes at the Mineral-Water and Particle-Water Interface in Natural Systems*, Wiley-Interscience, New York, NY, 1992.
88 J. N. Israelachvili and R. M. Pashley, *Nature*, 1983, **306**, 249–250.
89 L. Cheng, P. Fenter, K. L. Nagy, M. L. Schlegel and N. C. Sturchio, *Phys. Rev. Lett.*, 2001, **87**, 156103.
90 S. J. T. Brugman, E. R. Townsend, M. M. H. Smets, P. Accordini and E. Vlieg, *Langmuir*, 2018, **34**, 3821–3826.
91 S. Pintea, W. de Poel, A. E. F. de Jong, R. Felici and E. Vlieg, *Langmuir*, 2018, **34**, 4241–4248.
92 P. Fenter and N. C. Sturchio, *Prog. Surf. Sci.*, 2004, **77**, 171–258.
93 P. Fenter, C. Park, K. L. Nagy and N. C. Sturchio, *Thin Solid Films*, 2007, **515**, 5654–5659.
94 A. Tuladhar, S. M. Piontek and E. Borguet, *J. Phys. Chem. C*, 2017, **121**, 5168–5177.
95 S. Dewan, M. S. Yeganeh and E. Borguet, *J. Phys. Chem. Lett.*, 2013, **4**, 1977–1982.
96 E. T. Herruzo, H. Asakawa, T. Fukuma and R. Garcia, *Nanoscale*, 2013, **5**, 2678–2685.
97 T. Fukuma and R. Garcia, *ACS Nano*, 2018, **12**, 11785–11797.
98 H. Songen, B. Reischl, K. Miyata, R. Bechstein, P. Raiteri, A. L. Rohl, J. D. Gale, T. Fukuma and A. Kuhnle, *Phys. Rev. Lett.*, 2018, **120**, 116101.
99 K. Miyazawa, J. Tracey, B. Reischl, P. Spijker, A. S. Foster, A. L. Rohl and T. Fukuma, *Nanoscale*, 2020, **12**, 12856–12868.
100 H. Songen, R. Bechstein and A. Kuhnle, *J. Phys.: Condens. Matter*, 2017, **29**, 274001.
101 H. Songen, C. Marutschke, P. Spijker, E. Holmgren, I. Hermes, R. Bechstein, S. Klassen, J. Tracey, A. S. Foster and A. Kuhnle, *Langmuir*, 2017, **33**, 125–129.
102 A. F. Payam, D. Martin-Jimenez and R. Garcia, *Nanotechnology*, 2015, **26**, 185706.
103 D. Martin-Jimenez and R. Garcia, *J. Phys. Chem. Lett.*, 2017, **8**, 5707–5711.
104 S. Kerisit, C. Liu and E. S. Ilton, *Geochim. Cosmochim. Acta*, 2008, **72**, 1481–1497.
105 S. Kerisit, M. Okumura, K. Rosso and M. Machida, *Clays Clay Miner.*, 2016, **64**, 389–400.
106 Z. Shen, E. S. Ilton, M. P. Prange and S. N. Kerisit, *J. Phys. Chem. C*, 2017, **121**, 13692–13700.
107 S. N. Kerisit and J. J. De Yoreo, *J. Phys. Chem. C*, 2020, **124**, 5480–5488.
108 I. C. Bourg, S. S. Lee, P. Fenter and C. Tournassat, *J. Phys. Chem. C*, 2017, **121**, 9402–9412.
109 J. A. R. Willemsen, S. C. B. Myneni and I. C. Bourg, *J. Phys. Chem. C*, 2019, **123**, 13624–13636.
110 T. D. Li, H. C. Chiu, D. Ortiz-Young and E. Riedo, *Rev. Sci. Instrum.*, 2014, **85**.
111 E. Riedo, *Phys. Rev. Lett.*, 2008, **100**, 106102.
112 Y. Zhu and S. Granick, *Phys. Rev. Lett.*, 2001, **87**, 096104.
113 T.-D. Li, J. Gao, R. Szoszkiewicz, U. Landman and E. Riedo, *Phys. Rev. B: Condens. Matter Mater. Phys.*, 2007, **75**.

114 L. Fumagalli, A. Esfandiar, R. Fabregas, S. Hu, P. Ares, A. Janardanan, Q. Yang, B. Radha, T. Taniguchi and K. Watanabe, *Science*, 2018, **360**, 1339–1342.
115 M. Raju, A. C. van Duin and K. A. Fichthorn, *Nano Lett.*, 2014, **14**, 1836–1842.
116 P. Grochowski and J. Trylska, *Biopolymers*, 2008, **89**, 93–113.
117 L. Landau, *Electrodynamics of Continuous Media, Landau and Lifshitz Course of Theoretical Physics*, Elsevier Butterworth-Heinemann, Burlington, MA, 1984, vol. 8.
118 V. Parsegian, *Van der Waals Forces, A Handbook for Biologists, Chemists, Engineers, and Physicists*, Cambridge University Press, Cambridge, 2005.
119 V. Parsegian and T. Zemb, *Curr. Opin. Colloid Interface Sci.*, 2011, **16**, 618–624.
120 R. M. Pashley, *J. Colloid Interface Sci.*, 1981, **80**, 153–162.
121 R. M. Pashley, *J. Colloid Interface Sci.*, 1984, **102**, 23–35.
122 Y. Marcus, *Chem. Rev.*, 2009, **109**, 1346–1370.
123 R. Pashley, *J. Colloid Interface Sci.*, 1981, **83**, 531–546.
124 T. Baimpos, B. R. Shrestha, S. Raman and M. Valtiner, *Langmuir*, 2014, **30**, 4322–4332.
125 T. Colla, L. Nunes Lopes and A. P. Dos Santos, *J. Chem. Phys.*, 2017, **147**, 014104.
126 A. P. dos Santos and Y. Levin, *J. Chem. Phys.*, 2015, **142**, 194104.
127 V. M. Agranovich and V. L. Ginzburg, *Phys.-Usp.*, 1962, **5**, 323.
128 J. Chun, J.-L. Li, R. Car, I. A. Aksay and D. A. Saville, *J. Phys. Chem. B*, 2006, **110**, 16624–16632.
129 D. Saville, J. Chun, J.-L. Li, H. Schniepp, R. Car and I. A. Aksay, *Phys. Rev. Lett.*, 2006, **96**, 018301.
130 X. Zhang, Y. He, M. L. Sushko, J. Liu, L. Luo, J. J. De Yoreo, S. X. Mao, C. Wang and K. M. Rosso, *Science*, 2017, **356**, 434–437.
131 J. Lee, E. Nakouzi, M. Song, B. Wang, J. Chun and D. S. Li, *ACS Nano*, 2018, **12**, 12778–12787.
132 J. Lee, E. Nakouzi, D. Xiao, Z. G. Wu, M. Song, C. Ophus, J. Chun and D. S. Li, *Small*, 2019, **15**.
133 S. Kim and S. J. Karrila, *Microhydrodynamics: Principles and Selected Applications*, Courier Corporation, 2013.
134 W. B. Russel, W. Russel, D. A. Saville and W. R. Schowalter, *Colloidal Dispersions*, Cambridge university press, 1991.
135 D. Jeffrey and Y. Onishi, *J. Fluid Mech.*, 1984, **139**, 261–290.
136 D. Xiao, Z. Wu, M. Song, J. Chun, G. K. Schenter and D. Li, *Langmuir*, 2018, **34**, 1466–1472.
137 A. A. Gribb and J. F. Banfield, *Am. Mineral.*, 1997, **82**, 717–728.
138 M. R. Ranade, A. Navrotsky, H. Z. Zhang, J. F. Banfield, S. H. Elder, A. Zaban, P. H. Borse, S. K. Kulkarni, G. S. Doran and H. J. Whitfield, *Proc. Natl. Acad. Sci. U. S. A.*, 2002, **99**, 6476–6481.
139 V. M. Yuwono, N. D. Burrows, J. A. Soltis and R. L. Penn, *J. Am. Chem. Soc.*, 2010, **132**, 2163–2165.
140 C. Frandsen, B. A. Legg, L. R. Comolli, H. Z. Zhang, B. Gilbert, E. Johnson and J. F. Banfield, *CrystEngComm*, 2014, **16**, 1451–1458.
141 O. Söhnel, *J. Cryst. Growth*, 1982, **57**, 101–108.
142 E. Zhu, S. Wang, X. Yan, M. Sobani, L. Ruan, C. Wang, Y. Liu, X. Duan, H. Heinz and Y. Huang, *J. Am. Chem. Soc.*, 2019, **141**, 1498–1505.

143 J. Cassidy, D. Harankahage, J. Ojile, D. Porotnikov, L. Walker, M. Montemurri, B. S. L. Narvaez, D. Khon, M. D. E. Forbes and M. Zamkov, *Chem. Mater.*, 2022, **34**, 2484–2494.

144 R. Li, A. Smolyakova, G. Maayan and J. D. Rimer, *Chem. Mater.*, 2017, **29**, 9536–9546.

145 W. Qin, A. Agarwal, M. K. Choudhary, J. C. Palmer and J. D. Rimer, *Chem. Mater.*, 2019, **31**, 3228–3238.

146 N. A. Merrill, F. Yan, H. Jin, P. Mu, C. L. Chen and M. R. Knecht, *Nanoscale*, 2018, **10**, 12445–12452.

147 M. Monahan, B. Cai, T. Jian, S. Zhang, G. Zhu, C. L. Chen, J. J. De Yoreo and B. M. Cossairt, *Nanoscale*, 2021, **13**, 1273–1282.

148 L. Liu, D. Song, B. Jin, M. A. Sinnwell, J. Liu, J. J. De Yoreo and M. L. Sushko, *J. Phys. Chem. C*, 2020, **124**, 24215–24222.

149 C. Zhu, S. Liang, E. Song, Y. Zhou, W. Wang, F. Shan, Y. Shi, C. Hao, K. Yin and T. Zhang, *Nat. Commun.*, 2018, **9**, 1.

150 U. Anand, J. Lu, D. Loh, Z. Aabdin and U. Mirsaidov, *Nano Lett.*, 2016, **16**, 786–790.

151 F. Yan, L. Liu, T. R. Walsh, Y. Gong, P. Z. El-Khoury, Y. Zhang, Z. Zhu, J. J. De Yoreo, M. H. Engelhard, X. Zhang and C. L. Chen, *Nat. Commun.*, 2018, **9**, 2327.

152 B. Jin, F. Yan, X. Qi, B. Cai, J. Tao, X. Fu, S. Tan, P. Zhang, J. Pfaendtner, N. Y. Naser, F. Baneyx, X. Zhang, J. J. DeYoreo and C. L. Chen, *Angew. Chem., Int. Ed.*, 2022, **134**, e202201980.

153 Y. Cheng, J. Tao, G. Zhu, J. A. Soltis, B. A. Legg, E. Nakouzi, J. J. De Yoreo, M. L. Sushko and J. Liu, *Nanoscale*, 2018, **10**, 11907–11912.

154 N. Deng, A. G. Stack, J. Weber, B. Cao, J. J. De Yoreo and Y. Hu, *Proc. Natl. Acad. Sci. U. S. A.*, 2019, **116**, 13221–13226.

Faraday Discussions

ROYAL SOCIETY OF CHEMISTRY

PAPER

Solvent-mediated isotope effects strongly influence the early stages of calcium carbonate formation: exploring D_2O *vs.* H_2O in a combined computational and experimental approach†

Michael King, ‡[a] Jonathan T. Avaro,‡[ab] Christine Peter,[a] Karin Hauser [a] and Denis Gebauer *[c]

Received 26th October 2021, Accepted 7th December 2021
DOI: 10.1039/d1fd00078k

In experimental studies, heavy water (D_2O) is employed, *e.g.*, so as to shift the spectroscopic solvent background, but any potential effects of this solvent exchange on reaction pathways are often neglected. While the important role of light water (H_2O) during the early stages of calcium carbonate formation has been realized, studies into the actual effects of aqueous solvent exchanges are scarce. Here, we present a combined computational and experimental approach to start to fill this gap. We extended a suitable force field for molecular dynamics (MD) simulations. Experimentally, we utilised advanced titration assays and time-resolved attenuated total reflection Fourier transform infrared (ATR-FTIR) spectroscopy. We find distinct effects in various mixtures of the two aqueous solvents, and in pure H_2O or D_2O. Disagreements between the computational results and experimental data regarding the stabilities of ion associates might be due to the unexplored role of HDO, or an unprobed complex phase behaviour of the solvent mixtures in the simulations. Altogether, however, our data suggest that calcium carbonate formation might proceed "more classically" in D_2O. Also, there are indications for the formation of new structures in amorphous and crystalline calcium carbonates. There is huge potential towards further improving the understanding of mineralization mechanisms by studying solvent-mediated isotope effects, also beyond calcium carbonate. Last, it must be appreciated that H_2O and D_2O have significant, distinct effects on mineralization mechanisms, and that care has to be taken when experimental data from D_2O studies are used, *e.g.*, for the development of H_2O-based computer models.

[a]*Department of Chemistry, University of Konstanz, Universitätsstr. 10, 78457 Konstanz, Germany*
[b]*Empa, Lerchenfeldstrasse 5, 9014 St. Gallen, Switzerland*
[c]*Institute of Inorganic Chemistry, Leibniz University of Hannover, Callinstr. 9, 30167 Hannover, Germany. E-mail: gebauer@acc.uni-hannover.de*

† Electronic supplementary information (ESI) available. See DOI: 10.1039/d1fd00078k

‡ These authors contributed equally and retain the right to list their names first in their respective CVs.

Introduction

Deuteration of compounds is often employed to investigate reaction mechanisms based on the so-called kinetic isotope effect,[1] where doubling of the masses due to the change from hydrogen (H) to deuterium (D) reduces the vibrational frequency of the involved bonds.[2,3] This effect also allows shifting the vibrational modes of the aqueous solvent background in IR spectroscopic studies to explore, *e.g.*, protein dynamics.[4–7] In proton nuclear magnetic resonance spectroscopy, for instance, replacing H_2O by D_2O allows one to get rid of unwanted solvent resonance.[8] However, when changing the solvent from light to heavy water, also thermodynamic effects have to be considered.[9] Not only quantum nuclear effects but also electronic effects come into play and affect a variety of properties.[10] The O–D bond is shorter than the O–H bond, *i.e.*, 0.985(5) Å *vs.* 0.990(5) Å,[11,12] rendering the average number of hydrogen bonds within the bulk of D_2O slightly higher.[13] Also, the peaks in the radial distribution function are more pronounced in the case of D_2O.[14] Thus, D_2O is more ordered than H_2O.[14–17] This, in turn, causes different macroscopic and microscopic properties of these aqueous solvents. For example, the compressibility and polarizability of D_2O are higher than those of H_2O.[18] Some further selected differences can be found in Table 1, and more details in ref. 9, 19 and 20.

The different properties of D_2O also have consequences for solutes in heavy water. The neutral pH value of water increases from 7.0 to ~7.4 when changing from H_2O to D_2O.[21,22] Also, the pK_a value of acids shifts by an increment in the range of 0.5–0.7 in D_2O with respect to the value in H_2O, *e.g.*, for the second deprotonation step of carbonic acid $pK_a^D - pK_a^H = 0.60$,[23] or 0.741.[21] Even deuteration of aliphatic C-atoms near a carboxyl group (*i.e.*, of C-alpha or C-beta atoms) affects the pK_a value of the corresponding acids.[26]

The higher solubility of hydrocarbons and noble gases in D_2O[2,27,28] can be attributed to its higher compressibility (Table 1),[29] reflecting the ability to form nano-cavities more easily than light water. This also leads to a higher hydrophobicity of D_2O, even overcoming the opposing effect of stronger hydrogen bonds than in light water.[30] This effect, however, depends on solute size, shape and charge and can switch from attractive to repulsive.[29] The differences between light and heavy water can furthermore affect the stability of proteins, *i.e.*, the melting temperatures of some proteins in heavy water increase relative to H_2O.[31]

Table 1 Comparison of selected properties of light and heavy water

	H_2O	D_2O
Melting point (101.325 kPa) [°C][19]	0.00	3.82
Boiling point (101.325 kPa) [°C][19]	100.00	101.42
$C_{p,m}$ [J K^{-1} mol^{-1}][24]	75.23	84.67
$C_{V,m}$ [J K^{-1} mol^{-1}][24]	74.44	84.42
Density (25 °C)[a] [g cm^{-3}][19]	0.997048	1.1044
Temperature of maximum density [°C][19]	4.0	11.2
Compressibility [10^{-10} Pa^{-1}][18]	4.599	4.763
Diffusion coefficient [10^{-5} cm^2 s^{-1}][25]	2.272	2.109
Polarizability of vapor near 100 °C [cm^3 mol^{-1}][9]	58.5	61.7

[a] The measurements for H_2O were done at 1 atm, for D_2O at 1 bar.

However, other proteins are less stable in D_2O than in light water.[32,33] Also, the phase behaviour of solutes is different in D_2O. For instance, the critical micelle concentration is typically lower in D_2O than in H_2O.[2,34,35] Also, ion solubilities, as well as solvation strengths differ,[36–39] *e.g.*, the heat of transfer for $CaCl_2$ between the waters is $\Delta H_t(H_2O - D_2O) = -5.0$ kJ mol^{-1}.[40]

However, little is known about the effects of changing the aqueous solvent from light water to D_2O on mineralization. Lee *et al.*[41] explored thermodynamic effects of deuterium on the aqueous synthesis of inorganic materials, revealing new structural and magnetic properties of the obtained manganese and iron materials in D_2O. Lundager Madsen reported no significant effect of magnetic fields on calcium carbonate precipitation when using D_2O, in contrast to the observed effect in light water.[42] Raudino *et al.*[43] found that smaller crystals formed in D_2O compared to H_2O solutions in the case of alkaline earth carbonates. Furthermore, the conversion of calcite into aragonite in D_2O seemed to be less favourable at high temperatures. Recently, Morris *et al.*[44] studied the kinetics of amorphous calcium carbonate (ACC) precipitation in H_2O and D_2O, reporting very similar rates for the two aqueous solvents, suggesting that neither ion dehydration nor deprotonation of bicarbonate ions represented significant energetic barriers to the formation of ACC.

Considering the various effects of a solvent exchange from light to heavy water described above begs the question: which influences on the early stages of precipitation of calcium carbonate can be expected, in principle? In the following, we contemplate these possibilities within the frameworks of different existing nucleation theories.

Potential effects of D_2O *vs.* H_2O according to classical nucleation theory (CNT)

According to CNT, as schematically illustrated in Fig. 1 (top), the nucleation rate J can be expressed as:[46]

$$J = L\cdot\exp(-E_A/kT)\cdot\exp(-\Delta g_c/kT) \qquad (1)$$

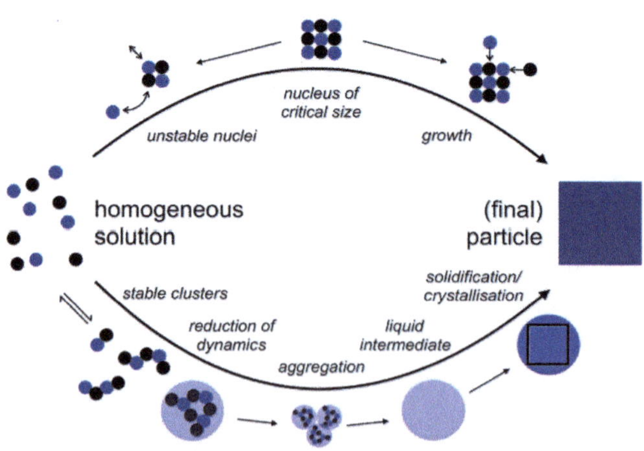

Fig. 1 Comparison of the pathways from homogeneous solution to crystals according to the notions of classical nucleation theory (CNT, top) and the pre-nucleation cluster (PNC) pathway (bottom). For explanation see the text; adopted from ref. 45.

where the pre-factor L can be derived from the shape and material-dependent bulk properties of the nuclei. The first exponent represents the so-called "kinetic barrier" characterized by the activation energy E_A (Boltzmann constant k, absolute temperature T), which is, however, probably negligible and difficult to assess *a priori* (also see below). The second exponent gives rise to the so-called "thermodynamic barrier" to nucleation characterized by the standard free energy associated with the formation of the critical nucleus, $\Delta g_c = B \cdot \alpha^3/\sigma^2$. Here, B is a constant that depends on the nucleus shape and density.[46] According to CNT, Δg_c is proportional to the cube of the interfacial free energy of the nuclei, α, which can be obtained from the macroscopically accessible interfacial tension (capillary assumption). Also, Δg_c is inversely proportional to the square of supersaturation σ given by:

$$\sigma = kT \cdot \ln(\mathrm{IAP}/K_{sp}) \tag{2}$$

where IAP is the actual ion activity product (in the supersaturated, metastable solution) according to the composition of the nascent mineral phase, and K_{sp} is its solubility constant.

The above considerations allow us to qualitatively analyze potential effects of the solvent change on classical nucleation behaviour. A change to heavy water is rather unlikely to affect the pre-factor, L, as this parameter is mainly associated with mineral bulk properties and nucleus shape. Regarding the kinetic barrier, in the case of calcium carbonate, computer simulations indeed suggested that no significant activation energies, *e.g.*, due to de-hydration of the ions, exist,[47] a notion that recent experiments seem to corroborate.[44] This then leaves the thermodynamic barrier, and it is indeed conceivable that the change of the aqueous solvent alters the surface free energy and/or the solubility constant, K_{sp}. As the latter enters the quantitative expressions in a logarithm (eqn (2)) that then enters the rate expression in the denominator of an exponent (eqn (1)), any potential effects of the solvent change on the interfacial free energy are expected to dominate, as this parameter is cubed and then enters the exponent for calculating the nucleation rate J.

The surface tension of D_2O is slightly less than that of H_2O,[48] which is attributed to the somehow larger molecular volume and, thus, a somewhat larger separation of the dipoles in heavy water, reducing the mutual attraction between the dipoles within the bulk of the solvent. In the absence of data on interfacial tensions between calcium carbonate and heavy water, it can thus be speculated that the solvent change to heavy water would also somewhat reduce the interfacial free energies of the different polymorphs and forms, potentially by increments that are similar in magnitude. Since the interfacial free energy is cubed for the calculation of Δg_c at a given supersaturation, and then enters the exponent for estimating the nucleation rate, corresponding effects could become large even for small changes in interfacial free energy, but would promote rather than inhibit "classical nucleation" of calcium carbonate in D_2O, with respect to the scenario in H_2O.

Potential effects of D_2O *vs.* H_2O according to the pre-nucleation cluster (PNC) pathway

The so-called PNC pathway (Fig. 1, bottom) provides an alternative perspective on the mechanisms underlying the early stages of mineralization.[45,49,50] Here, thermodynamically stable solute clusters (*i.e.*, standard free energy of formation, ΔG^0

< 0) form initially, independent of supersaturation. Their structural form is called a 'dynamically-ordered liquid-like oxyanion polymer' (DOLLOP).[47] A change towards higher coordination numbers than in the initially chain-like DOLLOPs in larger PNCs was suggested to significantly reduce the PNC dynamics and thereby lead to phase separation.[49,51] Indeed, the PNC size distribution depends on the actual IAP in solution, and a recently introduced quantitative framework allows predicting the specific IAPs, at which PNCs can and must transform into dense liquid calcium carbonate (nano)droplets, that is, of the corresponding liquid–liquid binodal and spinodal limits, respectively.[52] Input parameters of the experimentally verified model are, essentially, the ion association constant, K_{cluster}, and the solubilities of the different polymorphs, $K_{\text{sp,polym.}}$, accounting for the experimentally observed phenomenon of amorphous polymorphism in ACCs[53–55] in the binodal limit, given by the corresponding $\text{IAP}_{\text{binodal}}$:[52]

$$\text{IAP}_{\text{binodal}} = A_{\text{polym.}} \cdot K_{\text{sp,polym.}} \cdot \ln K_{\text{cluster}} \tag{3}$$

The unit-less constant A was determined to be $A_{\text{vaterite}} = \sim 0.4$, $A_{\text{aragonite}} = \sim 1.0$ and $A_{\text{calcite}} = \sim 1.3$.[52] The predicted spinodal limit, $\text{IAP}_{\text{spinodal}}$, on the other hand, is given by:[52]

$$\text{IAP}_{\text{spinodal}} = 1/(K_{\text{cluster}})^2 \tag{4}$$

$\text{IAP}_{\text{spinodal}}$ agrees with the commonly accepted value of the solubility of (disordered) ACC[56] that is typically precipitated from high levels of supersaturation. Indeed, according to the PNC pathway, the as-formed liquid intermediates dehydrate towards the formation of solid ACC, as observed *in situ* recently,[57,58] thus transferring structural characteristics from the PNCs, over the liquid intermediates, to the initial amorphous solids. The solid ACCs later transform into crystals, at which stage different mechanisms may operate.[59,60]

Water plays an important role in the non-classical nucleation of calcium carbonate.[54,61,62] For instance, the entropic contribution to the overall free energy change due to the release of hydration waters from ion hydration shells during ion association drives the formation of PNCs,[63] and it is thus conceivable that the change from H_2O to D_2O might affect the ion association constant. For instance, since D_2O is more ordered than H_2O (*cf.* above), the entropic gain upon releasing hydration waters to a more ordered environment during ion association might be somewhat reduced in heavy water, which would render PNCs in D_2O less stable than in H_2O. Since the solubilities of the different polymorphs and forms might be affected too, within the framework of the quantitative PNC model,[52] this would then shift the binodal and spinodal limits, eqn (3) and (4). With it, the critical point and the whole liquid–liquid coexistence region would be shifted, potentially giving rise even to the formation of distinct, potentially new proto-structures in ACC. In turn, new crystal polymorphs might become accessible. A changed locus of the miscibility gap would in any case alter the kinetics of liquid–liquid demixing from specific levels of supersaturation, opening up reaction channels for alternative pathways. For instance, in D_2O, calcium carbonate might behave more 'classically'. In our opinion, virtually anything is possible.

However, very little is known about the effects of changing the solvent from light to heavy water on calcium carbonate mineralization. Since strong effects are possible, which promise to shed new light on the nucleation mechanisms both from a kinetic and thermodynamic perspective, we studied the early stages of calcium carbonate formation in light and heavy water, and mixtures of light and heavy water, computationally and experimentally.

Results and discussion

Computer simulations of PNCs in light *vs.* heavy water

The development of a flexible heavy water model and its properties are described in the Experimental section on the computational methods (ESI, Experimental section and Fig. S1, S2, Tables S1 and S2†). In brief, the parameterization by scaling the charges of the flexible water model SPC/fw led to a good D_2O model (SPC/HW/fw) that correctly represents the structure of heavy water and that can be combined with the $CaCO_3$ model of Raiteri *et al.*[64]

The formation of PNCs in D_2O and H_2O was simulated utilizing a cubic simulation box with a side length of 55 Å, containing 10 Ca^{2+}- and 10 CO_3^{2-}-ions. The resulting concentration was 0.1 M with respect to Ca^{2+}-ions. 20 replicas with randomized ion positions were simulated per solvent. During the course of these simulations, clusters emerged and disappeared. These clusters were further analysed to search for differences in structure and kinetics between light and heavy water.

Cluster size distribution. The ions in solution were clustered with a cut-off distance of 3.6 Å, see the ESI for details.† The size distribution of the different clusters was calculated by counting the number of ions inside a cluster yielding distributions shown in Fig. 2. The change of solvent from H_2O to D_2O shifts the cluster size distribution to smaller structures with fewer ions.

In addition, only half so many clusters containing more than two ions, were found in D_2O with respect to H_2O. We thus visualized the internal structure of ion aggregates by creating 2D histograms of angle–distance combinations within the clusters. Here, we looked at all ion pairs within a cut-off distance of 4.0 Å. Then we

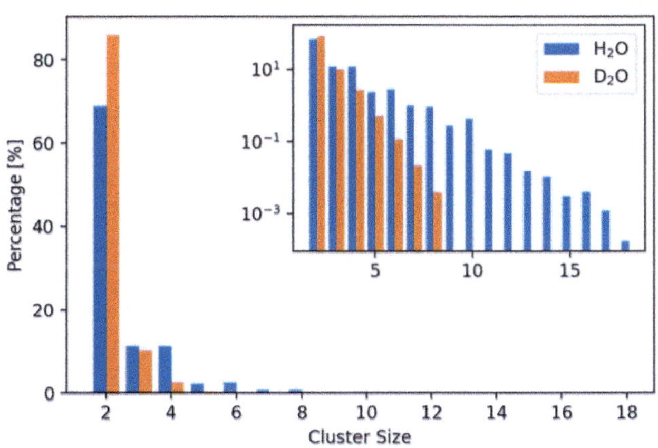

Fig. 2 Distribution of the number of ions per found cluster in light and heavy water. The distributions are normalized based on the total number of clusters found in the respective solvent. Two times more clusters (consisting of N(ions) > 2) were found in H_2O compared to D_2O.

measured the distance and angle of a third ion within 8.0 Å and binned the values in histograms. The resulting histograms (Fig. S3 and S4†) show that the overall features stay the same. However, the features, especially at large angles, are less pronounced in D_2O. This agrees with the observation of shorter and more compact clusters in D_2O and indicates that the structure of the clusters differs between the two aqueous solvents.

Lifetime of the clusters. The lifetime of different clusters was calculated to evaluate the kinetic stability of these species. The time while the cluster consists of three or more ions is thereby considered the lifetime of a cluster. The solvent change to D_2O leads to a reduction of the lifetime of pre-nucleation clusters (Fig. 3) along with a reduction of their total number (Fig. 2). Hence, D_2O reduces both their kinetic and thermodynamic stabilities with respect to the scenario in light water.

Discussion. The computational results show that a change from H_2O to D_2O leads to smaller, more compact PNCs which also occur less often and with shorter lifetimes than in light water. The potential cause of these changes remains unclear and has to be evaluated further. To this end, we first calculated the potential of mean force (PMF) between calcium and carbonate ions in the different aqueous solvents (Fig. 4). Again, it is obvious that ion association is weaker in D_2O than in H_2O, while the general features of the PMF stay the same. The resulting differences may partially be explained by the higher dipole moment of D_2O, leading to a higher dielectric constant, and, with it, to a lower resulting electrostatic force between ions due to screening. In order to study this effect further, the ion association constant was calculated using the PMF, following the approach used by Raiteri *et al.*[64] The ratio of the association constants in the distinct aqueous solvents should be rather robust to errors that might occur when analysing the PMF. Indeed, the association constant is 6 times higher in H_2O than in D_2O (Fig. S5†), explaining why fewer and smaller clusters are observed in heavy water than in light water.

The difference in mass between the two aqueous solvents is often considered to be the only difference, while changes of internal structure and polarizability are

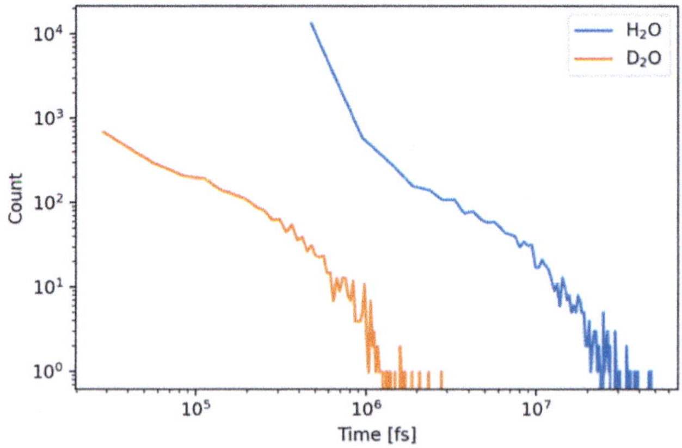

Fig. 3 Histogram of the lifetime of $CaCO_3$ clusters containing more than 3 ions for simulations in H_2O and D_2O as indicated.

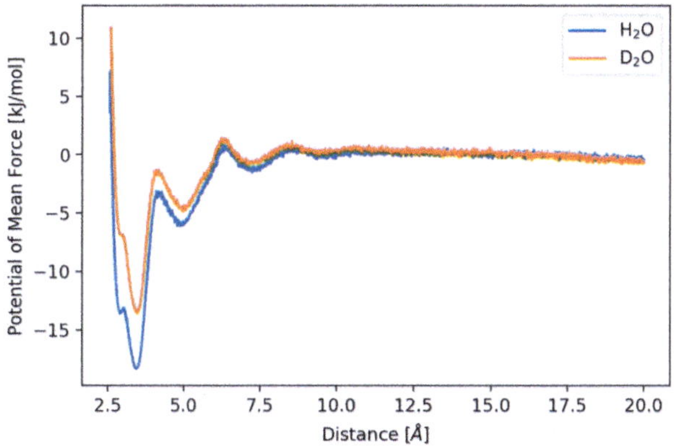

Fig. 4 Potential of mean force for the distance between calcium and carbonate ions in H_2O and D_2O as indicated.

often neglected. So as to investigate the over-simplified assumption further, an H_2O model with masses of D_2O was simulated. The corresponding 2D histogram of the radius of gyration *vs.* cluster size (Fig. S6†) reveals no structural changes between the clusters in H_2O and in H_2O with increased masses, that is, in "pseudo-D_2O". Also, the distribution of cluster sizes persists after the mass change (Fig. S7†). Only the lifetime of these clusters in "pseudo-D_2O" decreases (Fig. S8†). Thus, these simulations of "pseudo-D_2O" reveal that the above-described change in behaviour arises from the change of the charge distribution within the solvent rather than from the mass difference of the two aqueous solvents.

Titration assays of calcium carbonate formation

So as to elucidate the effects of a solvent change from light to heavy water experimentally, we performed titration experiments in different mixtures of the two aqueous solvents, as well as in pure H_2O and 99% D_2O. In brief, dilute calcium solution was added into dilute carbonate buffers, while the pH was maintained at a constant value by automatic counter-titration with dilute sodium hydroxide solution. The specific H_2O/D_2O ratio (v/v) in these solutions was the same, and adjusted to 100/0, 87.5/12.5, 75/25, 62.5/37.5, 50/50, 25/75 and 0/99 in separate repetitions. The pH of the respective mixtures was thereby adjusted to a value that corresponds to the same carbonate/bicarbonate ratio in the buffer in pure H_2O at pH 9.00 (for details see ESI,† Experimental section). This is the equivalent pH of these aqueous mixtures enabling direct comparison of the titration profiles.

Initially, the calcium carbonate ion activity product (IAP) increases linearly until it drops to a plateau during the course of the titration (Fig. 5). Due to the very slow titration, the linear parts of the curves represent equilibrated (meta)stable states.[65] The slope of the pre-nucleation part (*i.e.*, before the steep drop of the IAP) thus correlates with the equilibrium constant for ion association yielding PNCs, whereas a flatter slope points towards a larger equilibrium constant, *i.e.*, more stable ion associates. On the other hand, the post-nucleation plateau value (*i.e.*, after the steep drop of the IAP) represents the solubility of the initially precipitated phase. Note that the most soluble phase determines these solubility measurements, that is, ripening towards more stable (less soluble) forms is only visible in the titration profiles once the more soluble one has completely

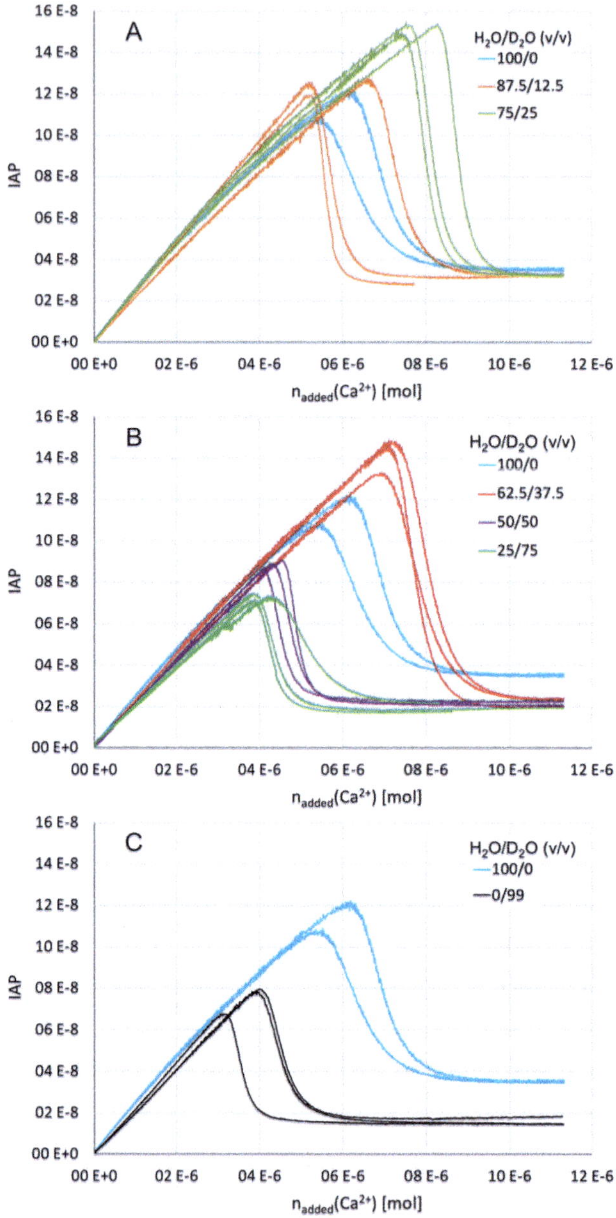

Fig. 5 Titration profiles for different H_2O and D_2O mixtures at an equivalent pH of 9.00 in pure water as indicated for (A) low heavy water contents; (B) high heavy water contents; (C) pure light *vs.* 99% heavy water. The data for pure light water is identical in panels (A), (B), and (C) so as to facilitate comparison. Individual repetitions are shown in the same respective colours, so as to illustrate reproducibility.

dissolved. In pure water, the IAP of $\sim 3.5 \times 10^{-8}$ corresponds to the solubility of proto-calcite amorphous calcium carbonate (pc-ACC).[53,65] At the same time, this IAP corresponds to the liquid–liquid binodal limit, $IAP_{binodal} = \sim 3.5 \times 10^{-8}$, as demonstrated by THz spectroscopy,[66] which can be quantitatively predicted based on eqn (3).[52] Indeed, the IAP development upon exceeding the binodal limit represents liquid–liquid coexistence, and strong evidence suggests that the initially formed solid pc-ACC is formed *via* dehydration of the dense liquid precursors.[52] In this sense, the delay between exceeding the binodal limit and the steep drop in IAP is associated with the kinetics of dehydration of phase-separated dense liquid rather than nucleation kinetics *per se*. In any case, it is obvious that the change from light to heavy water influences (i) the pre-nucleation slope, (ii) the point of nucleation, and (iii) the solubility of the initially precipitated phase. In the following, we discuss these observations individually.

Effects on pre-nucleation slope. At low D_2O contents (H_2O/D_2O ratios of 87.5/12.5 and 75/25, Fig. 5A), the pre-nucleation slope seems to be largely unaffected by the presence of heavy water, especially concerning the initial development before reaching the liquid–liquid binodal limit (the one for pure H_2O, *i.e.*, $\sim 3.5 \times 10^{-8}$). If anything, above $IAP_{binodal}$, the IAP pre-nucleation development in pure light water bends down earlier than in the solvent mixtures with low heavy water contents. This might indicate that dehydration of phase-separated dense liquid (nano)droplets is, by comparison, kinetically weakly inhibited (see below) in the presence of low amounts of heavy water.

On the other hand, there seem to be no distinct effects of the low heavy water contents on the ion association thermodynamics. However, with increasing heavy water contents (H_2O/D_2O ratios of 50/50 and 25/75, Fig. 5B), the pre-nucleation slope becomes noticeably flatter, with a minimum at 75% D_2O. In terms of standard free energy of ion pair formation, the ion associates are by *ca.* 2 kJ mol^{-1} more stable in a mixture of H_2O/D_2O of 25/75 than in pure light water. However, the trend appears to be reversed in 99% heavy water (Fig. 5C), where the pre-nucleation slope is essentially parallel to the one observed in pure light water, even though the IAP development in 99% heavy water is curved downwards in the very early stage of the experiment (*i.e.*, before the IAP values are reached that are established in the post-nucleation plateau), as opposed to pure light water. In this sense, the experimental data on ion association seems to contradict our results from computer simulations; if anything, ion associates are somewhat more stable in the presence of heavy water.

Effects on the point of nucleation. Above 12.5% D_2O, the steep drop in IAP is shifted to higher added amounts of calcium ions (and higher maximal IAPs) in comparison to pure light water as the reference (Fig. 5A), indicating that the presence of low amounts of D_2O in the solvent mixtures inhibits dehydration processes of the dense liquid precursors. A similar effect is observed in 37.5% D_2O (Fig. 5B). However, when an effect of the presence of heavy water on the stability of ion associates becomes noticeable in the 50/50 mixture (see above), the drop in the IAP occurs earlier, now indicating that dehydration of dense liquid precursors is facilitated at even higher heavy water contents—if liquid–liquid separation still occurs in these systems. Notably, this behaviour does not directly correlate with the development of the post-nucleation solubility thresholds (see below), as a distinctly more stable phase is formed initially already in 37.5% D_2O, where nucleation is still inhibited rather than promoted.

Effects on solubilities of initially precipitated phases. In the presence of low amounts of heavy water (H_2O/D_2O ratios of 87.5/12.5 and 75/25, Fig. 5A), the solubility of the initially precipitated phase decreases slightly and is now on the order of $\sim 3.3 \times 10^{-8}$. While the difference in solubility compared to the situation in pure light water is small, it is on the order of the difference of the solubilities of pc-ACC and proto-vaterite-ACC (pv-ACC),[53,65] and thus seems to be significant. Upon increasing the D_2O content from 25% to 37.5%, the solubility of the initially precipitated phase decreases further, and is then on the order of 1.7×10^{-8} to 2.3×10^{-8}, which is slightly above the solubility of vaterite in pure water at 25 °C (1.2×10^{-8}).[56] Considering the reproducibility of the measurements in different mixtures, it seems that there is no obvious trend, but rather a switch to a different behaviour. This could be due to a shift of the liquid–liquid binodal limit that is caused by a change in the solubility of the different polymorphs or the constant A (eqn (3)). Indeed, according to the theory (eqn (3)), a change in solubilities or A would have a significantly stronger effect on the locus of the liquid–liquid binodal limit than altered ion association thermodynamics. On the other hand, the dehydration and crystallization kinetics might be significantly faster in the presence of higher amounts of heavy water, too, leading to the accelerated formation of vaterite, thereby potentially missing solid ACC intermediates in the development of the IAP. Third, liquid–liquid separation as a precursor to ACC may not occur in the solvent mixtures at higher heavy water contents, and nucleation of vaterite might occur directly, *i.e.*, "classically". In order to elucidate these possibilities further, we attempted to quench the amorphous intermediates in an excess of ethanol, as established for the isolation of proto-structured ACCs in pure light water.[53,54] In H_2O, ethanol serves as a drying agent for the liquid–liquid separated dense calcium carbonate droplets, but at high contents of D_2O, unfortunately, the solubilities of sodium bi/carbonate become an issue for this isolation procedure, and we largely obtained crystalline sodium bi/carbonate in ethanol quenches in the presence of heavy water—with additional reflexes (Cu-Kα) at $\sim 24°$, $\sim 27°$ and $\sim 33°$ 2θ, which we could not assign to any compound that is in principle accessible in this system (Fig. S9†).

ATR-FTIR kinetic investigations

The titration experiments suggest that at relatively low levels of supersaturation (*i.e.*, close to the binodal limit), low amounts of heavy water may inhibit the dehydration kinetics of liquid–liquid separated states towards the formation of solid ACCs, while even higher amounts of heavy water may, in turn, accelerate precipitation and crystallisation in this region of the phase diagram—if the liquid–liquid miscibility gap still exists in D_2O containing solutions. In order to explore the precipitation kinetics from the putative spinodal regime of the phase diagram, *i.e.*, at high levels of initial supersaturation, we implemented stopped-flow mixing experiments monitored over time by ATR-FTIR (see the ESI for experimental details†). Calcium chloride (0.2 M) and sodium carbonate (0.2 M) solutions were prepared in different H_2O/D_2O mixtures (100/0; 87.5/12.5; 75/25; 50/50 and 0/100) and corresponding normalised time transients after mixing were extracted at 869 cm^{-1}, corresponding to the minimum of the second derivative of the ν_2 carbonate vibrational band. These concentrations—at least for the pure H_2O sample—lead to liquid–liquid demixing from the spinodal regime of the aqueous calcium carbonate phase diagram.[52] However, as per the theory (eqn (4)), the altered ion association thermodynamics would shift the spinodal regime to lower IAPs for intermediate heavy water contents, due to the somewhat higher ion association constant.

Indeed, the time transients (Fig. 6A) strongly differ for the different aqueous mixtures. In pure light water, we observe a steep initial increase, which gradually flattens, kinks, and then develops into a plateau. In an H_2O/D_2O mixture of 87.5/12.5, the initial increase is significantly flatter, and the transient does not seem to even come close to reaching a final plateau as observed in pure light water, within the observation time (note that due to the normalization to the plateau value, which is not reached in the experiment with 12.5% heavy water, the initial kinetics is probably the slowest in this mixture). However, a plateau is again

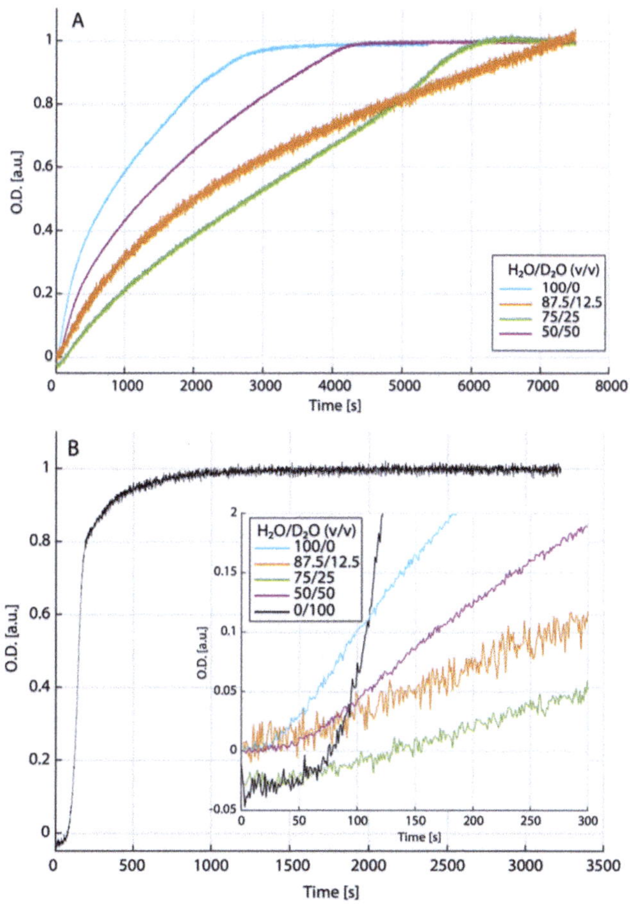

Fig. 6 Normalised transients extracted from the time development of the ν_2 carbonate vibrational band upon mixing 0.2 M calcium and carbonate solutions (liquid–liquid spinodal regime for pure light water) in different mixtures of light and heavy water as indicated (A); and in essentially pure (99.9%) heavy water (B). The inset in (B) shows a zoom into the early times of the transients shown in (A), together with the transient for heavy water.

reached in the 75/25 mixture within the observation time. In the 50/50 mixture, the kinetics are then again closest to the situation in pure water. Also qualitatively, the kinetics of calcium carbonate precipitation in the 50/50 mixture is similar to the pure light water system, while at low heavy water content (75/25), the transient exhibits rather distinct features. In almost pure heavy water (99.9%; Fig. 6B), on the other hand, the transient looks again completely distinct. Initially, there is a rather flat linear increase for *ca.* 50–70 s (inset of Fig. 6B) before the transient directly develops into a plateau value with the steepest increase and thus, fastest kinetics observed in all solvent mixtures. In the other solvent mixtures (see inset of Fig. 6B), but not pure light water, we find a similarly flat initial slope, however, the very steep increase as in 99.9% heavy water does not occur.

Only in the 50/50 mixture, there is a notable upwards kink after *ca.* 50 s, which then gradually levels off after *ca.* 150 s. In pure light water, a less steep increase than in heavy water occurs after *ca.* 25 s and we interpret this as the time required for a sufficient amount of calcium carbonate to sediment on the ATR crystal, rather than an induction time. Indeed, a similar "deadtime" is observed in all other transients. However, the extended, flat, initial linear regime in presence of heavy water could be interpreted as the presence of an induction time, which would then provide evidence that we did not probe the spinodal regime in presence of pure heavy water. Unfortunately, due to heavily convoluted carbonate and aqueous solvent bands, the single spectra (data not shown) cannot be evaluated towards an assignment of different calcium carbonate forms to the different kinetic stages in any solvent mixture, at this time.

Simulations of residence times of water molecules and water exchange

Our kinetic analyses provide evidence for distinct effects of the solvent mixtures in the binodal as well as putative spinodal regimes. In order to explore these observations further, computationally, we studied the residence time of water in the first solvation shell of the ions. The average time that a water molecule stays within the solvation shell was calculated utilizing a self-developed method by bookkeeping the ions and *via* a time correlation function (see ESI,† Experimental section, for details). The water exchange for calcium ions in heavy water is around two times slower than in light water (Table S4†). The errors observed for both methods show that the residence times of water vary substantially with a maximum value of 1.8 ns for calcium ions in light and 4.8 ns in heavy water, while the average times are on the order of experimentally determined values. Notably, previous calculations of the water residence time for calcium ions in light water with MD simulations yielded a large range of values, depending on the used force field,[67] or more precisely, on the size of ions and the water models. An influence of the box size was also found, with a smaller box leading to shorter residence times.[68] The difference between water and heavy water may occur due to the changes in mass and charge between the models. However, the change in charge from SPC to TIP3P is around +1.7%, while the change from SPC/fw to SPC/fw/HW is around +2.7%. In the case of SPC *vs.* TIP3P (GROMOS96 *vs.* GROMOS87), the residence time decreases by 10%. In the case of water to heavy water, it increases by over 100%. The increase in residence time for D_2O seems to depend on the change in mass and maybe slightly stronger ion–water interactions.

Moreover, in kinetic measurements, the mixing behaviour of the two aqueous solvents might play additional roles, *i.e.*, the dynamics of light water/heavy water exchange of the ions and potential memory effects. Mixtures of H_2O and D_2O were simulated while the issue of the formation or presence of HDO molecules was not covered. Negative controls of pure H_2O/D_2O mixtures without ions showed a rapid mixing of both previously separated phases without any form of aggregation present. The interactions of the ions with the mixed solvents were tested with two setups. On one hand, an ion was simulated in one of the solvents. Then, the ion and a sphere of water, consisting of the first and second solvation shell (Ca^{2+}: 5.4 Å; $CO_3{}^{2-}$: 6.2 Å), was cut out and placed in a box of the other solvent. On the other hand, the box was split into two equal compartments and the ion placed in either one of them. After an equilibration with position restrains, the coordination number of light and heavy water in the first shell was monitored, as shown in Fig. 7. At first glance, the behaviours of the systems seem similar, independent of the placement in H_2O or D_2O. Yet, there are minor differences between the two solvents. In the first setup, with a calcium ion in a sphere of H_2O placed in D_2O, the H_2O is completely dissolved in D_2O and the ion is exclusively surrounded by D_2O molecules during the second half of the simulation, from 5 ns after mixing onwards. When a sphere of D_2O is placed in H_2O, we observe a similar effect. However, at least one D_2O molecule detaches and reattaches repeatedly during the whole course of the simulation. In the case of the second setup, with an equal amount of both solvents present, there are on average 1.7 more D_2O molecules present around the ion than H_2O molecules.

The fluctuation of H_2O and D_2O around the ion is smaller in the case where the ion is placed directly in D_2O. Both setups show that the calcium ion is more attracted to D_2O and is thus more influenced by it in mixtures. The second observation indicates that D_2O could form a stable solvation shell upon time,

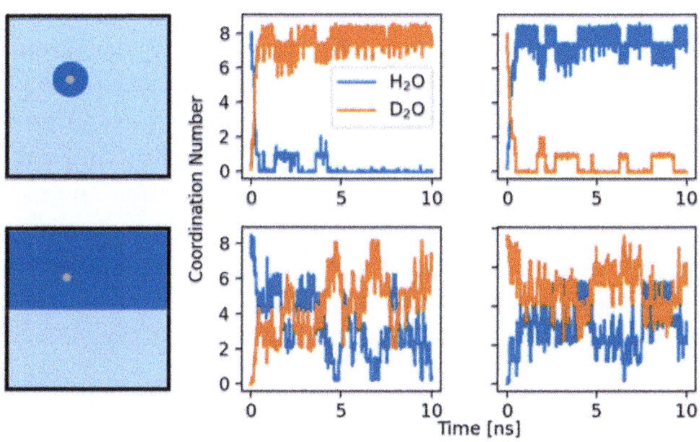

Fig. 7 Coordination number of H_2O and D_2O around a calcium ion. The rows show the results for two different setups. Top row: the ion with its first and second solvation shell is placed in the opposite solvent. Bottom row: the box is split into two compartments containing H_2O and D_2O and the ion is placed in one of them. Left: schemes of the initial setups. Middle: calcium ion initially surrounded by H_2O. Right: calcium ion initially surrounded by D_2O.

maybe with stable, chelating motifs in it. The light water exchange with this solvation shell appears to be slower. The formation of a more stable solvation shell after some induction time is a known problem in MD simulation of magnesium ions, especially in conjunction with RNA.[69–72] The better solubility of some ions and small molecules in D_2O is also known, assumed to be caused by its different compressibility and hydrophobicity.[29]

Discussion

We observed various, distinct effects of the aqueous solvents, D_2O *vs.* H_2O, on the early stages of calcium carbonate formation, both computationally and experimentally. In MD simulations of ion association in D_2O, to which end we developed a suitable force field that is also presented herein, the number of calcium carbonate clusters (PNCs) was reduced by a factor of two when compared to the situation in light water. Furthermore, the clusters in D_2O were smaller and more compact than the ones found in H_2O, and also shorter lived. This could indicate that nucleation processes in D_2O might be "more classical" as the role of PNCs would be reduced, and the cluster distribution shifted more towards single ions or ion pairs—as reflected by a *ca.* 6 times smaller ion association constant in D_2O than in H_2O. Simulations of a "pseudo-D_2O", *i.e.*, where only the mass of light water was altered, showed that rather than mass differences, the distinct charge distribution within the aqueous solvents is the key to the different behaviours. The residence time of water molecules on calcium ions is increased by a factor of 2.2–2.5 in D_2O, and there seems to be a somewhat higher attraction of calcium ions to D_2O molecules, indicating that the ion interactions are probably influenced already at low D_2O concentrations in H_2O solution. Furthermore, indications were found that the behaviour of the solvent mixtures was determined by its initial condition—*i.e.*, whether a calcium ion was dissolved in D_2O or H_2O prior to mixing. All of the computationally observed differences between the behaviours of calcium carbonate in D_2O and H_2O are eventually due to changes in water properties, ion–water interactions and cluster formation, which could even lead to different calcium carbonate structures and formation pathways.

Indeed, experimentally, we observed distinct effects of the solvent exchange as well, however, the influence on ion association has the opposite trend when compared to the computer simulations (*i.e.*, ion associates become somewhat more stable in the presence of heavy water in experiments), and the effect is only on the order of thermal energy in terms of standard free energies. Also, with increasing D_2O contents, there is no clear trend, and the behaviour somehow reverses again when approaching pure heavy water, in the binodal regime of calcium carbonate formation. This could indicate that in purest heavy water (note that for cost reasons, we employed 99% heavy water in the titration assays, but did use 99.9% heavy water in the ATR-IR-based kinetic analyses), there might actually be a (slight) decrease in stability of ion associates also in experiments. At low heavy water contents in the binodal regime of calcium carbonate formation, dehydration processes of dense liquids towards the formation of solid ACC seem to be weakly inhibited, which is in line with the computer simulations and the residence times of the water molecules on the ions. It is difficult to find information on the kinetics of the exchange of H and D between D_2O and H_2O in the literature. Mammoli *et al.*[73] assessed the lifetime of proton exchange to be on the order of $\sim$1 ms in pure water. If this holds true also for the D nucleus, at least for the order of magnitude, corresponding exchanges should be no issue at all with regards to our ATR-FTIR kinetic analysis (where timescales of seconds are discussed). Also, the exchange of H/D should be slow enough so that the computational analysis of the lifetimes of water around calcium-ions are not affected, because the interconversion should be slower than ion–water exchange.

The complex dependence of the calcium carbonate formation pathway on different light/heavy water mixtures is also reflected in kinetic measurements of calcium carbonate formation from high initial levels of supersaturation. While in pure light water, we probed the spinodal regime towards calcium carbonate liquid–liquid demixing, in the presence of heavy water, we observed initial rather flat slopes that might point towards the presence of an induction time for calcium carbonate formation at the same initial IAPs. This suggests that in the presence of heavy water, we might still have probed the kinetics of precipitation of calcium carbonate in the binodal regime. At this point, it remains unclear whether the locus of the liquid–liquid miscibility gap was altered in unpredictable ways, if it even still exists, or whether indeed, calcium carbonate nucleation proceeded "more classically". The latter might explain the observation of Lundager Madsen[42] that there was no effect of magnetic fields on calcium carbonate formation in heavy water (as opposed to light water), since Coey's theory[74] on the effects of magnetic fields on scaling relies on the role of PNCs. The found influence of the solvent and solvent mixture on the point of nucleation of the initial solid is interesting in this regard, and low amounts of heavy water in H_2O seem to inhibit the dehydration processes towards formation of solid ACC. The reverse observed behaviour at higher heavy water contents (*i.e.*, promoting nucleation) might indeed reflect that a distinct pathway is populated under these conditions. Thus, above *ca.* 40% D_2O, vaterite may form directly rather than *via* liquid–liquid separation and solid ACC. After all, this could occur even if liquid–liquid separation still happens in these solvent mixtures, *i.e.*, when the kinetics of dehydration of "non-classical" intermediates is reduced and the classical rate of direct nucleation is increased (potentially, due to lower interfacial free energies of crystals in heavy water, *cf.* the introductory section). This could be the case to an extent that the classical pathway becomes significantly more populated than the "non-classical" one *via* PNCs, once heavy water contents are high enough, *i.e.*, above *ca.* 40% heavy water. However, the kinetics of ACC formation might, on the other hand, be accelerated in these mixtures, so they are not observable in our experiments, and unfortunately, we were not able to isolate the intermediate states due to solubility issues of sodium carbonates in these aqueous solvent mixtures. Notably, it cannot be excluded that new, distinct forms of ACC or even crystalline calcium carbonate become accessible—for which there are at least some indications (Fig. S9†).

When comparing the results from experiment and computations, it has to be kept in mind that the latter did not account for HDO molecules, which might actually be an important species for explaining the experimentally observed behaviour, especially at low heavy water contents. Note that in these mixtures, no significant fractions of D_2O are expected to be present, but they would rather be mixtures of mostly H_2O and HDO. The changed trends observed in the titration assays may then correlate with the onset of the presence of significant D_2O contents within H_2O/HDO mixtures. Also, the unclear trends upon changing the compositions of the mixtures in a systematic fashion indicate that there might be a complex, underlying phase behaviour of the water systems, which seems to be rather complicated already in pure light water alone,[75] and which might in turn distinctly affect the calcium carbonate formation pathway.

Conclusions

In our opinion, further studies of the differences in the behaviours of aqueous calcium carbonate systems in D_2O and H_2O, and mixtures thereof, open up unprecedented possibilities for obtaining a better understanding of nucleation and crystallisation mechanisms in the future, and we believe that also the behaviour of other mineral systems in such different aqueous systems should be explored. However, our study provides a warning towards usage of different experimental data, too. Many experiments use D_2O as substitute for H_2O, for example, so as to get rid of unwanted spectral backgrounds, or for other reasons (also see the introduction section). However, in terms of mineralisation, our data unambiguously demonstrates that the pathway can be considerably altered, depending on the specific solvent (mixture). Notably, experimental data from heavy water systems are sometimes used for the creation of H_2O-based (!) models in computer simulations, where the original solvent may not even be mentioned in citations, and mechanisms are then erroneously treated to arise from light water properties. Thus, considering the distinct differences between the waters might even reconcile previous, seemingly contradicting results.

Author contributions

Conceptualization: DG, KH, CP; data curation: JTA, MK, CP; formal analysis: JTA, MK, DG; funding acquisition: CP, KH, DG; investigation: JTA, MK; methodology: JTA, MK, CP, KH, DG; project administration: CP, KH, DG; resources: CP, KH, DG; software: JTA, MK, CP; supervision: CP, KH, DG; validation: JTA, MK; visualization: JTA, MK, DG; writing – original draft: MK, DG; writing – review & editing: JTA, MK, CP, KH, DG.

Conflicts of interest

There are no conflicts to declare.

Acknowledgements

We gratefully acknowledge funding by the German Research Foundation (DFG) through SFB1214 (projects A02 and A04). The authors acknowledge support by the state of Baden-Württemberg through bwHPC and the German Research Foundation (DFG) through grant INST 35/1134-1 FUGG.

Notes and references

1 J. J. Katz, *Am. Sci.*, 1960, **48**, 544–580.
2 G. C. Kresheck, H. Schneider and H. A. Scheraga, *J. Phys. Chem.*, 1965, **69**, 3132–3144.

3 S. Scheiner and M. Čuma, *J. Am. Chem. Soc.*, 1996, **118**, 1511–1521.
4 G. Zuber, S. J. Prestrelski and K. Benedek, *Anal. Biochem.*, 1992, **207**, 150–156.
5 K. Hauser, C. Krejtschi, R. Huang, L. Wu and T. A. Keiderling, *J. Am. Chem. Soc.*, 2008, **130**, 2984–2992.
6 M. Jawurek, J. Dröden, B. Peter, C. Glaubitz and K. Hauser, *Chem. Phys.*, 2018, **512**, 53–61.
7 A. Krüger, A. Bürkle, K. Hauser and A. Mangerich, *Nat. Commun.*, 2020, **11**, 2174.
8 H. H. Mantsch, H. Saitô and I. C. P. Smith, *Prog. Nucl. Magn. Reson. Spectrosc.*, 1977, **11**, 211–272.
9 G. Némethy and H. A. Scheraga, *J. Chem. Phys.*, 1964, **41**, 680–689.
10 T. Clark, J. Heske and T. D. Kühne, *ChemPhysChem*, 2019, **20**, 2461–2465.
11 A. Zeidler, P. S. Salmon, H. E. Fischer, J. C. Neuefeind, J. M. Simonson and T. E. Markland, *J. Phys.: Condens. Matter*, 2012, **24**, 284126.
12 R. L. Cook, F. C. D. Lucia and P. Helminger, *J. Mol. Spectrosc.*, 1974, **53**, 62–76.
13 A. K. Soper and C. J. Benmore, *Phys. Rev. Lett.*, 2008, **101**, 353.
14 J. H. Root, P. A. Egelstaff and A. Hime, *Chem. Phys.*, 1986, **109**, 437–453.
15 B. Tomberli, C. J. Benmore, P. A. Egelstaff, J. Neuefeind and V. Honkimäki, *J. Phys.: Condens. Matter*, 2000, **12**, 2597–2612.
16 R. T. Hart, C. J. Benmore, J. Neuefeind, S. Kohara, B. Tomberli and P. A. Egelstaff, *Phys. Rev. Lett.*, 2005, **94**, 047801.
17 R. A. Kuharski and P. J. Rossky, *J. Chem. Phys.*, 1985, **82**, 5164–5177.
18 M. N. Rodnikova, *J. Mol. Liq.*, 2007, **136**, 211–213.
19 *CRC Handbook of Chemistry and Physics*, ed. D. R. Lide, 89th edn, Taylor & Francis Ltd, 2019.
20 M. Chaplin, https://water.lsbu.ac.uk/water/, accessed July 2, 2020.
21 R. A. Robinson, M. Paabo and R. G. Bates, *J. Res. Natl. Bur. Stand., Sect. A*, 1969, **73A**, 299.
22 M. Ceriotti, W. Fang, P. G. Kusalik, R. H. McKenzie, A. Michaelides, M. A. Morales and T. E. Markland, *Chem. Rev.*, 2016, **116**, 7529–7550.
23 J. Curry and Z. Z. Hugus, *J. Am. Chem. Soc.*, 1944, **66**, 653–655.
24 M. Nakamura, K. Tamura and S. Murakami, *Thermochim. Acta*, 1995, **253**, 127–136.
25 D. Eisenberg and W. Kauzmann, *The Structure and Properties of Water*, Oxford University Press, 2005.
26 A. Streitwieser and H. S. Klein, *J. Am. Chem. Soc.*, 1963, **85**, 2759–2763.
27 M. M. Lopez and G. I. Makhatadze, *Biophys. Chem.*, 1998, **74**, 117–125.
28 E. Wilhelm, R. Battino and R. J. Wilcock, *Chem. Rev.*, 1977, **77**, 219–262.
29 G. Hummer, S. Garde, A. E. García and L. R. Pratt, *Chem. Phys.*, 2000, **258**, 349–370.
30 G. Graziano, *J. Chem. Phys.*, 2004, **121**, 1878–1882.
31 B. Kuhlman and D. P. Raleigh, *Protein Sci.*, 1998, **7**, 2405–2412.
32 G. I. Makhatadze, G. M. Clore and A. M. Gronenborn, *Nat. Struct. Biol.*, 1995, **2**, 852–855.
33 R. Guzzi, L. Sportelli, C. L. Rosa, D. Milardi and D. Grasso, *J. Phys. Chem. B*, 1998, **102**, 1021–1028.
34 G. C. Kresheck, *J. Am. Chem. Soc.*, 1998, **120**, 10964–10969.
35 Y. R. E. Silva and J. R. Grigera, *RSC Adv.*, 2015, **5**, 70005–70009.
36 H. S. Taylor, E. R. Caley and H. Eyring, *J. Am. Chem. Soc.*, 1933, **55**, 4334–4335.

37 E. C. Noonan, *J. Am. Chem. Soc.*, 1948, **70**, 2915–2918.
38 A. A. Sunier and J. Baumbach, *J. Chem. Eng. Data*, 1976, **21**, 335–336.
39 M. Jelińska-Kazimierczuk and J. Szydłowski, *J. Solution Chem.*, 1996, **25**, 1175–1184.
40 J. Greyson and H. Snell, *J. Phys. Chem.*, 1969, **73**, 3208–3214.
41 J. Lee, N. D. Chasteen, G. Zhao, G. C. Papaefthymiou and S. M. Gorun, *J. Am. Chem. Soc.*, 2002, **124**, 3042–3049.
42 H. E. Lundager Madsen, *J. Cryst. Growth*, 2004, **267**, 251–255.
43 M. Raudino, F. Sarri, D. Tatini, M. Ambrosi, G. D. Aloisi, B. W. Ninham, L. Dei and P. Lo Nostro, *J. Solution Chem.*, 2020, **49**, 289–305.
44 P. D. Morris, I. J. McPherson, G. N. Meloni and P. R. Unwin, *Phys. Chem. Chem. Phys.*, 2020, **22**, 22107–22115.
45 D. Gebauer, *Minerals*, 2018, **8**, 179.
46 J. J. De Yoreo and P. G. Vekilov, *Rev. Mineral. Geochem.*, 2003, **54**, 57–93.
47 R. Demichelis, P. Raiteri, J. D. Gale, D. Quigley and D. Gebauer, *Nat. Commun.*, 2011, **2**, 590.
48 G. Jones and W. A. Ray, *J. Chem. Phys.*, 1937, **5**, 505–508.
49 D. Gebauer, M. Kellermeier, J. D. Gale, L. Bergström and H. Cölfen, *Chem. Soc. Rev.*, 2014, **43**, 2348–2371.
50 D. Gebauer and S. E. Wolf, *J. Am. Chem. Soc.*, 2019, **141**, 4490–4504.
51 A. F. Wallace, L. O. Hedges, A. Fernandez-Martinez, P. Raiteri, J. D. Gale, G. A. Waychunas, S. Whitelam, J. F. Banfield and J. J. De Yoreo, *Science*, 2013, **341**, 885–889.
52 J. T. Avaro, S. L. P. Wolf, K. Hauser and D. Gebauer, *Angew. Chem., Int. Ed.*, 2020, **59**, 6155–6159.
53 D. Gebauer, P. N. Gunawidjaja, J. Y. P. Ko, Z. Bacsik, B. Aziz, L. J. Liu, Y. F. Hu, L. Bergström, C.-W. Tai, T.-K. Sham, M. Edén and N. Hedin, *Angew. Chem., Int. Ed.*, 2010, **49**, 8889–8891.
54 M. Farhadi Khouzani, D. M. Chevrier, P. Zhang, N. Hedin and D. Gebauer, *Angew. Chem., Int. Ed.*, 2016, **55**, 8117–8120.
55 J. H. E. Cartwright, A. G. Checa, J. D. Gale, D. Gebauer and C. I. Sainz-Díaz, *Angew. Chem., Int. Ed.*, 2012, **51**, 11960–11970.
56 L. Brečević and A. E. Nielsen, *J. Cryst. Growth*, 1989, **98**, 504–510.
57 D. J. Kelly, N. Clark, M. Zhou, D. Gebauer, R. V. Gorbachev and S. J. Haigh, *Adv. Mater.*, 2021, **33**, 2100668.
58 P. J. M. Smeets, A. R. Finney, W. J. E. M. Habraken, F. Nudelman, H. Friedrich, J. Laven, J. J. De Yoreo, P. M. Rodger and N. A. J. M. Sommerdijk, *Proc. Natl. Acad. Sci. U. S. A.*, 2017, **114**, E7882–E7890.
59 J. Ihli, W. C. Wong, E. H. Noel, Y.-Y. Kim, A. N. Kulak, H. K. Christenson, M. J. Duer and F. C. Meldrum, *Nat. Commun.*, 2014, **5**, 3169.
60 E. Ruiz-Agudo, C. V. Putnis and A. Putnis, *Chem. Geol.*, 2014, **383**, 132–146.
61 P. Raiteri and J. D. Gale, *J. Am. Chem. Soc.*, 2010, **132**, 17623–17634.
62 H. Du and E. Amstad, *Angew. Chem., Int. Ed.*, 2020, **59**, 1798–1816.
63 M. Kellermeier, P. Raiteri, J. K. Berg, A. Kempter, J. D. Gale and D. Gebauer, *ChemPhysChem*, 2016, **17**, 3535–3541.
64 P. Raiteri, R. Demichelis and J. D. Gale, *J. Phys. Chem. C*, 2015, **119**, 24447–24458.
65 D. Gebauer, A. Völkel and H. Cölfen, *Science*, 2008, **322**, 1819–1822.

66 F. Sebastiani, S. L. Wolf, B. Born, T. Q. Luong, H. Cölfen, D. Gebauer and M. Havenith, *Angew. Chem., Int. Ed.*, 2016, **56**, 490–495.
67 D. Di Tommaso and N. H. de Leeuw, *J. Phys. Chem. B*, 2008, **112**, 6965–6975.
68 J. Anwar and P. K. Boateng, *J. Am. Chem. Soc.*, 1998, **120**, 9600–9604.
69 N. M. Fischer, M. D. Polêto, J. Steuer and D. van der Spoel, *Nucleic Acids Res.*, 2018, **46**, 4872–4882.
70 O. Allnér, L. Nilsson and A. Villa, *J. Chem. Theory Comput.*, 2012, **8**, 1493–1502.
71 J. P. Larentzos and L. J. Criscenti, *J. Phys. Chem. B*, 2008, **112**, 14243–14250.
72 N. Schwierz, *J. Chem. Phys.*, 2020, **152**, 224106.
73 D. Mammoli, N. Salvi, J. Milani, R. Buratto, A. Bornet, A. A. Sehgal, E. Canet, P. Pelupessy, D. Carnevale, S. Jannin and G. Bodenhausen, *Phys. Chem. Chem. Phys.*, 2015, **17**, 26819–26827.
74 J. M. D. Coey, *Philos. Mag.*, 2012, **92**, 1–9.
75 O. Mishima, *Proc. Jpn. Acad., Ser. B*, 2010, **86**, 165–175.

Faraday Discussions

ROYAL SOCIETY OF CHEMISTRY

PAPER

Multiple pathways in NaCl homogeneous crystal nucleation†

Aaron R. Finney * and Matteo Salvalaglio *

Received 15th November 2021, Accepted 16th December 2021
DOI: 10.1039/d1fd00089f

NaCl crystal nucleation from metastable solutions has long been considered to occur according to a single-step mechanism where the growth in the size and crystalline order of the emerging nuclei is simultaneous. Recent experimental observations suggest that significant ion–ion correlations occur in solution and that NaCl crystals can emerge from disordered intermediates which is seemingly at odds with this established view. Here, we performed biased and unbiased molecular dynamics simulations to analyse and characterise the pathways to crystalline phases from solutions far into the metastable region. We find that large liquid-like NaCl clusters emerge as the solution concentration is increased and a wide distribution of crystallisation pathways are observed with two-step nucleation pathways—where crystalline order emerges in dense liquid NaCl regions—being more dominant than one-step pathways to phase separation far into the metastable region. Analyses of cluster size populations and the ion pair association constant show that these clusters are transient, unlike the thermodynamically stable prenucleation cluster solute species that were suggested in other mineralising systems. A Markov state model was developed to analyse the mechanisms and timescales for nucleation from unbiased molecular dynamics trajectories in a reaction coordinate space characterising the dense regions in clusters and crystalline order. This allowed calculation of the committor probabilities for the system to relax to the solution or crystal states and to estimate the rate of nucleation, which shows excellent agreement with literature values. From a fundamental nucleation perspective, our work highlights the need to extend the attribute 'critical' to an ensemble of clusters which can display a broad range of structures and include sizeable disordered domains depending upon the reaction conditions. Moreover, our recent simulation studies demonstrated that carbon surfaces catalyse the formation of liquid-like NaCl networks which, combined with the observations here, suggests that alternative pathways beyond the single-step mechanism can be exploited to control the crystallisation of NaCl.

Thomas Young Centre and Department of Chemical Engineering, University College London, London WC1E 7JE, UK. E-mail: a.finney@ucl.ac.uk; m.salvalaglio@ucl.ac.uk

† Electronic supplementary information (ESI) available: Additional figures. See DOI: 10.1039/d1fd00089f

Introduction

Many pathways to crystals have been observed in a range of systems that *seemingly* deviate from the single-step crystal nucleation mechanism from a parent solution phase that is predicted by classical nucleation theory (CNT).[1,2] It was shown some time ago in simulations of spherical particles with short-ranged interactions that, under appropriate conditions, liquid-like intermediates precede the formation of crystals and the rate for crystal nucleation increases.[3] Experiments were later presented that support these observations in systems containing proteins[4,5] and small organic molecules.[6,7] The nucleation of crystals from supersaturated solutions was therefore described as 'two-step'[8] in light of findings such as these.

Evidence was presented for hydrous mineral dense liquid phases which precede the precipitation of amorphous solid and crystalline phases at room temperature,[9–14] demonstrating how multi-step crystallisation pathways are a common feature in a range of crystallising systems. In principle, this observation does not negate CNT as an adequate framework to predict the rates for phase separation in mineralising solutions.[15,16] Indeed, it should be noted at the outset that a two-step pathway can be described using thermodynamic concepts consistent with CNT[17] adopting a core–shell composite cluster model.[18,19] Furthermore, observation of amorphous intermediates alone does not confirm their direct role in crystallisation *i.e.*, crystal nuclei could form in a single-step *via* stochastic density fluctuations of the dissociated ions in the surrounding lean solution in quasi-equilibrium with amorphous states. An alternative to the pathway described by CNT posits that liquid-like microscopic associates are *thermodynamically stable* species with respect to dissociated ions and are present even in undersaturated solutions; these were described as 'prenucleation clusters' on the basis that they purportedly represent the solute species directly involved in nucleation.[20–23] The prenucleation cluster pathway was described as a paradigmatic example of 'nonclassical' nucleation because here, aggregation of the stable clusters is the limiting step to phase separation.[14,20,24]

In NaCl(aq), which is arguably the simplest mineralising solution, computer simulations have been and continue to be instrumental for our understanding of crystal nucleation.[25–35] Molecular dynamics simulations using the classical force field of Joung and Cheatham[36] identified that the nominal equilibrium molality for a saturated NaCl(aq) solution at ambient temperature and pressure was 3.7 m (ref. 37–42) (where m refers to mol kg^{-1} units) and the limit of solution stability was identified at 15 m.[26] This provides a relatively wide range of concentrations where the solution is metastable and crystal nucleation is favoured but does not occur spontaneously.

In metastable NaCl(aq) solutions at $T = 298$ K, crystal nucleation rates were determined using seeded molecular dynamics simulations and classical nucleation theory (CNT).[27,28,34] The rates calculated by Lamas *et al.* were found to match well with those calculated from forward flux sampling (FFS), with both sets of computed rates deviating from the experimentally determined values.[25,34] Zimmermann *et al.* showed that the rates from seeded MD simulations can be aligned to the experimental data by adopting suitable criteria for the identification of ions belonging to the crystal phase.[28] These results thus indicate that CNT provides a consistent model for determining the nucleation kinetics in metastable

NaCl(aq) solutions. At the limit of solution stability, however, simulations identified that, before crystal nucleation, NaCl clusters with minimal crystalline order display large fluctuations in size.[26] This informed a two-step description for nucleation of NaCl at and beyond the spinodal, which marks the transition for a change in the type of nucleation mechanism.[26,34] The recent experimental observation and characterisation of hydrated dense liquid NaCl domains in supersaturated solutions puts this schematic picture into question. In ref. 43, Hwang *et al.* measured long-lived liquid-like nanometre-sized ion domains in levitated droplets by *in situ* X-ray and Raman spectroscopy, and further supported their findings with results from molecular dynamics (MD) simulations. The formation of these domains at high supersaturations was described as a possible first step in a two-step nucleation process.

Direct observation of amorphous NaCl intermediates to crystal nucleation was obtained in studies of NaCl(aq) confined to aminated carbon nanotubes performed using transmission electron microscopy after removal of the solvent from the samples.[44] Before the nucleation of a nanocrystal, disordered NaCl clusters emerge and dissipate multiple times. The nucleation step involved the formation of a $(NaCl)_{50}$ nanocrystal from a semi-ordered precursor. While this process appears compatible with a two-step mechanism, the drying of the samples *in vacuo*, presence of the (vibrating) carbon nanotube wall and confinement means that the mechanism exposed here is not directly applicable to the case of crystal nucleation in metastable bulk solutions. Amorphous NaCl nanoparticles have been stabilised over long times by spray–drying;[45] the mechanism of nanoparticle formation here is not clear and, given the method of preparation, kinetically arrested amorphous structures could well result from spinodal decomposition.

Here, we employed swarms of MD simulations, seeded MD simulations and swarms of well-tempered metadynamics[46] simulations to study the pathways associated with crystal nucleation in highly supersaturated, metastable NaCl(aq) solutions at room temperature and pressure. We analyse and discuss speciation in solution and the role of liquid-like clusters in order to characterise nucleation mechanisms in the context of those described above and identify the signature for multi-step nucleation in this regime.

Computational methods

MD simulations of solutions

A series of MD simulations were performed to investigate the structure and dynamics of ions in bulk NaCl(aq) solutions. The molality of these solutions was 2, 3, 4, 5.1, 6.1, 7.1, 8.1, 9, 10, 11.1, 12.5, 13.7, 15 and 16 *m* which provided supersaturation ratios, $S = 0.5$–4.3, defined as $S = b(\mathrm{NaCl})/b^{\mathrm{eq}}(\mathrm{NaCl})$ and where $b^{\mathrm{eq}}(\mathrm{NaCl}) = 3.7\ m$.[42] At the lowest concentrations (2–5 *m*) ions and 4000 water molecules were randomly placed into a cubic simulation cell; the number of NaCl ion pairs increased from 145 to 370 over this molality range. In the remaining simulations, the number of ions was fixed at 370 NaCl and the number of water molecules decreased from 4000 to 1280 to achieve the desired molality. Ions and water molecules were assigned random initial positions (ensuring no overlap between atoms) in a cubic simulation cell. A short 0.5 ns simulation was then performed in the (N, p, T) ensemble to relax the simulation cell volume and system potential energy, before a minimum of 10 ns of production runs were performed at each *S*. In order to account for slower relaxation times at higher concentrations the simulation time was increased to 20 ns beyond 9 *m* and 50 ns beyond 11.1 *m*. In this set of simulations, 13.7 *m* represents a special case as it is at this molality where nucleation pathways were evaluated and, hence, the MD simulation time here was substantially extended, reaching 300 ns.

Simulations at $S = 3.7$

Two approaches were taken to prepare input configurations for simulations at $b(\mathrm{NaCl}) = 13.7\ m$ ($S = 3.7$) initiated from either a homogeneous solution phase (solution) or crystalline seeds (seeded) in solution. In the first case, cubic simulation cells were randomly populated with ions and water molecules as described above; ten configurations in total were prepared in this manner. For seeded simulations, crystalline seeds were extracted from a rock salt supercell. Pseudo-spherical crystals with diameter $d = 0.6$–2.4 nm (in 0.2 nm increments) and containing $N = 7$ to 341 ions were cut from the bulk crystal where no constraint was placed on the charge of the resulting nanocrystals. The smallest seed was comprised of an octahedral $[\mathrm{NaCl_6}]^+$ complex and, hence, carried a high charge density. The absolute excess charge density (ρ^{ex}) in seeds decayed rapidly as the number of ions in the seed (N) increased, and this can be reasonably approximated by $\rho^{\mathrm{ex}} = 1.4N^{-0.6}e$ per atom. The crystalline seeds were subsequently immersed in a solution such that $b(\mathrm{NaCl}) = 13.7\ m$ according to the total number of ions the simulation box. All systems were relaxed during 0.5 ns (N, p, T) MD simulations in which the ions in crystalline seeds were fixed to their initial positions. This was followed by 300 ns production runs, where all ions were mobile except for one cation at the centre of the crystal which was fixed to the centre of the simulation cell. These initial configurations (both solutions and seeds) were also used to perform well-tempered metadynamics simulations, as described in the following section.

Simulation details

In line with other recent simulation studies of NaCl nucleation,[25–28] ions in solution were modelled using the Joung and Cheatham force field[36] which adopts the SPC/E model[47] for water. The geometry of water molecules was constrained using the LINCS algorithm.[48] MD simulations were performed using GROMACS 2018.6 (ref. 49) and the leapfrog time integration algorithm with a 1 fs timestep. The temperature and pressure of the system were held constant at 298 K and 1 bar, respectively, within statistical fluctuations using the Bussi–Donadio–Parrinello thermostat[50] and the barostat of Berendsen *et al.*[51] Particle Mesh Ewald summation[52] was adopted to compute the energies and forces on atoms arising from electrostatic interactions. As three-dimensional periodic boundary conditions were used throughout, the real-space contributions to the Ewald sum were computed for atoms within 0.9 nm, and Lennard-Jones interactions were truncated at 0.9 nm with a dispersion correction added to the energies of short-range intermolecular interactions.

Collective variables

We analysed pathways from solutions to crystals on a two-dimensional reaction coordinate space characterised by the collective variables (CVs) n and $n(q6)$ which quantify the total size of high ion density regions and crystalline regions in NaCl clusters, respectively. These calculations were performed using PLUMED (version 2.5).[53] Here, we determined the coordination between ions according to a smoothly varying geometric criterion:

$$f = \frac{1 - \left(\frac{r_{ij}}{r_0}\right)^p}{1 - \left(\frac{r_{ij}}{r_0}\right)^q} \tag{1}$$

where $p = 6$, $q = 12$, r_{ij} is the distance between two ions i and j and $r_0 = 0.38$ nm. The total coordination number for i was then calculated by shifting and stretching f_i, such that the value of coordination goes smoothly to zero at $r_{\max} = 1$ nm:

$$c_i = \frac{1}{N_j} \sum_{j=1}^{N_j} \frac{f(r_{ij}) - f(r_{\max})}{1 - f(r_{\max})} \tag{2}$$

where the sum runs over all j ions with opposite charge to i. A variable s is then defined as

$$s_i = \begin{cases} 1, & \text{if } c_i \geq 5 \\ 0, & \text{otherwise} \end{cases} \tag{3}$$

where a smooth truncation, with similar functional form to eqn (1), is applied to the continually varying c_i to determine the value of s_i. The sum over s provides a measure of the number of ions which are significantly dehydrated, due to coordination to at least five other ions of opposite charge:

$$n = \left(\sum_{i=1}^{N_i} s_i + \sum_{j=1}^{N_j} s_j \right)^{1/3} \tag{4}$$

We take the cubic root over the sum in eqn (4) to obtain a CV which scales approximately linearly with the radii of clusters emerging in solution—assuming that clusters have a relatively uniform ionic density throughout their volume—and to expand the reaction coordinate space in the region where the smallest clusters are represented.

The spatial distribution of s, $\xi(s)$, was calculated in order to identify ions as a function of distance, r, from the centre of the cubic simulation cell:

$$s^{\text{sph}} = \xi(s)\left(1 - \tanh\left(\frac{r}{r_0}\right)\right) \tag{5}$$

where r_0 is 1.5 nm. n^{sph} was calculated according to this equation during biased sampling simulations (see below) to limit the number of nucleation events that occur simultaneously in the simulations.

Together with CVs n and n^{sph}, which inform about regions of high local ion density, we also determined the degree of crystallinity in these regions. To this aim, we calculated a local average of the sixth order Steinhardt bond-orientational order parameter[54] defined according to:

$$S_i = \frac{\sum_j f(r_{ij}) \sum_{m=-6}^{6} q_{6m}^*(i) q_{6m}(j)}{\sum_j f(r_{ij})} \tag{6}$$

where f is provided in eqn (1) and, as for s_i, i and j represent ions with opposite sign of charge. The components of the complex $q6$ vector are calculated using the spherical harmonics Y_{6m}:

$$q6(i) = \frac{\sum_j f(r_{ij}) Y_{6m}(\mathbf{r}_{ij})}{\sum_j f(r_{ij})} \tag{7}$$

A Gaussian filter was then applied to identify ions with local q_6 symmetry that matched the rock salt structure:

$$\sigma_i = \begin{cases} 1, & \text{if } S_i \geq 0.55 \\ 0, & \text{otherwise} \end{cases} \tag{8}$$

The cut-off of 0.55 was chosen in this function as it correctly identifies ions at the surface of a bulk NaCl crystal as belonging to a crystalline configuration. Finally, eqn (4) and (5) are applied to calculate $n(q6)$ and $n^{\text{sph}}(q6)$.

Well-tempered metadynamics

Well-tempered metadynamics[46] was adopted in this work to enhance the sampling of local fluctuations of the ion density by biasing n^{sph} on the fly during MD simulations using PLUMED.[53] During the simulations, an artificial bias potential, V, was evolved over time by the iterative deposition of Gaussian contributions to V every 1 ps with $\sigma_n = 0.01$, initial height, $w_n = 2.5$ kJ mol^{-1}, and bias factor, $\gamma = 200$. The bias in n was stored on a grid with limits -1 to 10 (bin width $= 0.002$). Metadynamics simulations were performed for around 500 ns initiated from twenty independent configurations obtained from either a solution or crystal seed in solution at $S = 3.7$, providing a total of $\sim$10 μs of simulation time.

Relative free energies between states can be calculated according to:

$$F = -k_\text{B} T \ln(\mathbf{p}) \tag{9}$$

where k_B is Boltzmann's constant, T is temperature and $\mathbf{p}$ is the probability density of states in the reaction coordinate space. Metadynamics enhances the sampling of n^{sph}, but translation of the NaCl particle outside of the biased sphere can lead to inaccurate estimates for the probability of n^{sph} states. It was therefore necessary to calculate the free energies between states in a space characterised by n and $n(q6)$. To do this, we first calculate the biased probability density ($\mathbf{p}^b(n^{\text{sph}}, n, n(q6))$) by post-processing the simulation trajectories. We also calculated $\mathbf{p}(n^{\text{sph}}, n, n(q6))$ by reweighting the biased probability density according to the weights, $\exp\left(\frac{V(n^{\text{sph}})}{k_\text{B}T}\right)$. Metadynamics simulations that failed to sample both the solution and crystalline states and transition states between them were neglected, and the remaining 14 data sets were aligned such that the average $\mathbf{p}(n^{\text{sph}}, n, n(q6))$ for the solution phase was equal within noise. The relative free energy was then calculated according to the weighted average:

$$F = \frac{\sum_i \mathbf{p}^b(-k_B T \ln(\mathbf{p}))}{\sum_i \mathbf{p}^b} \tag{10}$$

where the sum runs over all i simulations that met the sampling criteria described above. $F(n, n(q6))$ was established by taking a thermodynamic average according to:

$$F(n,\ n(q6)) = -k_B T \ln \int_{n^{\text{sph}}} \exp\left(\frac{-F(n^{\text{sph}},\ n,\ n(q6))}{k_B T}\right) \mathrm{d}n^{\text{sph}} \tag{11}$$

Umbrella Sampling

To investigate ion association, we performed Umbrella Sampling simulations in solutions approaching the limit of infinite dilution. Here, one Na^+, one Cl^- and 4000 water molecules were inserted into a cubic simulation cell which was relaxed in (N, p, T) simulations for 0.5 ns. A total of 20 (N, p, T) simulations (windows) followed, where the distance between ions was restrained at $r^0_{\text{Na-Cl}} = 0.25$–1 nm in 0.025 nm increments by the introduction of a time-independent harmonic bias potential:

$$V_{\text{US}} = \frac{k}{2}\left(r_{\text{Na-Cl}} - r^0_{\text{Na-Cl}}\right)^2 \tag{12}$$

where k was 900 kJ mol^{-1} for $r_{\text{Na-Cl}} < 0.4$ nm and 500 kJ mol^{-1}, otherwise. Simulations were performed for 5.5 ns, with the initial 0.5 ns discarded in any subsequent analyses. With this setup, a significant overlap of the probability densities in the reaction coordinate between adjacent windows was obtained, enabling a well-converged calculation of the potential of mean force, obtained with the weighted histogram analysis method.[55]

Ion clusters and diffusion

A depth first search algorithm[56] was used to determine the cluster size distribution according to a continuous but sharp definition of ion coordination in the first sphere using eqn (2) and (1) where $r_0 = 0.355$ nm, $r_{\max} = 0.36$ nm, $p = 6$ and $q = 12$. In addition, self-diffusion coefficients, D, for ions and water were calculated from the mean squared displacement of ions according to $D = \lim_{t\to\infty} \mathrm{d}\langle(\mathbf{r}(t) - \mathbf{r}(0))^2\rangle/6\mathrm{d}t$, where $\mathbf{r}(t)$ are the atom positions at time t. A correction was applied to account for the finite size of the simulation cell[57] and here the shear viscosity of the adopted water model was taken from ref. 58.

Markov state model construction

All unbiased simulations, totalling more than 20 μs, were used to inform Markov State Models (MSMs) in the reaction coordinate space defined by CVs n and $n(q6)$, as well as by their localised counterparts n^{sph} and $n^{\text{sph}}(q6)$. MSM construction and analyses were implemented using PyEMMA 2.5.6.[59] The two-dimensional reaction coordinate spaces were partitioned into discrete sets of states *via* regular space clustering,[60,61] leading to sets of approximately 100 states in both (n, $n(q6)$) and (n^{sph}, $n^{sph}(q6)$) which are fully connected to one another *via* intermediate states. MSMs constructed from these sets showed a convergent behavior for the slowest implied timescale for lag-times $\tau > 50$ ps (see Fig. S3 and S4 in ESI†).[62] Bayesian MSMs were therefore employed to estimate the average and 95% confidence interval of the slowest implied timescale using $\tau = 100$ ps, and to estimate the equilibrium probability distribution.[63] Using discrete transition path theory (TPT),[64,65] the MSMs were used to compute the committor probability, as well as the state-to-state probability flux and the mean first passage time associated with the nucleation process. This analysis, further described in the Results and discussion section, provides quantitative information on the pathways connecting the liquid and crystal states.

Results and discussion

Solution speciation

Before discussing NaCl crystallisation pathways from solution, we analyse the ionic species characterising a stable ($S \leq 1$) and metastable ($S > 1$) solution at the steady state, *i.e.* for timescales significantly shorter than the characteristic nucleation times. The solution achieves a quasi-equilibrium within such timescales; thus, ensemble averages pertaining to pre-critical species in solution can be directly computed *via* brute-force sampling. To this aim, a series of MD simulations were performed where NaCl molality (b) was 2–13.7 m and $S = 0.5$–3.7. We also performed simulations at and beyond the limit of solution stability ($S = 4.1$)[26] where $b = 15$ and 16 m ($S = 4.1$ and 4.3). Fig. S1 in ESI† indicates how the supersaturation levels targeted in this work compare to other computational studies and recent experiments from the literature. Fig. 1 shows how the non-ideality of the solution increases as the system is progressively brought further into the metastable region ($1 < S < 4.1$). The average ion coordination number in Fig. 1A increases from a value of around zero (*i.e.*, indicating that ions are completely solvated by water molecules in the first coordination sphere) to 0.75 when $S = 3.7$. Across the same range of molality, the maximum coordination number increases from 1.1 ± 0.2 to 3.2 ± 0.2. In the stable solution when $S = 0.5$, the maximum coordination number is 0.7 ± 0.2, while at $S = 4.3$, this is 3.6 ± 0.2 (albeit with a conservative definition of ion coordination; see Methods). The increased ion–ion coordination, and partial dehydration across the metastable region, leads to clustering and the formation of extended ionic networks. This observation is in very good agreement with results from recent experiments and simulations that showed the presence of extended Na–Cl ionic networks when S is greater than 1.3.[43]

The extent to which clusters can grow is highlighted by the number-weighted cluster size probability distributions in Fig. 1B, representing the probability of finding an ion within a cluster of size N. All distributions below the limit of solution stability show a rapid decay with respect to N; indeed, at the lowest concentration in the metastable region, an exponentially decaying function ($N \cdot p(N) \propto \exp(-1.8N)$) can be fitted to the data. The complexity of the solution structure increases with concentration and large clusters emerge (the largest of which contain 8–18% of the simulated ions at $S = 3.7$). We anticipate a system size dependence on the extent to which clusters can grow; both due to the number of

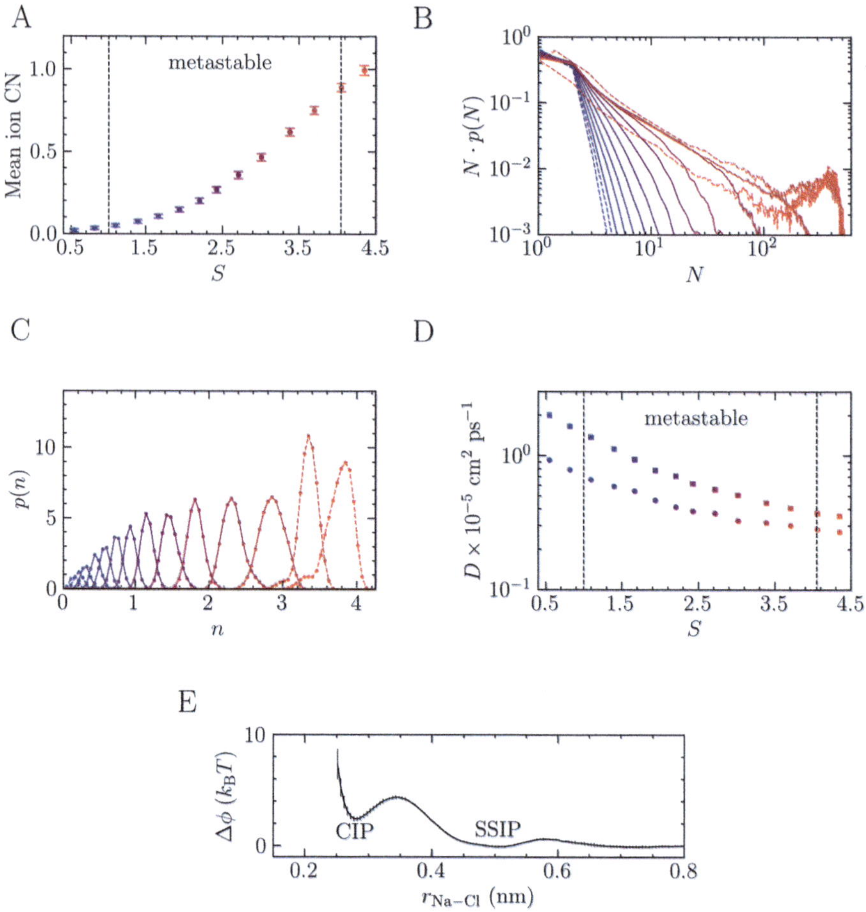

Fig. 1 Ion structure and dynamics in solution over a range of supersaturation ratios, S. (A) Provides the average ion coordination number in the first coordination sphere. (B) Shows the number weighted cluster size probability distribution ($p(N) \cdot N$) where N is the total number of ions in a cluster defined according to coordination in the first ionic sphere. (C) Provides the n probability densities. (D) Shows the cation (circles) and water (squares) self-diffusion coefficients; the uncertainties in the data are of the size of the data points. The blue → red colour scale in all panels indicates increasing supersaturation from $S = 0.5 \rightarrow 4.3$. The dashed lines in (A and D) indicate the boundaries for the metastable solution region with respect to crystallisation (a stable/unstable solution exists at lower/higher S) and the dashed curves in (B and C) highlight systems where S is outside of this region. (E) Shows the potential of mean force, ϕ, as a function of the separation distance between Na^+ and Cl^- in solution from Umbrella Sampling simulations which approach the limit of infinite dilution. The minima representing the contact ion pair (CIP) and solvent shared ion pair (SSIP) are highlighted. Error bars show the standard deviation in the data analysed from five, one nanosecond trajectories.

ions available to form clusters in finite sized simulations[66,67] and the size of the cluster domain with respect to the simulation cell geometry. While Lamas *et al.*[34] observed crystal nucleation in much larger brute force simulations when $S = 3.8$ ($b = 14\ m$ *i.e.*, close to the limiting case of $S = 3.7$ in the metastable region studied here), we did not observe spontaneous crystal nucleation (by monitoring $n(q6)$) in brute force simulations in the metastable region. To verify this, we performed nine additional 300 ns MD simulations at $S = 3.7$, where the initial solution configurations were generated independently, none of which crystallised. Based on the rate for nucleation calculated in this work (3.5×10^{31} s^{-1} m^{-3}, see below), the timescale for nucleation in this system is around 500 ns.

On crossing the spinodal, the cluster size distributions show markedly different behaviour (see the dashed curves in Fig. 1B where $S = 4.1$ and 4.3), with additional peaks in $N \cdot p(N)$ highlighting the spontaneous phase separation associated with large fluctuations of the local ion densities that occurs at these supersaturations. The probability for the smallest clusters at the highest concentrations tends to decrease over time, concomitant with an increase in the size of the largest clusters. This indicates that, beyond the limit of solution stability, the timescale for phase separation is comparable with the length of our simulations and it is likely that the curves in Fig. 1B for these cases would continue to evolve beyond the 50 ns used in our analyses. At the end of these trajectories we also observed crystalline regions which develop spontaneously in the clusters, in agreement with simulations reported elsewhere.[25,26,31,32,34]

Further insight into the coordination environment in clusters can be gained by considering $p(n)$ which is correlated to the number of ions in the clusters with particularly high local ion density. The variable n, defined in the Methods section, represents the cubic root of the number of ions possessing a five-fold or higher ion coordination. In the calculation of n, the function, f, describing coordination goes smoothly from one to zero over a distance 0.38–1 nm. Fig. 1C shows well-defined monomodal distributions for $p(n)$ which shift to larger values of n as the concentration of ions increases. Together with Fig. 1A and B, this shows that in a relatively lean metastable solution the ionic species present before nucleation are predominantly dissociated ions. As the system explores further into the metastable region, the large clusters which form contain regions of high ion density accounting for 0.1–0.3 of the total fraction of ions in the clusters. Phase separation, which occurs spontaneously at the spinodal point, increases the likelihood of these dense regions, which is shown by a shifting to larger n and the appearance of a shoulder in the peaks (see the dashed curves in Fig. 1C) averaged over the entire MD simulation.

Finally, we investigated how ion clustering affects the dynamics of ions and water in solutions crossing the spinodal. Fig. 1D shows a slowing of the diffusion of all solution species as the concentration increases. However, the observed decay is monotonic, as also shown elsewhere,[26] and the data does not provide clear evidence for a dynamic arresting of ions in the clusters. Furthermore, as for the mean ionic coordination number, no discernible discontinuity in the diffusion of solution species is detected when crossing the spinodal, at least on the timescale of $\sim 10^1$ ns. The self-diffusion coefficients, D, for Na^+ and water can be accurately determined as $7 \times 10^{-6} S^{-0.66}$ and $1.5 \times 10^{-5} S^{-0.96}$ cm^2 ps^{-1}, respectively for the range of S sampled here. With only limited reduction to the dynamics of ions, it is not surprising that, in experiments, dehydration and reorganisation of ions in dense regions to the crystal coordination geometry occur over very short timescales; hence, special techniques such as rapid drying must be used to avoid NaCl crystallisation in the production of amorphous NaCl nanoparticles.[45]

The presence of extended ionic networks with relatively high and low regions of ion density raises questions about their potential role in nucleation. In the context of characterising nucleation pathways, it is also useful to consider whether ion association is thermodynamically favoured, as this could indicate nucleation *via* the prenucleation cluster (PNC) pathway.[20] The mechanism for precipitation along this pathway involves the association of PNCs during a microscopic liquid–liquid phase transition with a slowing of the solution dynamics in the proceeding liquid intermediate which can undergo further transformations to amorphous solids or crystals.[14,20,22] A significant debate has surrounded the role of PNCs in nucleation and even their existence.[13,15,16,68,69] In the case of calcium carbonate, PNCs are described as dynamically ordered polymeric species.[21] The thermodynamic stability of the PNCs, relative to free ions in solution, is ascribed to a strongly exergonic ion association reaction where K_a^M—the molar equilibrium ion association constant—is independent of cluster size *i.e.*, for $(AX)_n + (AX) \rightleftharpoons (AX)_{n+1}$, where A, X and AX represent anions, cations and ion pairs, respectively, the free energy change for the forward reaction is a constant value independent of n. If $K_a^M > 1$, then according to a multiple binding model, it was argued that a (meta)stable PNC population should exist in solution, regardless of the value of S.[69]

To calculate K_a^M, we performed Umbrella Sampling simulations to evaluate the potential of mean force for ion pairing, ϕ, in a system that approached the limit of infinite dilution following the method described in detail in the work of Chialvo *et al.*[70] $\Delta\phi(r)$ is provided in Fig. 1E and shows excellent agreement with studies elsewhere.[71] This indicates that the contact ion pair (CIP; $r_{\text{Na–Cl}} = 0.28$ nm) is less stable than the solvent shared ion pair (SSIP; $r_{\text{Na–Cl}} = 0.5$ nm). Under these conditions,

$$K_a^M = \frac{4\pi}{C_0}\int_{r_0}^{r_1} \exp\left(\frac{-\Delta\phi(r)}{k_B T}\right) r^2 \mathrm{d}r \tag{13}$$

where r is the distance between ions, k_B is Boltzmann's constant and T is temperature. C_0 is a constant that ensures mol^{-1} dm^3 units for the equilibrium constant. The integration limit r_0 was defined as the minimum distance at which ions approach each other, while r_1 was 0.6 nm, *i.e.*, beyond the second maximum in $\Delta\phi(r)$. We evaluated $\log_{10}(K_a^M)$ as -0.41 ± 0.04 (mol^{-1} dm^3). This value is less negative than the prediction of $\log_{10}(K_a^M)$ by analytical models fitted to ion conductance measurements in solution;[72] although, the measurements were collected at temperatures beyond 400 K, and thus a direct comparison here should be treated with caution. A negative value of $\log_{10}(K_a^M)$ indicates thermodynamically favourable dissociation, even when all types of ion pair states described here are considered.

Perhaps a better indicator for ion association in solution is the equilibrium constant for the formation of a CIP from a SSIP, which can be calculated according to:

$$K_{\text{CIP}} = \frac{\int_{r_0}^{r_T} \exp\left(\frac{-\Delta\phi(r)}{k_B T}\right) r^2 \mathrm{d}r}{\int_{r_T}^{r_1} \exp\left(\frac{-\Delta\phi(r)}{k_B T}\right) r^2 \mathrm{d}r} \tag{14}$$

Here, r_T is 0.35 nm, *i.e.*, the first maximum in $\Delta\phi(r)$ which represents the transition state between the two ion pair types. $\log_{10}(K_{\text{CIP}})$ was -0.77 ± 0.02 in our calculations, indicating that, even in dilute solutions, a small fraction of ion pairs are contact ion pairs, as was already found using cluster population methods for determining the association constant.[73] Regardless of the choice of association constant, therefore, the equilibrium investigated will not lead to a thermodynamically stable population of clusters in the multiple binding model alluded to above. Furthermore, no stable cluster size population was detected throughout the metastable solution region, where the probability distributions in Fig. 1B indicate that monomers are the most stable species. In this regard, Ising lattice gas simulations showed that a broad distribution of cluster sizes in solution results from liquid–liquid demixing close to the critical point for the binodal curve for the liquid/liquid phase separation.[74] A dense liquid NaCl will likely be short lived; however, cluster populations from simulations could inform about the underlying physics which drives cluster formation, as was shown already in the case of, *e.g.*, disordered proteins.[75]

Sampling nucleation events using metadynamics

Well-tempered metadynamics (WTmetaD) simulations were initiated both from homogeneous supersaturated solutions and from crystals of various sizes immersed in solution (see Methods) to analyse the pathways for crystallisation when $S = 3.7$, *i.e.*, a supersaturation far into the metastable region. 10 μs total simulation time was generated from 20 individual simulations. The added bias potential ($V(n^{\mathrm{sph}})$) in WTmetaD enhances the sampling of the high ion density regions in the clusters within a spherical volume in the simulation cell (see Methods). Localising the enhancement of ion density fluctuations triggered by the bias is necessary at high S to limit the number of crystal nucleation events leading to the formation of structures stabilised by periodic boundary conditions. No constraint was applied to the position of the nuclei emerging from solution, however, and these can translate outside of the spherical volume. We therefore post-process the trajectories (see Methods) to calculate the probability density of states in the space n, $n(q6)$. Calculating the forces on atoms from the derivatives of the bias in $n(q6)$ is computationally very expensive, which makes biasing this CV in simulations intractable for the adopted system size and simulation times. This offers an advantage in that by introducing bias only to increase the fluctuations in local density, we avoid forcing the system to adopt any particular crystalline structure. However, since fluctuations in the CVs defining the energy landscape were not directly enhanced during the sampling, some regions of the reaction coordinate are less well converged than others; indeed, out of the 20 WTmetaD simulations, four failed to visit both the solution and crystalline states (and transition states between them) and were therefore discarded in subsequent analyses.

The free energy landscape in Fig. 2A indicates that a wide range of states with minimal crystalline order corresponding to small values of $n(q6)$ are accessible. The minimum accessible n is around 2.5 when $n(q6) = 0$, in agreement with our unbiased simulations. A value of $n = 2.5$ indicates that around 15 ions in solution have high ion coordination and form partially dehydrated ion clusters. The configuration in Fig. 2B provides an example of an ion cluster which emerged in MD simulations of NaCl(aq) solutions when $S = 3.7$; the colours for the ions indicate their relative coordination number for the case where $n = 2.8$, $n(q6) = 0$. Our MD simulations indicated that the average number of ions in the largest

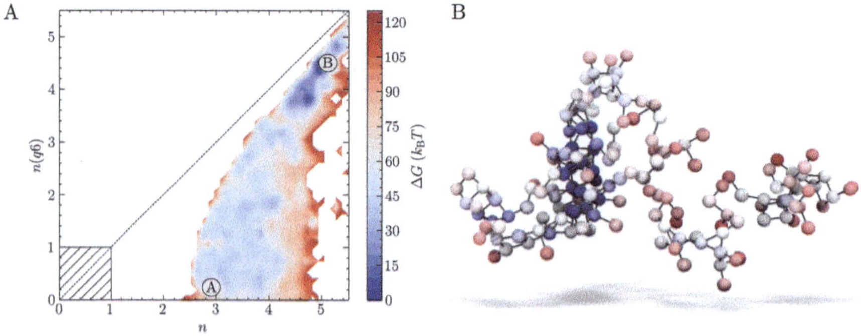

Fig. 2 Pathways from solutions to crystals from biased sampling calculations when $S = 3.7$. (A) Provides relative Gibbs free energies, ΔG, as a function of the reaction space variables n and $n(q6)$ characterising the size of dense and crystalline NaCl regions, respectively. A and B marked on the plot indicate the solution and crystal states. The hatched region indicates the range of n and $n(q6)$ where less than one ion displays five-fold ion coordination or a rock salt crystal geometry, respectively. (B) Shows a liquid-like NaCl cluster observed in MD simulations where $S = 3.7$. The grey lines emphasize first-sphere connections between the ion spheres which are coloured red ($c = 0.6$) → blue ($c = 6$) according to their value of c: the coordination environment which informs about the relative ionic density in the cluster and is used to calculate n (see Methods).

cluster at $S = 3.7$ was 100 ± 40. It is clear that non-uniform chemical ordering occurs in the ion clusters: high density regions, shown by the blue spheres in Fig. 2B, are surrounded by extended ionic networks where ion coordination is much lower, and where the geometry and connectivity of the cluster evolves rapidly during the simulations. The energy landscape generated using WTmetaD indicates that n states beyond the maximum found in MD simulations, up to $n \approx 4$, are accessible in the long time limit. The molality of ions in the high density regions when $S = 3.7$, therefore, is around 0.6 m, but this can increase by several m as the system explores n.

As the degree of crystalline order increases, the accessible region in n narrows significantly. Up to $n(q6) \approx 2$, the range of accessible n states is largely independent of $n(q6)$, but it is around this point that the accessible states converge towards a limiting case for crystallisation on the diagonal, where the size of the dense and crystalline regions increase simultaneously. The crystal basin is observed beyond $n = 5$ and $n(q6) = 4.5$. Although the definition of $n(q6)$ was able to correctly identify crystal ions at rock salt planar surfaces, the rough crystal nanoparticle surfaces found in solution limits the accessibility of states on the diagonal to systems which contain particularly large crystals. From the minimum corresponding to the metastable solution and labelled 'A', lowest energy pathways to the crystal, 'B', were calculated using nudged elastic band (NEB) calculations.[76] While these correctly identified the crystal as the most stable state, the profiles were noisy and no clear transition state for crystallisation was identifiable with confidence. Nevertheless, WTmetaD was instrumental to obtain an extensive sampling of configurations spanning the entire reaction coordinate space which are inaccessible to sub-μs seeded MD or solution simulations.

Nucleation pathways in unbiased simulations

MD simulations. To obtain a quantitative estimation of transition paths, we took advantage of the extensive sampling of configurations obtained from WTmetaD to initialise a comprehensive set of unbiased MD trajectories covering the entire reaction coordinate space. For the analysis of the unbiased trajectories, we focus our discussion on the n^{sph}, $n^{sph}(q6)$ reaction coordinate space: a localisation of the n, $n(q6)$ CVs in the centre of the simulation cell. Considering a subspace of the simulation cell reduces the noise associated with diffusion of the system in the reaction coordinate and, because the crystal seeds discussed below were anchored to the centre of this subspace, no translation of the nanoparticles occurs during the simulations, which would otherwise lead to inaccuracies in the relative probability density of states. Nonetheless, the general features we describe below were apparent regardless of this choice in the reaction coordinate. Our MD simulations indicated that the mean n^{sph} from extended simulations of the homogeneous solution phase is 1.61 ± 0.13 and the normalised probability density for this is shown in Fig. 3B.

Two approaches were taken to investigate diffusive fluxes in the reaction coordinate space. In the first of these, 10 crystalline seeds of various sizes were simulated in solution for 300 ns and their growth or dissolution was monitored by tracing trajectories in n^{sph}, $n^{sph}(q6)$. Fig. 2A shows that, depending on the size of the initial crystal seed, systems consistently relax to either the solution (A) or

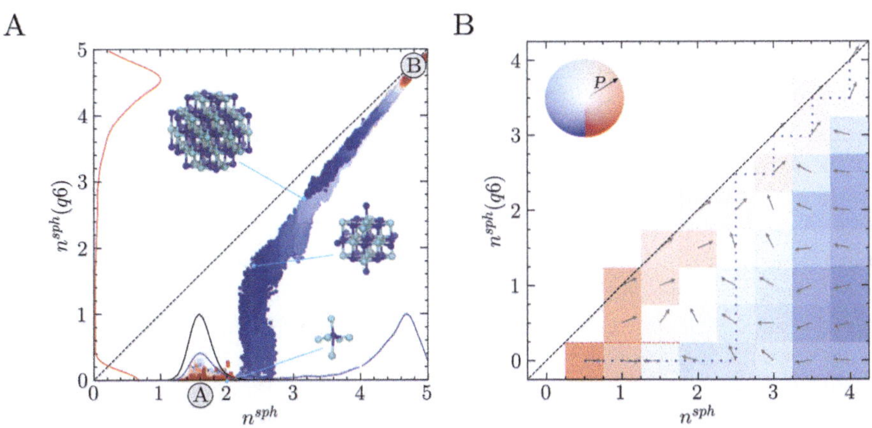

Fig. 3 (A) Seeded MD simulations initiated from crystal seeds with varying sizes; the colour scale here indicates the time in the simulation as blue ($t = 0$) → red ($t = 300$ ns). Also provided by the solid lines are the probability densities in the reaction space variables n^{sph} (blue) and $n^{sph}(q6)$ (red) calculated using all of the data from seeded MD and the probability density from MD simulations of the solution phase (black). Inset are configurations for three crystal seeds around $t = 0$, with blue and cyan spheres indicating Na^+ and Cl^- ions, respectively. The initial values of n^{sph}, $n^{sph}(q6)$ for these seeds are highlighted by the light blue points. (B) The mean flux of trajectories on a discretised n^{sph}, $n^{sph}(q6)$ reaction coordinate space sampled using swarms of MD simulations. The mean paths for trajectories from each bin are shown by the arrows and colours, with the opacity of the colour indicating the probability of transition to adjacent bins occurring on the timescale of the simulations. The red dashed line highlights the case where zero probability of transition to adjacent bins occurs.

crystal states (B). Overlapping multiple seeded trajectories provides a representation of the path between these states in the two-dimensional reaction coordinate space n^{sph}, $n^{sph}(q6)$. While the smallest and largest crystal seeds dissolved and grew, respectively, seeds with diameters in the range 0.8–1 nm relaxed to states which were opposite to this general trend, as shown in ESI Fig. S2.† A seed containing 19 ions where $n^{sph}(q6) = 1.3$, and which was highly faceted, grew in solution; whereas, a seed containing 27 ions with perfectly planar surfaces, and where $n^{sph}(q6) = 1.5$ (since in both seeds only one ion has complete rock salt coordination geometry), dissolved. Ion coordination between seeds and other liquid-like clusters occurred readily at the beginning of the simulations and ion exchange with the surrounding ionic networks ultimately consumed these less dense regions of the clusters when the seeds were post-critical. The probability densities shown in Fig. 2A indicate that the pathway to bulk crystals involves increasing the density in ion clusters in a first step to crystal nucleation: a departure from the mean n^{sph} in solutions at the steady state (see the black curve) occurs before crystalline order emerges, when $n^{sph}(q6) > 0$, in a second step. This is particularly significant as it indicates that crystallisation occurs in regions where relatively high levels of ion–ion coordination, and therefore low levels ion solvation, are observed compared to the homogeneous solution state. The minimum in the probability density in $n^{sph}(q6)$ is 0.5–0.6; here, $n = 2$ indicating at least eight ions—on average—in the subspace with an ion rich first coordination sphere. The rate limiting step to crystallisation, therefore, is the reordering of ions in the dense liquid domains to a geometry which begins to resemble that of the crystalline phase.

Further support for the two-step pathway to phase separation is gained from our second approach to investigate diffusive flux in the reaction coordinate space. Here, the n^{sph}, $n^{sph}(q6)$ space was discretised into 0.5×0.5 bins. Swarms of MD simulations were initiated in each bin where the initial configurations for the simulations were extracted from MD simulations of solutions and from the WTmetaD simulations already described. Initial configurations were selected randomly, ensuring that no two initial starting points were correlated. At least 40 independent 2.5 ns simulations were initiated in each bin which was increased to ~100 simulations in regions of the configuration space where diffusion is particularly slow. Fig. 3B shows the direction of flux for the average trajectory initialised in each bin. Using this coarse-grained representation of the flux of configurations in the reaction coordinate space, the two-step nucleation pathway spontaneously emerges, and is clearly apparent. This is highlighted by the most probable pathway which propagates the system beyond previously visited states in the solution phase to the crystal based on the transitions between adjacent bins, as shown by the blue dotted line in Fig. 3B. Here, $n^{sph}(q6)$ initially increases when $n^{sph} = 2.5$ until the system approaches and follows the diagonal beyond $n^{sph} = 2.5$, $n^{sph}(q6) = 2.5$, in close agreement with the pathway identified in Fig. 3A. Furthermore, the qualitative features here are in good agreement with the free energy landscape provided in Fig. 2A.

Systems where $n^{sph}(q6) = 0$ and n^{sph} is particularly small diffuse rapidly to states where the ion density in clusters increases and there was zero probability of transition from bins where $n^{sph} < 2$ and $n^{sph}(q6) = 0$ to those where $n^{sph}(q6) > 0$ during the 2.5 ns timescale. When $n^{sph} > 2.5$ and $n^{sph}(q6) = 0$, the system tends to diffuse towards $n^{sph} = 1.5$–2.5, with the probability of transitions (highlighted by the opacity of the colours in the figure) increasing with n^{sph}. This provides confidence to the qualitative assessment regarding the accessibility of states with increased ion coordination relative to the most probable state for a homogeneous solution. It is notable too that the direction of the mean path when $n^{\text{sph}} = 1$, $n^{\text{sph}}(q6) = 1$ indicates concerted growth in the reaction space variables. A pathway along the diagonal from $n^{\text{sph}} = 1$, $n^{\text{sph}}(q6) = 1$ could be consistent with single-step nucleation occurring *via* the addition of single monomers to growing crystal nuclei; although technically, these nuclei could still reside within a liquid-like cluster. Therefore, while the most likely pathway under these conditions appears consistent with two-step nucleation, multiple, coexisting pathways may contribute to phase separation.

Markov state model. To gain quantitative information about the nucleation pathways spontaneously emerging from unbiased MD, we used the data gathered from all unbiased simulations, amounting to a total simulation time of 20 μs, to construct a Markov State Model (MSM) for the nucleation process. Here, the reaction coordinate was partitioned into approximately 100 discrete states which are shown in Fig. S3A and S4A in ESI.† The MSM provides independent insight into the free energies associated with the nucleation process, the implied

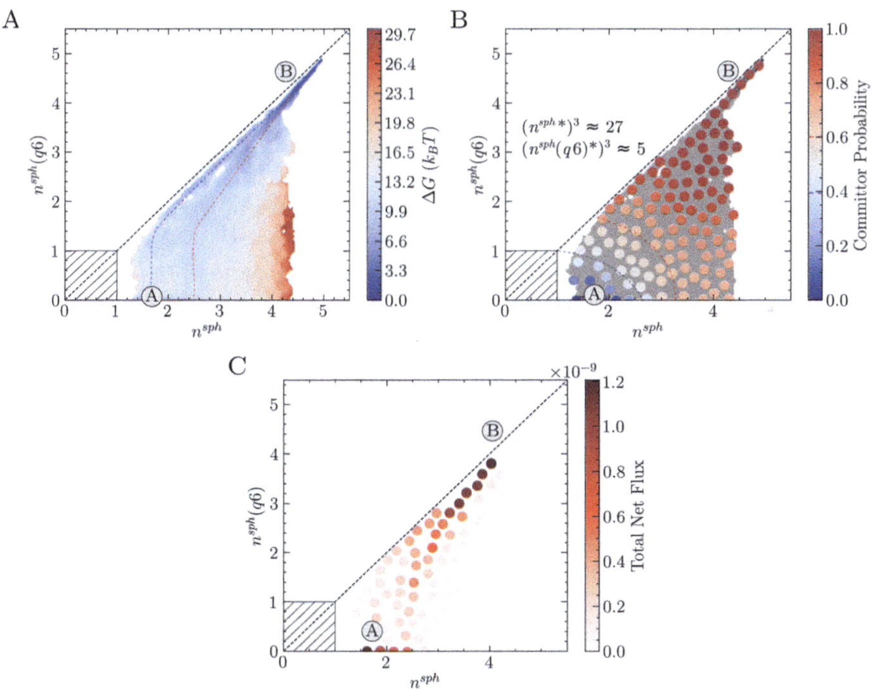

Fig. 4 (A–C) Provide the free energy landscape, committor probabilities and total net flux between partitioned states, respectively, in the n^{sph}, $n^{\text{sph}}(q6)$ reaction coordinate space calculated from the MSM described in the text. The hatched regions indicate the range of n^{sph} and $n^{\text{sph}}(q6)$ where less than one ion displays five-fold ion coordination or a rock salt crystal geometry, respectively. The blue and red dashed lines in A provide a guide to the eye for the lowest energy one- and two-step nucleation pathways, respectively. The blue and red dashed lines in B indicate the approximate range of the transition state ensemble of states where the committor probability is 0.4–0.6 (see key).

timescales associated with the relaxation of the system, as well as the information necessary to characterise the nucleation process using tools from discrete transition path theory.

The free energy landscape in Fig. 4A shows two basins for the solution and crystal states (labelled A and B, respectively) and again confirms the wide distribution of accessible n^{sph} states corresponding to clusters with negligible crystalline order. Following the two-step pathway already identified in Fig. 3, the barrier to crystallisation (extracted from nudged elastic band calculations along the most probably pathway) is 7 $k_{\mathrm{B}}T$, in reasonable agreement with the barrier determined elsewhere by interpolating the data from Umbrella Sampling simulations and mean first passage time in forward flux sampling.[26] An alternative pathway exists from the solution basin to $n^{\mathrm{sph}} = n^{\mathrm{sph}}(q6) = 1$, and which subsequently follows closely to the diagonal towards the crystal basin, is also evident in the energy landscape; this is captured due to the finer partitioning of states here compared to the coarse approach taken in Fig. 3A. The barrier for crystal nucleation along this pathway is 9 $k_{\mathrm{B}}T$, which further supports the notion that multiple pathways for nucleation are available to the system far into the metastable region. The energy landscape in Fig. 4A is compatible with those theorised using a general framework to describe two-step nucleation pathways (using CNT concepts) recently proposed by Kashchiev.[17] Both one- and two-step pathways to crystal nucleation are predicted by this framework when the maximum in the free energy landscape—characterised by cluster size and crystalline order—is associated with a large amorphous cluster (see ref. 17 and Fig. 3A). The distinction between the pathways identified in Fig. 4A tends to fade when the free energy in the space n, $n(q6)$ is analysed (see Fig. S4C in ESI†), due to the noise associated with n fluctuations in the entire simulation cell.

A committor analysis was performed to calculate the probability for partitioned states (associated with the MSM) in the reaction coordinate space to relax to either the solution or crystal basins. This analysis (see Fig. 4B and S4D†) revealed that the transition state ensemble, identified here by states with committor probability values of 0.4–0.6, spans a wide range of cluster configurations including both fully ordered and completely disordered ones. Indeed, even for the largest clusters with no apparent crystalline order (*i.e.*, when n^{sph} is beyond 2 and $n^{\mathrm{sph}}(q6)$ is zero), the committor is larger than 0.5, indicating their post-critical nature, *i.e.*, they display a greater likelihood to crystallise than to dissolve. This observation is evidence for the coexistence of multiple pathways from the solution to crystal states and introduces the necessity of generalising the concept of a critical nucleus, to that of a *critical ensemble of nuclei*. Our simulations indicate that, within the critical ensemble, the average cluster contains around 27 ions with five-fold coordination or more, 5 of which have rock salt crystal coordination geometries.

The MSM provides quantitative estimates for the kinetics associated with state-to-state dynamics. By interpreting the slowest implied timescale as the time for relaxation of the system between states A and B (Fig. 4A–C), we obtain a nucleation rate of 3.5×10^{31} m^{-3} s^{-1} (95% confidence interval: 2.7×10^{31} to 5.1×10^{31} m^{-3} s^{-1}). Alternatively, the nucleation rate has been computed from the mean first passage time between two metastable states identified using the robust Perron cluster analysis[77] as 6.0×10^{31} m^{-3} s^{-1}. Both estimates are in excellent agreement with the rates determined elsewhere using alternative methods.[26,34]

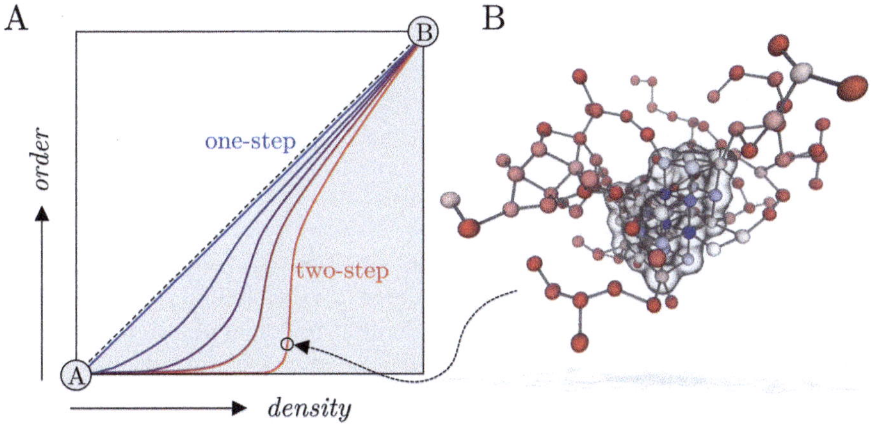

Fig. 5 (A) Shows a schematic of the two-dimensional reaction coordinate sampled in this work. Crystallisation pathways from a metastable solution phase (A) to states where a NaCl crystal is in equilibrium with a saturated solution (B) are highlighted. A pathway consistent with one-step nucleation is provided by the blue path on the diagonal. Paths which deviate from the diagonal indicate inaccuracies associated with the capillary approximation of CNT, while paths which go *via* extended, high density amorphous clusters (in which crystallisation subsequently occurs) suggest a two-step nucleation mechanism. (B) Provides a snapshot of a pre-critical cluster along the two-step pathway containing around 130 ions and where $n^{sph} = 2.4$, $n^{sph}(q6) = 0.8$. Ions are coloured according to their local value of $q6$ in the range 0 (red) to 0.6 (blue), where a value of 1 would indicate a perfect crystal coordination geometry for ions. The transparent surface highlights those ions which begin to adopt a pseudo-crystal rock salt lattice structure with (111) crystallographic orientation in the projection shown.

Finally, we project the total current of reactive trajectories in state space, defined for state i as $\frac{1}{2}\sum_j |F_{ij}|$, where F_{ij} is the current of reactive trajectories between states i and j. This provides a quantitative analysis of the transition mechanism between states A and B, which further supports the emergent picture for phase separation described so far. In particular, analysis of the high current trace between states, shown in Fig. 4C, indicates the presence of a main reaction channel corresponding to the most probable pathway identified by seeded simulations (Fig. 3A), and by analysing the flux of configurations directly obtained from swarms of short MD simulations (Fig. 3B). Interestingly, a secondary channel, closer to the definition of a one-step mechanism and associated to lower values of the net flux can be identified, quantitatively supporting the co-existence of multiple pathways from solution to crystals in the metastable region.

Jiang *et al.*[26] sampled similar reaction coordinate spaces using forward flux sampling simulations and brute force simulations. They find that the nucleation pathway below the spinodal (when $S = 2.7$ and 3.2) occurs in a single-step through correlated fluctuations in the density and crystallinity of emerging clusters. Beyond the spinodal (when $S = 4.3$), the phase separation mechanism was consistent with two-step nucleation, where a wide distribution in the size of amorphous clusters was observed before the formation of crystalline regions (as we also observed in MD simulations at this molality). It is possible then that, on moving the system far into the metastable region, there is a transition in the most probable pathway from one- to two-step nucleation and this may occur in the window $S = 3.2–3.7$. Multiple accessible pathways to crystals allow for greater control of the nucleation process, as was shown mechanistically for the case of urea.[78,79]

Conclusions

In this work, homogeneous NaCl nucleation mechanisms from solutions far into the metastable region were analysed in detail. To this aim, an extensive set of molecular trajectories obtained using different computational methods, including biased and unbiased dynamics, was generated. In our analyses we explicitly considered nucleation as a process evolving along a multidimensional reaction coordinate space where both the size for ion dense regions and crystalline order in emerging nuclei evolve along pathways connecting metastable solutions and crystalline phases. An analysis of the cluster size populations in this work shows that ion clusters are apparent in solutions at all concentrations in the metastable region at the steady state (before nucleation), in line with experiments;[43] although no special thermodynamic status can be attributed to these species, as was suggested in other mineralising systems.[69] In metastable solutions, the clusters emerge through density fluctuations and are unstable with respect to dissociated ions; however, beyond the spinodal, aggregation leads to extended amorphous clusters during a spontaneous phase separation. Both the average size for the largest clusters and the size of high ion density regions in the clusters (where there is a significant level of dehydration) increase as the system explores further into the metastable solution region.

Inspired by the recent approach of Kashchiev[17] to provide a thermodynamic and kinetic framework for two-step nucleation, we analysed free energy landscapes characterising the size of the dense cluster regions and the degree of crystalline order in clusters (see the schematic in Fig. 5A). These show that multiple nucleation pathways are available to the system when $b(\mathrm{NaCl(aq)}) = 13.7\ m$ and $S = 3.7$, and we identified the lowest energy one- and two-step routes to crystals from the range of possible pathways, beginning from the solution phase (see Fig. 5A). Two-step nucleation is the most probable of these pathways to phase separation under the conditions studied, and a wide range of states are accessible where clusters which contain regions of high ion density (compared to lower ion density clusters on average in solution) are completely amorphous; indeed, committor probabilities computed *via* the construction of a MSM showed that a critical ensemble of nuclei exist, and that even large amorphous clusters tend to relax to a crystal rather than dissolving. The activation energy for crystallisation along the one-step pathway we identify is greater than in the case of two-step nucleation by 2 k_BT, and a single thermodynamic barrier separates the solution and crystal basins. It is important to note that nucleation in a single step, defined in terms of the reaction space variables n and $n(q6)$, could occur by direct association of dissociated ions into a crystal nucleus or by concerted increases to the density and crystalline order of ions already present in liquid-like clusters in solution. From an experimental perspective, distinguishing these types of one-step nucleation mechanisms would be challenging, particularly at large values of S. Fig. 5B shows a pre-critical cluster along the two step pathway which contained around 100 ions. Crystal nucleation occurs in the high density region of the cluster shown by the blue spheres and highlighted by the transparent surface in the image. The kinetics of nucleation computed here, explicitly acknowledging the complexity of the configuration space, showed excellent agreement with those calculated elsewhere.[26,34]

Given the finding that crystal nucleation in NaCl(aq) solutions occurs *via* a single-step pathway when $b(\mathrm{NaCl(aq)}) = 12\ m$,[26] we hypothesise that the transition from predominantly one- to two-step nucleation occurs in the metastable region at $S = 3.2$–3.7; however, we predict that large amorphous clusters with high density ion regions could still provide pathways to crystals within this range of supersaturation. In other mineralising systems, amorphous mineral phases are observed during two-step nucleation from solutions.[9,10,13,14] On the other hand, amorphous NaCl phases are only stabilised using special preparation techniques.[45] The difference here may simply be due to undersaturation of the solutions with respect to a dense liquid (if one exists) or amorphous solid NaCl phase; nonetheless, this scenario could still lead to a two-step nucleation pathway,[17] where nascent amorphous clusters rapidly transform to crystals. It is possible that kinetic stabilisation of the amorphous phase occurs in other systems, where the energetic barriers associated with translating bulkier anions to their crystalline geometries results in longer-lived amorphous precipitates.

Recently, we showed that surfaces catalyse the formation of large liquid-like NaCl clusters in solutions where the chemical potential is held constant in CμMD simulations.[80,81] Both the size and density of ions in the clusters were increased in the presence of graphite surfaces. Within the context of CNT, interfaces reduce the surface energies of emerging nuclei and therefore the thermodynamic barriers to nucleation. Given the pivotal role of dense clusters in the two-step nucleation pathway, it may also be the case that surfaces catalyse a change in the nucleation mechanism from predominantly one-step to two-step in the metastable region, at lower supersaturations than those analysed here. This observation opens up new avenues to experimentally validate the mechanistic picture emerging from simulations, and to exploit surfaces to control nucleation.

Data availability

PLUMED input files used in this work are available *via* PLUMED-NEST (https://www.plumed-nest.org),[82] the public repository for the PLUMED consortium, using the project ID: plumID:21.044.

Conflicts of interest

There are no conflicts to declare.

Acknowledgements

The authors acknowledge funding from an EPSRC Programme Grant (Grant EP/R018820/1) which funds the Crystallization in the Real World consortium. The authors acknowledge the use of the UCL Myriad High Throughput Computing Facility (Myriad@UCL), and associated support services, in the completion of this work.

References

1 J. J. De Yoreo, P. U. P. A. Gilbert, N. A. J. M. Sommerdijk, R. L. Penn, S. Whitelam, D. Joester, H. Zhang, J. D. Rimer, A. Navrotsky, J. F. Banfield, A. F. Wallace, F. M. Michel, F. C. Meldrum, H. Cölfen and P. M. Dove, Crystallization by particle attachment in synthetic, biogenic, and geologic environments, *Science*, 2015, **349**, aaa6760.

2 J. De Yoreo, in *ACS Symposium Series*, ed. X. Zhang, American Chemical Society, Washington, DC, 2020, vol. 1358, pp. 1–17.

3 P. R. Ten Wolde and D. Frenkel, Enhancement of Protein Crystal Nucleation by Critical Density Fluctuations, *Science*, 1997, **277**, 1975–1978.

4 O. Galkin and P. G. Vekilov, Control of protein crystal nucleation around the metastable liquid-liquid phase boundary, *Proc. Natl. Acad. Sci. U. S. A.*, 2000, **97**, 6277–6281.

5 O. Galkin, K. Chen, R. L. Nagel, R. E. Hirsch and P. G. Vekilov, Liquid-liquid separation in solutions of normal and sickle cell hemoglobin, *Proc. Natl. Acad. Sci. U. S. A.*, 2002, **99**, 8479–8483.

6 B. A. Garetz, J. Matic and A. S. Myerson, Polarization Switching of Crystal Structure in the Nonphotochemical Light-Induced Nucleation of Supersaturated Aqueous Glycine Solutions, *Phys. Rev. Lett.*, 2002, **89**, 175501.

7 P. E. Bonnett, K. J. Carpenter, S. Dawson and R. J. Davey, Solution crystallisation via a submerged liquid–liquid phase boundary: oiling out, *Chem. Commun.*, 2003, 698–699.

8 P. G. Vekilov, The two-step mechanism of nucleation of crystals in solution, *Nanoscale*, 2010, **2**, 2346.

9 M. Faatz, F. Gröhn and G. Wegner, Amorphous Calcium Carbonate: Synthesis and Potential Intermediate in Biomineralization, *Adv. Mater.*, 2004, **16**, 996–1000.

10 S. E. Wolf, J. Leiterer, M. Kappl, F. Emmerling and W. Tremel, Early Homogenous Amorphous Precursor Stages of Calcium Carbonate and Subsequent Crystal Growth in Levitated Droplets, *J. Am. Chem. Soc.*, 2008, **130**, 12342–12347.

11 X. Wang, I.-M. Chou, W. Hu and R. C. Burruss, In situ observations of liquid–liquid phase separation in aqueous $MgSO_4$ solutions: Geological and geochemical implications, *Geochim. Cosmochim. Acta*, 2013, **103**, 1–10.

12 M. A. Bewernitz, D. Gebauer, J. Long, H. Cölfen and L. B. Gower, A metastable liquid precursor phase of calcium carbonate and its interactions with polyaspartate, *Faraday Discuss.*, 2012, **159**, 291–312.

13 P. J. M. Smeets, A. R. Finney, W. J. E. M. Habraken, F. Nudelman, H. Friedrich, J. Laven, J. J. De Yoreo, P. M. Rodger and N. A. J. M. Sommerdijk, A classical view on nonclassical nucleation, *Proc. Natl. Acad. Sci. U. S. A.*, 2017, **114**, E7882–E7890.

14 J. T. Avaro, S. L. P. Wolf, K. Hauser and D. Gebauer, Stable Prenucleation Calcium Carbonate Clusters Define Liquid–Liquid Phase Separation, *Angew. Chem., Int. Ed.*, 2020, **59**, 6155–6159.

15 Q. Hu, M. H. Nielsen, C. L. Freeman, L. M. Hamm, J. Tao, J. R. I. Lee, T. Y. J. Han, U. Becker, J. H. Harding, P. M. Dove and J. J. De Yoreo, The

thermodynamics of calcite nucleation at organic interfaces: Classical vs. non-classical pathways, *Faraday Discuss.*, 2012, **159**, 509.

16 A. Carino, A. Testino, M. R. Andalibi, F. Pilger, P. Bowen and C. Ludwig, Thermodynamic-Kinetic Precipitation Modeling. A Case Study: The Amorphous Calcium Carbonate (ACC) Precipitation Pathway Unravelled, *Cryst. Growth Des.*, 2017, **17**, 2006–2015.

17 D. Kashchiev, Classical nucleation theory approach to two-step nucleation of crystals, *J. Cryst. Growth*, 2020, **530**, 125300.

18 M. Iwamatsu, Free-energy landscape of nucleation with an intermediate metastable phase studied using capillarity approximation, *J. Chem. Phys.*, 2011, **134**, 164508.

19 M. Iwamatsu, Nucleation pathway of core-shell composite nucleus in size and composition space and in component space, *Phys. Rev. E*, 2012, **86**, 041604.

20 D. Gebauer, A. Völkel and H. Cölfen, Stable Prenucleation Calcium Carbonate Clusters, *Science*, 2008, **322**, 1819–1822.

21 R. Demichelis, P. Raiteri, J. D. Gale, D. Quigley and D. Gebauer, Stable prenucleation mineral clusters are liquid-like ionic polymers, *Nat. Commun.*, 2011, **2**, 590.

22 F. Sebastiani, S. L. P. Wolf, B. Born, T. Q. Luong, H. Cölfen, D. Gebauer and M. Havenith, Water Dynamics from THz Spectroscopy Reveal the Locus of a Liquid-Liquid Binodal Limit in Aqueous $CaCO_3$ Solutions, *Angew. Chem., Int. Ed.*, 2017, **56**, 490–495.

23 J. Scheck, B. Wu, M. Drechsler, R. Rosenberg, A. E. S. Van Driessche, T. M. Stawski and D. Gebauer, The Molecular Mechanism of Iron(III) Oxide Nucleation, *J. Phys. Chem. Lett.*, 2016, **7**, 3123–3130.

24 D. Gebauer, M. Kellermeier, J. D. Gale, L. Bergström and H. Cölfen, Pre-nucleation clusters as solute precursors in crystallisation, *Chem. Soc. Rev.*, 2014, **43**, 2348–2371.

25 H. Jiang, A. Haji-Akbari, P. G. Debenedetti and A. Z. Panagiotopoulos, Forward flux sampling calculation of homogeneous nucleation rates from aqueous NaCl solutions, *J. Chem. Phys.*, 2018, **148**, 044505.

26 H. Jiang, P. G. Debenedetti and A. Z. Panagiotopoulos, Nucleation in aqueous NaCl solutions shifts from 1-step to 2-step mechanism on crossing the spinodal, *J. Chem. Phys.*, 2019, **150**, 124502.

27 N. E. R. Zimmermann, B. Vorselaars, D. Quigley and B. Peters, Nucleation of NaCl from Aqueous Solution: Critical Sizes, Ion-Attachment Kinetics, and Rates, *J. Am. Chem. Soc.*, 2015, **137**, 13352–13361.

28 N. E. R. Zimmermann, B. Vorselaars, J. R. Espinosa, D. Quigley, W. R. Smith, E. Sanz, C. Vega and B. Peters, NaCl nucleation from brine in seeded simulations: Sources of uncertainty in rate estimates, *J. Chem. Phys.*, 2018, **148**, 222838.

29 J. Alejandre and J.-P. Hansen, Ions in water: From ion clustering to crystal nucleation, *Phys. Rev. E*, 2007, **76**, 061505.

30 F. Giberti, G. A. Tribello and M. Parrinello, Transient Polymorphism in NaCl, *J. Chem. Theory Comput.*, 2013, **9**, 2526–2530.

31 G. Lanaro and G. N. Patey, Birth of NaCl Crystals: Insights from Molecular Simulations, *J. Phys. Chem. B*, 2016, **120**, 9076–9087.

32 D. Chakraborty and G. N. Patey, How Crystals Nucleate and Grow in Aqueous NaCl Solution, *J. Phys. Chem. Lett.*, 2013, **4**, 573–578.

33 D. Zahn, Atomistic Mechanism of NaCl Nucleation from an Aqueous Solution, *Phys. Rev. Lett.*, 2004, **92**, 040801.

34 C. P. Lamas, J. R. Espinosa, M. M. Conde, J. Ramírez, P. Montero de Hijes, E. G. Noya, C. Vega and E. Sanz, Homogeneous nucleation of NaCl in supersaturated solutions, *Phys. Chem. Chem. Phys.*, 2021, **23**, 26843–26852.

35 T. Karmakar, P. M. Piaggi and M. Parrinello, Molecular Dynamics Simulations of Crystal Nucleation from Solution at Constant Chemical Potential, *J. Chem. Theory Comput.*, 2019, **15**, 6923–6930.

36 I. S. Joung and T. E. Cheatham, Determination of Alkali and Halide Monovalent Ion Parameters for Use in Explicitly Solvated Biomolecular Simulations, *J. Phys. Chem. B*, 2008, **112**, 9020–9041.

37 F. Moučka, I. Nezbeda and W. R. Smith, Molecular simulation of aqueous electrolytes: Water chemical potential results and Gibbs-Duhem equation consistency tests, *J. Chem. Phys.*, 2013, **139**, 124505.

38 F. Moučka, I. Nezbeda and W. R. Smith, Chemical Potentials, Activity Coefficients, and Solubility in Aqueous NaCl Solutions: Prediction by Polarizable Force Fields, *J. Chem. Theory Comput.*, 2015, **11**, 1756–1764.

39 Z. Mester and A. Z. Panagiotopoulos, Mean ionic activity coefficients in aqueous NaCl solutions from molecular dynamics simulations, *J. Chem. Phys.*, 2015, **142**, 044507.

40 Z. Mester and A. Z. Panagiotopoulos, Temperature-dependent solubilities and mean ionic activity coefficients of alkali halides in water from molecular dynamics simulations, *J. Chem. Phys.*, 2015, **143**, 044505.

41 A. L. Benavides, J. L. Aragones and C. Vega, Consensus on the solubility of NaCl in water from computer simulations using the chemical potential route, *J. Chem. Phys.*, 2016, **144**, 124504.

42 J. R. Espinosa, J. M. Young, H. Jiang, D. Gupta, C. Vega, E. Sanz, P. G. Debenedetti and A. Z. Panagiotopoulos, On the calculation of solubilities via direct coexistence simulations: Investigation of NaCl aqueous solutions and Lennard-Jones binary mixtures, *J. Chem. Phys.*, 2016, **145**, 154111.

43 H. Hwang, Y. C. Cho, S. Lee, Y.-H. Lee, S. Kim, Y. Kim, W. Jo, P. Duchstein, D. Zahn and G. W. Lee, Hydration breaking and chemical ordering in a levitated NaCl solution droplet beyond the metastable zone width limit: evidence for the early stage of two-step nucleation, *Chem. Sci.*, 2021, **12**, 179–187.

44 T. Nakamuro, M. Sakakibara, H. Nada, K. Harano and E. Nakamura, Capturing the Moment of Emergence of Crystal Nucleus from Disorder, *J. Am. Chem. Soc.*, 2021, **143**, 1763–1767.

45 E. Amstad, M. Gopinadhan, C. Holtze, C. O. Osuji, M. P. Brenner, F. Spaepen and D. A. Weitz, Production of amorphous nanoparticles by supersonic spray-drying with a microfluidic nebulator, *Science*, 2015, **349**, 956–960.

46 A. Barducci, G. Bussi and M. Parrinello, Well-Tempered Metadynamics: A Smoothly Converging and Tunable Free-Energy Method, *Phys. Rev. Lett.*, 2008, **100**, 020603.

47 H. J. C. Berendsen, J. R. Grigera and T. P. Straatsma, The missing term in effective pair potentials, *J. Phys. Chem.*, 1987, **91**, 6269–6271.

48 B. Hess, H. Bekker, H. J. C. Berendsen and J. G. E. M. Fraaije, LINCS: A linear constraint solver for molecular simulations, *J. Comput. Chem.*, 1997, **18**, 1463–1472.
49 B. Hess, C. Kutzner, D. van der Spoel and E. Lindahl, GROMACS 4: Algorithms for Highly Efficient, Load-Balanced, and Scalable Molecular Simulation, *J. Chem. Theory Comput.*, 2008, **4**, 435–447.
50 G. Bussi, D. Donadio and M. Parrinello, Canonical sampling through velocity rescaling, *J. Chem. Phys.*, 2007, **126**, 014101.
51 H. J. C. Berendsen, J. P. M. Postma, W. F. van Gunsteren, A. DiNola and J. R. Haak, Molecular dynamics with coupling to an external bath, *J. Chem. Phys.*, 1984, **81**, 3684–3690.
52 U. Essmann, L. Perera, M. L. Berkowitz, T. Darden, H. Lee and L. G. Pedersen, A smooth particle mesh Ewald method, *J. Chem. Phys.*, 1995, **103**, 8577–8593.
53 G. A. Tribello, M. Bonomi, D. Branduardi, C. Camilloni and G. Bussi, PLUMED 2: New feathers for an old bird, *Comput. Phys. Commun.*, 2014, **185**, 604–613.
54 P. J. Steinhardt, D. R. Nelson and M. Ronchetti, Bond-orientational order in liquids and glasses, *Phys. Rev. B*, 1983, **28**, 784–805.
55 A. Grossfield, *WHAM: the weighted histogram analysis method*, http://membrane.urmc.rochester.edu/wordpress/?page_id=126.
56 G. A. Tribello, F. Giberti, G. C. Sosso, M. Salvalaglio and M. Parrinello, Analyzing and Driving Cluster Formation in Atomistic Simulations, *J. Chem. Theory Comput.*, 2017, **13**, 1317–1327.
57 I.-C. Yeh and G. Hummer, System-Size Dependence of Diffusion Coefficients and Viscosities from Molecular Dynamics Simulations with Periodic Boundary Conditions, *J. Phys. Chem. B*, 2004, **108**, 15873–15879.
58 M. A. González and J. L. F. Abascal, The shear viscosity of rigid water models, *J. Chem. Phys.*, 2010, **132**, 096101.
59 M. K. Scherer, B. Trendelkamp-Schroer, F. Paul, G. Pérez-Hernández, M. Hoffmann, N. Plattner, C. Wehmeyer, J.-H. Prinz and F. Noé, PyEMMA 2: A software package for estimation, validation, and analysis of Markov models, *J. Chem. Theory Comput.*, 2015, **11**, 5525–5542.
60 J.-H. Prinz, H. Wu, M. Sarich, B. Keller, M. Senne, M. Held, J. D. Chodera, C. Schütte and F. Noé, Markov models of molecular kinetics: Generation and validation, *J. Chem. Phys.*, 2011, **134**, 174105.
61 J. A. Hartigan, *Clustering Algorithms*, John Wiley & Sons Inc., New York, NY, 1975.
62 G. R. Bowman, V. S. Pande and F. Noé, *An Introduction to Markov State Models and Their Application to Long Timescale Molecular Simulation*, Springer Science & Business Media, 2013, vol. 797.
63 B. Trendelkamp-Schroer, H. Wu, F. Paul and F. Noé, Estimation and uncertainty of reversible Markov models, *J. Chem. Phys.*, 2015, **143**, 174101.
64 P. Metzner, C. Schütte and E. Vanden-Eijnden, Transition path theory for Markov jump processes, *Multiscale Model. Simul.*, 2009, **7**, 1192–1219.
65 F. Noé, C. Schütte, E. Vanden-Eijnden, L. Reich and T. R. Weikl, Constructing the equilibrium ensemble of folding pathways from short off-equilibrium simulations, *Proc. Natl. Acad. Sci. U. S. A.*, 2009, **106**, 19011–19016.
66 J. Wedekind, D. Reguera and R. Strey, Finite-size effects in simulations of nucleation, *J. Chem. Phys.*, 2006, **125**, 214505.

67 M. Salvalaglio, P. Tiwary, G. M. Maggioni, M. Mazzotti and M. Parrinello, Overcoming time scale and finite size limitations to compute nucleation rates from small scale well tempered metadynamics simulations, *J. Chem. Phys.*, 2016, **145**, 211925.
68 K. Henzler, E. O. Fetisov, M. Galib, M. D. Baer, B. A. Legg, C. Borca, J. M. Xto, S. Pin, J. L. Fulton, G. K. Schenter, N. Govind, J. I. Siepmann, C. J. Mundy, T. Huthwelker and J. J. D. Yoreo, Supersaturated calcium carbonate solutions are classical, *Sci. Adv.*, 2018, **4**, eaao6283.
69 D. Gebauer, P. Raiteri, J. D. Gale and H. Cölfen, On classical and non-classical views on nucleation, *Am. J. Sci.*, 2018, **318**, 969–988.
70 A. A. Chialvo, P. T. Cummings, H. D. Cochran, J. M. Simonson and R. E. Mesmer, Na^{+}–Cl^{-} ion pair association in supercritical water, *J. Chem. Phys.*, 1995, **103**, 9379–9387.
71 C. Zhang, F. Giberti, E. Sevgen, J. J. de Pablo, F. Gygi and G. Galli, Dissociation of salts in water under pressure, *Nat. Commun.*, 2020, **11**, 3037.
72 P. C. Ho, D. A. Palmer and R. E. Mesmer, Electrical conductivity measurements of aqueous sodium chloride solutions to 600°C and 300 MPa, *J. Solution Chem.*, 1994, **23**, 997–1018.
73 I. S. Joung and T. E. Cheatham, Molecular Dynamics Simulations of the Dynamic and Energetic Properties of Alkali and Halide Ions Using Water-Model-Specific Ion Parameters, *J. Phys. Chem. B*, 2009, **113**, 13279–13290.
74 A. F. Wallace, L. O. Hedges, A. Fernandez-Martinez, P. Raiteri, J. D. Gale, G. A. Waychunas, S. Whitelam, J. F. Banfield and J. J. De Yoreo, Microscopic Evidence for Liquid-Liquid Separation in Supersaturated $CaCO_3$ Solutions, *Science*, 2013, **341**, 885–889.
75 U. Rana, C. P. Brangwynne and A. Z. Panagiotopoulos, Phase separation vs. aggregation behavior for model disordered proteins, *J. Chem. Phys.*, 2021, **155**, 125101.
76 G. Henkelman and H. Jonsson, Improved tangent estimate in the nudged elastic band method for finding minimum energy paths and saddle points, *J. Chem. Phys.*, 2000, **113**, 9978–9985.
77 S. Röblitz and M. Weber, Fuzzy spectral clustering by PCCA+: application to Markov state models and data classification, *Adv. Data Anal. Classif.*, 2013, **7**, 147–179.
78 M. Salvalaglio, M. Mazzotti and M. Parrinello, Urea homogeneous nucleation mechanism is solvent dependent, *Faraday Discuss.*, 2015, **179**, 291–307.
79 M. Salvalaglio, C. Perego, F. Giberti, M. Mazzotti and M. Parrinello, Molecular-dynamics simulations of urea nucleation from aqueous solution, *Proc. Natl. Acad. Sci. U. S. A.*, 2015, **112**, E6–E14.
80 A. R. Finney, I. J. McPherson, P. R. Unwin and M. Salvalaglio, Electrochemistry, ion adsorption and dynamics in the double layer: a study of NaCl(aq) on graphite, *Chem. Sci.*, 2021, **12**, 11166–11180.
81 A. R. Finney and M. Salvalaglio, Bridging the gap between mesoscopic and molecular models of solid/liquid interfaces out-of-equilibrium, *arXiv*, 2021, preprint, arXiv:2109.00568.
82 The PLUMED consortium, Promoting transparency and reproducibility in enhanced molecular simulations, *Nat. Methods*, 2019, **16**, 670–673.

Faraday Discussions

ROYAL SOCIETY OF CHEMISTRY

PAPER

Possible embryos and precursors of crystalline nuclei of calcium carbonate observed by liquid-cell transmission electron microscopy†

Yuki Kimura, * Hiroyasu Katsuno and Tomoya Yamazaki

Received 27th December 2021, Accepted 11th February 2022

DOI: 10.1039/d1fd00125f

Several different building blocks or precursors play an important role in the early stages of the crystallization of calcium carbonate ($CaCO_3$). Many studies have been conducted over a long period to elucidate the fundamental processes involved in this crystallization. Here, we report the role of an amorphous phase and embryo at the beginning of the nucleation of $CaCO_3$ from solutions of relatively low supersaturation. Prenucleation crystals formed in amorphous calcium carbonate (ACC) at a significantly large formation rate of 2×10^{22} m^{-3} s^{-1}, suggesting that a low interfacial energy exists between the ACC and crystals. Only one calcite crystal exceeded the size for a critical nucleus ($\sim10^4$ molecules) in 150 pre-nucleation crystals. Each pre-nucleation crystal might consist of a different polymorph, and ACCs have a similar composition and structure. A particle-detection algorithm, used in conjunction with machine learning, suggested that an embryo with a characteristic structure exists in solution and might play a crucial role in nucleation. No similar embryonic structure could be observed immediately after the dissolution of pre-nucleation particles, implying that their dissolution process is not simply the reverse process of their growth. This method should provide a new approach to understanding nucleation processes.

Introduction

Using a liquid cell, solution samples can now be observed by transmission electron microscopy (TEM), permitting the visualization of nanoscale

Institute of Low Temperature Science, Hokkaido University, Kita-19, Nishi-8, Kita-ku, Sapporo, 060-0819, Japan. E-mail: ykimura@lowtem.hokudai.ac.jp

† Electronic supplementary information (ESI) available: Video 1: *in situ* TEM observation of the formation of calcium carbonate crystals in ACCs. This video is 10× actual speed. The scale bar is 40 nm. Frames of the same event are presented in Fig. 1A. Video 2: *in situ* TEM observation of the formation of calcium carbonate crystals in ACCs. This video is 10× actual speed. The scale bar is 50 nm. Frames of the same event are presented in Fig. 1B. Video 3: *in situ* HRTEM observation of the precipitation of two different kinds of pre-nucleation crystals. This video is 4× actual speed. Frames of the same event are presented in Fig. 4. See DOI: 10.1039/d1fd00125f

processes in aqueous solution.[1–3] The technique has allowed us to observe the dynamic processes occurring in the early stages of crystallization that were previously unpredictable and this has stimulated an acceptance of nonclassical concepts in crystallization, including unexpected relationships between precursors and final crystals,[4,5] the relationship between dynamic behaviours of defects and the stability of the resulting crystal,[6] the timing of the dehydration process in crystallization,[7–9] and fluctuation-triggered or fluctuation-selected nucleation.[10,11]

Calcium carbonate ($CaCO_3$) attracts considerable attention because of the complexity of its nucleation due to the appearance of various polymorphs,[12] and because of its use as an industrial material,[13] its action as a reservoir for carbon dioxide,[14] and its role in biomineralization.[15] To discuss the early stages of crystallization, we use calcium carbonate as a model substance in view of the variety of polymorphs present in its nucleation process. In the case of nucleation of calcium carbonate from aqueous solutions of relatively high supersaturation, the formation of an amorphous phase before that of a crystalline phase has long been known.[16] Solution-mediated phase transition from an amorphous phase into a crystalline phase has been directly observed by optical microscopy.[17] In general, it is believed that Ostwald's rule of stages holds in many cases. A contribution of pre-nucleation clusters to the nucleation has also been reported.[18] An identification of the existence of pre-nucleation clusters using cryo-TEM has been claimed,[19] but this claim is controversial.[20]

Here, we have attempted to understand the roles of the amorphous phase and an embryo phase in the early stages of crystallization by two different approaches. The first was the observation of amorphous calcium carbonate (ACC) produced by mixing two solutions in a liquid cell TEM (LC-TEM), because ACC is a typical precursor of crystalline phases.[15,21] The second was the detection of an embryo during dissolution and precipitation of calcium carbonate by means of a machine-learning particle-detection algorithm.

Formation of ACC by mixing of two solutions

Although LC-TEM is a powerful tool for observing crystallization processes in aqueous solution, the experimental conditions are determined by the specific constraints of TEM, such as the thickness of the solution as controlled by spacers (typically 50–500 nm), the volume of the solution (0.1–1 nL), radiolysis by the electron beam,[22,23] the time required from preparation to observation (several minutes), and the temporal resolution of the recorded images (generally less than the video-frame rate).

Capturing the early stages of crystallization often requires a great deal of ingenuity. Mixing two different liquids can create a large driving force towards crystallization at certain times. Nevertheless, the direct observation of the moment of nucleation is still difficult, because the mixing initially takes place outside the area of the observation window, so that nucleation occurs before the mixed solution flows into the observable area. We therefore observed the produced ACC particles, which are believed to act as precursors of crystalline calcium carbonate, by LC-TEM.

Detection of embryos by machine learning

One aim of observing a dissolution process is to identify differences between nucleation and dissolution processes. We show, here, competing processes in the formation and dissolution of pre-nucleation particles/crystals of calcium carbonate with diameters of less than 10 nm. These processes were captured in high-resolution TEM (HRTEM) images. To understand the behaviours of embryos and pre-nucleation particles/crystals, we used a machine-learning particle-detection algorithm in conjunction with time-resolved sequential HRTEM images. We define an embryo as a possible cluster of calcium carbonate molecules that has no clear interfaces with the solution and is therefore difficult to recognize as a particle in an HRTEM image. If the nucleus maintains its composition and structure in both growth and dissolution, growing embryos should be detectable using a particle-detection algorithm, trained with ghost images recorded after dissolution of pre-nucleated particles/crystals as a training dataset.

Experimental procedure

TEM observations

A 3 mM aqueous solution of calcium chloride ($CaCl_2$) and a 3 mM aqueous solution of sodium carbonate (Na_2CO_3) were allowed to flow into a liquid cell in a Poseidon holder (Protochips, Morrisville, NC) at a rate of 2 μL min^{-1} so that the liquids mixed while under TEM observation at room temperature (23 °C). To permit liquid flow, the thickness of the solution in the liquid cell was controlled by means of four spacers, each with a thickness of 150 or 500 nm, situated at the corners of the liquid cell; the actual thickness of the liquid layer was somewhat larger due to bulging of the cell walls caused by the vacuum present in TEM.[24] The TEMs used in this study were a Hitachi H-8100 (200 kV, tungsten filament) with an XR-611B TEM camera (Advanced Microscopy Techniques Corp., Woburn, MA) connected to a conventional digital video recorder, and a Hitachi HF-3300 (300 kV, field-emission gun) with an ORIUS SC1000 CCD TEM camera (Gatan Inc., Pleasanton, CA). The *in situ* TEM images were recorded at ~8 and 10 frames s^{-1} (fps), respectively.

Machine learning

We have previously developed a low-dose-electron observation method that uses machine learning to reduce the effect of radiolysis by the electron beam,[25] together with a method for the early detection of nucleation events by machine learning that permits efficient observations on small solution volumes (or small observation areas).[26] On the basis of these previous studies, we have now developed a machine-learning program that uses a standard YOLOv5 algorithm (https://github.com/ultralytics/yolov5) to detect particles in a dataset of *in situ* TEM images. Annotation was performed using the open-source program, LabelImg (https://github.com/tzutalin/labelImg) for training and validation, we used different 40 s portions of a single 10 fps video recording. The YOLO algorithm for object detection works effectively with TEM images.[26] We prepared training datasets for the detection of pre-nucleation particles or embryos. The images were extracted from HRTEM observations captured as video recordings.

First, individual particles that had clear interfaces with the solution were annotated to confirm that the algorithm was capable of detecting particles. Secondly, we identified images that showed the formation of a pre-nucleation particle, and we annotated the same position in the images recorded up to 0.1, 0.3, 0.5, 0.7, and 1 s before the formation of the particle. If the embryo of calcium carbonate has a characteristic structure before nucleation, we should have been able to detect this embryo before a nucleation event occurred. Thirdly, we identified images in which a pre-nucleation particle dissolved, and we annotated the same position in the images recorded up to 0.1, 0.5, 1, and 1.5 s after the disappearance of the pre-nucleation particle by dissolution. If the dissolution process of calcium carbonate is the same as the formation process, the embryo for the formation process should be capable of being detected after the disappearance of the pre-nucleation particle.

Results and discussion

Formation of ACC

Approximately 15 min after starting the liquid flow, particles with a rounded shape and size of less than 100 nm in diameter were observed by TEM (0 s in Fig. 1A and Video 1†). Those particles did not fuse to each other like liquid droplets, suggesting the presence of a solid phase. In addition, because those solid particles exhibited no diffraction contrast, they were amorphous in nature. Of the two possible precursor forms, ACC,[15,21] or dense liquid,[27,28] these features suggest that the particles consist of ACC with a spherical shape rather than a dense liquid.

The ACC particles gradually expanded and their contrast became relatively weaker at the interior and stronger at the periphery. Then, intense contrast appeared from the inside of the ACC, as shown by the white arrowhead at 4 s in Fig. 1A or, more clearly, in Video 1.† The contrast of the particle became stronger at 10 s. This intense contrast corresponds to a diffraction contrast originating from a crystal. Soon after, the intense contrast disappeared. A similar phenomenon involving the appearance and disappearance of crystals has been observed for ACC. All ACC particles disappeared after four minutes (Fig. 1A and Video 1†). The electron dose rate had a constant value of approximately 4×10^3 electrons nm^{-2} s^{-1} and, therefore, the total dose was $\sim 1 \times 10^6$ electrons nm^{-2}. Electron irradiation at a dose rate of 4×10^4 electrons nm^{-2} s^{-1} for 300 s has been shown to cause decomposition of $CaCO_3$ to CaO and CO_2 (ref. 29) whereas a dose rate of 6×10^3 electrons nm^{-2} s^{-1} does not.[30] Our electron dose rate was below the second lower value.

All crystals were formed from the interior and/or inner surface of ACC. We did not observe direct nucleation of crystalline particles from the solution under our experimental conditions (Videos 1 and 2†). Instead, an amorphous particle formed from the solution in the central part of the aggregate in the second observation, as shown by the black arrowheads in Fig. 1B. This might be because the dissolution of ACCs results in a transient flow of the solution towards the area indicated by the black arrowheads in Fig. 1B. The inertia of the flow increases the concentration sufficiently to exceed the critical concentration required for nucleation of ACC. This suggests that there is a local inhomogeneity in concentration in the region of the solution where a cluster of ACC exists. In other words, the distribution of ACC in the solution enhanced any inhomogeneities in the concentration of the solution. Finally, all particles disappeared.

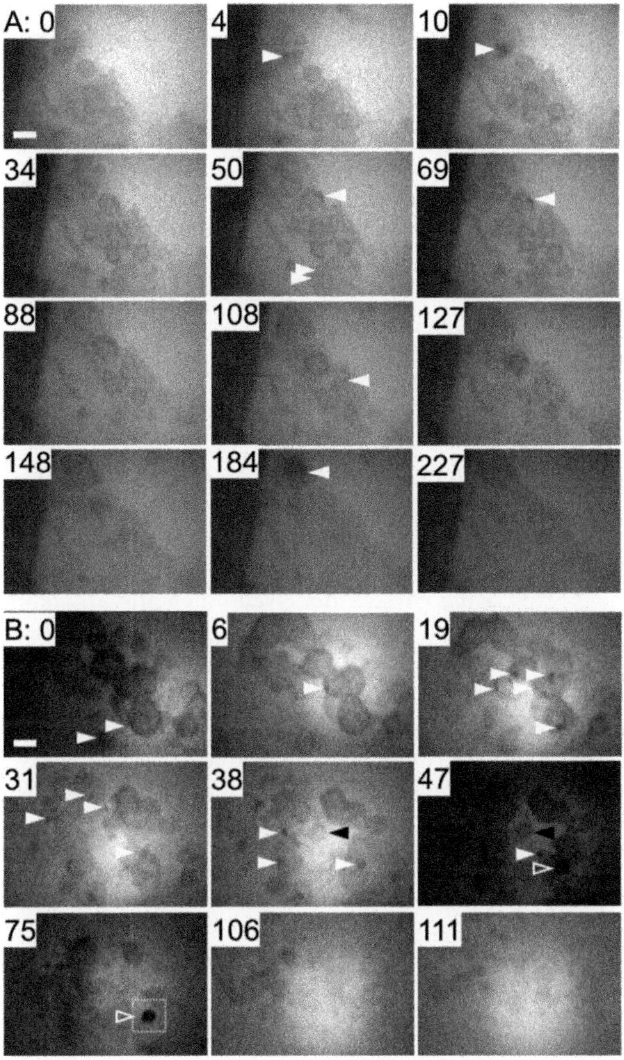

Fig. 1 Time-resolved bright-field TEM (Hitachi H-8100 TEM) images showing the formation of $CaCO_3$ crystals in ACCs. The spacer thickness of the liquid cell was 150 nm. The numbers are the observation times in seconds after the first image, which was recorded immediately after starting electron-beam irradiation. The white arrowheads indicate regions of stronger contrast than that of the initial particle, suggesting the formation of denser particles, *i.e.*, crystals. Scale bar: 50 nm. (A) Constant electron dose rate during the observation. (B) Variable electron dose rate during the observation. The black arrowheads in (B) indicate the formation of an amorphous particle. The open arrowheads indicate a crystal with longer lifetime. The dotted square region has been magnified in Fig. 2.

Effects of electron irradiation

In the second series of observations, shown in Fig. 1B, the electron dose rate was reduced from 7×10^3 electrons nm^{-2} s^{-1} at 2–40 s to 1×10^3 electrons nm^{-2} s^{-1} at 41–105 s to suppress the effects of electron irradiation. After 106 s, the electron dose rate was increased to 6×10^3 electrons nm^{-2} s^{-1} to confirm the dissolution of ACC from brighter images. Under the higher (7×10^3 electrons nm^{-2} s^{-1}) and lower (1×10^3 electrons nm^{-2} s^{-1}) electron dose rates, the maximum radius (8.8 $\pm$ 1.9 nm and 9.0 $\pm$ 4.0 nm, respectively), the growth rate (20.6 $\pm$ 10.8 nm s^{-1} and 15.7 $\pm$ 9.0 nm s^{-1}, respectively), and the dissolution rate (7.4 $\pm$ 3.1 nm s^{-1} and 9.5 $\pm$ 5.6 nm s^{-1}, respectively) in averaged values of the crystals were similar within a standard deviation (one sigma). All three values should be dependent on the saturation level. The similarity of these values therefore suggests that the effects of electron irradiation on the solution are not different at any of the dose rates that we used.

Formation of pre-nucleation crystals and calcite nuclei

In Fig. 1B, the lifetimes of the formed crystals at higher and lower electron dose rates also do not differ (1.6 $\pm$ 1.3 and 2.0 $\pm$ 1.4 s, respectively), apart from one long-lived particle with a lifetime of 49.4 s, as shown by the open white arrowheads in Fig. 1B. The face angle of the largest crystal was 75°, which corresponds to the characteristic face angle of calcite, as shown in the enlarged image in Fig. 2. The formation of calcite from the interior of ACC was not observed in a previous LC-TEM study.[12] This large crystal dissolved after the dissolution of the surrounding ACC. These results suggest that only this single calcite crystal exceeded the size of a critical nucleus, whereas others dissolved because the surface-free-energy disadvantage was greater than the bulk-free-energy advantage. All the smaller crystals were therefore pre-nucleation crystals.

The size dependence of the lifetime of the produced crystals shows that all pre-nucleation crystals smaller than 8 nm in radius had a lifetime of less than 5.5 s (Fig. 2). The average lifetime and its deviation have almost constant values, suggesting that they are independent of the size. If the pre-nucleation crystal

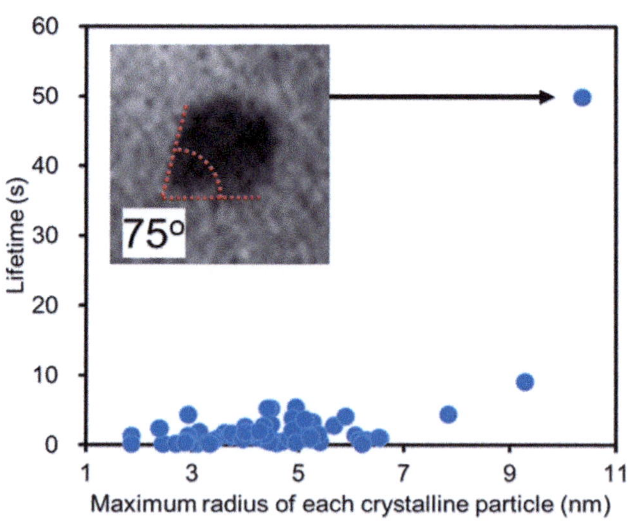

Fig. 2 Size dependence of the crystalline lifetime. The inset is an enlarged image of the largest crystal, indicated by the open white arrowheads, in Fig. 1B. This larger crystal has a face angle of 75°, corresponding to the typical face angle of calcite.

consisted entirely of calcite, its lifetime should have increased with increasing size. However, the results did not show such a trend. This can be explained by considering the formation of various polymorphs of pre-nucleation crystals with different bulk solubilities. Namely, pre-nucleation crystals probably do not only calcite, but other polymorphs, such as aragonite and vaterite, are present. These might form at differing rates in ACC. Another possibility is that the structure of the smaller crystals is still under construction and, therefore, contains many defects, resulting in differences in the stabilities of the smaller crystals. In other words, the relaxation time of the structure takes longer relative to the growth rate, as expected like for glass transitions.

The size of the critical nucleus of calcite in ACC is around 9–10 nm under our experimental conditions. Calcite nuclei of critical size therefore contain approximately 5×10^4 molecules of $CaCO_3$. To achieve or detect nucleation within a realistic timescale in a laboratory, experiments have generally been conducted under conditions of high supersaturation. The size of critical nuclei under such supersaturation conditions is less than 10^3 molecules.[31] The size of the critical nucleus of 5×10^4 $CaCO_3$ molecules in the current experiment was significantly larger than that under typical experimental conditions. The larger size of the critical nucleus suggests that the supersaturation of calcite to ACC is smaller than that to solution. Calcite does not form from the solution despite being supersaturated, probably because of the high interfacial energy.

The role of ACC as a precursor

We observed the formation of pre-nucleation crystals in the ACC at a rate of 1.3 s^{-1} (Fig. 1). If the particle shape is assumed to be spherical, the formation rate of pre-nucleation crystals was 2×10^{22} m^{-3} s^{-1} in the ACC. This apparently large value can be explained by the small interfacial energy between calcite (or other polymorphs) and ACC. Comparing ACC and solution, the structure of ACC is closer to the structure of calcite than to that of the solution. Therefore, the interfacial energy between calcite and ACC should be smaller than that between calcite and the solution. However, only one crystal of a size that exceeded that of a critical nucleus was observed among 150 pre-nucleation crystals. Consequently, the nucleation rate of calcite in the ACC is approximately 10^{20} m^{-3} s^{-1}.

It is surprising that a crystal consisting of as many as 10^3 molecules can form within a few seconds in an amorphous particle at a significantly large rate of 10^{22} m^{-3} s^{-1}. This might be explained by the diffusion coefficient of calcium and carbonate ions in ACC being so large that the ACC behaves in a fluid manner. However, the existence of high fluidity is contrary to the observed fact that ACC particles do not fuse to each other. Another possibility is that there are many water-filled voids in the ACC that facilitate the transport of molecules and induce a large inhomogeneity of concentration. Indeed, it has been reported that voids present in ACC contain mobile water.[30]

It has been proposed that ACCs have polymorphs with a structural similarity to such crystalline phases as calcite and vaterite.[32] Since the clusters of ACCs in Fig. 1 exist in the same place, their solubilities should be similar. Therefore, their composition and local structure should also be similar. Nevertheless, the results regarding the formation of various polymorphs imply that there is less correlation between the structure of the ACC and the polymorphism of $CaCO_3$.

Phenomena that possibly differ from the classical view

When the size of the crystal is smaller than a critical size, its growth rate should be smaller than the absolute value of its dissolution rate. This is because growth is a stochastic process based on thermal fluctuations, whereas dissolution is close to being deterministic. In our observation, the result was the opposite: the growth rate was larger than the absolute value of the dissolution rate. At the time of the formation of a crystal, the concentration should be relatively high and, therefore, the thermodynamic energy barrier could be more easily overcome. Once a crystal forms, the concentration surrounding the crystal becomes lower than the equilibrium concentration of the crystalline phase, and then, the crystal starts to dissolve more slowly than it grows, because the surrounding ACC also dissolves.

Observation of pre-nucleation processes in HRTEM images

Fig. 3 shows an aragonite crystal, which is one of the polymorphs of $CaCO_3$, used in dissolution and precipitation experiments in a liquid cell with a spacer of 500 nm, subject to HRTEM observations by an HF-3300 TEM with a field-emission gun, operated at an acceleration voltage of 300 kV. The solution did not flow, *i.e.*, the solution concentration was maintained by the solubility of the aragonite crystal.

When the magnification was increased from that shown in Fig. 3a to that shown in Fig. 4, the aragonite began to dissolve because the pH of the solution decreased as a result of the intense electron dose.[22] Because the solubility of $CaCO_3$ increases with decreasing pH,[33] aragonite dissolved until the solution condition reached equilibrium. Then, two different kinds of object precipitated near the crystal (Fig. 4 and Video 3†). One was a thin crystal that nucleated heterogeneously on the amorphous silicon nitride window of the liquid cell (Fig. 4A–C). The other was a particle that nucleated homogeneously around the crystal (Fig. 4D). Both these rapidly disappeared, like the pre-nucleation crystals

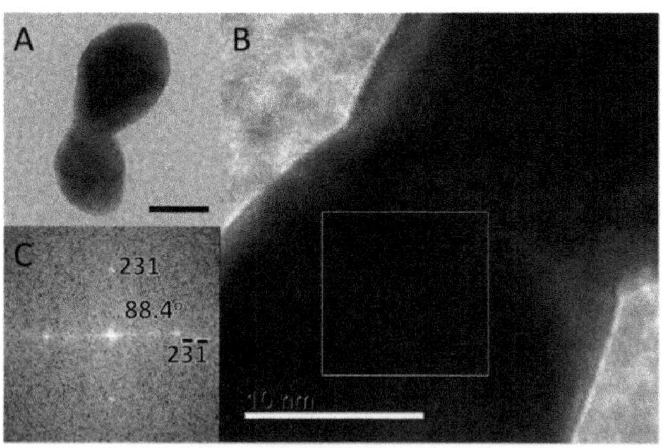

Fig. 3 Mother crystal for the dissolution and precipitation experiment. (A) Micrograph of the complete mother crystal. The scale bar is 20 nm. (B) Lattice image of the mother crystal. The scale bar is 10 nm. (C) Fast-Fourier-transform image corresponding to the red square in (B). The spots correspond to the {231} planes of aragonite (JCPDS card # 41-1475) crossing at an angle of 88.4°.

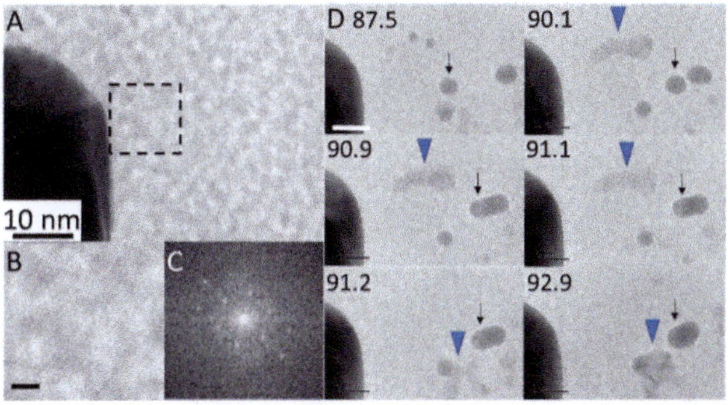

Fig. 4 Two different kinds of pre-nucleation crystals. (A) Thin crystal two-dimensionally precipitated on the window of amorphous silicon nitride. (B) Magnified image corresponding to the dashed square region. The scale bar is 2 nm. (C) Fast-Fourier-transform images corresponding to the magnified image. (D) Pre-nucleation crystals three-dimensionally precipitated in the solution. The numbers indicate the time after the observation area and magnification were fixed. The arrows show a particle that merged on contact with another particle. The blue arrowheads show an object that heterogeneously coagulated on the three tiny particles seen at 87.5 s, and then dissolved at 91.1–91.2 s. Simultaneously, another object with a similar contrast heterogeneously coagulated on the bottom particle at 91.2 s. This appears to correspond to dissolution and precipitation of the object.

formed from ACC shown in Fig. 1, suggesting that these tiny particles were also prenucleation crystals/particles.

Thin crystals formed throughout the view at a formation rate of 1.0×10^{15} m^{-2} s^{-1} and with a lifetime of less than 0.3 s. Assuming they were spherical, their average radius was 6.2 nm. They had various lattice constants and the polymorphs present were not identified as ikaite, CaO, or $Ca(OH)_2$. The thin crystal shown in Fig. 4A–C with lattice fringes of ~0.5 nm might consist of calcite viewed from the *a*-axis.

The formation rate of the three-dimensional particles, as shown in Fig. 4D, was 6.3×10^{14} m^{-2} s^{-1} and, assuming they were spherical, their average radius was 2.7 nm. Some of the pre-nucleated particles had diffraction contrasts originating from crystals. Their lifetime was less than 2 s, typically 0.3 s, suggesting these particles were also smaller than the size of a critical nucleus. The lifetime of these particles showed no size dependency, like the crystals formed from ACC shown in Fig. 1 and 2. The crystalline structure and/or crystallinity of the pre-nucleated particles might vary regardless of the system. Particles formed slightly further from the crystal, even though the concentration was expected to decrease in this region. We hypothesize that a hydrated layer is present on the crystal and might prevent the precipitation of particles in this region. The biggest difference from the case shown in Fig. 1 is that the crystals appeared from the solution directly and did not appear *via* ACC.

Detection of pre-nucleation particles by machine learning

To confirm the applicability of our particle-detection algorithm for HRTEM images, particles were detected after training with annotated data. Examples of annotated particles are shown in Fig. 5A, and results of the validation are shown

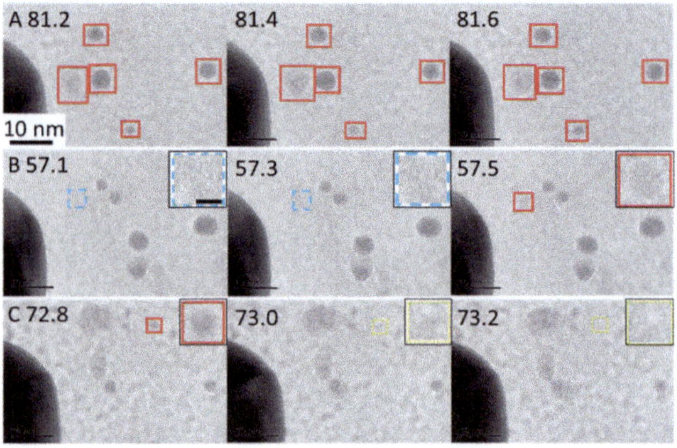

Fig. 5 Examples of annotation in HRTEM images. The numbers indicate the time after the observation area, magnification was fixed. (A) Annotation for particle detection. (B) An embryo has been annotated, as indicated by the blue dashed square, to detect embryos. The red square in the last panel at 57.5 s corresponds to a particle. The insets are enlarged images of the squares in each panel. The scale bar in the enlarged image at 57.1 s is 2 nm, and is the same for the enlarged images in the insets. (C) A ghost has been annotated, as indicated by the yellow dotted square, to detect embryos.

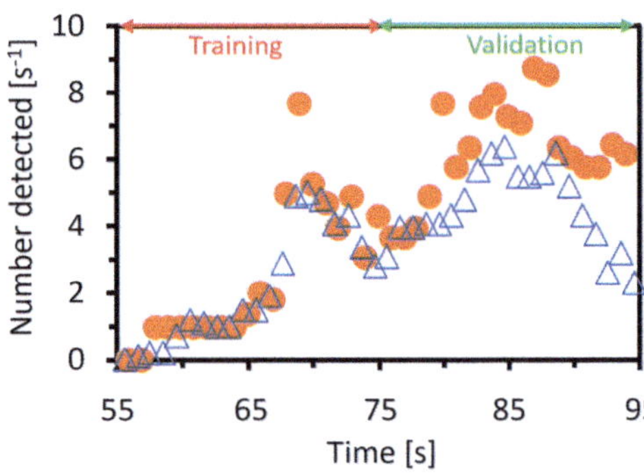

Fig. 6 Number of particles detected s^{-1} in Video 3.† The circles and triangles represent labels applied manually and those detected by machine learning, respectively. The datasets at 55–75 s and 75–93 s were used for training and validation, respectively.

in Fig. 6. The precision score and the recall score reached 0.82 and 0.73, respectively.

Detection of embryos by machine learning

To detect embryos and to predict nucleation events from a solution, potential embryos were annotated as shown in Fig. 5B for machine learning. An embryo is defined as a cluster, possibly consisting of $CaCO_3$, that has no clear interfaces with the solution, as shown by the blue and yellow squares in Fig. 5. Therefore, it is difficult to recognize as a particle in an HRTEM image. Pre-nucleation particles, on the other hand, have clear interfaces, as shown by the red squares in Fig. 5. Examples of results of training and validation for the periods 0.1, 0.5, and 1 s are

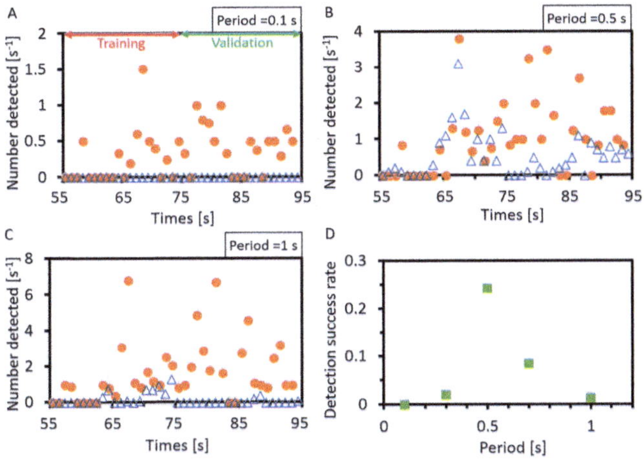

Fig. 7 Number of embryos detected s^{-1} in Video 3.† The circles and triangles represent labels applied manually and detected by machine learning, respectively. The datasets at 55–75 s and 75–93 s were used for training and validation, respectively. (A–C) Examples of detection for periods of 0.1, 0.5, and 1 s, respectively. (D) The ratio of the number of detections to the number of labels in the validation.

shown in Fig. 7A–C, respectively. The filled circles and open triangles represent labels applied manually and those detected by machine learning, respectively. Almost no items were detected in Fig. 7A and C. In Fig. 7B, machine-learning reproduces the labelled data in the training, and detected several objects during the validation. For a quantitative comparison, we plotted the number of detections over the number of labels (Fig. 7D). The abscissa represents the annotation period just before the nucleation event. The number of detections in the period 0.5 s is obviously larger than that in other periods. After detection using the model parameter for the period 0.5 s, nucleation sometimes occurred, as shown in Fig. 5B.

In ordinary machine learning, the larger the number of datasets, the better the accuracy of the machine learning. In the case of the 0.1 s period, no embryos were detected, possibly because the number of training datasets was insufficient to permit their detection (Fig. 7A). In the case of the 0.5 s period, 25% of embryos were detected by machine learning (Fig. 7B), suggesting that embryos with a characteristic structure related to that of the pre-nucleation crystals were present in the solution. In the case of the 0.7 s period, the ratio of the number of detections to the number of labels in the validation decreased to 10%. Finally, in the case of the 1 s period, no embryos were detected, despite a larger number of datasets. These results suggest that the dataset is incorrect when the period becomes too long. Consequently, the embryo must form ~0.5 s before the formation of the pre-nucleation crystal in this system.

Detection of embryos using an algorithm for detection of ghosts

Using a similar method to the above, we also attempted to detect embryos using the algorithm to detect ghosts present after the disappearance of particles. However, no embryos were detected in training or validation in any of the training

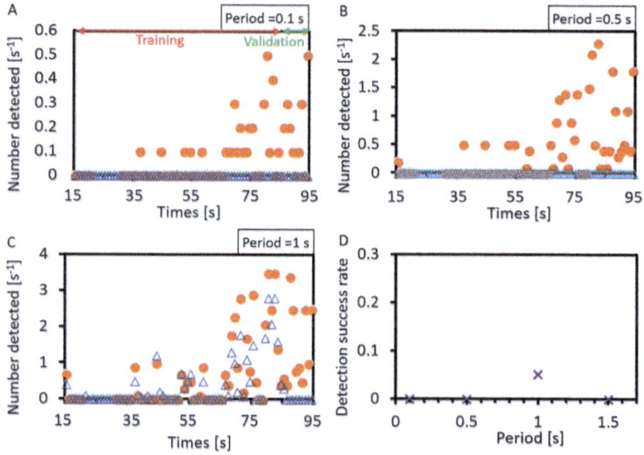

Fig. 8 Number of embryos detected s^{-1} in Video 3.† The circles and triangles represent labels applied manually and detected by machine learning, respectively. The datasets at 0–85 s and 85–93 s were used for training and validation, respectively. (A–C) Examples of detections for periods of 0.1, 0.5, and 1 s, respectively. (D) The ratio of the number of detections to the number of labels in the validation.

periods of 0.1, 0.5, 1, and 1.5 s after the dissolution of pre-nucleation crystals, as shown in Fig. 8. There are two possible explanations for this. The first is that the dissolution process involves different structural changes to the growth process, *i.e.*, dissolution is not the reverse process of growth. The other reason is a difference in the timescale, *i.e.*, the dissolution process might be terminated within 0.1 s. This is consistent with our observation that particles with sizes of the order of several nanometres disappear within 0.1 s. The dissolution of the particles is complete in a shorter timescale than that required for their growth. In addition, the fact that the algorithm cannot detect even ghosts suggests that no characteristic structure exists in the timescale longer than 0.1 s after dissolution. It is probably impossible to detect them at the current temporal resolution of our system.

Conclusions

Pre-nucleation processes in the crystallization of calcium carbonate were directly visualized under near-equilibrium conditions by LC-TEM. Several different building blocks or precursors, such as amorphous, pre-nucleation particles/crystals, and embryos, play a role near the beginning of nucleation of $CaCO_3$. Pre-nucleation crystals formed in ACC with a significantly large formation rate of $2 \times 10^{22}\ m^{-3}\ s^{-1}$, suggesting a lower interfacial energy between ACC and crystals. Among 150 prenucleation crystals, only one calcite crystal exceeded the size for a critical nucleus ($\sim 10^4$ molecules). The lifetime of the pre-nucleation crystals was independent of their size, implying that several different polymorphs formed in the ACC as pre-nucleation crystals. Since the ACC particles existed in the same area, their solubilities should be identical, and therefore their compositions and local structures should all be identical. These results imply that ACC has no polymorphs related to the various crystalline polymorphs, such as calcite or vaterite.

When a particle-detection algorithm based on machine learning was applied to a video captured by HRTEM observations, the results suggest the existence of embryos with a characteristic structure and lifetime of ~0.5 s in a solution that might play a crucial role in nucleation. No related embryonic structure remained immediately after the dissolution of a pre-nucleation particle/crystal, suggesting that the dissolution process does not follow the same structural path as the growth process, or that the timescale for the dissolution process is much shorter than that of the growth process.

Author contributions

Y. K. conceived the presented idea. Y. K. and T. Y. performed the TEM observations. H. K. designed the computational framework. Y. K. analysed the data and wrote the manuscript. All the authors contributed to the final manuscript.

Conflicts of interest

The authors declare no competing interests.

Acknowledgements

This work was partly supported by Grants-in-Aid for Scientific Research (S) from KAKENHI (15H05731 and 20H05657). The authors thank M. Shizuka Hirakawa of the Institute of Low Temperature Science, Hokkaido University, for supporting the production of annotation data for the machine learning, and Manabu Shirai and Hiroaki Matsumoto of Hitachi High-Tech Co. for HRTEM observations.

References

1 N. de Jonge and F. M. Ross, *Nat. Nanotechnol.*, 2011, **6**, 695.
2 J. M. Yuk, J. Park, P. Ercius, K. Kim, D. J. Hellebusch, M. F. Crommie, J. Y. Lee, A. Zettl and A. P. Alivisatos, *Science*, 2012, **336**, 61.
3 H.-G. Liao, L. Cui, S. Whitelam and H. Zheng, *Science*, 2012, **336**, 1011.
4 D. Li, M. H. Nielsen, J. R. I. Lee, C. Frandsen, J. F. Banfield and J. J. De Yoreo, *Science*, 2012, **336**, 1014.
5 L. Houben, H. Weissman, S. G. Wolf and B. Rybtchinski, *Nature*, 2020, **579**, 540.
6 T. Yamazaki, A. E. S. Van Driessche and Y. Kimura, *Soft Matter*, 2020, **16**, 1955.
7 T. Yamazaki, Y. Kimura, P. G. Vekilov, E. Furukawa, M. Shirai, H. Matsumoto, A. E. S. Van Driessche and K. Tsukamoto, *Proc. Natl. Acad. Sci. U. S. A.*, 2017, **114**, 2154.
8 F. Sebastiani, S. L. P. Wolf, B. Born, T. Q. Luong, H. Cölfen, D. Gebauer and M. Havenith, *Angew. Chem., Int. Ed.*, 2017, **56**, 490.
9 J. T. Avaro, S. L. P. Wold, K. Hauser and D. Gebauer, *Angew. Chem., Int. Ed.*, 2020, **59**, 6155.
10 K. Nakajima, T. Yamazaki, Y. Kimura, M. So, Y. Goto and H. Ogi, *J. Phys. Chem. Lett.*, 2020, **11**, 6176.

11 Y. Kimura, H. Niinomi, K. Tsukamoto and J. M. García-Ruiz, *J. Am. Chem. Soc.*, 2014, **136**, 1762.
12 M. H. Nielsen, S. Aloni and J. J. De Yoreo, *Science*, 2014, **345**, 1158.
13 J. Rieger, M. Kellermeier and L. Nicoleau, *Angew. Chem., Int. Ed.*, 2014, **53**, 12380.
14 J. A. Kleypas, R. W. Buddemeier, D. Archer, J.-P. Gattuso, C. Langdon and B. N. Opdyke, *Science*, 1999, **284**, 118.
15 L. Addadi, S. Raz and S. Weiner, *Adv. Mater.*, 2003, **15**, 959.
16 T. Ogino, T. Suzuki and K. Sawada, *Geochim. Cosmochim. Acta*, 1987, **51**, 2757.
17 J. Kawano, N. Shimobayashi, M. Kitamura, K. Shinoda and N. Aikawa, *J. Cryst. Growth*, 2002, **237–239**, 419.
18 D. Gebauer and H. Cölfen, *Nano Today*, 2011, **6**, 564.
19 E. M. Pouget, P. H. Bomans, J. A. C. M. Goos, P. M. Frederik, G. de With and N. A. Sommerdijk, *Science*, 2009, **323**, 1455.
20 P. J. M. Smeets, A. R. Finney, W. J. E. M. Habraken, F. Nudelman, H. Friedrich, J. J. J. LavenDe Yoreo, P. M. Rodger and N. A. J. M. Sommerdijk, *Proc. Natl. Acad. Sci. U. S. A.*, 2017, **114**, E7882.
21 M. P. Schmidt, A. J. Ilott, B. L. Phillips and R. J. Reeder, *Cryst. Growth Des.*, 2014, **14**, 938.
22 N. M. Schneider, M. M. Norton, B. J. Mendel, J. M. Grogan, F. M. Ross and H. H. Bau, *J. Phys. Chem. C*, 2014, **118**, 22373.
23 B. Ambrožič, A. Prašnikar, N. Hodnik, N. Kostevšek, B. Likozar, K. Ž. Rožman and S. Šturm, *Chem. Sci.*, 2019, **10**, 8735.
24 M. E. Holtz, Y. C. Yu, J. Gao, H. D. Abruna and D. A. Muller, *Microsc. Microanal.*, 2013, **19**, 1027.
25 H. Katsuno, Y. Kimura, T. Yamazaki and I. Takigawa, *Microsc. Microanal.*, 2022, **28**, 138.
26 H. Katsuno, Y. Kimura, T. Yamazaki and I. Takigawa, *Front. Chem.*, 2022, **10**, 818230.
27 A. F. Wallace, L. O. Hedges, A. Fernandez-Martinez, P. Raiteri, J. D. Gale, G. A. Waychunas, S. Whitelam, J. F. Banfield and J. J. De Yoreo, *Science*, 2013, **341**, 885.
28 J. T. Avaro, S. L. P. Wolf, K. Hauser and D. Gebauer, *Angew. Chem., Int. Ed.*, 2020, **59**, 6156.
29 C. Rodriguez-Navarro, E. Ruiz-Agudo, A. Luwue, A. B. Rodriguez-Navarro and M. Ortega-Huertas, *Am. Mineral.*, 2009, **94**, 578.
30 H. Du, M. Steinacher, C. Borca, T. Huthwelker, A. Murello, F. Stellacci and E. Amstad, *J. Am. Chem. Soc.*, 2018, **140**, 14289.
31 Q. Hu, M. H. Nielsen, C. L. Freeman, L. M. Hamm, J. Tao, J. R. I. Lee, T. Y. J. Han, U. Becker, J. H. Harding, P. M. Dove and J. J. De Yoreo, *Faraday Discuss.*, 2012, **159**, 509.
32 J. H. E. Cartwright, A. G. Checa, J. D. Gale, D. Gebauer and C. I. Sainz-Diaz, *Angew. Chem., Int. Ed.*, 2012, **26**, 11960.
33 P. W. Hart, G. W. Colson and J. Burris, *J. Sci. Technol. For. Prod. Processes*, 2011, **1**, 67.

Faraday Discussions

PAPER

Multistep nucleation compatible with a single energy barrier: catching the non-classical culprit

A. R. Lauer, [a] M. A. Durán-Olivencia, *[bcd] A. Fernandez-Martinez [a] and A. E. S. Van Driessche *[ae]

Received 22nd November 2021, Accepted 19th January 2022
DOI: 10.1039/d1fd00092f

In this work we link experimental results of $SrSO_4$ precipitation with a nucleation model based on mesoscopic nucleation theory (MeNT) to stride towards a cohesive view of the nucleation process that integrates both classical and non-classical views. When $SrCl_2$ and Na_2SO_4 are co-titrated at slow dosing rates, time-resolved turbidity, conductivity and ion-specific data reveal that the initial stage of the nucleation process is driven by neutral species, *i.e.* ion-pairs or larger, akin to the prenucleation cluster model. However, when co-titrations are conducted at higher rates, the onset of nucleation is dominated by the consumption of free ions, akin to the explanation provided by classical nucleation theory (CNT). The occurrence of both mechanisms for the same system is explained by a toy model that includes both the thermodynamics (consisting of a single energy barrier) and kinetics of cluster formation formally obtained from MeNT. This gives rise to an effective energy barrier exhibiting a local intermediate minimum, which does not originate from a minimum in the thermodynamic free energy. Rather, it is associated with an increased probability of observing a specific class (in terms of size/density) of precursor clusters due to their slower kinetics. At high supersaturations this minimum in the kinetics of cluster formation becomes less pronounced and the effective barrier is also significantly lowered. Consequently, the probability of observing an intermediate state is blurred and we recover a nucleation pathway more closely following the one envisaged by the classical model. Thus, our model is capable of capturing both single and multistep nucleation mechanisms observed experimentally considering only a single energy barrier.

[a] *Université Grenoble Alpes, Université Savoie Mont Blanc, CNRS, IRD, IFSTTAR, ISTerre, Grenoble, France*
[b] *Department of Computer Science, Universidad Pontificia Comillas, C/Alberto Aguilera 25, Madrid, 28015, Spain. E-mail: madolivencia@icai.comillas.edu; m.duran-olivencia@imperial.ac.uk*
[c] *Institute for Research in Technology (IIT), ICAI School of Engineering, Universidad Pontificia Comillas, C/ Santa Cruz de Marcenado 26, Madrid, 28015, Spain*
[d] *Department of Chemical Engineering, Imperial College London, London, UK*
[e] *Instituto Andaluz de Ciencias de la Tierra, IACT, CSIC – University of Granada, Granada, Spain. E-mail: alexander.vd@csic.es*

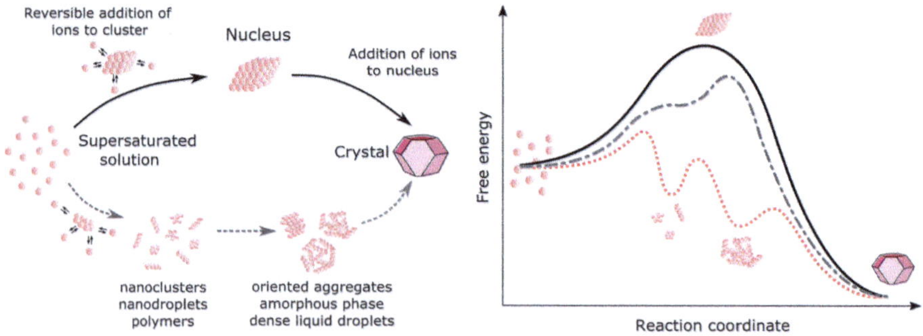

Fig. 1 Schematic representation of one- and multi-step mechanisms for the formation of crystals from supersaturated solutions and their associated energy barriers: a single barrier representative of classical nucleation (black curve), a multi-step barrier model proposed for non-classical nucleation (dotted pink curve) and a multislope single barrier considered by MeNT (grey curve).

1. Introduction

Nucleation is an activated process in which a system overcomes a free energy barrier during the first-order phase transition from a metastable to a stable phase. In the particular case of crystallization from solution, it is commonly accepted that the nucleation process starts by random aggregation of ions/species in solution, which form clusters that eventually may evolve into a crystal. When accounting for the contributions of bulk (favorable) and surface (unfavorable) interactions as the randomly formed aggregates grow in size, the overall free energy of a nucleating system first increases, thus posing a barrier. Consequently, many of the initially formed clusters will quickly re-dissolve and only a few, as a result of statistical fluctuations, reach a certain critical size (where they become an actual nucleus), at which the gain in volume free energy equals the costs of surface free energy (maximum in ΔG). Any further addition of ions/species will lead to a decrease in overall energy of the nucleus, which will continue to grow to a mature crystal. According to classical nucleation theory (CNT),[1] the formation of a nucleus is a one-step process that produces a microscopic particle with the same characteristics (*e.g.* order, density, composition, *etc.*) as a fully-grown macroscopic crystal. In CNT, this transition from dissolved ions/species to crystalline entities is associated with a single energy barrier (black curve, Fig. 1).

However, experimental results gathered during the last decades have challenged this classical picture by revealing that the route from a supersaturated solution to a solid (crystalline) material can involve multiple steps in which distinct precursors and/or intermediate phases occur and interconvert across different length and time scales.[2,3] It is generally assumed that such a multi-step (or so-called non-classical) pathway also implies a multi-barrier energetic landscape for nucleation[4–6] (pink curve, Fig. 1). For example, some of the precursors/intermediates correspond to local free energy minima separated by individual barriers that need to be overcome for the system to transform (note that each transformation step can involve a new nucleation event, either in the condensed phase – structural reorganization – or in solution after re-dissolution). Nonetheless, recent theoretical considerations within the context of the Mesoscopic

Nucleation Theory (MeNT)[7] postulate that energy minima are not a prerequisite for multi-step nucleation to occur, since precursor species could potentially emerge due to the competition between thermodynamics and kinetics during cluster formation on a continuous uphill energy profile[8,9] (grey curve, Fig. 1).

In this work we attempt to link experimental results of $SrSO_4$ precipitation with MeNT modeling to stride towards a cohesive view of the nucleation process. When $SrCl_2$ and Na_2SO_4 are co-titrated at slow dosing rates, nucleation occurs at relatively low supersaturations and time-resolved turbidity, conductivity and ion-specific data reveal that the initial phase separation is controlled by neutral species, *i.e.* ion-pairs or larger, akin to the pre-nucleation pathway.[10] When co-titrations are conducted at higher rates, nucleation occurs at considerably higher supersaturations and the data obtained from the different probes show that the initially nucleating phase is formed mainly through the consumption of free ions, akin to the CNT model. The occurrence of both nucleation mechanisms for the same system is also observed when using a toy model to describe the nucleation pathway, consisting of a single energy barrier and kinetics of cluster formation inspired by the ones formally obtained from MeNT. The potential implications of these preliminary findings and future endeavours are discussed.

2. The MeNT-based toy model

The classical nucleation theory (CNT) presents a series of inherent problems, such as the lack of consistency when expressing measurable quantities, such as the nucleation rate, as a function of one order parameter, *e.g.* the cluster molecular size N or the cluster radius R.[1,11,12] In addition, CNT entails other constraints that further limit its range of applicability. It assumes a spherically symmetric nucleation cluster, with an associated work of formation determined by the equilibrium thermodynamic potential (free energy in the case of closed systems, or Landau's potential in the case of open systems). Also, it imposes a very naïve nucleation pathway that fixes the structural properties of the clusters to be the same as the ones of the final stable bulk phase. This nucleation pathway can be represented as a straight line in the parameter space, connecting two points: zero and infinity. Nevertheless, it is reasonable to expect that clusters are not only characterised by their size, but by several other features (*e.g.* density/order). Thus, the assumption of clusters growing or shrinking along the cluster-size (R or N) axis with all their other properties being exactly the same as the macroscopic phase is in many cases an oversimplification of reality.

Indeed, the constraints of the CNT model come at a great cost. For example, it cannot account for the rich family of clusters that have been observed experimentally during the nucleation process of a variety of systems.[2,3,10] Of the main CNT constraints we have just pinpointed, the spherical symmetry and the local-equilibrium approximation do not affect the internal properties of nucleation clusters. These two assumptions are usually invoked to facilitate the mathematical expressions involved, and allow the formulation of an analytical theory in terms of manageable equations. As for the remaining constraint, the imposed nucleation pathway cancels out all clusters that are not perfect microscopic replicas of the final phase. For instance, according to CNT, if the final stable phase is a crystal then clusters must have the very same structure as the equilibrium configuration.

To overcome these inherent limitations of CNT, the description of the nucleation process used in this work relies upon a fluctuating-hydrodynamical picture of out-of-equilibrium systems, also known as MeNT.[7] MeNT starts from an atomistic description of out-of-equilibrium fluids, and finally recovers CNT-like expressions.[11,13,14] To reduce the number of independent variables of the MeNT theory and obtain an explicit solution of fundamental quantities, such as the nucleation rate or the cluster distribution function, a general framework for nucleation (GFN)[8] has been derived within the context of MeNT establishing the nucleation pathway as the core of the process. This led to a single-variable theory to describe nucleation. Thus, if we have a model for the nucleation pathway GFN will provide us with an exact expression of the cluster distribution, which can then be contrasted against experimental observations.

The key components of the nucleation model used in this work are the following:

(I) A parametric description of the density fluctuations (clusters), $\rho(\mathbf{r};t)$, in terms of measurable properties, such as cluster radius, inner density, order, *etc.* In order to get tractable analytical equations, clusters are assumed to be spherically symmetric, so that $\rho(\mathbf{r};t) = \rho(r;t)$. The parametric description of clusters is then accomplished by introducing a parameterisation in terms of those variables:

$$\rho(r;t) = \rho(r;x_1(t),x_2(t),\ldots,x_N(t)) \tag{1}$$

where $x_1(t),x_2(t),\ldots,x_N(t)$ are the so-called order parameters (or reaction coordinates). CNT is recovered by setting $x_1(t) = R(t)$, with R being the cluster radius, and maintaining all the other properties constant and equal to the values of the final equilibrium phase. A more general theory than CNT is reached by setting: $\rho(r;t) = \rho(x_1 = R(t),\ x_2 = \rho_0(t))$, with ρ_0 being the average inner density of the cluster, assumed constant inside the cluster, now free to change unlike in CNT. For the sake of brevity, we will drop the time-dependency of the order parameters, by simply referring to them as $x_1,x_2,\ldots,x_N$, bearing in mind the underlying dependency. In this study, we will consider a two-variable description of clusters, although more order parameters could be added if needed. However, it has been shown that a two-variable description is good enough to capture the most important aspects of nonclassical nucleation pathways whilst keeping the theory analytically tractable.[8,13]

(II) A model for the free energy functional, $\Omega[\rho] = \Omega(x_1,x_2,\ldots,x_N)$, which determines the equilibrium states of the system. For this we will use a standard fluid equation of state, based on the thermodynamic perturbation theory.[15,16] Once we have a model for the free energy of the system, we will be able to build the nucleation barrier by differencing the energy in the presence of a cluster, $\Omega[\rho]$, and the energy of the system in the absence of any cluster, $\Omega[\rho_\infty]$, where $\rho_\infty = \lim_{r\to\infty} \rho(r;t)$, which refers to the average density of the metastable phase. This has been shown to yield a prototypical volume-*vs.*-surface barrier.[13]

(III) A nucleation pathway, Γ, which represents a trajectory in the parameter space (*i.e.*, the space where the coordinate axes are the order-parameter variables mentioned above). As any trajectory in a geometric space, the pathway can be rewritten in terms of the natural arc-length parameter, s, or by any other variable which is one-to-one related to s. The average inner density of the cluster, ρ_0, is in a one-to-one relationship with s, as was shown elsewhere.[7,8] This is quite convenient given that ρ_0 is experimentally accessible, unlike the arc-length of the nucleation pathway. Hence, from here on we will not differentiate between s and ρ_0.

Now, based on the nucleation trajectory Γ where the clusters will be fluctuating over, and the cluster-density parameterisation, steady-state and equilibrium cluster distributions (*i.e.* the probability density functions to find a cluster at any point of the nucleation pathway, with the properties determined by the value of s) can be derived:[8]

$$P_{\rm st}(s) = A_{\rm st}\sqrt{g(s)}\,{\rm e}^{-\beta\Omega(s)}\int_s^{s_+}{\rm d}z\sqrt{g(z)}{\rm e}^{\beta\Omega(z)} \tag{2}$$

$$P_{\rm eq}(s) = A_{\rm eq}g^{\frac{1}{2}}(s){\rm e}^{-\beta\Omega(s)} = A_{\rm eq}\exp\left(-\beta\Omega(s)+\frac{1}{2}\log g(s)\right) \tag{3}$$

where $A_{\rm st}$ and $A_{\rm eq}$ are normalisation constants, $g(s)$ is the so-called metric function which recovers the monomer attachment rate in the case of a single parameter $x_1 = R$, $\beta = 1/k_{\rm B}T$ is the reciprocal temperature, and $s_+ = (1+\varepsilon)s_*$ is a boundary point in the vicinity of the critical cluster s_* separating the nucleation and growth stages, with $\varepsilon < 1$. The steady-state and equilibrium distributions of the clusters are nearly identical within the pre-critical region of the parameter space, as was shown recently.[8] For this reason, we focus our attention on the equilibrium distribution, as it is more straightforward and contains all the elements necessary to study a nucleation process.

The metric function $g(s)$ represents the kinetics of cluster formation, given that it recovers the monomer attachment rate from CNT when a single order-parameter description of the cluster is adopted.[11] This is worth mentioning because, if we have a careful look at the equilibrium cluster distribution (eqn (3)), the kinetics can be embedded in an effective energy barrier that allows us to rewrite $P_{\rm eq}$ as a standard Arrhenius-like equation:

$$P_{\rm eq}(s) = A_{\rm eq}\exp(-\beta\Omega_{\rm eff}(s)) \tag{4}$$

with

$$\beta\Omega_{\rm eff}(s) = \beta\Omega(s) - \frac{1}{2}\log g(s) \tag{5}$$

This is the essential part of this model, because it allows studying the properties of the cluster distribution by merely observing the behaviour of the effective energy barrier $\beta\Omega_{\rm eff}(s)$. The outcome of this approach is described in detail in the Results section.

3. Materials and methods

3.1. Co-titration experiments

Stock solutions of $SrCl_2$ and Na_2SO_4 were prepared by dissolving $SrCl_2\cdot 6H_2O$ (Acros organics, 99% extra pure) and Na_2SO_4 (Carl Roth, ≥99% pure) in deionized water. Using both stock solutions, equimolar potentiometric co-titrations were conducted using a Metrohm 905 Titrando equipped with two 800 Dosino devices both controlling a 20 mL dosing unit which has a minimum dosing speed of 0.015

mL min^{-1}. The titrants were added to 50 mL of deionized water in a 20–90 mL Titration Vessel purchased from Metrohm. A 20 × 6 mm Teflon-coated stir bar at 500 rpm agitated the reaction. The reaction was monitored simultaneously for transmittance (Metrohm optrode – part 6.115.000), conductivity (Metrohm 5-ring conductivity measuring cell $c = 0.7$ cm^{-1} with Pt1000 – part 6.0915.100), and cation concentration (Ion Selective Electrode (ISE) that consisted of two half-cells: a Mettler-Toledo DX337-Ba membrane with a Metrohm LL ISE reference electrode – part 6.0750.100).

ISE calibration was conducted to convert voltage readouts to actual free ion concentrations. These calibrations were carried out in two steps by titrating $SrCl_2$ (100 mM at 0.5 mL min^{-1}) and NaCl (200 mM at 0.5 mL min^{-1}) into deionized water. This allowed for the accounting of the effect of both cations on the measured potential. From each calibration curve, the Nernstian equation was fitted for the parameters U_0 and p (the sign of p has been changed from standard representation for clarity):

$$U = U_0 + p\log(c + c_0) \tag{6}$$

with c_0 assumed to be negligible. The fitting procedure first found the slope of $\log(c_{\text{dosed}})$ *vs.* U to fit for the parameter p. Then, the equation can be rearranged to find:

$$e^{U/p} = ce^{U_0/p} \tag{7}$$

and treating $e^{-U_0/p}$ as a constant K, we obtain an easy expression for treating the data where:

$$c_{\text{measured}} = e^{U_0/p}/K \tag{8}$$

K was determined by plotting the concentration of dosed ions against the function $e^{U/p}$ using the p parameter defined in the first step. For each titration, this calibration procedure was conducted two times with strontium chloride and one time with sodium chloride. With a new ISE, the s parameter was found to be quite consistent (standard deviation below 3%), so to determine the $1/K$ parameter for sodium, the average p for the strontium calibrations was used. Then the two $1/K$ parameters were summed, and applied to the experimental data alongside the s parameter obtained from the strontium calibration.

Different dosing rates were used to vary the saturation rate across several orders of magnitude (Table 1). At the point of nucleation, we know the concentration of ions that has been dosed into the solution and from this the critical saturation index, SI_{crit}, relative to the solid phase (*i.e.* celestite) was calculated using Phreeqc[17] and the BRGM Thermoddem geochemical database.[18] As expected, SI_{crit} increased with increasing dosing rates (Table 1).

3.2. Nucleation point determination

During co-titration experiments, three probes, *i.e.* optrode, ion selective probe and conductivity probe, were used to monitor *in situ* the nucleation process. From these time-resolved data the onset of phase transition was determined by an automated analysis routine based on the changes in the behavior of the first

Table 1 Dosing rates and the critical supersaturation index, SI_{crit}, at which nucleation is first detected. The error represents the standard deviation from three replicate experiments

Dosing rate (mol min^{-1})	SI_{crit}	Error
0.002	1.15	0.08
0.010	1.33	0.04
0.100	1.63	0.01
0.500	1.91	0.01
1.000	2.02	0.02

derivative, *i.e.* the time point of nucleation was defined when the first derivative reached 5% of its minimum value. For optical transmittance, this was the point at which the derivative became non-zero. Both the conductivity and free cation concentration curves had a downward concavity that required a more complex algorithm to precisely define the time of nucleation observed by those probes. To this end, the curves were smoothed, derived, smoothed again, and then the slope before the nucleation point was fit with a line. Nucleation was defined as the point at which the first derivative deviated significantly (5%) from that fit line.

4. Results and discussion

4.1. Co-titration experiments

In order to study the evolution of the ionic environment during the precipitation of strontium sulfate, equimolar co-titrations were conducted in which equal

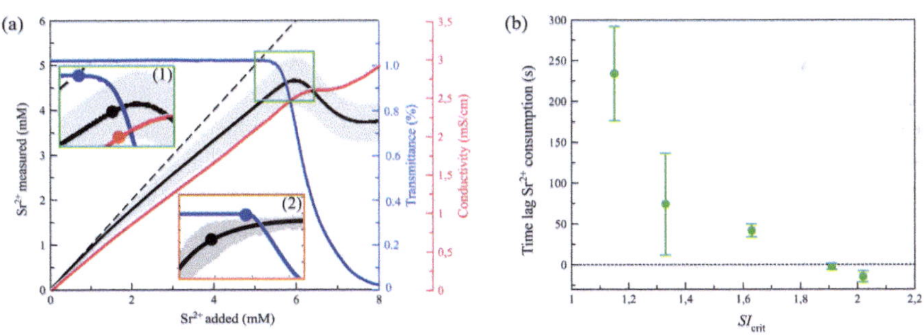

Fig. 2 $SrSO_4$ nucleation times monitored using co-titration experiments. (a) Example of an equimolar co-titration experiment at low dosing rate showing the time evolution of Sr^{2+} concentration (black), optical transmittance of the solution (blue), and the conductivity (red). For reference, the total added strontium concentration is shown (dashed line). Inset 1 (green square) shows an enlarged view of the nucleation event, which is defined as the point where transmittance begins to decrease or when the increases in Sr^{2+} concentration and conductivity slow. The nucleation points are indicated by large dots added to each curve. Inset 2 (orange square) shows a zoom in of the relevant area for the highest $SrSO_4$ dosing rate indicating that nucleation is detected first by the ISE. (b) Difference in measured nucleation time between the transmittance probe and the ISE as a function of the critical SI at which nucleation occurred. A positive difference means the optrode detected the phase transition before the ISE. Error bars represent the standard deviation of three replicate experiments.

amounts of Sr^{2+} and SO_4^{2-} were added at a steady rate into a reaction vessel containing 50 mL of deionized water and continuously stirred to assure a homogeneous mixing. These titrations were monitored for cation activity, transmittance and conductivity (Fig. 2a). The first stage of these titrations is characterized by a monotone increase of cation activity and conductivity with a steady state transmittance signal. Upon nucleation, a second stage is reached in which transmittance drops sharply, while the conductivity signal starts to flatten out and the free ion curve experiences a drop. The point at which these changes occur is defined as the onset of nucleation and was determined by an automated analysis routine based on the behavior of the first derivative.

A close examination of Fig. 2a (see insets) shows that there is an apparent difference between the nucleation times measured by the transmittance probe and those measured by the ISE and conductivity. Using the previously described method to determine the nucleation time from the different data sets it is possible to compare between the probes for experiments run with different dosing rates, from 0.002 to 1 mol min^{-1}. From these dosing rates and the known volume of the reaction solution volume the critical saturation index, SI_{crit}, at the nucleation time point can be calculated (Table 1).

At the three lowest dosing rates, and thus lowest SI_{crit}, the transmittance probe detected the formation of a new phase significantly before the ISE and conductivity probe registered the consumption of free ions (Fig. 2b). Importantly, this effect is most pronounced at the lowest dosing rate – where relatively low saturations would result in the lowest possible probe lags. This observation can be explained if we consider that the first step in strontium sulfate nucleation occurs *via* the consumption of neutral – *i.e.* bound – species that are not detected by the ISE. Indeed, Fig. 2a shows that the detected free Sr^{2+} is lower than the total added strontium, confirming the presence of bound ions that can be consumed to create this effect. Consequently, the particles detected by the optrode are formed through the aggregation of neutral particles, the smallest of which could be ion pairs. It is important to note here the error margin in the calibration of the ISE, which is represented by the grey shaded area surrounding the free ion curve in the main plot of Fig. 2a. This error represents the absolute maximum and minimum that the true reading for Sr^{2+} concentration could be considering 4 calibrations and two experiments, all conducted with a new ISE membrane on the same day (the error in the inset (2) comes from a different source – the standard deviation of three replicate experiments). While a significant portion of the shaded region coincides with the unity line (where free Sr^{2+} = dosed Sr), it is important to note that above the low concentration regime ($<\sim$2 mM Sr), the rate of change is less than one. In the low concentration regime, the results from this ISE are highly irreproducible and should not be considered as meaningful, and thus it is significant that the slope of the curve outside this regime indicates ion pairing or complexation even when considering the measurements least favorable to this hypothesis. Unfortunately, the inconsistency in the low concentration regime means we are unable to comment about the pairing and complexation of Sr^{2+} in this system below its saturation concentration ($\sim$0.7 mM under experimental conditions).

Now, we focus on the two highest dosing rates used for the co-titrations of $SrSO_4$. In these experiments, the early detection of nucleation by the transmittance probe vanishes and at the highest dosing rate (1.0 mol min^{-1}), the detection of nucleation by transmittance even lags behind the potentiometric detection of nucleation (Fig. 2a, inset 2). An example of typical data from an experiment at this dosing rate is shown in Fig. 2a. Hence, at high dosing rates, *i.e.* high supersaturations, the first stage of nucleation is mainly driven by ion consumption and not by bound species. The fact that at the highest supersaturation the transmittance probe lags behind is probably due to the very small size of the initial nuclei, which go undetected at the used wavelength (660 nm). In short, the above-discussed experimental observations reveal that as the supersaturation rate, and thus also the critical supersaturation, increases during $SrSO_4$ co-titrations, the preferred nucleation mechanism changes.

4.2. The effective nucleation barrier: thermodynamics meets kinetics

In this section we present the outcome of our theoretical modelling based on the equations of GFN[8] for the cluster distribution after adopting a model nucleation pathway inspired by the one obtained from a two-variable MeNT.[13] As we mentioned in Section 2, to obtain a nucleation barrier a model for the free-energy functional of the system is required. For this, we used a fluid equation of state (EOS) to build an energy barrier, $\Delta\Omega[\rho]$, based on the Barker–Henderson perturbation theory with the reference system given by the Carnahan–Starling EOS, which has been shown to be well-suited for the description of transitions from low-density to high-density states.[7,8,11,13,14] We used this model EOS to initially obtain the bulk densities at coexistence for a given temperature, *i.e.* $\rho_\infty^{(c)}$ for the low-density phase and $\rho_0^{(c)}$for the high-density state at a given β. Once these values have been determined, we place our simulated system into a supersaturated initial metastable state by imposing a density of $\rho_\infty = S \times \rho_\infty^{(c)}$, with $S = \rho_\infty/\rho_\infty^{(c)}$. Regarding the model of the nucleation barrier, $\beta\Omega$, we consider the surface-*vs.*-volume barrier:[1,7,11,13]

$$\Delta\Omega(R,\rho_0) = V(R,\rho_0)\Delta\omega + S(R,\rho_0)(\rho_0 - \rho_\infty)^2\sigma \tag{9}$$

with $\Delta\omega = \omega(\rho_0) - \omega(\rho_\infty)$ being the negative increment in the system's pressure in the presence of a cluster of volume and surface, $V = \frac{4\pi}{3}R^3$ and $S = 4\pi R^2$, respectively; and σ a constant which defines the planar surface tension.

Now that we have established the bulk system at a given temperature and a given initial density, ρ_∞, we need to introduce a nucleation pathway so that we can compute $\beta\Omega_{\text{eff}}(s)$ along such a trajectory. As mentioned above, our pathway is inspired by the one obtained from MeNT,[13] which selects the most probable pathway by following the steepest-descent path over the energy landscape. The resulting pathway has a characteristic U-shape (Fig. 3) and can be divided into three stages:[11] (i) the first part consists of large-size-low-density fluctuations (bottom branch in Fig. 3), followed (ii) by a densification at an almost constant size (between $R = 1$ and $R = 2$) and finally, (iii) the growth stage of the clusters (top branch in Fig. 3). Here, we assume a similar U-shaped nucleation pathway.

Having established the nucleation pathway (Fig. 3), we now apply the equations previously discussed (Section 2), and study the effects of kinetics on the pre-critical cluster distribution. Fig. 4a shows the outcome of a computation of the thermodynamic energy barrier along the natural parameter $s \sim \rho_0$ for a set of ten supersaturation values, $S \in [1.75, 3.25]$, which results in a nucleation barrier of approximately 80 k_BT at 1.75 and that decreases with increasing supersaturation

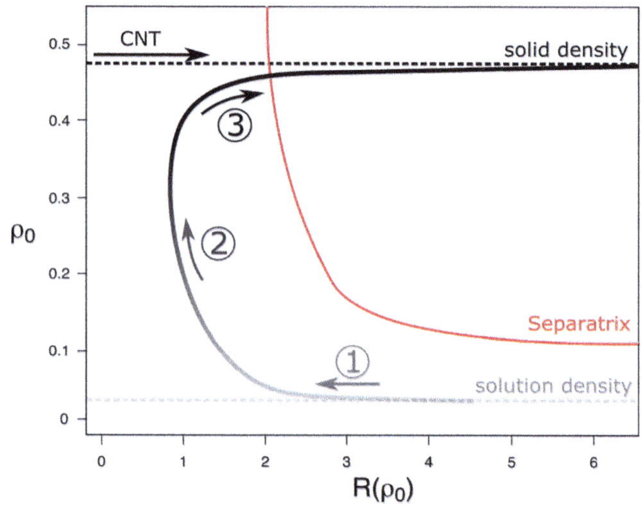

Fig. 3 Model nucleation pathways predicted by two-variable MeNT and CNT. Bottom and top dashed lines represent the initial and final stable-state densities, calculated using the EOS. To achieve nucleation the separatrix must be crossed. The critical cluster for each pathway is the intersection between the pathway and the separatrix. The MeNT pathway shows three main stages:[13] (1) contraction of the large-size-low-density fluctuations (bottom branch); (2) densification at an almost constant size (between $R = 1$ and $R = 2$, left branch); and, finally, (3) the growth stage of the clusters (top branch). During this final stage, the MeNT is similar to the CNT pathway, where the interior density is comparable to that of the bulk solid phase and only the size, *i.e.* R, changes.

(red arrow Fig. 4a). In Fig. 4b the computed effective energy barriers for the same set of supersaturation values, $S \in [1.75, 3.25]$, are shown. Simultaneously, Fig. 5a shows how this translates into the probabilities of observing a cluster of any given density between the values of the initial metastable and the final stable state. Remarkably, the effective energy barrier displays a minimum at an intermediate density between the mother phase and the critical-cluster density values. This has a direct impact on the cluster probability distribution (Fig. 5a), which, as we can observe, shows an intermediate local maximum of probability in pre-critical densities. Hence, the kinetic term does not affect all pre-critical clusters in the same way, since it induces a local minimum of the effective potential at an intermediate density, ρ_m, which is related to a maximum of probability. This tells us that such intermediate clusters will be observable before nucleation happens, therefore behaving as nucleation precursors.

Importantly, the local intermediate minimum observed in Fig. 4b does not come from a minimum of the thermodynamic free energy but emerges after taking into account the kinetics of cluster formation. This indicates that apparent multi-step energy barriers can be the result of the kinetics of cluster formation, while actually involving just a single energy barrier. But, Fig. 5a shows that with increasing supersaturation the likelihood of observing pre-critical clusters over a broad range of intermediate densities grows. This gradually decreases the importance of any specific region along the density axis (say the vicinity of an intermediate density, ρ_m), since any pre-critical density becomes more and more accessible as the supersaturation increases. Consequently, a separation between

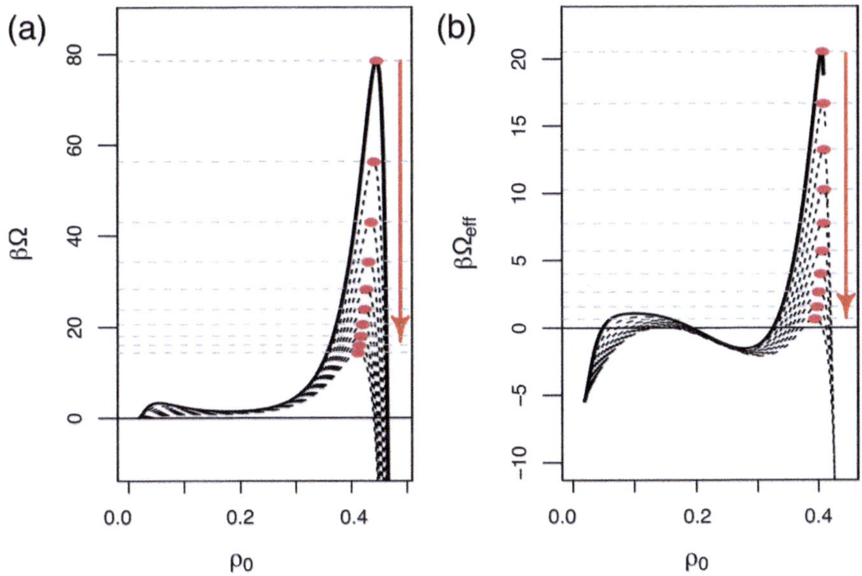

Fig. 4 Energy barrier (a) and effective energy barrier (b) for a range of supersaturation values $S \in [1.75, 3.25]$. Solid black line corresponds to $S = 1.75$, while all the other supersaturation values are represented as black-dashed lines. Red circles represent the critical clusters, and grey-dashed lines are drawn as a guide to the eye to highlight the nucleation barrier for each S value. Red arrows shows the direction of increasing S values.

a specific class of pre-nucleation clusters and the eventual nucleation event becomes less and less evident as the supersaturation value increases. Eventually, the effective energy barrier becomes of the order of thermal fluctuations so that all possible pre-critical densities, *i.e.* from monomers and onward, are all likely to be observed, making it harder to differentiate any specific class of precursor clusters.

Since a well-defined population of precursor clusters appears to be more evident as the supersaturation decreases we also studied a lower interval of supersaturation values $S \in [1.01, 1.75]$, which is shown in Fig. 5b. As can be observed, the results show that the emergence of precursor clusters is consistent

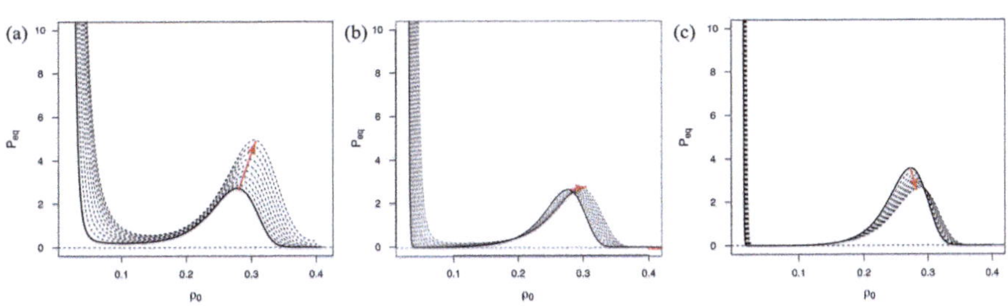

Fig. 5 Equilibrium probability distributions ($P_{eq} \sim e^{-\beta\Omega_{eff}}$) of clusters associated with the supersaturation values $S \in [1.75, 3.25]$ (a), $S \in [1.01, 1.75]$ (b) and $S \in [0.75, 1.0]$ (c). Solid black line corresponds to $S = 1.75$, $S = 1.01$ and $S = 0.75$, respectively. All the other supersaturation values are represented as black-dashed lines and the red arrow indicates increasing S values.

even for very low supersaturations. Provided that the cluster-formation pathway remains unaltered for undersaturated conditions, we could in principle also study whether these intermediate densities would still be there. The results for undersaturated conditions, *i.e.* $S < 1$, are shown in Fig. 5c. The results of our analysis for the supersaturation range $S \in [0.7, 1.0]$ are rather remarkable: (i) precursor clusters are also present for undersaturated conditions and (ii) gain importance (in terms of likelihood to be observed) as the density of the motherphase is reduced, *i.e.* as the system gets farther from the coexistence line.

It is worth noting that the precursor clusters described in the present model will appear in the mother phase almost instantaneously, and will persist throughout the whole nucleation process. This is because they appear as a bump of probability in the stationary cluster distribution. If anything, the time for them to be observed would be approximately the relaxation time of the cluster distribution, which is typically on the order of the diffusion time (which is negligible in most nucleation experiments). Regarding the characteristic size and density of such clusters, they are associated with the left branch (densification stage) of the nucleation pathway depicted in Fig. 3. Hence, their inner density is approximately the same as the critical cluster of the final equilibrium phase. Taking their associated range of densities, and checking the nucleation pathway, this means that these will be considerably smaller than the critical cluster, approximately on the order of one or two molecular radii.

Thus, on a qualitative level the characteristics of the precursor clusters observed in the model are comparable to the neutral species, *i.e.* ion pairs or larger, detected in the $SrSO_4$ co-titration experiments. Moreover, the observed evolution of the system with increasing supersaturation in the model and experimental results follow the same trend, *i.e.* with increasing supersaturation the role of a specific (in terms of size/density) class of precursor clusters in the nucleation process becomes less pronounced. The observed trend in the model at undersaturated conditions, *i.e.* increasing probability distribution of precursor clusters, could not be confirmed experimentally due to the large errors associated with the ISE measurements in the low concentration (undersaturation) regime of the $SrSO_4$ system. Notwithstanding, prenucleation clusters at undersaturated conditions have been observed experimentally for the $CaCO_3$ system.[10]

5. Implications

Up to now, classical nucleation has been associated with a single energy barrier (black curve, Fig. 1), while non-classical (multi-step) nucleation pathways are usually linked to an energy landscape including multiple individual barriers and/or local minima (pink profile, Fig. 1). Nonetheless, the preliminary experimental and model results presented in this study point towards the fact that such a complex energy landscape is not a prerequisite to warrant multi-stage crystallization, since precursor species can also emerge due to the competition between thermodynamics and kinetics during cluster formation on a continuous uphill energy profile. Interestingly, the resulting effective energy landscape significantly changes as a function of supersaturation and reveals that more than one nucleation mechanism is attainable for one system without changing the thermodynamic barrier of the system. It is important to note that the precursor particles observed in the model do not represent a thermodynamic phase, but rather transient particles that in a specific region of the U-shaped nucleation pathway have slower kinetics which renders them "observable". Stated differently, this specific low kinetic region of the effective barrier corresponds with a certain size/density of precursor clusters, which disguises them as a microscopic phase with a local energy minima. It is not inconceivable to consider that such kinetically induced precursor clusters have been frequently observed in recent times due to the advent of imaging, scattering and spectroscopy tools with high spatial and time resolutions.[3]

As in CNT, in our model nucleation also occurs through inherent thermal fluctuations that take the system over a nucleation barrier, which is largely defined by the chemical potential and mass-transport kinetics (opposed to the surface energy of the nucleus in CNT). The effective energy landscape that arises offers a plausible explanation for our data, as well as other *in situ* and *in silico* observations of systems where precursor clusters appear to play a decisive role, including $CaCO_3$,[10] $CaSO_4$,[19] NaCl,[20] proteins,[21] pharmaceutical compounds,[22] quantum dots[23] or metals.[24] Future challenges to support our model concern the U-shaped nucleation pathway derived by MeNT. In particular, the initial long-wavelength low-density fluctuations, which have been observed in Monte Carlo simulations using forward-flux sampling,[25] but still await experimental confirmation. Also, the current mono-component system should be extended to a multicomponent system, which will be a more realistic representation of *e.g.* salt nucleation. Additionally, the influence of the shape of the pathway on the probability distribution of precursor clusters needs to be addressed in detail.

6. Concluding remarks

This work represents a first attempt to merge experimental evidence of nonclassical mechanisms for nucleation and state-of-the-art theoretical advances that have been put forward in recent years. The application of such recent theories to experimental evidence has been severely hampered by the complexity of theoretical formalisms (for practitioners) and by the non-ideal experimental environment (for theoreticians). This has led to a parallel, yet largely disconnected, evolution of both theory and experiments. This needs to be resolved in order for the field to keep advancing. This work argues in favour of MeNT as a holistic nucleation theory capable of inherently capturing recent nonclassical observations without relying upon *ad-hoc* "designed" energy landscapes. Nonetheless, MeNT needs to further evolve in order to attain the capacity of fully describing multi-component systems. This is paramount to provide nucleation practitioners with a theoretical tool capable of predicting the outcome of a nucleation process.

Conflicts of interest

There are no conflicts to declare.

Acknowledgements

A. R. L., A. F.-M. and A. E. S. V. D. acknowledge the French national research agency, ANR, for funding through the project CATCH, ANR-18-CE05-0035-03.

References

1 D. Kashchiev, *Nucleation: Basic Theory with Applications*, Butterworth-Heinemann, Oxford, 2000.

2 J. J. De Yoreo, P. U. P. A. Gilbert, N. A. J. M. Sommerdijk, R. L. Penn, S. Whitelam, D. Joester, H. Zhang, J. D. Rimer, A. Navrotsky, J. F. Banfield, *et al.*, *Science*, 2015, **349**, aaa6760.

3 *New Perspectives on Mineral Nucleation and Growth: from Solution Precursors to Solid Materials*, ed. A. E. S. Van Driessche, M. Kellermeier, L. G. Benning and D. Gebauer, Springer International Publishing, Cham, 2017.

4 P. G. Vekilov, *Cryst. Growth Des.*, 2010, **10**, 5007–5019.

5 D. Gebauer, P. Raiteri, J. D. Gale and H. Colfen, *Am. J. Sci.*, 2018, **318**, 969–988.

6 J. De Yoreo, in *ACS Symposium Series*, ed. X. Zhang, American Chemical Society, Washington, DC, 2020, vol. 1358, pp. 1–17.

7 J. F. Lutsko, *J. Chem. Phys.*, 2012, **136**, 034509.

8 M. A. Durán-Olivencia, P. Yatsyshin, S. Kalliadasis and J. F. Lutsko, *New J. Phys.*, 2018, **20**, 083019.

9 J. F. Lutsko, *Sci. Adv.*, 2019, **5**, eaav7399.

10 D. Gebauer, A. Volkel and H. Colfen, *Science*, 2008, **322**, 1819–1822.

11 J. F. Lutsko and M. A. Durán-Olivencia, *J. Chem. Phys.*, 2013, **138**, 244908.

12 K. Kelton and A. L. Greer, *Nucleation in Condensed Matter: Applications in Materials and Biology*, Pergamon Materials Series, 2009.

13 J. F. Lutsko and M. A. Durán-Olivencia, *J. Phys.: Condens. Matter*, 2015, **27**, 235101.

14 M. A. Durán-Olivencia and J. F. Lutsko, *Phys. Rev. E: Stat., Nonlinear, Soft Matter Phys.*, 2015, **91**, 022402.

15 J. Hansen and I. McDonald, *Theory of Simple Liquids*, Academic, San Diego, CA, 1986.

16 N. F. Carnahan and K. E. Starling, *J. Chem. Phys.*, 1969, **51**, 635–636.

17 D. L. Parkhurst and C. A. J. Appelo, User's guide to PHREEQC - a computer program for speciation, reaction-path, 1D-transport, and inverse geochemical calculations, *U.S. Geol. Survey Water Res. Invest. Rep.*, 1999, **99**, 4259.

18 P. Blanc, A. Lassin, P. Piantone, M. Azaroual, N. Jacquemet, A. Fabbri and E. C. Gaucher, *Appl. Geochem.*, 2012, **27**, 2107–2116.

19 T. M. Stawski, A. E. S. Van Driessche, M. Ossorio, J. D. Rodriguez-Blanco, R. Besselink and L. G. Benning, *Nat. Commun.*, 2016, **7**, 11177.

20 H. Hwang, Y. C. Cho, S. Lee, Y.-H. Lee, S. Kim, Y. Kim, W. Jo, P. Duchstein, D. Zahn and G. W. Lee, *Chem. Sci.*, 2021, **12**, 179–187.

21 A. E. S. Van Driessche, P. H. H. Bomans, R. R. M. Joosten, D. Gil-Carton, N. A. J. M. Sommerdijk and M. Sleutel, *Nature*, 2018, **556**, 89–94.

22 Y. Tsarfati, I. Biran, E. Wiedenbeck, L. Houben, H. Cölfen and B. Rybtchinski, *ACS Cent. Sci.*, 2021, **7**, 900–908.

23 D. C. Gary, M. W. Terban, S. J. L. Billinge and B. M. Cossairt, *Chem. Mater.*, 2015, **27**, 1432–1441.

24 J. B. K. Cao, C. T. Stoppiello, R. L. McSweeney, T. W. Chamberlain, Z. Liu, K. Suenaga, E. Besley, S. T. Skowron, A. N. Khlobystov and U. Kaiser, *Nat. Chem.*, 2020, **12**, 921–928.

25 J. F. Lutsko and J. Lam, *Phys. Rev. E*, 2020, **101**, 052122.

Faraday Discussions

PAPER

The unexpected dominance of secondary over primary nucleation

Johannes Hoffmann,[ab] James Flannigan,[bc] Andrew Cashmore,[bc] Maria L. Briuglia,[bc] René R. E. Steendam,[ab] Charline J. J. Gerard,[d] Mark D. Haw,[c] Jan Sefcik[bc] and Joop H. ter Horst*[d]

Received 23rd November 2021, Accepted 5th January 2022
DOI: 10.1039/d1fd00098e

It is still a challenge to control the formation of particles in industrial crystallization processes. In such processes, new crystals can be generated either by primary or secondary nucleation. While in continuous stirred tank crystallization processes, secondary nucleation is thought to occur due to the shear or attrition of already present larger crystals; in antisolvent crystallization processes, where mixing at the inlets locally causes high supersaturations, primary nucleation is understood to be the main mechanism. We aim to show here that secondary nucleation is the dominant nucleation mechanism, even under conditions that are generally considered to be dominated by primary nucleation mechanisms. Measurements of primary and secondary nucleation rates under similar industrial crystallization conditions of sodium bromate in water, sodium chloride in water, glycine in water and isonicotinamide in ethanol show that the secondary nucleation rate is at least 6 orders of magnitude larger in all these systems. Furthermore, seeded fed-batch and continuous antisolvent crystallizations of sodium bromate under high local supersaturation, seeded with crystals of a specific handedness, result in a close to chirally pure crystalline product with the same handedness. This shows that indeed, enantioselective secondary nucleation is the dominant mechanism in these antisolvent crystallizations. It is even possible to use the enantioselective secondary nucleation mechanism to control the product chirality in such a process, making antisolvent crystallization a viable crystallization-enhanced deracemization technique, having a superior productivity compared to other crystallization-enhanced deracemization methods. Our finding of a dominant secondary nucleation mechanism, rather than primary nucleation, will have a strong impact on nucleation control strategies in industrial crystallization processes.

[a]*Strathclyde Institute of Pharmacy and Biomedical Sciences (SIPBS), University of Strathclyde, 161 Cathedral Street, Glasgow, G4 0RE, UK*
[b]*EPSRC Centre for Innovative Manufacturing in Continuous Manufacturing and Crystallisation (CMAC), Technology and Innovation Centre, University of Strathclyde, 99 George Street, Glasgow, G1 1RD, UK*
[c]*Department of Chemical and Process Engineering, University of Strathclyde, 75 Montrose Street, Glasgow, G1 1XJ, UK*
[d]*Laboratoire Sciences et Méthodes Séparatives, Université de Rouen Normandie, Place Emile Blondel, 76821 Mont Saint Aignan Cedex, France. E-mail: Joop.ter-Horst@univ-rouen.fr*

1. Introduction

Crystallization is a separation process, producing solid particles through the combination of primary crystal nucleation, crystal growth, secondary nucleation and agglomeration.[1] Primary nucleation is the formation of new crystals from a clear supersaturated solution.[2] Secondary nucleation is the process of forming new crystals in a supersaturated solution due to the presence of parent crystals of the same compound.[3–6] The simultaneous occurrence of all these subprocesses in a batchwise or continuous crystallization make it a challenge to recognize the nucleation mechanism and determine the primary or secondary nucleation rates and their relevance to the specific process. Whether primary or secondary nucleation is the dominant mechanism is very important for the control of industrial crystallization processes as the control strategies are very different.

In batchwise seeded cooling crystallizations, usually both primary and secondary nucleation must be avoided to ensure control over the final crystal size distribution.[7] In unseeded batchwise cooling crystallizations, the initial burst of nucleation is usually assumed to be primary heterogeneous nucleation. Even more so, in batchwise and continuous antisolvent crystallizations, due to the very high local supersaturations at the mixing point of the antisolvent and feed streams, primary nucleation is generally thought to be the dominant mechanism. However, in continuous evaporative and cooling crystallizations, in which crystals already are present, secondary nucleation is generally assumed to occur.[2,5] Conversely, continuous evaporative and cooling crystallization processes performed in continuously stirred tanks rely on secondary nucleation to provide a continuous source of secondary nuclei in the suspension.[3,5,8] The mechanism with which these parent crystals serve as a catalyst for the formation of small new crystals is still not well understood and it is suggested that attrition[9] or fluid shear[10] are involved.

An interesting process in which secondary nucleation would be favoured over primary nucleation is the deracemization of a racemic system into an enantiopure crystalline product.[11] Crystallization-enhanced deracemization is a hybrid process of a solution-phase racemization reaction and a crystallization that allows the formation of an enantiopure product from a racemic solution or suspension.[12] Viedma ripening is a particular deracemization process, where a racemic mixture of chiral crystals can be deracemized by attrition under isothermal conditions.[13] Enantioselective secondary nucleation in a process with an enantiopure seed suspension would generate only new crystals with the same handedness as that of the seed crystals, while unselective primary nucleation in such a process would result in the loss of product chirality, as crystals with both handednesses are formed. While deracemization by Viedma ripening is haunted by a relatively low productivity, in terms of units of product weight per units of time and volume, due to the relatively low supersaturations during the process, temperature cycling induced deracemization (TCID)[14,15] and second order asymmetric transformation (SOAT)[16] processes allow an improvement in productivity, as the supersaturations achieved for crystallization are higher, while primary nucleation can be avoided. In respect to the productivity, it is interesting to investigate antisolvent crystallization as a deracemization technique: the driving forces in this crystallization process are very high, leading to high productivities, as drastic drops in solubility can be achieved due to the addition of the antisolvent. However, generally it is thought that primary nucleation, rather than secondary, is occurring in such processes.

We aim to show that secondary nucleation is the dominant nucleation mechanism, even under conditions that are generally considered to be dominated by primary nucleation mechanisms (Fig. 1). We use the model compound sodium bromate ($NaBrO_3$), which is achiral in solution but crystallizes in a chiral space group,[17] making its crystals either left- or right-handed, like the crystals of a conglomerate forming chiral compound.[18] First, we determine the secondary nucleation rate, as well as the handedness of the crystallized product in single crystal seeded supersaturated solutions. Then, we investigate the preservation of handedness in fed-batch and continuous antisolvent crystallization processes of this compound. Finally, we determine the primary and secondary nucleation rates of several compounds to show the more general applicability of the results we find.

2. Experimental

Sodium bromate (VWR, 99%), sodium chloride (Sigma-Aldrich, ACS reagent, 99%), glycine (Sigma-Aldrich, 99%), double filtered water (Milli-Q gradient, Millipore SAS, solvent, prepared in-house) and ethanol (VWR, 99%, antisolvent) were used as received.

2.1 Nucleation rate measurements

2.1.1 Primary nucleation. The primary nucleation rates were obtained from the induction time distributions determined under isothermal conditions in agitated vials using the Crystal16 instrument (Technobis, The Netherlands) for $NaBrO_3$ in water and the Crystalline instrument (Technobis, The Netherlands) for isonicotinamide (INA) in ethanol, glycine in water and sodium chloride in water. The induction time was measured as the time taken from the beginning of the

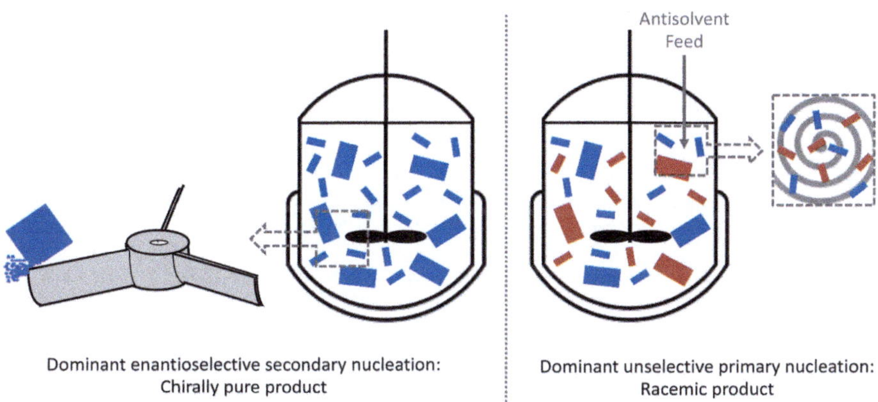

Fig. 1 Chirally unselective primary nucleation and enantioselective secondary nucleation in a highly productive antisolvent crystallization-enhanced deracemization lead to, respectively, a racemic and an enantiopure product.

isothermal conditions to the detection of crystals using light transmission measurements.

For the Crystal16 experiments, first, samples with various concentrations of $NaBrO_3$ were used to determine the clear and cloud point temperatures using heating and cooling rates of 0.5°C min^{-1} between −25.0 °C and 50.0 °C for $NaBrO_3$ in water.[18] The samples were stirred with a controlled stirring speed of 700 rpm, using a PTFE coated magnetic stirring bar. Then, a clear stock solution of 429 g l^{-1} at elevated temperature was made for $NaBrO_3$ to fill 8 vials for the induction time measurement series at each supersaturation. The induction times were measured at $T_x = 20.0$ and 25.0 °C, which, respectively, result in supersaturation ratios $S = 1.17$ and 1.08 for $NaBrO_3$.

For the crystalline experiments, 40, 24 and 10 vials were prepared, each containing 3 ml of solution with equal compositions of INA in ethanol, glycine in water and sodium chloride in water, respectively (Table 1). The solubilities of isonicotinamide (INA) in ethanol,[19] glycine in water[20] and sodium chloride in water[21] were taken from the literature, after a check of their validity. The vials were subjected to a temperature profile that allowed several induction time measurements per sample. The vials were initially held at a temperature T_{high} above their saturation temperature to ensure complete solute dissolution. Then, they were cooled at a rate of −5 °C min^{-1} to the temperature T_i, at which the induction time measurement took place. The vials were held at the low temperature T_i for 1.5, 4 and 3 hours for INA in ethanol, sodium chloride in water, and glycine in water, respectively. Throughout the experiments, the vials were agitated for INA in ethanol and glycine in water using a Teflon coated magnetic stirrer bar at 700 rpm. Agitation for sodium chloride in water was supplied by an overhead downflow short shaft propeller at 1250 rpm. The distribution of the induction times was then fitted against a model developed by Jiang and ter Horst[22] to obtain the primary nucleation rate J and growth time t_g:

$$P(t) = 1 - \exp(-JV(t - t_g))$$

2.1.2 Secondary nucleation. The secondary nucleation rates were determined using a recently proposed method.[19] The secondary nucleation rate was

Table 1 Experimental conditions used in the determination of the primary and secondary nucleation rates in 4 systems. For an induction time measurement at temperature T_i, the sample composition is equal to the concentration C, while the solubility is C^* at that temperature. The solubility and concentration have units of mg solute per gram of solvent. The secondary nucleation rate experiments of $NaBrO_3$ took place at slightly higher supersaturations S

Solute	Solvent	T_i [°C]	C^* [mg g^{-1}]	C [mg g^{-1}]	S [−]
$NaBrO_3$	Water	20	366.4	429	1.17
		25	397.5	429	1.08
NaCl	Water	25	359.3	367.1	1.02
Glycine	Water	25	249.5	294.4	1.18
INA	Ethanol	18	74.3	83.2	1.12

determined from the rate of increase in the number concentration of particles following seeding isothermal supersaturated solutions with a single crystal, using multiple reactors (Crystalline, Technobis Crystallization Systems B.V.) set up with 8 ml glass vials. All reactors were independent and monitored by a camera, from which the images were analyzed automatically to obtain the particle count.

For the $NaBrO_3$ experiments, the seed crystals were prepared through evaporative crystallization of a saturated solid-free solution, which was filtered beforehand. After about three days, the triangular shaped crystals were large enough to distinguish their chirality by polarization microscopy. The formed crystals were washed with cold water to remove any possible residual particles, which could otherwise result in initial breeding.[23] After the characterization through the polarized microscope, the single crystals were weighed and classified for their size.

For $NaBrO_3$ in water, two stock solutions at different supersaturation ratios were prepared (Table 1). The solutions were stirred for two hours at 20 °C above the saturation temperatures to ensure a clear solution. The stock solutions were quickly filtered and transferred into a pre-warmed bottle, ensuring the absence of heterogeneous particles, and avoiding any possible nucleation. Before use, the stock solutions were heated up to around 40 °C for 20 minutes, to ensure a complete dissolution, after which the vials were filled with 3 ml of the clear solution and heated to around 40 °C in the Crystalline to give a clear solution. Then, the solutions were cooled down to the desired temperature with a cooling rate of 1 °C min^{-1}. Then, two crystalline vials were taken out from the Crystalline. One vial was seeded with a single seed crystal, previously characterized for chirality and size, added from a small sample vial into the crystallization vial. The second vial did not receive any crystal, providing an unseeded test.

For each other system, three vials were filled with solutions having the same composition as in the primary nucleation measurements. These vials were subjected to a single hot–cold temperature profile under the same thermal and agitation conditions as in the primary nucleation rate measurements. Only for INA in ethanol, the cooling to 18 °C was performed at −1 °C min^{-1}. Once isothermal conditions were reached, each vial was seeded with a single washed crystal measuring approximately 3 mm × 3 mm × 2 mm for sodium chloride, with a length of 2.5 mm for α-glycine and with a projected area of around 3–6 mm^2 form II INA. Washing was performed using a small volume of the pure solvent. The vials were seeded at times shorter than the shortest induction time. At set intervals, images were recorded using a built-in camera in the Crystalline setup to allow the number of solid particles visible to be counted. A previously performed calibration process, where 50 μm diameter polystyrene spheres in suspension were used to determine the correlation between the number of particles visible and the overall volumetric number concentration, was used.[19] The number concentration associated with each image was calculated and plotted against time. The secondary nucleation rate B was determined from the slope of the fitted line to the initial rise in the particle density. Data where the images contained fewer than 10 or more than 160 identified particles were excluded from analysis to eliminate background effects and to allow each particle to be individually distinguished, respectively.[19]

A relatively low supersaturation was used for the experiments with sodium chloride as, at higher supersaturation levels, a significant proportion of vials nucleated before isothermal conditions were reached, while also crowning of a crystal layer just above the liquid level was observed in such samples using overhead stirring.[24] The supersaturations used for INA and glycine were chosen to allow the vials to be seeded to ensure that the washed seed was the source of the secondary nuclei and to reduce the likelihood of primary nucleation occurring during the experiments. Using higher supersaturations in combination with glycine would not have left enough time between reaching isothermal conditions and primary nucleation occurring to allow this procedure to be conducted.

2.2 Antisolvent deracemization of $NaBrO_3$

2.2.1 Fed-batch antisolvent crystallization of sodium bromate. A schematic overview of the fed-batch antisolvent crystallization set up is shown in Fig. 2, where the black lines represent the process streams in the fed-batch set up, consisting of a crystallizer and an antisolvent feed tank connected to the crystallizer through a peristaltic pump (Watson Marlow, 520u). Both jacketed vessels were temperature controlled at 20 °C through a Pt100 probe (Radleys) connected to a heater chiller (Lauda pro line).

An initial solution of 500 ml, containing water and 7 v% antisolvent (ethanol), at a concentration of 274 mg ml^{-1} $NaBrO_3$, was prepared in the crystallizer (Fig. 2, middle). The initial antisolvent fraction was chosen to prevent too large suspension densities. The initial content of the crystallizer was heated up to 30 °C and was kept at that temperature for 20 minutes to ensure a clear solution. After

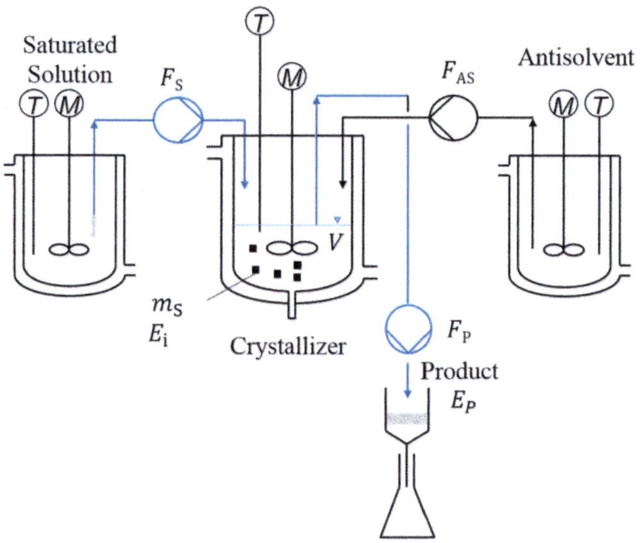

Fig. 2 Schematic overview of the fed-batch (black lines) and continuous (black and blue lines) antisolvent crystallization set ups with the saturated solution feed tank (left), the crystallizer (middle), suspension removal and filtration (bottom) and the antisolvent feed vessel (right). In the fed-batch configuration, the antisolvent was added with a feed flow rate F_{AS} to the crystallizer containing a suspension of crystals of a single handedness. In the continuous configuration, both a solution feed flow rate F_S and an antisolvent feed flow rate F_A were continuously added to the crystallizer, while a continuous suspension flow F_P was leaving the crystallizer.

cooling the clear solution back to its saturation temperature of 20.0 °C, a mass m_S of seed crystals of a specific handedness (enantiomeric excess E_i = 96.1%) was added.

The fed-batch antisolvent crystallization experiment was initiated by starting the flow of antisolvent F_{AS} into the crystallizer using a peristaltic pump (Watson Marlow, 520u). The experiment was run until a target antisolvent fraction of x_{AS} = 50 v% was reached. Then, the suspension was filtered and the solid product dried for analysis. During the experiment, suspension samples were taken to determine the handedness of the crystals (see Chiral product purity section). The solute concentration in the suspension liquid phase of the sample was determined using a gravimetric method, in which the solvent was evaporated from the solution sample using an oven at 40 °C. The antisolvent fraction development in time was determined through the antisolvent feed pump calibration.

The suspension was vigorously stirred at 800 rpm. To prevent fouling, the inlet tube of the antisolvent flow was placed above the suspension, with the antisolvent dropping into the stirred suspension. Baffles were used to prevent the formation of a vortex and to ensure good mixing of the suspension.

2.2.2 Continuous antisolvent crystallization of sodium bromate. The continuous antisolvent crystallization set up (Fig. 2) was realized by extending the fed-batch set up with a continuous solution feed and suspension outflow. The saturated solution was fed from a 5 l feed vessel into the crystallizer using a pump (Watson Marlow, 530). The inlets in the crystallization vessel were positioned such that the incoming solution and antisolvent did not get in direct contact with the baffles or walls of the crystallizer before entering the suspension, preventing fouling. The continuous product outflow was realised by a level control, fixating the outflow tubing (MasterFlex, d_i = 6.4 mm) to a baffle at a defined height, so that the suspension volume remained constant at 0.5 l under operating conditions. A peristaltic pump (MasterFlex, Pump Drive) at a high rotation rate was used to constantly extract slurry into the filter. During the crystallization process, high stirring rates of 800 rpm were used to prevent the accumulation of solids in the crystallizer.

For each experiment, 500 ml of a starting $NaBrO_3$ solution in water and antisolvent at a particular antisolvent fraction was prepared in the crystallizer at 30 °C, after which it was cooled down to its saturation temperature of 20.0 °C. Then, the saturated clear solution was seeded with seed crystals (enantiomeric excess E_i = 95.2%). A seed mass of 10 g l^{-1} was used for each experiment.

The continuous experiment was started by starting the pumps that controlled the crystallizer's ingoing feed flows of solution and antisolvent and outgoing suspension flow. The feed flow rates were chosen in such a way that the antisolvent fraction of the initial solution in the crystallizer was maintained. A $NaBrO_3$ solution at a concentration C_0 of 350 mg ml^{-1} at 20 °C was used as the feed, while pure ethanol was used as the antisolvent. The experiment duration was at least 7 residence times $\tau = V/F$, where V is the constant volume of the suspension in the crystallizer and F is the overall incoming flow rate. During each experiment, suspension samples were taken, similar to the procedure in the fed-batch antisolvent crystallization experiments, in order to determine the enantiomeric excess as a function of the antisolvent fraction x_{AS}.

For the experiments C1–C3 (Table 4), the overall feed flow $F = F_S + F_{AS}$ was kept constant (20 ml min^{-1}) and with that the residence time (between 24–26 min), where the differences were due to slight variations in the set overall flow rates. The antisolvent fraction was controlled by the relative flow rates of the saturated solution, F_S, and antisolvent, F_{AS}. For continuous experiments C4 and C5 (Table 4), the overall flow rate F was varied, while the overall antisolvent fractions $x_{AS} = 20$ v% were kept constant, investigating the influence of the residence time on the process, as at a higher flow rate F, the residence time $\tau = V/F$ is shorter.

2.2.3 Chiral product purity. $NaBrO_3$ is achiral in solution and chiral in the solid state, thus standardised methods, such as HPLC, cannot be used for the determination of the handedness of the crystalline product. $NaBrO_3$ crystals are sensitive to polarized light and therefore they can be analysed using a microscope equipped with polarized light. The crystals from the filtered and dried suspension samples were suspended in a clear saturated solution of $NaBrO_3$ in water and left to grow overnight through slow evaporation of water under conditions at which no further nucleation occurred. Under polarized light, crystals of one handedness appeared orange, while those of the other handedness appeared blue. By changing the polarization angles, the colour of the crystals of each handedness were reversed.

For all solid samples, the same microscope (Leica DM6000M) was used at the polarization angles of 100° and 70° and pictures were taken with both polarizations. For both pictures, the crystals of each handedness were then counted. Each picture rendered a set of 200–400 crystals. For each purity sample, 2 to 3 sets of crystal pictures were analyzed, resulting in 400–1000 crystals counted per solid sample to ensure a high statistical accuracy. The crystals with an orange (N_O) and blue (N_B) colour were counted for one of the two polarization angles in all pictures to determine the enantiomeric excesses E_1 and E_2 in the solid sample through eqn (1):

$$E = \frac{N_O - N_B}{N_O + N_B} \times 100 \quad (1)$$

Although the crystals usually changed colour upon the inversion of the angle of polarized light, typically 10–24 crystals did not, in a sample of at least 400 crystals. This results in a maximal variation of around 2.5–6% between the enantiomeric excesses E_1 and E_2. Therefore, the average of E_1 and E_2 was used as the final enantiomeric excess of the solid sample.

3. Results

3.1 Rate and chirality of sodium bromate nucleation in water

We chose to work with the model compound sodium bromate ($NaBrO_3$), which is a water soluble salt and achiral in solution, while it crystallizes as chiral crystals because of the chiral space group $P2_13$.[17] Therefore, each crystal has only one of the two possible handednesses. If chirally unselective primary nucleation occurs, an initial excess in handedness would be lost and a mixture of solid particles with both handednesses would develop. However, if enantioselective secondary nucleation would occur, the initial excess in handedness would be retained. We used the recently developed secondary nucleation rate measurement method based on a single crystal seeding procedure on a ml scale[19] that allows comparison to the primary nucleation rate measurements on a similar scale with a previously developed method.[22,24]

3.1.1 Primary nucleation. The solubility of $NaBrO_3$ in water ranges from about 300 g l^{-1} at around 10 °C to about 425 g l^{-1} at around 30 °C. For all these concentrations, the metastable zone width (MSZW) is roughly 25 °C, when using a cooling rate of 0.5 °C min^{-1}. It is therefore possible to measure relatively long induction times at a constant temperature below the saturation temperature of a sample solution, if the temperature at which the induction time is measured is within 25 °C of the saturation temperature.

Fig. 3 shows such induction time measurements at temperatures of 20.0 and 25.0 °C for $NaBrO_3$ solutions in water at supersaturation ratios of $S = 1.08$ and 1.17, respectively. The measurements per series are represented as induction time distributions, where the probability $P(t)$ to measure a certain induction time is plotted against time.[22,24] While the shortest induction time is more than 100 minutes, it takes a further substantial time to detect crystals in all samples: for a supersaturation of $S = 1.17$, the induction times range from 116 to 168 minutes.

The sample-to-sample variation in the induction time under the same conditions is due to the stochastic behaviour of primary nucleation and indicates that the primary nucleation rate is rather low:[22,24] in a small volume of a stirred clear solution only a single crystal forms, which, after growth, is responsible for the formation of the subsequent crystals, probably by undergoing collisions with the stirrer, which result in many attrition fragments. This single crystal nucleation mechanism is observed in 1 ml supersaturated solutions just before the suspension formation[22] and can be dominant up to volumes as large as 1 liter, as observed from the variations in induction times up to these volumes.[25] Under concomitant crystallization conditions of different polymorphs in solvent

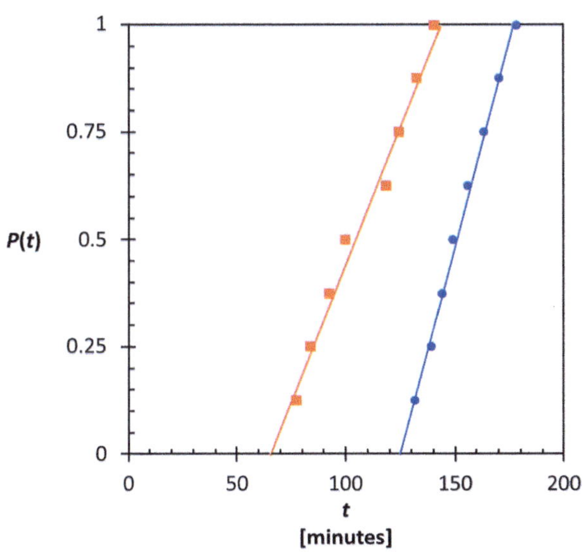

Fig. 3 The induction time distributions for $NaBrO_3$ in water at temperatures of 25 and 20 °C and supersaturation ratios $S = 1.08$ (blue) and 1.17 (orange) based on 8 measurements per series. The lines are a guide to the eye.

mixtures, pure polymorphic products appear randomly from experiment to experiment, indicating a secondary nucleation mechanism taking place, due to which the secondary nuclei have the same crystal structure as the first single crystal that is formed. The same principle applies to chiral compounds, where chiral pure crystals appear randomly from experiment to experiment, through crystallization from achiral solutions.[26]

Assuming that a single nucleus mechanism initiates crystallization, the nucleation rate can be roughly estimated from the slope of the fit of a straight line to the induction time distribution.[24] Due to the small number of measurements, a more accurate model description would not result in a more accurate nucleation rate estimate. In this way, the primary nucleation rate can be estimated for both series of experiments to be roughly $0.01 < J < 0.04$ ml^{-1} min^{-1}. This range of nucleation rates means that, on average, it takes as long as 25 to 100 minutes to form a single primary nucleus in the 1 ml vials. There is thus sufficient time to study secondary nucleation under these conditions.

3.1.2 Secondary nucleation rate and chirality. Under exact supersaturation conditions, within a time frame for which no crystals from primary nucleation are detected, secondary nucleation now can be studied. We chose to perform the secondary nucleation rate measurements by seeding a single crystal in the supersaturated solutions following a previously developed method.[19] Under the chosen conditions, the seed crystal is solely responsible for the generation of new crystals. By following the change in the number density of crystals in time in the suspension, it is possible to determine the secondary nucleation rate under well-defined conditions.

Fig. 4 shows the typical outcome of a pair of secondary nucleation rate experiments, in which the suspension number density N_ρ of particles is monitored after seeding in one of the vials is initialized at $t = 0$. The pair of

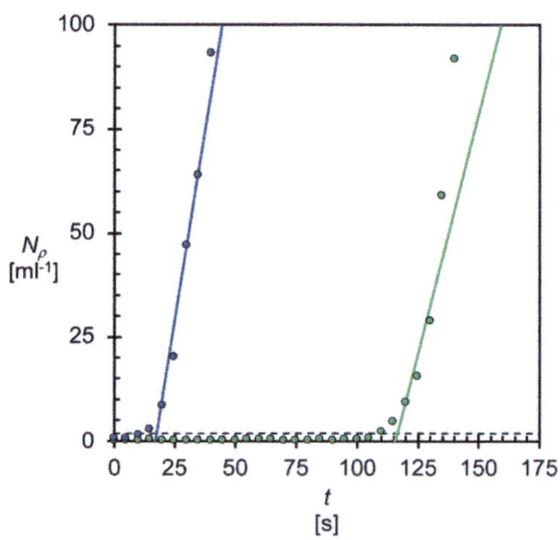

Fig. 4 The change in number-based suspension density N_ρ of supersaturated solutions at $S = 1.2$ and a temperature of 18.0 °C for single crystal seed experiments and unseeded experiments. Blue circle: single crystal self-seeding of a $NaBrO_3$ single crystal in a $NaBrO_3$ solution; green circle: an unseeded $NaBrO_3$ solution. The horizontal dashed line indicates the lower detection limit of $N_\rho = 2 \times 10^3$ ml^{-1}.

experiments includes a seeded single crystal seed experiment of a $NaBrO_3$ seed crystal in a $NaBrO_3$ supersaturated solution and an unseeded experiment, both performed at a supersaturation ratio $S = 1.2$. Around 10 seconds after the addition of the single crystal seed, the suspension density N_ρ starts to rise. This is substantially earlier than in the parallel unseeded experiment, showing that the single crystal is responsible for the detected suspension density increase. The roughly linear increase of the suspension density in the seeded experiment allows the secondary nucleation rate to be determined to be $B = (221 \pm 24).10^3$ ml^{-1} min^{-1}.

Table 2 shows the secondary nucleation rates B for several experiments, Br1–8, measured at a supersaturation of $S = 1.2$ using seed crystals of 2.4 to 14.4 mg to have values varying from 61×10^3 to 224×10^3 ml^{-1} min^{-1}. These values mean that around 1000 to 4000 secondary nuclei per second are formed in a 1 ml sample. Comparing these secondary nucleation rates $61 \times 10^3 < B < 224 \times 10^3$ ml^{-1} min^{-1} to the estimated primary nucleation rates $0.01 < J < 0.04$ ml^{-1} min^{-1}, shows that the secondary nucleation rates are orders of magnitude higher than the primary nucleation rates in similarly supersaturated solutions. Therefore, under these conditions, secondary nucleation is dominant.

The only prerequisite for secondary nucleation is the presence of a parent crystal, as this is the origin of the secondary nuclei. In that respect, it is interesting to see the behaviour of the unseeded experiment in Fig. 4, which is similar to that of the seeded experiment. This indicates that the underlying mechanism of suspension formation is similar in the unseeded and the seeded experiments, which suggests that, in the unseeded case, a single crystal is also formed, which is subsequently the origin of the secondary nuclei that are detected. It seems that also in the unseeded case, the majority of crystals are formed by secondary rather than primary nucleation.

We further investigated how the chirality of the seed crystal impacted on the chirality of the crystals in the formed suspension, by measuring the chirality of the seed crystal, as well as the dominant chirality and the enantiomeric excess in representative samples of the final product. A positive chirality (+) means that the crystals predominantly rotate the polarized light of the microscope clockwise; if the chirality is negative (−), the obtained crystals rotated the polarized light

Table 2 The secondary nucleation rate B and the chirality of the seed crystal with mass m_0 and the product chirality and enantiomeric excess E from single crystal seeded secondary nucleation rate measurements of $NaBrO_3$ single crystals in supersaturated solutions of $NaBrO_3$ at a supersaturation $S = 1.2$

Exp.	Seed mass [mg]	Seed chirality	B [$\times 10^3$ ml^{-1} min^{-1}]	Product chirality	E
Br1	2.4	−	224 ± 28	−	−99
Br2	5.5	+	136 ± 27	+	99
Br3	6.5	+	104 ± 16	+	100
Br4	9.5	−	61 ± 8	−	−97
Br5	9.8	−	113 ± 18	−	−100
Br6	10.6	+	175 ± 28	+	100
Br7	13.7	−	221 ± 24	−	−100
Br8	14.4	+	175 ± 28	+	97

counter-clockwise. Table 2 shows that the dominant chirality of the product mirrors the chirality of the seed crystal. Furthermore, the chiral purity of the product is extremely high, more than 97%. This indicates that all crystals in the suspension are formed as a result of the presence of the seed crystal, probably due to an attrition mechanism. This means that the dominant secondary nucleation allows reproducible control of the chirality of the formed product.

3.2 Antisolvent deracemization of $NaBrO_3$

Antisolvent crystallization is a widely applied process in the pharmaceutical industry, due to its high yields and short process duration. To the best of our knowledge, it has not been reported to be used for chiral resolution or deracemization as it is generally thought that, due to the high supersaturations produced during the process, especially at the mixing points, chirally unselective primary nucleation rather than enantioselective secondary nucleation would occur. However, the previous section showed that, at supersaturation conditions representing industrial crystallization, primary nucleation rates are very low, while the secondary nucleation rate in the presence of a single crystal is dominant. In addition, the chirality of the generated secondary nuclei is the same as that of the parent crystal. If dominance of secondary nucleation also prevails at higher supersaturation S, antisolvent crystallization can be used as a deracemization technique.

$NaBrO_3$ has a very high solubility in water, while being highly insoluble in most alcohols.[17,18] The solubility of $NaBrO_3$ at 20 °C as a function of antisolvent volume fraction, as obtained in another study,[27] is shown in Fig. 5. The $NaBrO_3$ solubility decreases strongly and non-linearly with the antisolvent fraction from $C^* =$

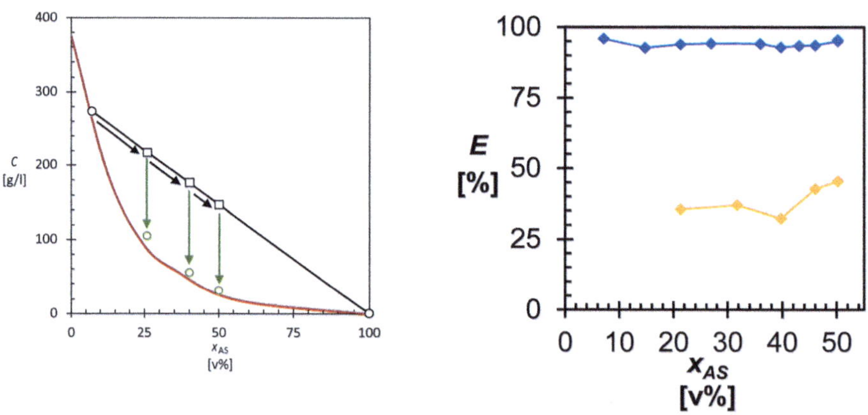

Fig. 5 Left: the solubility of $NaBrO_3$ at 20 °C (red line) as a function of the antisolvent fraction x_{AS} in water. The operating line (black line) of the fed-batch experiment FB1 runs from the composition (blue circle) of the saturated solution in the crystallizer having an antisolvent fraction $x_{AS} = 7$ v%, to the 100% pure antisolvent composition (red circle), which is added to the crystallizer with a constant rate until an antisolvent fraction $x_{AS} = 50.1$ v% (blue square) is reached. The concentrations measured during the process are indicated by green circles. Right: product enantiomeric excess (blue) measured over the course of fed-batch antisolvent crystallization experiment FB1 (blue) and of unseeded experiment FB4 (yellow).

370.5 g l^{-1} in water, down to $C^* = 19.5$ g l^{-1} in a solvent mixture with 55.8 v% ethanol, a more than tenfold decrease. Using this phase diagram, isothermal $NaBrO_3$ antisolvent crystallization experiments can be designed using ethanol as the antisolvent. First, we will discuss fed-batch antisolvent crystallization experiments and then continuous antisolvent crystallization experiments.

3.2.1 Fed-batch antisolvent crystallization. Fed batch experiment FB1 is exemplified in the phase diagram in Fig. 5. During fed-batch antisolvent crystallization experiment FB1, the antisolvent was added with a flow rate of $F_{AS} = 3$ ml min^{-1} to the saturated solution containing enantiopure seeds up to a specific antisolvent fraction. The fed-batch process starts with a saturated solution in the crystallizer ($C_0 = 273$ mg ml^{-1}, blue circle Fig. 5) at an antisolvent volume fraction $x_{AS} = 7$ v%. This initial antisolvent fraction lowers the solubility compared to that in pure water and allows us to start the experiment with a saturated solution containing 20 mg ml^{-1} $NaBrO_3$ crystal seeds of an almost single handedness (enantiomeric excess $E = 96.1\%$), while the suspension density towards the end of the process at $x_{AS} = 50.1\%$ remains sufficiently low, around 200 mg ml^{-1} or 17 w %, to suspend the crystals well. Upon starting the process, a constant flow of antisolvent is pumped into the crystallizer leading to a dilution of the initial concentration following the operating line in the direction of the blue arrow in Fig. 5 (left). At the same time, the solubility is decreasing non-linearly, following the solubility line. As the solubility becomes lower than the overall concentration, product immediately starts crystallizing. Experiment FB1 was stopped when arriving at an antisolvent fraction of $x_{AS} = 50.1$ v%.

At the end of experiment FB1, the solution concentration is measured to be $C = 31$ mg ml^{-1} (green circle at 50.1% x_{AS}, Fig. 5, left), slightly higher than the solubility $C^* = 21.8$ mg ml^{-1} at that antisolvent fraction. Using the measured concentration, solubility, and the overall composition at the operating line $C_M = 147$ mg ml^{-1} (Fig. 5, left, blue rectangle), the yield $X = (C_M - C)/(C_M - C^*)$ can be calculated as the obtained amount of product over the maximum amount obtainable at the antisolvent fraction 50.1%. For experiment FB1 a yield of $X = 92.4\%$ is obtained. As the produced amount (around 200 mg ml^{-1}) substantially supersedes the amount of seeds used (20 mg ml^{-1}), extensive nucleation must have taken place throughout fed-batch experiment FB1.

Fig. 5 (right) shows that the measured enantiomeric excess E of the crystals in the suspension remains high throughout the experiment. The process starts with an enantiomeric excess in the seed material of 96.1%, while the final value is determined at $E = 95.6\%$, showing that the handedness in the seeds can be retained in the product, even up to an antisolvent fraction of $x_{AS} = 50.1\%$. This means that the nucleation occurring during the fed-batch antisolvent crystallization is enantioselective and must follow a secondary nucleation mechanism.

The main parameter for control over the handedness of the crystals in the fed-batch experiments is the addition rate of antisolvent. If the antisolvent addition rate is increased, the local supersaturation at the mixing point is increased, which increases the likelihood of excessive primary nucleation of the undesired handedness. Additional fed-batch antisolvent crystallization experiments with higher antisolvent addition rates of $F_{AS} = 6$ and 20 ml min^{-1} (Table 3, experiment FB2 and FB3) show that the local supersaturations become sufficiently high to induce some primary nucleation, resulting in a reduced final enantiomeric excess of the crystalline product (respectively, 84.0 and 88.5%). However, all the conditions led

Table 3 Overview of the experimental conditions (antisolvent addition rate F_{AS}, seed mass m_S, enantiomeric excess E_S of the seed material) and results (enantiomeric excess E_P of product crystals and the yield X) of the fed-batch antisolvent crystallization experiments

	Process conditions			Experimental results	
Experiment	F_{AS} [ml min^{-1}]	m_S [mg ml^{-1}]	E_S [%]	E_P [%]	X [%]
FB1	3	20	96.1	95.6	92.4
FB2	6	20	88.5	86.5	84.0
FB3	20	20	89.3	78.5	88.5
FB4	6	0	—	44.8	88.0

to an enantiomeric excess only slightly lower than that of the seed material, indicating that still secondary nucleation is the dominant mechanism, even at more extreme local supersaturations.

Subsequently, we performed an unseeded fed-batch antisolvent crystallization experiment FB4, under otherwise similar conditions to those of FB2. Surprisingly, even in this unseeded experiment, no racemic solid product is obtained but rather a product with a substantial enantiomeric excess ($E_P = 44.8\%$, Fig. 5, right), at similar yields. Apparently, a limited number of primary nucleation events at the start creates a bias in the handedness of the nucleated crystals, which remains after enantioselective secondary nucleation takes over as the dominant nucleation mechanism, with the initially primary nucleated crystals serving as seeds for the secondary nucleation.

Overall, we showed that seeded fed-batch antisolvent crystallization processes can be used for the spontaneous resolution of $NaBrO_3$ to produce a desired handedness.

3.2.2 Continuous antisolvent crystallization. While the antisolvent fraction in fed-batch antisolvent crystallization increases over time, in continuous antisolvent crystallization in a stirred tank, the antisolvent fraction is constant over time. This allows us to investigate at which antisolvent fraction the local supersaturation becomes too high to maintain the handedness in the suspension.

In continuous antisolvent crystallization experiment C1, the crystallizer initially contained a starting solution with $x_{AS} = 50$ v%, saturated at $C^* = 21.7$ mg ml^{-1} (4, Fig. 6, left) and almost enantiopure seed crystals $E = 95.2\%$ at a suspension density of 10 g l^{-1} (Table 4). With these starting conditions, the process quickly arrives at the steady state value of the concentration $C(t)$, as shown in Fig. 6 (right, green circles). Upon starting the process, saturated solution (1, Fig. 6, left) and antisolvent (2, Fig. 6, left) are pumped into the crystallizer with flow rates of, respectively, $F_{AS} = 9.9$ and $F_S = 9.4$ ml min^{-1}, resulting in an antisolvent fraction $x_{AS} = 51.4$ v% and a residence time $\tau = 26$ min. The antisolvent feed and the saturated solution feed are mixed with the suspension in the crystallizer, represented through the mixing point (3, Fig. 6, left) on the dilution line in the phase diagram, while continuously a suspension flows out of the crystallizer so that the crystallizer volume remains constant. The overall composition in the crystallizer is therefore equal to that in the mixing point on the operating line ($x_{AS} = 51.4$ v%, $C_M = 170.3$ mg ml^{-1}).

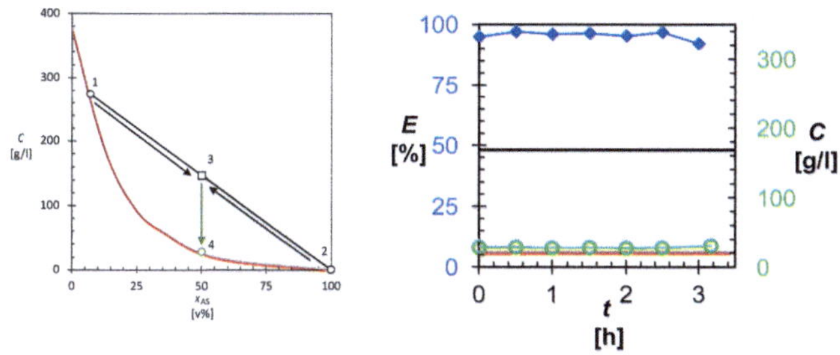

Fig. 6 Left: the phase diagram of $NaBrO_3$ in water–ethanol mixtures at 20 °C and the pathway of continuous antisolvent crystallization experiment C1 at $x_{AS} = 51.4$ v% and $\tau = 26$ min; solution feed flow composition (1), pure antisolvent feed flow composition (2), mixing point of both feed streams at $x_{AS} = 51.4\%$ (3), measured solution concentration within the crystallizer solution in steady state (4). Right: product purity (blue diamonds) and concentration (green circles) of continuous antisolvent crystallization experiment C1 at 20 °C, a residence time of $\tau = 26$ min and constant antisolvent fraction of $x_{AS} = 51.4$ v% (experiment C1, Table 4) with the overall concentration (black line) and the solubility (red line).

Table 4 Overview of the experimental conditions (antisolvent fraction x_{AS}, residence time τ, enantiomeric excess E_i of seed crystals) and results (enantiomeric excess E_P of product crystals, the productivity P, the yield X) of the continuous antisolvent crystallization experiments

	Experimental conditions			Results		
Experiment	x_{AS} [v%]	τ [min]	E_i [%]	E_P [%]	P [g l^{-1} h^{-1}]	X [—]
C1	51.4	26	95.2	92.3	321.9	93.9
C2	19.6	24	95.2	91.2	360.2	91.4
C3	80.3	26	95.2	95.4	153.3	99.9
C4	18.4	9	95.2	59.4	837.3	90.3
C5	20.4	5	95.2	45.4	1675	90.2

The enantiomeric excess measured during the process is plotted in Fig. 6 (right, blue diamonds) and remains high throughout the process, with a final value of $E = 92.3\%$, only slightly lower than the initial enantiomeric excess in the seed suspension. Comparing the mixing point concentration $C_M = 170.3$ mg ml^{-1} and solute concentration $C(t) = 29.4$ mg ml^{-1} indicates the crystallization of the substantial amount of 140.9 mg ml^{-1} each residence time of $\tau = 26$ minutes. This must coincide with a substantial amount of nucleation. Again, as the enantiopurity is maintained throughout the process of more than 6 residence times, we can conclude that the dominant mechanism must be an enantioselective secondary nucleation mechanism, even under these high local supersaturation conditions.

The local supersaturation conditions are determined by the antisolvent fraction and the residence time. First, additional continuous antisolvent crystallization experiments C2 and C3 were performed at antisolvent fractions of 19.6 and 80.3 v%, respectively, using a similar residence time of 24–26 min compared to that of experiment C1. The results are shown in Fig. 7, where the seed crystal purity $E = 95.2\%$ (blue line), the enantiomeric excess values, measured of C1 (blue rectangles) at an average $E = 95.8\%$, C2 (green rectangles) at an average of $E = 91.7\%$ and C3 (orange rectangles) at an average $E = 95.7\%$, are plotted against the number of residence times $N_\tau = t/\tau$. The results show that the enantiomeric excess of the crystalline product can be maintained close to that of the seed material at all 3 antisolvent fractions used. While at a higher antisolvent fraction, the local supersaturation at the solution feed mixing point is expected to increase as the solubility in the crystallizer solution is lower, the same residence time at a higher antisolvent fraction reduces the feed flow rate of the saturated solution, which reduces the amount of solution to be mixed and with it the local supersaturation. Apparently, the mixing effect is strong enough to avoid conditions for non-enantioselective primary nucleation. It is possible that, due to the high stirring rates, the antisolvent is mixed quickly with the bulk solution, preventing high local supersaturations.

Thus, a residence time of around 25 minutes allows control over the product handedness in a continuous antisolvent crystallization with an antisolvent fraction between $20 < x_{AS} < 80$ v%.

The effect of the residence time on the enantiomeric excess E of the product is investigated through experiments C2, C4 and C5. These experiments are performed at the same antisolvent fraction $x_{AS} = 20$ v%, varying the overall flow rate $F = F_S + F_{AS}$ from 20.8 up to 100 ml min^{-1}. The resulting enantiomeric excess E as a function of the number of residence times $N_\tau = t/\tau$ is shown in Fig. 8.

While the enantiomeric excess E for experiment C2 remains close to the purity of the seed crystals (blue line, Fig. 8), if the residence time is reduced to 9 min, the enantiomeric excess E of experiment C4 declines rapidly after 7 residence times. Decreasing the residence time further to 5 min in experiment C5 (orange, Fig. 8)

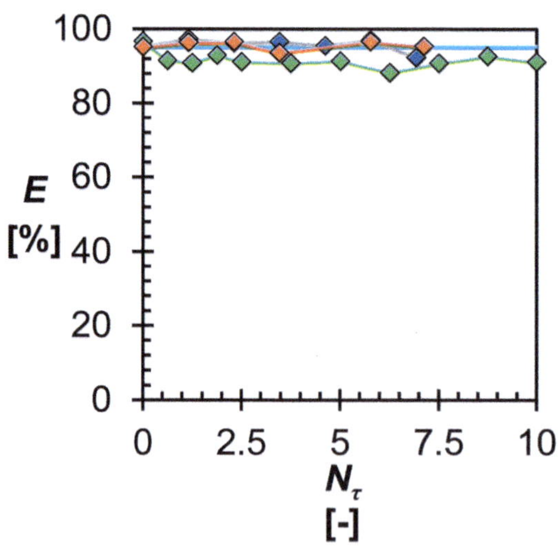

Fig. 7 Change in enantiomeric excess E of the product crystals over the number of residence times N_τ at different antisolvent fractions in continuous antisolvent deracemization experiments C1 ($x_{AS} = 51.4$ v%, blue rectangles), C2 ($x_{AS} = 20$ v%, green rectangles) and C3 ($x_{AS} = 80.3$ v%, orange rectangles) at a residence time of 24–26 minutes.

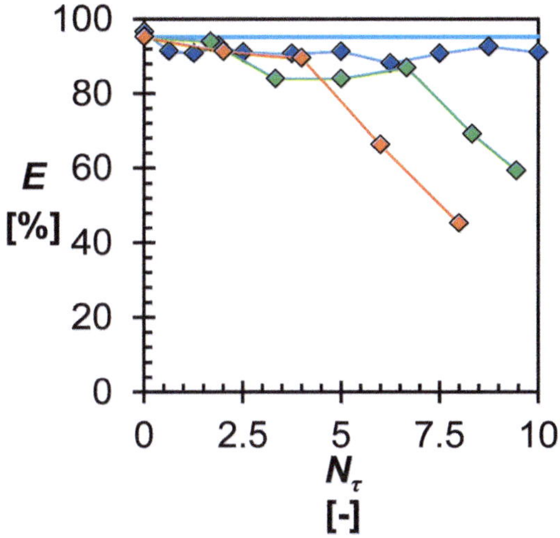

Fig. 8 Enantiomeric excess E of the product crystals over the number of residence times N_τ for experiment C2 (blue rectangles, $\tau = 24$ min), C4 (green rectangles, $\tau = 9$ min) and C5 (orange rectangles, $\tau = 5$ min). The enantiomeric excess in the seed material is indicated by the blue solid line.

leads to a loss of the handedness after 4.5 residence times. It is shown that the enantiomeric excess declines sooner with decreasing residence times, probably due to local high supersaturations inducing primary nucleation of crystals of the other handedness.

In order to compare all continuous experiments with regards to process stability, the excess productivity of the pure enantiomer crystals $P_E = E(C_M - C)/\tau$, is plotted against the overall process productivity $P = (C_M - C)/\tau$, in Fig. 9. Experiments C1–C3, represented by squares, show good steady state behaviour through a constant productivity of the overall and enantiopure product produced. Experiments C4 and C5 show a strong fluctuation between the overall and enantiopure productivity in C5, resulting in a clear diversion from the grey dotted line, as shown in Fig. 9 (red circles). First, the productivity P decreases because the suspension density in the crystallizer is decreasing due to the flushing out of crystals at the high overall flow rate, as the growth rate cannot keep up with the removal of crystals from the suspension. This increases the overall concentration and supersaturation in the bulk solution in the crystallizer. It is likely that, due to the smaller suspension density, the secondary nucleation rate also decreases, although that might be compensated by an increase in secondary nucleation due to an increase in bulk supersaturation as a result of the increase in solute concentration. Then, at some point, primary nucleation occurs of crystals of the other handedness, contaminating the suspension and reducing P_E, as now crystals of both handednesses are present. The points then deviate from the diagonal, as the enantiopure productivity P_E becomes much smaller than the productivity P.

Fig. 9 also makes clear that the productivity of this process is quite large, even though the process should be limited to conditions at which no unselective primary nucleation takes place that reduces the purity of the product. It seems that productivities up to 500 g l^{-1} h^{-1} could be achievable for this process. This

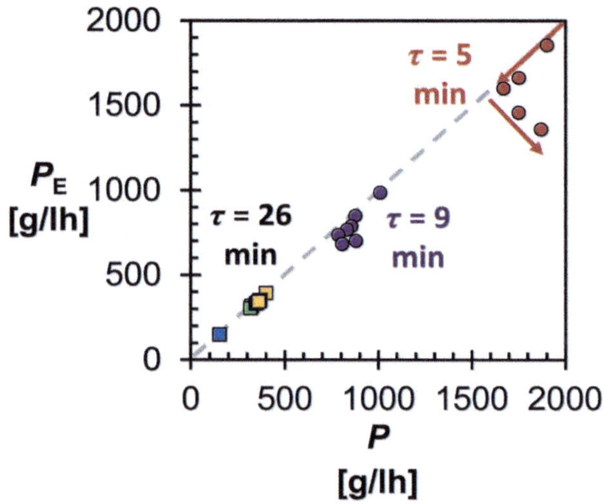

Fig. 9 Productivity of the pure enantiomer product P_E over the overall product productivity P for each continuous experiment conducted with C1 (green squares), C2 (blue squares), C3 (orange squares), C4 (purple circles) and C5 (red circles). The grey dashed line represents an ideal process, producing only pure enantiomer.

high productivity is a result of the high difference in concentration and solubility at higher antisolvent fractions and the relatively low residence time. For instance, at an antisolvent fraction $x_{AS} = 50\%$, the mixing point gives a concentration of about 170 g l^{-1}, while the solubility is about 22 g l^{-1}, giving a productivity $P =$ 296 g l^{-1} h^{-1} at a residence time $\tau = 30$ minutes. This high productivity could even be improved by optimizing the concentration difference and residence time and reducing the possibility for primary nucleation. Such large concentration differences and small residence times are usually not possible in continuous evaporative and cooling crystallizations, let alone in a batch-wise TCID (temperature cycling-induced deracemization) processes: usually TCID productivities are as low as 0.1–5 g l^{-1} h^{-1}.[16,28,29] As the process time for SOATs (second order asymmetric transformations) is shorter than that of TCID processes, as long as primary nucleation can be avoided, SOAT productivity is higher: a batchwise SOAT process for an atropisomeric naphthamide showed productivities of up to 140 g l^{-1} h^{-1}.[16] Antisolvent deracemization seems to be an interesting option to increase the productivity of a crystallization-enhanced deracemization with high yields and high chiral product purity due to the unexpectedly low primary nucleation rate in the process.

3.3 Competing primary and secondary nucleation

Up to now, we focused on the primary and secondary nucleation of $NaBrO_3$ in water, which showed dominant secondary nucleation, even at high local supersaturations in antisolvent crystallization. We investigate here, the primary and secondary nucleation rates at the same supersaturation of several other compounds to establish whether they behave in a similar way compared to $NaBrO_3$. We surveyed three solute/solvent systems, encompassing organic and

Table 5 Comparison of primary nucleation rate *J* and secondary nucleation rate *B* for sodium bromate in water, sodium chloride in water, glycine in water and INA in ethanol, at supersaturation ratio *S*

Solute	Solvent	S [—]	J [ml^{-1} min^{-1}]	B [ml^{-1} min^{-1}]
Sodium bromate	Water	1.08–1.17	10–40 × 10^{-3}	61–224 × 10^{3}[a]
Sodium chloride	Water	1.02	0.8 × 10^{-3}	3.5 × 10^{3}
Glycine	Water	1.18	20 × 10^{-3}	160 × 10^{3}
INA	Ethanol	1.12	40 × 10^{-3}	9.7 × 10^{3}

[a] Supersaturation of sodium bromate in water during the secondary nucleation rate measurements was $S = 1.2$.

inorganic solutes and solvents. The resulting primary and secondary nucleation rates at supersaturations representative of typical industrial crystallization process conditions are shown in Table 5.

For these systems, the primary nucleation rates measured were from 0.8×10^{-3} to 40×10^{-3} ml^{-1} min^{-1}, while the secondary nucleation rate showed values from 3.5×10^{3} to 160×10^{3} ml^{-1} min^{-1}. In all cases examined, the secondary nucleation rate was 5–7 orders of magnitude higher than the primary nucleation rate measured under the same conditions. This was also the case for lower supersaturations. This means that, in the presence of parent crystals, secondary nucleation tends to be dominant, also for compounds other than $NaBrO_3$.

4. Discussion

We showed that the secondary nucleation rates at industrial crystallization conditions are more than 6 orders of magnitude higher than the primary nucleation rates for several compounds. This huge difference makes it likely that, once a crystal has been finally formed by primary nucleation in a clear solution, secondary nucleation quickly takes over as the dominant mechanism. It thus might well be that the entire suspension in many crystallization processes originates, because of secondary nucleation, from a primary nucleated single crystal that happens to be the one that nucleated first.

The prevalence of secondary rather than primary nucleation has important implications for the design, operation and control of industrial crystallization processes in general, since the key controlling factors of these fundamental nucleation mechanisms differ. While primary nucleation is determined by interfacial energy and supersaturation, as well as the presence of active heterogeneous particles,[2,30] secondary nucleation is an empirical function of stirrer speed, suspension density and supersaturation ratio. Control methods for industrial crystallizers are therefore completely different for primary and secondary nucleation.

An example of this difference in control is given by the seeded fed-batch and continuous antisolvent crystallizations of sodium bromate. These processes, unexpectedly, result in spontaneous chiral resolution and a crystal product of high chiral purity. The high enantiomeric excess, which can be retained throughout the process in a very stable manner, is surprising, as antisolvent crystallization usually involves relatively high local supersaturations compared to other crystallization techniques, such as cooling or evaporative crystallization. Especially locally around the mixing point of the bulk solution and antisolvent inlet, the local antisolvent fraction can be high and, due to the low solubility of the compound in the mixture at high antisolvent fractions, locally high supersaturations occur. These high local supersaturations have a high likelihood to result in the primary nucleation of crystals of both handednesses, as primary nucleation is thought not to be enantioselective.

However, our results indicate that an enantioselective nucleation mechanism occurs during the performed antisolvent crystallization and deracemization. The more likely nucleation mechanism to provide enantioselectivity is a secondary nucleation mechanism, for which the secondary nuclei, generated from the crystals already present in the suspension, retain the handedness of the seed crystals during the fed-batch antisolvent crystallization. This can be either through an attrition-based or a fluid shear secondary nucleation mechanism.[31]

This dominance of enantioselective secondary nucleation demonstrates that antisolvent deracemization is a feasible deracemization technique to produce an enantiopure product. As productivity and yield can be high for antisolvent crystallization, fed-batch or continuous antisolvent deracemization can act as an alternative for deracemization techniques, such as Viedma ripening[32–34] and temperature cycling induced deracemization.[35]

5. Conclusions

This work reveals that, unlike as is generally thought, secondary, rather than primary, nucleation dominates in cooling, as well as antisolvent, crystallization processes, even with high local supersaturations occurring due to the mixing of antisolvent and bulk solution. Measurement data using $NaBrO_3$ in water, glycine in water, isonicotinamide in ethanol and sodium chloride in water shows that the secondary nucleation rate is at least 6 orders of magnitude faster than that of primary nucleation at typical industrial crystallization conditions. A dominant secondary nucleation rather than primary nucleation will strongly affect nucleation control strategies in industrial crystallization processes. We further demonstrate that, despite the very high local supersaturations, enantioselective secondary nucleation is the dominant nucleation mechanism in the seeded fed-batch and continuous antisolvent crystallization of $NaBrO_3$, as we obtain close to a single handedness crystal product. This creates the opportunity to use antisolvent crystallization as a crystallization-enhanced deracemization method, having a superior productivity compared to other crystallization-enhanced deracemization methods.

Conflicts of interest

There are no conflicts to declare.

Acknowledgements

Part of this research has received funding as part of the CORE project (October 2016–September 2020) from the European Union's Horizon 2020 Research and

Innovation Programme under the Marie Sklodowska-Curie grant agreement no. 722456 CORE ITN. We thank the EPSRC Centre for Innovative Manufacturing in Continuous Manufacturing and Crystallisation (http://www.cmac.ac.uk) for support (EPSRC funding under grant reference: EP/I033459/1).

References

1 J. H. Ter Horst, C. Schmidt and J. Ulrich, Fundamentals of industrial crystallization, in *Handbook of Crystal Growth*, Elsevier, 2015, pp. 1317–1349.

2 R. J. Davey, S. L. M. Schroeder and J. H. ter Horst, Nucleation of Organic Crystals—A Molecular Perspective, *Angew. Chem., Int. Ed.*, 2013, **52**(8), 2166–2179.

3 J. W. Mullin, *Crystallization*, Elsevier, 2001.

4 R. Ó. Meadhra, H. J. Kramer and G. M. Van Rosmalen, Model for secondary nucleation in a suspension crystallizer, *AIChE J.*, 1996, **42**(4), 973–982.

5 A. Myerson, *Handbook of Industrial Crystallization*, Butterworth-Heinemann, 2002.

6 P. J. Frawley, N. A. Mitchell, C. T. Ó'Ciardhá and K. W. Hutton, The effects of supersaturation, temperature, agitation and seed surface area on the secondary nucleation of paracetamol in ethanol solutions, *Chem. Eng. Sci.*, 2012, **75**, 183–197.

7 N. Kubota, N. Doki, M. Yokota and A. Sato, Seeding policy in batch cooling crystallization, *Powder Technol.*, 2001, **121**(1), 31–38.

8 J. Garside and R. Davey, *From Molecules to Crystallizers: an Introduction to Crystallization*, Oxford University Press, 2000.

9 C. Gahn and A. Mersmann, Brittle fracture in crystallization processes Part A. Attrition and abrasion of brittle solids, *Chem. Eng. Sci.*, 1999, **54**(9), 1273–1282.

10 C. J. Callahan and X.-W. Ni, An investigation into the effect of mixing on the secondary nucleation of sodium chlorate in a stirred tank and an oscillatory baffled crystallizer, *CrystEngComm*, 2014, **16**(4), 690–697.

11 W. L. Noorduin, T. Izumi, A. Millemaggi, M. Leeman, H. Meekes, W. J. Van Enckevort, R. M. Kellogg, B. Kaptein, E. Vlieg and D. G. Blackmond, Emergence of a single solid chiral state from a nearly racemic amino acid derivative, *J. Am. Chem. Soc.*, 2008, **130**(4), 1158–1159.

12 C. Viedma, Chiral symmetry breaking during crystallization: complete chiral purity induced by nonlinear autocatalysis and recycling, *Phys. Rev. Lett.*, 2005, **94**(6), 065504.

13 G. Coquerel and M. Hoquante, Spontaneous and Controlled Macroscopic Chiral Symmetry Breaking by Means of Crystallization, *Symmetry*, 2020, **12**(11), 1796.

14 K. Suwannasang, A. Flood, C. Rougeot and G. Coquerel, Using programmed heating–cooling cycles with racemization in solution for complete symmetry breaking of a conglomerate forming system, *Cryst. Growth Des.*, 2013, **13**(8), 3498–3504.

15 W. W. Li, L. Spix, S. C. De Reus, H. Meekes, H. J. Kramer, E. Vlieg and J. H. Ter Horst, Deracemization of a racemic compound via its conglomerate-forming salt using temperature cycling, *Cryst. Growth Des.*, 2016, **16**(9), 5563–5570.

16 R. Oketani, M. Hoquante, C. Brandel, P. Cardinael and G. Coquerel, Resolution of an Atropisomeric Naphthamide by Second-Order Asymmetric

Transformation: A Highly Productive Technique, *Org. Process Res. Dev.*, 2019, **23**(6), 1197–1203.
17 R. M. Pagni and R. N. Compton, Asymmetric Synthesis of Optically Active Sodium Chlorate and Bromate Crystals, *Cryst. Growth Des.*, 2002, **2**(4), 249–253.
18 R. R. E. Steendam and J. H. ter Horst, Continuous Total Spontaneous Resolution, *Cryst. Growth Des.*, 2017, **17**(8), 4428–4436.
19 M. L. Briuglia, J. Sefcik and J. H. ter Horst, Measuring secondary nucleation through single crystal seeding, *Cryst. Growth Des.*, 2019, **19**(1), 421–429.
20 D. Rowland, Thermodynamic properties of the glycine + H_2O system, *J. Phys. Chem. Ref. Data*, 2018, **47**(2), 023104.
21 O. Stern, Zur theorie der elektrolytischen doppelschicht, *Z. Elektrochem. Angew. Phys. Chem.*, 1924, **30**, 508–516.
22 S. Jiang and J. H. ter Horst, Crystal Nucleation Rates from Probability Distributions of Induction Times, *Cryst. Growth Des.*, 2011, **11**(1), 256–261.
23 R. R. E. Steendam and P. J. Frawley, Secondary Nucleation of Sodium Chlorate: The Role of Initial Breeding, *Cryst. Growth Des.*, 2019, **19**(6), 3453–3460.
24 C. Brandel and J. H. ter Horst, Measuring induction times and crystal nucleation rates, *Faraday Discuss.*, 2015, **179**, 199–214.
25 S. S. Kadam, S. A. Kulkarni, R. C. Ribera, A. I. Stankiewicz, J. H. ter Horst and H. J. Kramer, A new view on the metastable zone width during cooling crystallization, *Chem. Eng. Sci.*, 2012, **72**, 10–19.
26 R. R. E. Steendam, L. Keshavarz, M. A. R. Blijlevens, B. de Souza, D. M. Croker and P. J. Frawley, Effects of Scale-Up on the Mechanism and Kinetics of Crystal Nucleation, *Cryst. Growth Des.*, 2018, **18**(9), 5547–5555.
27 C. Mack, J. Hoffmann, J. Sefcik and J. H. ter Horst, Accurate Phase Diagrams for the Basic Design of Antisolvent Crystallization Processes, to be published, 2022.
28 R. Oketani, M. Hoquante, C. Brandel, P. Cardinael and G. Coquerel, Practical role of racemization rates in deracemization kinetics and process productivities, *Cryst. Growth Des.*, 2018, **18**(11), 6417–6420.
29 F. Breveglieri, G. M. Maggioni and M. Mazzotti, Deracemization of NMPA via temperature cycles, *Cryst. Growth Des.*, 2018, **18**(3), 1873–1881.
30 D. Kashchiev, *Nucleation: Basic Theory with Applications*, Butterworth-Heinemann, Oxford, 2000.
31 S. G. Agrawal and A. H. J. Paterson, Secondary Nucleation: Mechanisms and Models, *Chem. Eng. Commun.*, 2015, **202**(5), 698–706.
32 W. L. Noorduin, P. van der Asdonk, A. A. C. Bode, H. Meekes, W. J. P. van Enckevort, E. Vlieg, B. Kaptein, M. W. van der Meijden, R. M. Kellogg and G. Deroover, Scaling Up Attrition-Enhanced Deracemization by Use of an Industrial Bead Mill in a Route to Clopidogrel (Plavix), *Org. Process Res. Dev.*, 2010, **14**(4), 908–911.
33 M. W. van der Meijden, M. Leeman, E. Gelens, W. L. Noorduin, H. Meekes, W. J. P. van Enckevort, B. Kaptein, E. Vlieg and R. M. Kellogg, Attrition-Enhanced Deracemization in the Synthesis of Clopidogre l – A Practical Application of a New Discovery, *Org. Process Res. Dev.*, 2009, **13**(6), 1195–1198.
34 W. L. Noorduin, W. J. P. vanEnckevort, H. Meekes, B. Kaptein, R. M. Kellogg, J. C. Tully, J. M. McBride and E. Vlieg, The Driving Mechanism Behind

Attrition – Enhanced Deracemization, *Angew. Chem., Int. Ed.*, 2010, **49**(45), 8435–8438.
35 K. Suwannasang, A. E. Flood and G. Coquerel, A Novel Design Approach To Scale Up the Temperature Cycle Enhanced Deracemization Process: Coupled Mixed-Suspension Vessels, *Cryst. Growth Des.*, 2016, **16**(11), 6461–6467.

Faraday Discussions

ROYAL SOCIETY OF CHEMISTRY

PAPER

Influence of anisotropy on heterogeneous nucleation of gold nanorod assemblies†

Ann-Kathrin Göppert, Guillermo González-Rubio and Helmut Cölfen *

Received 9th November 2021, Accepted 13th December 2021

DOI: 10.1039/d1fd00087j

In this study, we analysed for the first time heterogeneous nucleation with anisotropic nanoparticles as a model system for non-spherical building units on the nanoscale. Gold nanorods were synthesised and assembled to investigate the phenomenon of heterogeneous nucleation. To determine the influence of the particle shape on heterogeneous nucleation, we utilised gold nanorods with varying aspect ratios, ranging from 3.00 and 2.25 to 1.75, while keeping the surface chemistry constant. First, the nucleation of the gold nanorod assemblies in solution and the process kinetics were analyzed with UV-vis-NIR spectroscopy followed by a microscopic examination of the gold nanorod-based superstructures formed heterogeneously on substrates. Here, positively charged cetyltrimethylammonium bromide (CTAB)-functionalized gold nanorods and negatively charged polystyrene sulfonate (PSS) functionalized substrates ensured the directed heterogeneous nucleation on the substrates. A combination of light microscopy with simultaneous UV-vis-NIR spectroscopy allowed us to observe the gold nanorod-based superstructure formation on the substrates *in situ* and to determine the nucleation rates of the process. We analysed the resulting data with the classical nucleation theory, which revealed a dominating kinetic term and a negligible thermodynamic term in contrast to ionic systems like calcium carbonate. Our studies consistently exhibit an influence of the aspect ratio on the nucleation behaviour resulting in faster nucleation of superstructures as the aspect ratio decreases. Hence our studies show unprecedented insight into the influence of particle anisotropy on the nucleation and growth of nanorod-based superstructures and reveal significant differences in the nucleation of nanoparticle building units compared to the nucleation of atoms or molecules as building units.

Introduction

Heterogeneous nucleation is a common phenomenon in nature, evident for example, in the formation of clouds or biominerals.[1,2] However, it is no less

Physikalische Chemie, Universität Konstanz, Universitätsstr. 10, D-78457 Konstanz, Germany. E-mail: helmut.coelfen@uni-konstanz.de

† Electronic supplementary information (ESI) available. See DOI: 10.1039/d1fd00087j

important in industrial applications, like the hardening of cement or the formation of mineral scale in pipes.[3,4] Despite its relevance, heterogeneous nucleation is not fully understood, probably due to the demanding analysis required to investigate it, which entails high spatial and temporal resolution at the same time.[5,6] To simplify the analysis of nucleation phenomena, micron-sized particles have been proposed and exploited as model systems for atoms or molecules building up to bigger superstructures.[7–10] Most analysed microparticle-based systems rely on isotropic hard spheres as the simplest system for nucleation. Today, we have a relatively good understanding of the nucleation of hard spherical particles into superstructures thanks to the use of a combination of computer simulations[11,12] and experiments utilizing, for example, spherical PMMA particles.[13–18]

Although most work on nucleation and crystallization focuses on isotropic particles as building units, systems with anisotropic particles or building units are of great importance to investigate many crystallization phenomena. For example, in concrete where nucleation of the anisotropic calcium silicate hydrate platelets to superstructures leads to hardening of the material.[19,20] Concrete is the second most widely used material on earth, and its vast carbon dioxide emission urges us to adapt greener alternatives like magnesium silicates.[21] An understanding and manipulation of the nucleation process in these anisotropic systems can thus potentially help to improve the properties of magnesium silicate-based or alternative concrete, allowing a partial replacement of calcium silicates. Proteins constitute another prominent example, as they can also exhibit anisotropic forms. Today, X-ray crystallography remains the gold standard for the determination of the macromolecular structure of proteins.[22] However, the crystallization of proteins, which is needed for analysis with X-ray crystallography, is extremely challenging. Also, in this example, a better understanding of the nucleation of anisotropic proteins can bring us to a more controlled crystallization of proteins and thus simplify their analysis enormously.

Moreover, anisotropic nanoparticles have direction-dependent properties like different shapes or surface chemistry. *Via* the shape of a particle, the properties of the single-particle such as electronic,[23] catalytic,[24] or magnetic,[25,26] can be influenced.[27] Beyond that, anisotropic particles can not only assemble to simple closest packing, like isotropic spherical particles, but furthermore, they can assemble into more complex structures.[28–31] Here, the nature of the resulting bulk material depends on the nanoparticle-based superstructure and determines its macroscopic properties. This is the field of non-classical crystallization, with mesocrystals as a prominent example.[28,32–34] Therefore, anisotropic building units from molecules to particles are not only common in nature but additionally open the possibility to design new material properties resulting from a variation in the assembly structures.[35] However, in many cases, it is yet unclear how the shape of the building units influences the nucleation events that eventually lead to the formation of nanoparticle-based materials.

In this study, we focused on investigating the phenomenon of heterogeneous nucleation with anisotropic particles as building units on the nanoscale. Nanoparticles show more similar behaviour to atoms and molecules than microparticles due to their similar dimensions and diffusion properties, and are therefore well suited as model systems. There are already a lot of examples for the (heterogeneous) nucleation of nanoparticles on substrates, which play an important role for the formation and analysis of nanoparticle assemblies. Typical subsequent characterizations of nanoparticle assemblies like transmission electron microscopy (TEM), scanning electron microscopy (SEM), atomic force microscopy (AFM), or small-angle X-ray scattering (SAXS) require the immobilization of the nanoparticle superstructures on a substrate. In retrospect, it can be difficult to estimate whether the superstructures nucleated heterogeneously on the substrate, if the substrate had any influence on the structures formed, or how big this influence was.[35] Here, theoretical investigations can be helpful, like the example of Luedtke *et al.* who simulated dodecanethiol capped gold nanoparticles adsorbing on the basal plane of graphite.[36] The simulations of the hydrocarbon chains while attaching to the substrate and with this an edge orientation of the nanoparticles further induced the resulting structure of the superstructures. Next to planar substrates textured substrates on the order of the nanoparticle diameter open the possibility of direct structural control during nucleation and growth. For instance, lithographically patterned templates can trap solvated nanoparticles over capillary forces,[37–39] or periodically structured templates allow the arrangement of gold nanorods to plasmonic supercrystals.[40] Further, spherical substrates like emulsion droplets[41] or microspheres[42] can be used to control the structuring of nanoparticle assemblies. Additionally, the assembly of cadmium selenide nanoparticles to monolayers at liquid–liquid interfaces give more insight into the formation mechanism.[43] Pendant drop tensiometry and *ex situ* TEM measurements showed three steps in this process: first free nanoparticle diffusion to the interface, second, a closer packing of the nanoparticles to clusters, and third, a thermally activated exchange between incoming and adsorbed nanoparticles resulting in a dense packed monolayer. Another example at liquid–liquid interfaces constitutes the assembly of tri-*n*-octylphosphine oxide-covered cadmium selenide nanorods.[44] The structure of these assemblies ranged from low density smectic packing over more dense columnar ordering to crystalline-like phases. These different structures could be induced by varying the interfacial energy between the liquids and the nanorods, the concentration of the nanorods, or the aspect ratio of the nanorods.

To gain better insight into how the shape of the building units or more precisely the particle anisotropy, influences the heterogeneous nucleation of nanoparticles into superstructures, we utilised gold nanorods with similar width and distinct aspect ratios while keeping their surface chemistry constant. First, the nucleation of the gold nanorod assemblies in solution was analysed followed by an examination of the gold nanorod superstructures formed heterogeneously on substrates. A combination of light microscopy with simultaneous UV-vis-NIR (ultraviolet-visible-near-infrared) spectroscopy allowed us to observe the gold nanorod-based superstructure formation on the substrates *in situ* and determine the nucleation rates of this process.

Results and discussion

Gold nanorods

Our project aimed to analyse the impact of nanoparticle anisotropy on heterogeneous nucleation into superstructures. Therefore, gold nanorods were chosen as building blocks due to their ease of synthesis, low size dispersity, tuneable anisotropy, and ability to self-assemble into (meso)crystalline superstructures.

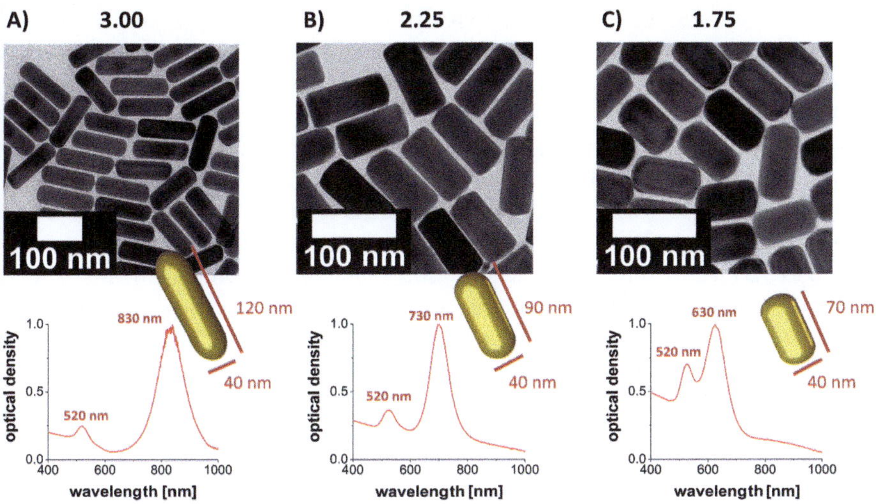

Fig. 1 Gold nanorods with various aspect ratios utilised in the heterogeneous nucleation experiments. The upper row shows HR-TEM images of the gold nanorods with aspect ratios of (A) 3.00, (B) 2.25, and (C) 1.75 and the bottom row shows the corresponding UV-vis-NIR spectra and a scheme with the edge lengths of the gold nanorods.

Moreover, their strong plasmonic activity, which is sensitive to the nanoparticle aggregation state, enables the use of UV-vis-NIR spectroscopy to investigate self-assembly and nucleation processes. The synthesis of colloidal gold nanoparticles with controlled dimensions and low size dispersity can typically be achieved by seed-mediated growth methods based on the spatial and temporal separation of nucleation and growth stages.[45,46] More specifically, in this work, we synthesised three gold nanorod samples with a different aspect ratio (detailed description in the experimental part).[47] Thereby, gold nanorods of 40 nm in diameter and lengths of 120, 90, and 70 nm, corresponding to aspect ratios of 3.00, 2.25, and 1.75, respectively, were successfully prepared. The resulting nanorods presented narrow plasmon bands of 830, 730 and 630 nm (Fig. 1).

Controlled nucleation in solution

The system of three identical gold nanorods only varying in their aspect ratio allowed the study of the influence of the particle shape on the heterogeneous nucleation of these gold nanorods into superstructures. To investigate the heterogeneous nucleation experiments in a reproducible manner, a methodology for the controlled destabilization of the gold nanorod dispersions was first required. This type of colloidal gold nanocrystal is stabilised through the adsorption of cationic cetyltrimethylammonium bromide (CTAB) micelles on their surface, which is the surfactant utilised during their synthesis as a stabilizing and shape directing agent. When aiming at the controlled self-assembly of the gold nanorods, destabilisation of such electrostatic forces through the addition of water-soluble organic solvents probably represents the most straightforward strategy.[35,48,49] In our case, we choose ethanol to trigger the gold nanorod self-assembly into superstructures. Then, to obtain information about the aggregation process, we employed UV-vis-NIR spectroscopy, as it is proven to be a helpful analytic in the case of plasmonic gold nanoparticles where their optical properties are sensitive to the presence of other plasmonic particles in the vicinity.[50] This fact opens access to characterise the self-assembly process *via* UV-vis-NIR spectroscopy directly in solution and receives fast insight into the destabilization process (ESI Fig. S1†).[51,52] The gold nanorods present two characteristic plasmonic bands: a transversal resonance mode located at approximately 530 nm, and a longitudinal mode at longer wavelengths that depends on the aspect ratio.[53] During the nucleation and self-assembly process, a shift and broadening of the longitudinal plasmon bands for all investigated nanorod colloids were observed, meaning new bands corresponding to the formation of aggregates emerged at longer wavelengths.[50,54,55] Overall, these results indicate the suitability of using ethanol to trigger the aggregation of CTAB-stabilised gold nanorods. Moreover, analysis of the destabilization kinetics showed that the reactions proceeded as a second-order reaction (ESI Table S1†). These findings suggest a mechanism based on bimolecular collisions.

Heterogeneous nucleation on substrates

After a methodology for the controlled destabilization of gold nanorods was established, in the next step, we aimed to accomplish heterogeneous nucleation and growth of the gold nanorod-based superstructures triggered by solid substrates. The utilization of CTAB as a colloidal stabilizing agent results in gold nanorods with a positively charged surface due to the quaternary ammonium cation of CTAB. To ensure a favourable interaction of the positively charged gold nanorods with the desired substrate, we functionalized the surface of mica substrates with negatively charged polystyrene sulfonate (detailed description in the experimental part). In addition, the use of mica substrates, which are atomically cleavable and thereby exhibit a very smooth surface, allows us to minimize the potential undesired nucleation triggered by cracks or the roughness on the surface. The interaction of the gold nanorods with the sulfonate-functionalized mica substrates was evaluated by destabilizing gold nanorods in the presence of these substrates *via* UV-vis-NIR spectroscopy. The self-assembly kinetics were then compared with that of samples destabilized in the absence of substrates (ESI Fig. S2†). The experiments showed that sulfonate-functionalized substrates can successfully enhance the destabilisation of the gold nanorod dispersions. Consistent with this finding, we obtained similar results with a favourable interaction for CTAB stabilized gold nanocubes with sulfonate-functionalized substrates in another project.[56]

In the next step, we examined if the sulfonate-functionalized substrates can effectively lead to a superstructure growth on their surface *via* scanning electron microscopy (SEM) (Fig. 2). Here, the rods form randomly organized assemblies which is not surprising given the short building time of 40 min. The rods with an aspect ratio of 3.00 formed large superstructures up to a size of around 5 μm, which tend to be elongated and tattered. However, lower aspect ratios (2.25 and 1.75) lead to more and denser-packed gold nanorod superstructures with larger dimensions of up to 10 μm for the aspect ratio of 1.75 (ESI Table S2 and Fig. S3†). Overall, these results show that sulfonate-functionalized substrates not only enhanced the destabilization of the gold nanorods but further led to the heterogeneous nucleation and growth of gold nanorod-based superstructures on

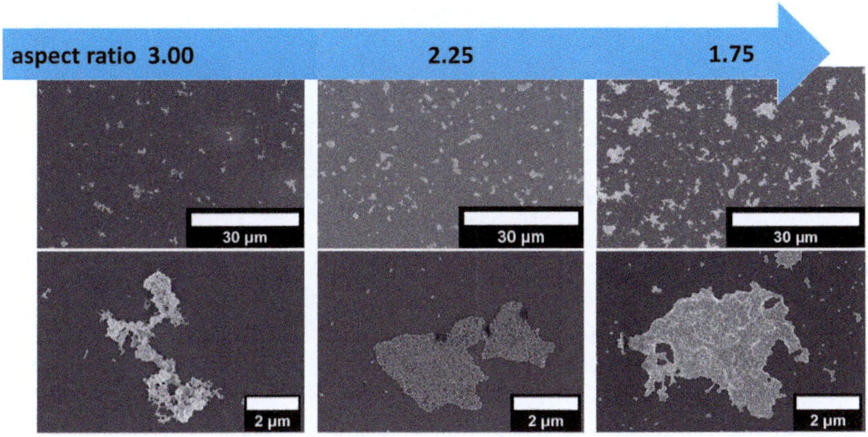

Fig. 2 Gold nanorod-based superstructures. SEM images of gold nanorod-based superstructures on sulfonate-functionalised substrates. The gold nanorods have various aspect ratios: 3.00 (left column), 2.25 (middle column), and 1.75 (right column). The density of superstructures on the substrate is less for gold nanorods with a high aspect ratio (3.00) and the superstructures seem to be more tattered. Going to lower aspect ratios (1.75), the density of superstructures increases, the superstructures seem to become more compact, and increase in height.

the substrate surface. Thus, the combination with sulfonate-functionalized substrates was chosen for the heterogeneous nucleation experiments.

Analysing heterogeneous nucleation kinetics

The heterogeneous nucleation of the gold nanorod-based superstructures on sulfonate-functionalised mica substrates was analysed with light microscopy and simultaneous UV-vis-NIR spectroscopy. We explained this general approach in detail in a previous publication and refer the reader to more detailed information about the experimental set-up and the evaluation of the data in that publication.[56] For this project, a hydrophobised quartz cuvette was used as a vessel for the heterogeneous nucleation experiments. The gold nanorods came into contact with the sulfonate-functionalised mica substrates through a hole in the cuvette. With light microscopy, it was possible to observe the superstructure growth on the mica surface, which opened the possibility to analyse the kinetics of the heterogeneous nucleation. The simultaneous UV-vis-NIR measurement gave information about the gold nanorod concentration and potential undesired homogeneous nucleation processes. The gold nanorods were then destabilized in a controlled manner with 20 vol% ethanol to ensure the colloidal stability of the nanoparticles in solution while enabling the superstructure growth on the substrate (ESI Fig. S4†). A detailed comparison of the results obtained from the different investigated nanorod aspect ratios (Fig. 3) showed that the number of superstructures detected on the substrate *via* light microscopy, their formation speed and with it the slope, correlate with their anisotropy degree. For instance, gold nanorods with a high aspect ratio (3.00) tended to form fewer superstructures and exhibited lower nucleation rates. However, those gold nanorods with

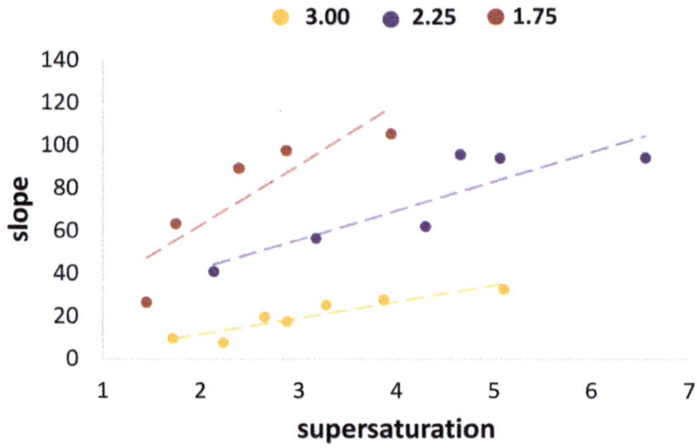

Fig. 3 Slopes of superstructure growth dependent on the aspect ratio. The determined slopes are the number of counted superstructures against the experiment time, which is equivalent to the nonnormalized nucleation rate. These slopes are plotted against the supersaturation for gold nanorods with varying aspect ratios: 3.00 (yellow), 2.25 (purple), and 1.75 (red).

a lower aspect ratio (1.75) nucleated faster and formed a higher number of superstructures. These findings are consistent with the results obtained from SEM characterisation experiments (Fig. 2) where gold nanorods with a lower anisotropy degree (1.75) formed more, denser-packed and bigger superstructures.

A logarithmic plot of the nucleation rates J_n against the inverse square of the supersaturations σ (Fig. 4) allowed us to determine the interfacial energies α for the heterogeneous nucleation process according to eqn (1).[5]

$$\ln(J_n) = -\frac{f\Omega^2}{(k_B T)^3}\alpha^3\left(\frac{1}{\alpha^2}\right) + \ln\left(A \times e^{-\frac{E_A}{k_B T}}\right) \quad (1)$$

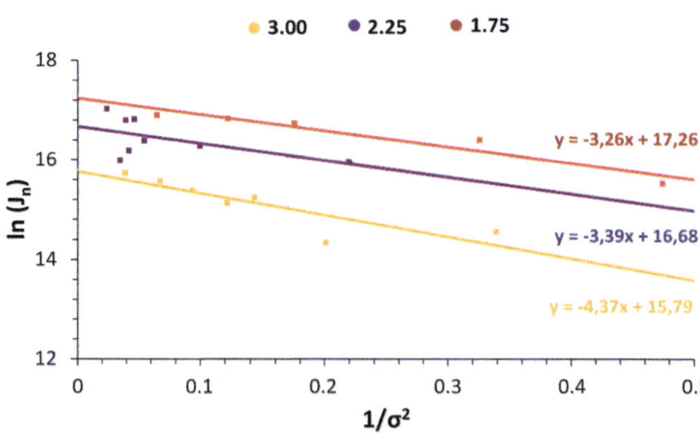

Fig. 4 Nucleation rates *versus* supersaturation. The logarithms of the determined nucleation rates are plotted against the inverse square of the supersaturation for gold nanorods with aspect ratios of 3.00 (yellow), 2.25 (purple), and 1.75 (red). The nucleation rates correlate with the aspect ratio of the gold nanorods.

With Ω being the volume per building unit of the solid phase, f being a numerical factor depending on the geometry of the nucleus, k_B being the Boltzmann constant, T being the temperature, A being a complicated function of the molecular-level parameters,[57] and E_A being the effective activation barrier. The calculated interfacial energies are summarized in ESI Table S3.† Herein, the interfacial energies increased with the decreasing gold nanorod aspect ratio. Furthermore, the nucleation barriers ΔG for the heterogeneous nucleation of gold nanorod-based superstructures on sulfonate-functionalized mica were calculated according to eqn (2).[5]

$$\Delta G = f\alpha^3\left(\frac{V_M}{k_B T\sigma}\right)^2 \tag{2}$$

Fig. 5 shows the nucleation barriers obtained for different gold nanorod supersaturations. In general, the obtained nucleation barriers were low compared to the thermal energy, thereby favouring nucleation without a thermodynamic barrier at $\sigma > 2$ for all analysed aspect ratios. This result is not surprising if the opposite charges of particles and substrates are considered. Furthermore, the nucleation barriers augmented with increasing aspect ratio: gold nanorods with an aspect ratio of 3.00 exhibited the highest nucleation barriers (1.09 k_BT for $\sigma = 2$), while those of 2.25 and 1.75 hardly differed in terms of their magnitude (0.85 k_BT and 0.82 k_BT, respectively for $\sigma = 2$).

Finally, the kinetic and thermodynamic terms for the heterogeneous nucleation of the investigated gold nanorod superstructures were calculated (Table 1). The thermodynamic term constitutes the first addend in eqn (1) and the kinetic term constitutes the second addend. A low correlation was observed between the magnitude of the kinetic term and the nanorod anisotropy. For instance, the kinetic term increased when lowering the aspect ratios, indicating a decrease of E_A consistent with the higher nucleation rates depicted in Fig. 3. However, the obtained values hardly differ and are of the same magnitude. In general, the kinetic terms dominate over the thermodynamic ones. The magnitude of the thermodynamic term related to the nucleation barrier depends on the supersaturation σ and we determined the values for a supersaturation of 1.0 and two typical

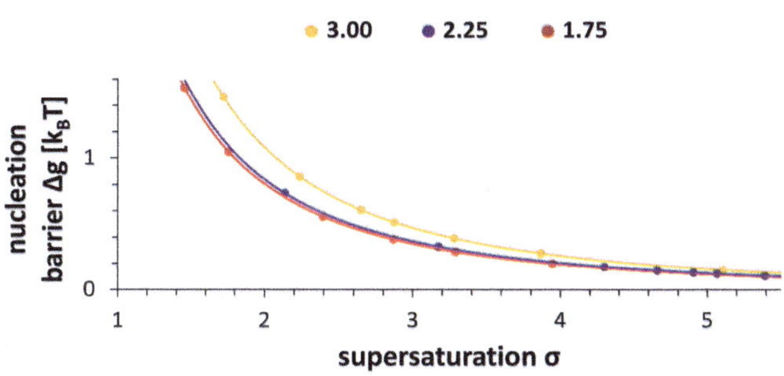

Fig. 5 Nucleation barrier. The nucleation barrier determined for the heterogeneous nucleation of gold nanorods is plotted against the supersaturation. The gold nanorods differ in their aspect ratio: 3.00 (yellow), 2.25 (purple), and 1.75 (red). The particles with the highest aspect ratio exhibit the highest nucleation barrier (3 : 1), while the nucleation barriers for the other two aspect ratios hardly differ.

Table 1 Kinetic and thermodynamic terms. The kinetic and thermodynamic terms determined for the heterogeneous nucleation of gold nanorod-based superstructures on sulfonate-functionalized mica substrates. Since the thermodynamic term is dependent on the supersaturation σ, we determined it for supersaturations of 1.0, 1.5, and 2.0

	3.00	2.25	1.75
Kinetic term	15.79	16.68	17.26
Thermodynamic term with σ = 1.0	−4.37	−3.39	−3.26
Thermodynamic term with σ = 1.5	−1.94	−1.51	−1.45
Thermodynamic term with σ = 2.0	−1.09	−0.85	−0.82

Table 2 Growth conditions for gold nanorod synthesis

Localized surface plasmon resonance [nm]	Volume of silver nitrate 0.01 M solution [ml]	Volume of hydrochloric acid 1 M solution [ml]	Volume of small gold nanorod solution [μl]
630	0.4	—	50
730	2.0	2.0	50
800	2.5	6.0	40

supersaturations from our experiments: 1.5 and 2.0 (Table 1). The thermodynamic terms decrease in absolute numbers with decreasing aspect ratio indicating a decreasing nucleation barrier, which makes nucleation with a high aspect ratio thermodynamically less favourable than nucleation with a low aspect ratio. This finding is similar to the decreasing E_A with decreasing aspect ratio reflected in increasing kinetic terms. The values of the kinetic terms are dominating even for the lowest supersaturation of 1.0. This order of magnitude of thermodynamic and kinetic terms stands in contrast to typical molecular or ionic systems, for example, the nucleation of calcium carbonate on organothiol self-assembled monolayers.[5] As a rough estimate from the published graph, these examined systems showed thermodynamic terms of about −550, for a supersaturation σ of 1.0, which are significantly larger than the kinetic terms of around 40. The difference between the nanoparticle systems to ionic and molecular systems could be explained by a higher impact of diffusion for the nucleation with larger nanoparticles compared to small ions or molecules. This impact would make diffusion a crucial aspect for the heterogeneous nucleation of nanoparticle building units due to the more limited supply of building units to the nucleation site as compared to fast diffusing small ions or atoms. The increasing kinetic term with decreasing aspect ratios could support this idea (Table 1). Furthermore, a stronger diffusion is observed for the rods with a lower aspect ratio compared to those with a larger aspect ratio, due to the larger hydrodynamic radii of the latter. As the dominating kinetic term for the nucleation of gold nanorod-based superstructures stands in great contrast to the usually analysed atomic or molecular systems, further studies with nanoparticle systems would be of interest to verify whether this applies to the nucleation of nanoparticle building units in general.

Conclusions

It was possible to quantitatively investigate the heterogeneous nucleation phenomenon of superstructures based on anisotropic nanoparticles using a set of gold nanorods with different aspect ratios as a model system. Time-resolved light microscopy and UV-vis-NIR spectroscopy allowed for the determination of nucleation rates as well as the nanoparticle concentration in solution at each time. The UV-vis-NIR spectra allowed for a quantitative evaluation of the nucleation process by showing the time-dependent aggregation of the nanorods *via* spectral shifts of the nanorod longitudinal plasmonic band to longer wavelengths.

The shape of nanoparticles was found to strongly influence the heterogeneous nucleation rate of superstructures from anisotropic building blocks, as faster nucleation was noticed for nanorods with lower aspect ratios. In addition, evaluation of the nucleation rates in dependence of the supersaturation revealed a much larger kinetic nucleation barrier than the thermodynamic one. This is the opposite of the situation reported for the nucleation of ionic crystals like calcium carbonate.[5] The reason might be the much faster diffusion of ions than the nanorods and, therefore, a larger kinetic hindrance for the nucleation of the nanorod building units. On the other hand, the thermodynamic barrier did not practically exist for the investigated nanorods, as it was found to be $<1\ k_BT$ already for supersaturations >2, and only reached a few k_BT at its maximum at supersaturations approaching 1. The reason could be the electrostatic attraction between the negatively charged mica substrates and the positively charged nanorods, which almost eliminates the thermodynamic barrier, while the kinetic one is still significant.

Overall, our data set on heterogeneous nucleation of monodisperse nanorods with different anisotropy degrees into superstructures, provides important insight into the formation of superstructures from anisotropic nanoparticles. Moreover, the reported experiments can also serve as experimental data for theoretical descriptions of nucleation using non-spherical building units. The herein presented first results are limited, but a more systematic future investigation using the established experimental methodology can potentially yield larger data sets to systematically reveal the influence of the building unit shape on nucleation processes.

That a theory describing nucleation with anisotropic building units is important can already be seen from the fact that molecules are rarely spherical and also many particles are anisotropic. Therefore, understanding their nucleation and crystallization better could be truly beneficial for several applications ranging from nanoparticle superstructures like mesocrystals *via* macromolecular crystals in protein crystallization to the crystallization of small molecules in pharmaceuticals.

Experimental

Materials

All the starting materials were obtained from commercial suppliers and used without further purification: hexadecyltrimethylammonium bromide (CTAB, ≥99%), 1-decanol (*n*-decanol, 98%), hydrogen tetrachloroaurate trihydrate ($HAuCl_4 \cdot 3H_2O$, ≥99.9%), silver nitrate ($AgNO_3$, ≥99.0%), hydroquinone

(≥99.0%), L-ascorbic acid (≥99%), sodium borohydride ($NaBH_4$, 99%), *N*,*N*-diisopropylethylamine, and 3-aminopropyltriethoxysilane were purchased from Aldrich. Heptadecafluoro-1,1,2,2-tetrahydrodecyltriethoxysilane was purchased from ABCR. Ethanol (EtOH, ≥ 99.8%) and hydrochloric acid (HCl, 37.0%) were purchased from VWR. Nitric acid (HNO_3, ≥ 65%) was purchased from Carl Roth. MilliQ grade water (resistivity 18.2 MΩ cm at 25 °C) was used in all experiments. Mica sheets were obtained from micro to nano in different shapes with a thickness of 0.15 to 0.21 mm and the highest grade V-1 quality. Quartz cuvettes from VWR were used.

Synthetic procedures

Gold nanorods were synthesized according to a recently reported procedure.[47]

Synthesis of 1–2 nm gold seeds. 200 μl of a 0.05 M chloroauric acid solution and 200 μl of a 0.1 M ascorbic acid solution were added to 20 ml of a 50 mM CTAB and 13.5 mM *n*-decanol solution (prepared by dissolving 9.111 g of CTAB and 1.068 g in a 500 ml glass beaker in 500 ml of warm water, ~60 °C, under stirring until complete dissolution). After 1–2 min, 800 μl of a freshly prepared 0.02 M sodium borohydride solution was injected under vigorous stirring (1000 rpm using a polytetrafluorethylene plain magnetic stirring bar: 30 × 6 mm) at 25–27 °C. The seed solution was aged for 30–60 min at room temperature prior to use.

Synthesis of small gold nanorods (21 nm in length and 7.5 nm in width). In a typical synthesis, 2400 μl of 0.01 M silver nitrate, 21 ml of 1 M hydrochloric acid, 3000 μl of 0.05 M chloroauric acid, and 3900 μl of 0.1 M ascorbic acid were added to 300 ml of a 50 mM CTAB and 11 mM *n*-decanol solution at exactly 25 °C (prepared by following the same procedure described for the 1–2 nm gold seeds). The growth was initiated by adding 18 ml of the seed solution under stirring. The mixture was left undisturbed at exactly 25 °C for at least 4 h. The obtained small gold nanorods, with a localized surface plasmon resonance longitudinal band located between 725–730 nm, were centrifuged at 14 000–15 000 rpm for 45–60 min in 2 ml tubes. The precipitate was redispersed with 4 ml of a 10 mM CTAB solution. The concentrated gold nanorods were centrifuged twice under the same conditions, and the final Au^0 concentration was fixed to 4.65 mM (Abs 400 nm: 1, optical path: 0.1 cm).

Synthesis of gold nanorods with distinct dimensions. In a typical synthesis, a certain amount of a 0.01 M silver nitrate solution (Table 2), 1 ml of 0.05 M chloroauric acid, 0.8 ml of a 0.1 M ascorbic acid solution (for nanorods with localized surface plasmon resonances at 730 nm and 800 nm), or 5 ml of a 0.1 M hydroquinone solution (for gold nanorods with localized surface plasmon resonances at 630 nm) were added under stirring, to 100 ml of a 50 mM CTAB and 11 mM *n*-decanol (for gold nanorods with localized surface plasmon resonances at 730 nm and 800 nm) solution or a 50 mM CTAB and 5.5 mM *n*-decanol (for gold nanorods with localized surface plasmon resonances at 630 nm) solution at 28 °C (Table 2). A certain volume of 1 M hydrochloric acid was subsequently added under stirring, and the growth was initiated *via* the addition of a given amount of small gold nanorod solution (Table 2). The mixture was left undisturbed for 4 h, and the obtained overgrown nanorods were washed by precipitation at 4000 rpm for 10 min and redispersed in 5 ml of a 1 mM CTAB solution.

Functionalization of mica substrates. Mica substrates were freshly cleaved using cello tape and treated with an oxygen plasma for 10 min at 80 W to generate free hydroxyl groups. The surface functionalization was adapted from Crampton *et al.* and was carried out using a vapour phase diffusion under a protective gas atmosphere.[58] A 4 L desiccator was two times evacuated and flooded with nitrogen. The substrates were placed in the desiccator, followed by evacuation and flooding with nitrogen for a third time to reach a humidity under 25%. 60 µl of 3-aminopropyltriethoxysilane and 20 µl of 1,1-diisopropylethylamine were added to the desiccator in separate beakers. The desiccator was evacuated for 30 s, sealed afterwards and left for 2 h at RT. Afterwards, the functionalized substrates were baked for 2 h at 150 °C under ambient conditions to ensure the homogeneous distribution of the siloxanes on the surface and a complete crosslinking of the siloxane network. For functionalization with polystyrene sulfonate, the 3-aminopropyltriethoxysilane functionalized substrates were given into a solution of 1 wt% polystyrene sulfonate in water for 1 h. Afterwards, the surfaces were cleaned thoroughly with water and dried at 50 °C. All functionalized substrates were used in a period of one to two days after the functionalization.

Hydrophobisation of quartz cuvettes. The quartz cuvettes were cleaned in aqua regia for 1 h and afterwards cleaned thoroughly with water and treated with an oxygen plasma for 10 min at 80 W. The coating of quartz cuvettes with (heptadecafluoro-1,1,2,2-tetrahydrodecyl)methyldichlorosilane was performed similarly to the functionalization of mica substrates with 3-aminopropyltriethoxysilane except for the coating time, which amounted to 12 h instead of 2 h.

Nucleation experiments

Preparation of gold nanorod dispersions. The concentrated gold nanorods from the synthesis were washed with a 2 mM CTAB aqueous solution, centrifuged at 6000 rpm for 20 min, and the supernatant was removed. The washing procedure was repeated two times. A solution of 0.02 mM CTAB in water was prepared. The concentrated gold nanorods were added to the CTAB solution to reach the desired optical density. Before using the dispersion of gold nanorods in experiments, the stability of the gold nanorods was checked with UV-vis-NIR spectroscopy.

Destabilisation of gold nanorod dispersions. To destabilize the dispersion of gold nanorods, ethanol was added to an amount of 20 vol% (for the microscopic analysis). During or after the destabilization experiments, the formed aggregates and remaining single particles were stabilized again using an excess amount of CTAB. This allowed freezing and analysing aliquots at their actual destabilization stage. Therefore, a CTAB solution was added to adjust the amount of CTAB to 2.00 mM. To analyse the structures formed on mica surfaces, the mica platelets were washed thoroughly with water and dried with a nitrogen stream at RT.

Image analysis. During the experiments, light microscope pictures were taken every 30 s and processed with ImageJ respective to Fiji. A Trainable Weka Segmentation was applied, differentiating between nuclei on the surface, background and moving aggregates in solution. The classifier results were given out as results and the number of nuclei species was counted. Therefore, a threshold (MaxEntropy) was applied and the number of nuclei was counted using the analyse particles function (size = 10 − infinity pixel). The number of counted species of the first picture (only background, structure growth did not start yet) was subtracted from all following results.

Determination of supersaturation. For the determination of the gold nanorod concentrations, the optical density at 400 nm was taken from UV-vis-NIR spectra. After Hendel *et al.*, we assumed an extinction coefficient of 2.685 l mol^{-1} cm for our gold nanorods.[59] With the Beer–Lambert law (eqn (3)) the Au(0) concentration c was calculated since we knew the extinction E_λ from the UV-vis-NIR measurements and the depth of the used cuvette d.[60]

$$E_\lambda = \varepsilon_\lambda c d \tag{3}$$

With this, the Au(0) concentration of the solution was known. However, in the system, the monomer was not the gold atom but the gold nanorods. Therefore, the gold nanorod concentration had to be calculated from the Au(0) concentration. The volume for each gold nanorod aspect ratio was calculated: 6.02×10^{-22} m^3 (3.00), 4.52×10^{-22} m^3 (2.25), and 3.52×10^{-22} m^3 (1.75). Gold only forms a face-centred cubic package with a lattice parameter of 0.4078 nm resulting in the volume of the unit cell of 0.07 nm^3.[61] With this, the number of gold unit cells fitting into one gold nanorod was calculated. One gold unit cell contains 4 atoms of gold which gave us the number of gold atoms in one gold nanorod: 3.56×10^7 (3.00), 2.67×10^7 (2.25), and 2.08×10^7 (1.75).[62] With this quantity, we converted the Au(0) concentration into a gold nanorod concentration.

To calculate the supersaturation σ, the concentration of gold nanorods in a saturated solution x^* was determined for each particle surface system. Therefore the Au(0) concentration after the destabilization of the nanorods (when a steady state was reached) was determined with UV-vis-NIR spectroscopy and the corresponding gold nanorod concentration was calculated. Finally, the resulting supersaturation for each destabilization experiment at a gold nanorod concentration x was calculated with eqn (4).

$$\sigma = \frac{x}{x^*} \tag{4}$$

Instruments

HR-TEM images were recorded on a JEOL JEM-2200FS microscope at an accelerating voltage of 200 kV. The samples were applied on Quantifoil carbon-coated Cu 400 mesh grids.

Light microscopy images were recorded with an AxioImager from Zeiss with an LD Epiplan 50×/0.50 HD DIC objective using transmitted light, bright field illumination, a condenser numerical aperture at 0.9, an Axiocam 506 bw as an imaging device, and an exposure time of 10 ms.

SEM images were recorded with a Gemini500 by Zeiss operating at 3 kV equipped with an Inlens and a secondary electron detector for secondary and backscattered electrons. Samples were coated with a 2.5 nm platinum layer, mounted on aluminium stubs, and attached by carbon conductive layers.

UV-vis-NIR measurements were performed with a Varian Cary 50 spectrometer in quartz cuvettes. The UV-vis-NIR measurements carried out simultaneously with light microscopy measurements were performed with a modular USB2000+ spectrometer from Ocean Optics equipped with a USB-DT miniature deuterium tungsten halogen lamp.

Author contributions

A. G. performed the investigation, formal analysis, methodology, visualization and project administration for this project. G. G.-R. provided resources (gold nanorods) and conceptualized this project. H. C. secured the funding, conceptualized and administrated this project. All authors contributed to writing of this manuscript.

Conflicts of interest

There are no conflicts to declare.

Acknowledgements

This work was supported by the Deutsche Forschungsgemeinschaft (SFB 1214, Project B3). G. G.-R. gratefully acknowledges the Alexander von Humboldt Foundation for financial support. We thank the BIC of the University of Konstanz for providing the light microscope instrumentation and assistance with the imaging and analysis of the pictures. Furthermore, we thank the PAC and the Nano Lab of the University of Konstanz for providing the SEM and TEM.

Notes and references

1 D. W. Oxtoby, *J. Phys.: Condens. Matter*, 1992, **4**, 7627.
2 J. J. De Yoreo and P. G. Vekilov, *Rev. Mineral. Geochem.*, 2003, **54**, 57–93.
3 K. Sangwal, *Nucleation and Crystal Growth: Metastability of Solutions and Melts*, 2018.
4 S. A. Walling and J. L. Provis, *Chem. Rev.*, 2016, **116**, 4170–4204.
5 M. H. Nielsen, J. R. I. Lee, Q. N. Hu, T. Y. J. Han and J. J. De Yoreo, *Faraday Discuss.*, 2012, **159**, 105–121.
6 R. P. Sear, *J. Phys.: Condens. Matter*, 2007, **19**, 033101.
7 L. Assoud, F. Ebert, P. Keim, R. Messina, G. Maret and H. Löwen, *J. Phys.: Condens. Matter*, 2009, **21**, 464114.
8 H. J. Schöpe and T. Palberg, *J. Phys.: Condens. Matter*, 2002, **14**, 11573.
9 V. de Villeneuve, D. Verboekend, R. Dullens, D. Aarts, W. Kegel and H. Lekkerkerker, *J. Phys.: Condens. Matter*, 2005, **17**, S3371.
10 S. Auer and D. Frenkel, in *Advanced Computer Simulation*, Springer, 2005, pp. 149–208.
11 B. J. Alder and T. E. Wainwright, *J. Chem. Phys.*, 1957, **27**, 1208–1209.
12 W. W. Wood and J. D. Jacobson, *J. Chem. Phys.*, 1957, **27**, 1207–1208.
13 P. N. Pusey and W. Van Megen, *Nature*, 1986, **320**, 340–342.
14 L. Antl, J. Goodwin, R. Hill, R. H. Ottewill, S. Owens, S. Papworth and J. Waters, *Colloids Surf.*, 1986, **17**, 67–78.
15 S.-E. Phan, W. B. Russel, Z. Cheng, J. Zhu, P. M. Chaikin, J. H. Dunsmuir and R. H. Ottewill, *Phys. Rev. E: Stat., Nonlinear, Soft Matter Phys.*, 1996, **54**, 6633–6645.
16 R. S. Jardine and P. Bartlett, *Colloids Surf., A*, 2002, **211**, 127–132.
17 U. Gasser, *J. Phys.: Condens. Matter*, 2009, **21**, 203101.
18 J. R. Savage and A. D. Dinsmore, *Phys. Rev. Lett.*, 2009, **102**, 198302.

19 A. Nonat, *Cem. Concr. Res.*, 2004, **34**, 1521–1528.
20 P. Nadeau, *Clay Miner.*, 1985, **20**, 499–514.
21 I. Soroka, in *Portland Cement Paste and Concrete*, Springer, 1979, pp. 28–45.
22 J. Holcomb, N. Spellmon, Y. Zhang, M. Doughan, C. Li and Z. Yang, *AIMS Biophys.*, 2017, **4**, 557–575.
23 C. Lizandara-Pueyo, S. Siroky, M. R. Wagner, A. Hoffmann, J. S. Reparaz, M. Lehmann and S. Polarz, *Adv. Funct. Mater.*, 2011, **21**, 295–304.
24 Y. Kang, M. Li, Y. Cai, M. Cargnello, R. E. Diaz, T. R. Gordon, N. L. Wieder, R. R. Adzic, R. J. Gorte and E. A. Stach, *J. Am. Chem. Soc.*, 2013, **135**, 2741–2747.
25 M. Aslam, R. Bhobe, N. Alem, S. Donthu and V. Dravid, *J. Appl. Phys.*, 2005, **98**, 074311.
26 M. Darques, L. Piraux and A. Encinas, *Appl. Phys. Lett.*, 2005, **86**, 072508–072511.
27 S. Polarz, *Adv. Funct. Mater.*, 2011, **21**, 3214–3230.
28 E. V. Sturm and H. Cölfen, *Chem. Soc. Rev.*, 2016, **45**, 5821–5833.
29 D. V. Talapin, *MRS Bull.*, 2012, **37**, 63.
30 Z. W. Quan and J. Y. Fang, *Nano Today*, 2010, **5**, 390–411.
31 S. Y. Zhang, M. D. Regulacio and M. Y. Han, *Chem. Soc. Rev.*, 2014, **43**, 2301–2323.
32 H. Cölfen and M. Antonietti, *Mesocrystals and Nonclassical Crystallization*, John Wiley & Sons, 2008.
33 H. Cölfen and M. Antonietti, *Angew. Chem., Int. Ed.*, 2005, **44**, 5576–5591.
34 M. Niederberger and H. Cölfen, *Phys. Chem. Chem. Phys.*, 2006, **8**, 3271–3287.
35 M. A. Boles, M. Engel and D. V. Talapin, *Chem. Rev.*, 2016, **116**, 11220–11289.
36 W. D. Luedtke and U. Landman, *J. Phys. Chem.*, 1996, **100**, 13323–13329.
37 L. Jiang, X. D. Chen, N. Lu and L. F. Chi, *Acc. Chem. Res.*, 2014, **47**, 3009–3017.
38 Y. Zhou, X. Z. Zhou, D. J. Park, K. Torabi, K. A. Brown, M. R. Jones, C. Zhang, G. C. Schatz and C. A. Mirkin, *Nano Lett.*, 2014, **14**, 2157–2161.
39 Y. Cui, M. T. Bjork, J. A. Liddle, C. Sonnichsen, B. Boussert and A. P. Alivisatos, *Nano Lett.*, 2004, **4**, 1093–1098.
40 C. Hanske, E. H. Hill, D. Vila-Liarte, G. González-Rubio, C. Matricardi, A. Mihi and L. M. Liz-Marzán, *ACS Appl. Mater. Interfaces*, 2019, **11**, 11763–11771.
41 X. Wang, J. Feng, Y. Bai, Q. Zhang and Y. Yin, *Chem. Rev.*, 2016, **116**, 10983–11060.
42 A. D. Dinsmore, M. F. Hsu, M. G. Nikolaides, M. Marquez, A. R. Bausch and D. A. Weitz, *Science*, 2002, **298**, 1006–1009.
43 Y. Lin, A. Boker, H. Skaff, D. Cookson, A. D. Dinsmore, T. Emrick and T. P. Russell, *Langmuir*, 2005, **21**, 191–194.
44 J. He, Q. Zhang, S. Gupta, T. Emrick, T. R. Russell and P. Thiyagarajan, *Small*, 2007, **3**, 1214–1217.
45 Y. Xia, K. D. Gilroy, H. C. Peng and X. Xia, *Angew. Chem., Int. Ed.*, 2017, **56**, 60–95.
46 M. N. O'Brien, M. R. Jones, K. A. Brown and C. A. Mirkin, *J. Am. Chem. Soc.*, 2014, **136**, 7603–7606.
47 G. Gonzalez-Rubio, V. Kumar, P. Llombart, P. Diaz-Nunez, E. Bladt, T. Altantzis, S. Bals, O. Pena-Rodriguez, E. G. Noya, L. G. MacDowell, A. Guerrero-Martinez and L. M. Liz-Marzan, *ACS Nano*, 2019, **13**, 4424–4435.
48 M. Grzelczak, J. Vermant, E. M. Furst and L. M. Liz-Marzán, *ACS Nano*, 2010, **4**, 3591–3605.

49 J. Liao, Y. Zhang, W. Yu, L. Xu, C. Ge, J. Liu and N. Gu, *Colloids Surf., A*, 2003, **223**, 177–183.

50 V. Myroshnychenko, J. Rodríguez-Fernández, I. Pastoriza-Santos, A. M. Funston, C. Novo, P. Mulvaney, L. M. Liz-Marzán and F. J. G. De Abajo, *Chem. Soc. Rev.*, 2008, **37**, 1792–1805.

51 L. M. Liz-Marzán, *Langmuir*, 2006, **22**, 32–41.

52 G. Haran and L. Chuntonov, *Chem. Rev.*, 2018, **118**, 5539–5580.

53 R. Yu, L. M. Liz-Marzán and F. J. G. de Abajo, *Chem. Soc. Rev.*, 2017, **46**, 6710–6724.

54 G. González-Rubio, J. González-Izquierdo, L. Bañares, G. Tardajos, A. Rivera, T. Altantzis, S. Bals, O. Peña-Rodríguez, A. Guerrero-Martínez and L. M. Liz-Marzán, *Nano Lett.*, 2015, **15**, 8282–8288.

55 D. J. Schupp, J. Angst, E. A. Schaefer, S. M. Schupp and H. Cölfen, *J. Phys. Chem. C*, 2021, **125**, 20343–20350.

56 A.-K. Göppert, G. González-Rubio and H. Cölfen, *J. Phys. Chem. A*, 2020, 5657–5663.

57 O. Galkin and P. G. Vekilov, *J. Phys. Chem. B*, 1999, **103**, 10965–10971.

58 N. Crampton, W. A. Bonass, J. Kirkham and N. H. Thomson, *Langmuir*, 2005, **21**, 7884–7891.

59 T. Hendel, M. Wuithschick, F. Kettemann, A. Birnbaum, K. Rademann and J. r. Polte, *Anal. Chem.*, 2014, **86**, 11115–11124.

60 A. Beer, *Ann. Phys.*, 1852, **162**, 78–88.

61 F. M. Mirabella, *Internal reflection spectroscopy: theory and applications*, CRC Press, 1992.

62 F. M. Fowkes, *Ind. Eng. Chem. Res.*, 1964, **56**, 40–52.

Faraday Discussions

PAPER

Mechanism of ice nucleation in liquid water on alkali feldspars†

Alice Keinert,[a] Kathrin Deck,‡[a] Tilia Gaedeke,[a] Thomas Leisner[ab] and Alexei A. Kiselev*[a]

Received 5th December 2021, Accepted 11th January 2022
DOI: 10.1039/d1fd00115a

Crystallization of supercooled liquid water in most natural environments starts with heterogeneous nucleation of ice induced by a nucleation site. Mineral surfaces, which form the majority of aqueous interfaces in Earth's ecosystem, possess a plethora of surface morphological and chemical features that can serve as ice nucleation sites. The nature of surface sites responsible for ice nucleation from supersaturated water vapor have been recently identified for alkali feldspar, a family of rock-building minerals constituting 60% of the Earth's crust. It was demonstrated that ice preferentially forms upon the patches of crystalline surface with (100) orientation, exposed in the surface defects such as cracks, pores, and pits arising due to chemically induced stress and further enhanced by hydrothermal alterations of natural feldspars. However, whether the same sites were responsible for nucleation from liquid water, remained to be shown. Here, we investigate the mechanism of heterogeneous ice nucleation in a layer of aqueous sucrose solution on top of thin sections of feldspar prepared along the (010) crystalline plane. We observe a preferential orientation of ice crystals defined by an epitaxial relationship between feldspar and ice, with ice crystals growing on the crystalline surfaces of feldspar with (100) orientation. We thus conclude that the ice nucleating sites active in deposition freezing mode are also active in the immersion freezing regime. This conclusion is further supported by the enhancement of ice nucleation active site density observed for the thin sections of feldspar prepared sub-parallel to the (100) plane as compared to sections prepared along (010) and (001) crystallographic orientations.

Introduction

Although water in crystalline form constitutes less than 2% of the total water reservoir on Earth, almost 70% of the world's fresh water exists in the form of

[a]*Karlsruhe Institute of Technology, Institute of Meteorology and Climate Research, Karlsruhe, Germany. E-mail: alexei.kiselev@kit.edu*
[b]*Institut für Umweltphysik, Universität Heidelberg, Heidelberg, Germany*

† Electronic supplementary information (ESI) available. See DOI: 10.1039/d1fd00115a

‡ Now at TU Delft, Faculty of Aerospace Engineering, Department of control and operations, Delft, The Netherlands.

snow and ice.[1] Crystallization of supercooled water is one of the main mechanisms responsible for precipitation formation in atmospheric clouds, and as such is important for the Earth's water cycle and for the cloud-feedback as part of the global radiation budget.[2–4] Thus, the importance of water–ice transformation for the Earth's ecosystem cannot be overestimated.

In the atmosphere, cloud droplets can be supercooled to a temperature as low as approximately −38 °C before spontaneous nucleation of an ice embryo leads to the crystallization of water. At temperatures below −25 °C formation of ice may happen directly from supersaturated water vapor *via* deposition on the surface of aerosol particles. This latter mechanism was shown to be efficient if mineral dust particles are present in the air, acting as ice nucleating particles (INP).

Even though the composition of mineral dust is highly heterogeneous,[5] not all minerals are equally efficient INPs. It was recently demonstrated that certain types of alkali feldspar are capable of triggering ice nucleation at temperatures just below the melting point,[6–8] with the mineralogy of feldspar apparently being a decisive factor for its ice nucleating efficacy. The highly ice nucleation (IN) active types of alkali feldspars feature micron-scale lamellar intergrowth of more K-rich and more Na-rich alkali feldspar, known as perthitic intergrowth. It was suggested that ice nucleates preferentially at the sites associated with microstructural defects (the so-called nanochannels and pull-aparts) forming due to the tensile stress near the intergrowth boundaries.[9]

In our previous study of ice crystals nucleation and growth on natural perthitic alkali feldspars, the preferential epitaxial nucleation of ice by the alignment of the ice $(10\bar{1}0)$ prism plane with (100) faces of the feldspar was identified as the mechanism associated with high IN activity of alkali feldspar in deposition freezing mode.[10] Similar results have been reported in a later experimental study conducted with different feldspar samples under similar conditions,[11] but could not be reproduced by molecular dynamics (MD) simulation performed using the (100) surfaces of an idealized K-feldspar crystal.[12] To explain how the non-cleavable crystalline (100) face becomes accessible for ice nucleation, we hypothesized that small patches of (100) become exposed in the cracks and steps that are characteristic for feldspar cleavage surfaces (001) and (010). This hypothesis has received experimental support in the more recent study demonstrating enhancement of the IN activity of K-rich feldspar caused by chemically-induced fracturing.[13] Preferential nucleation of ice in surface pits and cracks has been reported repeatedly for immersion[14,15] and deposition freezing,[10,11] however, no direct evidence that ice nucleation is associated with (100) surfaces in *both freezing mechanisms* has been provided so far.

Recent studies of deposition freezing suggest that the presence of liquid water either adsorbed on the mineral surface[16] or condensing in pores and cavities *via* capillary condensation is required for initial ice nucleation,[17,18] which would imply that the same IN active sites should be responsible for freezing under both mechanisms. And yet, the direct evidence that the patches of (100) are responsible for ice nucleation in liquid water were missing.[19] In this communication we report the results of ice nucleation experiments in concentrated sucrose solution in contact with feldspar samples, prepared as thin sections parallel to the three main crystallographic orientations (001), (010) and (100). We observe a preferential orientation of ice crystals nucleating heterogeneously on feldspar allowing us to deduce the nature of the IN active sites. We address the question, if the frequency of appearance of the (100) patches is associated with the feldspar microtexture, by comparing the ice nucleating efficacy of two distinctly different feldspars: a perthite microcline feldspar with rich microtexture and an almost pure K-rich adularia orthoclase feldspar, devoid of any visible microtexture.

Materials and methods

Sample preparation

Specimens of two natural feldspars have been used in this study. The first one is a natural perthitic alkali feldspar from Pakistan (samples prepared from this specimen are labelled FS06). Visually, the centimetre-sized specimen with partially developed facets is pale beige and turbid (see ESI Fig. S1A†). The X-ray powder diffraction analysis (XRD, Panalytical, Cu K-alpha 1&2) yielded a phase content of 41% orthoclase, 39% microcline, and 20% albite (see Table ST1† for details). FS06 has a complex microperthitic structure overlain by polysynthetic twinning according to the pericline law[20] and porosity, very likely related to hydrothermal or deuteric alteration[21] (see Fig. S2†). The fully coherent exsolution lamellae are characteristic for slowly cooled microcline feldspar[21] and are oriented so that the lattice misfit between the more K-rich and more Na-rich regions is minimized. This condition defines the so-called *Murchison plane*,[22,23] which has non-rational Miller indices between ($\bar{6}01$) and ($\bar{8}01$). We use this fact for identification of crystallographic orientations in the samples prepared from the FS06 specimen.

The second specimen (labelled FS07) is a K-rich sanidine feldspar (adularia) from Austria. The FS07 contains over 90% pure orthoclase phase (see ESI Table ST1†) and is almost transparent. Apart from few macroscopic cracks, FS07 does not exhibit any microtexture at any scale, neither appearing in polarization microscope nor in the scanning electron microscope imaging.

For this study, both specimens were cut along the main crystallographic plains (001), (010) and (100) and the thin sections parallel to these planes were prepared following standard petrographic sample preparation techniques. The crystallographic orientations of the thin sections parallel to the cleavage planes (010) and (001) were later confirmed by backscattered electron diffraction (Hikari Super BSED system, AMETEK EDAX) and analysed using TSL OIM 8 software. The orientation of the (100) surface was deduced from the known geometric relationship to the cleavage planes. The deviation between the actual crystallographic plane and the orientation of the thin section was estimated at below 2° for the cleavage planes but could be as large as ±10° for the thin sections prepared along (100).

For comparison, the fragments of both specimens were ground in a ceramic mortar and sieved through a 20 μm sieve. The powders of FS06 and FS07 were then used for preparation of aqueous suspensions with various concentrations of solid material for the droplet freezing assay (DROFA) experiment.

Sucrose solution experiment

Nucleation and growth of ice in supercooled water is too fast to be observed with the usual means and requires high-speed video-recording techniques to capture the initial stage of ice formation.[14] However, the growth rate of ice crystals can be significantly slowed down in a concentrated solution where water diffusivity is significantly reduced. In 1957, Bigg[24] introduced a method of counting airborne ice crystals by allowing them to settle onto the surface of a supercooled sugar solution. We use this approach to investigate the kinetics of crystal nucleation and growth on the surface of thin sections of feldspar.

In this study, a small droplet (approximately 1 μL volume) of 60 wt% (weight%) aqueous solution of sucrose (Merck) was placed on the surface of a sample thin section and covered with a microscope cover slide to avoid evaporation (Fig. S3†). The sample was positioned on the Peltier stage mounted on the rotation table of the polarization microscope (Leica DM4P), and the whole microscope was enclosed into an air-tight flexible glove box filled with water-free synthetic air to avoid water condensation and frost formation on the cold surfaces. The Peltier stage was slowly cooled down to −30 °C or until several ice crystals became visible in the field of view of the 5× objective piece. The temperature was then manually adjusted to the apparent melting point of ice in the sucrose solution (between −13 °C and −15 °C) to slow down or completely stop the growth of ice crystals. In several cases the temperature was slightly raised to achieve partial melting of ice crystals allowing for observation of the dendrite geometry and identification of crystallographic axes of individual ice crystals. We refer to such crystals as being "optically thinned".

DROFA measurements

Thin sections. The ice nucleating efficacy of the thin sections and powder suspensions of feldspar samples was measured in the DROFA setup as described in our earlier work.[8,13] Briefly, a 10 × 10 mm thin section was placed into a temperature-controlled cold stage setup consisting of a Linkam MDBCS-196 motorized cold stage, a piezo-driven drop-on-demand dispenser (GeSIM, model A010-006 SPIP) and a video camera with a wide-field objective allowing for detection of individual freezing events with 0.125 s time resolution. In the experiments described in this work, 50 to 150 droplets of pure water each with a volume of approximately 1.4 nL were deposited onto the polished surface of a thin section in a square pattern with 400 μm centre-to-centre separation between the droplets. During the freezing experiments the cold stage was cooled with a rate of 3 K min^{-1}. Time and temperature of freezing of every individual droplet were recorded automatically and stored in the form of the temperature dependent fraction of frozen droplets (freezing spectra) for further analysis.

Suspension droplets. Approximately 70 droplets with a volume of 21.6 nL containing 0.01 wt%, 0.1 wt%, or 1 wt% suspensions of feldspar particles were deposited in a 9 × 9 array with a PipeJet Nano dispenser (BioFluidix GmbH) on a clean 10 × 10 mm silicon wafer (Ted Pella, Inc.). Freezing of the suspension droplets was measured using the DROFA technique as described above. The specific surface area (SSA) of the feldspar powders was determined using an Autosorb iQ model 7 gas sorption system (Quantachrome Instruments, a brand of Anton Paar QuantaTec Inc.). Using argon as a sorbent gas at the boiling point of 87 K and applying Brunauer–Emmett–Teller (BET) theory,[25] the SSA of (2.5 ± 0.2) m^2 g^{-1} was found for the FS06 and (1.8 ± 0.2) m^2 g^{-1} for the FS07 powder samples.

The heterogeneous freezing efficacy of a substrate was expressed in terms of Ice Nucleation Active Site (INAS) surface density,[26] $n_s(T)$, as a function of temperature T using the fraction of frozen droplets $f_{ice}(T)$ and the total area of ice nucleating particles contained in a single droplet S_d. In the DROFA experiments with suspension droplets the S_d is calculated from the SSA of the powder and the weight concentration of feldspar material in a suspension droplet. For droplets deposited on a thin section, S_d is the area of contact between the pure water droplet and the substrate. In this way, the IN efficacy of the powder samples and thin sections can be compared in a quantitative way.

ESEM IN experiments. An environmental scanning electron microscope (ESEM, FEI Quattro FEG, Thermo Fisher Scientific) was used to investigate the surface morphology of feldspar samples and to conduct a limited series of deposition ice nucleation experiments following the method described in our previous work.[10] Using the modified cooling specimen stage of the microscope, the feldspar sample was cooled down to −28 °C with the rate of $-1\ \mathrm{K\ min^{-1}}$ to achieve ice nucleation and growth. The overall pressure in the ESEM specimen chamber was kept at 450 Pa (humidified N_2 gas) with the partial water pressure set to about 50 Pa, which is just above the saturation pressure with respect to ice but below the water saturation pressure at this temperature. Sample cooling was terminated once ice crystals were visually detected on the surface, thus allowing them to develop well-defined pristine crystal habits.

Results and discussion

Ice crystals growing in sucrose solution

Thin section FS06. The main result of this study is that the ice crystals, heterogeneously nucleating in aqueous sucrose solution in contact with the (010) thin section of FS06 feldspar, would grow in a preferential orientation defined by the epitaxial relationship between the crystal structures of feldspar and ice (Fig. 1). The apparent habit of the ice crystals (rectangular envelope filled with branching dendrites) requires some explanation which is given below.

At very small supercooling (below 1 K), ice crystals in aqueous solutions grow preferentially along the basal plane, forming first platelets and then flat dendritic crystals as growth proceeds.[27] At moderate supercooling the dendrites become inclined to the basal plane, with the inclination angle being a function of supercooling temperature and solution concentration.[28,29] At a supercooling of 5 K below the melting point of ice, the primary dendrites split into two symmetric segments, forming three-dimensional hollow hexagonal pyramids joined at their apices.[29–31] The secondary dendrites would then branch off at angles defined by the crystal structure of ice and the splitting angle. Similar shapes have been reported by Pruppacher,[27] Macklin and Ryan,[31] and Knight and DeVries,[29] where a thorough discussion of the observed growth behaviour can also be found.

The sucrose solution in our experiments forms a thin layer (about 0.2 mm) on the surface of feldspar, and thus the growth of ice crystals is constrained in the direction perpendicular to the sample surface. This constraint is not only due to the confinement, but also because the solution is cooled through the substrate, and thus the growth of ice away from the surface is less favourable. In our experiments we observe ice crystals assuming two stable orientations: with the basal plane either parallel or perpendicular to the plane of confinement. If the ice crystals grow with their basal plane parallel to the substrate, their habit exhibits clearly hexagonal symmetry, although the dendrite branches become curved in

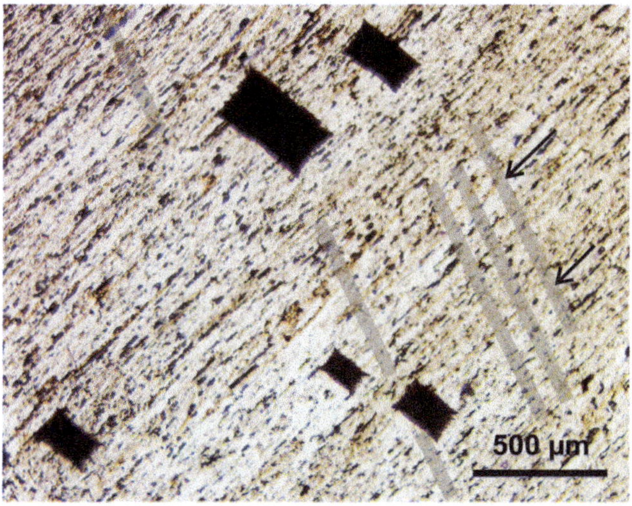

Fig. 1 Ice crystals growing in sucrose solution on top of FS06 thin section along the (010) crystallographic plane of feldspar. The grey bands are manually overlaid on top of the (001) cleavage cracks to guide the eye. The rectangular shape of the crystals is explained in the text. Note the identical orientation of the ice crystals.

response to the asymmetrical temperature gradients in the liquid,[32] as shown by the ESI Fig. S7.† In the latter case, the major axis of ice (*C*-axis) lies in the plane of the substrate, but the full 3D pyramid cannot develop in the lateral plane because the growth of a basal plane is constrained in the direction perpendicular to the sample surface. As a result, the ice crystals have the habit of inverse double pyramid cut along the prismatic plane. Fig. 2 shows a structure of the ice crystal

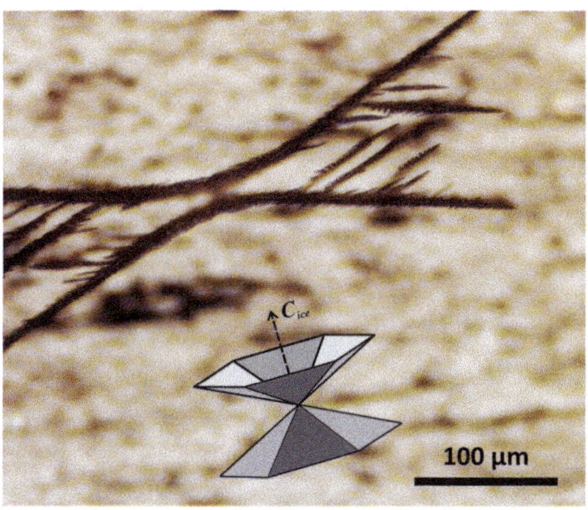

Fig. 2 3D-dendrite structure of an ice crystal in equilibrium with sucrose solution at −14.2 °C. The overlay shows the structure of the inverse double pyramid of ice redrawn after Knight and DeVries,[29] with the orientation of the *C*-axis of the hexagonal ice crystal indicated by the arrow.

stabilized shortly after nucleation by adjusting the temperature to the melting point of ice in the sucrose solution at −14.2 °C. The idealized structure of the inverse double pyramid is overlaid on the image to illustrate the spatial arrangement of the crystallographic axes of the ice. Interestingly, Holden and co-authors[14,33] recently published similar images of rectangular ice crystal shapes appearing in pure water in contact with a thin section of microcline feldspar polished along the (010) cleavage plane.

In order to establish the geometrical relationship between the orientation of ice and crystalline structure of feldspar, we notice two morphological features visible on the surface of the thin section: (a) the cleavage cracks and (b) the exsolution lamellae. Feldspar has two perfect cleavage planes along (010) and (001) with approximately 90° angle between them,[34] and in a bulk crystal of natural feldspar cleavage cracks would appear along these planes if mechanical stress (as a result of thermal expansion or chemical alteration) is applied. On a thin section with orientation (010) as the one shown in Fig. 1, cracks would be aligned with the orthogonal cleavage plane (001). To aid the eye, the cleavage cracks are overlaid with transparent bands in Fig. 1. ESI Fig. S4A† gives a clearer view of a series of cracks on the same substrate.

Orientation of cleavage cracks together with the orientation of the thin section (010) allow for unambiguous identification of the crystallographic orientation of the FS06 feldspar specimen (010). A projection of the unit cell of feldspar linked to the orientation of cleavage cracks is shown in Fig. 3.

Comparing the orientation of ice with the orientation of the feldspar unit cell in the (010) thin section (Fig. 3), it becomes obvious that the primary prismatic plane ($10\bar{1}0$) of ice crystals must be aligned with the (100) plane of feldspar in such a way that the directions of the main crystallographic axes of both crystals coincide (see also ESI Fig. S4B†). This is only possible if ice grows epitaxially on

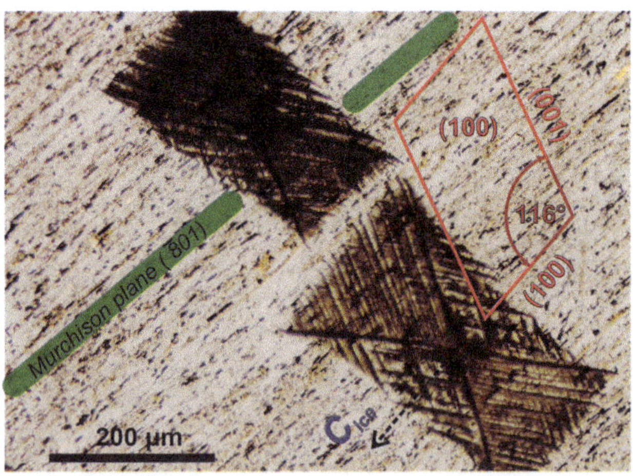

Fig. 3 Oriented ice crystals growing in sucrose solution in contact with FS06 thin section with orientation of (010). Projection of feldspar unit cell identified from the orientation of cleavage cracks and exsolution lamellae is shown in red. Orientation of Murchison plane ($\bar{6}01$) is highlighted in green, while the direction of the main crystallographic axis of the ice is indicated by the broken arrow line.

the (100) surfaces of feldspar, implying that these surfaces must be the nucleation sites of ice. This is in agreement with our previous finding[10] that ice crystals growing from water vapor would be nucleating on the patches of (100) plane exposed in cracks and steps present on the surface of natural perthitic feldspars. Such surface defects are present in abundance on the surface of the (010) thin section, and are associated with the boundaries between exsolution lamellae, as illustrated by the SEM image of the FS06 (010) sample shown in Fig. S5a.† In contrast, the surface of FS07 samples is almost devoid of any surface topography, as illustrated by ESI Fig. S5b.†

The exsolution lamellae are stretched along the Murchison plane having the Miller indices between ($\bar{8}01$) and ($\bar{6}01$) and is thus sub-parallel to (100) with the angle between 8° and 11°. The Murchison plane is described by non-rational Miller indices and does not, in general, coincide with any of the primary atomic planes in feldspar.[22,23] In practice that means that the orientation of exsolution lamellae can be anywhere between planes with Miller indices of ($\bar{6}01$) and ($\bar{8}01$). We investigate the relationship between the ice crystal orientation (which we assume is aligned with the (100) plane of feldspar) and the orientation of exsolution lamellae. Fig. 4 shows the frequency distribution of the angles. From the 87 crystals observed in all IN experiments with the sucrose solution in contact with FS06 thin section (010), approximately 60 were in the range between 5° and 11° with the maximum at 8.1°. For the microcline feldspar with the lattice parameters defined in the ESI Table ST1,† this value is exactly equal to the angle between the planes with Miller indices ($\bar{8}01$) and (100) (which is equal to 8.13°), thus nicely confirming our hypothesis that ice crystals are epitaxially nucleated and grown on the sites with (100) orientation. The variability of ±10° (Fig. 3) comes mostly from the irregular form of exsolution lamellae, which complicates the identification of the Murchison plane.

Interestingly, we could not observe the oriented growth of ice in sucrose solution in contact with other thin sections prepared from the same specimen. Ice crystals growing in all possible orientations were observed on the thin sections with orientations parallel to the (001) and sub-parallel to the (100). However, in most cases ice was growing with its basal face parallel to the substrate. One

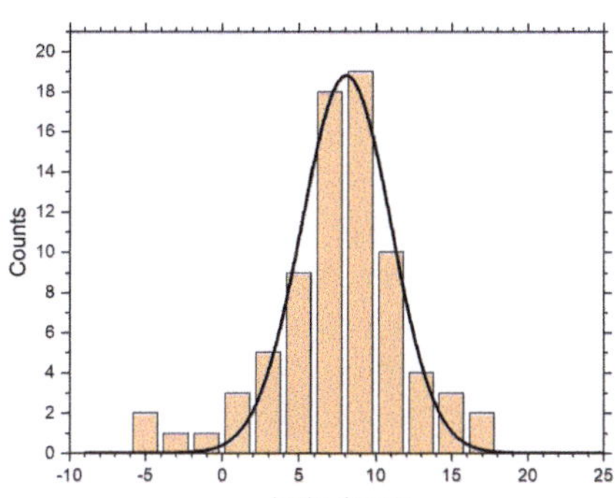

Fig. 4 Distribution of angles between the main crystallographic axis of ice and the preferential orientation of exsolution lamellae within the (010) plane of FS06. The distribution is fitted with a Gauss curve (the black line) with maximum at 8.1°.

possible reason for such behaviour could be that the only sample where ice crystals would have their *c*-axis exactly parallel to the surface is the thin section parallel to the (010). In all other cases, the ice crystals would have to grow with their basal plane tilted with respect to the sample surface. Such geometry, which is easily achievable for deposition growth of ice (as demonstrated in ESI Fig. S6†), could be unfavourable for ice growth in a confined liquid layer. The same argument applies to the thin section of FS06 which was prepared sub-parallel to (100): since the alignment is not perfect, the patches with (100) orientation would be inclined with respect to the surface resulting in an inclination between the basal face of ice and the sucrose solution layer.

Furthermore, no orientational growth of ice could be observed on any of the thin sections of the adularia feldspar FS07. We explain this observation by the lack of morphological features where the patches of (100) could have been exposed. This is consistent with our previous observation of absence of orientational growth of ice on the sanidine feldspar in the deposition mode.[10]

Deposition IN experiments. This study was focused on immersion freezing and therefore only a limited series of experiments with the FS06 sample has been performed. These experiments, conducted in the ESEM in an atmosphere of humidified nitrogen gas, have revealed preferential orientation of ice crystals growing upon the cleaved (001) surface of the FS06 specimen (see ESI Fig. S6†). As in our earlier work,[10] pristine ice crystals with identical orientation were nucleating at a temperature of −28 °C at the vapor pressure below saturation with respect to water, which is traditionally described as the deposition freezing regime. The tilt angle and the orientation of ice crystals with respect to the exsolution lamellae and the cleavage cracks suggest that the nucleation of ice crystals occurred epitaxially on the patches with (100) orientation. We thus conclude that the ice nucleating sites active in deposition and in the immersion freezing modes have the same crystallographic nature. In both freezing mechanisms, patches of feldspar surface with (100) orientation must be responsible for ice nucleation.

In a very recent study Holden and co-authors[33] compared ice nucleation in immersion and deposition mode by using high-speed video imaging. They observed oriented growth of ice crystals on a polished section of feldspar from water vapor and explained it by an epitaxial relationship between the crystalline structure of feldspar and ice. As the orientation of ice crystals was apparently in agreement with that observed in immersion freezing experiments, a common mechanism of ice nucleation in both modes has been suggested. However, no correlation between the sites of initial ice nucleation in the two freezing regimes could be established.

Droplet freezing assay measurement results. The DROFA measurements have been performed with the thin sections polished along the two cleavage planes of feldspar specimens (001) and (010), and a thin section that was prepared along the (100) plane according to geometry considerations. The results of these measurements are shown in Fig. 5 and 6, together with the results for the aqueous suspensions of powder samples in three different weight concentrations. For comparison, the parameterization of the INAS density of a reference feldspar[6] is shown as a straight dashed line.

The first glance at the freezing behaviour of both specimens reveals that droplets in contact with FS06 freeze at 10 K higher temperature than for the FS07,

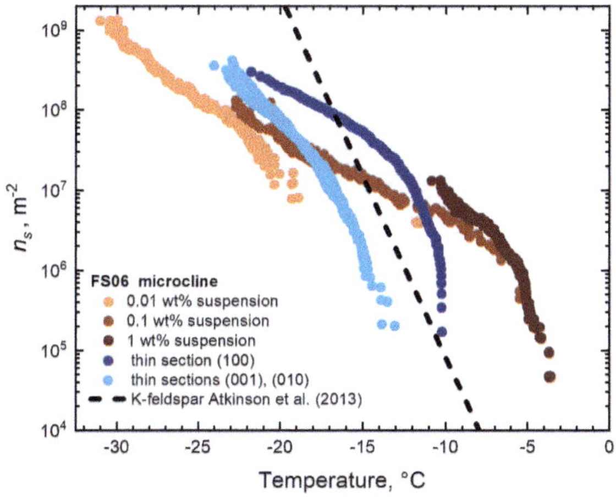

Fig. 5 IN active sites surface densities on the thin sections and for the suspensions of FS06 feldspar.

which we do not find surprising given the well-established empirical relationship between the feldspar microtexture and its IN efficacy.[7–9,13] What is indeed noteworthy is that the droplets are freezing at almost 5 K higher temperature on the surfaces of a thin section polished along the (100) plane as compared to other orientations. This result is somewhat unexpected given the absence of oriented ice growth in sucrose solution in contact with the same surfaces. As the preparation of a thin section along a non-cleavable plane is connected with certain practical difficulties (starting with but not limited to the orientation uncertainty), we have not expected such a striking difference in IN efficacy of the thin sections with different crystallographic orientations. The plausible explanation of this result is that the patches of surface with (100) orientation should appear more frequently in the steps and pits present on the surface of this thin section. The

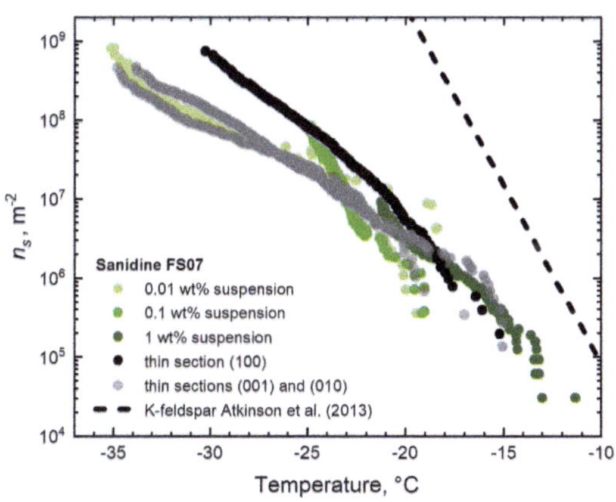

Fig. 6 IN active sites surface densities on the thin sections and for the suspensions of FS07 sanidine feldspar.

absence of apparent epitaxial growth in the sucrose solution experiment is likely due to the mismatch between the preferential plane of ice crystal growth and the plane of confinement defined by the solution layer, as discussed above.

The thin section with (100) orientation shows the highest IN efficacy for both FS06 and FS07 specimens, although the relative enhancement in the case of FS07 is smaller and only pronounced below −20 °C. Assuming that the mechanism responsible for the enhanced ice nucleation is the same (exposure of the surface patches with (100) orientation), the smaller variation of the $n_s(T)$ is expected, given the relative homogeneity and reduced density of surface topography features on the polished surfaces of the FS07 samples as compared to FS06.

We also note a good general agreement between $n_s(T)$ values obtained for pure water droplets in contact with the thin sections parallel to (001) and (010) and for aqueous suspensions. For FS06, however, the number of high-temperature active sites (which are active above freezing temperature of −10 °C) in the powder sample is significantly higher than on the thin section sample with highest IN efficacy (the one sub-parallel to (100)). We hypothesize that such high-temperature sites might have been created by grinding the bulk specimen, thus allowing access to more IN active sites per unit surface area. Similar conclusions have been reached by Holden and co-authors.[14] Indirectly, this hypothesis goes along with our recent observation that chemically induced cracks in the initially texture-free sanidine feldspar (similar to FS07) indeed increase its IN efficacy.[13] No such high-temperature IN active sites have been found in the samples prepared from the FS07 specimen.

Finally, the similarity of $n_s(T)$ values for the thin sections with (001) and (010) orientation observed for both samples FS06 and FS07 should be noted. The values for both orientations are identical within the experimental uncertainties on the order of 100%, typical for droplet freezing assay experiments. Within the suggested mechanistic explanation, this is not surprising for high-temperature IN efficacy of the FS06 specimen given the similar surface density of exsolution lamellae exposed on both the cleavage planes. This, however, cannot explain the identity of $n_s(T)$ values observed for the whole temperature range from −35 to −10 °C for both the FS06 and FS07 samples, as patches of crystalline plane parallel to (100) are rare in FS06 and absent in FS07. We thus have to assume that the reason for identical IN efficacy of the orthogonal cleavage planes lies in the similarity of lattice constants and surface energies of these planes.[34] For FS07 samples, the close match between the $n_s(T)$ values of cleavage planes and aqueous suspensions suggests that mechanical grinding of the crystalline specimen exposes predominantly surfaces coinciding with the cleavage planes, whereas in the ground FS06 sample crystalline surface parallel to the (100) dominates the freezing behaviour at higher temperature.

Conclusions

In summary, we have applied a well-established experimental technique initially devised 70 years ago for detection and counting of airborne ice-crystals to investigate the nature of ice nucleating active sites associated with the microcline feldspar. The technique relies on the drastically reduced growth rates of ice in the concentrated (60 wt%) aqueous solution of sucrose, allowing for observation of crystal habits evolution in real time without the necessity to use high-speed recording methods. By observing the ice crystal growth in a thin layer of sucrose solution in contact with the (010) thin section of microperthitic feldspar, we were able to establish a relationship between the lattice structure of feldspar and the crystalline structure of ice. The ice crystals were preferentially oriented in a way that their main crystal axis is parallel to the main crystallographic axis of feldspar. This is only possible if ice crystals nucleate and grow epitaxially with their prismatic crystal planes $(10\bar{1}0)$ attached to the feldspar crystal surface with (100) orientation. Such surfaces must be exposed in the cavities, steps, and pits which are characteristic for the exsolved alkali feldspars. The evolution of a perthite microstructure in alkali feldspar, often enhanced by hydrothermal/deuteric alteration, fosters the exposure of (100) crystal surfaces and thus enhances the IN activity of exsolved alkali feldspars. This conclusion is strongly supported by the results of the droplet freezing assay study. Here, we have shown that the IN efficacy of the microperthitic microcline FS06 is significantly higher than that of the sanidine sample FS07, which is devoid of cracks, inclusions or perthitic structure. The IN efficacy of the thin section prepared sub-parallel to the (100) orientation was further enhanced in comparison to other orientations, supposedly due to the higher surface density of the crystal surfaces with the true (100) orientations.

Comparing the results of ice nucleation experiments in the aqueous sucrose solutions with the previous experiments conducted with feldspar samples in deposition freezing mode, we arrive at the conclusion that the nature of the most active ice nucleating sites in both processes must be the same. In both cases, the crystal surfaces with (100) orientation must be responsible for the onset of ice nucleation. Considering recent doubts of the validity of the deposition freezing concept, the identity of IN active sites in both types of experiments reported here supports the conclusion that the so-called deposition freezing always requires formation of liquid water in contact with active sites to initiate heterogeneous freezing. The fact that deposition freezing is observed below water saturation, should be explained in view of the recently suggested pore condensation and freezing (PCF) mechanism. The surface topography of the samples investigated in this study fully justify this conclusion.

Author contributions

AAK conceived the idea of the study and led the manuscript preparation with contributions from all authors. AK conducted the droplet freezing assay experiment, evaluated the results and wrote the manuscript. KD and TG performed sucrose solution experiments, analysed the results and prepared the graphics for the manuscript. TL contributed to the conception of the experiment and provided the general supervision. All authors contributed to the discussion of the results.

Conflicts of interest

The authors declare no conflicts of interest.

Acknowledgements

The authors acknowledge financial support by the Helmholtz Association under Atmosphere and Climate Programme (ATMO). The authors also greatly acknowledge XRD analysis of feldspar powder samples performed by Dr Jörg Göttlicher (Institute of Photonic Science KIT-IPS). AAK acknowledges Prof. Rainer Abart from the Vienna University for his valuable help in interpretation of crystallographic relationships in feldspars.

References

1 S. H. Schneider, T. L. Root and M. D. Mastrandrea, *Encyclopedia of Climate and Weather*, Oxford University Press, 2011.

2 S. Bony, R. Colman, V. M. Kattsov, R. P. Allan, C. S. Bretherton, J.-L. Dufresne, A. Hall, S. Hallegatte, M. M. Holland, W. Ingram, D. A. Randall, B. J. Soden, G. Tselioudis and M. J. Webb, *J. Clim.*, 2006, **19**, 3445–3482.

3 O. Boucher, D. Randall, P. Artaxo, C. Bretherton, G. Feingold, P. Forster, V.-M. Kerminen, Y. Kondo, H. Liao, U. Lohmann, P. Rasch, S. K. Satheesh, S. Sherwood, B. Stevens and X.-Y. Zhang, in *Climate Change 2013: the Physical Science Basis. Contribution of Working Group I to the Fifth Assessment Report of the Intergovernmental Panel on Climate Change*, ed. T. F. Stocker, D. Qin, G.-K. Plattner, M. Tignor, S. K. Allen, J. Boschung, A. Nauels, Y. Xia, V. Bex and P. M. Midgley, Cambridge University Press, Cambridge, 2014, ch. Clouds and Aerosols, pp. 571–658.

4 B. J. Murray, K. S. Carslaw and P. R. Field, *Atmos. Chem. Phys.*, 2021, **21**, 665–679.

5 N. A. Marsden, R. Ullrich, O. Möhler, S. Eriksen Hammer, K. Kandler, Z. Cui, P. I. Williams, M. J. Flynn, D. Liu, J. D. Allan and H. Coe, *Atmos. Chem. Phys.*, 2019, **19**, 2259–2281.

6 J. D. Atkinson, B. J. Murray, M. T. Woodhouse, T. F. Whale, K. J. Baustian, K. S. Carslaw, S. Dobbie, D. O'Sullivan and T. L. Malkin, *Nature*, 2013, **498**, 355–358.

7 A. D. Harrison, T. F. Whale, M. A. Carpenter, M. A. Holden, L. Neve, D. O'Sullivan, J. Vergara Temprado and B. J. Murray, *Atmos. Chem. Phys.*, 2016, **16**, 10927–10940.

8 A. Peckhaus, A. Kiselev, T. Hiron, M. Ebert and T. Leisner, *Atmos. Chem. Phys.*, 2016, **16**, 11477–11496.

9 T. F. Whale, M. A. Holden, A. N. Kulak, Y.-Y. Kim, F. C. Meldrum, H. K. Christenson and B. J. Murray, *Phys. Chem. Chem. Phys.*, 2017, **19**, 31186–31193.

10 A. Kiselev, F. Bachmann, P. Pedevilla, S. J. Cox, A. Michaelides, D. Gerthsen and T. Leisner, *Science*, 2017, **355**, 367–371.

11 E. Pach and A. Verdaguer, *J. Phys. Chem. C*, 2019, **123**, 20998–21004.

12 A. Soni and G. N. Patey, *J. Chem. Phys.*, 2019, **150**, 214501.

13 A. A. Kiselev, A. Keinert, T. Gaedeke, T. Leisner, C. Sutter, E. Petrishcheva and R. Abart, *Atmos. Chem. Phys.*, 2021, **21**, 11801–11814.

14 M. A. Holden, T. F. Whale, M. D. Tarn, D. O'Sullivan, R. D. Walshaw, B. J. Murray, F. C. Meldrum and H. K. Christenson, *Sci. Adv.*, 2019, **5**, eaav4316.

15 R. W. Friddle and K. Thürmer, *Atmos. Meas. Tech.*, 2020, **13**, 2209–2218.

16 A. Laaksonen and J. Malila, *Atmos. Chem. Phys.*, 2016, **16**, 135–143.

17 R. O. David, C. Marcolli, J. Fahrni, Y. Qiu, Y. A. Perez Sirkin, V. Molinero, F. Mahrt, D. Brühwiler, U. Lohmann and Z. A. Kanji, *Proc. Natl. Acad. Sci. U. S. A.*, 2019, **116**, 8184–8189.

18 C. Marcolli, *Atmos. Chem. Phys.*, 2020, **20**, 3209–3230.
19 B. J. Murray, *Science*, 2017, **355**, 346–347.
20 I. Parsons, P. Thompson, M. Lee and N. Cayzer, *J. Sediment. Res.*, 2005, **75**, 919–940.
21 I. Parsons, *Mineral. Mag.*, 2010, **74**, 529–551.
22 J. D. Fitz Gerald, I. Parsons and N. Cayzer, *Am. Mineral.*, 2006, **91**, 772–783.
23 J. V. Smith and W. L. Brown, *Feldspar Minerals*, Springer, Berlin Heidelberg, 2nd edn, 1988.
24 E. K. Bigg, *Tellus*, 1957, **9**, 394–400.
25 S. Brunauer, P. H. Emmett and E. Teller, *J. Am. Chem. Soc.*, 1938, **60**, 309–319.
26 G. Vali, *Atmos. Chem. Phys.*, 2014, **14**, 5271–5294.
27 H. R. Pruppacher, *J. Glaciol.*, 1967, **6**, 651–662.
28 C. Knight, *Cryst. Growth Des.*, 2020, **20**, 580–589.
29 C. A. Knight and A. L. DeVries, *Phys. Chem. Chem. Phys.*, 2009, **11**, 5749–5761.
30 W. C. Macklin and B. F. Ryan, *J. Sediment. Res.*, 1965, **22**, 452–459.
31 W. C. Macklin and B. F. Ryan, *Philos. Mag.*, 1966, **14**, 847–860.
32 C. A. Knight, *J. Appl. Phys.*, 1962, **33**, 1808–1815.
33 M. A. Holden, J. M. Campbell, F. C. Meldrum, B. J. Murray and H. K. Christenson, *Proc. Natl. Acad. Sci. U. S. A.*, 2021, **118**, e2022859118.
34 P. H. Ribbe, in *Feldspars and Their Reactions*, ed. I. Parsons, Springer Netherlands, Dordrecht, 1994, pp. 1–49.

Faraday Discussions

ROYAL SOCIETY OF CHEMISTRY

PAPER

A zeolite crystallisation model confirmed by *in situ* observation†

Nick Pellens, [a] Nikolaus Doppelhammer, [ac] Karel Asselman, [a] Barbara Thijs, [a] Bernhard Jakoby, [c] Erwin K. Reichel, [c] Francis Taulelle, [ab] Johan Martens, [ab] Eric Breynaert [ab] and C. E. A. Kirschhock *[a]

Received 22nd November 2021, Accepted 14th December 2021

DOI: 10.1039/d1fd00093d

Probing nucleation and growth of porous crystals at a molecular level remains a cumbersome experimental endeavour due to the complexity of the synthesis media involved. In particular, the study of zeolite formation is hindered as these typically form in multiphasic synthesis media, which restricts experimental access to crystallisation processes. Zeolite formation from single phasic hydrated silicate ionic liquids (HSiL) opens new possibilities. In this work, HSiL zeolite crystallisation is investigated *in situ* using a specifically designed conductivity measurement set-up yielding access to crystallisation kinetics. Based on the conductivity data and final yields, a crystallisation model explaining the results based on a surface growth mechanism was derived. The excellent agreement between experiment and theory indicates zeolite crystallisation from highly ionic media proceeds *via* a multi-step mechanism, involving an initial reversible surface condensation of a growth unit, followed by incorporation of that unit into the growing crystal. The first step is governed by the liquid phase concentration and surface energy, while the final step shows a correlation to the mobility of the cation involved.

1. Introduction

Classical nucleation theory (CNT) describes the stochastic condensation of atoms, ions, and molecules into an ordered lattice.[1] While it successfully explains the formation of many simple crystals, a score of crystalline materials crystallise *via* more complex multi-step mechanisms involving metastable intermediates, such as oligomers, complexes, the initial formation of an emulsion of a dense liquid in a solvent, or amorphous and crystalline nanoparticles.[2] Indeed, experimental observations clearly show that molecular-scale processes during crystal

[a]KU Leuven, COK-KAT, Leuven, Belgium. E-mail: christine.kirschhock@kuleuven.be

[b]KU Leuven, NMRCoRe, Leuven, Belgium

[c]JKU Linz, Institut für Mikroelektronik und Mikrosensorik, Linz, Austria

† Electronic supplementary information (ESI) available. See DOI: 10.1039/d1fd00093d

nucleation and growth need to be accounted for. Magnetite, for example, can nucleate both *via* ordered or non-ordered intermediates, depending on the relative thermodynamic (meta)stability of these units in the system.[3] Interestingly, condensation of these precursors follows a kinetic regime reminiscent of colloidal assembly,[4] disregarding differences between surface energies of crystal facets. Also, calcium carbonates and calcium phosphates can nucleate *via* the condensation of solute pre-nucleation clusters,[5–7] which for calcium phosphate can be related to ion–association complexes.[7] To account for such complex multistep mechanisms, CNT has been extended with modern nucleation theory. But even here, perceived initial states and growth units are often described with thermodynamic properties of the bulk, owing to the limited mechanistic insight into crystal growth.[1] Crystallisation kinetics usually are evaluated *via* DSC, XRD, TEM or NMR, and data interpretation is often quite limited by the signal-to-noise ratio.[8] Accurate description of crystallisation, however, critically depends on models integrating the chemistry and energetics of the specific system in detail.[8,9] Unfortunately, molecular-level observation of growth processes is difficult and requires advanced *in situ* diagnostics.[10,11]

Zeolite nucleation is a striking example. Insight into early zeolite formation is hindered by the sol–gel nature of common zeolite synthesis media.[10] Despite these limitations, several attempts have been made to formulate general scenarios,[10,12–15] hypotheses ranging from classical amorphous densification to non-classical crystallisation *via* precursor attachment involving oligomers up to nanocrystals.[2,16] To experimentally access molecular steps of zeolite formation, simplified zeolite synthesis media and/or a step change in diagnostics are required.

The discovery of zeolite formation in monophasic hydrated silicate ionic liquids (HSiLs) provides opportunities to study zeolite crystallisation in more detail.[17] HSiLs are entirely inorganic, homogenous liquids composed of hyposolvated silicate oligomers and alkali hydroxides. They allow the study of zeolite formation on a molecular level[18] and thereby may aid the advance of modern nucleation theory in general.[19] Within HSiLs, zeolite formation is initiated by pre-nucleation cluster consisting of ion pairs of soluble aluminosilicate oligomer anions with alkali cations.[19] When the chemical conditions in the liquid favour formation of such ion pairs, ordered assembly into a zeolite phase is observed.

Insight in kinetics and rate-limiting steps of crystal nucleation and growth is essential to control crystal morphology, size, and eventually topology. In hypohydrated conditions, negatively charged deprotonated silanol functions prevent disordered silicate condensation into amorphous phases, and inner sphere ion-pairing of aluminosilicate oligomers with alkali cations into pre-nucleation clusters is observed. Ordered assembly of these pre-nucleation clusters into a zeolite phase implies the removal of the negative charge by nucleophilic substitution reactions, generating hydroxide ions as leaving groups.[19] These hydroxide ions, along with charge-compensating cations, are released into the crystallisation medium. This increases the conductivity of the synthesis liquid as zeolite crystallisation proceeds, providing experimental access to the progressing crystallisation *in situ*. In the absence of amorphous phases or dissolving framework sources, the increase in conductivity must show a linear relationship with every additional siloxation in the medium, providing a direct correlation between zeolite crystallisation and accurate *in situ* conductivity measurements. Discovery of ion-paired pre-nucleation clusters in zeolite crystals is in full agreement with the theory describing magnetite precipitation, a model entirely based on thermodynamic arguments.[3]

In this work, it will be shown that the interaction strength of the cation with the pre-nucleation cluster and crystal surface determines crystal morphology and size, allowing rationalisation and control of the formation of zeolites, and by extension, complex materials in general. With the discovery of zeolite pre-nucleation clusters, modern nucleation theory can now also adequately describe zeolite formation. Adding zeolites to the list of minerals for which nucleation is explained with a similar level of detail as has been possible for calcium carbonate and phosphate,[5] holds promise that the documented approach here will inspire researchers to add many more complex porous crystals to the list of crystallisation systems explained by modern nucleation theory.

2. Experimental details

HSiL synthesis liquids with molar composition 0.5 $Si(OH)_4$:0.03 $Al(OH)_3$:1 MOH:5 H_2O were prepared in the presence of sodium or cesium hydroxide ($M = Na^+$ or Cs^+) and stirred for 24 h at room temperature (details in ESI†). Zeolites were crystallised from the synthesis liquids at temperatures of 60, 70, 80, and 90 °C. During synthesis, conductivity was measured *in situ* by MEEIS, using a custom measurement cell (Fig. S1†). HSIL zeolite synthesis liquids were loaded into the preheated cell after being stirred for 24 h at room temperature, achieving fast thermal equilibration times of <10 min with $\Delta T < 0.01$ °C. To inhibit evaporation, the moveable electrode was tightly sealed by a silicone cap. Impedance spectra were measured at 13 electrode distances between 5 and 9 cm. For each electrode distance, impedance values at 30 logarithmically spaced frequencies were recorded in the range of 10 Hz to 1 MHz, in two-electrode potentiostatic mode, using a peak-to-peak amplitude of 0.1 V. Conductivity sampling times of 7 and 30 minutes were chosen for fast and slow crystallising samples, respectively. Samples were monitored until the conductivity curve plateaued. Zeolite syntheses were replicated in synthesis liquid from the same batch as used for *in situ* conductometry (details in ESI†). The recovered solids were collected, washed *via* repeated dispersion–centrifugation (15 min, 35.000*g*), and studied with powder X-ray diffraction (PXRD) and scanning electron microscopy (SEM). PXRD patterns ($CuK_{\alpha1}$) were recorded on a STOE STADI P Combi diffractometer in high-throughput mode, with a curved image plate PSD and focusing Ge(111) monochromator. High resolution SEM images were recorded with a Nova NanoSEM450 (Fei, Hillsboro, OR).

3. Results

3.1 Product characterisation

According to PXRD measurements (Fig. 1) and SEM images (Fig. 2) all synthesis products are zeolitic, being of ANA topology for all Cs-containing samples. The Na syntheses yielded mixtures of FAU, GIS and SOD, with increasing GIS fractions at higher temperature. SEM revealed significant differences in crystal size and morphology of the synthesis products. All Cs samples show nearly identical isotropic morphologies of 100–200 nm crystal sizes. Note that during synthesis,

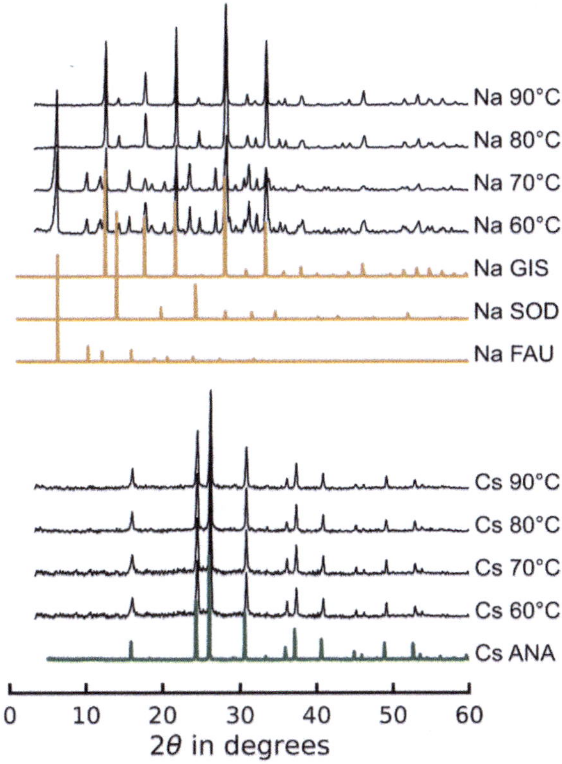

Fig. 1 X-ray diffraction measurements of the synthesis products. Synthesis liquids containing Na cations yielded mixtures of the GIS/FAU/SOD zeotypes. Zeolite synthesis in the presence of Cs cations yielded a phase-pure ANA zeotype in all cases. The sample labels indicate a synthesis liquid either containing Na or Cs cations, synthesized at temperatures of 60, 70, 80 or 90 °C.

Fig. 2 Scanning electron microscopy imaging of the synthesis products. The white markers represent 2 μm in all figure frames. Images were selected to be representative of the studied samples.

these spheres are isolated and aggregate only during product recovery. Crystals of the Na samples are more than an order of magnitude larger and crystal morphology strongly depends on synthesis temperature. At low temperatures, the crystals show emerging crystal facets, but are partly spherical. Increasing synthesis temperature defines the facets more clearly, while overall crystal size increases. An earlier study on the crystallisation kinetics of MER in K systems yielded a similar result.[20] These observations suggest zeolite nucleation and crystal growth to be strongly affected by crystallisation kinetics.

3.2 Modelling of the conductivity curves

All *in situ* conductivity curves show sigmoidal behaviour, strongly dependent on temperature and cation type (Fig. 3). The plateaus of the curves were used to define the synthesis time (t_{final}, Table S1†) as the time where no further formation of zeolite material is detected. To analyse the measured conductivity in terms of crystallisation, a model describing the global zeolite content of the liquids as a function of time is necessary. For this, the growth of all observed zeolite phases, irrespective of topology, is considered to proceed *via* ion-paired pre-nucleation clusters involving surface condensation and release of metal hydroxide to the synthesis liquid, which relates to the measured change of conductivity,[19]

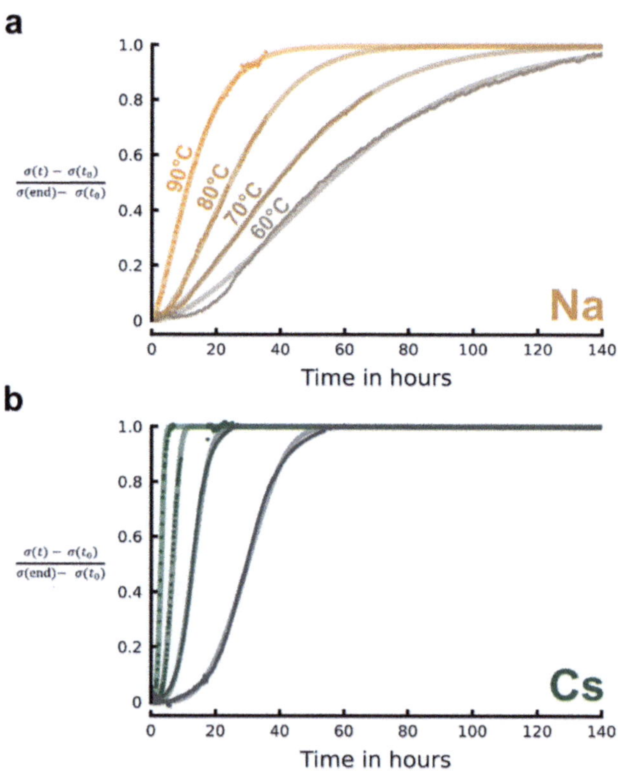

Fig. 3 MEEIS conductivity measurements. (a and b) Conductivity curves of samples containing Na and Cs cations, respectively. Measured data is given by point markers. The best fit is given by the line markers (details in ESI†).

$$\frac{\text{Solid product }(t)}{\text{Total solid product }(t_{\text{final}})} \propto \frac{\sigma(t)-\sigma(t_0)}{\sigma(t_{\text{final}})-\sigma(t_0)}$$

3.2.1 Liquid phase chemistry. Composing the synthesis mixture, aluminate is added to a purely siliceous HSIL, followed by stirring at room temperature for 24 h. Aluminate is consequently incorporated into aluminosilicate oligomers which are all in chemical equilibrium. As a result, Al can be assumed to always be homogeneously distributed over the silicate oligomers in the liquid, including the pre-nucleation clusters.[18] In the liquid, negative charges arising from deprotonated silanol and aluminol groups are compensated by alkali cations, the nature of the cation determining the interaction strength in the ion pair. In the hypo-hydrated conditions governing the synthesis medium, cations complete their coordination sphere with aluminosilicate oligomers. This results in a dense network of interlinked coordination polyhedra reminiscent of what is observed in highly concentrated sub-hydrated alkalihydroxides.[18,21] Assuming rapid interaction kinetics in the polyhedral network, each species, including viable pre-nucleation clusters, are available in any volume element of the synthesis medium. Therefore, upon heating, nucleation occurs at random within the total volume of the network.

3.2.2 Model of increasing zeolite fraction. The conductivity data acquired reflects the increasing fraction of zeolite, assuming the volume and the chemical state of the liquid fraction does not change significantly. This leads to a quasi-linear relation between conductivity increase and solid phase formation. Since silica-gelation is prevented by the high charge density of the clusters and the absence of a bulk solvent phase, at the start of the crystallisation the studied systems do not contain any solids. During the experiment, only the formation of crystalline zeolitic material is observed, indicating zeolite formation occurs as result of the limited solubility of the aluminosilicate oligomers in the concentrated alkali metal hydroxide liquid. As zeolite formation is correlated to the removal of pre-nucleation clusters (PNC) from the liquid, the relation between solid and liquid phase can be described with a very simple microscopic picture.

Considering a set of assumptions, a statistical model describing the probability for a PNC to have crystallised can be formulated. First it is assumed that any PNC in the vicinity of a growing crystal will attach, as long as the concentration c exceeds the liquid-state solubility limit c_S: $(c - c_s > 0)$ implies PNC attachment. Based on this assumption, the focus can be shifted to the number of liquid-state PNCs, which, at the end of the reaction, have been incorporated into zeolite crystals. This number is a direct function of the initial supersaturation $(N_0 - N_s = V_{\text{liquid}}(c(t_0) - c_s))$. Inspecting the morphology of the products (Fig. 2), 3D growth is evident, *i.e.*, it is assumed that the volume of the crystals can be described by a cubic function of their effective radius. With these considerations, the description of these systems can be simplified further. If the number density of PNC exceeds solubility, zeolite will form with nucleation and growth rates characteristic for the growing crystals.[1] This allows the product fraction at a time t to be used as an expression describing crystallisation progress. To allow comparison between systems, irrespective of cation and formed topology, several additional simplifications are made: (1) at each time, the system consists of a small fraction of growing crystals in the presence of abundant liquid growth medium, such that the liquid volume can be considered constant. (2) Kinetics and dynamics in the liquid medium are fast, implying chemical equilibrium in the liquid throughout the crystallisation. This implies that every species relevant for crystal growth is always available within any volume fraction of the liquid medium, and crystal growth is rate-limited by processes occurring on the crystal surface. (3) Crystal growth proceeds until the solubility of the growing phases (c_s) is reached, *i.e.* $c(t_{\text{final}}) - c_s = 0$. (4) For a given cation and synthesis temperature, the growth mechanism does not change. (5) Surface processes are similar for all surfaces and phases growing at the same moment, and total yield is defined by the total solid obtained at t_{final}.

Considering N_0 to be the total number of PNC present at time zero in a constant liquid volume V_{liquid} (assumption 1), at t_{final}, N_c ($=N_0 - N_s$) PNCs have transformed into the solid. N_c thus represents the number of PNC exceeding the solubility in V_{liquid} and $(c(t_0) - c_s) > 0$ (assumption 3). This averages supersaturation over all growing phases so that phase transformations are currently not explicit in this model. Instead, only the time-dependent increase of total zeolitic material is considered. Assuming fast liquid-phase kinetics and dynamics (assumption 2), it is now possible to determine a probability for the progress of crystallisation. In the liquid, all PNCs in the population N_c are in constant exchange with all oligomeric species, so that the statistical model can be reduced to the conversion of N_c PNCs into crystalline solid. The increasing crystal fraction in time thus determines the probability that a PNC of population N_c has already crystallised. Identifying the volume increase of the crystal phases per added PNC as,

$$V_{\text{PNC}} = \frac{V_{\text{c,final}}}{N_c}$$

and considering the total obtained product to be irrespective of the framework composition and density, V_{PNC} reflects the average increase of solid volume per condensed PNC. In terms of already present crystal volume, $V_c(t) = V_{\text{PNC}}(N_0 - N(t))$, the probability that at time t one of the N_c PNCs in the final crystals has already crystallised, can thus be expressed as:

$$P(t) = \frac{V_c(t)}{N_c V_{\text{PNC}}}$$

The formed crystal volume consists of all crystals present at time t. The volume of a single growing crystal, φ, is a cubic function of its characteristic radius, multiplied with a geometric factor ϑ (*e.g.* $4\pi/3$ for a sphere or 1 for a cube),

$$\varphi = \vartheta r^3$$

A crystal nucleated at time τ, which increases its radius in time, with a function $g(t,\tau)$ describing the temporal evolution of the crystal radius yields the volume of a crystal born at τ, as a function of time,

$$\varphi(t) = \vartheta[g(t,\tau)]^3$$

Defining $j(\tau)$ as the nucleation rate, the crystal volume arising from all crystals nucleated at τ in an interval $[\tau, \tau + \Delta\tau]$ is:

$$\phi(t,\tau,\Delta\tau) = j(\tau)\Delta\tau\vartheta[g(t,\tau)]^3$$

The probability that a PNC within N_c is already part of any crystal nucleated in $\Delta\tau$, therefore relates to the fraction of N_c in $\phi(t,\tau,\Delta\tau)$,

$$P(t,\tau,\Delta\tau) = \frac{\phi(t,\tau,\Delta\tau)}{N_c V_{PNC}}$$

The total probability for crystallised PNC, is the product of probabilities for all possible $\Delta\tau$:

$$P_{crystal}(t,\tau) = \prod_{all\ \Delta\tau}\left[\frac{\phi(t,\tau,\Delta\tau)}{N_c V_{PNC}}\right] = \exp\left[\sum_{all\ \Delta\tau}\ln\left[\frac{\phi(t,\tau,\Delta\tau)}{N_c V_{PNC}}\right]\right]$$

Expansion of the logarithm leads to,

$$P_{crystal}(t,\tau) = 1 - \exp\left[\sum_{all\ \Delta\tau} - \frac{\phi(t,\tau,\Delta\tau)}{N_c V_{PNC}}\right]$$

If nucleation is instantaneously occurring during a time interval $\Delta\tau$ at $\tau = 0$, *i.e.* only occurring at the start of crystallisation, and being rapidly exhausted afterwards, $P_{crystal}(t,\tau)$ can be approximated by:

$$P_{crystal}(t,\tau) = 1 - \exp\left[-\frac{n_\phi\vartheta(g(t))^3}{N_c V_{PNC}}\right],$$

with n_ϕ being the number of nuclei formed in an induction period τ. In the case of continuous random nucleation, the summation is transformed into an integral,

$$P_{crystal}(t) = 1 - \exp\left[\sum_{all\ \Delta\tau}\left[-j(\tau)\Delta\tau\frac{\vartheta[g(t,\tau)]^3}{N_c V_{PNC}}\right]\right] \xrightarrow{\Delta\tau\to 0} 1 - \exp\left[-\frac{\vartheta}{N_c V_{PNC}}\int_0^t j(\tau)[g(t,\tau)]^3 d\tau\right]$$

This distribution function resembles the often used JMAK approach, which is based on a statistical model derived in 1937 by Kolmogorov, to specifically describe volume to volume transformations from liquid to solid, or solid to solid state, *e.g.* crystallisation of glass.[22] However, in the present case, incongruent transformation of only a fraction of the initial system with different composition of solid and liquid phase is considered. The possibility to use similar statistics in both cases is rooted in the fast liquid dynamics and kinetics in the homogeneous liquid phase, compared to the surface processes. Furthermore, assuming a large and constant liquid volume, no troublesome impingement of nucleating crystals needs to be considered as the crystal volume is formally not part of the liquid volume.

3.2.3 Model of surface reactions. The first test for the derived kinetic model is the option to fit it to the measured data, which requires a radial growth function $g(t,\tau)$. Such a function can be obtained from the kinetics of the surface reactions, if known. The measurement principle, as well as the statistical model focus on the increase of zeolite product with time. Therefore, the suggested surface reactions also emphasise growth reactions, in a first approach ignoring surface dissolution. Fast dynamics in the liquid phase during nucleation and growth, defined as a prerequisite for derivation of the statistic model, imply the surface processes are rate-limiting and determine the increasing product fraction. We first derive rate laws for our proposed surface reactions and then discuss these in view of radial growth $g(t,\tau)$, to deduce which analytical forms $P_{\text{crystal}}(t)$ can take. Fast liquid-phase dynamics ensure identical PNC concentration at each point and time within the growth medium, also when a liquid volume is in contact with any growing crystal. The radial growth rate of the surface must be related to the number of crystallised PNC per time and area. In the herein studied system the pre-nucleation clusters consist of aluminosilicate oligomers, ion-paired to alkali-ions. PNCs react with the surface by nucleophilic substitution, releasing alkali cations, and hydroxide ions as leaving groups into the liquid volume. In a liquid supersaturated with PNC, this process can be described by a precipitation dissolution equilibrium, followed by the release of MOH from the surface (Fig. 4).

Two cases need to be considered for determination of the rate laws. In the case of fast MOH desorption ($k_\text{d} \gg k_\text{r}$, $k_{-\text{r}}$), the transition of liquid-state PNC to the active surface state, $\overline{\text{(C-PNC)*MOH}}$, is growth rate-limiting, indicating a steady state approximation of the crystal growth rate:

$$\frac{\text{dPNC}}{\text{d}t} = -k_\text{r}\text{PNC} + k_{-\text{r}}\overline{\text{(C-PNC)*MOH}}$$

$$\overline{\text{(C-PNC)*MOH}} = \frac{k_\text{r}}{k_{-\text{r}} + k_\text{d}}\text{PNC}$$

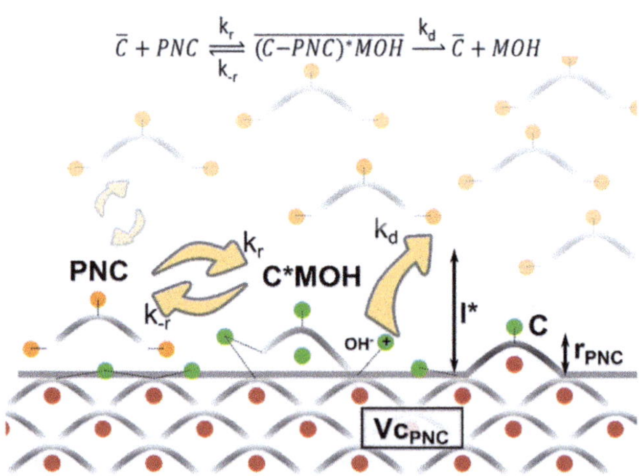

Fig. 4 Suggested mechanism of growth reactions. PNC are in equilibrium with the liquid. Once attached to the surface, the formed cation and hydroxide (not shown) have to move to a distance $l > l^*$ from the surface to prevent re-dissolution. Also shown: volume taken by 1 PNC in the solid (V_PNC) and added height per attached PNC (r_PNC). The active surface state $\overline{\text{(C-PNC)*MOH}}$ is symbolised by C*MOH. Circles represent cations attached to PNC (grey bows) in liquid volume (orange), growing surface (green) and crystal (red).

$$\frac{\mathrm{dPNC}}{\mathrm{d}t} = \left(-k_\mathrm{r} + k_{-\mathrm{r}}\frac{k_\mathrm{r}}{k_{-\mathrm{r}} + k_\mathrm{d}}\right)\mathrm{PNC}$$

$$k_\mathrm{d} \gg k_\mathrm{r}, k_{-\mathrm{r}} : \frac{\mathrm{dPNC}}{\mathrm{d}t} \approx -k_\mathrm{r}\mathrm{PNC}$$

$$\mathrm{PNC} \propto \exp(-k_\mathrm{r}t)$$

When desorption is slow and rate-limiting, the surface states are in equilibrium with liquid-state PNC, as defined by the formal solubility product, $K_{\mathrm{sp}} = k_{-\mathrm{r}}/k_\mathrm{r}$. With the pre-equilibrium approximation, the rate of growth is determined by k_d,

$$-\frac{\mathrm{dPNC}}{\mathrm{d}t} = k_\mathrm{d}\overline{(\mathrm{C\text{-}PNC})^*\mathrm{MOH}} = k_\mathrm{d}\frac{1}{K_{\mathrm{sp}}}\mathrm{PNC}$$

To dissolve $\overline{(\mathrm{C\text{-}PNC})^*\mathrm{MOH}}$ back into the liquid as PNC, a nucleophilic attack by the hydroxide ion on the formed siloxane bond is necessary. Therefore, the rate of dissolution is linked to the mobility of MOH in the system. Considering preservation of local charge neutrality, the hydroxide moves with the cation, so that the cation mobility determines the rate of crystallisation. Therefore, the reaction $\overline{(\mathrm{C\text{-}PNC})^*\mathrm{MOH}} \rightarrow \mathrm{C} + \mathrm{MOH}$ is defined by a rate that reflects the time needed to remove M^+ from the site of condensation by a critical distance l^*. In a first approximation, the cation mobility in the system can be described as a random change of coordination partners, *i.e.* the negative charges on the liquid-state oligomers and on the growing surface. With assumption 1, a constant concentration of cations in the growth medium, an average distance for each ligand switch and an average residence time with each partner can be presumed. Such a process is closely related to chemical exchange. In this picture, the cation mobility can be described as a random walk with a hopping rate D, related to the square of the average step distance, which depends on the interaction strength between cation and ligand. The probability that a freshly condensed PNC still resides on the surface ($0 < x < l^*$) after a time t, is obtained by integration of the probability distribution function describing the random walk process,

$$\mathrm{d}P(x,t) = \frac{\mathrm{d}\overline{(\mathrm{C\text{-}PNC})^*\mathrm{MOH}}(x,t)}{\overline{(\mathrm{C\text{-}PNC})^*\mathrm{MOH}}} = \frac{1}{\sqrt{\pi D t}}\exp\left(-\frac{x^2}{4Dt}\right)\mathrm{d}x\mathrm{d}t$$

Integration for x in the limits 0 to l^*, using the expanded error function, results in the probability function,

$$\mathrm{d}P(t) = \frac{\mathrm{dC^*MOH}}{\mathrm{C^*MOH}} \approx \frac{l^*}{\sqrt{\pi D t}}\mathrm{d}t$$

Considering fast dynamics compared to sampling times, only the first term of the expansion is kept since the standard deviation of the Gaussian distribution ($\sigma = 4Dt$) is large compared to l^*. This probability function reflects the changing number of active surface states resulting from MOH release. Therefore, the radial growth rate is related to,

$$\frac{\mathrm{dPNC}}{\mathrm{d}t} = \mathrm{PNC}\frac{-l^*}{K_{\mathrm{sp}}\sqrt{\pi D t}}$$

$$\mathrm{PNC} \propto \exp k_{\mathrm{d}}/K_{\mathrm{sp}}\sqrt{t}$$

To derive the global evolution of solid product, $P_{\mathrm{crystal}}(t)$, a function $g(t,\tau)$ has been introduced, describing the crystal radius at time t for a crystal nucleated at τ. In the following section, the derived surface reactions are linked to $g(t,\tau)$ for the cases of steady-state and pre-equilibrium conditions.

3.2.4 Model of conductivity data. The attachment of a PNC on a surface element of a crystal with radius r leads to a radius increase by the average height of a PNC in the crystal in the growth direction, r_{PNC} (Fig. 4). For crystal nucleation at time τ and a crystal growth time of $(t - \tau)$, the crystal radius increase depends on $g(t,\tau)$. When step 1 of the surface processes (Fig. 4 and 6) is rate-limiting (steady-state approximation), this results in a crystal growth rate of,

$$g_{\mathrm{r}}(t,\tau) = r_{\mathrm{PNC}}(\mathrm{PNC}(\tau) - \mathrm{PNC}(t)) \propto r_{\mathrm{PNC}}(\mathrm{e}^{-g\tau} - \mathrm{e}^{-gt}) \approx r_{\mathrm{PNC}}k_{\mathrm{r}}(t - \tau)$$

If the growth is limited by the desorption of MOH from the crystal surface (Fig. 4 and 6), the resulting crystal growth rate $g_{\mathrm{d}}(t,\tau)$ becomes,

$$g_{\mathrm{d}}(t,\tau) = r_{\mathrm{PNC}}(\mathrm{PNC}(\tau) - \mathrm{PNC}(t)) \propto r_{\mathrm{PNC}}\left(\mathrm{e}^{-k_{\mathrm{d}}\sqrt{\tau}} - \mathrm{e}^{-k_{\mathrm{d}}\sqrt{t}}\right) \approx r_{\mathrm{PNC}}k_{\mathrm{d}}\left(\sqrt{t} - \sqrt{\tau}\right)$$

Therefore, the probability of PNC crystallisation, $P_{\mathrm{crystal}}(t,\tau)$, can be determined for the two limiting cases, assuming a nucleation rate of the form $j(\tau) = J\tau^y$,

$$P_{\mathrm{crystal}}(t,\tau) = 1 - P_{\mathrm{crystal}}(t) = 1 - \exp\left[-\frac{J\vartheta(r_{\mathrm{PNC}}k)^3}{N_{\mathrm{c}}V_{\mathrm{PNC}}}\int_0^t \tau^y[(t^x - \tau^x)]^3\mathrm{d}\tau\right]$$

with $x = 1$ and $k = k_{\mathrm{r}}$ for surface reaction limitation, and $x = 1/2$ and $k = k_{\mathrm{d}}$ in the case of desorption limitation. Assuming a constant nucleation rate upon reaching crystallisation conditions ($\tau = 0$), integration yields

$$P_{\mathrm{crystal}}(t) = 1 - \exp\left[-\frac{J\vartheta(r_{\mathrm{PNC}}k_{\mathrm{r}})^3}{N_{\mathrm{c}}V_{\mathrm{PNC}}}t^{3x+y}\left(\frac{1}{y+1} - \frac{3}{x+y+1} + \frac{3}{2x+y+1} - \frac{1}{3x+y+1}\right)\right] = 1 - \exp\left[-G_{\mathrm{n}}t^{3x+y}\right]$$

In the case of instantaneous nucleation, terminated before crystal growth proceeds, $P_{\text{crystal}}(t)$ can be expressed as:

$$P_{\text{crystal}}(t) = 1 - \exp\left[-\frac{n_\phi \vartheta t^{3x}}{N_c V_{\text{PNC}}}\right] = 1 - \exp\left[-G_{\text{in}} t^{3x}\right]$$

For the here studied systems, crystal growth occurs by the addition of pre-nucleation clusters consisting of aluminosilicate-oligomers ion-paired to alkali-cations. Growth proceeds by siloxation reactions under release of the corresponding ion-paired cations and hydroxide into the medium. The charge release into the synthesis liquid, driven by the condensation of PNC as a function of time, can therefore directly be related to the derived distribution function,

$$P_{\text{crystal}}(t) = \frac{\text{solid product } (t)}{\text{total solid product } (t_{\text{final}})} = \frac{\sigma(t) - \sigma(t_0)}{\sigma(t_{\text{final}}) - \sigma(t_0)}$$

The derived probability as a function of time can take the form,

$$P_N(t) = 1 - \exp[-Gt^z]$$

All conductivity data could be described with such a function, suggesting the derived model indeed describes crystal growth as shown in Fig. 3. The fits along with their residuals are supplied in Fig. S4.† The conductivity curve is determined by two parameters, being the exponent, and multiplication factor of the variable time. The former indicates the radial growth of the crystals and the absence or presence of a nucleation rate. The latter contains the rate constants G_{in} and G_{n} accounting for growth and initially present nuclei or a combination of growth and nucleation for the scenarios of instantaneous *vs.* constant nucleation, respectively. As shown in Fig. 3, where the lines represent the fit of the data, the derived model is in excellent agreement with experimental conductivity data. The derived exponents (Fig. 5) do not significantly vary for the same cation with temperature, a clear indication that the global mechanism did not change. The slight variation of z for sodium with temperature will be addressed later in this article. Furthermore, the growth constants G show approximately linear behaviour in an Arrhenius plot, which indicates the rate-limiting steps are thermally activated processes, respecting the Boltzmann distribution, valid for surface condensation (step 1) as well as for chemical exchange (step 2). Therefore, G is linked to the activation energies for surface condensation, when reaction-limited, or the cation–ligand interaction strength for desorption limitation.

If the surface reactions suggested in 3.2.2 describe the surface growth and the reaction kinetics do not change during growth, then x can either take a value of $x = 1$, for reaction limitation, or $x = 1/2$ for desorption limitation. Therefore, the experimentally determined exponent z takes the form $z = 3 + y$ for the former, and $z = 3/2 + y$ for the latter. If constant or instantaneous nucleation is considered as limiting cases, this means any exponent z must either be between $3 < z < 4$ or $3/2 < z < 5/2$, respectively. Comparing these values to the experimentally obtained parameters, it is concluded that crystallisation in the sodium-containing systems proceeds with $z \sim 3/2$, and zeolite growth is controlled by the desorption of MOH from the growing surface ($x = 1/2$). Nucleation is quickly exhausted, *i.e.* y is small

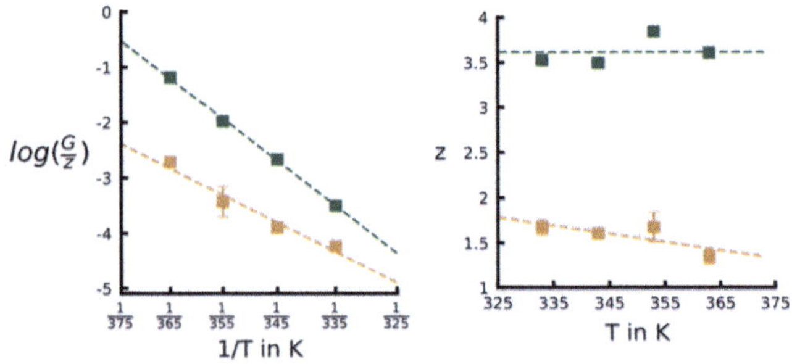

Fig. 5 Fitting parameter trends as given by an Arrhenius plot of the growth constant, G, and the determined values for the exponent, z. The values for Cs and Na are given in green and yellow, respectively.

or zero ($y \sim 0$). Therefore, the crystal surface is observed to be in (meta)equilibrium with the growing surface. In the case of Cs as the counterion, the experimental value for z is close to $z \sim 4$, the maximum value possible for the here proposed growth and surface processes, implying quasi-constant nucleation over the whole crystallisation process. In this case, zeolite growth is limited by the surface reaction ($x = 1$) and MOH release is so fast that no equilibrium between the liquid and the solid surface can be established.

4. Discussion

4.1 Model justification and rationale

The surprisingly good agreement with the predicted exponents for surface reaction limitation and constant nucleation for Cs ($x = 1, y \sim 1$ and $z \sim 4$), and MOH desorption limitation for Na ($x = 1/2$, $y \sim 0$, and $z \sim 3/2$) strongly suggests the proposed model, despite extensive simplification, adequately describes the here observed kinetics of zeolite crystallisation (Fig. 6). This model directly relates the state of the liquid to the growing surface and increasing crystal volume. It needs to be pointed out that the derived model is necessarily incomplete because it only accounts for the global increase of zeolite product.

Considering the combination of a fairly invariable liquid phase speciation, fast liquid state dynamics and low number density of hydroxide ions in the system, a critical distance for irreversible removal of the hydroxide leaving group from the active surface state can be introduced. At this distance, a nucleophilic attack by the hydroxide will occur on liquid-state species, rather than on the active surface state (Fig. 4 and 6). Therefore, reversing the second step in the reaction mechanism would require generation of a free hydroxide in the liquid phase at a distance to the surface closer than l^*. As removal of a fully incorporated growth unit would imply multiple of these unlikely de-siloxation events, the model only considers the initial, most important, siloxation in step 1 (Fig. 6). Attachment on the surface with multiple bonds and the option that a PNC can be optimised according to the growing surface structure is not explicit in the proposed model. However, once a PNC is irreversibly attached by the first removal of MOH, any following reorganisation and/or surface siloxation can be assumed to be fast, so that the kinetics should remain dominated by k_r and k_d.

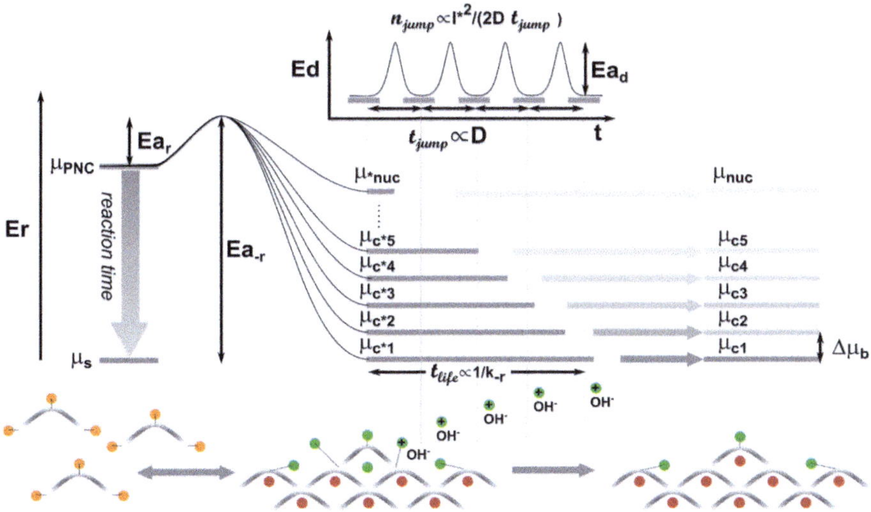

Fig. 6 Illustration of energetics and timeline of proposed surface reactions. Supersaturated liquid state PNC (left) has a chemical potential $\mu_{PNC} > \mu_S$ and condenses on the surface upon thermal activation by Ea. Consumption of PNC results in lowering its energy until saturation is reached $\mu_{PNC} > \mu_S$. The different active surface states C_i^* (middle), have energies μ_{c*i}, (μ_{nuc*} indicates the energy of a PNC in emerging nucleus), which determines the lifetime t_{life} of C_i^*. Cation and hydroxide generated in the condensation need to be removed to consolidate C_i^* into the final crystal C (right). This process is related to the chemical exchange rate of the cation (green circle), characterised by activation energy Ea_d. Note that (C-PNC)*MOH cations that will be included as extra-framework cations (red circles) still participate in chemical exchange, driven by Ea_d. After n_{jump} of the cation, the associated hydroxide is no longer available to return C* to PNC. If t_{life} of C_i^* is larger than it takes the cation to perform n_{jump}, the active surface state turns into crystal, or else the attached PNC returns to the liquid.

Upon association of a PNC with the growing crystal surface, generation of the initial bond between surface and PNC also generates a hydroxide ion. In the first approximation, the activation energy (Fig. 6, Ea_r) for this reaction, yielding the intermediate $\overline{(\text{C-PNC})}^*$MOH, should be similar, if not identical to siloxation reactions in the liquid. This assumption is justified by the reversible and fast bond breaking and making, needed for (meta)-equilibration in the liquid, a prerequisite to allow the derivation of the statistical model. The backwards reaction $\overline{(\text{C-PNC})}$*MOH → PNC, re-dissolving the active surface state, however, requires activation energy Ea_{-r}, the sum of Ea_r and the gain in energy $\Delta\mu_{C*} = \mu_{PNC} - (\mu_s + \mu_{C*})$ (Fig. 6). In the case of high PNC supersaturation, *i.e.* high aluminate content in the liquid, the chemical potential (μ_{PNC}) of PNC is high, resulting in a high activation energy (Ea_{-r}) for the backwards reaction. Consequently, the lifetime (t_{life}) of $\overline{(\text{C-PNC})}$*MOH is high compared to the time needed to remove the generated hydroxide to a distance l^*, inhibiting the backward reaction (Fig. 6).

This second step, $\overline{(\text{C-PNC})}$*MOH → $\overline{\text{C}}$ + MOH, was modelled as a function of the average interaction time between a cation and its coordination partner in the ion pairs. This process is related to the random walk continuously occurring in the system, where cations jump between coordination partners. Each jump involves breaking the existing bond between cation and coordination partner, reforming a similar bond with the next partner. This occurs with a hopping rate D and associated activation energy (Ea_d), characteristic for the cation type in the growth medium. Given that the nucleophilic attack of the hydroxide ion is needed to re-dissolve PNC from the surface, the hopping rate D of a cation in the growth medium decides on the (ir)reversibility of PNC attachment to the growing crystal. If cation hopping is slow, because interaction energy with partners is high, removal of hydroxide is slow. The resulting long residence time of $\overline{\text{(C-PNC)*MOH}}$ allows reversal of the PNC attachment to the surface by the pre-equilibrium. Only long-lived surface sites, with high stability and long lifetime ($t_{life} > n_{jump} \times t_{jump}$, Fig. 6) can transform into crystallised material.

During crystallisation, PNC will be consumed, until the concentration reaches the solubility and the PNC energy in the liquid is equal to μ_s (Fig. 6). The closer μ_{PNC} approaches μ_s, the higher the rates of the re-dissolution of the surface sites become, and the more strongly their respective energy differences will affect their relative lifetimes. PNC concentration should however not affect the intrinsic cation hopping frequency, so that the time to remove cation and hydroxide from $\overline{\text{(C-PNC)*MOH}}$ should remain largely unaffected. Therefore, supersaturation should not affect the time needed to make an active surface state a part of the crystal. This implies, that for low supersaturation levels and slow cation dynamics, those surface condensation sites with the lowest energy will be preferred, leading to increasingly faceted crystals of the most stable crystal phase throughout the reaction. On the other hand, when the cation is extremely mobile, *i.e.* only weakly interacting, then $\overline{\text{(C-PNC)*MOH}}$ changes into a crystal almost as soon as it is formed, and the resulting crystal shape should have an isotropic, spherical character.

As is evident from the Arrhenius plots in Fig. 5, all crystallisation processes proceed faster with increasing T, because all reactions (k_r, k_{-r} and k_d) are thermally activated processes. Furthermore, temperature also affects the supersaturation, because of temperature dependency of the solubility of the solid phases.[20] This readily explains obeisance or ignorance of the Ostwald step rule, as the latter only can be followed when the possible surface states on all solid phases can achieve (quasi)equilibrium distribution. In other words, a dissolution–reprecipitation process can only proceed when the dissolution reaction (k_{-r}) is not inhibited by the second reaction step. Therefore, one should expect more prominent facets at higher T, and increasing fractions of the thermodynamically more stable phases for slow cation dynamics, as observed for the Na system.

4.2 Nature of cations

All results point to the critically important role of the cation in zeolite crystallisation from homogeneous, highly ionic media. The cation is involved in finalising the individual pre-nucleation cluster attachment by removal of the hydroxide nucleophile from the surface and inhibiting immediate re-dissolution. On a chemical level, Na and Cs cations strongly differ. Sodium is a small, kosmotropic cation and is highly stable in a hydrated state. It only forms ion pairs with (alumino)silicate oligomers, in the absence of or with limited water molecules or other favourable ligands.[21] Cesium, on the other hand, is a large chaotropic cation, known to weakly interact with water, and even less with hard charges. This difference is well-known,[24,25] and reflects the affinity for oxygen ligands. The hydration energy of the cations can serve as a measure of oxophilicity. For Na^+ and Cs^+, hydration energies of 409 *vs.* 264 kJ mol^{-1} are reported.[26] This marked contrast also results in vastly different ionic mobility in water, where sodium moves very slowly, compared to cesium (50 *vs.* 77 m^{2}S mol^{-1}).[27] Admittedly, aqueous solutions are not an ideal comparison for the here discussed hypohydrated ionic liquids. Nonetheless, similar principles of cation mobility should apply. Microscopically, the alkali-cations, with low covalent bonding character, move through the medium, while changing available coordination partners. In aqueous systems, this is a function of the average residence time of water molecules next to the cation. This time is one order of magnitude larger for Na than for Cs.[28] Interestingly, the here measured ratio between conductivities in Na and Cs systems is similar to the ratio of reported ionic mobilities of Na^+ and Cs^+ (Table S2†). The observed conductivity is mainly driven by the mobility of Na^+ and Cs^+, because hydroxide ions are mostly depleted. Therefore, the conductivity of comparable ionic liquids reflects the mobility of the cations. In view of this, it is no surprise that the cation migration from the surface of the growing zeolite is rate-limiting in the case of sodium, but not for cesium, even though direct experimental data for chemical exchange rates are not yet available.

4.3 Topology and morphology of products

Inspecting the morphology of the obtained zeolites, a number of observations can readily be explained. In the case of Cs, many small unfaceted, almost perfectly spherical crystals are obtained. The proposed model predicts such crystal morphology based on a continued nucleation and indiscriminate PNC attachment, ($y = 1$, $x = 1$; $z = 4$). Both imply the energy of $\overline{\text{(C-PNC)*MOH}}$ on the crystal surface is much lower compared to its ion-paired liquid state, PNC, which also is reflected in a very low solubility and a high yield. Therefore, it is concluded, that the initial supersaturation in the studied Cs-systems is always extremely high, and/or the mobility of Cs in the medium is very fast in comparison. Both should lead to many small spherical crystals.

For Na, very different conclusions can be drawn. The exponent close to $z \sim 3/2$ indicates that the transport of the cation from the surface is rate-limiting. PNC attachment and dissolution is fast and reversible, in comparison. The obtained large crystals are faceted and different zeolite phases are observed during the crystallisation, with the most stable GIS dominating at higher temperatures. This suggests that the energy of the liquid-state PNC is close to its energy on the surface(s), which enables faceted crystals and obeisance of the Ostwald step rule, both gaining importance with increasing temperature.[23] The suggested growth model only accounts for the global increase of zeolite crystals. In accordance with modern nucleation theory, the first appearing solid should be the one where the energy difference between growth units in liquid and solid is the smallest.[5] This represents the solid with the lowest thermodynamic stability. Over time, this solid is then transformed into the thermodynamically more stable structure(s). In the case of Na, the initial formation of FAU is faster than for GIS, so the fraction of FAU is decreasing with crystallisation temperature. The general conclusion for sodium that the liquid-state energy of PNC (μ_{PNC}) is of the same order of magnitude as in the forming crystals, possibly also explains the slightly too high exponents observed at lower temperatures. Initially, at a given temperature the supersaturation is higher so that nucleation should slow down with reaction time. Therefore, the possibility of decreasing continuing nucleation cannot be discounted, which leads to a higher number of smaller crystals and slightly increased exponents ($x = 1/2$, $0 < y < 1$), as observed at lower temperatures for sodium systems (Fig. 5).

In this work, the transition from Na-FAU to Na-GIS with increasing temperature is observed. While not studied *in situ*, high temperatures or extended crystallisation times lead to the formation of analcime, also with sodium. This allows direct comparison of the crystal morphology of Cs-ANA and Na-ANA formed at the same temperature. An experiment with a prolonged synthesis time of 1 week at 90 °C was performed for zeolite synthesis liquids in the presence of Na and Cs cations, both yielding the ANA framework. SEM images of the synthesized ANA zeolites (Fig. 7) indicate that even for a prolonged synthesis time, and for the synthesis of an identical zeolite framework, the crystal morphologies are highly dissimilar. As expected from the observed crystallisation kinetics, the formation of the ANA framework in the presence of sodium cations results in highly faceted crystals, indicating the active surface sites are sufficiently long-lived to achieve equilibrated populations, and the removal of NaOH from the surface is slow and rate-limiting. For Cs synthesis, even after a highly prolonged synthesis time (168 h instead of 4 h), when the liquid should have reached saturation, Cs-ANA crystals are still isotropic spheres, indicating comparably fast removal of CsOH. Following, these general conclusions will be put in perspective of generally accepted concepts of (zeolite) crystal growth.

While zeolite nucleation and growth is considered highly complex, Anderson *et al.* demonstrated that crystal growth, even of zeolites, needs to respect simple rules.[29,30] The surface of a growing crystal has a predictable energy landscape that can be explored *via* a unified three-dimensional partition model.[29] For a given zeolite framework, crystal growth can be approximated by partitioning the zeolite in its cage structures, mathematically represented by the natural tiles concept.[31] It needs to be stressed that these natural tiles do not necessarily relate to precursor structure and molecular processes on a growing crystal surface. Instead, the concept identifies the energy of different surface structures from the perspective of the growing surface.[29] As most stable and desirable structures, states with high connectivity are favoured, implying surface silicate or aluminate to be oxolated to three, in some cases two, neighbours. This directly results in closed tiles, present on the surface and requires each growth step to close the next set of tiles, as demanded by topology. This way, the topology of the zeolite defines a set of active

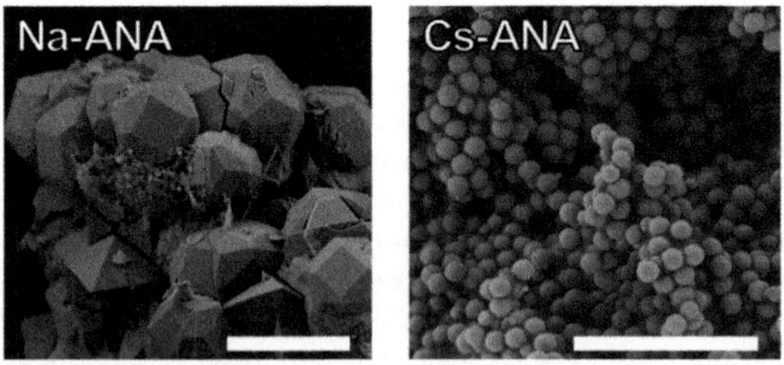

Fig. 7 SEM images of ANA zeotype, synthesized in the presence of Na^+ (Na-ANA) or Cs^+ (Cs-ANA). The white line represents 4 μm in both picture frames.

surface states with differing energies, analogous to kinks, steps and terraces in the simple TLK model. Energy differences between these sites are defined as multiples of $\Delta\mu_b$. The CrystalGrower software utility[30] explores the resulting crystal morphology based on kinetic Monte Carlo simulations, evaluating the crystal shape, entirely based on $\Delta\mu_b$, *versus* the supersaturation. In the case of low supersaturation, $\Delta\mu_b$ is large in comparison, facets are expressed as precipitation and dissolution can occur with similar rates, preferring the most stable states. If supersaturation is high, the energetic differences between surface states are negligible, compared to the very high activation energy necessary for dissolution, and isotropic crystal morphology is predicted. Comparing our proposed model to the CrystalGrower, it is obvious that we should obtain the same result of spherical growth, when initial supersaturation is very high, the energy differences between active surface states $\overline{\text{(C-PNC)*MOH}}$ is comparably small, and/or the cation dynamic in the system is fast.

CrystalGrower simulations were performed for the observed majority phases, being the ANA, FAU and GIS topologies. Crystallisation parameters were varied to match the observed morphologies (Fig. 8). As expected, isotropic crystal growth, observed for all Cs-ANA cases irrespective of T and even crystallisation time, is predicted by CrystalGrower for extremely low energy separations $\Delta\mu_b$. However,

Fig. 8 Kinetic Monte Carlo simulations were performed in the CrystalGrower software utility. Synthesis parameters were chosen to represent the observed crystal morphologies (details in ESI†). Values of $\Delta\mu_b$ are given in kcal mol^{-1}.

even after very long crystallisation times, when supersaturation after initial solidification can be expected to be low, and $\Delta\mu_b$ large in comparison, Cs-ANA still does not show any faceting. This can be explained with a very fast second crystallisation step because of the high mobility of Cs. Fast removal of the cation from the surface freezes a purely statistical distribution of active surface states. In the case of sodium, we always obtain faceted crystals, corresponding to large $\Delta\mu_b$ in CrystalGrower, suggesting a low supersaturation throughout, and comparably slow cation dynamics. Considering, our proposed model reveals transport limitation for Na ($x = 1/2$), the resulting morphology mirrors the (quasi)equilibrium distribution over the available surface sites as predicted by CrystalGrower for large $\Delta\mu_b$.

5. Conclusions

In summary, crystal nucleation and growth often displays behaviour that cannot be explained by classical nucleation theory.[5,32] Zeolite formation forms no exception.[10] In this work, we perform *in situ* conductivity measurements during zeolite synthesis from highly ionic homogeneous liquids. A suggested mechanism for the observed growth assumes ion-paired pre-nucleation clusters reversibly attach to and dissolve from the growing surface, only locked in place by desorption of metal cation and hydroxide removal. The cation release increases the conductivity of the synthesis liquid, which is directly accessible by state-of-the-art MEEIS.[33,34] A kinetic model of zeolite crystallisation was developed in excellent agreement with the measurements. This model proposes zeolite crystallisation can be limited either by the condensation of a pre-nucleation cluster on the surface or by transport of the resulting cation and hydroxide into the crystallisation medium. This model is verified by the variation of the alkali metal cation (Cs^+ *vs.* Na^+) present during crystallisation. The systems result in different zeolite topologies, but both can be described adequately, and reveal reaction limitation for the case of Cs^+ and transport limitation for Na^+. Even though the suggested model is still very rudimental, this success may be taken as an indication that with an improved version of the model and accurate measurement of zeolite growth rates, a generally valid description of zeolite growth might become possible.

Conflicts of interest

There are no conflicts to declare.

Acknowledgements

This work is funded by the bilateral FWO/FWF ZeoDirect project (I3680-N34), the Austrian COMET-K2 program, and FWO (1.5.061.18N and G083318N). This work received long-term funding from the Flemish Methusalem program and is funded by the European Research Council (ERC) under grant 371 agreement no. 834134 (WATUSO).

Notes and references

1 J. J. d. Yoreo, Principles of Crystal Nucleation and Growth, *Rev. Mineral. Geochem.*, 2003, **54**, 57–93.

2 J. J. d. Yoreo, *et al.*, Crystallization by particle attachment in synthetic, biogenic, and geologic environments, *Science*, 2015, **349**, 498–508.

3 J. Baumgartner, *et al.*, Nucleation and growth of magnetite from solution, *Nat. Mater.*, 2013, **12**, 310–314.

4 G. Mirabello, *et al.*, Crystallization by particle attachment is a colloidal assembly process, *Nat. Mater.*, 2020, **19**, 391–396.

5 D. Gebauer, M. Kellermeier, J. D. Gale, L. Bergström and H. Cölfen, Pre-nucleation clusters as solute precursors in crystallisation, *Chem. Soc. Rev.*, 2014, **43**, 2348–2371.

6 D. Gebauer, A. Völkel and H. Cölfen, Stable Prenucleation Calcium Carbonate Clusters, *Science*, 2008, **322**, 1819–1822.

7 W. J. E. M. Habraken, *et al.*, Ion-association complexes unite classical and non-classical theories for the biomimetic nucleation of calcium phosphate, *Nat. Commun.*, 2013, **4**, 1507.

8 E. D. Dill, J. C. W. Folmer and J. D. Martin, Crystal Growth Simulations To Establish Physically Relevant Kinetic Parameters from the Empirical Kolmogorov–Johnson–Mehl–Avrami Model, *Chem. Mater.*, 2013, **25**, 3941–3951, DOI: 10.1021/cm402751x.

9 M. Fanfoni and M. Tomellini, The Johnson-Mehl- Avrami-Kohnogorov model: A brief review, *Il Nuovo Cimento D*, 1998, **20**, 1171–1182.

10 C. S. Cundy and P. A. Cox, The hydrothermal synthesis of zeolites: Precursors, intermediates and reaction mechanism, *Microporous Mesoporous Mater.*, 2005, **82**, 1–78.

11 M. J. van Vleet, T. Weng, X. Li and J. R. Schmidt, In Situ, Time-Resolved, and Mechanistic Studies of Metal-Organic Framework Nucleation and Growth, *Chem. Rev.*, 2018, **118**, 3681–3721.

12 R. M. Barrer, J. W. Baynham, F. W. Bultitude and W. M. Meier, Hydrothermal Chemistry of the Silicates. Part VIII, *J. Chem. Soc.*, 1959, 195–208.

13 D. W. Breck, Crystalline molecular sieves, *J. Chem. Educ.*, 1964, **41**, 678–689.

14 S. P. Zhdanov, Some Problems of Zeolite Crystallization, *Adv. Chem.*, 1971, **101**, 20–43.

15 E. M. A. Flanigen, Review and New Perspectives in Zeolite Crystallisation, *Adv. Chem.*, 1973, **121**, 119–139.

16 M. Kumar, M. K. Choudhary and J. D. Rimer, Transient modes of zeolite surface growth from 3D gel-like islands to 2D single layers, *Nat. Commun.*, 2018, **9**, 2129.

17 L. van Tendeloo, *et al.*, Zeolite synthesis in hydrated silicate ionic liquids, *Faraday Discuss.*, 2015, **179**, 437–449.

18 K. Asselman, *et al.*, Super-ions of sodium cations with hydrated hydroxide anions: inorganic structure-directing agents in zeolite synthesis, *Mater. Horiz.*, 2021, **8**, 2576–2583.

19 N. Pellens *et al.*, DOI: 10.1021/acs.chemmater.2c00418.

20 M. Houlleberghs, *et al.*, Evolution of the crystal growth mechanism of zeolite W (MER) with temperature, *Microporous Mesoporous Mater.*, 2019, **274**, 379–384.

21 M. Hellström and J. Behler, Structure of aqueous NaOH solutions: insights from neural-network-based molecular dynamics simulations, *Phys. Chem. Chem. Phys.*, 2016, **19**, 82–96.

22 A. N. Kolmogorov, On the statistical theory of the crystallization of metals, *Bull. Acad. Sci. USSR*, 1937, **1**(3), 355–359.

23 M. Maldonado, M. D. Oleksiak, S. Chinta and J. D. Rimer, Controlling crystal polymorphism in organic-free synthesis of Na-zeolites, *J. Am. Chem. Soc.*, 2013, **135**, 2641–2652.

24 S. Koneshan, J. C. Rasaiah, R. M. Lynden-Bell and S. H. Lee, Solvent Structure, Dynamics, and Ion Mobility in Aqueous Solutions at 25 °C, *J. Phys. Chem. B*, 1998, **102**, 4193–4204.

25 R. M. Lynden-Bell and J. C. Rasaiah, From hydrophobic to hydrophilic behaviour: A simulation study of solvation entropy and free energy of simple solutes, *J. Chem. Phys.*, 1997, **107**, 1981–1991.

26 D. W. Smith, Ionic hydration enthalpies, *J. Chem. Educ.*, 1977, **54**, 540.

27 D. Lide, *CRC Handbook of Chemistry and Physics*, CRC Press, 96th edn, 2015.

28 L. Helm and A. E. Merbach, Water exchange on metal ions: experiments and simulations, *Coord. Chem. Rev.*, 1999, **187**, 151–181.

29 M. W. Anderson, *et al.*, Predicting crystal growth via a unified kinetic three-dimensional partition model, *Nature*, 2017, **544**, 456–459.

30 A. R. Hill, *et al.*, CrystalGrower: a generic computer program for Monte Carlo modelling of crystal growth, *Chem. Sci.*, 2020, **12**, 1126–1146.

31 V. A. Blatov, A. P. Shevchenko and D. M. Proserpio, Applied Topological Analysis of Crystal Structures with the Program Package ToposPro, *Cryst. Growth Des.*, 2014, **14**, 3576–3586.

32 S. Karthika, T. K. Radhakrishnan and P. Kalaichelvi, A Review of Classical and Nonclassical Nucleation Theories, *Cryst. Growth Des.*, 2016, **16**, 6663–6681.

33 N. Doppelhammer, *et al.*, Moving Electrode Impedance Spectroscopy for Accurate Conductivity Measurements of Corrosive Ionic Media, *ACS Sens.*, 2020, **5**, 3392–3397.

34 N. Doppelhammer, *et al.*, Using moving electrode impedance spectroscopy to monitor particle sedimentation, *IEEE Sens.*, 2021, **21**, 9636–9641.

Faraday Discussions

ROYAL SOCIETY OF CHEMISTRY

PAPER

Nucleation in sessile saline microdroplets: induction time measurement *via* deliquescence–recrystallization cycling†

Ruel Cedeno, [ab] Romain Grossier, [*a] Mehdi Lagaize, [a] David Nerini, [c] Nadine Candoni, [a] Adrian Flood [*b] and Stéphane Veesler [*a]

Received 19th November 2021, Accepted 28th January 2022

DOI: 10.1039/d1fd00090j

Induction time, a measure of how long one will wait for nucleation to occur, is an important parameter in quantifying nucleation kinetics and its underlying mechanisms. Due to the stochastic nature of nucleation, efficient methods for measuring large numbers of independent induction times are needed to ensure statistical reproducibility. In this work, we present a novel approach for measuring and analyzing induction times in sessile arrays of microdroplets *via* deliquescence/recrystallization cycling. With the help of a recently developed image analysis protocol, we show that the interfering diffusion-mediated interactions between microdroplets can be eliminated by controlling the relative humidity, thereby ensuring independent nucleation events. Moreover, possible influence of heterogeneities, impurities, and memory effect appear negligible as suggested by our 2-cycle experiment. Further statistical analysis (*k*-sample Anderson–Darling test) reveals that upon identifying possible outliers, the dimensionless induction times obtained from different datasets (microdroplet lines) obey the same distribution and thus can be pooled together to form a much larger dataset. The pooled dataset showed an excellent fit with the Weibull function, giving a mean supersaturation at nucleation of 1.61 and 1.85 for the 60 pL and 4 pL microdroplets respectively. This confirms the effect of confinement where smaller systems require higher

[a]*CNRS, Aix-Marseille University, CINaM (Centre Interdisciplinaire de Nanosciences de Marseille), Campus de Luminy, Case 913, F-13288 Marseille Cedex 09, France. E-mail: stephane.veesler@cnrs.fr; romain.grossier@cnrs.fr*

[b]*Department of Chemical and Biomolecular Engineering, School of Energy Science and Engineering, Vidyasirimedhi Institute of Science and Technology, Rayong 21210, Thailand. E-mail: adrian.flood@vistec.ac.th*

[c]*Aix Marseille University, University Toulon, Mediterranean Institute of Oceanography, CNRS INSU, IRD, UM 110, Marseille, France*

† Electronic supplementary information (ESI) available: Video of a full %RH cycle (deliquescence/recrystallization). Details of chemical products. Details of instrumentation. Humidification control and dispatching module. Distribution of sigma-curves characteristic points *versus* time. Cycle-to-cycle nucleation times shift. Anderson–Darling *k*-sample test for different aggregation subsets. Modeling of DITs with a Weibull distribution function. Supersaturation calculation. See DOI: 10.1039/d1fd00090j

supersaturations to nucleate. Both the experimental method and the data-treatment procedure presented herein offer promising routes in the study of fundamental aspects of nucleation kinetics, particularly confinement effects, and are adaptable to other salts, pharmaceuticals, or biological crystals of interest.

Introduction

Nucleation in solutions has been the subject of numerous investigations due to its significance in materials synthesis,[1] pharmaceutical purification,[2] and biomineralization.[3] In general, nucleation is the step that determines how long we must wait before the appearance of a stable crystal cluster in a supersaturated solution.[4] This "waiting time", referred to as induction time, is a function of the nucleation rate and the system size. Most induction time measurements are carried out at constant supersaturation for the sake of "simplicity" of data interpretation and modeling.[4] However, in reality, most nucleation processes occur at varying supersaturation, either by cooling, antisolvent, or evaporative crystallization.[5] Thus, a thorough understanding of the nucleation kinetics of such systems is important.

In this context, due to the stochastic nature of nucleation, droplet-based microfluidic systems have been widely used as experimental tools as they permit numerous independent experiments using very small quantities of material. Moreover, their small sizes promote homogeneity in temperature and composition. Despite these advantages, there remain some drawbacks in the use of droplet-based microfluidics in the context of nucleation studies. First, since induction time scales inversely with the system size due to the kinetic confinement effect, high supersaturation level must be achieved in order to observe nucleation events in small volumes within reasonable time scales. Consequently, in constant supersaturation experiments, not all of the microdroplets (denoted as μDs) will nucleate, thereby reducing the statistical quality of the data (data is censored). Another issue is that the time for a critical cluster to grow to detectable size depends on the instrument sensitivity, which can also affect the robustness of the measurements.[6] Regarding these limitations, evolving supersaturation experiments where the solvent is allowed to evaporate seem promising:[7] μDs rapidly reach high supersaturations where growth time to detectable size is negligible, and every μD will give rise to a nucleation event (uncensored data). Thus, in our previous work, we have developed an experimental setup allowing facile generation of monodisperse arrays of nanoliter to femtoliter droplets immersed in an oil film, which can serve as evaporative microcrystallizers.[8,9] We have also shown that a simple and efficient digital-image processing method based on the standard deviation of the pixels' grey-levels of a single μD and its immediate vicinity (σ) is useful in probing the microdroplet dynamics, particularly the onset of nucleation.[9,10] Interestingly, we have observed that in these experiments, microdroplets can interact with each other *via* water diffusion dynamics, *i.e.* when one μD nucleates, water can diffuse to its closest neighbors. Unfortunately, these interactions would complicate the analysis of induction times obtained from μD arrays.

In this contribution, we develop a new approach to measure induction time in sessile arrays of microdroplets using deliquescence–recrystallization cycles. With the help of our in-house developed microdroplet generation system with humidity control module, we show that by controlling the relative humidity, the interfering diffusion-mediated interactions between microdroplets can be eliminated. Then, by using appropriate statistical tests, we show that our measurement method is robust and reproducible. Finally, our evaporative nucleation experiments on 60 pL and 4 pL microdroplets (at equilibrium) confirm the confinement effect[11] where smaller systems reach higher supersaturations.

Experimental section

I. Production of microdroplets with microinjectors

We employed a mechanically controlled micropipette to generate arrays of monodisperse aqueous NaCl sessile microdroplets (1.7 M) on the surface of a poly(methyl methacrylate) (PMMA)-coated glass slide immersed in a 0.4 mm-thick layer of polydimethylsiloxane (PDMS) oil (10 cSt) (Fig. 1a). A detailed description of this procedure has been presented previously[8] while the details of the chemical products and equipment used are shown in ESI 1 and 2.†

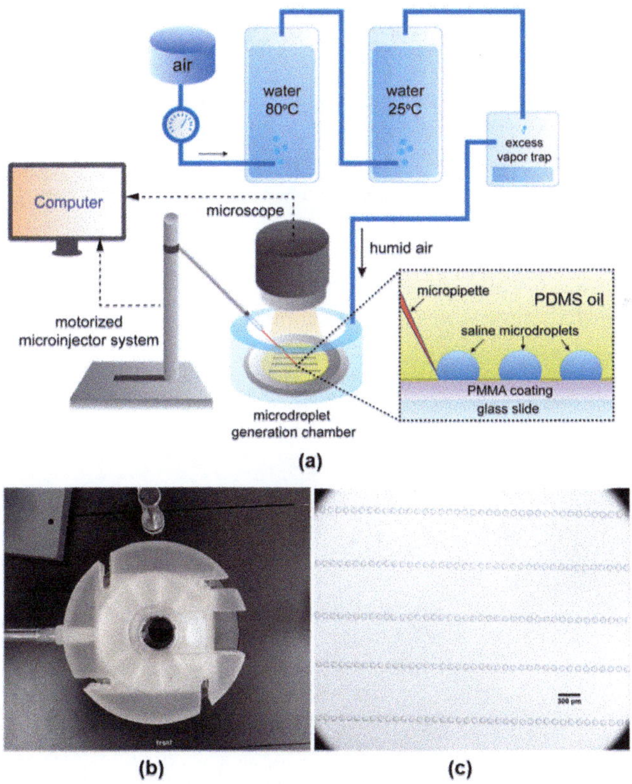

Fig. 1 (a) Schematic diagram of the microdroplet generation system with humidity control module. (b) Picture of the confined microdroplet generation chamber that contains PMMA-coated glass plate covered with PDMS oil. (c) Sessile microdroplets printed on the surface of the PMMA using a motorized microinjector.

II. %RH control system

A home-made, 3D printed (Formlabs printer, PMMA resin) %RH controller module (Fig. 1b) allows us to control the %RH of an air flow on the oil surface, while still having physical access to μDs (not a closed box, but an open-to-air design). Humid air is obtained through consecutive bubbling of a dry air-flow: first in hot-water, then in room-temperature water. A liquid trap ends the process to remove undesired water droplets generated by the bubbling process and conveyed in tubing. Mixing humid-air with dry-air allows fine control of % RH, from a maximum humidity condition (95% RH) to a dry condition (10% RH). The flow is then distributed uniformly above the oil layer. Calibration and stability of %RH were conducted with a %RH sensor (Sensirion SHT85) plugged onto a RaspberryPi3, and a home-made Python3 script to record %RH measurements (see ESI 3†).

III. (σ,t) Measurements

The dynamics of each individual μD is followed through an already developed image-processing technique. In a Region-Of-Interest (ROI), centered on each μD and its immediate vicinity as shown in Fig. 2, we extract, for each time step (image), a single scalar value: the standard-deviation σ of ROI-pixels' grey-levels. This is a post-processing technique: images of the whole μDs-array are acquired at a desired frequency until the end of the experiment, then we process the image-stack with ImageJ (FIJI). First, we extract each μD's edges on first images, and plot a ROI around each one. We then propagate σ measurement in each ROI through the whole image-stack to obtain the (σ,t)-curves which reflect the spatial and temporal dynamics of each μD. (σ,t)-Curves are then processed with home-made Python3 scripts (available on request).

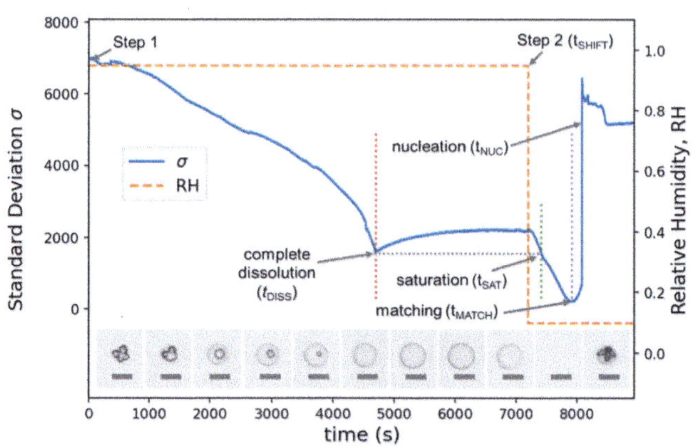

Fig. 2 Typical (σ,t)-curve, during a full %RH cycle (deliquescence/recrystallization), associated with each Region-Of-Interest (ROI). An illustrative ROI is provided in the lower part of the graph to show typical evolution of μDs. Characteristic time points are reported on both %RH curve and (σ,t)-curve. %RH cycle (dotted orange) goes from step 1 (crystal deliquescence up to t_{DISS} followed by μD dilution) to step 2 (μD contraction) at time t_{SHIFT}. During contraction, μD is at saturation at time t_{SAT}, optically disappears at time t_{MATCH}, and crystals nucleate at time t_{NUC}.

IV. Experimentation timeline

The experiments can be divided into 3 consecutive steps. Step 0 – this step is dedicated to obtaining an array of sessile crystals. A μD array is generated (Fig. 1c), following a previously developed method. %RH is here not controlled, and just needs to be lower than $\%RH_{eq}$ (the %RH corresponding to the equilibrium water partial vapor pressure above a saturated solution), so that μDs will contract (water diffuses to the oil and above-oil atmosphere) until nucleation and growth of a unique crystal in each μD. This step ends when an array of fully-dried crystals is obtained. Step 1 – the %RH above the oil layer is increased so that $\%RH > \%RH_{eq}$. Crystals will then, until complete dissolution, absorb water through the oil-layer. Step 1 ends after every crystal has been completely dissolved and the size of the resulting μDs seems constant. Step 2 – the %RH above the oil layer is decreased so that $\%RH < \%RH_{eq}$. Water then selectively diffuses from μDs, through the oil to the above-oil atmosphere; this contraction ultimately leads to nucleation and growth of a single crystal in each μD. Step 2 ends when we recover the initial conditions of step 1: an array of fully dried sessile crystals. Cycling – the sequence Step 1 and Step 2 can then be repeated at will.

Results and discussion

The dynamical behavior of the generated μDs depends on the relative humidity % RH above the oil in comparison with the equilibrium $\%RH_{eq}$ above the saturated solution (for NaCl in water, $\%RH_{eq} = 75\%$ at 25 °C). Generally, if $\%RH > \%RH_{eq}$, the crystal would absorb moisture leading to deliquescence, dissolution, and dilution. If $\%RH < \%RH_{eq}$, water would evaporate from the μD leading to an increase in salt concentration until nucleation. By controlling the prevailing % RH, we demonstrate the use of our image analysis protocol to track the dissolution and nucleation of μDs through (σ,t)-curves.

I. Analysis of (σ,t)-curves

Fig. 2 presents the typical time evolution during a nucleation experiment of σ measured for each ROI, *i.e.*, for each μD. The analysis of (σ,t)-curves shows characteristic points which are useful datapoints giving information on the dynamics of each μD, on both dissolution (deliquescence of the crystal), contraction and nucleation (see the video in the ESI† for the whole process). Their sequence and characteristics are discussed hereafter:

t_0–t_{DISS}. Dissolution time (t_{DISS}) is the first encountered characteristic point, appearing as a local minimum on the curve. From t_0 (at time $t = 0$) to t_{DISS}, the crystal is dissolving through water absorption, and t_{DISS} is the ultimate point of this process, when the crystal was last seen before total dissolution. It is noted that, on the path to t_{DISS}, the signal can present large and erratic oscillations, due to small movements of the crystal (jiggling) along the process leading to its total dissolution.

t_{DISS}–t_{SHIFT}. This is the μD dilution step, where, after the crystal completely dissolves, we allow μDs to continue to absorb water, thereby decreasing their NaCl concentration: μDs are undersaturated. At some time (t_{SHIFT}), when μDs do not visually evolve anymore, we shift %RH above the oil-layer, from above $\%RH_{eq}$ to a lower value, here from 95% to 10% RH.

t_{SHIFT}–t_{MATCH}. From the time we shift %RH, each μD will contract by selective diffusion of water through the oil-layer to the controlled atmosphere. Initially, the μDs have a refractive index lower than the oil one, but as contraction occurs, their refractive index will increase, and will at some time (t_{MATCH}) match the oil refractive index: at this point μDs optically disappear.

t_{MATCH}–t_{NUC}. The refractive index increases, linked to the increase in NaCl concentration, until crystal nucleation. Therefore, sigma will gradually re-increase after t_{MATCH} and crystal nucleation will appear as a large instantaneous (<1 s) jump in sigma values at t_{NUC}. The growth of the single nucleated crystal will continue to occur as long as the contraction of the μD continues, until no more water is available: we recover the initial state, a dry sessile crystal.

II. Estimating t_{SAT}

For experiments with evolving supersaturation, the time at which the micro-droplet becomes saturated (t_{SAT}) must be subtracted from the nucleation time t_{NUC} to calculate the induction time.[12] Otherwise, the measured induction time would depend on the arbitrary initial concentration. In principle, t_{SAT} can be estimated by computing the evaporation rate as a function of %RH and system geometry[13] but this could introduce another layer of uncertainty. Alternatively, we can extract t_{SAT} from (σ,t)-curves by making two assumptions. First, we consider that during the dissolution step of the crystal (t_0 to t_{DISS}), the solution is in equilibrium with the crystal,[14] *i.e.* at saturation concentration. Second, we consider that along the expansion stages (t_{DISS} to t_{SHIFT}) and contraction stages (t_{SHIFT} to t_{MATCH}), the μD shape has the same symmetric evolution, *i.e.* its shape follows the same smooth reversible path. Given that σ is a function of only shape and refractive index (so NaCl concentration), we can use the value of σ at t_{DISS} (where μD is saturated) as reference to find t_{SAT} during the contraction stages (Fig. 2). Thus, the μDs are assumed to be at saturation when their σ is equal to that of t_{DISS}. This is a direct consequence of our symmetric shape path hypothesis.

III. Eliminating interactions *via* %RH control

When the droplet contraction is done at %RH close to the %RH$_{eq}$, the μDs can interact with each other *via* water diffusion dynamics. This was previously reported and explained in a previous paper.[10] To summarize, when a crystal nucleates in a microdroplet, it suddenly increases the chemical potential of water inside the microdroplet. As a result, water diffuses from the nucleated microdroplet to its closest neighbors and thereby decreases their solute concentration. This is shown in Fig. 3, where selected (σ,t)-curves show clear oscillations, typical of nucleation induced μD-to-μD interactions. This data was obtained in contracting μDs at 55% RH (room conditions, without using the %RH module). Unfortunately, these interactions would make it impossible to consider each μD as an independent micro-crystallizer. To perform experiments in conditions where one can rule-out any possibility of μD-to-μD interactions, the %RH must be lowered such that the driving force for diffusion towards the atmosphere is high enough to prevent diffusion to neighboring μDs.

For this purpose, we developed two versions of an air-flow dispatcher module (v1 and v2, see ESI 3†). If the interactions are fully eliminated, there should be no oscillations in the σ-curve prior to nucleation in contrast to Fig. 3. Although both

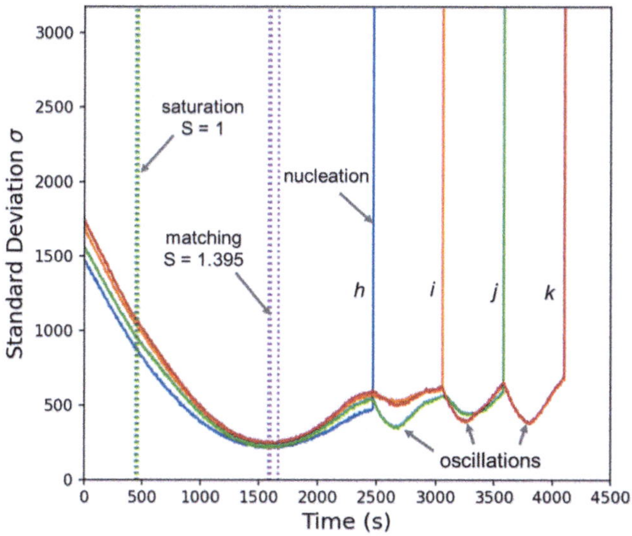

Fig. 3 (σ,t) Curves for 4 neighboring μDs, under conditions (room conditions at 55% RH, without using %RH module) which promote μD-to-μD interactions, which appear as oscillations in the σ signal. While a μD nucleates a crystal at time 'h', water is released to other neighboring μDs, thus diluting them. Their associated σ temporarily decreases before μD contraction re-occurs, thereby re-increasing their associated σ. Amplitude and time shift of the interactions are driven by both μD and nucleating-μD parameters such as size, supersaturation, and also by their separation distance. The spatial arrangement of μDs, named by their nucleation time, is the following: i, k, j, h. It is notable that these interactions are unobservable on images while looking at μD diameter changes under optical resolution.

designs were able to eliminate such oscillations, the v2 module design provides a much better spatial homogeneity of hygrometric conditions as revealed by the statistical-momentums analysis of t_{DISS} distributions (see ESI 3†). This led us to choose the v2 design for the experiments conducted and analyzed here.

IV. Small and big-droplets (SD and BD) experiments

To study the effect of confinement on nucleation, we performed experiments involving small droplets SD (4 pL) and big droplets BD (60 pL). First, we generated lines of monodisperse aqueous NaCl sessile microdroplets (1.7 M). Big μDs (total number of imaged μDs: 191, in 5 lines: 38, 36, 39, 39 and 39 μDs) were generated first. μDs are left free to contract (Step 0), at room conditions (55% RH), until we obtain lines of totally dried sessile crystals. Then, as the %RH above the oil is increased to 95% RH (Step 1), we start image acquisitions at a 1 Hz rate for BD and

Fig. 4 Representative lines of small μDs and large μDs.

4 Hz rate for SD. Following this experiment, we generated smaller μDs (total number of imaged μDs: 170, in 3 lines: 56, 56 and 58 μDs), and conducted the same experiment, except that for these small μDs we cycle them twice through Step 1/Step 2. Images of representative lines of small and large μDs are provided in Fig. 4.

Datasets

$S(t)$ is the cumulative probability that nucleation has not occurred as a function of time t, and is called the survival function. In experiments presented here, the time evolution of the fraction of droplets in which nucleation has not occurred is considered a good estimator of the underlying $S(t)$. Each μDs line is treated as an ensemble (with associated $S(t)$), and constitute a dataset. We consider datasets relative to the μDs sizes they are associated to: BD datasets (big μDs) and SD datasets (small μDs). These datasets are in fact different uncensored (every generated μD gives rise to a detected crystal) samplings of the nucleation times survival function, if existing.

Costs and benefits of increasing supersaturation

For experimentations conducted at constant supersaturation, there are two major drawbacks, sources of uncertainties in interpreting obtained nucleation data: t_{GROWTH} and data censoring. t_{GROWTH} is the time to grow from thc critical size (the nucleation event in itself) to a detectable size, and needs additional hypothesis to be modeled. Data censoring is the fact that not every experiment (supersaturated μDs in our case) will lead to a nucleation event, because nucleation rates may span over a large bandwidth, so that there's missing information. Here, conducted experiments provide access to uncensored data: every single μD leads to a detectable nucleation event, due to constantly increasing supersaturation. For t_{GROWTH}, we also benefit from the supersaturation constantly increasing, combined with a kinetic confinement effect,[15] as the probability to nucleate scales with the inverse of the volume, μDs reach high supersaturations before nucleation occurs. As a consequence of these high supersaturations, growth from critical cluster to detectable size is fast enough to be ignored: in less than a second, where experiment spans over hundreds to thousands, a large crystal is detected. These benefits in experimentations translate to a cost relative to the increased complexity to model nucleation in this evolving supersaturation framework.

Checking for heterogeneities and memory effect

A common source of uncertainties in nucleation experiments are heterogeneities, impurities, and their impact on measured nucleation rates (see the discussion on thermal and quenched disorder in ref. 4). The cycling can be viewed as an interesting approach[16] to detect a subset (some μDs in the experiments) where nucleation is mainly due to heterogeneities, which translates to a population of μDs that "statistically always" nucleates first. For SD datasets, that's what Fig. 5 is about: with two cycles, we can compare the nucleation ranks of every μD from one cycle to the following. If a population of μDs presenting heterogeneities leading to faster nucleation rates exists, it could appear as a pattern; for first ranks: they would statistically nucleate faster than other μDs, so in lower rank positions in both cycles. Fig. 5 does not show any kind of patterning, the distribution of ranks

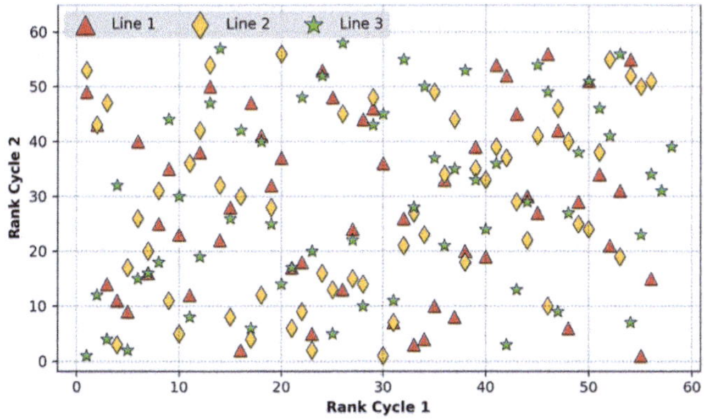

Fig. 5 μDs nucleation rank comparison in the two consecutive cycles for SD datasets. "Rank Cycle 1" is μD nucleation rank position in cycle 1 SD datasets, while "Rank Cycle 2" is the μD corresponding rank position where it nucleates in cycle 2 SD datasets.

is spread homogeneously in the whole available space. Moreover, it also indicates lack of historical contamination or memory effect from a cycle to the following, that is, nucleation is independent of previous cycle behavior. Nevertheless, further analyses and a larger number of cycles would be needed to confirm this and will be the subject of another study.

Distribution of characteristic time points

Due to the stochastic nature of nucleation, a distribution of times t_{NUC} is expected. But we also see, in Fig. 6, that contraction characteristic time points t_{SAT} and t_{MATCH} also distribute: despite optimization in hygrometric conditions, μDs dynamics still show individual characteristics. Such discrepancies can find their source in varying local conditions such as, without being an exhaustive list,

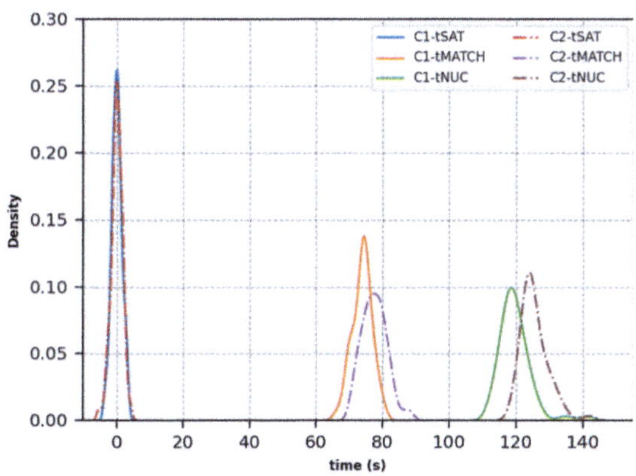

Fig. 6 Distributions (kernel density estimation of probability density function) of characteristic time points (t_{SAT}, t_{MATCH} and t_{NUC}), for a typical dataset (here SD line 1 dataset), for both deliquescence/nucleation cycles (C1 for cycle 1, C2 for cycle 2). Plots are against a shifted time where the mean t_{SAT}, respective to the cycle, is taken as the origin of times.

surface roughness or chemistry on which µDs rely, local water concentration in oil, small differences or variations in imposed above-oil %RH. Also, small differences in initial diameter, lower than the resolution of our optical system, cannot be ruled out. However, we see in Fig. 6 the astonishing similarity of the diverse characteristic time points distributions, in terms of both shape and scale. All µDs lines, whatever the cycle for SDs, show such resemblance (see ESI 4†). Ultimately, these distributions of t_{SAT} and t_{MATCH} will be reflected in nucleation time t_{NUC} distributions. Also, for the case of SDs, which experienced two deliquescence/nucleation cycles, Fig. 6 shows a cycle-to-cycle times location shift in distributions of times t_{MATCH} and t_{NUC}, similar for every SD line. It seems that, statistically, µDs are slower to reach t_{MATCH} and t_{NUC} during cycle 2 as compared to cycle 1: µDs contract slower in cycle 2.

Dimensionless induction time τ (DIT)

To compare µDs and associated t_{NUC}, we have to take into account differences that may only arise from their individual behavior (discrepancies in contracting dynamics), and also from their "cycle" behavior (the location shift shown from cycle 1 to cycle 2 for both t_{MATCH} and t_{NUC} distributions). For this purpose, we non-dimensionalized individual induction times t_{NUC} relative to t_{SAT} and t_{MATCH}. The individual µDs times to nucleation t_{NUC} are converted to a dimensionless induction time, as specified by eqn (1), here:

$$\tau = \frac{t_{NUC} - t_{SAT}}{t_{MATCH} - t_{SAT}} \quad (1)$$

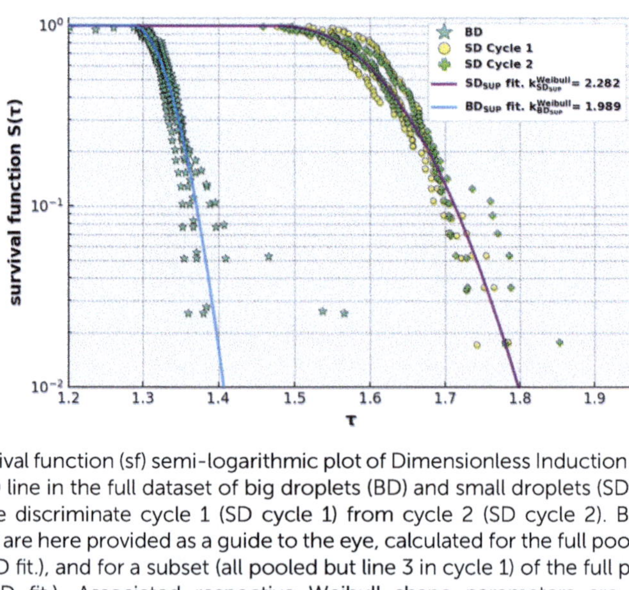

Fig. 7 Survival function (sf) semi-logarithmic plot of Dimensionless Induction Times (DITs) of every µD line in the full dataset of big droplets (BD) and small droplets (SD). Among SD datasets we discriminate cycle 1 (SD cycle 1) from cycle 2 (SD cycle 2). Best fits from a Weibull sf are here provided as a guide to the eye, calculated for the full pooling of all BD datasets (BD fit.), and for a subset (all pooled but line 3 in cycle 1) of the full pooling of SD datasets (SD fit.). Associated respective Weibull shape parameters are reported as $k^{Weibull}_{BD}$ and $k^{Weibull}_{SD}$.

DITs seem similar

DITs semi-logarithmic plots of their survival functions are reported in Fig. 7. At first sight, such nucleation dimensionless time allows comparison of μDs with individual contraction dynamics; the first benefit of which, for the SDs case, is to make cycle-to-cycle t_{NUC} location shifts vanish: there is no evidence of a location shift (see ESI 5†), from cycle 1 to cycle 2, so that nucleation times as reported through the use of DITs compare well. It seems, to the eye, that lines in BD datasets and SD datasets can, respectively, be fully pooled to be treated as representative of a unique experiment, thereby increasing the statistical representation. This will be addressed hereafter in the "Datasets pooling" section. Also evident in Fig. 7, for both BD and SD datasets, the log(sf) of their respective DITs do not follow a purely exponential decay, which would appear as a straight line on the plot: lowest values of DITs draw a plateau. This plateau is both typical of experimentation inaccuracies (t_{GROWTH}, heterogeneities), and of compressed exponential decays. As a reason for this plateau, we can here eliminate inaccuracies that arise from a shift between nucleation and detection of the crystal as growth is almost instantaneous to a visible and consequent size: growth time does not pollute these measurements. Also, as shown in Fig. 5, the lack of patterns (at least for SDs, the only μDs that underwent two cycles) in their cycle-to-cycle ranks do not promote evidence of a μD population (which nucleates first) subject to a higher nucleation rate due to nucleation promoting heterogeneities. In doing so, this plateau seems to have an existence independent of experimentation inaccuracies, and will be treated as such, promoting datasets fittings (addressed hereafter in a specific part) with the help of a typical stretch/compressed distribution: the Weibull distribution, where shape parameter k discriminates between stretching ($k < 1$) and compression ($k > 1$) of an exponential decay. The Weibull survival function, in its standardized form, can be written as follows (eqn (2)), with k being the shape parameter:

$$S(t) = \exp(-t^k) \tag{2}$$

Datasets pooling

Up to now, we implicitly considered a μD line as the scale of the sampling of the nucleation times distribution, each line forming a dataset, and kept comparing line-to-line similarities. There is high temptation to pool all datasets toward a "superdataset" that could be treated as a whole, but the problem would be to justify such data pooling. Such authorization for datasets pooling could rely on the use of a non-parametric statistical test: the "k-sample Anderson–Darling test" (kADt).[17] The kADt tests the null hypothesis "H0: the k-samples are drawn from the same population" *versus* the alternative hypothesis "H1: the k-samples are drawn from different populations". In other words, can here presented DITs associated to each μD line be considered as samples representative of the same distribution, *i.e.*, could we model these samples with the same distribution function, and pool datasets? For the 5 BD datasets (see ESI 6†), with a 5% significance level, kADt fails to reject the null hypothesis H0: the 5 μD lines and associated DITs distributions can be modeled with the same distribution function, and we pool datasets in BD_{SUP} superdataset. Regarding (see ESI 6†) the 6 SD datasets (considering each cycle – 3 lines each – as independent sampling as lines could be), the kADt rejects null hypothesis H0: the 6 datasets cannot be considered to be drawn from the same distribution. We then test every combination of 5 among 6 lines with *k*ADt, and one SD dataset seems to be the source of the null hypothesis rejection: line 3 in cycle 1. In removing this dataset, *k*ADt cannot reject the null hypothesis H0 anymore, and we can then construct a SD superdataset SD_{SUP} constituted of 5 pooled SD datasets: lines 1 and 2 in cycle 1, and all 3 lines in cycle 2.

Datasets fittings

Complementary to the analysis of survival function (sf) semi-logarithmic plots of Dimensionless Induction Times (DITs) presented in Fig. 7, the contracting process (where, as volume decreases, supersaturation increases, imposing also nucleation rate to do so upon nucleation of a crystal) encompassed by μDs also justifies the modeling of DITs with a Weibull distribution function, which would correspond to a compressed exponential function of class III as suggested by Sear.[4] A fitting procedure was applied to newly constructed SD_{SUP} and BD_{SUP} datasets, and results are presented in Fig. 8. It is of note that sensitivity of fittings algorithm to both outliers and initial fitting guesses was addressed in two steps: we first fit "roughly", with guessed outliers removed, then we re-incorporate 98% of the full dataset to refine fit. R^2 and mean squared errors were calculated on the full dataset. Despite such precautions, as seen in Fig. 8, there is departure of the fit from data at the distribution tail (highest DITs). This is not an artefact, the tail is not well fitted, but using a semi-logarithmic scale here does not render justice to the quality of fittings, as measured with both R^2 and mean squared errors (see ESI 7†). Moreover, Fig. 8 confirms confinement effects where the induction time shifts due to system (μD) size: the smaller the system, the longer the induction time.

Supersaturations at nucleation times

To translate DITs in supersaturations, one needs a contraction model: how μD volume evolves with time. An exact and precise model is complex and out of the scope of this paper, and will be the object of specific studies later on. But from

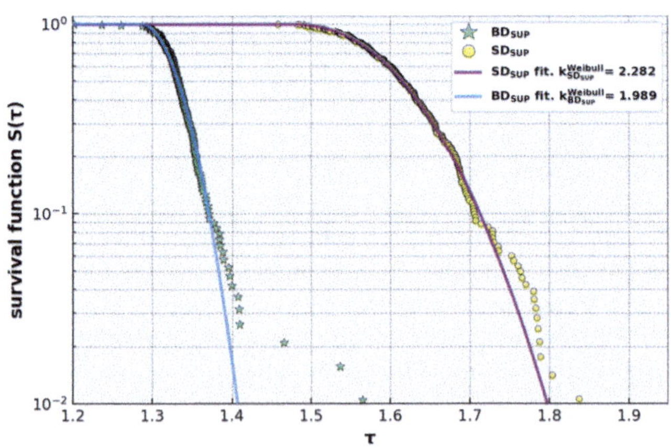

Fig. 8 Survival function (sf) semi-logarithmic plot of Dimensionless Induction Times (DITs) for both BD_{SUP} and SD_{SUP} datasets. Best fits with a Weibull function are shown as 'SD_{SUP} fit.' and 'BD_{SUP} fit.' with reported associated distribution shape parameters as $k_{SD}^{Weibull}$ and $k_{BD}^{Weibull}$.

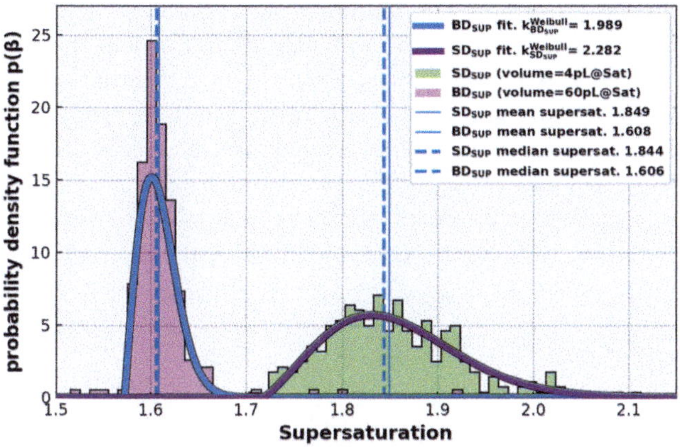

Fig. 9 Supersaturation histograms and best fits with a Weibull probability density function (pdf), using a linear decrease of droplets volume with time model. SD_{SUP} and BD_{SUP} are the pooled data as determined by Anderson–Darling *k*-sample tests. 'SD_{SUP} fit.' and 'BD_{SUP} fit.' are, respectively, SD_{SUP} and BD_{SUP} Weibull pdf best fits. Median and mean times of each fit are reported on the graph for both SD and BD fits.

previous works, and for the sake of simplicity, we can choose a linear evolution of volume with time, which allows (see ESI 8†) calculation of supersaturation β_{NUC} from DITs and supersaturation β_{MATCH} at matching time t_{MATCH}, following eqn (3) here:

$$\beta_{NUC} = \frac{1}{1 - \tau \times \left(1 - \frac{1}{\beta_{MATCH}}\right)} \quad (3)$$

In Fig. 9, we report calculated supersaturation distributions for both BD_{sup} and SD_{sup} superdatasets, with corresponding Weibull function best fits for both. For BD_{SUP}, nucleation occurs at a mean supersaturation of 1.61 and 95% spans from 1.58 to 1.65. For SD_{SUP}, nucleation occurs at a mean supersaturation of 1.85 and 95% spans from 1.74 to 2. There is high correspondence with supersaturation values already reported in the literature for comparable system sizes.[18–20]

Conclusions

In this communication, we present a microdroplet generation system with humidity control module and a novel approach for measuring and analyzing induction times in lines of saline sessile microdroplets undergoing a deliquescence/recrystallization cycling process. With the help of a recently developed image analysis protocol, we first assert microdroplets act independently: there are no microdroplet-to-microdroplet interactions. Second, using a dimensionless time, dynamics of microdroplets along deliquescence/recrystallization cycles are specified, allowing to obtain precise estimation of the nucleation induction times survival function $S(t)$, for every microdroplet line. Considered as samples of the underlying (unknown) nucleation probability density function, line datasets that satisfy a *k*-sample Anderson–Darling test are pooled, and resultant superdatasets are fitted with a Weibull function. Moreover, evaporation experiments on 60 pL and 4 pL microdroplets at saturation confirm the confinement effect that smaller systems reach higher supersaturations. Both the experimental method and the data-treatment procedure seem promising for the study of fundamental aspects of nucleation kinetics, confinement effects in particular, and is adaptable to other salts, pharmaceuticals, or biological crystals of interest.

The modeling of evaporation and the use of a modified Poisson distribution which considers the time-dependence of the nucleation driving force[5] will be a subject of future work.

Abbreviations

SD	Small droplets
BD	Big droplets
μD	Microdroplet

Author contributions

Conducted the experiments and collected data: R. C.; designed experiments: all but D. N.; analyzed datasets: all but M. L.

Conflicts of interest

There are no conflicts of interest to declare.

Acknowledgements

R. Cedeno acknowledges the financial support of Vidyasirimedhi Institute of Science and Technology (VISTEC) and the Eiffel Excellence Scholarship (No. P744524EP744524E) granted by the French Government.

References

1 N. T. Thanh, N. Maclean and S. Mahiddine, *Chem. Rev.*, 2014, **114**, 7610–7630.
2 S. Zhang, C. J. J. Gerard, A. Ikni, G. Ferry, L. M. Vuillard, J. A. Boutin, N. Ferte, R. Grossier, N. Candoni and S. Veesler, *J. Cryst. Growth*, 2017, **472**, 18–28.
3 X. Zhong, J. Ren and F. Duan, *J. Phys. Chem. B*, 2017, **121**, 7924–7933.
4 R. P. Sear, *CrystEngComm*, 2014, **16**, 6506–6522.
5 L. Goh, K. Chen, V. Bhamidi, G. He, N. C. Kee, P. J. Kenis, C. F. Zukoski, 3rd and R. D. Braatz, *Cryst. Growth Des.*, 2010, **10**, 2515–2521.
6 D. Kashchiev, D. Verdoes and G. M. Van Rosmalen, *J. Cryst. Growth*, 1991, **110**, 373–380.
7 A. I. Toldy, A. Z. M. Badruddoza, L. Zheng, T. A. Hatton, R. Gunawan, R. Rajagopalan and S. A. Khan, *Cryst. Growth Des.*, 2012, **12**, 3977–3982.
8 R. Grossier, Z. Hammadi, R. Morin, A. Magnaldo and S. Veesler, *Appl. Phys. Lett.*, 2011, **98**, 091916.
9 I. Rodríguez-Ruiz, Z. Hammadi, R. Grossier, J. Gómez-Morales and S. Veesler, *Langmuir*, 2013, **29**, 12628–12632.

10 R. Grossier, V. Tishkova, R. Morin and S. Veesler, *AIP Adv.*, 2018, **8**, 075324.
11 R. Grossier and S. Veesler, *Cryst. Growth Des.*, 2009, **9**, 1917–1922.
12 L. Goh, K. Chen, V. Bhamidi, G. He, N. C. S. Kee, P. J. A. Kenis, C. F. Zukoski and R. D. Braatz, *Cryst. Growth Des.*, 2010, **10**, 2515–2521.
13 S. Talreja, P. J. A. Kenis and C. F. Zukoski, *Langmuir*, 2007, **23**, 4516–4522.
14 R. McGraw and E. R. Lewis, *J. Chem. Phys.*, 2009, **131**, 194705.
15 Z. Hammadi, R. Grossier, S. Zhang, A. Ikni, N. Candoni, R. Morin and S. Veesler, *Faraday Discuss.*, 2015, **179**, 489–501.
16 P. Laval, A. Crombez and J.-B. Salmon, *Langmuir*, 2009, **25**, 1836–1841.
17 F. W. Scholz and M. A. Stephens, *J. Am. Stat. Assoc.*, 1987, **82**, 918–924.
18 R. Grossier, Z. Hammadi, R. Morin and S. Veesler, *Phys. Rev. Lett.*, 2011, **107**, 025504.
19 J. Desarnaud, H. Derluyn, J. Carmeliet, D. Bonn and N. Shahidzadeh, *J. Phys. Chem. Lett.*, 2014, **5**, 890–895.
20 A. Utoft, K. Kinoshita, D. L. Bitterfield and D. Needham, *Langmuir*, 2018, **34**, 3626–3641.

Nucleation in sessile saline microdroplets: induction time measurement *via* deliquescence–recrystallization cycling

Ruel Cedeno, Romain Grossier, Mehdi Lagaize, David Nerini, Nadine Candoni, Adrian Flood and Stéphane Veesler

Faraday Discuss., 2022, DOI: 10.1039/D1FD00090J. ***Amendment published 20 April 2022.***

On the 3rd page of the article, first paragraph of the Experimental section, the thickness of the polydimethylsiloxane layer should be amended from 0.4 mm to 0.8 mm so that the sentence reads "We employed a mechanically controlled micropipette to generate arrays of monodisperse aqueous NaCl sessile microdroplets (1.7 M) on the surface of a poly(methyl methacrylate) (PMMA)-coated glass slide immersed in a 0.8 mm thick layer of polydimethylsiloxane (PDMS) oil (10 cSt)".

The Royal Society of Chemistry apologises for these errors and any consequent inconvenience to authors and readers.

Faraday Discussions

ROYAL SOCIETY OF CHEMISTRY

PAPER

Studying the impact of the pre-exponential factor on templated nucleation

Vivek Verma, *[a] Hamish Mitchell,[a] Mingxia Guo,[a] Benjamin K. Hodnett [b] and Jerry Y. Y. Heng [ac]

Received 25th November 2021, Accepted 13th December 2021

DOI: 10.1039/d1fd00101a

Traditionally, the enhancement of nucleation rates in the presence of heterogeneous surfaces in crystallisation processes has been attributed to the modification of the interfacial energy of the system according to the classical nucleation theory. However, recent developments have shown that heterogeneous surfaces instead alter the pre-exponential factor of nucleation. In this work, the nucleation kinetics of glycine and diglycine in aqueous solutions have been explored in the presence and absence of a heterogeneous surface. Results from induction time experiments show that the presence of a heterogeneous surface increases the pre-exponential factor by 2-fold or more for both glycine and diglycine, while the interfacial energy remains unchanged for both species. This study suggests that the heterogeneous surface enhances the nucleation rate *via* hydrogen bond formation with both glycine and diglycine. This is verified by hydrogen bond propensity calculations, molecular functionality analysis, and calculation of the time taken for a solute molecule to attach to the growing nucleus, which is an order of magnitude shorter than the estimated lifetime of the hydrogen bond. The effect of the heterosurface is of greater magnitude for diglycine than for glycine, which may be due to the heightened molecular complementarity between the hydrogen bond donor and acceptor sites on diglycine and the heterosurface.

1. Introduction

The universal nature of crystallization has been explored in every aspect of science, from organics to inorganics to bio-organics, and from atmospherics to minerals. Crystallisation is a two-step process, with nucleation being the initial step of forming a stable new phase from a supersaturated medium. Nucleation is subsequently followed by growth as well as ripening, agglomeration, aggregation,

[a]*Department of Chemical Engineering, Imperial College London, South Kensington Campus, London, SW7 2AZ, UK. E-mail: v.verma@imperial.ac.uk*

[b]*Synthesis and Solid State Pharmaceutical Centre, Department of Chemical Sciences, Bernal Institute, University of Limerick, Limerick, V94 T9PX, Ireland*

[c]*Institute of Molecular Science and Engineering, Imperial College London, South Kensington Campus, London, SW7 2AZ, UK*

and phase transformation. In both steps, supersaturation is the driving force, caused by the difference between the actual concentration of the crystallizing solute and the saturation concentration (solubility). There are a variety of methods for generating supersaturation, such as cooling, anti-solvent addition, evaporation, and the addition of precipitants, amongst others. Since the physical properties of the crystals, such as shape, structure, defects, and polymorphism are determined at the nucleation stage, this step is usually rate limiting for the crystallisation process.[1,2]

The nucleation process occurring at interfaces is both profoundly important and widely encountered in different areas of science, such as material synthesis,[3–5] biomineralisation,[6] battery operation,[7,8] geochemistry,[9,10] geologic CO_2 sequestration,[11] cement hardening,[12] and drug production.[13,14] It is essential to analyse, predict, control, and fine-tune these systems by comprehending nucleation and quantifying the relevant thermodynamic and kinetic parameters. Heterogeneous nucleation occurs in 'impure' liquids, for example, where surface walls or nucleating agents (not crystal seeds) are present, rather than in an idealized supersaturated liquid, where internal fluctuations in density trigger the crossing of the kinetic barrier to nucleation, resulting in homogeneous nucleation. Ward and colleagues laid a modern scientific foundation for the study of heterogeneous nucleation in organic molecules.[15–17] Their work on benzoic acid nucleation on the freshly cleaved surface of β-succinic acid or L-valine delivered by sublimation led to the concept of ledge-directed epitaxy (LDE). It was postulated that benzoic acid nucleation at a pair of intersecting cleavage planes on the heterosurface was a result of the favourable lattice matching.[16]

Over the past few decades, the 'state-of-the-art' approaches responsible for studying heterogeneous nucleation have been used for the non-specific adsorption of molecules on the heterosurface,[18] specific interactions such as functional group matching,[19] or selective interactions such as lattice matching.[20] The materials that have successfully served the function of the heterosurface for heterogeneous nucleation are pharmaceutical excipients,[14,21–23] silanised glass substrates,[13,24,25] amino acids,[26] self-assembled monolayers,[27] bio-compatible[28] and synthetic polymers,[29,30] and porous substrates.[31] Protein crystallization has also been achieved with chemically modified surfaces.[32] These surfaces were used either to induce nucleation and/or study the nucleation kinetics of the drug molecules. Parambil *et al.*[33] have reviewed the existing literature discussing both experimental and simulation studies directed towards deeper understanding of the underpinning mechanisms. Despite the significant number of publications in this area, it remains unclear how heterogeneous nucleation occurs and how the thermodynamic and kinetic factors are influenced by heterogeneous nucleation.

In recent years, functional group matching between the nucleating solute and the heterosurface has emerged directly or indirectly as the leading responsible factor for the occurrence of heterogeneous nucleation. Diao *et al.*[34] showed that the nucleation density of aspirin increased dramatically in the presence of two polymers, in particularly poly(4-acryloylmorpholine) and poly(2-carboxyethyl acrylate). Aspirin's hydrogen bond donor (HBD) sites can interact most efficiently within these polymers since they have hydrogen bond acceptor (HBA) sites. Later, Diao *et al.*[35] also showed that nanopores of different shapes (spherical, square and hexagonal), specifically angular nanopores (15–120 nm), promoted the nucleation rate of aspirin. Although the porosity in polymers with specific wall angles enhanced the nucleation rate, pointing towards lattice matching as the potential reason, this was only evident for the polymers with HBA properties. Polymers lacking HBA properties did not influence the nucleation rate of aspirin.

The concept of hydrogen bonding complementarity in the template induced nucleation of pharmaceutical ingredients was further explored by Chadwick *et al.*[19,36] for the heterogeneous nucleation of acetaminophen (AAP) in the presence of graphite, δ-mannitol, α-lactose monohydrate, L-histidine, poly(-methylmethacrylate), and poly-*n*-butylmethacrylate. Interestingly, graphite was among the poorest in accelerating the heterogeneous nucleation of AAP. Later, Bueno *et al.*[22] studied the influence the range of different heterosurfaces such as δ-mannitol, silica, microcrystalline cellulose (MCC), carboxymethyl cellulose, α/β-lactose, and polycaprolactone (PCL) on the heterogeneous nucleation of fenofibrate (FF). They found that FF nucleated slowest in the presence of the PCL heterosurface. This is because except for PCL, all the heterosurfaces studied possessed HBDs which could easily form hydrogen bonds with the HBAs of fenofibrate. Therefore, this study shows the influence of functional group complementarity in template induced nucleation systems.

Recently, Ouyang *et al.*[37] were able to successfully nucleate and stabilise the carbamazepine (CBMZ) metastable form (FII) at lower supersaturations, at which normally only the stable form of CBMZ (FIII) crystallises, when crystallised in the presence of a phenyl functionalised silica heterosurface. The possible mechanism for this observation was the stabilisation of CBMZ molecules on the heterosurface due to the aromatic–aromatic interaction between the surface of phenyl-functionalized silica and the CBMZ molecule. Later, Ouyang *et al.*[38] expanded the work of using functionalized silica (SiO_2, SiO_2–NH_2, SiO_2–COOH) as a heterosurface in the nucleation of vanillin. The functionalised silica was able to significantly enhance the nucleation rate of vanillin, which is evident by the reduction in the induction time from 4 hours to 20 minutes. The heterosurface also increased the growth rate of the nucleated vanillin crystal. These increases in the nucleation and growth rates were potentially due to the formation of hydrogen bonds between the vanillin molecules and the SiO_2–NH_2 and SiO_2–COOH heterosurfaces. Recently Li *et al.*[39] discovered that the use of silica particles as a heterosurface improved the crystallisation rate of lysozyme by reducing the induction time even in the presence of an impurity protein, thaumatin.

A recent study by Verma *et al.*[14] studied the influence of a heterosurface (microcrystalline cellulose (MCC)) on the nucleation kinetic parameters of seven different active pharmaceutical ingredients (APIs), namely carbamazepine (CBMZ), acetaminophen (AAP), caffeine (CAF), clozapine (CPB), risperidone (RIS), phenylbutazone (PBZ), and FF. Crystallisation of the latter five APIs was accelerated in the presence of MCC compared to the former two, CBMZ and AAP. The maximum acceleration was observed for FF, which was 16 times faster than what was observed for homogeneous nucleation. This acceleration was possibly because of the formation of a stable hydrogen bond between the HBAs of the latter five APIs with the HBD sites on MCC. This study confirmed that the functional group complementarity between the API and heterosurface is an important factor in improving the API nucleation rate. This confirmation was supported by the hypothesis that the lifetime of the single adsorbed molecule on the surface of a hetero substance is sufficiently long to allow for the adsorption of other API molecules or clusters to attach to the adsorbed molecule, thereby leading to the formation of stable nuclei that remain attached to the heterosurface. This hypothesis was recently proved by Cazade *et al.*[40] using Monte Carlo (MC) and molecular dynamics (MD) simulations. The MD simulation results for the adsorption of RIS onto MCC surfaces concluded that the hydrogen-bonded lifetime of 32 ns is comparatively longer than that of the RIS–RIS attachment lifetime in the solution. It is also an order of magnitude longer than the time to add a single molecule of RIS to a growing crystal at a moderate supersaturation.

Verma *et al.*[14] also reported that the heterogeneous nucleation of APIs in the presence of MCC had minimal influence on the interfacial energy of APIs in the nucleating solution. This is contradictory to the reported literature claiming that heterosurfaces decrease the interfacial energy, thereby reducing the nucleation energy barrier for nucleation, resulting in an increased nucleation rate.[2,41] Instead, they claimed that the increase in the nucleation rate in the presence of MCC was due to the increase in the pre-exponential factor (A) in the presence of MCC. The value of A increased more than 2-fold for the API, exhibiting an increase in the nucleation rate in the presence of MCC. This signifies that the rate of subcritical size cluster formation is at least 2-fold greater in the presence of MCC for most of the APIs, thereby promoting heterogeneous nucleation. This is consistent with their previous findings on the nucleation of clozapine–methanol solvate in the presence of MCC, confirming that the pre-exponential factor is the major contributing factor to heterogeneous nucleation.[42]

Glycine and diglycine were selected as molecules of interest for this work. The reason for this is that glycine is the simplest and most studied amino acid, consisting of a non-heavy hydrogen side chain, which eliminates the need to consider the effect of the side chain of the amino acid when analysing the effect of the heterosurface. Glycine exhibits a total of six polymorphs (α, β, γ, δ, ε, and ζ),[43] but the latter three only form under extreme pressures.[43] At atmospheric conditions, only α, β, and γ will form.[43] The γ polymorph is the thermodynamically stable form of glycine, while α- and β-glycine exist as metastable and unstable polymorphs, respectively.[44] However, the nucleation of γ-glycine is kinetically hindered when compared to that of α-glycine, meaning that under agitation at atmospheric conditions, α-glycine is the initially forming polymorph. The solution-mediated phase transformation of glycine occurs slowly, taking around 20 hours to begin.[45] There have been numerous studies into the effects of process conditions on the polymorphic outcome of glycine crystallisation, which have identified factors such as humidity,[46] pH,[47] the usage of microdroplets,[48] and the addition of precipitants such as sodium chloride.[45] It is possible to distinguish between the α and γ forms visually – α-glycine forms as rod-like monoclinic crystals belonging to the space group $P2_1/n$,[49] whereas γ-glycine forms trigonal pyramidal crystals belonging to $P3_1$ or $P3_2$.[50]

Recently, Vesga *et al.*[51] showed that for the cooling crystallisation of aqueous solutions of glycine, agitation was a critical factor in determining the polymorphic outcome – the presence of agitation led to the preferential formation of α-glycine, while γ-glycine was able to nucleate under quiescent conditions.[51] They also showed that the degree of supersaturation influenced the polymorphic outcome, with higher concentrations leading to the co-existence of both α-glycine and γ-glycine.

Diglycine was selected for a similar reason as it is the simplest dipeptide, consisting of two glycine residues. There have been comparatively fewer studies on the crystallisation of diglycine, but it has been shown to exhibit three polymorphs, α, β, and γ.[52] However, the β and γ polymorphs are notoriously elusive, requiring many recrystallisations to form.[52] As well as this, the α-polymorph is the thermodynamically stable form, meaning that diglycine does not suffer from the same polymorphic issues as glycine.[53] As a result, the α-polymorph is the most extensively studied polymorph as it forms most readily. The α-polymorph forms plate-like crystals belonging to the space group $P2_1/c$.[54]

This work further explores the hypothesis that the heterosurface (glass beads) will enhance the nucleation rates of the amino acid (glycine) and dipeptide (diglycine) through heterogeneous nucleation. Previous work from the authors revealed that functional group complementarity between the heterosurface and the nucleating solute encouraged hydrogen bond formation between them, resulting in the heterogeneous nucleation of the solute. However, their work was only restricted to small organic drug molecules. The aim of this work is to advance the hydrogen bond complementarity and hydrogen bond lifetime hypotheses to small biomolecules, such as amino acids and dipeptides. The goal is to crystallise glycine and diglycine in the absence and presence of silica at different supersaturations to calculate heterogeneous nucleation parameters and to use the classical nucleation theory (CNT) to deduce the kinetic parameters of nucleation for each. The possible knowledge of hydrogen bonding behaviour between heterosurfaces and the crystallising solute, alongside appropriate crystallisation conditions, will result in the generation of crystals with the desired size, morphology, and polymorphic outcome.

2. Experimental section

2.1. Materials

Glycine (Gly, >99%, suitable for electrophoresis) and diglycine (Digly, >99% by titration) were supplied by Sigma-Aldrich and used without further purification. Deionized water was generated by a PURELAB Chorus 1 water purification system (ELGA LabWater (High Wycombe, UK)). Glass beads with an average diameter of 50 μm were supplied by Sigmund Linder GmbH (Warmensteinach, Germany). The molecular structures of the heterosurface and the two organic compounds are presented in Fig. 1.

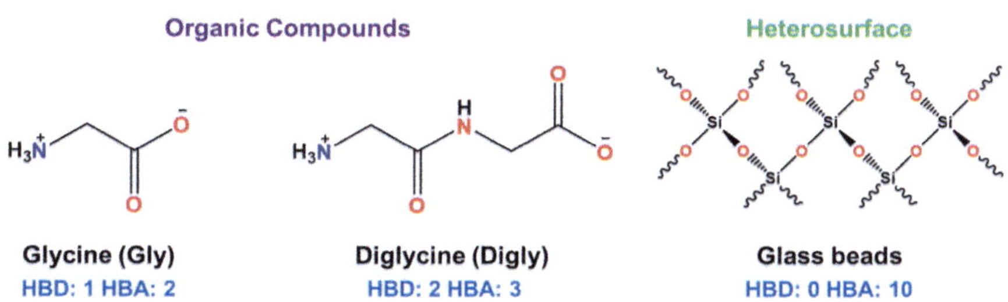

Fig. 1 Chemical structures of the heterosurface (glass beads) and the two organic compounds (glycine and diglycine) used in this study with the number of hydrogen bond donor (HBD) and hydrogen bond acceptor (HBA) groups in each structure shown.

2.2. Determination of the metastable zone width (MSZW) for the glycine and diglycine solutions in water in the absence of silica

The MSZW for both glycine and diglycine were determined using the following procedure. A known amount of glycine/diglycine was added to a lidded bottle containing 200 mL of deionised water. The solution was placed in a water bath controlled by a Grant GP200 circulator (Royston, UK) at 45 °C for an hour to allow the solids to dissolve. After this, the solutions were checked visually to ensure the solids had dissolved and the solution was filtered *via* a Sartorius 0.22 μm pore size PTFE syringe filter (Göttingen, Germany) and transferred to a 250 mL glass bottle submerged in a water bath connected to a JULABO F32-GB thermoregulator (Seelbach, Germany). The thermoregulator was programmed to follow a set heating profile. Firstly, the system was held at 30 °C (50 °C for diglycine) for one hour to allow the system to equilibrate and to ensure solid dissolution. The temperature was then decreased to 25 °C (45 °C for diglycine), and the system was held for a further 30 minutes to ensure thermal equilibrium. From here, a cooling ramp of 0.1 °C min^{-1} was imposed until the system reached 5 °C. At this point, the system was maintained at 5 °C for 1 hour. The onset of nucleation was determined *via* a Nikon D90 camera fitted with an AF-S 18–105 mm lens, which took regularly timed images of the solution, and the metastable zone widths were calculated from these images.

2.3. Determination of the induction time of glycine and diglycine in the absence of glass beads at different supersaturations

All induction time experiments in the absence of glass bead particles were carried out *via* the methodology outlined below (see Fig. 2). A known amount of glycine/diglycine was added to a lidded bottle containing 100 mL of deionised water. The solution was placed in a water bath controlled by a Grant GP200 circulator (Royston, UK) at 45 °C for an hour to allow the solids to dissolve. After this, the solutions were checked visually to ensure that the solids had dissolved and 50 mL of solution was then filtered *via* a syringe filter and transferred to a lidded 100 mL bottle. The bottle was then placed back in the water bath at 45 °C for an hour to further ensure that no crystallisation or precipitation had occurred. The concentration of the filtrate was verified *via* a Thermo Scientific NanoDrop One^{C} UV-Vis spectrophotometer (Waltham, Massachusetts). The bottle was then placed in a second water bath controlled by a JULABO F32-GB thermoregulator, which was set at a lower temperature to generate supersaturation. The onset of nucleation was determined *via* a Nikon D90 camera fitted with an AF-S 18–105 mm lens,

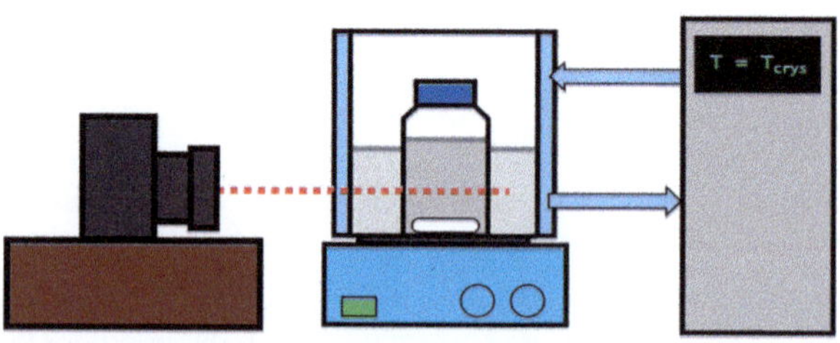

Fig. 2 Schematic representation of the experimental set-up.

Table 1 Complete list of crystallisation temperatures and associated supersaturations, along with the solubility and metastable zone width information used in this study, $n \geq 3$; MSZW: metastable zone width

Compound	Supersaturation (S)	Saturation temperature, T_{sat}/°C (solubility (mg mL^{-1}))	Supersaturation temperature, T_{cry}/°C (solubility (mg mL^{-1}))	Solubility data	MSZW (°C)
Glycine	1.17	26 (234.35)	17.5 (201.10)	56	9.0
(Gly)	1.22		15.0 (192.25)		(ref. 55)
	1.33		10.0 (175.70)		
Diglycine	1.30	43 (298.24)	30.0 (228.76)	56	17.5
(Digly)	1.40		26.5 (213.00)		(this work)
	1.50		23.0 (198.32)		

which took regularly timed images of the solution. The images were then analysed to determine the onset of nucleation. After nucleation had occurred, the solution was removed from the bath and the entire procedure was repeated at least three times to obtain an average value for the induction time, alongside an indication of the standard deviation. This was then repeated for two more different crystallisation temperatures, allowing for induction time measurements for three different supersaturations for both glycine and diglycine. The conditions for each experiment are given in Table 1.

2.4. Crystallisation of glycine and diglycine in the presence of glass beads at different supersaturations

The procedure of determining the induction time for the presence of glass beads is similar to that outlined in Section 2.3 (see Fig. 2). However, there are two key differences. Firstly, once the solutions were filtered, a known mass of glass beads corresponding to 50% of the maximum theoretical yield of glycine and diglycine (calculated as the difference in concentration between the initial supersaturation and the solubility) was added to the bottles containing filtered solution. Secondly, it was not possible to monitor crystallisation *via* visual observation methodologies, as the presence of glass beads rendered the solutions cloudy from the offset. Instead, 150 μL of the solution was collected *via* a syringe into an Eppendorf tube, and then filtered using 200 nm PTFE membrane syringe filters. From the supernatant, 100 μL was then diluted in 500 μL of deionised water in a second Eppendorf tube, to prevent any further crystallisation from occurring and to bring the concentration to within the limits of calibration. The concentration of the diluted sample was then characterised *via* a Thermo Scientific NanoDrop One^{C} UV-Vis spectrophotometer (Waltham, Massachusetts) by collecting absorbance data at 220 nm (glycine) and 230 nm (diglycine), and the concentration of the original sample was back-calculated *via* a linear correlation between concentration and absorbance. This correlation was calculated by dissolving known quantities of glycine and diglycine at concentrations well below the solubility limit and measuring their absorbance at 220 nm (glycine) and 230 nm (diglycine) to form a linear calibration curve. From this, it was possible to plot desupersaturation profiles for each experiment, from which the induction time could be estimated. These experiments were again repeated a minimum of three times for each of the conditions described in Table 1.

2.5. Solid state characterisation of the glycine/diglycine–glass bead composites

2.5.1. Powder X-ray diffraction (PXRD). Powder X-ray diffractograms (PXRDs) were recorded on a PANalytical Empyrean diffractometer using a Cu radiation source ($\lambda = 1.541$ nm) at 40 mA and 40 kV. Scans were performed between 5 and 35° 2θ at a scan rate of 0.013° 2θ min^{-1}.

2.5.2. Microscopy imaging. Visual analysis of the systems was performed using an Olympus CX-41 microscope (Essex, UK) at magnifications of 5× and 20×. The microscope was connected to a GT Vision GXCAM HiChrome Met display (Suffolk, UK), which allowed for digital imaging and postprocessing of the images taken. A sample of around 50 µL was taken at the end of each experiment *via* a syringe and dispensed onto a clear glass slide, on top of which a cover slip was placed to prevent solvent evaporation. Images were obtained to help verify the polymorphic outcome of each crystallisation experiment.

2.6. Hydrogen bond propensity calculations

To find the most probable HBAs and HBDs in the glycine and diglycine, hydrogen bond propensities were calculated[57] using the program Mercury 3.10.1 as follows. The training dataset for the statistical models was composed of molecules extracted from the Cambridge Structural Database (CSD) that contain all the functional groups present in the target APIs. A logistic regression was then applied to the training dataset that allowed the predictions in the form of H-bond propensities upon consideration of the environmental variables for the functional groups (*e.g.*, aromaticity, steric density) of the target API. The H-bond propensity for a donor–acceptor pair can take a value between 0 and 1, where 0 indicates no likelihood of H-bond formation and 1 indicates that a hydrogen bond will always be found.

3. Results and discussion

3.1. Determination of the MSZW

Table 1 summarizes the key process parameters used during the various crystallizations of glycine and diglycine from their respective saturated aqueous solutions in the presence of the heterosurface, glass beads. The MSZW determined during this study for Digly is included in Table 1, along with the previously reported MSZW for Gly.[55] Bonnin-Paris *et al.*[55] showed that the MSZW of glycine aqueous solution is approximately 9 °C at a cooling rate of 0.1 °C min^{-1}. Recently, Ramakers *et al.*[58] studied the influence of the addition of antisolvents such as ethanol, methanol, and dimethyl formamide to the aqueous glycine solution on the MSZW of glycine. The MSZW of glycine is narrower in aqueous solution but wider when the initial glycine solution has a greater antisolvent mass fraction. The MSZW of Digly ($T_{sat} = 43$ °C) was determined to be 17.5 °C at a cooling rate of 0.1 °C min^{-1}. Table 1 further indicates that most of the crystallization temperatures used to generate the desired supersaturations reside inside or within 0.5 °C

Table 2 Summary of the induction times, the nucleation rates, and the nucleation rate ratios for the crystallization of glycine and diglycine from supersaturated aqueous solutions in the absence and presence of glass beads at various supersaturations

			Homogeneous		Heterogeneous		
Compounds	Supersaturation (S)	$RT \ln S$ (J mol^{-1})	T_{ind} (min) ± SD (min)	Nucleation rate (J^*, nuclei per m^3 per s)	T_{ind} (min) ± SD (min)	Nucleation rate (J, nuclei per m^3 per s)	Nucleation rate ratio (J/J^*)
Glycine (Gly)	1.17	379	211 ± 34	1.58	120 ± 30	2.78	1.76
	1.22	346	50 ± 11	6.67	45 ± 15	7.41	1.11
	1.33	671	17 ± 2	19.61	10 ± 5	33.33	1.70
Diglycine (Digly)	1.30	661	188 ± 44	1.77	60 ± 15	5.56	3.13
	1.40	838	80 ± 26	4.17	30 ± 15	11.11	2.67
	1.50	998	44 ± 9	7.58	15 ± 5	22.22	2.93

of the MSZW based on the data presented, thereby favouring heterogeneous nucleation in the presence of glass beads at the applied supersaturations.

3.2. Induction time of glycine and diglycine in the absence and presence of glass beads

Table 2 summarises the induction time of glycine and diglycine in both the presence and absence of glass beads. The extent of the change in induction in the absence and the presence of the heterosurface between both compounds varied considerably. As such, the glass beads display a positive yet discriminating influence over the nucleation of Gly and Digly. Table 2 also summarises the nucleation rate for both Gly and Digly in the absence and presence of glass beads (J^* and J, respectively), calculated according to eqn (1), previously reported by Mealey *et al.*,[59] along with the corresponding nucleation rate ratios (J/J^*).

$$J \text{ or } J^* = \frac{1}{t_{\text{ind}} V}, \tag{1}$$

where J or J^* is the nucleation rate (nuclei m^{-3} s^{-1}), t_{ind} is the induction time (s), and V is the volume of the crystallization solution (m^3).

When comparing the nucleation behaviour of Gly and Digly, the nucleation rate ratio clearly indicates that glass beads have a pronounced effect on the nucleation of Digly compare to Gly. One of the possible reasons for this change in the nucleation rate in presence of glass beads is the abundant presence of hydrogen bond accepting oxygen atoms on the silibead molecules, resulting in hydrogen bond formation between the heterosurface and the crystallising solutes. Both Gly and Digly are rich in hydrogen bond donors (HBDs), which present complementarity to the HBA of glass beads, thereby favouring the formation of hydrogen bonds. These hydrogen bonds facilitate the formation of clusters of Gly and Digly on the surface of the glass beads, resulting in the faster nucleation. Similar results were observed by Verma *et al.*[40] for the nucleation of various APIs in the absence and presence of microcrystalline cellulose (MCC). It was observed that the APIs with only HBAs could easily form hydrogen bonds with the HBD sites of MCC, which resulted in the stable cluster of API molecules on the surface of MCC, therefore enhancing the nucleation rate of the APIs possessing only HBAs. Fig. 3 presents the desupersaturation profile of Gly and Digly in the presence of glass beads. Irrespective of the supersaturation, the rates of the

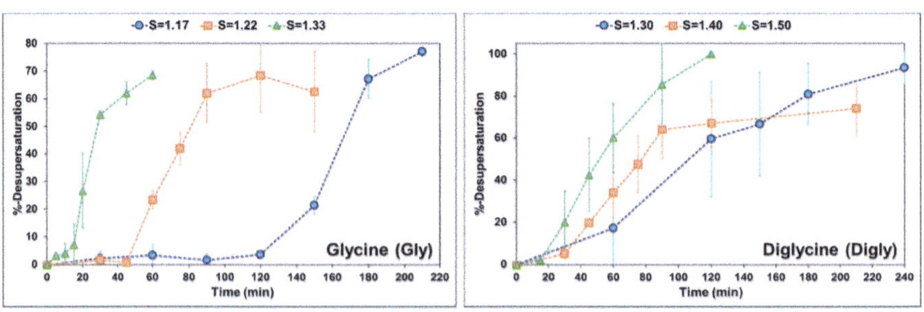

Fig. 3 %-Desupersaturation of glycine (left) and diglycine (right) in the presence of glass beads at the supersaturations indicated; volume = 50 mL.

desupersaturation of Gly and Digly aqueous solutions in the presence of the heterosurface are similar, as observed by the slope of the desupersaturation profiles. The standard deviations are slightly bigger due to the stochastic nature of the solution, as the volume is only 50 mL.

Despite similar molecular functionalities on Gly and Digly – amide and carboxylic groups – the extent of nucleation is different in each case. For example, the nucleation rate of Gly increased by more than 1.5 times at the lowest S of 1.17 in the presence of glass beads, whereas the corresponding increase in the case of Digly at its lowest S of 1.30 is more than 3-fold. This change in the nucleation rate of crystallising solute in the presence of the heterosurface, as shown in Table 2, is a clear indication that heterogeneous nucleation can be influenced by the interplay of supersaturation and the presence or absence of a heterosurface.

3.3. Solid state characterisation of isolated solids

Fig. 4 presents the powder X-ray diffraction (PXRD) patterns of the isolated solids after the full desupersaturation of Gly and Digly aqueous solutions in the presence of glass beads at the desired supersaturations. The PXRD patterns confirm the presence of the metastable α-form for Gly and the stable α-form for Digly in the presence of glass beads.

The solids isolated after the homogeneous and heterogeneous nucleation of Gly and Digly were observed under an inverted light microscope. Crystals (100 μm) of the metastable α-form of Gly were observed in the presence and absence of glass beads, while crystals (100 μm) of the stable α-form of Digly were crystallised in the absence and presence of the heterosurface, as seen in Fig. 5. Most of the Gly and Digly crystals seem to be associated with the silibead micro particles, supporting the arguments made in Section 3.2. Similar finding of drug crystals attached to the heterosurface have previously been reported extensively in the literature.[14,21,22,42]

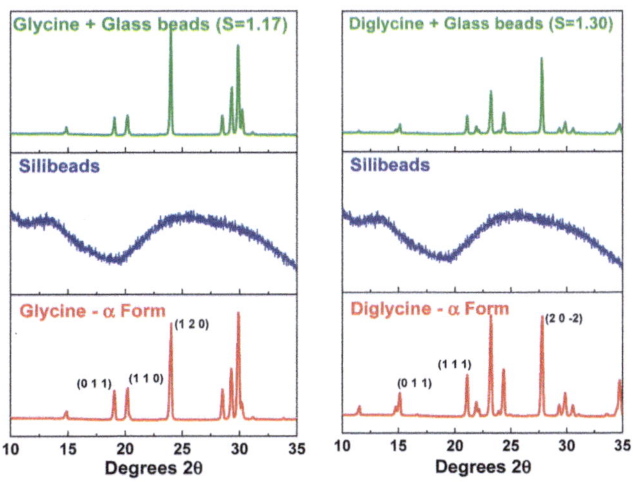

Fig. 4 Powder X-ray diffraction patterns of the composite solids isolated after the complete desupersaturation of both glycine (left) and diglycine (right) in the presence of glass beads at the indicated supersaturations, along with the glass beads and glycine and diglycine patterns.

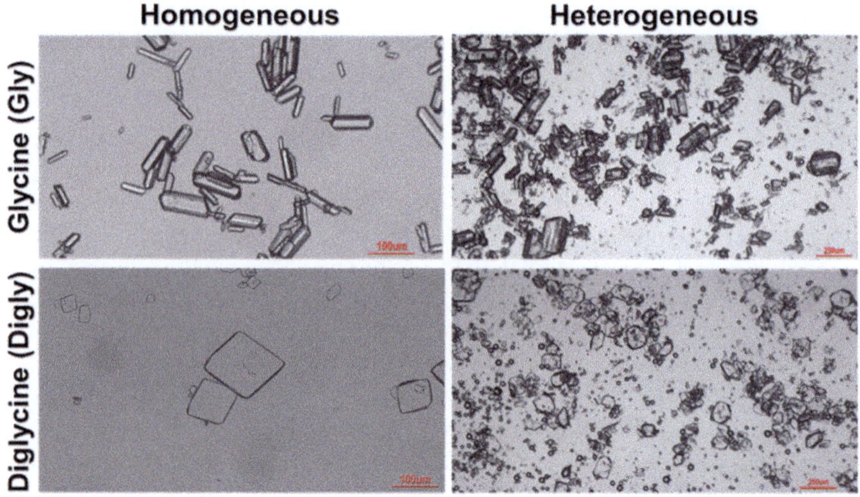

Fig. 5 Microscope images of the solids isolated after the homogeneous and heterogeneous nucleation of glycine and diglycine in the absence and presence of glass beads.

3.4. Effect of glass beads on the kinetic parameters of glycine and diglycine

The energy barrier associated with this desupersaturation is called the free energy barrier to nucleation (ΔG_c^*), which is mainly dependent on the temperature, supersaturation, and the interfacial energy at the crystal–solution interface. An equation for ΔG_c^*, which is based upon the classical nucleation theory, is shown in eqn (2).

$$\Delta G_c^* = \frac{16\pi\gamma^3 v_m{}^2 N_a}{3k^2T^2\ln^2 S}. \quad (2)$$

Here, γ is the interfacial energy, v_m is the molecular volume, N_a is Avogadro's constant, k is the Boltzmann constant, T is temperature (in Kelvin), and S is the supersaturation.

According to eqn (2), ΔG_c^* is inversely proportional to the supersaturation (S), therefore at low S, the ΔG_c^* is relatively large, and thus the possibility of the conversion of the solution to crystals is a rare event. A sufficiently large S is required to cross this energy barrier for nucleation to occur. The formation of nuclei has been postulated to occur either through a single step process of exchanging monomer units with other growing clusters in a structured way, or *via* a two-step process, whereby an unstructured aggregation of solute molecules is followed by the defined arrangement to form a crystal.[60]

The knowledge of supersaturation and the induction time of Gly and Digly in the absence and presence of a heterosurface were used to calculate the interfacial energy (γ^* and γ) and the pre-exponential factor (A^* and A) in the absence and presence of glass beads, respectively, for each compound. The kinetic and the thermodynamic parameters were derived using eqn (3) published by Kashchiev *et al.*[61] and used by the authors in the past.[14,42]

$$J(S) = AS\exp(-B/\ln^2 S) \quad (3)$$

where $J(S)$ = nucleation rate (nuclei per m^3 per s) at a given supersaturation, S, A = pre-exponential factor (nuclei per m per s), S = supersaturation, and

$$B = 16\pi v_0^{\,2}\gamma_{ef}^{\,3}/3(kT)^3, \tag{4}$$

where ν_0 = molecular volume (m^3) (for Gly: 9.94×10^{-29} m^3 and Digly: 1.45×10^{-28} m^3),[62] γ_{ef} = interfacial energy of the cluster/solution interface for heterogeneous nucleation (J m^{-2}), k = Boltzmann constant (J K^{-1}) (1.38×10^{-23} J K^{-1}) and T = crystallization temperature (K)

Eqn (3) can be re-written as follows:

$$\ln(J/S) = \ln A - (B/\ln^2 S) \tag{5}$$

Best fit linear equations, as presented in Fig. 6(A) and (B), were used to calculate γ^* and γ, and A^* and A for Gly and Digly crystallization systems in the absence and presence of glass beads, respectively, plotted using eqn (5). According to previously published literature, an interface "promotes nucleation by lowering the interfacial energy of aggregates",[32,63] but interestingly the data presented in Fig. 6(C) suggests that the presence of a heterosurface does not appreciably influence the interfacial energy of either the Gly or Digly crystallisation systems. Surprisingly, the major contributing factor to the heterogeneous nucleation of Gly and Digly is the pre-exponential factor (A^*). As reported earlier, we interpret A as the number of clusters of a size less than the critical radius which form per cubic metre of supersaturated solution each second.[14] Fig. 6(C) suggests that the presence of glass beads coincides with an increase in the pre-exponential (A^*) factor of *ca.* 2-fold for the crystallization of Gly and Digly.

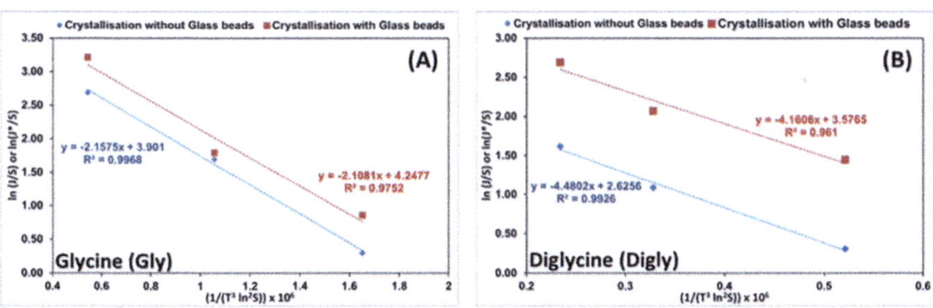

(C)

Compound	Crystallisation in the absence of Glass beads		Crystallisation in the presence of Glass beads		Nucleation rate ratio (J/J^*), (S)
	γ^* (mJ.m^{-2})	A^* (nuclei.m^{-3}.s^{-1})	γ^* (mJ.m^{-2})	A^* (nuclei.m^{-3}.s^{-1})	
Glycine (Gly)	3.25	49	3.1	51	1.76 (1.17)
Diglycine (Digly)	3.16	14	2.7	26	3.13 (1.30)

Fig. 6 (A and B) Plots of ln(J/S) or ln(J^*/S) against $(1/(T^3 \ln^2 S)) \times 10^6$, which illustrate the dependence of the nucleation rate on supersaturation for the crystallization of glycine (A) and diglycine (B) in the absence (blue diamond) and presence (red square) of glass beads at different supersaturations; the lines shown are the best linear fits and include the respective line equations and R^2 values. (C) Interfacial energies and pre-exponential factors for the crystallization of glycine and diglycine from aqueous solutions in the absence and presence of glass beads, and the related nucleation rate ratios at the lowest supersaturations.

These results are in support of the previous findings from the group, where the presence of MCC in the crystallisation system resulted in a minimal change to the interfacial energy of the crystallising solute but instead resulted in an increase of the pre-exponential factor by 2-fold or more.[14,42] Similar results were obtained by Heffernan *et al.*,[64] where the presence of structurally similar impurities (demethoxycurcumin (DMC) and bisdemethoxycurcumin (BDMC)) in the curcumin solution did not influence the interfacial energy of the crystallising solute, but instead changed the pre-exponential factor.[64] Another report by Bodnár *et al.*[65] also presents similar findings, where the presence of docusate sodium (DOSS) in a mefenamic acid solution in dimethylacetamide (DMA)–water mixtures resulted in an approximately 50% increase in the pre-exponential factor, while the interfacial energy remained uninfluenced. Hence, the nucleation of organics in the presence of heterosurfaces in either a dissolved or undissolved state is influenced by the change in the pre-exponential factor, while the interfacial energy remains uninfluential.

3.5. Analysis of the intermolecular interactions

The presence of hydrogen bond donor (HBD) or acceptor (HBA) sites on the organic molecules allow them to form inter- or intra-molecular hydrogen bonds. This hydrogen bonding network has been proven to influence the crystallisation kinetics,[14,22] as well as providing thermodynamic stability to the growing crystals.[66] The crystallising molecules in this study, glycine and diglycine, possess 1 and 2 HBDs, respectively, and 2 and 3 HBAs, respectively, as shown in Fig. 1 and Table 3. To quantify the contribution of HBAs and HBDs to the formation of the crystal structures of Gly and Digly, the logit hydrogen-bonding propensity (LHP) model was used, based on the statistical analysis of hydrogen bonds in the Cambridge Structural Database (CSD).[57] The hydrogen bond propensity (π) can be defined as a probability measure of an intermolecular hydrogen bond forming between two molecules within a crystal structure, with the assumption that the strongest possible donor–acceptor pairs will form this bond.[57] The values of π for Gly (0.98) and Digly (0.87) are reported in Table 3, suggesting that Gly has a slightly higher affinity to form hydrogen bonds with itself compared to Digly. This result is consistent with other work on glycine polymorphism in aqueous solutions, which has shown that α-glycine is constructed from zwitterionic dimers,[47] and that the presence of glycine dimers in solution (as a result of hydrogen bonding) is conducive to the preferential formation of α-glycine over γ-

Table 3 Summary of the number of hydrogen bond donor and acceptor groups of glycine and diglycine with their hydrogen bond propensity (π), the time required to add one molecule to a growing crystal (picoseconds), and nucleation rate ratio

Compound	Number of HBDs	Number of HBAs	Hydrogen bond propensity, π	Time required to add one molecule (ps)	Nucleation rate ratio (J/J^*), (S)
Glycine (Gly)	1	2	0.98	0.66	1.76 (1.17)
Diglycine (Digly)	2	3	0.87	2.01	3.13 (1.30)

glycine.[67] Comparing the values of π with the nucleation rate ratio (J/J^*), suggests that the solute with a larger value of π resulted in a lower nucleation rate ratio, and *vice versa*. Both Gly and Digly can form hydrogen bonds with the glass beads due to hydrogen bonding complementarity between the crystallising solute and the heterosurface. Considering the cumulative number of HBDs and HBAs on Gly and Digly, and the value of π, suggests that the amide ($-NH_2$) and carboxylic groups ($-C{=}O$) of Digly have a higher propensity to form hydrogen bonds with the oxygen molecules (–O–) of the glass beads compared to Gly. This, therefore, results in the comparatively faster nucleation rate of Digly to that of Gly in the presence of glass beads.

Heterogeneous nucleation is governed heavily by the lifetimes of the hydrogen bond interactions, as reported in past literature.[14,22,40,42] Previous literature has reported that the average lifetime of single molecules attached to a surface *via* hydrogen bonding ranges from <10 ns to <70 ns. A previous paper from the authors suggests that the lifetime of the hydrogen bond between a risperidone molecule and the surface of microcrystalline cellulose is about 30 ns, as computed by molecular dynamics (MD) simulations.[40] Table 3 also presents the time required (t_m) to add a single molecule to growing crystals of either Gly or Digly. These values were calculated from the growth phases of Fig. 3 for each of the APIs using eqn (6).[14]

$$t_m = \frac{t_g M_w}{\rho_{solid} V_p N_A}, \tag{6}$$

where t_g is the time required for a single crystal to grow to a certain size (s) (obtained from the growth phases of Gly and Digly in Fig. 3), M_w is the molecular mass of the crystallising solute (g mol^{-1}), ρ_{solid} is the density of the crystallising solute (g m^{-3}) (for Gly: 1.3×10^6 g m^{-3} and Digly: 1.5×10^6 g m^{-3}),[62] V_p is the volume of the solute particle (m^3) calculated using the particle diameter from the microscope images (particle diameter for Gly: 100 μm and Digly: 100 μm), and N_A is Avogadro's constant (6.023×10^{23} mol^{-1}).

The times required to add a single molecule to a growing crystal of Gly and Digly are 0.66 ps and 2.01 ps, respectively, which are much smaller than the average lifetime of the hydrogen bond (30 ns) between a molecule of the crystallising solute and the heterosurface. This also confirms that Gly is a fast-nucleating molecule compared to Digly. Alternatively, the adsorbed solute molecule can exist attached to a surface for a time scale which will allow the attachment of multiple solute molecules from the solution phase and facilitate the formation of stable nuclei and eventually fully grown crystals. However, in the solution phase, solute–solute interactions are much shorter-lived, increasing the difficulty of nucleus formation, as reported previously.[22] Therefore, the modest enhancement in the nucleation rate of Gly could be explained using the "lifetime effect", where once a Gly molecule adheres to the silibead surface *via* hydrogen bond formation, it remains adhered for a long enough period for other Gly molecules or clusters to attach to this silibead-bound Gly molecule. This would result in stable nuclei which subsequently grow into a well-formed crystal. Therefore, the heterogeneous nucleation observed for Gly in the presence of glass beads is postulated to be due to the "lifetime effect", whereas for Digly it is the combination of the "lifetime effect" and molecular complementarity.

4. Conclusions

The homogeneous and heterogeneous nucleation of glycine and diglycine in the absence and presence of glass beads were performed from aqueous solutions. It was observed that there was a modest increase in the nucleation rate of Gly compared to Digly. This confirms that glass beads displayed a positive yet discriminating effect on the heterogeneous nucleation of Gly and Digly. PXRD and microscope images confirmed the presence of the α-forms of both Gly and Digly after both the homogeneous and heterogeneous crystallisation processes. Experimental data suggests that the presence of glass beads did not dramatically influence the interfacial energies of Gly and Digly, but instead increased the pre-exponential factor by a factor of at least 2. Furthermore, molecular functionality analysis, hydrogen bond propensity calculations, and the time required to add a single molecule to the growing crystal of Gly and Digly all suggest that the heterogeneous nucleation of Gly is the consequence of the hydrogen bond "life-time effect", whereas for Digly it is a combination of this effect as well as molecular complementarity.

Conflicts of interest

There are no conflicts to declare.

Acknowledgements

The results incorporated in this paper have received funding from the European Union's Horizon 2020 research and innovation programme under the Marie Sklodowska-Curie grant agreement no. 101026339. KH thanks the Science Foundation Ireland (SFI) for support (award grant numbers 12/RC/2275_P2). JH and MG thank the EPSRC for funding SCoBiC (EP/N015916/1). HM is grateful for the financial support from the Engineering and Physical Science Research Council (EP/T518207/1 and EP/T005556/1).

References

1 Q. Li and Y.-S. Jun, The apparent activation energy and pre-exponential kinetic factor for heterogeneous calcium carbonate nucleation on quartz, *Commun. Chem.*, 2018, **1**(1), 56.

2 S. Karthika, T. K. Radhakrishnan and P. Kalaichelvi, A Review of Classical and Nonclassical Nucleation Theories, *Cryst. Growth Des.*, 2016, **16**(11), 6663.

3 C. Lindberg, A. Whiticar, K. A. Dick, N. Sköld, J. Nygård and J. Bolinsson, Silver as Seed-Particle Material for GaAs Nanowires—Dictating Crystal Phase and Growth Direction by Substrate Orientation, *Nano Lett.*, 2016, **16**(4), 2181.

4 Y. Xia, X. Xia and H.-C. Peng, Shape-Controlled Synthesis of Colloidal Metal Nanocrystals: Thermodynamic *versus* Kinetic Products, *J. Am. Chem. Soc.*, 2015, **137**(25), 7947.

5 J. Zeng, J. Huang, C. Liu, C. H. Wu, Y. Lin, X. Wang, S. Zhang, J. Hou and Y. Xia, Gold-Based Hybrid Nanocrystals Through Heterogeneous Nucleation and Growth, *Adv. Mater.*, 2010, **22**(17), 1936.

6 P. J. M. Smeets, K. R. Cho, R. G. E. Kempen, N. A. J. M. Sommerdijk and J. J. De Yoreo, Calcium carbonate nucleation driven by ion binding in a biomimetic matrix revealed by *in situ* electron microscopy, *Nat. Mater.*, 2015, **14**(4), 394.
7 A. Pei, G. Zheng, F. Shi, Y. Li and Y. Cui, Nanoscale Nucleation and Growth of Electrodeposited Lithium Metal, *Nano Lett.*, 2017, **17**(2), 1132.
8 S. Lau and L. A. Archer, Nucleation and Growth of Lithium Peroxide in the Li–O_2 Battery, *Nano Lett.*, 2015, **15**(9), 5995.
9 Y. Hu, B. Lee, C. Bell and Y.-S. Jun, Environmentally Abundant Anions Influence the Nucleation, Growth, Ostwald Ripening, and Aggregation of Hydrous Fe(III) Oxides, *Langmuir*, 2012, **28**(20), 7737.
10 Q. Li, C. I. Steefel and Y.-S. Jun, Incorporating Nanoscale Effects into a Continuum-Scale Reactive Transport Model for CO2-Deteriorated Cement, *Environ. Sci. Technol.*, 2017, **51**(18), 10861.
11 Y.-S. Jun, L. Zhang, Y. Min and Q. Li, Nanoscale Chemical Processes Affecting Storage Capacities and Seals during Geologic CO2 Sequestration, *Acc. Chem. Res.*, 2017, **50**(7), 1521.
12 J. J. A. Thomas, New Approach to Modeling the Nucleation and Growth Kinetics of Tricalcium Silicate Hydration, *J. Am. Ceram. Soc.*, 2007, **90**(10), 3282.
13 J. V. Parambil, S. K. Poornachary, R. B. H. Tan and J. Y. Y. Heng, Template-induced polymorphic selectivity: the effects of surface chemistry and solute concentration on carbamazepine crystallisation, *CrystEngComm*, 2014, **16**(23), 4927.
14 V. Verma, J. Zeglinski, S. Hudson, P. Davern and B. K. Hodnett, Dependence of Heterogeneous Nucleation on Hydrogen Bonding Lifetime and Complementarity, *Cryst. Growth Des.*, 2018, **18**(11), 7158.
15 P. W. Carter, A. C. Hillier and M. D. Ward, Nanoscale Surface Topography and Growth of Molecular Crystals: The Role of Anisotropic Intermolecular Bonding, *J. Am. Chem. Soc.*, 1994, **116**(3), 944.
16 P. W. Carter and M. D. Ward, Topographically directed nucleation of organic crystals on molecular single-crystal substrates, *J. Am. Chem. Soc.*, 1993, **115**(24), 11521.
17 P. W. Carter and M. D. Ward, Directing Polymorph Selectivity During Nucleation of Anthranilic Acid on Molecular Substrates, *J. Am. Chem. Soc.*, 1994, **116**(2), 769.
18 I. Weissbuch, L. Addadi, M. Lahav and L. Leiserowitz, Molecular Recognition at Crystal Interfaces, *Science*, 1991, **253**(5020), 637.
19 K. Chadwick, J. Chen, A. S. Myerson and B. L. Trout, Toward the Rational Design of Crystalline Surfaces for Heteroepitaxy: Role of Molecular Functionality, *Cryst. Growth Des.*, 2012, **12**(3), 1159.
20 V. López-Mejías, A. S. Myerson and B. L. Trout, Geometric Design of Heterogeneous Nucleation Sites on Biocompatible Surfaces, *Cryst. Growth Des.*, 2013, **13**(8), 3835.
21 V. Verma, R. V. G. Peddapatla, C. M. Crowley, A. M. Crean, P. Davern, S. Hudson and B. K. Hodnett, Experimental Study on the Influence of Excipients on the Heterogeneous Crystallization and Dissolution Properties of an Active Pharmaceutical Ingredient, *Cryst. Growth Des.*, 2018, **18**(1), 338.

22 R. Arribas Bueno, C. M. Crowley, P. Davern, B. K. Hodnett and S. Hudson, Heterogeneous Crystallization of Fenofibrate onto Pharmaceutical Excipients, *Cryst. Growth Des.*, 2018, **18**(4), 2151.
23 R. Arribas Bueno, C. M. Crowley, B. K. Hodnett, S. Hudson and P. Davern, Influence of Process Parameters on the Heterogeneous Nucleation of Active Pharmaceutical Ingredients onto Excipients, *Org. Process Res. Dev.*, 2017, **21**(4), 559.
24 J. V. Parambil, S. K. Poornachary, S. J. Hinder, R. B. H. Tan and J. Y. Y. Heng, Establishing template-induced polymorphic domains for API crystallisation: the case of carbamazepine, *CrystEngComm*, 2015, **17**(33), 6384.
25 H. Yang, C. L. Song, Y. X. S. Lim, W. Chen and J. Y. Y. Heng, Selective crystallisation of carbamazepine polymorphs on surfaces with differing properties, *CrystEngComm*, 2017, **19**(44), 6573.
26 F. J. Link and J. Y. Y. Heng, Enhancing the crystallisation of insulin using amino acids as soft-templates to control nucleation, *CrystEngComm*, 2021, **23**(22), 3951.
27 J. Aizenberg, A. J. Black and G. M. Whitesides, Control of crystal nucleation by patterned self-assembled monolayers, *Nature*, 1999, **398**(6727), 495.
28 H. B. Eral, V. López-Mejías, M. O'Mahony, B. L. Trout, A. S. Myerson and P. S. Doyle, Biocompatible Alginate Microgel Particles as Heteronucleants and Encapsulating Vehicles for Hydrophilic and Hydrophobic Drugs, *Cryst. Growth Des.*, 2014, **14**(4), 2073.
29 V. López-Mejías, J. W. Kampf and A. J. Matzger, Polymer-Induced Heteronucleation of Tolfenamic Acid: Structural Investigation of a Pentamorph, *J. Am. Chem. Soc.*, 2009, **131**(13), 4554.
30 L. Y. Pfund, C. P. Price, J. J. Frick and A. J. Matzger, Controlling Pharmaceutical Crystallization with Designed Polymeric Heteronuclei, *J. Am. Chem. Soc.*, 2015, **137**(2), 871.
31 J. Ling and K. Chadwick, Heterogeneous Crystallization Inside Microporous Polymer Particles as a Process Intensification Technology for the Manufacture of Drug Formulations, *Org. Process Res. Dev.*, 2017, **21**(6), 827.
32 Y.-X. Liu, X.-J. Wang, J. Lu and C.-B. Ching, Influence of the Roughness, Topography, and Physicochemical Properties of Chemically Modified Surfaces on the Heterogeneous Nucleation of Protein Crystals, *J. Phys. Chem. B*, 2007, **111**(50), 13971.
33 J. V. Parambil, S. K. Poornachary, J. Y. Y. Heng and R. B. H. Tan, Template-induced nucleation for controlling crystal polymorphism: from molecular mechanisms to applications in pharmaceutical processing, *CrystEngComm*, 2019, **21**(28), 4122.
34 Y. Diao, A. S. Myerson, T. A. Hatton and B. L. Trout, Surface Design for Controlled Crystallization: The Role of Surface Chemistry and Nanoscale Pores in Heterogeneous Nucleation, *Langmuir*, 2011, **27**(9), 5324.
35 Y. Diao, T. Harada, A. S. Myerson, T. Alan Hatton and B. L. Trout, The role of nanopore shape in surface-induced crystallization, *Nat. Mater.*, 2011, **10**(11), 867.
36 K. Chadwick, A. Myerson and B. Trout, Polymorphic control by heterogeneous nucleation - A new method for selecting crystalline substrates, *CrystEngComm*, 2011, **13**(22), 6625.

37 J. Ouyang, J. Chen, W. Chen, I. Rosbottom, M. Guo and J. Y. Y. Heng, Application of Phenyl-Functionalized Porous Silica for the Selective Crystallization of Carbamazepine Metastable Form II, *Ind. Eng. Chem. Res.*, 2021, **60**(2), 939.

38 J. Ouyang, X. Xing, J. Chen, L. Zhou, Z. Liu and J. Y. Y. Heng, Effects of solvent, supersaturation ratio and silica template on morphology and polymorph evolution of vanillin during swift cooling crystallization, *Particuology*, 2022, **65**, 93.

39 X. Li and J. Y. Y. Heng, Protein crystallisation facilitated by silica particles to compensate for the adverse impact from protein impurities, *CrystEngComm*, 2021, **23**(47), 8386.

40 P.-A. Cazade, V. Verma, B. K. Hodnett and D. Thompson, Extended Lifetime of Molecules Adsorbed onto Excipients Drives Nucleation in Heterogeneous Crystallization, *Cryst. Growth Des.*, 2021, **21**(4), 2101.

41 R. J. Davey, S. L. M. Schroeder and J. H. terHorst, Nucleation of Organic Crystals—A Molecular Perspective, *Angew. Chem., Int. Ed.*, 2013, **52**(8), 2166.

42 V. Verma, P. Bannigan, M. Lusi, C. M. Crowley, S. Hudson, B. K. Hodnett and P. Davern, The heterogeneous crystallization of a novel solvate of clozapine base in the presence of excipients, *CrystEngComm*, 2018, **20**(31), 4370.

43 C. L. Bull, G. Flowitt-Hill, S. de Gironcoli, E. Kucukbenli, S. Parsons, C. H. Pham, H. Y. Playford and M. G. Tucker, [zeta]-Glycine: insight into the mechanism of a polymorphic phase transition, *IUCrJ*, 2017, **4**(5), 569.

44 G. L. Perlovich, L. K. Hansen and A. Bauer-Brandl, The Polymorphism of Glycine. Thermochemical and structural aspects, *J. Therm. Anal. Calorim.*, 2001, **66**(3), 699.

45 X. Yang, J. Lu, X.-J. Wang and C.-B. Ching, Effect of sodium chloride on the nucleation and polymorphic transformation of glycine, *J. Cryst. Growth*, 2008, **310**(3), 604.

46 K. Srinivasan, Crystal growth of α and γ glycine polymorphs and their polymorphic phase transformations, *J. Cryst. Growth*, 2008, **311**(1), 156.

47 C. S. Towler, R. J. Davey, R. W. Lancaster and C. J. Price, Impact of Molecular Speciation on Crystal Nucleation in Polymorphic Systems: The Conundrum of γ Glycine and Molecular 'Self Poisoning', *J. Am. Chem. Soc.*, 2004, **126**(41), 13347.

48 S. K. Poornachary, J. V. Parambil, P. S. Chow, R. B. H. Tan and J. Y. Y. Heng, Nucleation of Elusive Crystal Polymorphs at the Solution–Substrate Contact Line, *Cryst. Growth Des.*, 2013, **13**(3), 1180.

49 G. Albrecht and R. B. Corey, The Crystal Structure of Glycine, *J. Am. Chem. Soc.*, 1939, **61**(5), 1087.

50 Y. Iitaka, The crystal structure of γ-glycine, *Acta Crystallogr.*, 1961, **14**(1), 1.

51 M. J. Vesga, D. McKechnie, P. A. Mulheran, K. Johnston and J. Sefcik, Conundrum of γ glycine nucleation revisited: to stir or not to stir?, *CrystEngComm*, 2019, **21**(13), 2234.

52 S. A. Moggach, D. R. Allan, S. Parsons and L. Sawyer, Effect of pressure on the crystal structure of [alpha]-glycylglycine to 4.7 GPa; application of Hirshfeld surfaces to analyse contacts on increasing pressure, *Acta Crystallogr., Sect. B: Struct. Sci.*, 2006, **62**(2), 310.

53 S. M. Clarke, B. A. Steele, M. P. Kroonblawd, D. Zhang, I. F. W. Kuo and E. Stavrou, An Isosymmetric High-Pressure Phase Transition in α-

Glycylglycine: A Combined Experimental and Theoretical Study, *J. Phys. Chem. B*, 2020, **124**(1), 1.
54 E. W. Hughes and W. J. Moore, The Crystal Structure of β-Glycylglycine, *J. Am. Chem. Soc.*, 1949, **71**(8), 2618.
55 J. Bonnin-Paris, S. Bostyn, J.-L. Havet and H. Fauduet, Determination of the Metastable Zone Width of Glycine Aqueous Solutions for Batch Crystallisations, *Chem. Eng. Commun.*, 2011, **198**(8), 1004.
56 M. Guo, H. Z. Chang, E. Liang, H. Mitchell, L. Zhou, Q. Yin, J. E. Guinn and Y. Y. J. Heng, The effect of chain length and side chains on the solubility of peptides in water from 278.15K to 313.15K: A case study in glycine homopeptides and dipeptides, *J. Mol. Liq.*, 2021, **352**, 118681.
57 P. T. A. Galek, L. Fabian, W. D. S. Motherwell, F. H. Allen and N. Feeder, Knowledge-based model of hydrogen-bonding propensity in organic crystals, *Acta Crystallogr., Sect. B: Struct. Sci.*, 2007, **63**(5), 768.
58 L. A. I. Ramakers, J. McGinty, W. Beckmann, G. Levilain, M. Lee, H. Wheatcroft, I. Houson and J. Sefcik, Investigation of Metastable Zones and Induction Times in Glycine Crystallization across Three Different Antisolvents, *Cryst. Growth Des.*, 2020, **20**(8), 4935.
59 D. Mealey, J. Zeglinski, D. Khamar and Å. C. Rasmuson, Influence of solvent on crystal nucleation of risperidone, *Faraday Discuss.*, 2015, **179**, 309.
60 J. De Yoreo James, U. P. A. Gilbert Pupa, A. J. M. Sommerdijk Nico, R. L. Penn, S. Whitelam, D. Joester, H. Zhang, D. Rimer Jeffrey, A. Navrotsky, F. Banfield Jillian, *et al.*, Crystallization by particle attachment in synthetic, biogenic, and geologic environments, *Science*, 2015, **349**(6247), aaa6760.
61 D. Kashchiev and G. M. van Rosmalen, Review: Nucleation in solutions revisited, *Cryst. Res. Technol.*, 2003, **38**(7–8), 555.
62 http://www.chemspider.com/, 2021.
63 V. López-Mejías, J. L. Knight, C. L. Brooks and A. J. Matzger, On the Mechanism of Crystalline Polymorph Selection by Polymer Heteronuclei, *Langmuir*, 2011, **27**(12), 7575.
64 C. Heffernan, M. Ukrainczyk, J. Zeglinski, B. K. Hodnett and Å. C. Rasmuson, Influence of Structurally Related Impurities on the Crystal Nucleation of Curcumin, *Cryst. Growth Des.*, 2018, **18**(8), 4715.
65 K. Bodnár, S. P. Hudson and Å. C. Rasmuson, Promotion of Mefenamic Acid Nucleation by a Surfactant Additive, Docusate Sodium, *Cryst. Growth Des.*, 2019, **19**(2), 591–603.
66 G. R. Desiraju, Crystal Engineering: From Molecule to Crystal, *J. Am. Chem. Soc.*, 2013, **135**(27), 9952.
67 Y. Yani, P. S. Chow and R. B. H. Tan, Glycine Open Dimers in Solution: New Insights into α-Glycine Nucleation and Growth, *Cryst. Growth Des.*, 2012, **12**(10), 4771.

Faraday Discussions

ROYAL SOCIETY OF CHEMISTRY

DISCUSSIONS

Understanding crystal nucleation mechanisms: where do we stand? General discussion

Michael W. Anderson, Matthew Bennett, Ruel Cedeno, Helmut Cölfen, Stephen J. Cox, Aurora J. Cruz-Cabeza, James J. De Yoreo, Rik Drummond-Brydson, Marta K. Dudek, Kristen A. Fichthorn, Aaron R. Finney, Ian Ford, Johanna M. Galloway, Denis Gebauer, Romain Grossier, John H. Harding, Alan Hare, Dezső Horváth, Liam Hunter, Joonsoo Kim, Yuki Kimura, Christine E. A. Kirschhock, Alexei A. Kiselev, Weronika Kras, Christian Kuttner, Alfred Y. Lee, Zhiyu Liao, Lucia Maini, Sten O. Nilsson Lill, Nick Pellens, Sarah L. Price, Ivo B. Rietveld, Jeffrey D. Rimer, Kevin J. Roberts, Jutta Rogal, Matteo Salvalaglio, Ilaria Sandei, Gábor Schuszter, Jan Sefcik, Wenhao Sun, Joop H. ter Horst, Marko Ukrainczyk, Alexander E. S. Van Driessche, Stéphane Veesler, Peter G. Vekilov, Vivek Verma, Thomas Whale, Helen P. Wheatcroft and Jacek Zeglinski

DOI: 10.1039/d2fd90021a

Kristen Fichthorn opened the discussion of the introductory lecture by James De Yoreo: I noted that in your potential of mean force (PMF) plots for various particle orientations, fluctuations occur such that different orientations could be preferred at a fixed distance. Could this affect the aggregation outcome?

James De Yoreo responded: That's a good point that I had not considered. The gradient of the PMF must impact the average trajectory at any given point. However, at least for the mineral systems we have simulated, both the gradients and the height of the barriers between adjacent minima favor trajectories in which the particles are correctly aligned. The implication is that those particles that happen to be correctly aligned at distances where the interparticle attraction begins to be felt (~100 nm) are most likely to reach the point of contact. Presumably, this also amounts to a torque that rotates particles into alignment as they approach.

Stephen J. Cox enquired: Thanks for the very nice talk. My question concerns the mechanism of dipolar interactions between these nanoparticles driving assembly. It seems you are looking at ZnO nanoparticles exposing their basal faces, which gives rise to the large dipole you have mentioned. How does the system support such a large dipole without incurring a large electrostatic energy penalty?

The basal face of ZnO is a classic type III polar surface in Tasker's classification,[1] and I would have thought that either the surface reconstructs to compensate the dipole, or ions from solution adsorb to the surface and screen the electrostatic interaction between particles.

1 P. W. Tasker, *J. Phys. C: Solid State Phys.*, 1979, **12**, 4977.

James De Yoreo replied: I do not know the answer to the question.

Wenhao Sun addressed James De Yoreo and Stephen J. Cox: Although polar surfaces and slabs are electrostatically unstable (based on Tasker's criterion of a Tasker Type 3 surface[1]), this is based on the argument of semi-infinite slabs, where the electrostatic energy diverges with slab thickness. Claudine Noguera has shown that the electrostatic energy of a nano-object does not necessarily diverge if the nanoparticle is small;[2] since I believe the electric field lines can close in on themselves.

1 P. W. Tasker, The stability of ionic crystal surfaces, *J. Phys. C: Solid State Phys.*, 1979, **12**(22), 4977.
2 C. Noguera and J. Goniakowski, Polarity in oxide nano-objects, *Chem. Rev.*, 2013, **113**(6), 4073–4105.

James De Yoreo replied: I agree. Another point to consider is the extent to which, in the case of particles in solution, electrolytes act to stabilize polarized surfaces.

Stephen J. Cox responded: Thanks for pointing out review article by Noguera and Goniakowski.

It is true that the divergence of the electrostatic energy refers to the limit of large crystals. The review highlights that the case for finite size particles is an even more complex affair than for semi-infinite slabs, for example, the electrostatic potential difference across the polar direction can either grow with the thickness of the crystal, or its width, depending on the geometry.

The review does not preclude polarity compensation at the nanoscale, and indeed, states that while polarity compensation mechanisms relevant to semi-infinite slabs may still be relevant, other mechanisms may also be at play.

Peter Vekilov asked: You referred to the several hundred debye dipole moment of some of the particles that attach to crystals during their growth as very large. Sickle cell hemoglobin (HbS) has dipole moment of about 450 D (ref. 1 and 2) and a size of about 6.5 nm. HbS molecules freely rotate and translate in solution driven by Browning collisions and they assemble into both sickle cell polymers and crystals. Do these considerations affect in any way the interpretation of the role of the dipole moment of the attaching crystallites in oriented attachment (OA)?

1 D. W. Rodgers, R. H. Crepeau and S. J. Edelstein, *Proc. Natl. Acad. Sci. U. S. A.*, 1987, **84**, 6157–6161.

2 O. Galkin, W. Pan, L. Filobelo, R. E. Hirsch, R. L. Nagel and P. G. Vekilov, *Biophys. J.*, 2007, **93**, 902–913.

James De Yoreo answered: I agree that, when compared to dipole moments of all other objects of similar size, several hundred Debeye is not anomalously large. In this case, my characterization of the dipole moment as being "large" was in respect to the magnitude of the interparticle force and torque it produces compared to the other interparticle forces at work in this system.

Peter Vekilov queried: In your very comprehensive talk, you discuss several examples of nonclassical crystal growth by the association of crystalline, liquid or amorphous precursors. Is this a common theme in the cases that you discussed? Can all these observations be summarized with one or a few sentences? What are the most significant recent achievements in the field of nonclassical growth mechanisms? What questions are still open and may be addressed, with our current tools, in the immediate future?

James De Yoreo responded: I do see some common themes. For the sake of brevity I will summarize them at a rather cursory level. The first is that one cannot divorce the observation of what I will generally refer to as crystallization by particle attachment (CPA) from supersaturation. This is, of course, pretty obvious, but, nonetheless, often ignored in the literature. The reason it matters is that the probability of forming clusters or nucleating particles scales strongly with supersaturation (exponential of the inverse supersaturation squared in the classical theory). Thus, one general theme is that strongly driven systems are more likely to exhibit CPA. The second general theme is that solubility matters. Solubilities can span nine orders of magnitude from molar-levels for salts like NaCl to nanomolar levels for oxides. However, the rate of growth by ion addition is proportional to the excess concentration ($C - C_e$), where C_e is the equilibrium concentration. The proportionality constant (the kinetic coefficient) only varies by approximately an order of magnitude over the full range of solubility and is generally smaller for systems with lower solubility. The consequence is that, even if the supersaturation $\sim\ln(C/C_e)$ is set to the same level (*i.e.*, $(C - C_e)/C_e$ approximately constant), the growth rate for a highly soluble phase will be greater by a factor that scales with the product of C_e and the kinetic coefficient. To take a concrete example, the ratio of the growth rate of KDP to that of hydroxyapatite is about 1010. That means that particle attachment rates would have to be 1010 times faster for KDP than for hydroxyapatite for the relative contribution of CPA to the growth rate to be similar. I've done the calculation for KDP at a typical supersaturation and found that particle attachment rates for ~5 nm particles would have to be many times per second, whereas, for hydroxyapatite, the required rate is 1010 times lower. The third theme is that size matters. Because the barriers to attachment tend to be elimination of the hydration layers, the larger the interface, the less likely that thermal energy alone will enable the particles to reach contact.

As far as a summary of all the observations in one or two sentences, I would say that is not possible at any useful level, because, even if one can state a general principle, as I have done above, the details matter and the number of possible outcomes is enormous. The situation is not very different than with nucleation.

The answer to what are the most significant recent achievements in studies of non-classical nucleation would take a great deal of writing to detail. I will highlight two. The first is the direct observation of non-classical pathways by high resolution methods, demonstrating: (i) order can develop gradually, (ii) amorphous clusters can be the stable form below a certain size and thus the system has two critical sizes, one for the amorphous phase and another for the crystalline phase that nucleates in the amorphous clusters, (iii) interfacial gradients drive near surface nucleation, thus biasing systems toward CPA. Some relevant references include ref. 1–3.

1 K. C. Cao, J. Biskupek, C. T. Stoppiello, R. L. McSweeney, T. W. Chamberlain, Z. Liu, K. Suenaga, S. T. Skowron, E. Besley, A. N. Khlobystov and U. Kaiser, Atomic mechanism of metal crystal nucleus formation in a single-walled carbon nanotube, *Nat. Chem.*, 2020, **12**(10), 921–928, DOI: 10.1038/s41557-020-0538-9.
2 L. Houben, H. Weissman, S. G. Wolf and B. Rybtchinski, A mechanism of ferritin crystallization revealed by cryo-STEM tomography, *Nature*, 2020, **579**(7800), 540–543, DOI: 10.1038/s41586-020-2104-4.
3 G. Zhu, M. L. Sushko, J. S. Loring, B. A. Legg, M. Song, J. A. Soltis, X. Huang, K. M. Rosso and J. J. De Yoreo, Self-similar mesocrystals form via interface-driven nucleation and assembly, *Nature*, 2021, **590**(7846), 416–422, DOI: 10.1038/s41586-021-03300-0.

Jeffrey Rimer asked: Regarding the observation of high Fe^{3+} concentrations in solution in close proximity to crystal surfaces, is this observation consistent with the Boltzmann distribution of counterions near negatively-charged interfaces? Are the concentrations higher than what would be predicted by this theory, which is typically used (along with DLVO theory) to predict the density of ions in the diffuse double layer surrounding charged particles.

James De Yoreo replied: The largest contributor to the distribution is the ion–ion correlation term, which I do not believe is included in the DLVO prediction. The details are contained in the supplementary information of ref. 1.

1 G. Zhu, *et al.*, Self-similar mesocrystals form via interface-driven nucleation and assembly, *Nature*, 2021, **590**(7846), 416–422, DOI: 10.1038/s41586-021-03300-0.

Jan Sefcik commented: It is reasonable to expect that dispersion (vdW) interactions between the solid interface and molecules in solution may be an important factor behind concentration profiles at solid–liquid interfaces. Specific effects of organic ligands may be important but there are always dispersion interactions between solution molecules and the solid phase behind the interface. Dispersion interactions would favour one mixture component over another, leading to concentration enhancement of one of the solution components at the interface compared to the bulk concentration far from the interface (see ref. 1).

1 D. McKechnie, *et al.*, *J. Phys. Chem. Lett.*, 2020, **11**(6), 2263–2271.

James De Yoreo replied: Yes, it is absolutely the case that dispersion forces impact concentration gradients. Moreover, the related ion–ion correlation forces, which arise from density fluctuations in electrolyte solutions, impact these concentration gradients and, in a number of the simulations we have done, come to dominate the interparticle interactions as well. However, the observation that organic functionalization of the surfaces is present in all three cases where near-surface nucleation or nucleation at a surface in undersaturated conditions is present, suggests that the addition of charged surfaces is an important factor.

Stéphane Veesler questioned: My question concerns your slide 27. You spoke about modifying the surface energy and/or supersaturation (*via* a solubility decrease). I did not understand what you meant and how you managed to do that.

James De Yoreo answered: What we know is that the nucleation rate is greatly enhanced in the region just near the interface – say within 2 nm. As we do not see any clusters aggregating to form the new particles, we assume that they nucleate classically, that is, ion-by-ion. There are two factors that could enhance the nucleation rate over that of the bulk. The first is the surface energy and the second is the supersaturation. With regard to the surface energy, because the chemistry of the solution and the structure of the solvent are different near the interface, we can imagine that the surface energy is also different. With regard to the supersaturation, we cannot assume that the chemical potential with respect to bulk hematite is different, because the chemical potential must be the same everywhere. It is the need to have uniform chemical potential that leads to non-uniform ion distributions in the first place. However, the solubility of hematite may not be the same in this region of structured solvent and distinct chemistry. So even if the chemical potential of Fe^{3+} relative to bulk hematite in bulk solution is uniform, the chemical potential relative to hematite in that interfacial region may be different. We do not know which of these is correct, or if there is another explanation. These are hypotheses.

Christian Kuttner enquired: You explained that DLVO theory is fundamental to crystallization. Could you please elaborate on the importance of the secondary minimum and how well it is understood about how it can be manipulated in order to control the process of aggregation?

James De Yoreo responded: My comment about DLVO theory was in recognition that CPA is largely a manifestation of colloidal physics with the added complexity of an atomically heterogeneous surface, which imposes structure on the solvent itself. Thus, DLVO theory is a reasonable starting point. The secondary minimum that emerges from DLVO theory provides a place where the approaching particle will have some residence time until it overcomes the intervening barrier to the primary minimum (*i.e.*, contact). During that residence, the particle may have time for Brownian rotation to take it to the minimum energy orientation, which the MD simulations predict is the one for which the particles are crystallographically aligned. From the form of the potential in DLVO theory, one would conclude that the way to manipulate the minimum and the barrier is through the electrolyte concentration and/or strength to alter the electrostatic repulsion, and the solvent to alter the dielectric constant and therefore the van der Waals attraction.

Rik Drummond-Brydson asked: In terms of the gaps between nucleating/growing particles (ZnO I think?) and the suggestion that the formation of necks between particles indicates higher supersaturation in that region, how does that square with the indication the structure of the solvent may be more ordered in the interparticle region? Could the necks form as a result of surface diffusion of say a hydrated surface layer (similar to that seen in particle sintering in the absence of liquid)?

James De Yoreo replied: The necking is not a consequence of surface diffusion causing sintering. We know this because we track the particle shape and the conclusion is that the particles are growing during this process. If they were deforming due to surface diffusion of atoms into the gap, the volume would not increase.

John Harding opened discussion of the paper by Denis Gebauer: The dipole moment of D_2O (1.8506 debye) is marginally greater than that of H_2O (1.8479 debye) – see ref. 1. However the dielectric constant of D_2O (78.06) is less than that of water (78.39) at 25 °C.[2] This suggests that the screening of electrostatic forces by the solvent would be (slightly) less for D_2O than for H_2O and so the stability of ion associates should be slightly greater in D_2O as the experiments suggest. Also, although the fact that the rates of amorphous calcium carbonate (ACC) precipitation are similar for D_2O and H_2O could be explained by saying that ion dehydration and deprotonation have negligible energetic barriers, an alternative explanation (and to my mind more probable one) is that the behaviour in D_2O and H_2O is very similar and the energetic barriers (whatever their absolute value may be) are very similar also.

1 S. A. Clough, *et al.*, *J. Chem. Phys.*, 1973, **39**, 2254.
2 G. A. Vidulich, *et al.*, *J. Phys. Chem.*, 1967, **71**, 656.

Denis Gebauer responded: It is indeed peculiar that some physical chemical parameters for H_2O and D_2O are only marginally different, while the behaviour of various solutes, proteins, and also minerals in the two solvents appear to be strongly affected (see the Introduction of our paper (https://doi.org/10.1039/d1fd00078k) for various references). Having said that, I agree that based on the values for the dielectric constants, ion association should be somewhat weaker in heavy than in light water. However, this viewpoint seems to neglect, *e.g.*, entropic effects arising from distinct solvent structures, as discussed in our paper. Moreover, values for the distinct polarizabilities of the waters seem to suggest a different trend in terms of screening effects.[1]

We did not determine the rates for ACC precipitation in the distinct solvents, however, the varying times of nucleation (Fig. 5 in our paper; https://doi.org/10.1039/d1fd00078k) suggest that there is a considerable effect on barriers. Since previous studies revealed that the ACC forms *via* the dehydration and solidification of dense liquid precursors,[2] this observation suggests that the solvent change affects dehydration barriers rather than nucleation barriers though.

1 M. N. Rodnikova, *J. Mol. Liq.*, 2007, **136**, 211–213.
2 J. T. Avaro, S. L. P. Wolf, K. Hauser and D. Gebauer, *Angew. Chem., Int. Ed.*, 2020, **59**, 6155–6159.

Romain Grossier enquired: In describing ACC, you use the term "amorphous polymorphism in ACCs", but polymorphism relates to solid crystalline phases, not amorphous ones, which also share the same chemical composition. As "ACCs" neither have crystalline structure, nor share the same chemical composition(?), using "polymorphism" for such phases is confusing. Would there be another word that could be use? Just "amorphous phases"?

Denis Gebauer replied: I agree that the term amorphous polymorphism (or polyamorphism) may be confusing.

The term was introduced by the Ukrainian scientists Palatnik *et al.*,[1,2] who wrote: "By analogy with the phenomenon of polymorphism, well known for crystalline materials, we will refer to the phenomenon of the existence of several such varieties of the amorphous state of the same substance as polyamorphism".

We have discussed the details of ACC polyamorphism in a previous review article.[3] Since the composition of ACC does not vary for the different forms of ACC, the use of the term polyamorphism seems justified.

In previous papers, we called the distinct forms "proto-structured" ACCs.[4] In the case of biogenic ACCs, the occurrence of distinct structures has been known for decades, using the notion of, *e.g.*, "vateritic" ACC.[5]

1 L. S. Palatnik, Y. A. Bykovskii, P. A. Panchekha, A. G. Dudoladov, V. I. Verchenko and S. V. Marun'ko, *Sov. Phys. Dokl.*, 1980, **25**, 770–772.
2 L. S. Palatnik, Y. A. Bykovskii, P. A. Panchekha, A. G. Dudoladov, V. I. Verchenko and S. V. Marun'ko, *Dokl. Akad. Nauk SSSR*, 1980, **254**, 632–635.
3 J. H. E. Cartwright, A. G. Checa, J. D. Gale, D. Gebauer and C. I. Sainz-Díaz, *Angew. Chem., Int. Ed.*, 2012, **51**, 11960–11970.
4 D. Gebauer, P. N. Gunawidjaja, J. Y. P. Ko, Z. Bacsik, B. Aziz, L. J. Liu, Y. F. Hu, L. Bergström, C.-W. Tai, T.-K. Sham, M. Edén and N. Hedin, *Angew. Chem., Int. Ed.*, 2010, **49**, 8889–8891.
5 L. Addadi, S. Raz and S. Weiner, *Adv. Mater.* 2003, **15**, 959–970.

Ivo B. Rietveld addressed Denis Gebauer and Romain Grossier: This is a remark in reference to the previous question asked by Romain Grossier, in which he referred to poly-amorphism and the issues related to its definition.

The semantic part of the problem is that amorphism means "no form", polyamorphism is therefore already a contradiction in terms as you cannot count something that is not.

But the scientific issue here goes deeper and is definitely not resolved. Polymorphs are different crystalline phases, with different energetic fingerprints, therefore it will always be possible to pinpoint a transformation of one form into another. When it comes to amorphous or glassy systems, each single one of them will be different from another depending on its history, the point being that to go from one amorphous system to another one can imagine the existence of a continuum of slightly different amorphous layouts connecting all of the amorphous appearances by slightly shifting molecules around into different conformations. This will simply not be possible between crystalline phases, because with a shift of a molecule the unit cell breaks down and a phase change must occur. For an amorphous system a slight shift of a molecule just gives another amorphous system and so on until one "specific" (I would like to say "specific" within our imagination) amorphous phase has been reached. This does not mean that within an amorphous system shifts cannot occur with visible or measurable volume and energy changes, because also volume and energy will be continuums and for example ageing of glassy systems is known to show "stabilisation" or shifts towards lower energy states. But as long as these states are all interconnected by slight – infinitesimal if you like – changes (of energy, volume, molecular positions), I don't see the necessity to speak of poly-amorphism; it's a continuum.

There may be an exception though, and that is where the species making up the amorphous material changes, such as changes in charge distributions, hydrogen positions, molecular changes *etc.* (which brings us into another contentious area as we can judge from the salt/co-crystal discussion). In that case, if the identity of the species changes, thermodynamically it will not be possible to speak of the same phase. But should we call this poly-amorphism? I'm not sure.

Denis Gebauer replied: There are many open questions in the rather young field of amorphous polymorphism. From a crystallographic point of view, atomic long-range orders give rise to well-defined crystal shapes and facets on the macroscopic scale. This correspondence principle of atomic structure and macroscopic shape does not occur in amorphous structures; however, short-range structures can certainly be distinct. There is no "absence of structure" in any matter. While the proposed notion of a continuum of amorphous structures may be true in some cases, there is strong evidence speaking against continuous changes of amorphous (short-range) structures for several examples. In amorphous ices, abrupt phase transitions between distinct forms are observed at well-defined pressures and temperatures.[1–3] There is no continuum, but rather a sudden change in properties (*i.e.*, density), and thus, in the structures, even though they are "amorphous". A related observation can be made in ACCs. For the different proto-structured forms (proto-aragonite,[4] proto-calcite,[5] proto-vaterite[5]), the solubilities depend on temperature, as is the case for crystalline forms. However, ACC polyamorphism is controlled by temperature and pH, whereby the solubilities of the distinct polyamorphic forms do not change continuously. They change discontinuously and abruptly between characteristic values,[6,7] as is also the case for crystals.

1 T. Bartels-Rausch, V. Bergeron, J. H. E. Cartwright, R. Escribano, J. L. Finney, H. Grothe, P. J. Gutiérrez, J. Haapala, W. F. Kuhs, J. B. C. Pettersson, *et al.*, *Rev. Mod. Phys.*, 2012, **84**, 885–944.
2 O. Mishima, *Proc. Jpn. Acad., Ser. B*, 2010, **86**, 165–175.
3 T. Loerting and N. Giovambattista, *J. Phys.: Condens. Matter*, 2006, **18**, R919–R977.
4 M. Farhadi Khouzani, D. M. Chevrier, P. Zhang, N. Hedin and D. Gebauer, *Angew. Chem., Int. Ed.*, 2016, **55**, 8117–8120.
5 D. Gebauer, P. N. Gunawidjaja, J. Y. P. Ko, Z. Bacsik, B. Aziz, L. J. Liu, Y. F. Hu, L. Bergström, C.-W. Tai, T.-K. Sham, M. Edén and N. Hedin, *Angew. Chem., Int. Ed.*, 2010, **49**, 8889–8891.
6 D. Gebauer, A. Völkel and H. Cölfen, *Science*, 2008, **322**, 1819–1822.
7 J. T. Avaro, S. L. P. Wolf, K. Hauser and D. Gebauer, *Angew. Chem., Int. Ed.*, 2020, **59**, 6155–6159.

Marta K. Dudek asked: Can you comment on the possibility to use other deuterated solvents to redirect/influence crystallization? Do you think it is possible to use deuterated solvents as a route to elusive polymorphs?

Denis Gebauer answered: In my opinion, it is likely that other deuterated solvents can also have distinct effects on crystallization pathways, and it is certainly possible that their utilisation may open up novel routes to elusive polymorphs. For calcium carbonate formation in light *vs.* heavy water, we do have some indications for the precipitation of some novel form in the latter solvent (see https://doi.org/10.1039/d1fd00078k). However, as we have not studied such effects beyond this system, this is certainly merely speculation on the general, potential use of deuterated solvents in crystallisation.

Wenhao Sun remarked: To comment on polyamorphism: there is work in thin-film deposition that shows that you can measure the pair distribution function of amorphous films deposited under different conditions; and that there can be two distinct forms of amorphous regions (indicated by different pair distribution patterns). Then, when this is annealed, it anneals into two different polymorphs. See ref. 1.

I have a question for Denis Gebauer; in your paper, you wrote that $CaCO_3$ nucleation is 'more classical' in D_2O than it is in H_2O. Can you explain what you mean by it being 'more classical', and more generally speaking, can you give us your perspective on what non-classical nucleation means? I think of all the people in the world, your perspective on this matter is of greatest interest.

1 K. H. Stone, *et al.*, Influence of amorphous structure on polymorphism in vanadia, *APL Mater.*, 2016, **4**(7), 076103.

Denis Gebauer responded: Thanks for your comment on polyamorphism in thin-films.

For a clarification of what we mean by "more classically" in the context of our paper (https://doi.org/10.1039/d1fd00078k), I will answer later.

I actually believe that the term "non-classical" nucleation should be avoided, as it is better to describe phenomena by what they are, rather than by what they are not. I thus apologize for using this poorly defined term in our paper (https://doi.org/10.1039/d1fd00078k), and will try to elucidate what I actually mean by this in the following. Generally speaking, in my opinion, "non-classical" nucleation means that the pathway from solution to solids does not depend on the formation of a critical nucleus (*i.e.*, as defined within Classical Nucleation Theory, CNT), as a required, fundamental transition state. Thus, "non-classical" processes can involve growth units that are (significantly) larger than monomers, as opposed to CNT and two-step nucleation theory (2S). In other words, "non-classical" nucleation precursors, pre-nucleation clusters (PNCs), are thermodynamically stable (*i.e.*, negative standard free energy of formation) and exist at significant populations as solute species in homogeneous solutions. From the point of view of CNT, it is then a conundrum how these species can participate in phase separation, as they are energetically further away from the required transition state than the monomers,[1] and they are thus neglected. In "non-classical" nucleation, there is an alternative reaction channel that explains the role of stable PNCs, as the event of phase separation does not depend on overcoming a certain size of the clusters but rather on a change in cluster dynamics.[2] Recently, these notions have allowed the development of a quantitative model for the so-called pre-nucleation cluster pathway,[2] which offers both predictive and explanatory powers for the complex phase behavior of the aqueous calcium carbonate system.[3]

1 D. Gebauer, P. Raiteri, J. D. Gale and H. Cölfen, *Am. J. Sci.*, 2018, **318**, 969–988.

2 D. Gebauer, M. Kellermeier, J. D. Gale, L. Bergström and H. Cölfen, *Chem. Soc. Rev.*, 2014, **43**, 2348–2371.
3 J. T. Avaro, S. L. P. Wolf, K. Hauser and D. Gebauer, *Angew. Chem., Int. Ed.*, 2020, **59**, 6155–6159.

Wenhao Sun commented: I want to emphasize that we, as a community, should be more careful when we use the word 'stable'. As Gibbs defines it, stable phase and equilibrium phase are *not* the same thing. Stable means that the internal energy surface, $U(X)$, is convex; meaning that the second derivative of U with respect to extensive variables (volume, entropy, composition) is positive. On the other hand, the equilibrium phase means that the phase is the lowest free-energy phase under those conditions. Metastability means that you can be 'stable', but not the lowest free-energy equilibrium phase. Sometimes I see people use the terminology 'stable' to indicate persistence of metastable phases, this usage would not be correct. Denis: My question to you is that under D_2O, we see that the titration curves are slightly flatter than for H_2O. Does this indicate fewer free nucleation clusters?

Denis Gebauer responded: I fully agree with the comment. In addition, there is also kinetic stability (perhaps, better, stabilization), which can add to the confusion. When it comes to thermodynamic stability of pre-nucleation clusters (PNCs), we have used the definition of Gibbs within the constraints of a homogeneous system (as PNCs are solute species that form independent of supersaturation, given by a negative standard free energy of formation ΔG_0). Please also remember that for the definition of the stability of phases, suitable boundary conditions need to be applied. As a hyperbole example, remember that all matter is eventually metastable with respect to some iron isotope.

Regarding your question on the pre-nucleation slopes of the titrations, a flatter slope indicates that fewer free ions are present, *i.e.*, more ions are bound, giving rise to a larger equilibrium constant K of cluster formation, and with $\Delta G_0 = -RT \ln K$ a more negative standard free energy of the clusters. Thus, the flatter slopes show that the PNCs become more stable (given a constant pH).

Aaron R. Finney enquired: Can the authors expand on the description 'more-classical'? If the nucleation of $CaCO_3$ in light water is non-classical, how do the authors reconcile their description of the nucleation in heavy water? Are they suggesting that both CNT-type and non-classical nucleation mechanisms contribute to phase separation according to a relative shifting of the binodal? (They write, "and nucleation of vaterite might occur directly, *i.e.*, 'classically'.") The non-classical nature of $CaCO_3$ nucleation was based partly on a multiple-binding model and the fact that the equilibrium constant for ion association is much bigger than 1 in light water; this is also very much the case in heavy water from the calculations presented. In that case, the model still predicts a population of cluster species before nucleation – are these simply spectator species if the system phase separates more classically, or is it the case that the calculations are performed at conditions that don't necessarily match with the experiments? In 'On classical and non-classical views on nucleation'[1] the authors write, "CNT is not a universal framework that can be employed for solute speciation, be it in under- or supersaturated solutions: It already fails at the ion pair" and "In our opinion, this idea – that is, that the CNT-type critical nucleus is not essential as a transition state for nucleation – is a major criterion rendering nucleation theories truly 'non-classical'." In light of this, it would be helpful to clarify the use of terms used to describe the pathways.

1 D. Gebauer, P. Raiteri, J. D. Gale and H. Cölfen, *Am. J. Sci.*, 2018, **318**, 969–988.

Denis Gebauer replied: First, I would like to emphasize that we have phrased the statement that calcium carbonate formation might proceed "more classically" in heavy water very carefully. A lot of additional work has to be done to actually show this, and we did mention also alternative explanations for our observation, *e.g.*, that ACC may be simply more transient and was thus not detected by us, but might still occur as an intermediate, formed from PNCs, to vaterite in presence of heavy water.

As we have discussed in the introduction section of our paper (https://doi.org/10.1039/d1fd00078k), it might be expected that the classical barrier towards nucleation may be reduced in heavy water. On the other hand, our computer simulations indicate that ion association becomes weaker in heavy water, whereas the cluster size distribution decays quicker than in light water. Taken together (lower barrier for classical nucleation, weakened ion association towards an alternative reaction channel), this could mean that, indeed, stable ion associates might be merely spectators, and might not participate in the nucleation of crystals in heavy water. Still, CNT fails to explain the thermodynamics of stable ion associates, but this would be no issue if these actually did not participate in nucleation. In light of this, the quote from ref. 1 actually addressed contrary claims of Henzler *et al.*[2] that CNT would accurately describe populations of stable ion pairs and clusters. As demonstrated in ref. 1, this notion is not sustainable. Please also see our earlier reply to the question from Wenhao Sun when it comes to a clarification of the use of terms, *i.e.*, "classical" *vs.* "non-classical".

1 D. Gebauer, P. Raiteri, J. D. Gale and H. Cölfen, *Am. J. Sci.*, 2018, **318**, 969–988.
2 K. Henzler, E. O. Fetisov, M. Galib, M. D. Baer, B. A. Legg, C. Borca, J. M. Xto, S. Pin, J. L. Fulton, G. K. Schenter, N. Govind, J. I. Siepmann, C. J. Mundy, T. Huthwelker and J. J. De Yoreo, *Sci. Adv.*, 2018, **4**, eaao6283.

Peter Vekilov remarked: You emphasized that CNT does not account for the role of solvent in nucleation. In fact, it does. Surface tension, the most important parameter in CNT, is strongly solvent dependent. The interactions between the solute molecules, which determine the supersaturation, the other crucial parameter of CNT, also strongly depend on the solvent chemistry. The activation barrier for attachment of molecules to the nucleus, a part of the kinetic part of CNT, is governed by the detailed structures of the solvent that erect around both the nucleus and the incoming solvent molecules.

Denis Gebauer responded: We fully agree. We have described the effects of the solvent on CNT parameters in exactly the same manner in our paper (https://doi.org/10.1039/d1fd00078k) and in our talk as Prof. Vekilov summarizes here. Perhaps, the hyperbole statement made in our talk towards the lack of an explicit treatment of solvation in CNT has caused this misunderstanding.

Ruel Cedeno questioned: In the experimental measurement of the ion association constant, the pH was set to basic conditions, while in simulations, the solvent essentially assumed a neutral pH (using the flexible SPC model).

Given the sensitivity of calcium carbonate to pH,[1] do you think the disagreement in terms of ion association behavior between the computational and experimental data might have also originated from this discrepancy in pH? Would it be useful to employ reactive forcefields which allows bond-breaking and bond-forming of water to account for this?[2,3]

1 D. J. Tobler, J. D. Rodriguez Blanco, H. O. Sørensen, S. L. S. Stipp and K. Dideriksen, *Cryst. Growth Des.*, 2016, **16**, 4500–4508, DOI: 10.1021/acs.cgd.6b00630.
2 J. D. Gale, P. Raiteri and A. C. T. van Duin, *Phys. Chem. Chem. Phys.*, 2011, **13**, 16666–16679, DOI: 10.1039/C1CP21034C.
3 W. Zhang and A. C. T. van Duin, *J. Phys. Chem. B*, 2017, **121**, 6021–6032, DOI: 10.1021/acs.jpcb.7b02548.

Denis Gebauer answered: We agree that pH is an important parameter in calcium carbonate formation, and the difference in pH between simulations and experiments may in part be a reason for the disagreement. However, please note that the pH in the simulations can be assumed to be basic rather than neutral, as the excess of carbonate (without added bicarbonate) would correspond to a pH value of around 14. Having said that, we did not add OH^- ions, which would probably start to interfere with calcium carbonate formation at this pH level.

Please note that we also performed simulations at pH 8.5 (again, as given by the carbonate/bicarbonate ratio) to investigate the system's pH dependence, resulting in, *e.g.*, more elongated clusters than at pH 14 and a shift towards smaller clusters.

A reactive force field, if it represents the aqueous calcium carbonate system correctly, would be great to study these systems. However, the slow speed of these force fields heavily counters their usefulness. While we already struggle to reach realistic solution conditions with an atomistic model, these seem to be yet out of reach employing a reactive force field.

Sten O. Nilsson Lill asked: In the MD-simulations, did you allow the O–H or O–D bonds to change upon interactions or were they kept fixed? Did you use any particular algorithm, for example SHAKE or something similar?

Denis Gebauer replied: We used the SPC/fw model.[1] This means the bonds are unbreakable and not interchangeable. However, they are not kept at a fixed length like in the SPC/E model, but can oscillate freely following a harmonic potential. The same holds true for the angle. So, no SHAKE algorithm, *etc.*, was used.

Details about the force field can be found in ref. 2 in the supplementary information.

1 Y. Wu, H. L. Tepper and G. A. Voth, *J. Chem. Phys.*, 2006, **124**, 024503.
2 P. Raiteri, R. Demichelis and J. D. Gale, *J. Phys. Chem. C*, 2015, **119**, 24447–24458.

Alexei Kiselev enquired: The effects of H_2O/D_2O mixtures on the crystallization of calcium carbonate presented in your paper (https://doi.org/10.1039/d1fd00078k) are really fascinating and open a whole new world of interesting problems. My initial question was going to be about the specific role of HDO in the crystallization process, but now I realize that this issue could require another paper or a series of papers to answer this question. So I reformulate my question in the following way: when the mixtures of H_2O/D_2O are prepared, did you check if the concentrations of the intermediate species (HDO, H_2O, D_2O, DH_2O^+, HD_2O^+, DO^-, HO^-, *etc.*) have reached equilibrium? I am asking because the Bernal and Fowler model of water expects the lifetime of a water molecule to be on the timescale of several hours, so one would expect that the equilibrium concentration of HDO could require a very long time to achieve.

1 J. D. Bernal and R. H. Fowler, A theory of water and ionic solution, with particular reference to hydrogen and hydroxyl ions, *J. Chem. Phys.*, 1933, **1**, 515–548, DOI: 10.1063/1.1749327.

Denis Gebauer answered: Thanks for your positive comment. To answer your question: no, we did not verify whether or not the light/heavy water mixtures had equilibrated prior to our measurements (though they likely were equilibrated, as we used stock mixtures and did not prepare the solutions for each experiment freshly). While I think that equilibration of the mixtures and corresponding potential effects have to be taken into account in future studies, I am not sure whether or not the Bernal and Fowler model is consistent with the (later?) measurements of proton diffusion coefficients and H–D exchanges, which show that these processes are very fast.

Peter Vekilov addressed Denis Gebauer and James De Yoreo: Professor Gebauer stated that he avoids using the term nonclassical nucleation because all nucleation complies with the general framework of Gibbs' theory. Gibbs proposed a thermodynamic framework applicable to the nucleation of liquid droplets in supersaturated vapors based on the Second law of thermodynamics, which he reformulated in terms of thermodynamic potentials so that it can provide guidance for systems held under isothermal and isobaric constraints, common in the laboratory, nature, and industry. As all processes in the universe comply with the Second law, it is no mystery that all nucleation processes fit Gibbs' general framework. Classical theory of crystal nucleation, however, includes a thermodynamic part, based on adapting Gibbs' ideas to ordered solids, and a kinetic part developed by L. Farkas, L. Szilard,[1] M. Vomer,[2] R. Becker, W. Döring,[3] and Y. B. Zeldovich[4] in the 1930s and 40s. The main thermodynamic assumptions of classical theory are that both the structure and the faceting of the nucleus replicate those of large crystals. In consequence, the surface free energy of the nucleus equals that of large crystals (and is thus size-independent) and its contribution to the nucleation barrier scales with the nucleus surface area. These assumptions are referred to as the "capillary approximation" and they are readily violated at high supersaturations, at which the nucleus is so small that both its structure and faceting diverge from those of large crystals.[5] More recently, deviations from the classical thermodynamic assumption have been recorded at moderate supersaturations. These include non-equilibrium nucleus shapes[6,7] or nucleus sizes of one molecule that lead to barrier-free nucleation,[8] akin to spinodal decomposition. The central assumption of the kinetic part of classical theory of crystal nucleation is that molecules join the nucleus individually. Whereas the value of the nucleation barrier does not depend on the pathway by which the molecules assemble into a nucleus, the rate of nucleation must. A cluster comprised $n + m$ molecules can assemble in multiple ways from the collision of two clusters of any combination of sizes n and m (Fig. 1a). Clusters of more than one molecule, however, possess larger surface and carry greater surface free energy. As a result, their concentration is low and the probability of a collision of any cluster with another cluster of size m is insignificant if m is greater than one. Guided by the low concentration of clusters with $m > 1$, classical theory presumes that nucleation proceeds as successive assembly of single molecules (Fig. 1b), as suggested in the first comprehensive model of crystal nucleation, by L. Farkas,[1] where this argument was attributed to L. Szilard. The "Szilard postulate" is perhaps the most consequential assumption of the classical theories of both nucleation and growth of crystals. Numerous recent examples of growth by association of ordered and disordered mesoscopic and microscopic precursors constitute violations of the Szilard postulate during crystal growth.[9–12] The most heavily discussed nonclassical mode of crystal nucleation involves violation of a corollary of the Szilard postulate, that the molecules join the nucleus directly from the solution. Crystals have been found to nucleate in liquid droplets, amorphous aggregates, mesoscopic clusters, and other solute assemblies.[13–35] A possible explanation of why two-step nucleation hosted by a precursor with high solute concentration is preferred to direct nucleation in the dilute solution centers on the lower surface free energy between the nucleus and its environment during two-step nucleation.[36] For a more detailed discussion of nonclassical nucleation scenarios, see our recent review.[37]

1 L. Farkas, *Z. Phys. Chem.*, 1927, **125**, 236–242.
2 M. Volmer and W. Schultze, *Z. Phys. Chem., Abt. A*, 1931, **156**, 1–22.
3 R. Becker and W. Döring, *Ann. Phys.*, 1935, **416**, 719–752.
4 Y. B. Zel'dovich, *Acta Physicochim. URSS*, 1943, **18**, 1–22.
5 A. Milchev, *Contemp. Phys.*, 1991, **32**, 321–332.
6 S.-T. Yau and P. G. Vekilov, *Nature*, 2000, **406**, 494–497.
7 U. Gasser, E. Weeks, A. Schofield, P. Pusey and D. Weitz, *Science*, 2001, **292**, 258–262.
8 L. F. Filobelo, O. Galkin and P. G. Vekilov, *J. Chem. Phys.*, 2005, **123**, 014904.
9 O. Gliko, N. Neumaier, W. Pan, I. Haase, M. Fischer, A. Bacher, S. Weinkauf and P. G. Vekilov, *J. Am. Chem. Soc.*, 2005, **127**, 3433–3438.
10 A. I. Lupulescu and J. D. Rimer, *Science*, 2014, **344**, 729–732.

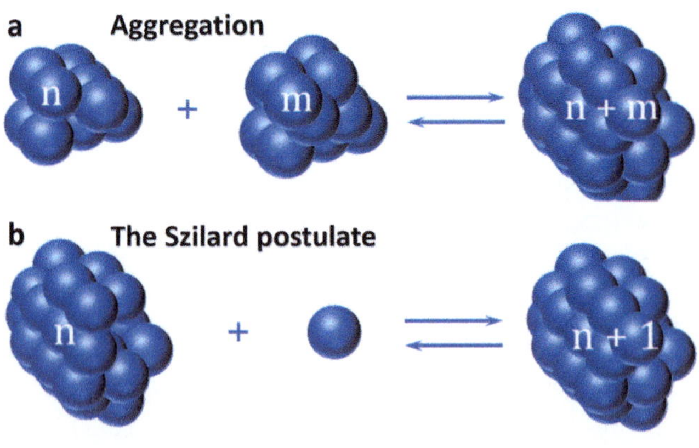

Fig. 1 Models of molecular assembly into nuclei. (a) Aggregation of two clusters comprised of *n* and *m* molecules, respectively. (b) Cluster growth by association of single molecules.

11 J. J. De Yoreo, P. U. P. A. Gilbert, N. A. J. M. Sommerdijk, R. L. Penn, S. Whitelam, D. Joester, H. Zhang, J. D. Rimer, A. Navrotsky, J. F. Banfield, A. F. Wallace, F. M. Michel, F. C. Meldrum, H. Cölfen and P. M. Dove, *Science*, 2015, **349**, aaa6760.
12 Y. Jiang, M. Kellermeier, D. Gebauer, Z. Lu, R. Rosenberg, A. Moise, M. Przybylski and H. Cölfen, *Nat. Commun.*, 2017, **8**, 15933.
13 Y. G. Kuznetsov, A. J. Malkin and A. McPherson, *J. Cryst. Growth*, 2001, **232**, 30–39.
14 T. Yamazaki, Y. Kimura, P. G. Vekilov, E. Furukawa, M. Shirai, H. Matsumoto, A. E. S. Van Driessche and K. Tsukamoto, *Proc. Natl. Acad. Sci. U. S. A.*, 2017, **114**, 2154–2159.
15 M. A. Vorontsova, D. Maes and P. G. Vekilov, *Faraday Discuss.*, 2015, **179**, 27–40.
16 D. Maes, M. A. Vorontsova, M. A. C. Potenza, T. Sanvito, M. Sleutel, M. Giglio and P. G. Vekilov, *Acta Crystallogr., Sect. F: Struct. Biol. Commun.*, 2015, **71**, 815–822.
17 P. G. Vekilov, *Nanoscale*, 2010, **2**, 2346–2357.
18 Y. Liu, L. Porcar, J. Chen, W.-R. Chen, P. Falus, A. Faraone, E. Fratini, K. Hong and P. Baglioni, *J. Phys. Chem. B*, 2010, **115**, 7238–7247.
19 A. Sauter, F. Roosen-Runge, F. Zhang, G. Lotze, A. Feoktystov, R. M. J. Jacobs and F. Schreiber, *Faraday Discuss.*, 2015, **179**, 41–58.
20 A. Sauter, F. Roosen-Runge, F. Zhang, G. Lotze, R. M. J. Jacobs and F. Schreiber, *J. Am. Chem. Soc.*, 2015, **137**, 1485–1491.
21 D. Soraruf, F. Roosen-Runge, M. Grimaldo, F. Zanini, R. Schweins, T. Seydel, F. Zhang, R. Roth, M. Oettel and F. Schreiber, *Soft Matter*, 2014, **10**, 894–902.
22 F. Zhang, F. Roosen-Runge, A. Sauter, R. Roth, M. W. A. Skoda, R. M. J. Jacobs, M. Sztucki and F. Schreiber, *Faraday Discuss.*, 2012, **159**, 313–325.
23 F. Zhang, R. Roth, M. Wolf, F. Roosen-Runge, M. W. A. Skoda, R. M. J. Jacobs, M. Stzucki and F. Schreiber, *Soft Matter*, 2012, **8**, 1313–1316.
24 R. Schubert, A. Meyer, D. Baitan, K. Dierks, M. Perbandt and C. Betzel, *Cryst. Growth Des.*, 2017, **17**, 954–958.
25 L. Houben, H. Weissman, S. G. Wolf and B. Rybtchinski, *Nature*, 2020, **579**, 540–543.
26 J. E. Aber, S. Arnold and B. A. Garetz, *Phys. Rev. Lett.*, 2005, **94**, 145503.
27 B. Garetz, J. Matic and A. Myerson, *Phys. Rev. Lett.*, 2002, **89**, 175501.
28 M. Warzecha, R. Guo, R. M. Bhardwaj, S. M. Reutzel-Edens, S. L. Price, D. A. Lamprou and A. J. Florence, *Cryst. Growth Des.*, 2017, **17**, 6382–6393.
29 J. R. Savage and A. D. Dinsmore, *Phys. Rev. Lett.*, 2009, **102**, 198302.
30 T. H. Zhang and X. Y. Liu, *J. Phys. Chem. B*, 2007, **111**, 14001–14005.
31 T. H. Zhang and X. Y. Liu, *J. Am. Chem. Soc.*, 2007, **129**, 13520–13526.
32 L. B. Gower, *Chem. Rev.*, 2008, **108**, 4551–4627.
33 E. M. Pouget, P. H. H. Bomans, J. A. C. M. Goos, P. M. Frederik, G. de With and N. A. J. M. Sommerdijk, *Science*, 2009, **323**, 1455–1458.
34 D. Gebauer, A. Volkel and H. Cölfen, *Science*, 2008, **322**, 1819–1822.
35 J. Baumgartner, A. Dey, P. H. H. Bomans, C. Le Coadou, P. Fratzl, N. A. J. M. Sommerdijk and D. Faivre, *Nat. Mater.*, 2013, **12**, 310.
36 M. Kaissaratos, L. Filobelo and P. G. Vekilov, *Cryst. Growth Des.*, 2021, **21**, 5394–5402.
37 P. G. Vekilov, in *Crystallization via Nonclassical Pathways Volume 1: Nucleation, Assembly, Observation and Application*, American Chemical Society, 2020, vol. 1358, ch. 2, pp. 19–46.

James De Yoreo commented: This is an excellent exposition of the issue of "non-classical" nucleation.

Denis Gebauer added: I would like to make two additions. Firstly, it has recently been demonstrated that CNT is able to reproduce quintessential features of two-step nucleation pathways,[1] suggesting that they should not be labeled as "non-classical nucleation". Secondly, the theories of the pre-nucleation cluster (PNC) pathway and two-step nucleation (2S) are fundamentally distinct. The former considers thermodynamically stable solute clusters (*i.e.*, negative standard free energy of formation) as the species that are fundamental to phase separation. According to the PNC pathway, phase separation is then not based upon overcoming a certain critical size, but on dynamical changes within stable solute clusters.[2] There is thus an alternative reaction channel to solids that does not rely on critical nuclei as transition states. The PNC theory does not rely on assigning interfacial free energies to nanoscopic species and can thus explain significant populations of nucleation precursors (as opposed to CNT and 2S),[3] and with it, also growth modes that rely on species larger than monomers. Please also consider our earlier replies to the questions from Wenhao Sun and Aaron R. Finney.

1 D. Kashchiev, *J. Cryst. Growth*, 2020, **530**, 125300.
2 D. Gebauer, M. Kellermeier, J. D. Gale, L. Bergström and H. Cölfen, *Chem. Soc. Rev.*, 2014, **43**, 2348–2371.
3 D. Gebauer, P. Raiteri, J. D. Gale and H. Cölfen, *Am. J. Sci.*, 2018, **318**, 969–988.

Christine Kirschhock asked: You observe a stepwise behavior in function of increasing D_2O content. Considering that H–D exchange is very fast, is it possible that the crystallising phase is affecting the distribution of isotopes and therewith influencing the speciation in the solvent?

Denis Gebauer answered: I think that it is possible, however, we can to a large extent only speculate about how the distribution of isotopes is affected during the early stages of calcium carbonate formation. Having said that, our simulations do indicate that heavy water molecules might have a stronger attraction to calcium cations than light water. This might lead to a bias towards heavy water hydration of mineral constituents already at rather low heavy water contents, which may be a clue to understand the stepwise behavior that we have observed in our experiments.

Romain Grossier opened discussion of the paper by Aaron R. Finney: Question 1: you show that CNT agrees well with your simulations for low supersaturations, but differs for "higher" supersaturations. As CNT was developed only for low supersaturation, it's not surprising it fails at higher supersaturations. So, in a way, could we see your simulation as a measure of the supersaturation bandwidth on which CNT is usable?

Question 2: you asked for experimentation data that could be used as force fields benchmarkings. Would solubility (NaCl here) be a pertinent benchmarking to test force fields?

Aaron R. Finney responded: Question 1: Classical Nucleation Theory (CNT) can be applied to predict nucleation rates at any level of supersaturation below the limit of solution stability (when considering solutions), *i.e.*, throughout the metastable zone, beyond which spinodal decomposition occurs. At high levels of supersaturation in macroscopic metastable solutions, many nuclei may be present and determining the rates for nucleation in order to compare with CNT predictions may be challenging. In simulations, the number of nucleation events is limited due to finite size effects and following the time evolution of the largest cluster is sufficient to estimate nucleation rates even at the high end of supersaturation in the metastable zone.[1,2]

For the system and force field under investigation in our paper, several other simulation groups established the supersaturation limits for the metastable solution zone at room temperature and pressure ($S = 1$ and 4.1 at b(NaCl) = 3.7 and 15 mol kg^{-1}, respectively); see, *e.g.*, ref. 3 and 4. At $S = 3.2$, simulations indicated that crystal nucleation occurs in a single step *i.e.*, there is concerted growth in the size of crystalline regions in emerging clusters and the total size of the clusters.[4] At $S = 3.7$, far into the metastable zone, our investigation of an extended reaction coordinate space, where we do not presuppose the pathway for the emergence of crystalline order in this space, indicates that a range of nucleation pathways are available for phase separation. Analysis of the free energy landscape presented in Fig. 4A of our paper (https://doi.org/10.1039/d1fd00089f) shows that the most probable pathway (highlighted by the red dashed line) involves first an enriching of the ion density in liquid-like transient clusters – which are apparent in solution throughout the entire metastable zone – before restructuring of the ions occurs to form a crystalline lattice. Given the evolution of the system along orthogonal degrees of freedom in the reaction coordinate space at early times during nucleation, we describe this as a two-step pathway.

We speculate that in a macroscopic system, a myriad of possible pathways could contribute to phase separation in the metastable zone and that a shift in the most likely pathway from one- to two-step could occur in the range $S = 3.2$–3.7. In addition, we suggest that the notion of a critical cluster be extended to an ensemble of critical clusters which can display a broad range of structures and include sizeable, disordered domains depending upon the reaction conditions.

If one considers the capillarity approximation as a central tenet of CNT, then all the pathways we identify deviate from the theory predictions because the smallest clusters emerging in solution do not have the same properties as the final bulk crystal phase with well-defined facets and chemical ordering. Nonetheless, if one allows for the fact that the smallest clusters on the pathway to crystals can have some intermediate structure and energetic properties between dispersed ions in solution and the final bulk crystal phase, we can reconcile the pathways we observe with the concepts associated with established nucleation theories which follow, *e.g.*, Becker–Döring statistics and the application of (single barrier) transition state theory to a phase separation process. See, *e.g.*, ref. 5. One crucial aspect here is that the lowest energy pathway is represented by an arc length on a multi-dimension manifold which appropriately defines the free energies of states in a reaction coordinate characterising nucleation.

Recent thermodynamic and kinetic frameworks were proposed to describe the driving forces and pathways to nucleation in an extended reaction coordinate, similar to those adopted in our study. One such framework from Kashchiev[6] adopts a composite cluster model which includes an inner crystalline phase surrounded by a metastable amorphous phase embedded in a solution. Depending upon the relative supersaturations between the solution, metastable and crystalline phases, different crystallisation pathways may be observed. In this framework, there is no size dependence on the interfacial tension or chemical potential difference between phases; hence, the author provides theorised energy landscapes for one- and two-step pathways (see Fig. 3 in ref. 6) making full use of CNT. While the energy landscapes we compute are consistent with those theorised elsewhere,[6] we cannot definitively say that the mechanisms we observe are "classical" or "nonclassical" not least because these terms are used somewhat subjectively; indeed, this was evident during this *Faraday Discussions* meeting. In our paper, we opted to refer to one- or two-step pathways as these terms better describe the structural evolution of the system. Nonetheless, we do not find evidence that the mechanisms for NaCl nucleation are inconsistent with the *essence* of established nucleation theories.

Question 2: for studies of crystallisation, or dissolution for that matter, measurements of solubilities and rates for these processes are extremely useful to compare simulation results to and to gauge the range of environmental conditions where the models perform adequately.

For NaCl, the experimental solubility of 6.15 mol kg^{-1} seems well established; however, there is still limited experimental data on the NaCl(s) homogeneous nucleation rate. A comparison of the rates from seeding simulations to experimental data was provided by Zimmermann *et al.*[1,2] Heterogeneous nucleation rates covering a range of supersaturations are also hard to come by and any input from experimental groups on this front would be most welcome.

1 N. E. R. Zimmermann, *et al.*, *J. Am. Chem. Soc.*, 2015, **137**, 13352–13361.
2 N. E. R. Zimmermann, *et al.*, *J. Chem. Phys.*, 2018, **148**, 222838.
3 A. L. Benavides, *et al.*, *J. Chem. Phys.*, 2016, **144**, 124504.
4 H. Jiang, *et al.*, *J. Chem. Phys.*, 2019, **150**, 124502.
5 I. J. Ford, *Proc. Inst. Mech. Eng.*, 2004, **218**, 883–898.
6 D. Kashchiev, *J. Cryst. Growth*, 2020, **530**, 125300.

Peter Vekilov opened discussion of the paper by Yuki Kimura: The nucleus is a cluster at the top of an energy barrier. You stated that there is not a free energy barrier for the formation of a nucleus during dissolution. What do you mean by nucleus during dissolution?

Yuki Kimura answered: You are right. To avoid confusion, it might be better to refer to the cluster with the maximum energy as the critical nucleus. That is why we chose our terminology with care in our paper (https://doi.org/10.1039/d1fd00125f). If I used the term nucleus for the dissolution process, that was my mistake. Perhaps I may have said nucleus in my talk, but I usually refer to it as particles.

In the paper, we defined particles as objects that have a clearly distinguishable interface with the solution and clusters as objects that do not. I use the terms nucleus and nucleation only for the formation processes that may have critical nuclei due to density fluctuation.

Gábor Schuszter asked: During your experiments applying TEM as the observation technique, you are bombarding water with high energy electrons which can cause radiolysis just like in the case of cloud chambers used for alpha particle detection. How can you make sure that this does not affect nucleation? Did you observe any change in the mechanism while varying the accelerating voltage?

Yuki Kimura replied: Radiolysis of water by electron irradiation leads to the formation of ions and radicals, and changes in pH. These reach equilibrium on a millisecond time scale. The most effective way to reduce the effect of electron irradiation on nucleation is to lower the electron dose rate. This can be done by observation at low magnification, image processing to brighten dark images, *etc.* However, the effects of the electron beam cannot be completely eliminated, and observation at high magnification is also necessary to visualize small nuclei. It should be considered in conjunction with other techniques such as optical microscopy.

We did not find any significant differences during the observations with two different acceleration voltages of 200 and 300 kV.

Rik Drummond-Brydson queried: As far as I understand the dose limits quoted in the paper are for damage of solid (dry) calcium carbonate I think. In the liquid cell, radiolysis of water is the key factor and at a dose rate of 40 $e^- Å^{-2} s^{-1}$ (this is 40–150 MGy s^{-1}) you would expect big pH changes for pH 7 and above – with (amongst other things) pH reducing and causing increased solubility of all calcium carbonate polymorphs and also ACC. I think you see this at longer times in your experiments and the behaviour and eventual dissolution is caused by the electron beam. However doesn't this mean that conditions in solution are changing during observation. I think the dose rate needs to be lowered by roughly two orders of magnitude to make conditions more stable.

Yuki Kimura responded: You are right. The pH of the solution is decreased by the radiolysis of water by electron beam irradiation. As a result, the solubility of calcium carbonate increases. If we can reduce the electron dose rate from 1000–10 000 $e^- nm^{-2} s^{-1}$ to 10–100 $e^- nm^{-2} s^{-1}$, the pH decrease will be very small, making the experiment more ideal. To achieve this, we recently developed an algorithm of machine learning to enhance the image clarity. This will lower the electron dose rate by 3–4 orders of magnitude.[1] Please look forward to our next report.

1 H. Katsuno, Y. Kimura, T. Yamazaki and I. Takigawa, *Microsc. Microanal.*, 2022, **28**, 138–144, DOI: 10.1017/S1431927621013799.

Michael Anderson questioned: Can you do electron diffraction on the particles in real time?

Yuki Kimura answered: Unfortunately, it is not possible to perform particle imaging and electron diffraction simultaneously in real time. However, it is possible to acquire (observe *in situ*) electron diffraction alone in real time.

Michael Anderson asked: Can you distinguish between an amorphous solid state and a liquid state or is this too difficult?

Yuki Kimura responded: For example, it can be inferred from whether two particles merge when they are in contact, or from the difference in contrast when comparing particles of the same size.

Stéphane Veesler enquired: In the video you presented, sometimes fringes appeared and disappeared. Are they artefacts or are they real?

Yuki Kimura replied: They are real. There are two different kinds of objects formed. One is two dimensionally nucleated thin crystal on the membrane of the liquid cell. Another one is a three dimensionally nucleated tiny particle. I assume that you are talking about the fringes of the thin crystal.

Jutta Rogal addressed Aaron R. Finney and Denis Gebauer: Concerning the PMF for ion dissociation, it was shown[1] that the distance between ions is not a good reaction coordinate and that the solvent degrees of freedom need to be considered in the free energy barrier. How confident are you about the PMF? (This actually also applies to the PMF for $CaCO_3$ in Denis Gebauer's paper).

1 P. L. Geissler, C. Dellago and D. Chandler, *J. Phys. Chem. B*, 1999, **103**, 3706.

Aaron R. Finney replied: We agree that a reaction coordinate defined using a single variable quantifying the distance between ions (r) is not always sufficient to understand mechanisms for ion association or to evaluate the free energy barriers for such processes. Often an additional collective variable, *e.g.*, quantifying the hydration level of the cation, is at least required to investigate these reactive features, as was shown in other biased simulation studies.[1,2] This is particularly necessary if the timescale for solvent exchange in the first ion coordination spheres is slow relative to the simulation times.

From the potential of mean force (PMF) calculation, our principle objective was to determine the equilibrium constants for ion association (K_a) which can be compared with experiments and used in the multiple-binding model for speciation[3] before nucleation, as we discussed. To calculate K_a, an integral over $\exp(\mathrm{PMF}(r)/k_BT)$ is performed and a $4\pi r^2$ factor is included to account for the increasing phase space volume as a function of the distance between ions, r. So long as the PMF profile is converged in the reaction coordinate space, the value of K_a is accurately evaluated and a single variable, r, is sufficient to establish the relative probabilities of bound and unbound ion pair states. In this regard, we performed 5 ns of simulations for each of the umbrella windows sampled. We are very confident that these are converged, not least because the uncertainty bars shown in Fig. 1E of our paper (https://doi.org/10.1039/d1fd00089f) are very small on the energy scale provided. Furthermore, the PMF is a very close match to those evaluated elsewhere in the literature.[4]

1 P. Raiteri, *et al.*, *J. Phys. Chem. C*, 2015, **119**, 24447–24458.
2 A. R. Finney, *et al.*, *Chem.–Eur. J.*, 2019, **25**, 8725–8740.
3 D. Gebauer, *et al.*, *Science*, 2008, **322**, 1819–1822.
4 C. Zhang, *et al.*, *Nat. Commun.*, 2020, **11**, 3037.

Denis Gebauer added: We are aware of this problem. That is why the original PMF for the force field was measured as a 2D PMF of ion-distance and cation–water coordination number, see Fig. 3 in ref. 1. This was found to be a better representation for a PMF between ions.[2,3]

We compared the reduction of the 2D PMF to 1D with a PMF that was computed in 1D and found them to be very similar. Thus, we decided to limit our calculations for D_2O to 1D PMFs. Please also note that we use these 1D PMFs only for comparison and for an assessment of relative binding strengths. We do not investigate the underlying dissociation mechanism. This would indeed need a 2D PMF.

1 P. Raiteri, R. Demichelis and J. D. Gale, *J. Phys. Chem. C*, 2015, **119**, 24447–24458.
2 A. J. Ballard and C. Dellago, *J. Phys. Chem. B*, 2012, **116**(45), 13490–13497.
3 R. Gotchy Mullen, J.-E. Shea and B. Peters, *J. Chem. Theory Comput.*, 2014 **10**(2), 659–667.

Jutta Rogal continued discussion of the paper by Aaron R. Finney: Do you have an idea of how the relative path probability of the different nucleation mechanisms depends on, for example, the supersaturation?

Aaron R. Finney responded: We have not explicitly evaluated the pathways for nucleation at lower levels of supersaturation, although this is perhaps the focus of future work. From the Markov state model we present, at $S = 3.7$ the thermodynamic barrier for the most-probable two-step pathway is $2k_BT$ lower in energy than the one-step pathway, but both of these types of mechanisms contribute to the phase separation and there is a myriad of possible routes from supersaturated solutions to crystals.

Previous computational studies using the same force field have shown that at $S = 3.2$, the growth in the size of crystalline regions in clusters during nucleation linearly correlates with the size of the clusters, and nucleation was described as one-step.[1] It is important to note that close inspection of the cluster population distribution at small cluster sizes in those studies (see middle panel of Fig. 7 of ref. 1) appears to indicate that a relatively wide range of (largest) cluster sizes are accessible to the system before significant crystalline order emerges. In support of the one-step mechanism at $S = 3.2$, the results from seeding simulations indicated that the rates for crystal nucleation – determined within a CNT framework from the rates at which crystal seeds grow or shrink in simulations – can be aligned to the NaCl crystal nucleation rates determined in experiments when an appropriate definition for ions within crystalline embryos is adopted.[2]

In light of these results, we speculate that there is a change in the most likely pathway from predominantly one- to two-step between $S = 3.2$ and 3.7 in the metastable zone. However, we also speculate that in a macroscopic system, two-step pathways could contribute to phase separation – albeit with often lower probability than one-step pathways (below the proposed transition region in S) – across the entire metastable zone. This is because, as shown in Fig. 1B of our paper (https://doi.org/10.1039/d1fd00089f), extended ionic networks are observed with increasing probability as S is increased, matching the results from recent experiments.[3] Our earlier work,[4] considering NaCl(aq) solutions at carbon surfaces, indicates that interfaces promote the formation of extended liquid-like networks, which raises the possibility that two-step pathways dominate the phase separation at lower levels of S than in the bulk solutions in the case of heterogeneous nucleation.

1 H. Jiang, *et al. J. Chem. Phys.*, 2019, **150**, 124502.
2 N. E. R. Zimmermann, *et al.*, *J. Chem. Phys.*, 2018, **148**, 222838.
3 H. Hwang, *et al.*, *Chem. Sci.*, 2021, **12**, 179–187, DOI: 10.1039/D0SC04817H.
4 A. R. Finney, *et al.*, *Chem. Sci.*, 2021, **12**, 11166–11180, DOI: 10.1039/D1SC02289J.

Sarah Price remarked: This is a comment on the earlier question about deuterating other solvents.

The effects of deuterating other solvents could be even larger. The thermodynamically stable form of isotopically normal pyridine (pyridine-h5) between the melting point (231 K) and 5 K is the complex structure of form I ($Pna2_1$, $Z' = 4$), and it can be cooled or heated within this temperature range without undergoing any phase transitions. In contrast the thermodynamically stable form of pyridine-d5 below 215 K is the simpler phase II ($P2_12_12_1$, $Z' = 1$).[1,2] Such a difference in crystal structures seems likely to affect the structure of pyridine as a solvent.

I would like to comment on force-field validation.

It would be good if all computational papers were very explicit about the evidence for their force-field or functional and simulation method being appropriate for the study. The force-fields that can be used in most molecular dynamics simulations are usually inadequate for crystal structure prediction studies. So knowing whether the force-field used for nucleation studies can reproduce the known polymorphs, and how the calculated relative stability compares with experiment, can be important in drawing conclusions from the simulation results. It can be harder to test the solute–solvent force-field, for example by comparing the calculated and observed solubility. However, defining the supersaturation relative to the calculated solubility with the force-field, as was done in this paper (https://doi.org/10.1039/d1fd00089f), can go a long way towards compensating for the limited accuracy of the force-field.

1 S. Crawford, M. T. Kirchner, D. Blaser, R. Boese, W. I. F. David, A. Dawson, A. Gehrke, R. M. Ibberson, W. G. Marshall, S. Parsons, *et al.*, Isotopic polymorphism in pyridine, *Angew. Chem., Int. Ed.*, 2009, **48**(4), 755–757.
2 N. Giordano, C. M. Beavers, B. J. Campbell, V. Eigner, E. Gregoryanz, W. G. Marshall, M. Peña-Álvarez, S. J. Teat, C. E. Vennari and S. Parsons, High-pressure polymorphism in pyridine, *IUCrJ*, 2020, 7, 58–70, DOI: 10.1107/S2052252519015616.

Kevin Roberts added: I have no comments to add regarding the issue of solvent deuteration.

Regarding the issue of force-field validation, I would comment that existing forcefields are broadly speaking quite appropriate for modelling organic systems. Whether crystal structure prediction (CSP) is an optimal test of forcefields validation is perhaps an open question as it predicts an energy landscape for bulk crystal structures governed by the periodic boundary conditions for a fully formed microscopic crystal structure. In contrast, the structural chemistry at nucleation is though more likely to be governed by the surface free energies of molecular clusters. In this, many of their constituent molecules will be under-coordinated and hence translationary displaced with respect to their bulk crystallographic structures and where, for flexible molecules, conformational variability molecule-to-molecule can also be expected. As the clusters grow the variability can be expected to decrease as the longer-range forces from the bulk structure dominate, see *e.g.* ref. 1. Hence, it is a challenge for CSP to predict structures with forcefields without some allowance for the size-dependent simulation of structural energetics.

1 R. B. Hammond, K. Pencheva and K. J. Roberts, Structural variability within, and polymorphic stability of, nano-crystalline molecular clusters of L-glutamic acid and D-mannitol, modelled with respect to their size, shape and 'crystallisability', *CrystEngComm*, 2012, **14**, 1069–1082.

Matteo Salvalaglio commented: Comparing experimentally measurable quantities, such as solubility, with estimates obtained from molecular simulations is a complex and often challenging task. This challenge reflects an inherent tension between two complementary aspects of molecular simulations that must simultaneously be satisfied to obtain precise and accurate estimates of macroscopic observables. These two aspects are the accuracy of the model used to represent the molecular system of interest and the quality of the sampling of the configurations accessible to that system in the relevant thermodynamic conditions for the problem.

The tension between these two aspects originates from the fact that the accuracy in the model used to represent a system drastically affects its computational cost and, thus, its application in calculations that require extensive sampling. This means that expensive and accurate models are impractical for simulating large systems for long timescales. The dual aspect of this issue is that often we can converge ensemble properties *via* sampling only for simplified and inherently inaccurate models. This tension naturally leads to identifying acceptable compromises every time simulations are deployed, especially in the context of crystallization. Such compromises depend on the system studied, the process investigated, and ultimately, the research question at the heart of a simulation campaign. Studying crystallization from complex liquids with molecular dynamics requires large systems, long timescales and often complex sampling algorithms to overcome activated events. In this case, favouring sampling under controlled conditions (*i.e.* known solubility of the model solute in the model solvent) is a typical choice, see, for instance, Aaron R. Finney's paper (https://doi.org/10.1039/d1fd00089f) and the discussion of it and Jutta Rogal's paper (https://doi.org/10.1039/d1fd00099c). The calculation of accurate lattice energies sits on the opposite end of the spectrum, where minute energy differences can drastically change conclusions. Thus, expensive electronic structure methods are needed to answer specific research questions. Awareness of this tension between model accuracy and sampling allows one to formulate better research questions, find more effective compromises in the choice of models, and ultimately extract usable insight on length and timescales inaccessible to direct experimental observation.

Joonsoo Kim continued discussion of the paper by Yuki Kimura: In the video, the brightness of the big crystal in the bottom left corner is fluctuating. Is that an artefact or change in the crystallinity?

Yuki Kimura replied: Diffraction contrast is the significant factor affecting crystal contrast in TEM images. Electron beams are diffracted when they are incident parallel to the crystal plane. If the diffracted electron is blocked by an aperture, the contrast becomes stronger (the crystal becomes darker). When the image is formed without being blocked by an aperture, the electron beam returns to a position slightly displaced from the crystal by the focal point. This location becomes brighter. The main reason for the fluctuating brightness of the crystal in the lower left is probably that the orientation of the crystal has changed slightly because it is in aqueous solution.

Wenhao Sun continued the general discussion: To address the earlier questions on validating theoretical models; I would just like to add that if we are using molecular dynamics potentials, it is important that the potentials are programmed to have the proper physics. For example, if you are studying a chemical reaction involving bond breaking, your potential should be able to account for that. If you are hoping to capture stericity or bond angles, your potential should account for that. There has been this discussion about theorists *vs.* experimentalists in this session thus far, which from my perspective, has mostly been experimentalists hoping that theorists can provide insights that are more actionable than we have so far. I have a criticism for the experimentalists, and perhaps for our community in general. **It would be best if our community decreased our reliance on the "Energy Landscape" picture**, which envisions a chemical reaction as a ball rolling on an energy landscape with hills and valleys; where valleys indicate the metastable or stable states (thermodynamics) and the hills represent the activation energy barriers (kinetics). This picture was developed for changes in molecular conformations, but **it is not proper to apply the energy landscape picture to solid-state transformations such as nucleation**, where the number of atoms are changing throughout the course of the reaction. There are two major arguments for this: (1) the *y*-axis is a generic 'energy' term, but energy is an extensive quantity, meaning it depends on the mol number of atoms being considered. For the 'valleys', which represent stable phases, we might have units of kJ mol^{-1}. However, for the barriers, the units of the barriers depend on the mechanism. If the barrier is a nucleation barrier, it should be energy per nucleus. If it is a diffusion barrier, the units would be energy per atomic hop. **It is not meaningful to put on a single figure both the bulk energies and the barriers at the same time.** (2) The reaction coordinate is often very vaguely defined. It sometimes refers to time, sometimes applied reaction conditions, sometimes a conformation. As a theorist, this is very frustrating, because **if you can't say what it is, we cannot calculate it.**

The problem with the energy landscape picture is that it is conceptually lazy and theoretically non-rigorous. Overall, it would promote theory/experimental collaboration if instead of using the energy landscape cartoon, we altogether moved to more precise descriptions of both the *y*-axis and the *x*-axis when we discussed thermodynamics and kinetics. It is frustrating for me to see a beautiful experimental observation in the literature, that has a discussion section with an 'energy landscape' rationalization of the mechanisms. Remember that scientific theory should be falsifiable. If the energy landscape picture can describe everything, then it cannot predict anything. The energy landscape picture is not a robust foundation to build a theory of nucleation or crystallization.

Denis Gebauer responded: I am not sure if I can generally agree with your comment. Even though nucleation theories consider barriers per nucleus, the barrier itself corresponds to a transition state that we can formally assign a standard free energy, and with it, also an equilibrium constant (in analogy to the valleys in free energy landscapes). It should thus always be possible, within the framework of a theory, to convert or normalize the energetics so as to illustrate barriers and valleys in the same diagram. Having said that, in my opinion, the free energy landscape illustrations are often used for the lack of a better description or quantitative expression, and I have seen this in both experimental and theoretical papers. I still believe that these illustrations are useful for demonstrating a principle idea, and they should be taken as what they are – cartoons (as you have rightfully pointed out). This cartoon perspective is then also often useful to illustrate basic differences between theories, without claiming any explanatory or predictive power going beyond that intention. I am sorry to have caused frustration by using these cartoons in our own works, however, the actual purpose always was to spark discussions, *e.g.*, on proper reaction coordinates, and to catalyse the development of new models. Finally, please note that for the illustration in ref. 1, we can now actually say what it is in terms of a quantitative model,[2] and look forward to theoreticians calculating it.

1 D. Gebauer, P. Raiteri, J. D. Gale and H. Cölfen, *Am. J. Sci.*, 2018, **318**, 969–988.
2 J. T. Avaro, S. L. P. Wolf, K. Hauser and D. Gebauer, *Angew. Chem., Int. Ed.*, 2020, **59**, 6155–6159.

Kristen Fichthorn remarked: Having accurate force fields is, of course, important in modelling experiments. However, force fields usually have deficiencies and even quantum calculations are not always accurate. As we listen to the talks presented here, we will see that experimentalists do not always know what they are measuring and they often have to do many different types of experiments to reach a conclusion. Theory is valuable in suggesting new mechanisms, new possibilities, revealing trends, and helping experimentalists tell their stories, not just predicting data to high accuracy. After all, theoreticians are not just accountants.

Ruel Cedeno opened discussion of the paper by Alexander Van Driessche: Sulfate ions have been shown to form large solvation shells in water.[1] To some extent, this also occurs with strontium ions.[2] These shells may cause charge shielding which could reduce the sensitivity of the probes. This effect might be more pronounced at lower dosing rate since such solvation shells are more well defined in dilute solutions. Conversely, the hydration numbers decrease in concentrated solution.[3] Could this phenomenon play a role in the apparent consumption of neutral species (prenucleation clusters) during nucleation at low dosing rate which is not detected by the ion selective electrode (ISE)?

1 J. T. O'Brien, J. S. Prell, M. F. Bush and E. R. Williams, *J. Am. Chem. Soc.*, 2010, **132**, 8248–8249, DOI: 10.1021/ja1024113.
2 P. D'Angelo, V. Migliorati, F. Sessa, G. Mancini and I. Persson, *J. Phys. Chem. B*, 2016, **120**, 4114–4124, DOI: 10.1021/acs.jpcb.6b01054.
3 A. V. Dighe and M. R. Singh, *Proc. Natl. Acad. Sci. U. S. A.*, 2019, **116**, 23954–23959, DOI: 10.1073/pnas.1910691116.

Alexander Van Driessche replied: It is unlikely that the solvation shells described in the above works play a significant role in our measurements. The solvation shell for Sr^{2+} ions described in the source provided[1] was derived from a XANES measurement taken of a bulk liquid (volume unspecified) which would have averaged the absorption spectra across all Sr atoms present in the solution. Thus, the absorption spectrum must be primarily providing information on what we, in our work, considered "free" Sr^{2+} ions as they are the majority species – in other words, we are measuring the hydrated Sr^{2+} ions with the ISE. It is also important to note that the particularly stable hydration shells of $n \sim 40$ for $SO_4{}^{2-}$ ions found in ref. 2 may change with supersaturation in a manner not dissimilarly to the 41<n<47 solvation shells simulated for glutamic acid.[3] It is a much larger leap in logic to presume that the measured and simulated hydration shell of $n = 8$ described in the Sr XANES study would necessarily be subject to the same effects. It is not theoretically impossible that there are some strongly hydrated Sr ions that are shielded from the ISE and participate in the early stages of nucleation.

However, there is no data in our experiment, or any that we know of in literature, that would support this assertion.

1 P. D'Angelo, V. Migliorati, F. Sessa, G. Mancini and I. Persson, *J. Phys. Chem. B*, 2016, **120**, 4114–4124, DOI: 10.1021/acs.jpcb.6b01054.
2 J. T. O'Brien, J. S. Prell, M. F. Bush and E. R. Williams, *J. Am. Chem. Soc.*, 2010, **132**, 8248–8249, DOI: 10.1021/ja1024113.
3 A. V. Dighe and M. R. Singh, *PNAS*, 2019, **116**, 23954, DOI: 10.1073/pnas.1910691116.

Aaron R. Finney questioned: In the paper (https://doi.org/10.1039/d1fd00092f), you indicate that at the lowest dosing rates, the transmittance probe detected the formation of a *new phase* before the consumption of dispersed ions indicated by the change in the ISE signal (*i.e.*, the inflection point in the LaMer diagram). You suggest that this phase emerges *via* consumption of neutral species *e.g.* ion pairs. What is the nature of this phase and can you comment on the possible comparison between the bound species (prenucleation clusters) in $SrSO_4$(aq) *cf.* $CaCO_3$(aq)? Furthermore, how does your model differentiate between the nanoclusters and dense liquids/amorphous phases (see Fig. 1 in the paper; https://doi.org/10.1039/d1fd00092f) that could be involved in the nucleation pathway?

Alexander Van Driessche responded: Neither the probes used in the titration experiments, nor the model used in this work have the implicit capacity to precisely determine the nature of the (pre)nucleation species present during the nucleation process. In fact, the model is completely independent of any chemical information about what is nucleating. It provides only information about the relative density of nucleating phase as it traverses a free energy landscape from a perfectly mixed fluid state to a thermodynamically stable solid phase. Similarly, the potentiometric and optic techniques used in the titrations serve primarily to characterize the solution, *i.e.* the fluid that is not nucleating. For example, the ion selective electrode characterizes the quantity of unbound Sr^{2+} ions in the solution, but does not provide a direct counting or characterization of the ions consumed in the reaction – any measures of those atoms are derived through subtraction from measurements carried out in the absence of any nucleation reaction (or pre-nucleation events). The most information gleaned about the nature of the phases formed during the titrations comes from the photometric sensor, and that is limited to a few minor details. We know that the phase is different enough from the solution surrounding it to form a light-scattering interface, and we can recover a limited qualitative measure of the concentration of those interfaces (higher turbidity = more interfaces). From the ISE data, and the information gleaned about bound species concentration, we could also attempt to extract some binding constants for the bound species and their dependence on time and supersaturation, but this analysis would create estimations blind to the actual structure and number of atoms contained in the bound species. Thus, we do not wish to speculate too extensively on the nature of the first formed phases. To obtain more insight on the nature of the transitory precursor cluster species other tools are necessary. For example, the structural properties of $CaSO_4$ precursor clusters have been derived by combining *in situ* high-energy X-ray diffraction experiments and molecular dynamics (MD) simulations.[1]

The pathway monitored by ISE and the photometric sensor is comparable to the pathway observed for $CaCO_3$ (ref. 2) using an identical setup and similar to the pathway revealed for $CaSO_4$ using *in situ* X-ray scattering techniques.[3,4] The only aspect that awaits experimental confirmation is the existence of $SrSO_4$ and $CaSO_4$ prenucleation clusters in the undersaturated regime, as has been shown for $CaCO_3$ prenucleation clusters. In any case, our model does predict the existence of a prenucleation population for undersaturated solution conditions. The model used in this work (initially introduced in ref. 5) reveals that before crossing the nucleation barrier a kinetically induced cluster population appears; yet this does not represent a true thermodynamic phase. Eventually, the system will cross the barrier and a new phase will nucleate with a density close the final one. Noteworthy, after crossing the nucleation barrier the system will continue to (slowly) evolve approaching the ideal density of the crystalline phase. Taking into account this pathway, the population of kinetically induced clusters would correspond to nanoclusters/PNCs shown in Fig. 1 in the paper (https://doi.org/10.1039/d1fd00092f), and are clearly differentiated in our model because these occur on the pathway before crossing the nucleation barrier. The first nucleated phase could potentially correspond to an amorphous phase because the density of this phase has not yet reached unity. Of course, this is one interpretation of the results and cannot be unequivocally confirmed from the current model. To do so, the simulations need to be run considering crystallinity as the variable, and not density as was done in the current work. In any case, the behavior observed in our model has been observed experimentally using *in situ* scattering during $CaSO_4$ precipitation at room and high temperature.[3,4] In these experiments, nanosized clusters formed first and subsequently aggregated into an amorphous phase (which is the first phase detected by an optical sensor). Eventually, a crystal structure develops through the reorganization of the nanosized clusters inside the disordered phase. Noteworthy, this reorganization continues long after the apparent precipitation reaction has finished. This is akin to what we observed in our model, where once the new phase has nucleated it will continue to evolve approaching the final density.

1 T. M. Stawski, A. E. S. Van Driessche, R. Besselink, E. H. Byrne, P. Raiteri, J. D. Gale and L. G. Benning, *J. Phys. Chem. C*, 2019, **123**, 23151–23158, DOI: 10.1021/acs.jpcc.9b04268.
2 P. I. Schodder, M. B. Gindele, A. Ott, M. Rückel, R. Ettl, V. Boyko and M. Kellermeier, *Phys. Chem. Chem. Phys.*, 2022, **24**, 9978–9989, DOI: 10.1039/D1CP05606A.
3 T. M. Stawski, A. E. S. Van Driessche, M. Ossorio, J. D. Rodriguez-Blanco, R. Besselink and L. G. Benning, *Nat. Commun.*, 2016, **7**, 11177, DOI: 10.1038/ncomms11177.
4 T. M. Stawski, R. Besselink, K. Chatzipanagis, J. Hövelmann, L. G. Benning and A. E. S. Van Driessche, *J. Phys. Chem. C*, 2020, **124**, 8411–8422, DOI: 10.1021/acs.jpcc.0c01041.
5 M. A. Durán-Olivencia, P. Yatsyshin, S. Kalliadasis and J. F. Lutsko, *New J. Phys.*, 2018, **20**, 083019, DOI: 10.1088/1367-2630/aad170.

Alan Hare enquired: Given that in Fig. 3 in the paper (https://doi.org/10.1039/d1fd00092f) the U-shaped curve ρ_0 tends towards a CNT limit, are you now able to draw any inference concerning the cluster shape? (I realise that this question could have a binary answer.)

Alexander Van Driessche answered: Clusters (or density fluctuations) are assumed to be spherically symmetric in mesoscopic nucleation theory (MeNT), consequently we only observe a spherical shape. Recently, a further generalization of MeNT has been put forward by Lutsko[1] allowing also for non-spherically symmetric shapes to be considered. However, the resultant formalism equations are considerably more complicated than the ones involved in the spherically-symmetric MeNT.

1 J. F. Lutsko, *New J. Phys.*, 2018, **20**, 103015, DOI: 10.1088/1367-2630/aae174.

Stéphane Veesler queried: Nucleation is a localized phenomena there is a resolution issue by using a global measure such as a conductimetric one to detect it. You do not know where nucleation will occur. Cannot we say that the Δt you observed is (by analogy) the growth time to a detectable size, this growth time being classically used in the interpretation of induction time measurement experiments?

Alexander Van Driessche responded: Indeed, it is important to consider growth when describing the nucleation events in the titration experiments. We do consider that the photometric sensor detects nucleation not at the formation of the first light scattering interface, but at the moment when the concentration/size of particles (and thus total amount of interfaces) present is significant to overcome the resolution limits of our sensor. Similarly, the ion selective electrode only detects nucleation when the change in free ion counts is significant enough to generate measurable change in the potential on the probe membrane – this would theoretically happen after the concentration of bound ions is greater than the detection limit for Sr^{2+} ions in solution. If we consider that the sensitivity threshold of both probes are comparable, then even if we do not detect only nucleation, but a combination of nucleation and growth, our main observation is still valid because when the photometric sensor is detecting the formation of a new phase, the ion selective electrode does not detect any change. Hence, the nucleation (and growth) of this new phase is mainly consuming bound species. We can also conduct a thought experiment where there is a distinct difference in detection limits for the probes. In this thought experiment, the nucleation process is assumed to be identical (classical) across all concentration ranges. We imagine classical nuclei that grow in size and number until the amount of interface present is enough to be detected by the photometric sensor. Some time later the concentration of nuclei increases until enough ions have been removed from the solution to be measure potentiometrically. This is a perfectly reasonable explanation for the results of a single titration with one supersaturation rate. However, if this were the complete picture for how nucleation happens in this system, we would expect the sequence of particle detection to remain unchanged regardless of the rate of ion addition. We would anticipate the same probes to have the highest sensitivity regardless of nucleation and growth rates. This is not what we see in the experiments.

Joonsoo Kim requested: Intuitively, larger nanoclusters should be more present at higher supersaturation but based on the observation in this work, larger nanoclusters are involved in the nucleation process when the concentration is low. May I ask your perspective?

Alexander Van Driessche replied: The involvement of large clusters in the nucleation pathway (step 1, Fig. 3 in the paper; https://doi.org/10.1039/d1fd00092f) mentioned in this work requires us to first reconsider the definition of "cluster". Within the classical realm, namely Classical Nucleation Theory (CNT), clusters are thought to be small replicas of the final stable phase. This definition (also referred to as "capillary approximation") leads to one of the main sources of problems for CNT, since it imposes the simplest possible nucleation pathway. Within the Mesoscopic Nucleation Theory (MeNT) framework, the concept of cluster is defined as the excess density with respect to the mother phase. This general definition, which lies at the core of MeNT, allows for the construction of a much more detailed theory of nucleation with enough flexibility to consider a much richer family of cluster and nonclassical nucleation pathways, and not just the classical kind. Within this context, the larger clusters (*i.e.* large-size–low-density fluctuations) that appear at the onset of the nucleation pathway obtained from MeNT, and are discussed in this work, are much larger in size than one molecular radius, but have an inner density that is much lower than the final phase and close to the mother phase. In this sense, the large clusters reported in this work are not replicas of the final phase. Instead, they are density fluctuations which extend several molecular radii in size, but whose intensity (or simply termed, average inner density) is very low. These initial large "clusters" are present in the nucleation pathway predicted by MeNT irrespective of the supersaturation.

Matteo Salvalaglio asked: The MeNT allows to describe nucleation using a multidimensional reaction coordinate space function of density and radius of the nucleus. Would it be possible to extend this approach to introduce additional parameters, such as measure of the order in the clusters/nuclei? Do you think there is a practical limit in the number of descriptors considered in this framework?

Alexander Van Driessche responded: Indeed, MeNT allows including as many order parameters as one might consider relevant to model nucleation. Some examples of other reaction coordinates considered under MeNT can be found, *e.g.*, in a previous work by Durán-Olivencia and Lutsko.[1] Additionally, a further extension of MeNT to consider other order parameters was carried out by Lutsko[2] in recent years. This new extension provides a roadmap to systematically develop nucleation theories considering all types of cluster geometries and reaction coordinates. Although there is no easy rule of thumb, or practical limit, to decide the optimal number of descriptors, we have observed that the complexities of the resultant equations grow exponentially with the number of order parameters.

1 M. A. Durán-Olivencia and J. F. Lutsko, *Phys. Rev. E*, 2015, **91**, 022402, DOI: 10.1103/PhysRevE.91.022402.
2 J. F. Lutsko, *New J. Phys.*, 2018, **20**, 103015, DOI: 10.1088/1367-2630/aae174.

Ivo B. Rietveld opened discussion of the paper by Joop ter Horst: It is clear that the detection rates of the crystals are very different in the case of seeding and in the case that no seeding is taking place. But how can we be sure that the seeded solutions are not simply showing crystal growth of the seeds; how can we prove that secondary nucleation is taking place (and how is secondary nucleation defined exactly)? And to go a bit further still, is there a possibility to decide whether CNT is valid in these nucleation experiments? Or are other processes possible, or even more likely (besides secondary nucleation)?

Joop ter Horst answered: As we operate the continuous antisolvent crystallization process for over 6 residence times and each residence time a lot of product mass is formed compared to the seed mass used, substantial secondary nucleation must be happening in order to create new crystals of the same chirality. Secondary nucleation follows a completely different mechanism compared to Classical Nucleation Theory or other primary nucleation mechanisms.

Aurora Cruz-Cabeza enquired: I really enjoyed this contribution Joop, thank you. I was wondering if the way the anti-solvent is added to the crystalliser matters to the resulting nucleation mechanisms (primary *versus* secondary).

Joop ter Horst responded: Thanks Aurora. I showed in the paper that secondary nucleation is dominant over primary nucleation. The secondary nucleation rate, whether it is shear- or attrition-based, is determined by the stirrer speed, number of crystals present and the bulk supersaturation, among others. The primary nucleation rate in continuous antisolvent crystallization is determined by the local supersaturation at the points where solution or antisolvent is mixed with the bulk solution. In order to increase the importance of primary nucleation one indeed can try to increase the local supersaturation in the continuous process by changing the local mixing. One way to create extreme local supersaturations is by pre-mixing the antisolvent and solution inflows before they enter the crystallizer. This might then result in a dominance of primary over secondary nucleation. There are many different ways to mix process streams and it would be very interesting to investigate the relative dominance of the nucleation mechanisms based on the local mixing conditions (and on parameters influencing secondary nucleation rate).

Ian Ford questioned: Are the crystals sufficiently fragile to break up to produce seeds for secondary nucleation under the typical shear stresses in the flow or impact with the mixer? The fragility must depend on crystal size as well as intrinsic strength. Is there experimental evidence for this?

Joop ter Horst replied: We find experimental evidence for secondary nucleation in the continuing enantiopurity of the product. As we operate the continuous antisolvent crystallization process for over 6 residence times and each residence time a lot of product mass is formed compared to the seed mass used, substantial enantioselective secondary nucleation must be happening in order to create new crystals of the same chirality.

Weronika Kras asked: I was wondering whether it is possible for the two enantiomers to have formed a solid solution with a crystal structure resembling one of the enantiomers?

Joop ter Horst responded: There are three main possibilities for enantiomers to crystallize from a racemic solution. (1) As a racemic compound, which is a 1 : 1 co-crystal of the two enantiomers. This happens for about 90% of the chiral compounds. (2) As a conglomerate, a physical mixture of chirally pure crystals, which happens for about 10% of the chiral compounds. When a chiral compound forms a conglomerate preferential crystallization and crystallization-enhanced deracemization techniques such as Viedma ripening can lead to an enantiopure crystalline product. (3) As a solid solution, for which the two enantiomers can both (more or less) occupy the same lattice position. This happens only occasionally.

In our paper (https://doi.org/10.1039/d1fd00098e) we used sodium bromate which is a compound that forms achiral ions in solution, but crystallizes in a chiral space group and therefore can form left- and right-handed crystals. So in effect it is a conglomerate system.

Alfred Y. Lee enquired: A single crystal seed strategy is described. Can the author comment on the potential impact of the shape and particle size of the seeds on the crystallization outcome and the solid-state properties of the end product? Does it not matter since it is an attrition-based secondary nucleation mechanism?

Joop ter Horst replied: The main parameters influencing secondary nucleation rate, whether it is shear- or attrition-based, are stirrer speed, crystal number and size, and the bulk supersaturation, among others. The size is important in attrition-based secondary nucleation as above a certain crystal size the crystals will have difficulty following the flow patterns close to the turbulently mixed stirrer region. Large particles then have a higher chance to collide with the stirrer blades and be attrited. The shape can also have an effect: as an example, needle crystals might be easier to break in the case of attrition-based secondary nucleation. In the paper (https://doi.org/10.1039/d1fd00098e) we focused on the enantiomeric excess of the product in comparison to the seed crystals. We did not yet look further into the effect of seed size and shape on the antisolvent crystallization of sodium bromate.

Jan Sefcik remarked: Both mechanical impact and fluid shear may be important in inducing secondary nucleation. While it is often assumed that mechanical impact and related crystal attrition is the main cause of secondary nucleation, it has been observed that fluid flow around a suspended crystal can also lead to secondary nucleation,[1–3] so attrition is not necessary. When antisolvent crystallisation is used, the manner of antisolvent addition can influence primary nucleation kinetics.[4] There may be intermittent heterogeneities in local composition due to differences in local mixing environments, although significant supersaturation overshoots over values corresponding to fully mixed conditions are not expected in antisolvent crystallisation systems.[5]

1 C. Y. Sung, *et al.*, *AIChE J.*, 1973, **19**, 957–962.
2 J. Wang and J. Estrin, *Chem. Eng. Commun.*, 1996, **152–153**, 275–286.
3 M. Yousuf and P. J. Frawley, *Cryst. Growth Des.*, 2018, **18**, 6843–6852.
4 L. A. I. Ramakers, *et al.*, *Cryst. Growth Des.*, 2020, **20**, 4935–4944.
5 R. Miller, *et al.*, *Cryst. Growth Des.*, 2022, **22**, 2192–2207.

Dezső Horváth queried: Based on Kondepudi's findings[1] in the crystallization of sodium chlorate, the suppression of primary nucleation is essential in maintaining the asymmetry during the competition between L- and D-crystals, besides enhancing secondary nucleation. Does this mean that in your system the residence time in the tank reactor has to be kept shorter than the induction time associated with primary nucleation?

1 D. K. Kondepudi, R. J. Kaufman and N. Singh, Chiral symmetry breaking in sodium chlorate crystallization, *Science*, 1990, **250**(4983), 975–976, DOI: 10.1126/science.250.4983.975.

Joop ter Horst answered: I should indeed have added a references to Kondepudi's nice paper.[1] In their batch-wise evaporative crystallization experiments it is not so much the suppression of unselective primary nucleation but rather the enhancement of chirally selective secondary nucleation that is responsible for the single chirality product from stirred solutions. The primary nucleation of the first crystal with a specific chirality gives rise to secondary nucleation of all other crystals having the same chirality.

In our continuous antisolvent crystallization experiments we use single chirality seed crystals to steer the product outcome to the same chirality as the seed crystals. This actually proves that a secondary nucleation mechanism is at play in this process, as only such a mechanism can be chirally selective due to the presence of the parent seed crystals. In the continuous process the seed material is gradually removed with the outgoing suspension flow. New crystals are created by secondary nucleation since chirality is maintained more than 6 residence times. These secondary nuclei have grown to undergo attrition or shear induced secondary nucleation to provide new crystals in the crystallizer.

While the bulk supersaturation is relatively low, in antisolvent crystallization at the local mixing points of antisolvent and solution with the bulk the local supersaturation is quite high. So it is actually quite surprising that no unselective primary nucleation takes place, as proven by the single chirality product, also after more than 6 residence times. It indicates that the primary nucleation rate in this process is very low, despite the high local supersaturations. The small number of primary nuclei of the other chirality that may form may be removed from the crystallizer before they can induce large amounts of secondary nuclei of the other chirality. Therefore it appears that such a continuous process can be operated to produce an enantiopure product much longer than any induction time measured in a batch-wise manner.

1 D. K. Kondepudi, R. J. Kaufman and N. Singh, Chiral symmetry breaking in sodium chlorate crystallization, *Science*, 1990, **250**(4983), 975–976, DOI: 10.1126/science.250.4983.975.

Helen Wheatcroft asked: You state that the secondary nucleation rate is at least 6 orders of magnitude greater than the primary nucleation rate at typical industrial crystallisation conditions. How much variation is expected in the ratio of the primary and secondary nucleation rates as a function of supersaturation?

Joop ter Horst responded: It is conventionally thought that with increasing supersaturation, primary nucleation will become dominant over secondary nucleation. What we see in our experiments is that if the flow rates increase at the same antisolvent fraction, at some point we lose control over enantiopurity and the product moves towards racemic composition (Fig. 8 in the paper; https://doi.org/10.1039/d1fd00098e). However, this point is well beyond what we had expected. Fig. 9 of the paper tells us that this loss of control happens as the suspension density in the crystallizer (related to productivity) goes down with time. This coincides with an increase in supersaturation. Due to the build up of supersaturation in the crystallizer, at some point nucleation occurs, which is probably unselective primary nucleation that compromises the product chirality.

Helen Wheatcroft enquired: Following up on the discussion of whether the secondary nuclei are generated by shear in the slurry or attrition by impact with the vessel and agitator, I wondered how large a secondary nucleus needs to be to grow? Does the critical nucleus size for primary nucleation also apply to secondary nucleation?

Joop ter Horst answered: A secondary nucleus formed by attrition might contain stress or strain, increasing somewhat the solubility of those small attrition fragments. Another effect might be that for small attrition fragments (<1 micron) indeed their total interfacial energy becomes a contributing factor and would also increase the solubility of the small fragments, similar to the primary nucleus in classical nucleation theory. Due to both causes the needed supersaturation to grow the attrition fragments must exceed a certain lower boundary limit. This lower boundary limit of the supersaturation is therefore indeed partly related to the primary nucleus size in classical nucleation theory.

Kristen Fichthorn opened discussion of the paper by Helmut Cölfen: Theoretical studies of the ordering of hard rods and nanowires reveal that various ordered phases can occur, depending on the aspect ratio.[1–3] This may also affect nucleation. Have you looked at the ordered phases that occur once aggregates have formed?

1 R. Blaak, D. Frenkel and B. M. Mulder, *J. Chem. Phys.*, 1999, **110**, 11652.
2 K. Zhao, C. Harrison, D. Huse, W. B. Russel and P. M. Chaikin, *Phys. Rev. E: Stat., Nonlinear, Soft Matter Phys.*, 2008, **76**, 040401(R).
3 D. Triplett and K. A. Fichthorn, *Phys. Rev. E: Stat., Nonlinear, Soft Matter Phys.*, 2008, **77**, 011707.

Helmut Cölfen responded: It is well known for liquid crystals that ordered phases form once a critical aspect ratio of the particles is reached. A preordering in solution may affect nucleation. However, in our system, we did not observe any order in the heterogeneously nucleated patches (scanning electron microscopy) while the particles themselves showed ordering on a TEM grid upon drying. The reason for the missing order for our heterogeneously nucleated particle superstructures is most likely the countercharged particles/surface, which have an attractive potential so that particles immediately stick to the surface once nucleation takes place without the possibility for orientational corrections. However, in other systems than the one with the attractive particle–surface interaction we have used in our study, I can well imagine that also ordered phases will form after heterogeneous nucleation. This is for example the case, when mesocrystals are formed.

Johanna Marie Galloway queried: Would it be possible to look at binary charged systems? Either a mixture of oppositely charged particles, or particles displaying different charges on different faces? Most colloidal self-assembly systems focus on equidimensional spherical or cube like particles, it would be really interesting to see what could be assembled with nanorods.

Helmut Cölfen answered: We would be able to also look at binary charged systems with our combined light microscopy/UV-Vis spectroscopy technique. Nucleation rates could be determined for these binary charged systems as for the presented gold nanorods since we detect nucleated assemblies when they have a size about a micrometer and become visible in light microscopy. I agree that it is interesting to observe the structure of the nanorod self-assemblies. In our case of the countercharged gold nanorods and surfaces, no order could be observed in the self-assembled structures, which we attribute to the charge–charge attraction, hindering orientational corrections of the nanorods on the surface to increase the order in the self-assembled structures.

Christian Kuttner enquired: Mechanistically, do the gold nanorods exclusively assemble at the surface or are there any indications to assume that the assembly takes place in the liquid phase as well? Could the aggregation of nanorods be considered reversible to some extent or is it a classical hit-and-stick mechanism?

Helmut Cölfen replied: We can be sure that the observed nanoparticle assembly only takes place at the particle surface because we place our functionalized mica surface on top of the nanoparticle dispersion. This avoids erroneously detecting nanoparticle assemblies, which may have formed in solution and sedimented onto the mica surface as heterogeneously nucleated particle assembly. Due to the charge–charge attraction between nanorods and functionalized mica surface, we can assume the aggregation of the nanorods at the surface to be irreversible. Support for this assumption comes from the observed lack of order in the nanoparticle assemblies in Fig. 2 in our paper (https://doi.org/10.1039/d1fd00087j) as a result of a lack of orientational correction possibilities for the nanorods in the assemblies.

Alan Hare asked: I realise that we are looking at 2D pictures of the nanorod, but if in 3D I allow myself the luxury of imagining a cylinder topped by a hemisphere at either end, then in nm^2 its surface area is $4\pi r^2$ times the aspect ratio; or, for a radius of 20 nm, a multiple of exactly 1600π nm^2. Given that we are concerning ourselves with heterogeneous nucleation, can you comment on the relationship between aspect ratio and surface area?

Helmut Cölfen answered: The surface area of the nanorods is not really relevant for the consideration of their attachment at a surface during heterogeneous nucleation since for the contact with a surface, only one face will be relevant and its contact area is proportional to the aspect ratio as long as the nanorods attach with a side face, which is the case here.

Alan Hare queried: In Fig. 3 in the paper (https://doi.org/10.1039/d1fd00087j), "slope" *vs.* supersaturation, we see non-linearity: a flattening-off of the curve, apparently; so that when σ has reached 4 or 5, there is little or no further increase in the nucleation rate. Could there be a secondary effect at work, perhaps?

Helmut Cölfen replied: Yes, this indeed seems to be the case, especially for the lowest aspect ratio of 1.75. The nature of a secondary effect is yet unknown.

Rik Drummond-Brydson questioned: For nanorods you have to consider rotational as well as translational diffusion coefficients. Could the differences in kinetics observed for different aspect ratio rods be due to the respective mobilities of the different aspect ratio rods on the surface as they self-assemble to form their superstructure?

Helmut Cölfen responded: On the surface, rotational diffusion is limited due to the limited degrees of freedom of a nanorod to rotate. Since we have charge attraction between the surface and the nanorods, also the translational mobility should be decreased in our case. But in principle, the higher diffusion coefficients of the shorter aspect ratio rods and thus their higher mobility on the surface support the self-assembly.

Joonsoo Kim asked: In Fig. 2 in the paper (https://doi.org/10.1039/d1fd00087j), the surface is not completely covered with AuNRs. If surface–AuNR interaction is based on charge, shouldn't the NPs cover the surface completely and later form aggregates? Is that really happening but just not visible in the SEM image?

Helmut Cölfen answered: The detected self-assembled aggregates are predominantly particle monolayers. If sufficient particles would be added, one could imagine a monolayer coverage with the nanorods because of their charge attraction to the surface but the number of added particles was not yet high enough to reach a full surface coverage in our case. If the surface would be fully covered by the nanorods, we would see this in SEM, because we are able to see single nanorods.

Christine Kirschhock enquired: You use surfactants to stabilise the gold particles and use surfactant desorption to initiate their assembly. Do you expect transport phenomena related to surfactant ad- and de-sorption to affect the particle–particle interactions in a size dependent manner?

Helmut Cölfen replied: Yes, partial surfactant desorption initiates the particle assembly and their heterogeneous nucleation on the negatively charged surface. It is unknown if and in how far the surfactant ad/desorption correlates with the particle size with respect to aspect ratio. If that would be the case, the particle–particle interactions would be size dependent.

Matthew Bennett asked: Surfactants affect the viscosity of the solvent. Does this have an effect upon crystallisation?

Helmut Cölfen answered: If the viscosity of the solvent changes for whatever reason, this will have a direct influence on the diffusion of the nanoparticles as can be quantified *via* the Stokes–Einstein equation. This can then have an effect on crystallization, as the supply of building units is changed, which can even change the growth mode for example to diffusion controlled if the particle diffusion is significantly decreased. A slow particle supply would allow for orientational corrections of the nanoparticles in the nucleated crystal superstructure and can lead to an ordered superstructure while a fast supply can have the opposite effect.

Alan Hare opened discussion of the paper by Alexei Kiselev: Given the beautiful Fig. 2 in your paper (https://doi.org/10.1039/d1fd00115a), is it true to say that the joint pyramidal apex has physicochemical significance and that ice nucleation begins at that point? If not, then what has been drawn as an inverse double pyramid could be drawn as a hexagonal prism, instead; in which case, could nucleation perhaps begin at a vertex of the hexagon?

Alexei Kiselev replied: As discussed in the paper (https://doi.org/10.1039/d1fd00115a), the ice nucleation indeed initiates at the apex of a double pyramid; so this point (or rather, the underlying feature on the feldspar surface) has physicochemical significance: it is the nucleation site. The shape of the double pyramid is a result of the ice growth process in the confined geometry of the sucrose layer. Due to the slow rate of crystal growth, we could visually follow the evolution of the crystal shape right after the nucleation event.

Peter Vekilov enquired: Your discussion of how feldspar, a common aluminosilicate, enhances ice crystal nucleation from vapor is fascinating. I have a small technical question. Even though feldspar may be main mineral in the earth's crust, is it indeed heavily represented in atmospheric dust, where it may interact with emerging ice nuclei?

Alexei Kiselev answered: Feldspars are indeed widely represented in the atmospheric mineral dust aerosol. The major components of atmospheric mineral dust are clay minerals (approx. 60%), quartz (approx. 25%), and feldspar (approx. 12%), but only feldspars have been shown to be ice nucleation active. For details, please refer to ref. 1

1 B. J. Murray, D. O'Sullivan, J. D. Atkinson and M. E. Webb, Ice nucleation by particles immersed in supercooled cloud droplets, *Chem. Soc. Rev.*, 2012, **41**(19), 6519–6554.

Peter Vekilov asked: Do all faces of feldspar crystal equally boost ice crystal nucleation? Are there correlations between the structures of the feldspar faces that support ice crystal nucleation and the ice faces that line them? Do you have an idea about the chemical interaction between water molecules in ice and the feldspar moieties that help the observed nucleation effect?

Alexei Kiselev responded: Not all crystalline faces perform equally well in nucleating ice. As you can see in Fig. 5 of our paper (https://doi.org/10.1039/d1fd00115a), the ice nucleation (IN) efficiencies of cleavage planes (001) and

(010) are almost identical, whereas ice nucleation on the substrate prepared along the (100) crystal plane is enhanced by almost 5 °C. Neither face has a noteworthy match to the crystal structure of ice, if taken in its ideal "undisturbed" state. However, the atomistic modeling has shown[1] that the presence of OH^- groups on the surface of the (100) face creates a flexible layer that can reduce the mismatch between the feldspar framework and prismatic face of ice. A similar layer would be also present at the other surfaces of feldspar, but the highest energy gain (and thus, the lowest free energy barrier for ice nucleation) is achieved on the (100).

1 A. Kiselev, *et al.*, Active sites in heterogeneous ice nucleation—the example of K-rich feldspars, *Science*, 2017, **355**(6323), 367–371.

Thomas Whale enquired: The work of Holden *et al.*[1] demonstrates that different active sites on feldspar nucleate ice in the deposition and immersion modes and provided a range of ideas regarding what might lead to this. Does this work provide any insight to what the differences might be?

1 M. A. Holden, J. M. Campbell, F. C. Meldrum, B. J. Murray and H. K. Christenson, Active sites for ice nucleation differ depending on nucleation mode, *Proc. Natl. Acad. Sci. U. S. A.*, 2021, **118**, e2022859118, DOI: 10.1073/pnas.2022859118.

Alexei Kiselev replied: This is an excellent comment showing that there are still gaps in our understanding of heterogeneous ice nucleation mechanisms. Our work was strictly focused on immersion freezing so I can only offer some thoughts in this respect. The deposition and immersion modes of ice nucleation differ in many ways, but the most important difference is in the availability of water for ice nucleation and growth. In deposition (or pore condensation and freezing (PCF)) mode, some water has to be adsorbed on the surface or form liquid pockets in pores and cracks prior to nucleation onset; whether the sites of condensation coincide with the most active ice nucleation sites is not guaranteed. On the contrary, in immersion mode, all surface defects are equally accessible to water, making nucleation on the most active IN sites highly probable. The other aspect could be that liquid water present in abundance (in immersion mode) would modify the surface properties in general or the properties of the IN sites in particular, by *e.g.* cation exchange or formation of a macroscopic electric double layer (EDL). In deposition mode, such a thing would not happen or could have a different effect on ice nucleation due to confinement effects.

Thomas Whale remarked: It is worth noting that not all ice nucleation sites on feldspar are active in only one mode – Holden *et al.*[1] showed that some sites are effective in both modes.

1 M. A. Holden, J. M. Campbell, F. C. Meldrum, B. J. Murray and H. K. Christenson, Active sites for ice nucleation differ depending on nucleation mode, *Proc. Natl. Acad. Sci. U. S. A.*, 2021, **118**, e2022859118, DOI: 10.1073/pnas.2022859118.

Sarah Price asked: How does the nucleation of ice on feldspars lead to snowflakes?

Alexei Kiselev answered: Formation of a snowflake is a multistage process. In a mixed-phase tropospheric cloud, it starts with the nucleation of ice in a super-cooled cloud droplet. This is where the ice nucleation particle (INP) is needed to facilitate freezing, and feldspar can be a very efficient INP. In a water-saturated environment, the frozen droplet would grow rapidly *via* the water vapor diffusion (Wegener–Bergeron–Findeisen mechanism) and at a certain point too heavy to be kept in an updraft, falling down as a result. On its way to the ground, the ice particle travels through the regions of various ambient temperature and of high and low vapor saturation, where the growth rate of crystal faces of ice varies strongly. This is how the typical dendrite structure of a snowflake appears. So, in principle, feldspar by itself is not responsible for formation of snowflakes, but its unique ability to facilitate freezing of water at low super-cooling is beneficial for formation of large snowflakes in the moderately supercooled convective tropospheric clouds.

Ilaria Sandei enquired: Feldspars and other minerals have been proven to be efficient ice nucleating materials and the paper (https://doi.org/10.1039/d1fd00115a) adds new and interesting insights about the nature of the active planes of feldspars and their effect on ice nucleation and growth. However, overall, the main characteristics that make a material a good ice nucleator are still unclear. In most cases, it has been suggested that a crystallographic match between ice and a material is necessary for this nucleation activity but, in other cases, it has also been shown that less oxidised materials can nucleate ice better than more oxidised ones[1] or that differences in surface roughness can be other important factors in this behaviour.[2] This leads to questions whenever the match between the crystalline lattices of ice and the mineral is the only property that makes a material a good ice nucleator.[3] Would the investigation of the ice nucleating activity of engineered substrates (obtained, for example, by modifying the morphology or the charge distribution of an otherwise neutral material) help to understand better, even in the case of feldspars or other minerals, if their activity results from a contribution of different surface features?

1 T. F. Whale, *et al.*, *J. Phys. Chem. Lett.*, 2015, **6**, 3012–3016.
2 L. Lupi, *et al.*, *J. Am. Chem. Soc.*, 2014, **136**(8), 3156–3164.
3 M. Fitzner, *et al.*, *J. Am. Chem. Soc.*, 2015, **137**(42), 13658–13669.

Alexei Kiselev replied: This is an excellent question. The problem with ice nucleation experiments involving natural substrates is that so many factors are affecting their IN behavior at the same time: surface topography, crystal structure, surface chemistry, charge distribution. In a real world experiment, disentangling the individual contributions of these factor is almost impossible. If we were able to manufacture an artificial substrate with fixed properties where we could vary only one thing at once, the role of individual factors would become more clear. To some extent, this is what we have tried recently by chemically modifying pure “gem-quality” potassium-rich feldspar to induce a series of cracks where certain morphological features become exposed, see ref. 1.

1 A. A. Kiselev, *et al.*, *Atmos. Chem. Phys.*, 2021, **21**(15), 11801–11814.

Jan Sefcik opened discussion of the paper by Nick Pellens: How would the proposed mechanism lead to the spherical shape of the submicron particles observed? What could explain the differences between the sodium and caesium outcomes? Perhaps the spherical shape of particles hints that surface tension plays a role in their formation. This would indicate the presence of liquid–liquid phase transitions which may be involved in particle formation pathways.[1,2] Liquid–liquid phase transitions can also be sensitive to the nature of cations present in solutions.[3,4]

1 J. Šefčik and A. V. McCormick, *Catal. Today*, 1997, **35**, 205–223.
2 K. Z. Gaca and J. Sefcik, *J. Colloid Interface Sci.*, 2013, **406**, 51–59.
3 S. J. Taylor, *et al.*, *Langmuir*, 2014, **30**, 10231–10240.
4 S. J. Taylor, *et al.*, *Langmuir*, 2015, **31**, 13571–13580.

Nick Pellens responded: The CrystalGrower simulation software presents the opportunity to gain deeper insights in the stability of the crystal surface during zeolite crystallisation *via* kinetic Monte Carlo simulations.[1] Comparison of the synthesis experiments with these simulations suggests that zeolite crystallisation in the presence of sodium or cesium cations results in a different crystal surface energy landscape. In the presence of sodium, crystal growth proceeds preferably *via* specific surface sites (large $\Delta\mu b$), while for cesium all surface sites seem to have a comparable stability (small $\Delta\mu b$). The proposed crystallisation mechanism explains such behaviour. Only in the presence of cesium cations, reaction-limited crystallisation kinetics are observed and crystal growth proceeds in pre-equilibrium conditions. In other words, due to the high mobility of Cs cations, fast removal of the cation from the surface freezes a purely statistical distribution of active surface sites. Because redissolution is hindered by the high mobility of Cs cations, nanosized isotropic crystals are perceived to be stable in the Crystalgrower simulations (small $\Delta\mu b$). Even for the synthesis of an identical zeolite framework, crystal morphologies remain dissimilar depending on the cation that is present. Because these observations are closely related to presence or absence of dissolution–reprecipitation chemistry during and after crystal growth, at this moment, it seems unlikely that a cation-dependent surface tension during the phase transition preceding nucleation plays a direct role in determining the crystal morphology of HSIL zeolite crystallisation. Our recent study for HSIL zeolite crystallisation in the presence of sodium, potassium, and cesium cations shows that ion pairs in aluminosilicate-alkali synthesis liquids determine the zeolite aluminium content and topology, further suggesting an invariable nucleation pathway.[2]

1 M. Anderson *et al.*, Predicting crystal growth via a unified kinetic three-dimensional partition model, *Nature*, 2017, **544**, 456–459; A. R. Hill *et al.*, Crystalgrower. A generic computer program for Monte Carlo modelling of crystal growth, *Chem. Sci.*, 2021, **12**, 1126–1146.
2 K. Asselman, *et al.*, Ion-pairs in aluminosilicate-alkali synthesis liquids determine aluminium content and topology of crystallizing zeolites, *Chem. Mater.*, 2022, DOI: 10.1021/acs.chemmater.2c00773.

Jeffrey Rimer queried: My question pertains to the properties of the clusters: is the chemical formula or general size and composition (Si, Al, and alkali content) known, and is the population uniform or is there a distribution of oligomeric species? This leads to the second question involving the chemical reaction used to model pre-nucleation cluster (PNC) addition: if there is a distribution of clusters, is this reflected in the chemical formula? Is the number of alkali metals released per PNC addition known or is there variability?

Nick Pellens answered: The chemical formula, size and composition of the identified ion-pair pre-nucleation are determined by the combination of several characterization tools. Based on conductivity measurements, ^{23}Na NMR and chemical modelling, it is feasible to quantify the fraction of sodium cations engaging in ion-pairing. Hydrogen electrode pH measurements provide an accurate probe to determine the global extent of (alumino)silicate deprotonation, measured *via* the depletion of hydroxide ions in the synthesis liquid. We performed ^{27}Al and ^{29}Si NMR measurements on the synthesis liquids to determine the (alumino)silicate speciation. Combined, this information led to the identification of ion-paired aluminosilicate 4 ring–alkali cations complexes. These 4 rings contain one or two aluminate centers based on the chemical composition of the synthesis liquid, as reflected in the zeolite framework Si/Al ratio. Based on the assumptions of crystallization *via* condensation of solute pre-nucleation clusters, a structural and compositional relation is implied between the liquid-state pre-nucleation clusters and the forming crystal nucleus. Therefore, it is not directly observed, but implied by the wealth of our experimental observations that this relation between the cluster properties and the crystallizing zeolite stands. The derived model is based solely on the first condensation step between a liquid-state PNC and a growing crystal nucleus, as this step should be rate-limiting. Therefore, the model is based on a single alkali metal and hydroxide ion release. However, our experimental observations show that a single ion-paired aluminosilicate 4 ring, on average and depending on the sample composition, ion pairs to 3–7 alkali metals due to the full deprotonation of silicate centers. We are in the progress of publishing these experimental results and, as a result, the related experimental evidence will become available in the near future.[1]

1 N. Pellens *et al.*, Nucleation of porous crystals from ion-paired pre-nucleation clusters, *Chem. Mater.*, 2022, DOI: 10.1021/acs.chemmater.2c00418; K. Asselman *et al.*, Ion-pairs in aluminosilicate-alkali synthesis liquids determine aluminium content and topology of crystallizing zeolites, *Chem. Mater.*, 2022, DOI: 10.1021/acs.chemmater.2c00773.

Michael Anderson asked: Your work is on very well-defined systems that are not the same as a gel synthesis or a templated synthesis of a zeolite. Is it possible to broaden out such studies to these systems?

Nick Pellens responded: Currently, we do not have experimental observation regarding gel or templated zeolite synthesis. Therefore, it is only possible to speculate regarding the general applicability of the observed crystallization kinetics. The main limitation is enabling a molecular investigation that is necessary to construct surface chemistry approximations of the crystallization kinetics. The well-defined physicochemical properties of HSIL zeolite synthesis liquids enable performing such research. To the best of our knowledge and at this moment, it is not possible to directly apply this research strategy to more complex zeolite synthesis systems. However, we are working on enabling *in situ* conductivity measurements for gel synthesis, that is inorganic zeolite crystallisation at a higher supersaturation compared to crystallisation from HSIL synthesis liquids.

Michael Anderson enquired: In the Na system you saw 3 or 4 different phases – do you know if these phases were successive or appeared at the same time?

Nick Pellens replied: In an earlier experiment, we performed an *ex situ* crystallization experiment to gain more insight into the stability of various zeolite frameworks. For the here-studied Na sample, at a synthesis temperature of 60 °C and slightly lower supersaturation (Si/Al = 18), it was observed that the three crystalline phases (FAU, GIS, SOD) appeared at the same time. The three different phases remained stable up to 120 h at 60 °C, but after 168 h, GIS formation was preferred over the FAU and SOD frameworks. In addition, at a higher synthesis temperature of 90 °C (t = 168 h), a mixture of the GIS/ANA frameworks was observed in the solid product. Pushing the synthesis temperature to 160 °C (t = 168 h) resulted in a full conversion into an ANA zeolite product. Based on these arguments, we conclude that dissolution–reprecipitation processes are responsible for our observations of zeolite polymorphism in Na-based HSIL zeolite synthesis.

Alan Hare asked: On seeing a curve like the conductivity curve in Fig. 3 in your paper (https://doi.org/10.1039/d1fd00093d), it is almost intuitive to think "this is Avrami", but did you make this assumption? Alternatively, were you able to *infer* Avrami, and *then* fit the curve to the data? I note that you acknowledge Kolmogorov's mathematics.

Nick Pellens responded: In the case of HSIL zeolite crystallisation, only a small fraction of the liquid solidifies into a crystalline phase. Therefore, the Avrami model was not inferred as is difficult to motivate the direct implementation of the crystallisation kinetics that was derived for the case of a total liquid to solid conversion. Instead, a reasonable set of assumptions was formulated to derive the crystallisation kinetics in the case of the incongruent transformation of only a fraction of the initial system with different compositions of solid and liquid phases. The fact that the resulting equation ($P_N(t) = 1 - \exp[-Gt^z]$) is of the same form as the Avrami model is rooted in the assumption of fast liquid dynamics and kinetics in the homogeneous liquid phase, compared to the surface processes.

Alan Hare enquired: In the model of increasing zeolite fraction (Section 3.2.2; https://doi.org/10.1039/d1fd00093d), your simplification (2) leapt off the page at me because many years ago Sam Levine (R.I.P.), when trying to teach us colloidal stability theory, made the very assertion: "chemical equilibrium does exist [in the liquid, throughout crystallisation]". Nearly 50 years on, my question today has mutated only slightly: not "does equilibrium exist" but "does it *have* to exist"? If equilibrium *doesn't* exist, how does this alter the system description? Does the problem become intractable; or does it now have a different solution?

Nick Pellens answered: For the derivation of the kinetic model, fast dynamics in the liquid state compared to the growth reactions is assumed. With this assumption the liquid phase has a homogeneous composition throughout and the *in situ* conductivity measurements report the accurate state of the growth medium. This way, no depletion effects or change of liquid state speciation needed to be accounted for, which would result in more complex crystallization kinetics, also accounting for the pre-equilibria reactions within the liquid. For example, if the liquid growth medium would be far from chemical equilibrium, the crystallization kinetics might be limited by the supply of suitable ion-paired pre-nucleation clusters, and the lifetime of these clusters must be longer than the time needed to pass the depletion zone. However, considering the nature of the used aluminosilicate fluids, a fast interchange between cation-paired oligomeric species can safely be assumed, so that the probability that a viable pre-nucleation cluster is formed is the same everywhere in the liquid, including the surface of a growing crystal. It also needs to be pointed out that a very low aluminium content was used for our study, so that the global change of the liquid state concentrations, especially the hydroxide content did not change significantly during crystal growth. This allowed focusing on the processes on the surface of the growing crystal, as the chemical state in the liquid only changed marginally.

Alan Hare asked: Regarding the pre-nucleation clusters: do you think that the nucleation mechanism could be consistent with a single energy-barrier model (such as that mentioned by Alexander Van Driessche in the context of $SrSO_4$; https://doi.org/10.1039/d1fd00092f)? I am wondering if there might be a continuous distribution of local energy minima.

Nick Pellens responded: In Na HSILs, the liquid zeolite crystallization medium has an experimentally determined liquid density of 1.3–1.5 g ml^{-1} and laboratory XRD measurements show that these liquids already exhibit local order. The ion-paired pre-nucleation clusters, thus, exist in a liquid medium that has similar properties to the crystallizing solid (average density $\approx$ 1.4 g ml^{-1}). The nucleation pathway proposed by Alexander Van Driessche is based on model systems with a large density gap between the solution and solid phases. Therefore, it is not possible to state whether or not this pathway can apply to HSIL zeolite nucleation.

Matteo Salvalaglio opened discussion of the paper by Romain Grossier: In your paper (https://doi.org/10.1039/d1fd00090j) you present new exciting experimental results on nucleation rate measurement within small sessile droplets. At such small size, the measured nucleation rates are affected by the droplet size, as you demonstrate experimentally and report in the manuscript. In the past, you have provided a general theoretical framework for understanding the nucleation of crystalline particles in droplets that don't exchange mass with a surrounding reservoir.[1] Can you use your own theoretical description to interpret the data gathered in this paper in order to extrapolate to the macroscopic limit the rates measured in small volumes?

1 R. Grossier and S. Veesler, *Cryst. Growth Des.*, 2009, **9**(4), 1917–1922.

Romain Grossier replied: Before interpretation of the here presented data, determining rates J as defined by CNT would imply to model more accurately the contraction dynamics, *i.e.* getting droplet volume evolution with time $V(t)$ (as stated in the paper: "An exact and precise model is complex and out of the scope of this paper, and will be the object of specific studies later on."). This is an ongoing, unpublished up to now, work.[1] It would then be possible to accurately calculate J, based on the here presented data. Then, before trying to extrapolate to larger volumes, I would first use this approach (and maybe collect more data on many other droplet sizes) to answer the following question: should depletion during nucleation be taken into account (thereby proving a thermodynamical aspect of confinement as developed in ref. 2)?

1 R. Cedeno, R. Grossier, N. Candoni, A. Flood and S. Veesler, *Evaporation Dynamics of Sessile Saline Microdroplets in Oil*, 2022, https://hal.archives-ouvertes.fr/hal-03440976.
2 R. Grossier and S. Veesler, Reaching one single and stable critical cluster through finite-sized systems, *Cryst. Growth Des.*, 2009, **9**(4), 1917–1922, DOI: 10.1021/cg801165b.

Peter Vekilov asked: You have collected an enormous set of data on rock salt crystal nucleation. What new insights on the nucleation of sodium chloride from aqueous solutions does this data set provide?

Romain Grossier answered: The aim here is first to present a novel approach for measuring highly time-resolved induction times (without interference of a time to grow to detectable size), able to provide large (statistically representative) reproducible datasets.

And second, to propose first steps, in analyzing data (such as removing outliers) without any nucleation mechanism inference.

Still, if the direct goal was not to learn about NaCl nucleation process, we measured, quantitatively, the (statistical) extent of confinement effects: smaller systems sustain higher supersaturations before nucleation occurs. This large dataset could be useful in simulations addressing such situations for "benchmarking" purposes, or to test different nucleation theories.

Sarah Price requested: Please can you comment on the location of the nucleation within the droplet. From the coffee-ring effect, I would have expected nucleation to tend to occur at the edge of the boundary between the droplet and the surface.

Romain Grossier responded: The exact location of nucleation is not attainable to us regarding spatial and temporal resolution, and also because of the large Marangoni flows that appear when nucleation occurs. Still, exact location (which interface? – contact line, PMMA surface or oil/droplet interface) could be studied in addressing more droplet sizes and see how J scales with diameter and its different exponents.

Liam Hunter enquired: During the generation of your droplet arrays do you assume that the micropipette is not in contact with the PMMA surface or does the size of the generated droplets make it necessary to be in contact? Can you comment on the effect of any pipette induced scratching of the PMMA substrate on the subsequent nucleation behaviour of the arrays? I assume that any surface defects generated this way would become apparent in your heterogeneity/memory tests. The low image acquisition rate may make it difficult to determine the point of nucleation; in your opinion does the nucleation generally seem to preferentially occur at one of the interfaces or is it homogeneous?

Romain Grossier replied: The micropipette needs to be in contact with the surface to generate droplets, and details on the generation mechanism can be found in ref. 1. Regarding potential scratching of the PMMA surface by the micropipette, under the same experimental conditions (PMMA on glass coverslip, spin coating conditions, "femtotip I" glass capillary), we never had any sign of a potential effect, and while observing resulting crystals in SEM[1] (Fig. 4 for example) no apparent scratching was seen. Despite this, we cannot rule out sub-resolution scratches, but one must have in mind the micropipettes have a small wall thickness (#200 nm), making them highly flexible (and fragile) and not prone to apply large forces so that to scratch PMMA. Regarding detection of such potential surface defects with our heterogeneity/memory tests, it may not help in detecting a "heterogeneity" that is uniformly distributed in droplets (or on the surface they rely on): if the micropipette uniformly scratches the PMMA surface during its course, each droplet would have the same "amount" of scratched surface, thereby no reasons to induce a particular behavior in only a subset of droplets. The tests we implemented are more effective to detect embedded heterogeneities ("impurities") which should not uniformly distribute among droplets, or PMMA surface defects localized only in some droplets, in fact any heterogeneity that would always give lower nucleation barrier to specific droplets.

You are right our image acquisition rate does not allow to determine the nucleation point, but higher acquisition rates would not give us such opportunity: spatial resolution to detect "critical cluster" is unattainable and growing crystals in the first seconds are subject to a lot of jittering due to large Marangoni flows induced by large concentration gradients inside the droplets when nucleation occurs, *i.e.* they could travel anywhere else in the droplet before being detected.

For your last question, I would prefer not to give an opinion, but if I had to, I would certainly preferentially bet on heterogeneous nucleation process. Now, it still has many "interfaces" on which nucleation could occur, and information could be obtained by studying a larger set of droplet diameters to see if induction time scales with their diameter (contact line), or with diameter squared (PMMA surface and droplet/oil interface) or diameter cubed (volume embedded heterogeneities – or homogeneous).

1 R. Grossier, Z. Hammadi, R. Morin, A. Magnaldo and S. Veesler, Generating nanoliter to femtoliter microdroplets with ease, *Appl. Phys. Lett.*, 2011, **98**, 091916, DOI: 10.1063/1.3560453.

Jan Sefcik commented: The term "induction time" is normally used to represent the time it takes from the point of reaching a given degree of supersaturation until the point of observing the first sign of crystal formation, while remaining at constant composition and temperature (see ref. 1, 2 and the paper of Joop ter Horst presented earlier (https://doi.org/10.1039/d1fd00098e)). Therefore it may be potentially misleading to use induction times to describe nucleation outcomes while supersaturation is time-dependent, for example in a cooling experiment. It would be useful to clarify terminology used to precisely describe crystallisation processes and their (stochastic) outcomes.

1 N. Javid, *et al.*, *Cryst. Growth Des.*, 2016, **16**, 4196.
2 M. L. Briuglia, *et al.*, *Cryst. Growth Des.*, 2019, **19**, 421.

Alexei Kiselev remarked: About the possibility of multiple nucleation events in the microscopic solution droplets: there are two arguments against it. First, decreasing the observation volume reduces the probability of having an impurity in the droplet which could induce nucleation at low supersaturation. Second, the probability of nucleation (which is a product of volume-specific nucleation rate and volume) is reduced in small droplets compared to larger volumes. Both factors lead to higher supersaturation than can be achieved in small droplets. On the other hand, if nucleation starts at higher supersaturation, the rate of crystal growth immediately after nucleation would also be higher, rendering the probability of a second nucleation event inside the same microdroplet negligible. In fact, this chain of arguments is exactly the reason why homogeneous nucleation should rather be studied in the system of many possibly small identical sample volumes. In view of this argument, the system presented by Dr Grossier is very well suited for such studies.

Several similar systems were reported recently, based on the principle of subdivision of sample volume.[1,2]

1 N. Reicher, L. Segev and Y. Rudich, The WeIzmann Supercooled Droplets Observation on a Microarray (WISDOM) and application for ambient dust, *Atmos. Meas. Tech.*, 2018, **11**, 233–248, DOI: 10.5194/amt-11-233-2018.
2 A. Peckhaus, A. Kiselev, T. Hiron, M. Ebert and T. Leisner, A comparative study of K-rich and Na/Ca-rich feldspar ice-nucleating particles in a nanoliter droplet freezing assay, *Atmos. Chem. Phys.*, 2016, **16**, 11477–11496, DOI: 10.5194/acp-16-11477-2016.

Lucia Maini asked: In your experiments you've got nice cube-like crystals in the first crystallization, but in the second crystallization the shape is totally different. Is it possible to obtain cube-like crystals also in the second or third crystallization?

Romain Grossier replied: Nice cube-like crystals were obtained at step 0, where droplets were left (still under oil layer) at ambient %RH, higher than 10% as used in step 2. Nucleation had time to occur before reaching too high supersaturations which promotes dendritic growth. Also, there is enough supersaturated solution left to complete smoothening to the cubic habit. So, cube-like crystals could certainly be obtained in larger droplets (images of the videos here are extracted from 60 pL droplets array) and/or in less dry conditions during step 2.

Stéphane Veesler noted: In the experiment presented here, relative humidity (RH) cycling (deliquescence–recrystallization), at the end of the crystallization step we observed one dendritic single crystal per droplet. The dendrite is the growth form, and there is no more solution (complete drying) at the end of the experiment to return to the equilibrium form, *i.e.* the cube. In other experimental conditions, in which complete drying was slower, we observed this return to the equilibrium form.[1]

1 Z. Hammadi, N. Candoni, R. Grossier, M. Ildefonso, R. Morin and S. Veesler, Small-volume nucleation, *C. R. Phys.*, 2013, **14**, 192–198.

Romain Grossier added: I agree.

Gábor Schuszter enquired: You mentioned that a mass flow controller will be purchased to more precisely adjust the RH of the air in contact with the covering oil layer. How thick is the oil layer above the droplets? What is the distance separating the droplets? Since the droplets are open to each other as well and not only to the air, do you see any communication (*i.e.*, collective behavior) between them?

Romain Grossier answered: Here, the oil layer thickness is 0.8 mm, as stated in the supplementary information file (note there is a typo error in the paper; https://doi.org/10.1039/d1fd00090j).

At step 0, while generating droplets to obtain the initial crystal array, droplet-to-droplet distance is a function of pressure applied in the micropipette and its velocity on the surface, as described in ref. 1. Resulting crystals, then have positions distributed around the initial droplets' centers. Step 1 (deliquescence of crystals) will then place the resulting droplets roughly centered on the crystals' initial positions.

You're right in saying droplet-to-droplet communications can occur, and this was the purpose of a previous study[2] where we explained their nature and developed the image processing technique we here use. This image processing technique is able to show droplet-to-droplet crystal nucleation mediated interactions: Fig. 3 in the paper clearly shows such individual interactions. Knowing how to measure such interactions, and parameters (oil chemistry, thickness, viscosity, above oil layer %RH) to either enhance or avoid them, we here developed and characterized two %RH controller module versions, which both revealed efficient removal of droplet-to-droplet interactions. Plus, other individual dynamics (see section "Distribution of characteristic time points"), not occurring through crystal nucleation, are addressed through the use of a dimensionless time. Finally, since we're able to both detect and get rid of droplets' interactions, we see no reason to focus on droplet-to-droplet distances.

1 R. Grossier, Z. Hammadi, R. Morin, A. Magnaldo and S. Veesler, Generating nanoliter to femtoliter microdroplets with ease, *Appl. Phys. Lett.*, 2011, **98**, 091916, DOI: 10.1063/1.3560453.
2 R. Grossier, V. Tishkova, R. Morin and S. Veesler, A parameter to probe microdroplet dynamics and crystal nucleation, *AIP Adv.*, 2018, **8**, 075324, DOI: 10.1063/1.5034443.

Sten O. Nilsson Lill opened discussion of the paper by Vivek Verma: For the hydrogen bond propensity (HBP), did you model both Gly and Digly as zwitterions? If yes, they both would interact *via* an ion–ion interaction and this is much stronger than H-bonding. I would suggest to look into the HBP for Digly again if this was modeled as a zwitterion, the value should be very similar to that found for Gly.

Vivek Verma responded: Yes, HBP was modelled for both Gly and Digly as zwitterions. Since Digly has two oxygens and two nitrogens compared to only the one nitrogen and one oxygen of glycine, the HBP of the terminal N and O on both Gly and Digly would be similar while the central N and O of Digly will have different HBP. Now, I am not sure which N and O of Digly are participating in the H-bonding with silica and I guess this will have to be varied with molecular dynamics simulations. We have started working on the MD simulations of the same systems to gain more insight on the heterogeneous nucleation mechanism and see which groups are participating in the H-bonding.

Ruel Cedeno enquired: My question is on the possible effect of glass beads on the hydrodynamics of the system. Studies have shown that as the shearing rate increases, the pre-exponential factor can also increase in certain hydrodynamic regimes due to the increased advective transport towards the nucleus.[1] The presence of glass beads during agitation (as in grinding experiments) can induce more turbulence in the system leading to higher shear rates. This is supported by your crystal images where more fine particles are seen in the case with glass beads.

Is it possible that the observed increase in prefactor is related to the increase in shearing rate rather than the H-bonding capability of the beads? To eliminate the interference of hydrodynamics, it would be interesting to compare the functionalized glass beads against inert beads in order to see the templating effect at identical hydrodynamic conditions.

1 F. Mura and A. Zaccone, *Phys. Rev. E*, 2016, **93**, 042803, DOI: 10.1103/PhysRevE.93.042803.

Vivek Verma replied: The increased shear rate could potentially increase the nucleation rate: $J = A\exp(-\Delta G_c{}^*/RT)$
Based on the dimensional analysis of the above equation, nucleation rate (J) and pre-exponential factor (A) should have the same unit. For J this is usually expressed as the number of nuclei of a size greater than the critical nucleus size generated per unit volume and per unit time (no. of stable nuclei $m^{-3}\ s^{-1}$). On this basis, A may be defined as the total number of nuclei (or clusters) of any size generated per unit volume and per unit time (no. of pre-critical and stable nuclei $m^{-3}\ s^{-1}$). The exponential factor is the fraction of these pre-critical nuclei which can advance to the critical size. So, it could be possible that shear would be leading to more nucleation sites. But, I still believe that H-bonding is responsible for the nucleation to start despite the increase in nucleation sites. H-bonding provides the driving force for the crystal nucleus to form leading to the crystal propagation.

I agree, that this could be cross verified by using the inert templates and we have already started working on using the inert beads along with the different functionalised beads to see the effect of different functional groups on the heterogeneous nucleation.

Aurora Cruz-Cabeza asked: Your microscopic image (Fig. 5 in the paper; https://doi.org/10.1039/d1fd00101a) shows very small fines for the samples obtained by crystallisation with glass beads. Glass beads will have a significant impact on breakage and attrition thus enhancing secondary nucleation. How can you ascribe your results to heterogeneous nucleation on beads rather than secondary nucleation?

Vivek Verma answered: I agree that the microscopic image shows some fines and they could potentially be due to the secondary nucleation from the impact of glass beads happening after the nucleation event. The main aim of the study was to observe the change in the induction time in the absence and the presence of glass beads. And the experimental data suggests that there was reduction in induction time in the presence of glass beads confirming the effect of the heterosurface and hence the propagation of heterogeneous nucleation. The induction time value was later used to obtain the nucleation rate and subsequently thermodynamic and kinetic parameters. The next phase of the study will investigate the secondary nucleation phenomenon.

Helen Wheatcroft questioned: Was the pH controlled in the experiments with and without the glass beads? As you mention in the introduction to the paper (https://doi.org/10.1039/d1fd00101a), the pH is known to affect the rate of crystallisation of glycine (ref. 47 in the paper; ref. 1 here). How was pH controlled or monitored in your experiments and did the addition of the glass beads cause a change in the solution pH?

1 C. S. Towler, R. J. Davey, R. W. Lancaster and C. J. Price, *J. Am. Chem. Soc.*, 2004, **126**, 13347.

Vivek Verma responded: We did not measure the pH before and after the experiments but I think the pH did not change much during the experiments. From a different study in the group we know it is difficult to change the pH of Gly and Digly due to the zwitterion effect. And secondly, the crystallisation of Gly and Digly will be influenced if the pH variation is significant. But having said that, I will measure the pH to see the change in pH with and without glass beads. Also I will use a different amount of glass beads to see the change in overall pH and if that has any effect on the nucleation behaviour.

Jan Sefcik said: Primary nucleation kinetics can depend on both glass–solution surface area[1] and agitation/fluid shear.[2] It would therefore be interesting to investigate varying the surface area of the beads and stirring in the nucleation experiments involving suspended beads.

1 C. Forsyth, *et al.*, *Cryst. Growth Des.*, 2016, **16**, 136.
2 M. J. Vesga, *et al.*, *CrystEngComm*, 2019, **21**, 2234.

Vivek Verma answered: Yes, I agree that nucleation kinetics will depend on both glass bead size and agitation rate. I am currently using a Design of Experiment approach to look at the effects of different glass bead size and agitation rate on the nucleation rate of glycine and diglycine.

Aaron R. Finney continued discussion of the paper by Nick Pellens by communicating: During the discussion, you listed criteria for species in solution before nucleation to be labelled prenucleation clusters. One of these was "stability". What specifically is meant by stability in this context and how is this determined in your work? Clusters are also predicted from classical nucleation theory and a key difference between these clusters and those proposed in the prenucleation cluster pathway is that the probability of "classical" clusters decays rapidly as a function of increasing size below some critical size, beyond which growth occurs. Why can't the clusters be classical?

Nick Pellens communicated in response: The paper (https://doi.org/10.1039/d1fd00093d) and discussion of the work refer to the stability of nm-scale ion-paired pre-nucleation clusters by the observation of their room-temperature stability in HSIL zeolite synthesis liquids. By ^{27}Al and ^{29}Si NMR measurements, it is verified in a nearly identical sample with a slightly lower aluminate content (Si/Al = 18) that liquid-state (alumino)silicate oligomer speciation remains unchanged, independent of whether the speciation is probed directly after sample preparation, or after resting the sample for 14 days at room temperature. The ion-paired aluminosilicate 4 ring oligomers, acting as pre-nucleation clusters, thus are stable observables, as probed by quantitative high-resolution NMR spectroscopy. Within the time period of weeks, these pre-nucleation clusters are kinetically stable and only condense to crystallize zeolites when exposed to a hydrothermal treatment. While we refer to ion-paired pre-nucleation clusters, these consist of aluminosilicate four ring oligomer anions, ion-paired to alkali metal cations. Classical nucleation theory states that these clusters should be highly unstable due to their small size. However, we show that these oligomers are stabilized by ion-pairing, something classical nucleation theory cannot account for. In addition, the impact of crystal surface stability on HSIL zeolite crystallisation, evidenced by the presented work, depends on the ion that is present, not solely on the minimization of the interfacial surface. Based on these considerations, while the observed experimental data and crystallization kinetics could be compatible with the classical theory it seems more appropriate to tackle the crystallization behaviour from the viewpoint of molecular (in)stability.

Zhiyu Liao continued discussion of the paper by Yuki Kimura by communicating: In the section 'The role of ACC as a precursor', it says 'Comparing ACC and solution, the structure of ACC is closer to the structure of calcite than to that of the solution', but later it is suggested that 'the ACC behaves in a fluid manner'. This is kind of contradicting; do you have evidence to support the former statement? Or is it an assumption?

Yuki Kimura communicated in reply: Because the existence of high fluidity is contrary to the observed fact that ACC particles do not fuse to each other, we reject that the ACC behaves in a fluid manner. Instead, another possibility has been suggested: that there are many water-filled voids in the ACC. Part of the supporting data indicating the structure of ACC is closer to the structure of calcite than to that of the solution is very low interfacial energy between ACC and calcium carbonate crystals, which has been described in the first paragraph of "The role of ACC as a precursor" section.

Zhiyu Liao continued discussion of the paper by Romain Grossier by communicating: In the paper, as well as during the discussion, it's stressed that only a single crystal is formed within each microdroplet. However, from Fig. 2 (bottom right; https://doi.org/10.1039/d1fd00090j), it seems there is more than one crystal formed in the droplet. What does it really imply when you have a single crystal or multiple crystals in each droplet, does it indicate heterogeneous or homogeneous crystallisation?

Romain Grossier communicated in response: In Fig. 2 in the paper, bottom right, there is only one single crystal: its fast growth implies a dendritic growth mechanism with possible subsequent smoothening if enough supersaturated solution is still available. Due to both droplet size and high supersaturation (fast growth), it is nearly impossible for a second nucleation event to occur before the first has entirely depleted the solution. At least, experimentally, under the here presented conditions, we never had more than a single crystal.

Equivalent droplets have been used previously where we did induce nucleation with a "controlled heterogeneity": a sharp tungsten tip.[1] In these experiments, yes, it was possible to obtain several crystals due to the heterogeneous nucleation mechanism. But identifying emergence of multiple crystals with a strict signature of a heterogeneous nucleation mechanism is a step I won't take.

1 R. Grossier, Z. Hammadi, R. Morin and S. Veesler, Predictive nucleation of crystals in small volumes and its consequences, *Phys. Rev. Lett.*, 2011, **107**, 025504, DOI: 10.1103/PhysRevLett.107.025504

Marko Ukrainczyk continued discussion of the paper by Joop ter Horst by communicating: Secondary nucleation mechanism: primary *vs.* secondary nucleation is investigated in small scale vial experiments and nicely exemplified with seeded fed-batch and continuous antisolvent crystallisations of sodium bromate deracemisation in this paper (https://doi.org/10.1039/d1fd00098e). The secondary nucleation mechanism responsible for the dominance over primary nucleation is briefly mentioned in this work, namely due to attrition/shear. However, it is hard to expect attrition is the dominant mechanism in the performed seeding crystallisation experiments using a single crystal to induce secondary nucleation. When a crystal seed is introduced into supersaturated solution a pronounced secondary nucleation event is observed. As discussed after the talk, this was also apparent after seeding supersaturated solution with a single crystal without stirring, where no significant attrition/shear should be expected. Recent work published by BMS[1] focused on the shear and attrition contribution of secondary nucleation to particle size distribution (PSD) control *via* population balance modeling (PBM). In contrast, this work investigated scale-dependency of secondary nucleation, by quantifying shear forces *via* impeller power number and impeller pumping number and introducing them in the PBM model to describe the secondary nucleation contribution, in order to model and gain control over PSD over different scales (lab and pilot/manufacturing). However, recent work shed light on the secondary nucleation mechanism (supported by both computational and experimental evidence[2]) and pointed to a surface catalytic activity mechanism *via* "loose" 2D nucleation on a crystal surface followed by detachment of the loosely bound nucleus from a surface into the solution. The energy barrier for such a surface 2D nucleation is significantly lower as opposed to primary (3D) nucleation in the bulk of a solution. Also such a 2D surface (crystal seed templated) nucleation could also be the dominant reason for nucleation control in terms of preferential polymorphism or chiral handedness deracemisation *via* crystallisation, as opposed to uncontrolled primary nucleation in the bulk of the solution where other undesired polymorphs/enantiomers can form. Such a surface catalytic nucleation mechanism can offer a better explanation for the dominant secondary nucleation event (studied/discussed in Prof. Joop ter Horst's paper) and thus the polymorph or chiral purification outcome (facilitated by templated nucleation on the crystal seed surface of the desired polymorph or enantiomer). It is expected (and in fact observed in our experimental reality) that such a secondary nucleation mechanism will be proportional to (beside supersaturation and temperature) the crystal seed-bed surface area and shear mixing, rather than solely attrition caused by the impeller or crystal/crystal impacts (dependent on agitation rate and crystal brittleness). Industrial relevant secondary nucleation after seeding is well established in industrial crystallization, where crystal seed is introduced in supersaturated solution to trigger (secondary) nucleation where often about 30% of the solute/batch is nucleated/crystallized at this early stage of the process (seeding stage followed by aging period). Such a secondary nucleation strategy upon seeding is often utilized to gain polymorph control (ensuring the desired form throughout the crystallisation process), and product PSD and morphology control. Therefore, fundamental knowledge of secondary nucleation is important, but it is less studied in the literature, in comparison to primary nucleation, due to experimental complexity and decoupling the problem of primary *vs.* secondary nucleation.

The application note on enantiomeric purification control (addressed in this paper) is indeed novel and much welcomed to be able to increase throughput/productivity of the deracemisation process, in comparison to Viedma ripening. How is the product purity affected by such a seeded process at relatively high supersaturations (to induce secondary nucleation after seeding), in comparison to the aging near equilibrium Viedma ripening process?

1 T. Rosenbaum, L. Tan and J. Engstrom, Advantages of utilizing population balance modeling of crystallization processes for particle size distribution prediction of an active pharmaceutical ingredient, *MDPI Processes*, 2019, 7, 355, DOI: 10.3390/pr7060355.
2 J. Anwar, S. Khan and L. Lindfors, Secondary crystal nucleation: nuclei breeding factory uncovered, *Angew. Chem., Int. Ed.*, 2015, **54**, 14681, DOI: 10.1002/anie.201501216.

Joop ter Horst communicated in reply: In our experiments we see enantioselective secondary nucleation occurring. We indeed do not make definite conclusions about the secondary nucleation mechanism. However, the fast rotating tip of the stirrer and the large seed crystal combined with the supersaturation create optimal conditions for attrition-based secondary nucleation. Anyway, we leave the door open to other enantioselective secondary nucleation mechanisms such as shear-induced secondary nucleation.

For deracemization processes the main product purity aspect is the enantiopurity, often quantified in the enantiomeric excess. The paper shows that under the right conditions the enantiomeric excess is maintained at close to 100% in a continuous antisolvent crystallization process. The impurity rejection of other impurities is not investigated. A major source for poor impurity rejection[1] can be agglomeration but this was not observed. Another source might be inclusion formation. Anyway, while impurity rejection in continuous antisolvent crystallization might be compromised by the high local supersaturation, in Viedma ripening the harsh grinding conditions might lead to incorporation of all kinds of impurities into the chiral product as well.

1 S. J. Urwin, G. Levilain, I. Marziano, J. M. Merritt, I. Houson and J. H. Ter Horst, A structured approach to cope with impurities during industrial crystallization development, *Org. Process Res. Dev.*, 2020, **24**(8), 1443–1456, DOI: 10.1021/acs.oprd.0c00166.

Jacek Zeglinski continued discussion of the paper by Aaron R. Finney by communicating: Can you see in your simulations that solvent/water molecules are entrapped in those transient disordered clusters of NaCl ions. If this is the case, is this observed in low density clusters only, or maybe also in the high density assemblies? Would you consider such entrapped water molecules to be highly impactful on the energy barriers to nucleation/crystallisation of NaCl? Can those solvent-induced energy barriers be quantified with MD simulations?

Aaron R. Finney communicated in response: The low-density ionic networks are highly hydrated and the water molecules – as well as ions for that matter – in these liquid-like clusters are in dynamic pseudo-equilibrium with the surrounding environment; therefore, while water is associated in the clusters, these can exchange readily with dispersed water molecules on the simulation timescales. Water molecules cannot be described as 'trapped' on this basis, in contrast to water molecules in amorphous solids or crystals with extremely slow diffusion. In brute force simulations, the solvent-exchange dynamics determines the efficiency of sampling different ion solvation states, and that was not an issue in the case of NaCl(aq) simulated for 10^1–10^2 ns.

As the size of transient clusters increases, we observe significant heterogeneities in the local chemical ordering of ions, and dense ion regions emerge which are internally dehydrated. Characterisation of the size of the dense regions in the clusters (see Fig. 1C of our paper; https://doi.org/10.1039/d1fd00089f) makes use of a first-coordination sphere criterion, such that n here represents the scaled number of ions with five-fold or more counter-ions in direct contact. We do not observe, therefore, entrapped water molecules in the high-density regions where crystallisation occurs.

Exchange of water molecules with counter-ions must contribute to the energy barriers associated with nucleation. After all, in the assembly of ions into an ion-rich amorphous precursor or directly into a crystal, solvent removal is a crucial step. In the case of the two-step pathways we identified in our work, the energy landscape in Fig. 4A (see the red dashed line; https://doi.org/10.1039/d1fd00089f) indicates first an increase in the size of the high ion density region in clusters, as n^{sph} increases when $n^{\text{sph}}(q6) = 0$. This is directly associated with solvent removal and carries an energy penalty. Though not presented explicitly in our paper, we used nudged elastic band calculations to determine the free energy profiles along the most likely two-step pathway, and while the barrier for the reordering of ions in the dense regions dominates the nucleation barrier, the removal of water in a first-step significantly contributes to the free energy change for nucleation (*i.e.* beyond thermal energy).

It is possible, therefore, to compute barriers for solvent removal in simulations and these are correlated with increased ion–ion coordination. In our approach, we performed a range of enhanced sampling and seeding simulations to provide an ensemble of initial configurations to initiate trajectory swarms using unbiased molecular dynamics; from these we constructed a Markov state model to determine the free energy landscape and the barriers for nucleation. We also note that in Fig. 1E of our paper we computed the potential of mean force for ion-pairing in the dilute limit, where the thermodynamic energy barrier for ion desolvation during the formation of a contact ion pair is around $4k_BT$. While this is only an approximate estimate for the barrier – an accurate estimate here requires sampling of an extended reaction coordinate – it offers a framework to compute the energy barriers for solvent removal around single ions. Additional computational methods for enhanced sampling and calculation of energy barriers, and challenges therein, are discussed in ref. 1 and 2.

1 G. C. Sosso, *et al.*, *Chem. Rev.*, 2016, **116**, 7078–7116.
2 K. E. Blow, *et al.*, *J. Chem. Phys.*, 2021, **155**, 040901.

Sten O. Nilsson Lill continued discussion of the paper by Joop ter Horst by communicating: Did you try any other antisolvents than ethanol, for example a chiral alcohol, and explore its impact on the kinetics of the control of enantiomeric excess/deracemization?

Joop ter Horst communicated in reply: We did not try other antisolvents for sodium bromate. It is an interesting idea to try though. Solubility measurements of chiral compounds in chiral solvents by the group of Heike Lorenz, however, did not indicate large changes (see ref. 1). I expect that the creation of supersaturation by the antisolvent is more important than its chirality in determining the outcome of the antisolvent crystallization.

1 S. K. Tulashie, H. Lorenz and A. Seidel-Morgenstern, Solubility of mandelic acid enantiomers and their mixtures in three chiral solvents, *J. Chem. Eng. Data*, 2010, **55**(11), 5196–5200, DOI: 10.1021/je1006955.

Alfred Y. Lee continued discussion of the paper by Vivek Verma by communicating: In Fig. 5 in the paper (https://doi.org/10.1039/d1fd00101a), there is significant particle size differences between the homogeneous and heterogeneous cases for both compounds. Was there agitation in each system? If yes, how can you conclusively rule out that secondary nucleation is not driving the enhanced nucleation and contributing as the dominant factor? In terms of the fines shown in both heterogeneous examples, is this akin to media milling?

Vivek Verma communicated in response: The overall particle sizes of glycine and diglycine crystals are almost the same, but they appear different due to different scale bars on each image. Yes, both the homogeneous and heterogeneous solutions were agitated and that could potentially lead to some fines especially in the heterogeneous case as visible in Fig. 5 in the paper (https://doi.org/10.1039/d1fd00101a) and that could be due to the secondary nucleation from the impact of glass beads happening after the nucleation event. Though we are not negating the secondary nucleation effect, the main aim of the study was to observe the change in the induction time in the absence and the presence of glass beads. And the experimental data suggests that there was reduction in induction time in the presence of glass beads confirming the effect of the heterosurface and hence the propagation of heterogeneous nucleation. The induction time value was later used to obtain the nucleation rate and subsequently thermodynamic and kinetic parameters. The next phase of the study will investigate the secondary nucleation phenomenon.

Alfred Y. Lee continued discussion of the paper by Romain Grossier by communicating: The RH cycling process for deliquescence–recrystallization is analogous to metastable zone width determination *via* heating and cooling. In the latter, it has been described in the literature[1] that the thermal history of the solution may influence the metastability. Curious to know if there is a similar effect noted in the RH cycling process – perhaps an effect with elevated RHs (akin to overheating), or possibly a crystallization memory effect due to the repetitive RH cycling?

1 K. Hussain, G. Thorsen and D. Mathe-Sorenssen, *Chem. Eng. Sci.*, 2001, **56**, 2295.

Romain Grossier communicated in reply: Fig. 5 (cycles nucleation rank comparison) in the paper (https://doi.org/10.1039/d1fd00090j) was designed for that purpose. If one has in mind that we here have only two cycles in interpreting it, it does not show any memory effects. More cycling may have to be done to specifically address the question.

Conflicts of interest

There are no conflicts to declare.

Faraday Discussions

ROYAL SOCIETY OF CHEMISTRY

PAPER

The influence of iodide on the solution-phase growth of Cu microplates: a multi-scale theoretical analysis from first principles†

Junseok Kim[a] and Kristen A. Fichthorn *[ab]

Received 21st November 2021, Accepted 14th December 2021

DOI: 10.1039/d1fd00091h

We use first-principles density functional theory (DFT) to quantify the role of iodide in the solution-phase growth of Cu microplates. Our calculations show that a Cu adatom binds more strongly to hcp hollow sites than fcc hollow sites on iodine-covered Cu(111) – the basal facet of two-dimensional (2D) Cu plates. This feature promotes the formation of stacking faults during seed and plate which, in turn, promotes 2D growth. We also found that iodine adsorption leads to strong Cu atom binding and prohibitively slow diffusion of Cu atoms on Cu(100) – a feature that promotes Cu atom accumulation on the {100} site facets of a growing 2D plate. Incorporating these insights into analog experiments, in which we initiated the growth of Cu plates from small seeds consisting of magnetic spheres, we confirmed that two or more stacking faults are required for lateral plate growth, consistent with prior studies. Moreover, plates can take on a variety of shapes during growth: from triangular and truncated triangular to round and hexagonal – consistent with experiment. Using absorbing Markov chain calculations, we assessed the propensity for 2D *vs.* 3D kinetic growth of the plates. At experimental temperatures, we predict plates can grow to achieve lateral dimensions in the 1–10 micron range, as observed in experiments.

Introduction

Understanding the growth of two-dimensional (2D) fcc metallic nanomaterials is both fundamentally and technologically important. From a technological perspective, the unique structure and properties (mechanical, plasmonic, electrical, and catalytic) of 2D metal nanoplates and microplates makes them promising for a range of applications,[1–5] including catalysis and electrocatalysis,[6–11] sensing,[12–15] conductive inks,[16] theranostics,[17–20] and multi-functional

[a]*Department of Chemical Engineering, The Pennsylvania State University, University Park, PA 16802, USA. E-mail: fichthorn@psu.edu*

[b]*Department of Physics, The Pennsylvania State University, University Park, PA 16802, USA*

† Electronic supplementary information (ESI) available. See DOI: 10.1039/d1fd00091h

property enhancement of polymer nanocomposites.[21] Many of the beneficial properties of 2D nanoplates, microplates, and nanoprisms stem from their surface plasmon resonance (SPR) and surface enhanced Raman scattering (SERS) properties, which depend on the plate dimensions.[22–24] Thus, it is of interest to obtain precise control of plate dimensions and shapes when they are synthesized.

One way to synthesize 2D metallic nanostructures is through solution-phase methods.[25–30] In these syntheses, a metal salt is reduced by solvent and/or solution-phase additives. Metal atoms and ions aggregate to form nuclei, which evolve to seeds. It has been suggested that 2D nanocrystals originate from plate-like seeds.[29,31] Capping agents, or additives present in the synthesis, can facilitate the formation of 2D structures from seeds.

As shown in Fig. 1, fcc metal nanoplates and their putative seeds possess relatively large, basal {111} facets on the top and bottom, while the sides contain a mix of {111}, {100}, and possibly higher-index facets. Many studies have emphasized the need for plates to have stacking faults and/or twin boundaries to maintain their 2D geometry.[32–36] Recently, Tan *et al.*[35] showed that the number of the twins affects the symmetry of the seed, which determines the final structure of the plate – or whether a plate will form at all.

In this study, we use multi-scale theoretical methods to examine the role of iodide (I^-) in producing Cu microplates in an experimental solution-phase synthesis.[37] In a previous study, we used first-principles, density-functional theory (DFT) to show that iodide stabilizes Cu(111) facets (lowers their surface energy) to a greater degree than it stabilizes Cu(100) facets, providing a thermodynamic driving force for plate growth from a seed.[37] While this thermodynamic driving force can be helpful in creating large basal {111} facets, it does not explain plate shapes or why, for example, plate seeds do not transform to other {111}-faceted shapes, such as octahedra, to maximize the bulk energy and minimize the surface energy. To address these questions, we use DFT calculations to quantify energies and kinetics relevant for the growth of plate-like Cu seeds to 2D crystals. We construct plate-like seeds containing different numbers of stacking faults from magnetic spheres and, guided by DFT energies, we grow the plates by adding spheres to the most energetically preferred sites on the sides of the plates. Incorporating the DFT rates into absorbing Markov chain calculations, we quantify net rates for inter-facet transport, which allows us to predict the size of

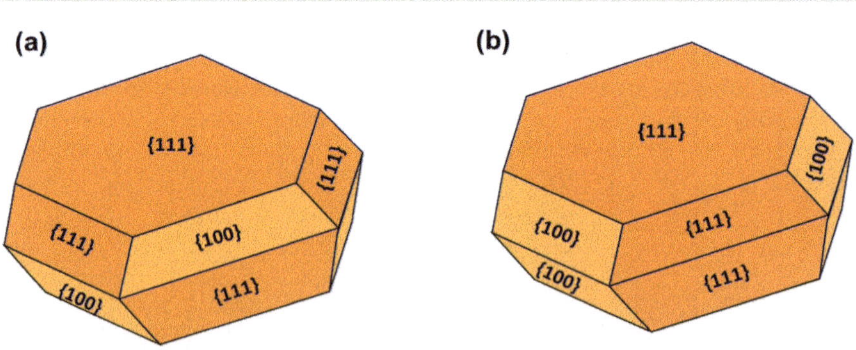

Fig. 1 Schematic of two popular models for fcc metal nanoplate geometries. (a) The alternating facet model; and (b) the mirror symmetry model.

the basal {111} facets. Our studies agree with experiment and confirm the important role of iodide in facilitating plate growth.

Results and discussion

Cu atom binding on I-covered Cu surfaces

In previous work, we used DFT to study likely surface environments for Cu microplates to form and grow from $CuCl_2$ salt in aqueous solution containing hexadecylamine (HDA) and NaI.[37] These studies showed that solution-phase I^- displaces Cl^- (from the $CuCl_2$ salt) and HDA from both Cu(111) and Cu(100) under the synthetic conditions that were the most conducive to plate formation and growth.[37] The predicted structures of Cu(100) and Cu(111) under these conditions are shown in Fig. 2.

For Cu(111), iodine has the $(\sqrt{3} \times \sqrt{3})R30°$ lattice structure at 1/3 monolayer (ML) coverage,[37] where a ML is defined as the ratio of adsorbed I atoms to the number of Cu atoms in the topmost surface layer. Iodine atoms can be placed on either fcc or hcp hollow sites of Cu(111), as shown in Fig. 2(a) and (b), respectively. Interestingly, both structures show similar binding energies for iodine [−1.90/

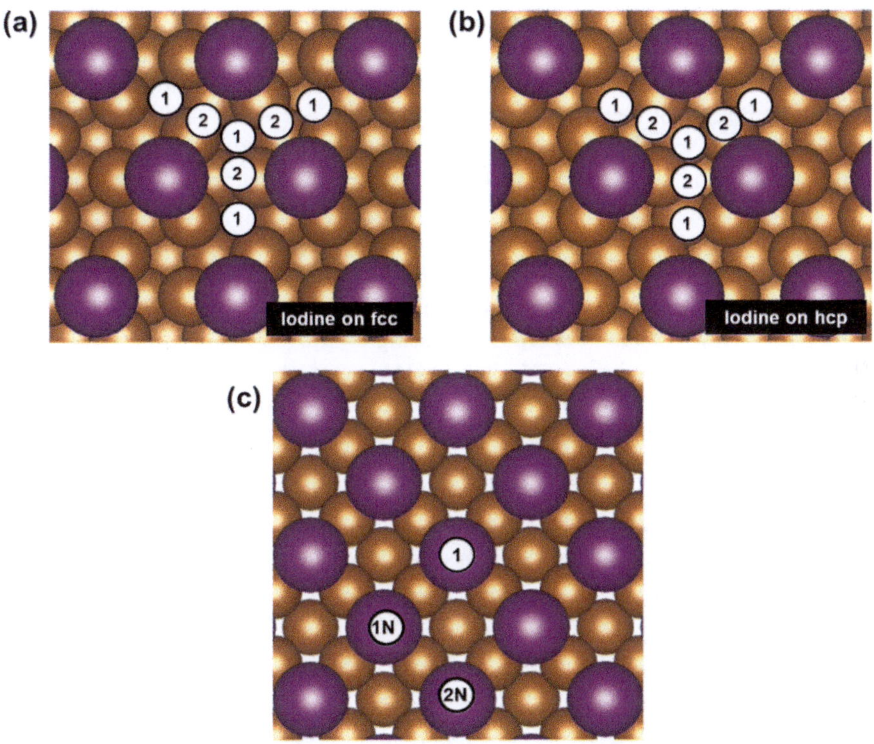

Fig. 2 Binding sites for Cu adatom on Cu surfaces with adsorbed I for: (a) 1/3 ML-I on fcc hollow sites of Cu(111); (b) 1/3 ML-I on hcp hollow sites of Cu(111); and (c) 1/2 ML-I on hollow sites of Cu(100). A Cu atom has two different binding sites on the {111} facets – fcc and hcp three-fold hollow sites – and one binding site on {100} – in a four-fold hollow site beneath an adsorbed I atom. First (1N) and second neighbor (2N) sites are indicated on the {100} surface. Cu is shown in orange, and I is purple.

Table 1 Binding energies and surface diffusion energy barriers for a Cu adatom on iodinated Cu surfaces

Surface	Binding position	Binding energy (eV)	Forward (reverse) diffusion barrier (eV)
Fig. 2(a)	fcc ①	−3.51	① → ②: 0.12 (<0.01)
	hcp ②	−3.39	
Fig. 2(b)	hcp ①	−3.53	① → ②: 0.14 (<0.01)
	fcc ②	−3.39	
Fig. 2(c)	Subsurface-hollow ①	−5.03	① → ① (1N) exchange: 2.01 ① → ① (2N) hop: 1.33

1.91 eV for Fig. 2(a) and (b)]. For Cu(100), a $c(2 \times 2)$ lattice structure with 1/2 ML coverage was found to be the most favorable structure,[37] as shown in Fig. 2(c).

Fig. 2 also shows possible binding positions for a Cu adatom on the iodinated Cu surfaces, with corresponding binding energies listed in Table 1. As we see in Fig. 2(a) and (b), a Cu adatom has two possible binding sites on the Cu(111) surfaces. Cu adatoms exhibit the strongest binding in three-fold fcc or hcp hollow sites ① and they coordinate with three neighboring iodine atoms. The binding in site ② is relatively weaker and the Cu adatom establishes two-fold coordination with its neighboring I atoms. Interestingly, a Cu adatom binds more strongly to the hcp hollow site of Cu(111) in the presence of iodine [Fig. 2(b)], unlike bare Cu(111). This binding preference promotes the formation of stacking faults which have been linked to microplate growth.[32–36]

Though we tried several possible initial sites on Cu(100), we found only one favorable binding site for a Cu adatom ①, in which the atom resides below an adsorbed I on a fourfold hollow site. As we see in Table 1, the Cu binding energy is strong in this site compared to the sites on Cu(111).

We investigated the surface diffusion of a Cu adatom on each of the surfaces in Fig. 2 using the climbing-image nudged elastic band (CI-NEB) method.[38] The diffusion energy barriers between neighboring binding sites are listed in Table 1. Here, we see that diffusion on 1/3 ML I–Cu(111) proceeds relatively rapidly. On 1/2 ML I–Cu(100), we investigated diffusion to two neighboring binding sites – the first (1N) and second (2N) neighbor subsurface hollow sites. We found diffusion to the first-neighbor binding site occurs *via* an exchange process, in which the Cu adatom takes the place of a Cu surface atom, which then becomes the adatom. We find hopping is preferred for diffusion to the second-neighbor site. As we see in Table 1, both diffusion pathways require prohibitively high energy barriers so that the diffusion of a Cu adatom is virtually impossible on 1/2 ML I–Cu(100) under experimental conditions. Thus, the adsorption of iodine leads to a stronger binding and slower diffusion of a Cu adatom on Cu(100) than Cu(111), which indicates a preferred accumulation of Cu on Cu(100). As we will elaborate below, this binding preference contributes to the lateral growth of Cu plates by promoting the accumulation of Cu on the side {100} facets of the plates.

Preferred microplate shapes

As shown in Fig. 1, two types of plate structures have been identified in previous studies: the alternating facet model [Fig. 1(a)] and the mirror-symmetry facet model [Fig. 1(b)]. As has been discussed previously,[35,37] a plate with alternating facets occurs when the number of stacking faults is even, while a plate with mirror symmetry is formed with an odd number of stacking faults. Here, we investigate these plate-shape models in the context of plate growth with the DFT binding energies in Table 1.

We first consider a seed plate with no stacking faults – the single crystal shown in Fig. S1 in the ESI.† When a plate seed has no stacking faults, it has alternating side

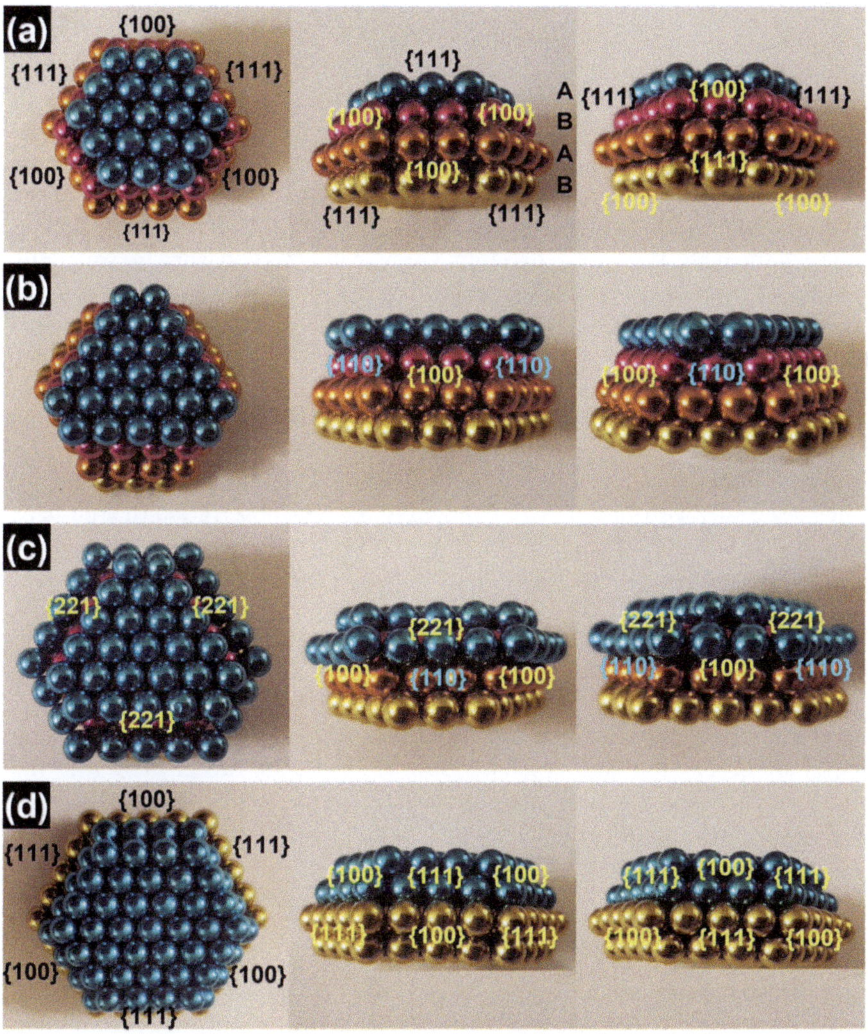

Fig. 3 Growth of an hcp plate seed, as a model of a larger plate containing two stacking faults. A top-down view of a basal facet is shown in the left panel and two different side views are shown in the center and right panels. In (a) we show three views of the initial seed, with the alternating facet structure in Fig. 1(a). The center panel indicates the hcp layer stacking. In (b) we show the structure after adding an atom to each of the four-fold {100} sites. After filling in all the {100} and {110} sites, we generate three rows of {221} sites and {100} sites in the middle, as shown in (c). Subsequent to filling in the {221} and {100} sites, we achieve the hexagonal structure in (d), which regenerates the structure in (a).

facets, as shown in Fig. 1(a), and an octahedron was obtained with addition to {100} sites only (Fig. S1†). When a plate has one stacking fault, it has the mirror-symmetry facet model shown in Fig. 1(b) and preferential Cu atom accumulation on {100} resulted in a triangular plate with all {111} surfaces (Fig. S2†). With further growth of the {111} facets, the final shape would be a 3D trigonal bipyramid and not a 2D plate.

We find lateral growth of the plate seeds is possible when there are two or more stacking faults. First, we grow an hcp plate as a model for a plate containing two stacking faults in the center. The growth progression is shown in Fig. 3. After adding an atom to each {100} site of the initial seed in Fig. 3(a), we achieve the structure in Fig. 3(b). We see in the side views to Fig. 3(b) that this creates {110} sites. We regard {110} (and sites on facets with higher indices) to be the most energetically preferred binding sites because of the relatively higher surface energies of such facets. After filling in all the {110} and {100} binding sites in Fig. 3(b), we attain the structure in Fig. 3(c), where we generated three rows of {221} and {100} sites in the middle. After filling in all the {221} and {100} sites, we achieve the hexagonal structure in Fig. 3(d), which creates a larger version of the structure in Fig. 3(a). This process can be repeated to achieve continuous lateral growth of the plate. We observed a similar growth progression when we varied the positions of the two stacking faults, as can be seen in Fig. S3,† and when we used a plate containing three stacking faults, as we see in Fig. S4.† In all cases with two or more stacking faults, we could create a larger version of our initial seed by adding Cu atoms to the sites with the highest binding energies.

Truncated triangular and hexagonal plate shapes have been observed in many experiments.[25,34–37,39] In an attempt to understand if or how these shapes depend on the number of stacking faults or the filling sequence of the side facets, we grew plates beginning with various seed configurations. The observation that perhaps best summarizes our attempts in this regard is shown in Fig. 4.

In Fig. 4, we show a sequence of snapshots as we grew a plate from a triangular seed containing two stacking faults. We see that the plate evolves through a progression of shapes as we fill in the energetically preferred binding sites on the sides of the plates – from a triangle in Fig. 4(b) to a round shape in Fig. 4(c) to a truncated triangle in Fig. 4(f) to a near hexagon in Fig. 4(h). Thus, we see that shape evolution to equilateral and truncated triangles, hexagons, and round shapes are all possible during lateral growth if there are two or more stacking faults, which is also observed in experiment.[35] Nevertheless, we found that most of the plate shapes during the lateral growth are hexagonal and truncated triangular structures, consistent with experiment.[37]

Cu atom inter-facet transport on I-covered Cu surfaces

It is clear from Table 1 that there is a strong thermodynamic driving force for Cu atoms to stick to the {100} sides of a growing Cu plate. By following an evolution sequence in which Cu atoms always bind to the sites with the highest binding energies, we showed that lateral plate growth could be maintained. However, plate growth is a kinetic process. If the Cu atom deposition rate (or reduction rate) is large compared to diffusion rates, atoms deposited on the large, basal facets of the plates will not reach the sides of the plates on the time scale of deposition. In this case, Cu atoms will accumulate on the basal facets and grow the plates vertically instead of laterally.

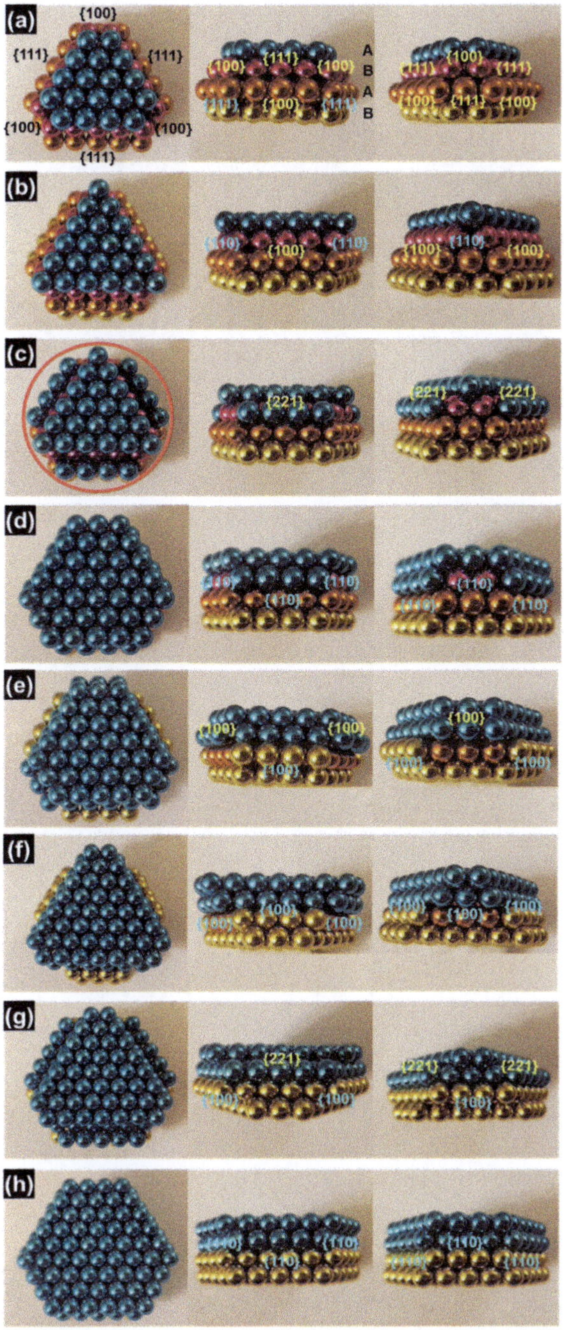

Fig. 4 Sequence of shapes that evolve as we grow a plate by adding spheres to sites with the highest binding energies on the sides of the plate. A top-down view of a basal facet is shown in the left panel and two different side views are shown in the center and right panels. The initial shape in (a) is a truncated triangular seed containing two stacking faults. A triangle is formed in (b), which transforms to a round shape in (c). After transforming between various truncated triangular and hexagonal shapes in (d)–(g) a near hexagon emerges in (h).

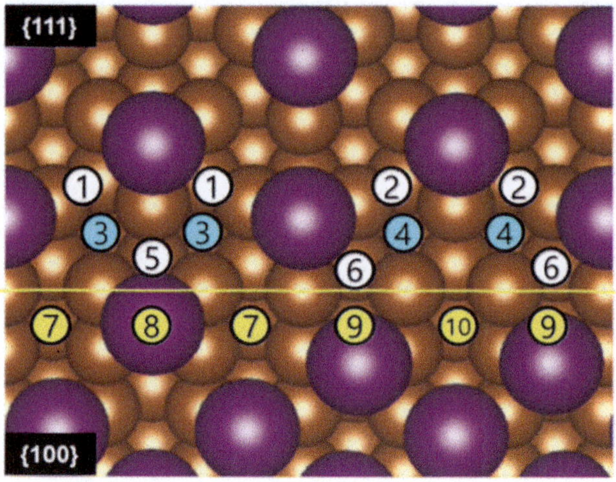

(a)

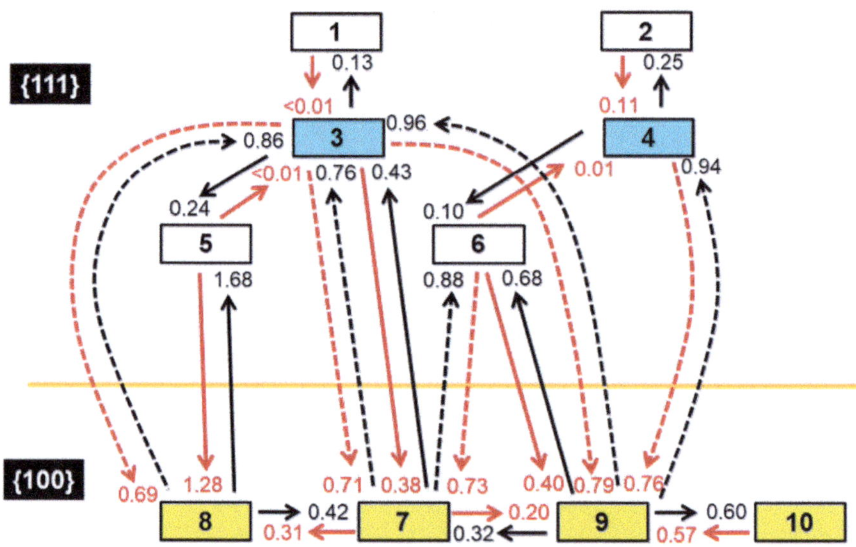

(b)

Fig. 5 (a) Binding sites for Cu adatom near the {111}–{100} edge when iodine is placed on fcc hollow sites on {111} – viewed from the perpendicular direction to {111} (Cu atoms are orange and I atoms are purple); and (b) the inter-facet diffusion pathways. Numbers at the end of each arrow denote the diffusion energy barrier (in eV) for that pathway. The red arrows denote lower barrier pathways in the forward–reverse pair. Arrows with solid lines represent hopping, while arrows with dashed lines represent exchange diffusion. Hcp{111} binding sites are shown in white, fcc{111} binding sites are shown in turquoise, and four-fold hollow sites on {100} are shown in yellow.

To assess the propensity of the plates to exhibit lateral growth, we examined the elementary steps underlying inter-facet transport on an iodine-covered Cu plate. We investigated two facet boundaries suggested by the two typical models for the sides of fcc plates, shown in Fig. 1.[33,35,36,39–41] Both models contain {111}–{100} and {111}–{111} facet boundaries, and we employed two different Cu slabs to characterize inter-facet diffusion. These two slabs are shown in Fig. S5.†

We first consider diffusion between the {100} and {111} facets. Iodine can be adsorbed to either fcc or hcp hollow sites on Cu(111) and we consider both of these scenarios. Fig. 5 shows Cu binding sites and diffusion pathways for the case that iodine is adsorbed on fcc hollow sites on Cu(111). As shown in Fig. 5(a), we identified four different hollow sites on {100} and six fcc and hcp hollow sites on {111} near the {100}–{111} edge. The corresponding binding energies are listed in Table S1.†

In Table S1,† we see the binding energies on sites ① and ② on the inter-facet slab are similar to those on sites on the flat 1/3 ML I–Cu(111) surface (compare Tables 1 and S1†). This result implies the inter-facet slab captures binding on a flat Cu surface away from the edge. On the other hand, Cu adatom binding on sites near the {100} edge is much weaker than on flat Cu(100), even though it is stronger than binding on {111} (see Table S1†). In addition, subsurface binding is not preferred near the {100}–{111} edge.

In Fig. 5(b), we show our calculated diffusion pathways and energy barriers among the various sites in Fig. 5(a). Here, we see that most of the diffusion pathways from {111} to {100} have lower energy barriers than those for the reverse direction, such that {111} → {100} transport is preferred to {100} → {111} transport. This aspect, combined with the strong binding we observe on the flat Cu(100) surface (*cf.*, Table 1), leads to our expectation that it would be difficult for Cu atoms to escape from {100} once they reach that facet.

We also investigated a {100}–{111} inter-facet slab in which iodine is adsorbed on hcp hollow sites on {111} (see Fig. S6 and Table S2†). Although there are differences between this system and the system in Fig. 5, we still find that most of the inter-facet diffusion pathways from {111} to {100} possess lower energy barriers than the reverse direction, as seen in Fig. S6(b).† Therefore, we can say that whether iodine is placed on fcc or hcp hollow sites on {111}, transport of Cu atoms from {111} to {100} is preferred over the opposite direction.

There are three different configurations for I at the {111}–{111} edge: (1) I adsorbed on fcc hollow sites on both facets; (2) I adsorbed on fcc hollow sites on one facet and hcp hollow sites on the other facet; and (3) I adsorbed on hcp hollow sites on both facets. Fig. 6(a) shows the first case, with possible Cu binding sites. In this case, we confirmed that a Cu adatom has similar binding energies for the same binding site on both {111} facets, as listed in Table S3.† In addition, we found that a fcc site far from the edge has a comparable binding energy (−3.48 eV) to that on flat 1/3 ML I–Cu(111), which shows the reliability of this inter-facet slab. Less states are involved in {111}–{111} diffusion than in {111}–{100} diffusion. We also noted that different symmetries of I are possible when it is placed on fcc sites of both facets. Fig. S7 and Table S4† summarize this case. Fig. S8 and S9† summarize the cases for which we have I on fcc sites on one facet and hcp on the other and when we have I in hcp sites on both facets. These calculations showed there is no significant difference in the binding energies and inter-facet diffusion energy barriers compared to those in Fig. 6. Thus, we expect that whether iodine is

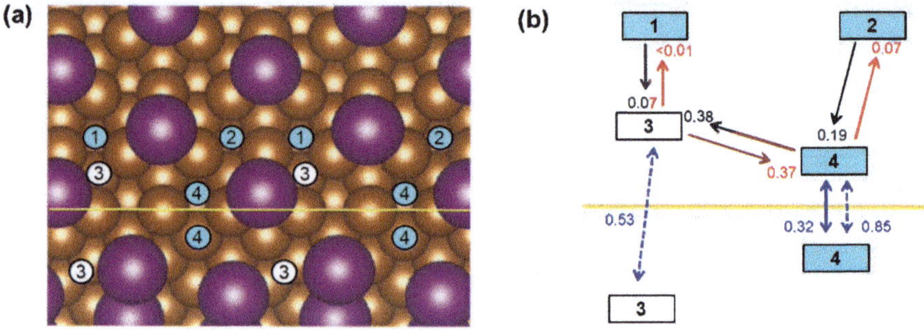

Fig. 6 (a) Binding sites for a Cu adatom on the {111}–{111} facet edge when iodine is placed on fcc hollow sites (Cu atoms are orange and I atoms are purple) and (b) inter-facet diffusion pathways. Numbers at the end of each arrow denote the diffusion energy barrier (in eV) for that pathway. The red arrows denote lower barrier pathways in the forward–reverse pair. Arrows with solid lines represent hopping, while arrows with dashed lines represent exchange diffusion. Hcp{111} binding sites are shown in white and fcc{111} binding sites are shown in turquoise.

absorbed on fcc or hcp hollow sites on {111} we would see similar inter-facet diffusion.

Lateral *vs.* vertical growth of Cu microplates

Using our calculated diffusion-energy barriers, we assessed the propensity for lateral *vs.* vertical growth of the Cu microplates. We expect that if the plate dimensions are sufficiently small, atoms can rapidly access the plate sides and accumulate on the energetically favored {100} (or higher index) sites to grow the plates laterally. However, as the plates grow, the mean-first passage time (MFPT) for a deposited atom to reach the sides of a plate will increase, the inter-facet transit rate will decrease, and we will eventually reach a critical plate size at which accumulation begins to occur on the basal facets. Below, we estimate that critical plate size.

To estimate the MFPT for a deposited atom on a basal plate facet to reach the sides of the plate, we use the theory of the absorbing Markov chains.[42,43] We used (and explained) this method in prior work.[44–46] Briefly, we solve for the average time, considering all possible deposition sites, for a Cu atom deposited on a basal {111} facet to diffuse to a side facet. This involves constructing a Markov matrix to characterize the network of diffusion rates for site-to-site hops/exchanges. There are two kinds of states in the kinetic network: transient states, which can be accessed and vacated an indefinite number of times as the Cu atoms diffuses around on the basal facet, and absorbing states at the sides of the plate. Absorbing states are occupied permanently once they are accessed. Two relevant sub-matrices of the Markov matrix are the transient matrix **T** and the recurrent matrix **R**.[42–44] The MFPT t_M is obtained from the transient matrix as

$$t_M = \bar{p}_0^T \cdot (\mathbf{I} - \mathbf{T})^{-1} \cdot \bar{\tau} \quad (1)$$

where $\bar{p}_0^T$ is the transpose of the probability vector for each transient state to be initially occupied. For N sites (transient states) on the basal facet, each element of $\bar{p}_0$ is $1/N$, assuming each site is equally likely to be occupied initially. **I** is the identity matrix and $\bar{\tau}$ is a vector containing the average time to exit each of the transient states. It is also important to determine where exits from the basal facets occur. This information is contained in

$$\bar{\pi}^{A} = \bar{p}_0^{T} \cdot (\mathbf{I} - \mathbf{T})^{-1} \cdot \mathbf{R} \tag{2}$$

where $\bar{\pi}^{A}$ contains the probability that each absorbing state will be occupied at the end of the process.

To construct **T** and **R**, we used $r_{ij} = \nu_0 \exp(-E_{B,ij}/kT)$ to calculate the transition rate between each pair of transient states or transient and absorbing states i and j. We calculated the rates at the experimental temperature of 363 K.[37] The values of $E_{B,ij}$ are taken from the DFT calculations summarized in Table 1, Fig. 5 and 6.

Fig. 7 shows an example lay-out for a small unit cell in our MFPT calculations and indicates the transient states on the basal {111} facet, as well as the absorbing states on the side {111} and {100} facets. We consider the situation in which iodine atoms are placed on fcc sites on the basal plane and the {111} side facet. Our unit cell represents one third of a hexagonal basal facet, as can be seen more clearly in Fig. S10,† and we apply reflecting boundary conditions for diffusion to transient states on sites that would cross a boundary indicated in Fig. 7. Table 2 shows average transit rates (t_M^{-1}) as a function of the plate dimension. Fig. S11† shows t_M as a function of the number of transient states (plate dimension). Since we consider every possible Cu binding site on (one-third of) the basal plane of a Cu plate, calculations for plate sizes beyond 200 nm become computationally expensive. We extrapolated our results to obtain MFPTs for large plates. Fig. S11† shows that the extrapolation model fits well to our calculated MFPTs.

To assess the lateral dimensions of the plates, we need to compare the MFPT to the Cu deposition rate in the syntheses. In similar experiments to those for plate synthesis,[37,47,48] the initial deposition rate in a synthesis solution of Cu nanowires including Cl^- and HDA was estimated to be around 10^4 to 10^5 s^{-1} and we expect the addition of I^- to the solution for growing Cu plates to decrease the reduction (deposition) rate.[37] Thus, we approximate the initial deposition rate to be 10^4 s^{-1} in the synthesis of Cu plates. This rate decreases over time as Cu salt is depleted to grow the plates, so we expect a reasonable range of deposition rates to be 10^3 to 10^4 s^{-1}. Using

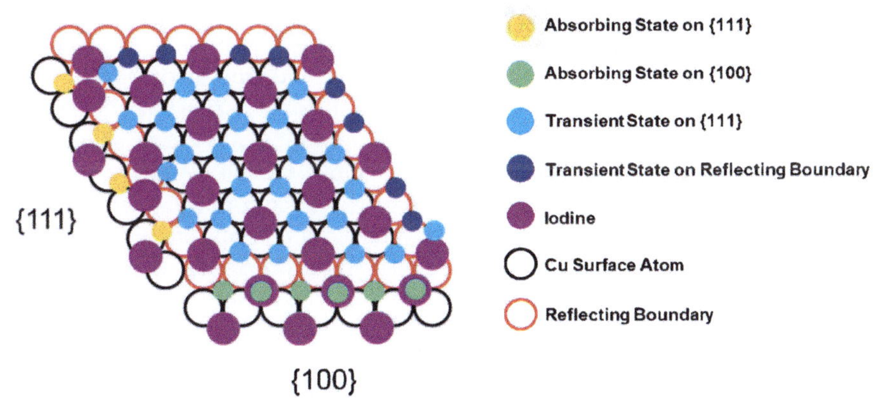

Fig. 7 Lay-out for a unit cell in MFPT calculations.

Table 2 Calculated inter-facet diffusion rates from basal {111} to side {111} and {100} facets in a hexagonal plate

Plate width (nm)	Inter-facet diffusion rate (1/s)
20	4.93×10^6
80	1.14×10^6
180	5.06×10^5
500	1.82×10^5
2000	4.52×10^4
9080	1.00×10^4

our model, the inter-facet diffusion rate becomes equal to a deposition rate of 10^4 s^{-1} when the plate dimension is 1.5 μm and the plates can achieve a lateral dimension of around 9 μm when the deposition rate is 10^4 s^{-1}. Plates with dimensions in the 1–10 micron range were observed experimentally,[37] in agreement with our results.

In addition to calculating the MFPT and inter-facet transit rates, we assessed the probability for a Cu atom on a basal $\{111\}$ plate facet to access either a $\{100\}$ or $\{111\}$ facet using eqn (2). These calculations indicate that for a hexagonal plate, the probability of a Cu atom to go from the basal plane to a $\{111\}$ side is 0.04 and the probability of the atom to go to a $\{100\}$ side is 0.96. In addition to their strong binding preference, this diffusion preference also contributes to the preferred accumulation of Cu atoms on the $\{100\}$ side facets. However, even with such a strong binding and transport preference to Cu(100), we still expect the plates will undergo a series of different shapes as they grow laterally, as we see in Fig. 4.

Conclusion

In summary, we quantified several aspects that facilitate the I-mediated growth of Cu microplates. First, we showed that a Cu adatom binds more strongly to hcp hollow sites than fcc hollow sites on iodine-covered Cu(111). This feature promotes the formation of stacking faults during seed and plate growth which, in turn, promotes 2D growth. We also found that iodine adsorption leads to strong Cu atom binding and prohibitively slow diffusion of Cu atoms on Cu(100). Incorporating these insights into analog experiments, in which we initiated the growth of Cu plates from small seeds consisting of magnetic spheres, we found that two or more stacking faults are required for lateral plate growth. Moreover, plates can take on a variety of shapes during growth: from triangular and truncated triangular to round and hexagonal. Though the diffusion of Cu atoms is slower on I-covered Cu(111) than on the bare surface,[46] it is still relatively rapid and, as we showed in our absorbing Markov chain calculations, it is rapid enough to promote lateral growth of microplates with dimensions in the 1–10 micron range. Therefore, we conclude the adsorption of iodine on Cu surfaces favors the anisotropic growth of thin Cu plates by influencing on both seed formation and growth stages.

Computational methods

All DFT calculations were carried out using the Vienna *Ab initio* Simulation Package (VASP) with the projector augmented-wave (PAW) method.[49–53] The generalized gradient approximation (GGA) by Perdew, Burke, and Ernzerhof (PBE) was used for the exchange–correlation functional.[54] 400 eV was used as an optimal value for the energy cut-off for the plane-wave basis set. We also used Monkhorst–Pack grids for integration over the first Brillouin zone.[55] A $(12 \times 12 \times 12)$ k-point mesh was used for calculations for bulk Cu. A cubic unit cell with a side length of 20 Å and a single k point was used to calculate the energy of an isolated Cu atom and a single I_2 molecule in the gas phase.

We used a periodic Cu slab with six Cu layers to describe the flat Cu surfaces. The bottom three layers of the slab were fixed to describe the bulk positions and the top three layers and adsorbed species were allowed to fully relax during structural optimization. Structural optimization was performed within a value of 10^{-6} eV for the energy convergence criterion and 0.01 eV Å^{-1} was used for the force convergence criterion. The Cu lattice constant was calculated to be 3.64 Å, which is well matched with other theoretical values using the GGA functional[56–58] and has less than 0.6% difference from the observed experimental values of 3.615 Å.[59] A (4×6) surface unit cell size was used for both Cu(100) and Cu(111). The optimal k-point mesh was determined to be $(6 \times 4 \times 1)$ for Cu(100) and $(7 \times 5 \times 1)$ for Cu(111). A vacuum space of 15 Å was included to separate periodic slabs and a dipole correction along the surface normal was also employed to avoid interactions between periodic slabs.

To calculate the binding energy for I and Cu atoms on the Cu surfaces, we used,

$$E_{\text{bind,X}} = \frac{1}{N_{\text{X}}}[E_{\text{slab+X}} - (E_{\text{slab}} + N_{\text{X}}E_{\text{X}})], \tag{3}$$

where X can be either Cu or I absorbed on the Cu surface, E_{X} is the energy of an isolated X in vacuum (here, $E_{\text{X}} = \frac{1}{2}E_{\text{X}_2}$ for X = I), N_{X} is the number of X on the Cu surface, E_{slab} is the energy of the Cu slab – either a bare or I-covered Cu slab, and $E_{\text{slab+X}}$ is the energy of an optimized Cu slab with X. Table S5 in the ESI† lists the convergence test for the binding of a Cu adatom on Cu surfaces, with regard to the energy cut-off, k-point mesh, and vacuum spacing.

To include long-range van der Waals (vdW) interactions in our DFT calculations, we used the DFT-D2 method of Grimme.[60] Especially for the screening effect of Cu, we employed the dispersion coefficient and the vdW radius suggested by Ruiz *et al.*,[61] while the default values of these parameters are those suggested by Grimme for the other elements. The cutoff radius for the vdW interactions was 40.0 Å.

To identify a transition state and energy barrier for the surface diffusion of a Cu adatom on the Cu surfaces, we employed the climbing nudged elastic band (CI-NEB) method using 3–6 images for an individual diffusion pathway.[38] Vibrational frequencies were calculated to confirm a single imaginary frequency for each identified transition state. The spring constant was set to 5.0 eV Å^{-2} and all CI-NEB calculations are converged when the force on each atom was less than 0.05 eV Å^{-1}.

To quantify diffusion between facets on plate-like Cu crystals, we created two Cu inter-facet slabs. One has {111} on the top and {111} and {100} on each side – Fig. S5(a),† while the other has {100} on the top and {111} on both sides – Fig. S5(b).† The former was used for calculations of the {111}–{111} edge, and the latter was used for the {111}–{100} edge. The inter-facet slabs consisted of six Cu layers. To optimize the slab configurations, we allowed the outer two Cu layers and all absorbed I atoms to fully relax, while fixing the bottom two Cu layers, their associated I atoms, and Cu atoms in the middle of the slab. Slabs were continuous in one dimension only – parallel to the inter-facet edges. We included a vacuum spacing in the *z* direction normal to the top facet and in the direction perpendicular to the inter-facet edge. We used a $(4 \times 1 \times 1)$ and $(1 \times 4 \times 1)$ *k*-point mesh for the slabs in Fig. S5(a) and (b),† respectively. Convergence tests with respect to *k*-points and vacuum space for the inter-facet slabs are summarized in Tables S5 and S6.†

Conflicts of interest

There are no conflicts to declare.

Acknowledgements

This work is funded by the Department of Energy, Office of Basic Energy Sciences, Materials Science Division, Grant DE FG02-07ER46414 (J. K.). J. K. acknowledges training provided by the Computational Materials Education and Training (CoMET) NSF Research Traineeship (DGE-1449785). This work used the Extreme Science and Engineering Discovery Environment (XSEDE), which is supported by National Science Foundation grant number ACI-1548562.

References

1 I. Pastoriza-Santos and L. M. Liz-Marzan, *J. Mater. Chem.*, 2008, **18**, 1724–1737.
2 Y. Chen, Z. Fan, Z. Zhang, W. Niu, C. Li, N. Yang, B. Chen and H. Zhang, *Chem. Rev.*, 2018, **118**, 6409–6455.
3 R. D. Neal, R. A. Hughes, A. S. Preston, S. D. Golze, T. B. Demille and S. Neretina, *J. Mater. Chem. C*, 2021, **9**, 12974–13012.
4 T. Wang, M. Park, Q. Yu, J. Zhang and Y. Yang, *Mater. Today Adv.*, 2020, **8**, 100092.
5 Q. Yu and Y. Yang, *ChemNanoMat*, 2020, **6**, 1683–1711.
6 H. Tian, Y. Yu, Q. Wang, J. Li, P. Rao, R. Li, Y. Du, C. Jia, J. Luo, P. Deng, Y. Shen and X. Tian, *Int. J. Hydrogen Energy*, 2021, **46**, 31202–31215.
7 F. Lv, B. Huang, J. Feng, W. Zhang, K. Wang, N. Li, J. Zhou, P. Zhou, W. Yang, Y. Du, D. Su and S. Guo, *Natl. Sci. Rev.*, 2021, **8**, nwab019.
8 F. Qin, Z. Yan, J. Fan, J. Cai and X. Zhu, *Macromol. Mater. Eng.*, 2021, **306**, 2000607.
9 X. Huang, S. Tang, X. Mu, Y. Dai, G. Chen, Z. Zhou, F. Ruan, Z. Yang and N. Zheng, *Nat. Nanotechnol.*, 2011, **6**, 28–32.
10 Y. Shi, Z. Lyu, M. Zhao, R. Chen, Q. N. Nguyen and Y. Xia, *Noble-metal Nanocrystals with Controlled Shapes for Catalytic and Electrocatalytic Applications*, 2021, vol. 121.
11 J. Yin, J. Wang, Y. Ma, J. Yu, J. Zhou and Z. Fan, *ACS Mater. Lett.*, 2021, **3**, 121–133.
12 P. K. Kalambate, J. Noiphung, N. Rodthongkum, N. Larpant, P. Thirabowonkitphithan, T. Rojanarata, M. Hasan, Y. Huang and W. Laiwattanapaisal, *TrAC, Trends Anal. Chem.*, 2021, **143**, 116403.

13 Y. Zhang, H. Lu, F. Yang, Y. Cheng, W. Dai, X. Meng, H. Dong and X. Zhang, *Anal. Chim. Acta*, 2020, **1139**, 164–168.

14 T. A. Mai Ngoc, D. T. D. Nguyen, V. K. T. Ngo, P. P. Nguyen Thi, D. H. Nguyen and M. T. Nguyen-Le, *Plasmonics*, 2020, **15**, 2185–2194.

15 C. Gao, Z. Lu, Y. Liu, Q. Zhang, M. Chi, Q. Cheng and Y. Yin, *Angew. Chem., Int. Ed.*, 2012, **51**, 5629–5633.

16 C. L. Lee, K. C. Chang and C. M. Syu, *Colloids Surf., A*, 2011, **381**, 85–91.

17 M. Chen, S. Tang, Z. Guo, X. Wang, S. Mo, X. Huang, G. Liu and N. Zheng, *Adv. Mater.*, 2014, **26**, 8210–8216.

18 B. Pelaz, V. Grazu, A. Ibarra, C. Magen, P. Del Pino and J. M. De La Fuente, *Langmuir*, 2012, **28**, 8965–8970.

19 I. S. Raja, M. S. Kang, K. S. Kim and Y. J. Jung, *Cancers*, 2020, **12**, 1657.

20 L. Cheng, X. Wang, F. Gong, T. Liu and Z. Liu, *Adv. Mater.*, 2020, **32**, 1–23.

21 Z. Zhang, J. Du, J. Li, X. Huang, T. Kang, C. Zhang, S. Wang, O. O. Ajao, W. J. Wang and P. Liu, *Prog. Polym. Sci.*, 2021, **114**, 101360.

22 Q. Zhang, L. Long, G. Zhang, Z. Y. Li and Y. Zheng, *CrystEngComm*, 2020, **22**, 173–177.

23 Q. Zhang, Y. Hu, S. Guo, J. Goebl and Y. Yin, *Nano Lett.*, 2010, **10**, 5037–5042.

24 F. Qin, T. Zhao, R. Jiang, N. Jiang, Q. Ruan, J. Wang, L. D. Sun, C. H. Yan and H. Q. Lin, *Adv. Opt. Mater.*, 2016, **4**, 76–85.

25 T. Yu, Z. Wu and W. S. Kim, *RSC Adv.*, 2014, **4**, 37516–37521.

26 J. Wang, X. Guo, Y. He, M. Jiang and R. Sun, *RSC Adv.*, 2017, **7**, 40249–40254.

27 G. S. Métraux and C. A. Mirkin, *Adv. Mater.*, 2005, **17**, 412–415.

28 S. Xu, X. Sun, H. Ye, T. You, X. Song and S. Sun, *Mater. Chem. Phys.*, 2010, **120**, 1–5.

29 Y. Xiong, I. Washio, J. Chen, M. Sadilek and Y. Xia, *Angew. Chem., Int. Ed.*, 2007, **46**, 4917–4921.

30 Y. Xiong, J. M. McLellan, J. Chen, Y. Yin, Z. Y. Li and Y. Xia, *J. Am. Chem. Soc.*, 2005, **127**, 17118–17127.

31 B. Y. Xiong and Y. Xia, *Adv. Mater.*, 2007, **19**, 3385–3391.

32 C. Lofton and W. Sigmund, *Adv. Funct. Mater.*, 2005, **15**, 1197–1208.

33 T. C. R. Rocha and D. Zanchet, *J. Phys. Chem. C*, 2007, **111**, 6989–6993.

34 S. D. Golze, R. A. Hughes, S. Rouvimov, R. D. Neal, T. B. Demille and S. Neretina, *Nano Lett.*, 2019, **19**, 5653–5660.

35 T. Tan, S. Zhang, J. Wang, Y. Zheng, H. Lai, J. Liu, F. Qin and C. Wang, *Nanoscale*, 2021, **13**, 195–205.

36 D. Aherne, D. M. Ledwith, M. Gara and J. M. Kelly, *Adv. Funct. Mater.*, 2008, **18**, 2005–2016.

37 M. J. Kim, M. A. Cruz, Z. Chen, H. Xu, M. Brown, K. A. Fichthorn and B. J. Wiley, *Chem. Mater.*, 2021, **33**, 881–891.

38 G. Henkelman, B. P. Uberuaga and H. Jónsson, *J. Chem. Phys.*, 2000, **113**, 9901–9904.

39 J. Goebl, Q. Zhang, L. He and Y. Yin, *Angew. Chem.*, 2012, **124**, 567–570.

40 A. I. Kirkland, D. A. Jefferson, D. G. Duff, P. P. Edwards, I. Gameson, B. F. G. Johnson and D. J. Smith, *Proc. R. Soc. London, Ser. A*, 1993, **440**, 589–609.

41 J. Zeng, J. Tao, W. Li, J. Grant, P. Wang, Y. Zhu and Y. Xia, *Chem.–Asian J.*, 2011, **6**, 376–379.

42 J. G. Kemeny and J. L. Snell, *Finite Markov Chains*, Springer-Verlag, New York, 1976.
43 M. A. Novotny, *Phys. Rev. Lett.*, 1995, **74**, 1–5.
44 K. A. Fichthorn and Y. Lin, *J. Chem. Phys.*, 2013, **138**, 164104.
45 X. Qi, Z. Chen, T. Yan and K. A. Fichthorn, *ACS Nano*, 2019, **13**, 4647–4656.
46 J. Kim, J. Cui and K. A. Fichthorn, *ACS Nano*, 2021, **15**, 18279–18288.
47 M. J. Kim, S. Alvarez, T. Yan, V. Tadepalli, K. A. Fichthorn and B. J. Wiley, *Chem. Mater.*, 2018, **30**, 2809–2818.
48 M. J. Kim, S. Alvarez, Z. Chen, K. A. Fichthorn and B. J. Wiley, *J. Am. Chem. Soc.*, 2018, **140**, 14740–14746.
49 G. Kresse and J. Hafner, *Phys. Rev. B: Condens. Matter Mater. Phys.*, 1993, **47**, 558–561.
50 G. Kresse and J. Hafner, *Phys. Rev. B: Condens. Matter Mater. Phys.*, 1994, **49**, 14251–14269.
51 G. Kresse and J. Furthmüller, *Phys. Rev. B: Condens. Matter Mater. Phys.*, 1996, **54**, 11169–11186.
52 P. E. Blöchl, *Phys. Rev. B: Condens. Matter Mater. Phys.*, 1994, **50**, 17953.
53 G. Kresse and D. Joubert, *Phys. Rev. B: Condens. Matter Mater. Phys.*, 1999, **59**, 1758–1775.
54 J. P. Perdew, K. Burke and M. Ernzerhof, *Phys. Rev. Lett.*, 1996, **77**, 3865–3868.
55 H. J. Monkhorst and J. D. Pack, *Phys. Rev. B: Solid State*, 1976, **13**, 5188.
56 Z. Xiong, S. Shi, C. Ouyang, M. Lei, L. Hu, Y. Ji, Z. Wang and L. Chen, *Phys. Lett. A*, 2005, **337**, 247–255.
57 J. L. F. Da Silva, C. Stampfl and M. Scheffler, *Surf. Sci.*, 2006, **600**, 703–715.
58 H. Y. Xiao, X. T. Zu, X. He and F. Gao, *Chem. Phys.*, 2006, **325**, 519.
59 K. Hermann, *Crystallography and Surface Structure: an Introduction for Surface Scientists and Nanoscientists*, Wiley-VCH, Weinheim, Germany, 2011.
60 S. Grimme, *J. Comput. Chem.*, 2006, **27**, 1787–1799.
61 V. G. Ruiz, W. Liu, E. Zojer, M. Scheffler and A. Tkatchenko, *Phys. Rev. Lett.*, 2012, **108**, 146103.

Faraday Discussions

ROYAL SOCIETY OF CHEMISTRY

PAPER

Significance of atomic-scale defects in flexible surfaces on local solvent and ion behaviour†

Veselina Marinova,* Colin L. Freeman and John H. Harding

Received 5th November 2021, Accepted 14th December 2021
DOI: 10.1039/d1fd00082a

Many factors can affect the course of heterogeneous nucleation, such as surface chemistry, flexibility and topology, substrate concentration and solubility. Atomic-scale defects are rarely investigated in detail and are often considered to be unimportant surface features. In this work, we set out to investigate the significance of atomic-scale defects in a flexible self-assembled monolayer surface for the behaviour of clusters of Ca^{2+} and $CO_3{}^{2-}$ ions in water. To this end, we use molecular dynamics simulations to estimate the diffusion coefficients of ion clusters at different topological surface features and obtain ionic radial distribution functions around features of interest. Well-tempered metadynamics is used to gain insight into the free energy of ions around selected surface defects. We find that certain defects, which we refer to as active defects, can impair ionic surface diffusion, as well as affect the diffusion of ions in close proximity to the surface feature in question. Our findings suggest that this effect can result in an ability of such topological features to promote ion clustering and increase local ionic concentration at specific surface sites. The work reported here shows how the presence of small atomic-scale defects can affect the role of a surface in the process of heterogeneous nucleation and contributes towards a rational definition of surfaces as effective nucleating agents.

1 Introduction

Crystallisation is a process of fundamental importance in both nature and industry. Examples include the formation of exoskeletons in marine species at ambient temperature and pressure,[1–3] the synthesis of active pharmaceutical ingredients with high purity[4,5] and the prevention of the formation of scale in industrial environments.[6–8] An in-depth understanding of the mechanisms which promote crystallisation is necessary to synthesise specific polymorphs, prevent

Department of Materials Science and Engineering, The University of Sheffield, Sheffield, S1 3JD, UK. E-mail: v.marinova@sheffield.ac.uk

† Electronic supplementary information (ESI) available. See DOI: 10.1039/d1fd00082a

the formation of crystals where it is unwanted, and tailor crystalline materials for a vast range of applications.

In an uncontrolled environment, crystallisation might be expected to occur on any available surface. However, although classical nucleation theory (CNT) suggests that many surfaces should promote nucleation and crystal formation, in fact surfaces with effective nucleating properties are often hard to identify.[9] It is generally acknowledged that a nucleant can operate in several ways. Some surfaces promote nucleation through the strength of the binding affinity of crystallising species to them, giving the convention that "good binders are good nucleators".[10,11] A particular example is epitaxial matching,[12–15] where a close lattice match between the growing crystal and substrate is believed to indicate strong binding between them. The presence of a surface may increase local supersaturation at the substrate (particularly if the surface is charged), decreasing nucleation times.[16] Nucleants may also exhibit surface features in the form of defects, pits or cracks, which can increase localised ion concentrations and so promote the formation of a nucleus.[14,17,18] Understanding the surface characteristics which promote and direct nucleation is key to controlling surface-aided crystallisation.

Self-assembled monolayers (SAMs) provide a diverse platform for studying surface effects on substrate behaviour, as their surface chemistry, charge and flexibility are easily varied, both in a computational and in an experimental setting. Nucleation of $CaCO_3$ on alkanethiol SAMs has been extensively studied as a model system for biomineralisation processes in living organisms.[19] Head-group charge,[16,20] changing functional groups,[10] epitaxial matching,[20] flexibility and disorder[21,22] have all been demonstrated to play a role in the crystal formation of $CaCO_3$ at the SAM surface. While little is known about the role of nanoscale surface defects on the local solution and ionic behaviour, previous work has revealed that the non-uniform surface chemistry of SAMs can affect the strength of the interactions and behaviour of certain biomolecular systems.[23,24] Specific studies on SAM surface defects suggest that these confined environments can affect the growth of nanoparticles.[25] We therefore believe that there is more to be discovered about the role of imperfections on flexible surfaces in crystal growth.

In this paper, we investigate the effect of relatively small (atomic-scale) defects in the surface of a self-assembled monolayer of 16-mercaptohexadecanoic acid (16-MHDA) on the behaviour of ionic clusters of Ca^{2+} and CO_3^{2-} ions using molecular dynamics (MD) simulations. We generate several surface topographies, for which we calculate the diffusion coefficient associated with the movement of ions out of the surface defect. In the most compelling case, we also obtain the free energy associated with the movement of a single Ca^{2+} and a single CO_3^{2-} ion in and out of the defect to understand more about the observed cluster behaviour. For the latter, we perform additional simulations at high ionic concentrations in solution. Our work shows how a better understanding of the way surface features affect surface-aided crystallisation can help us design and tailor effective nucleating agents.

2 Methods

In this work, we investigate the ion cluster behaviour of Ca^{2+} and CO_3^{2-} ions in a 50 : 50 ratio at 5 different SAM surface defects in water. Generating surface

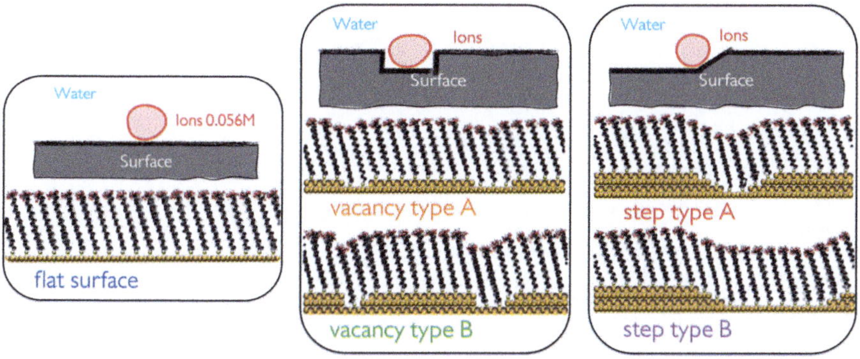

Fig. 1 Schematic representations of all of the surface configurations used in this study. From left to right: flat surface where no surface defects are present, vacancy types A and B with an offset in the surface of the depth of 1 and 2 Au-atoms, respectively, and step types A and B with a few consecutive steps in the underlying Au surface achieving a slope-like character in the surface, where the separation between the steps is altered between the two configurations.

defects in the monolayers was achieved through the removal of one or several surface atoms from the underlying Au(111) surface. This procedure creates nm-scale steps and vacancies which promote interactions between the functional groups of neighbouring chains through the offset produced. We generate a total of 5 configurations, which we refer to as flat surface, vacancy type A, vacancy type B, step type A and step type B, as displayed in Fig. 1.

We calculate the mean squared displacement (MSD) for each ion within the clusters to obtain average cluster diffusion coefficients. Solvent behaviour at the SAM surface was investigated using residence time analysis. A combination of MD and well-tempered metadynamics (WTmetaD) was used to obtain a profile of the single ion movement towards and away from the surface labelled vacancy type B for each ion type. We also perform large-scale simulations at much higher ionic concentrations to obtain an RDF profile between the surface and the solute.

2.1 Simulation setup

Self-assembled monolayer surfaces were generated using 16-MHDA as a building block on an 80 × 80 Å grid of an Au(111) surface using a total of 340 alkanethiol molecules. In this system, the 30° angle of the chains with respect to the normal to the surface plane (marked as the *z*-axis on Fig. 2a) was maintained throughout.[25,26] A spacing of 4.98 Å between the sulphur atoms in the *xy*-plane was maintained in all configurations with the exception of the surface labelled vacancy type B, where distances between the sulphurs within the defect had to be adjusted marginally to avoid overlap between the chains and to maintain consistency in the angle of the 16-MHDA molecules with respect to the underlying surface. More details on the spacing between building blocks can be found in the ESI.†

Surface defects were generated by removing layers of underlying Au atoms. The cross-section of the defect in the *xz*-plane is displayed in Fig. 1, while the defect itself runs parallel to the *y*-axis, a schematic of which can be seen in Fig. 2a.

2.1.1 Forcefield parameters. We have combined several potential models to achieve a comprehensive representation of the interactions between different

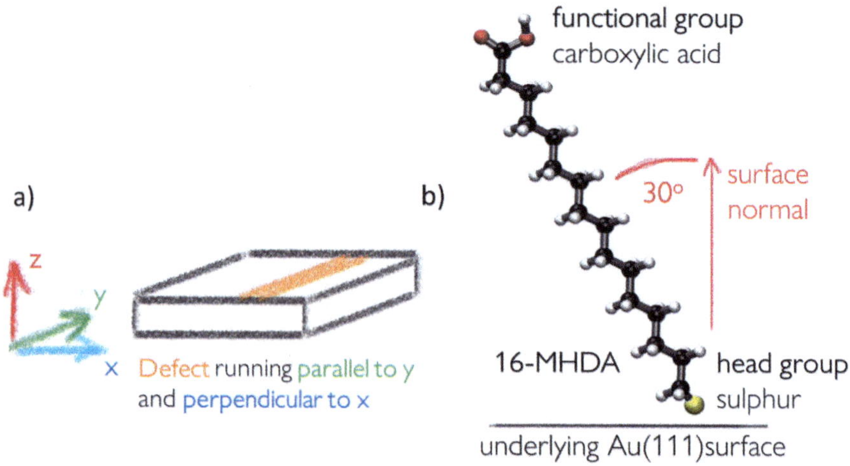

Fig. 2 (a) Schematic representation of how defects have been positioned within the SAM surface. The *xy*-plane is the plane of the SAM surface, while the *z*-axis is the normal to the surface plane (b) structure of 16-MHDA, used as a building block for the SAM surface, where the functional group and the head group have been labelled. The characteristic 30° angle with respect to the surface normal is displayed.

system components. The organic component, the chains of 16-MHDA, was parametrised according to the AMBER forcefield following the standard procedure with ANTECHAMBER.[27,28] The chains were parametrised as neutral and so the functional carboxylic groups are uncharged (–COOH). At the pH of most experimental systems we would expect the functional groups to be ionised ($-COO^-$). However, in this study we are focusing our attention on the effects of surface flexibility in combination with varying surface topography. By simplifying the surface chemistry, we can rigorously test the effect of the factors this study focuses on and avoid strong coulombic interactions that could confuse the interpretation of the results. Interactions involving $CaCO_3$ and water were parametrised with the Raiteri *et al.*[29] potential model. Non-bonded Au interactions[30] were used to prevent overlap between 16-MHDA and the Au under-layer at defects. These interactions were calculated following the standard Lorenz–Berthelot mixing rules.[31,32] The Lorenz–Berthelot mixing rules were also used to generate the interactions between the 16-MHDA chains and water and between 16-MHDA and the oxygen on the carbonate ion. The carboxylic C-atom interactions from AMBER were used to simulate the non-bonding interactions between the SAM chains and carbonate C-atom. Finally, non-bonded interactions between the carboxylic O-atoms in the SAM functional group and Ca^{2+} ions were parametrised according to the procedure outlined by Freeman *et al.*[33] A full list of all bonded and non-bonded interactions can be found in the ESI.†

2.1.2 Molecular dynamics setup. Molecular dynamics (MD) simulations were carried out using LAMMPS[34,35] with a time step of 1 fs and a PPPM[36] long range coulombic interactions solver with an accuracy of 1.0×10^{-5}. A cut-off of 9 Å was applied to all non-bonded interactions. Simulations were performed in the *NVT* ensemble using a Nosé–Hoover thermostat[37,38] at 300 K.

2.1.2.1 SAM–water simulations. Simulations of each of the 5 surface configurations in water were carried out. Production run lengths for these were 2 ns, with an additional initial 200–300 ps of system relaxation and water density equilibration.

2.1.2.2 Ion clusters at SAM defects in water. MD simulations of $CaCO_3$ ionic clusters at the SAM surface and within defects (where present) were carried out. The ionic clusters were generated using Packmol 20.010.[39] They contained 10 Ca^{2+} and 10 CO_3^{2-} ions, making up a total system concentration of 0.056 M. Clusters were positioned close to the surface defect and the cluster structure alone was relaxed with a short simulation of 100 ps. Three different cluster configurations per surface topography type were generated, which we will refer to as set 1, set 2 and set 3. Set 1 and set 3 were equilibrated by relaxing the ions onto the surface for 50 ps, followed by water relaxation for another 100 ps. On the other hand, set 2 was set up by relaxing the water onto the surface, followed by relaxation of the ions onto the surface. This variation in the setup enables us to evaluate whether the presence of water on the surface before relaxing the ions affects cluster behaviour on the surface. Each of the 3 setups for each surface type was run for a total of 56 ns.

2.1.2.3 Single ion at vacancy type B in water. In the case of surface vacancy type B, we perform further analysis by setting up simulations of a single ion at the surface in water. We perform a simulation for each ion type: Ca^{2+} and CO_3^{2-}. These simulations were configured by removing the extra ions from the starting configuration for set 1 of vacancy type B.

2.1.2.4 Large-scale simulations at vacancy type B in water. We perform additional simulations in the presence of a SAM surface which has a vacancy of type B. We refer to them as large-scale as the underlying surface grid is in this case 170 × 80 Å. Two different simulation sets were carried out, where Ca^{2+} and CO_3^{2-} ions are dispersed in solution, away from the surface, at the beginning of the simulation at concentrations of 0.41 M and 0.69 M. Starting configurations were generated using Packmol 20.010.

2.1.3 Well-tempered metadynamics setup. We perform MD with WTmetaD[40,41] in two simulations – vacancy type B with water and a Ca^{2+} ion and vacancy type B with water and a CO_3^{2-} ion. We sample the movements of the ions across the surface as well as up to 10 Å away from the surface. Metadynamics works by depositing an external biasing potential as a function of selected degrees of freedom, referred to as collective variables (CVs). By summing the deposited bias over time, a free energy profile as a function of those CVs can be recovered, giving information on energetically favourable states. The CVs we use are based on the distance between the centre of the lowest layer of Au atoms and the centre of the ion, where bias is deposited along the x-component and z-component of the distance, as shown in Fig. 3. We therefore sampled along the cross-section of the defect in x and normal to the surface in z. For computational efficiency, a wall was placed at 10 Å away from the lowest surface point.

The biasing potential was applied every 500 steps in the form of Gaussian functions of width 0.2 Å in the x-axis and 0.4 Å in the z-axis and a height of 12 kJ mol^{-1}. The bias factor for implementing WTmetaD was set to 10. Metadynamics was implemented by using the open source community-developed PLUMED library, version 2.6.[42,43]

Free energy surfaces were recovered using the sum_hills functionality in PLUMED. Simulations were run until the average positional error in the resulting

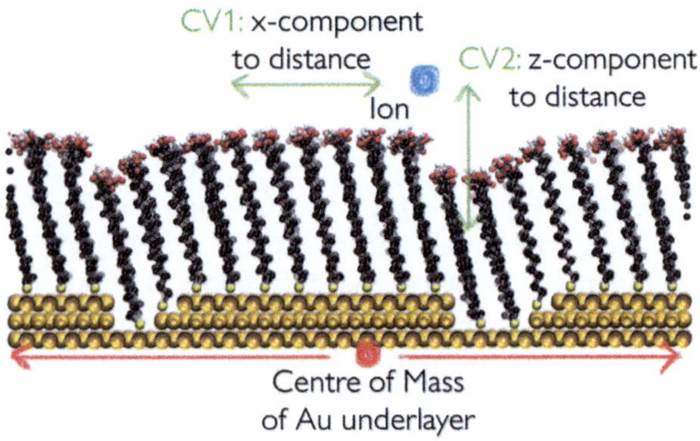

Fig. 3 Schematic representation of the collective variables used to apply WTmetaD. The distance between the centre of the outermost Au layer and the centre of the ion is broken down into *x*-,*y*- and *z*-components. Bias is applied on the *x*- and *z*-components.

free energy profile was within 1.5 k_BT units. More detail on our convergence analysis can be found in the ESI.†

2.2 Water analysis

The structure and dynamics of water were analysed in the absence of $CaCO_3$ ions in order to understand the effects of surface defects on local solvent behaviour. Water residence times at the SAM surface were calculated by counting the number of frames an O-atom of a water molecule spends within 2.45 Å of the H-atom on the –COOH SAM functional group (based on the RDF calculated between the two atom types as shown in Fig. 4). At each frame where a water molecule goes beyond that distance, the counting is stopped and only restarted when the molecule is back within the given distance range.

This procedure allows us to calculate an array of individual uninterrupted residence times of a water molecule on each SAM functional group, from which

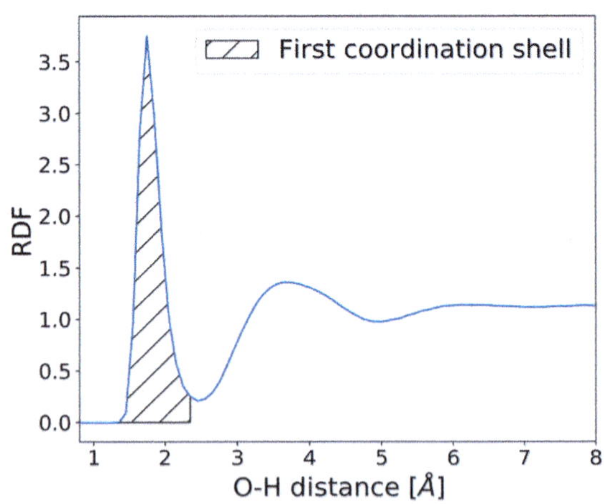

Fig. 4 Radial distribution function of water O-atoms and –COOH H-atoms.

we obtain the average. These values are further averaged across the *y*-axis, as shown in Fig. 2, and residence times of water at the SAM functional groups along the *x*-axis of the surface are reported.

2.3 Functional group flexibility analysis

The degree of movement in the SAM –COOH functional groups was recorded as a mean squared displacement of the carboxylic C-atom with respect only to the initial surface configuration, taken after equilibration at the start of the SAM-water production run. Similarly to the residence time analysis, we average the time-averaged functional group MSD across the *y*-axis of the surface and report the values along the cross-section of the defect in the surface *x*-direction.

2.4 Diffusion coefficient calculations

A diffusion coefficient for each ion cluster was obtained by calculating a mean squared displacement (MSD) for each ion within a cluster. The diffusion coefficient was extracted from the slope of a corresponding MSD plot for the time step fragment between 2 ns and 20 ns in each production run, as shown in Fig. 5a.

To obtain an average cluster diffusion coefficient which is consistent with the overall cluster behaviour, we perform an interquartile range (IQR) analysis on the 3D MSD values across the cluster and exclude any outliers. An example of this analysis is shown in Fig. 5b. The plot shows all individual ion diffusion data points (in blue) and the mean value of the cluster diffusion coefficient (in red). The mean cluster diffusion is skewed by the diffusion of several ions which are orders of magnitude higher than the median of the dataset (shown in green). These points correspond to individual ions which have detached from the cluster and therefore need to be excluded from the diffusion coefficient calculation of the ionic cluster. We use the IQR (in brown) to determine which points are to be excluded and report the corrected mean (in orange) in our results. All interquartile statistical analysis carried out when determining the mean cluster diffusion is reported in the ESI.†

All simulation images in this paper have been produced using VMD.[44] Trajectory analysis was carried out using Python 3.[45]

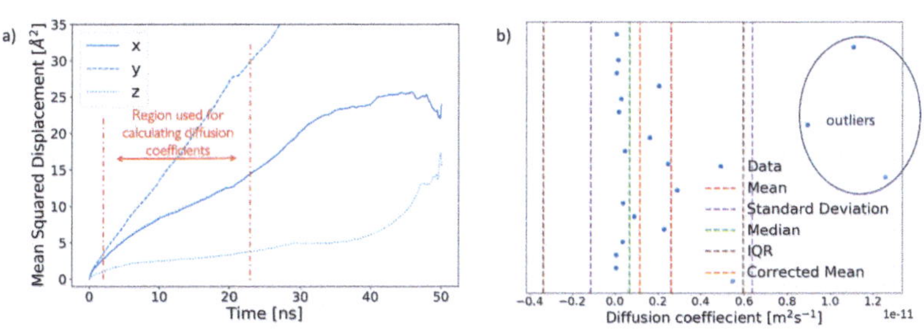

Fig. 5 (a) Example of an MSD plot of an ion broken into *x*-, *y*- and *z*-components. The region used for the calculation of diffusion coefficients is marked in red. (b) Interquartile range analysis on the diffusion coefficients of ions within a cluster. The analysis identifies three outliers, as marked on the plot, which are removed from the mean cluster diffusion coefficient to obtain a corrected mean.

3 Results

The results reported in this section are split into four main parts. We first take a look at the water behaviour in the presence of the different surface defect types on the SAM. In the second part, we discuss ionic cluster behaviour at the surface and how that varies at and around the different defect sites. Next, we report our work on the free energy calculations associated with an ion moving in and out of the surface defect of type vacancy type B. Finally, we discuss large-scale simulations where we start with $CaCO_3$ ions dispersed in solution and examine how the presence of a defect affects their behaviour using RDF analysis.

3.1 Water behaviour at the SAM surface

We start our investigation into the surface defects of a self-assembled monolayer of 16-MHDA by performing molecular dynamics simulations of each of the surface configurations, as described in Fig. 1, in pure water. Our analysis is focused on evaluating the degree of flexibility in the SAM functional groups, as well as recording the residence time of a water molecule at each individual chain, noting any particular features associated with the presence of a defect. A summary of this is reported in Fig. 6a. The figure shows the five different surface types we consider, where the x-axis of the plot corresponds to the functional group position along the x-axis of the SAM and the y-axis of the plot corresponds to the position in the z-direction of the box. As discussed in the Methods section, metrics on flexibility and residence times have been averaged across the y-axis of the surface, to which the defects run parallel, as shown in Fig. 2.

In Fig. 6a, each functional group is displayed as a circle, the colour of which represents the residence time of a water molecule at that particular chain using

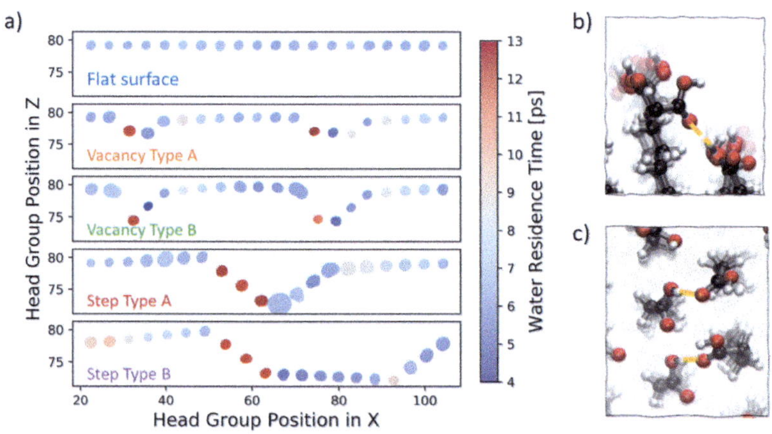

Fig. 6 (a) Summary of water residence times and –COOH functional group flexibility at the SAM surface in five configurations – flat surface, vacancy type A and B and step type A and B. Water residence times are represented by the colour, while the flexibility in the head groups is proportional to the size of the circles. (b) Image showing how the offset in the underlying Au layer in the case of vacancy type B creates space for functional groups to twist and interact with their offset neighbour. (c) Image of right-hand side steps in step type A and B where distances between –COOH are smaller and flexibility is high.

the colour bar on the right. The degree of movement in the –COOH group, on the other hand, is represented by the size of the circle; larger circles correspond to greater movement. For more information on the calculation of the residence times and flexibility of the SAM functional groups, the reader should refer to the Water analysis section in the Methods section.

Our analysis shows that, on a perfectly ordered defect-free region within a surface of a SAM which we refer to as a flat surface, the residence time of a water molecule at an individual chain is consistently around 6 ps. Equally, the variability in the degree of movement in the surface is minimal, as shown by the consistent size of the circles across the surface.

Moving to vacancy type A and B, we observe some variation in water residence times compared to that of a non-defective surface. At the first offset –COOH group, which is immediately next to the aliphatic chain of the previous, non-offset chain, the water residence time is 13 ps, which is more than double the average residence time we observe for the flat surface case. Additionally, we also note a slightly higher residence time of around 8–9 ps immediately next to the defect on the right. These results demonstrate that, due to the presence of a vacancy in the underlying Au layer, regions of higher water residence time emerge as a direct consequence of the disrupted surface pattern. The degree of movement in the –COOH functional group around vacancy type A is of a similar magnitude to that of the flat surface case, while for the chains immediately next to vacancy type B the flexibility is greater. Because of the large offset of approximately 5 Å, the part of the chain exposed to the solution has enough room to move more freely. In fact, as shown in Fig. 6b, the space is sufficient for the carboxylic acid group to twist and interact with the neighbouring offset chains through hydrogen bonding.

Step type A has three stepped chains of one offset Au-atom each, followed by three steps bringing the surface to its original level, forming a bowl-shaped defect. Step type B is similar, with the exception of having a larger separation in-between the two step defects, as shown in Fig. 1. An important feature of these stepped defects is the –COOH group spacing. The left-hand step defect has carboxylic acid groups which are marginally further apart, whereas at the right-hand going up the step the functional groups are a little closer together. This effect naturally emerges after equilibration and is a consequence of the characteristic 30° angle of the 16-MHDA aliphatic chain with respect to the normal to the plane (see Fig. 2b), in combination with the presence of steps in the underlying Au(111) surface. As a result, water residence times at the left-hand side steps are twice to three times larger that those observed on a defect-free SAM. Conversely, on the right-hand side of the steps, lower water residence times are observed, combined with a slightly higher degree of surface mobility. Looking at the trajectories, we find that in the latter regions we encounter coordination between the functional groups *via* hydrogen bonding, as shown in Fig. 6c.

In this section, we show that the presence of the defects affects both the residence times of water at the SAM functional groups and the flexibility of the SAM functional groups. The altered solvent behaviour at such defects can have important consequences for the solute. Longer water residence times will reduce solvent–solute exchange rates at the surface, increasing solute binding strengths and lowering surface diffusion rates. These effects, whether individually or combined, affect the kinetics of solute cluster formation, and therefore nucleation at surfaces.

3.2 Behaviour of the ionic clusters at the SAM surface

In order to understand how defects can affect ionic cluster behaviour at the SAM/water interface, we first set up simulations with each of the given SAM surface configurations and an ion cluster of 10 Ca^{2+} and 10 CO_3^{2-}, placed at a defect. A reference case using the defect-free surface was also included. In our system, this number of ions is equivalent to an overall concentration of 0.056 M. However, in this setup we are starting from a non-equilibrium system configuration, where all ions are placed in one particular place on the surface. We are interested in the rate of system evolution (*i.e.* ion surface diffusion) towards reaching equilibrium, which will give us insight into the role of defects on ionic behaviour. The setup procedure is performed 3 times, each with a different ion cluster configuration, for each surface type, producing three data points in the form of simulations when studying the ionic cluster behaviour. We refer to these simulations as set 1, set 2 and set 3 where applicable. For more information on the setup, the reader is referred to the section Simulation setup in the Methods section.

In Fig. 7a, we report the mean diffusion coefficient for the cluster, calculated as per the procedure outlined in the Diffusion coefficient calculations section in the Methods section. Additionally, we report all mean diffusion values, along with the lower and upper half averages of the diffusion values within each cluster, in Table 1. The latter numbers indicate whether both slow and fast diffusing ions are present within the cluster (and hence the likelihood that the cluster will fragment).

The mean diffusion coefficient of an ion cluster on a defect-free flat surface ranges between 1 and 2×10^{-12} m^2 s^{-1}. In this case, there are no topographical surface features and so we take this range as the benchmark for ionic surface diffusion when discussing the rest of the surface configurations.

In the case of vacancy type A (shown in orange), we can see that in two of the cases, the mean ionic diffusion is comparable to that in the flat surface case, indicating that the presence of one Au offset vacancy defect does not affect the

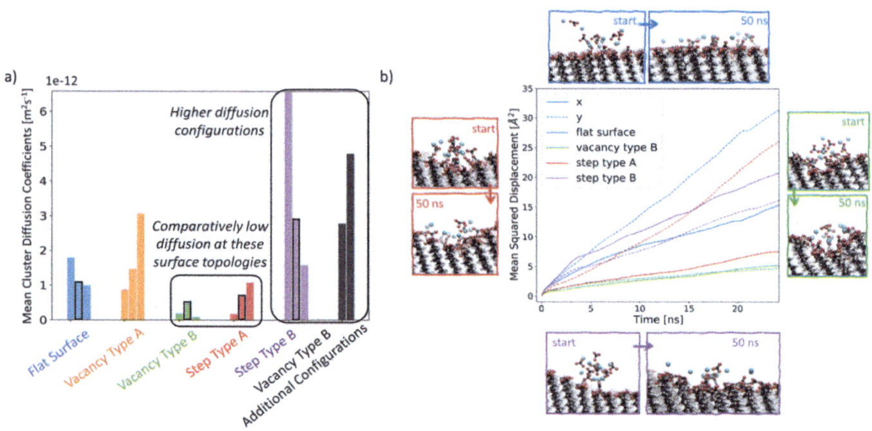

Fig. 7 (a) Mean diffusion coefficient of each ionic cluster for each simulation set, as described in the Methods section. Four of the simulations have been highlighted in black rectangles as they are shown in more detail in the second part of this figure. (b) Plot of the mean squared displacement of ions for selected clusters along the *x*- and *y*-system dimensions. Corresponding visual representations of the starting and the ending cluster configurations are displayed around the plot.

Table 1 Table of mean cluster diffusion coefficients, along with the lower (LH) and upper half (UH) average for each, representative of the spread in diffusion across the cluster. The data is shown in units of 10^{-12} m^2 s^{-1}

	LH	Set 1	UH	LH	Set 2	UH	LH	Set 3	UH
Flat surface	0.2	**1.8**	3.3	0.5	**1.1**	1.7	0.2	**1.0**	1.8
Vacancy type A	0.4	**3.1**	5.9	0.4	**1.5**	2.6	0.2	**0.9**	1.5
Vacancy type B	0.06	**0.2**	0.3	0.1	**0.5**	0.8	0.06	**0.1**	0.2
Step type A	0.09	**0.7**	1.5	0.1	**1.1**	2.2	0.06	**0.2**	0.4
Step type B	1.5	**6.6**	12.7	0.9	**2.9**	5.3	0.3	**1.6**	3.2
Vacancy type B shifted cluster	1.2	**4.2**	8.0	0.7	**2.8**	4.9	—	—	—

cluster behaviour much. An exception can be seen in set 1 of this group, where much higher average cluster diffusion is observed. In this case, the spread of diffusion coefficients is also large, showing that the cluster has fragmented.

Significantly smaller values of the mean cluster diffusion coefficients are found when ions are placed at vacancy type B (green) or step type A (red). In several of these cases, this is accompanied by an extremely compact cluster behaviour. These defect types create an environment where the ionic clusters are restrained and diffuse very little during our simulations. This implies that even relatively small surface features of the order of magnitude of tens of Å can substantially affect ionic cluster diffusion by providing a barrier to their natural surface mobility. Reducing cluster mobility, on the other hand, can produce more rapid growth of clusters around this defect as passing ions become trapped in its vicinity.

Ionic clusters placed at step type B (in purple) generally display quite dynamic cluster behaviour with mean diffusion coefficients between 2×10^{-12} m^2 s^{-1} and 6×10^{-12} m^2 s^{-1}, which are an order of magnitude higher than what we see in the flat surface case. We selected several of the configurations, highlighted by a black rectangular border in Fig. 7a, for more detailed analysis. Fig. 7b displays a visual representation of the starting and the ending configurations for each of the selected cases, together with a corresponding mean squared displacement plot along the x- and y-surface axes. Visual analysis of the ions on the flat surface SAM shows how the cluster has transformed to resemble an ionic layer across the surface. The corresponding MSD analysis shows that this spread occurs across both the x- and y-axes of the surface, being more substantial for the latter. This phenomenon is due to the packing arrangement of the underlying Au(111) surface, resulting in larger spacing between –COOH groups in the y-direction of the surface than in the x-direction. In the case of step type B, shown in purple, the slopes of the x- and y-component of the MSD are comparable to each other, indicating that the ions are diffusing both across and along the surface. Additionally, as shown in Fig. 7b in purple, the diffusion along the x-axis occurs preferentially away from the stepped region. This observation is consistent across all three simulations of this group, suggesting that the equilibration process of the ionic cluster is more favourable across a defect-free environment, rather than "climbing" the steps, and indicates a presence of a barrier associated with diffusion across the defect.

For the cases where a more compact cluster behaviour is observed, namely vacancy type B and step type A, a corresponding shift in the MSD, shown respectively in green and red in Fig. 7b, can be seen. In the case of step type A in set 2, the movement associated with spread across the *y*-axis (along the defect) is much larger than that observed in the *x*-axis (perpendicular to the defect), correlating with the general argument that has emerged from our simulations that defected surface features present a barrier to ionic surface diffusion.

3.2.1 A detailed look into vacancy type B. Additional configurations of ion clusters positioned on either side of defect vacancy type B, as shown in Fig. 8, were set up to understand better the behaviour of ionic clusters next to defects which have shown to be able to immobilise them. Once again, we start with a non-equilibrium ionic configuration to see whether such active defects can immobilise ions around them, as well as within them. In both cases, after a 50 ns simulation, ions diffuse preferentially and readily across the flat region away from the defect, with a few ions being trapped within the defect. Diffusion coefficients for these cases are reported in Fig. 7a in black and are 2–3 times higher than the diffusion coefficients for ions moving across a defect-free flat surface.

Analysis of the ionic cluster behaviour indicates that some surface defects will impair ion mobility across a surface, while others will not. We will refer to the former as active defects. Moreover, we see a general trend that in the process of equilibration, ion clusters diffuse much less readily across the active defect than across a defect-free feature, suggesting that topographical defects create a barrier to diffusion, reducing the available surface space. In our simulations, when starting with a local enhanced concentration, this effect results in higher diffusion rates away from a defect compared to the flat surface reference case, as shown in Fig. 7a (purple and black cases). However, as discussed, these simulations have artificially placed ion clusters as starting configurations and so the higher diffusion is due to the non-equilibrium starting conditions and the relaxations away from this state.

3.3 Free energy calculations on single ion movements with respect to vacancy type B

We take a closer look at the free energy associated with the movement of Ca^{2+} and CO_3^{2-} ions in and out of vacancy type B to gain insight into the solute–surface

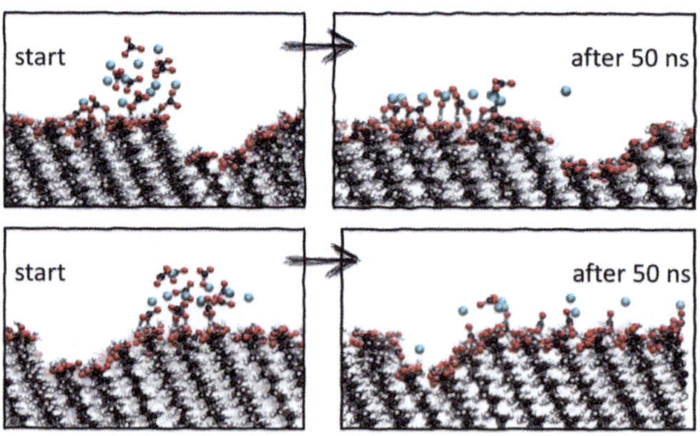

Fig. 8 Starting and final configurations of simulations of ion clusters placed next to vacancy type B, showing the preferential diffusion of ions across the defect-free surface side.

interactions promoting the observed ion cluster behaviour. Simulations were performed in combination with WTmetaD and the full details of the setup are outlined in the Well-tempered metadynamics setup section in the Methods section. Each simulation was carried out until the average positional standard error was within 1.5 k_BT units. All convergence analysis is reported in the ESI.†

Fig. 9 displays the free energy profiles recovered from the metadynamics runs. The free energy reported is projected onto the 2-dimensional space of the *x*- and *z*-distance components between the ion of interest and the centre of mass of the underlying Au surface. In this representation, we are making the assumption that, since the defect runs parallel with the *y*-axis of the box, ionic movements along that axis are degenerate. The free energy profile of the movement associated with a Ca^{2+} ion, reported on the left of Fig. 9, shows small variations across the system as a consequence of the presence of surface defects. In particular, we see a more stable state within the right-hand defect, which is the slightly wider of the two (refer to Fig. 1). This effect is relatively long-ranged, stretching as far as 10 Å away from the well of the defect. On either side of the defect, we see higher energy states of the Ca^{2+} ion. Despite the fact that the free energy difference is not large, energetically varied ionic states at and a few Å away from the surface can be clearly identified.

Looking at the right-hand side free energy surface of the counter CO_3^{2-} ion, we see a substantial shift in the energetics of the system, where the solution is now much less favourable than the surface states. This trend is closely related to the fact that the functional groups at the SAM surface are uncharged and CO_3^{2-} ions are associating strongly with the surface through hydrogen bonding. We can speculate that if the chains were charged and had a –COO^- functionality instead, the Ca^{2+} ions would be the ones with the stronger surface–solute interaction. Nonetheless, we believe that in the context of surface topography effects, our observations are valid and give insight into general mechanisms by which small surface defects can operate on nucleating species. The free energy surface (FES), corresponding to the movement of CO_3^{2-}, also shows a high energy region in the solution right above the defect, particularly for the wider defect on the right-hand side. This observation explains why we see little diffusion of ions out of and into

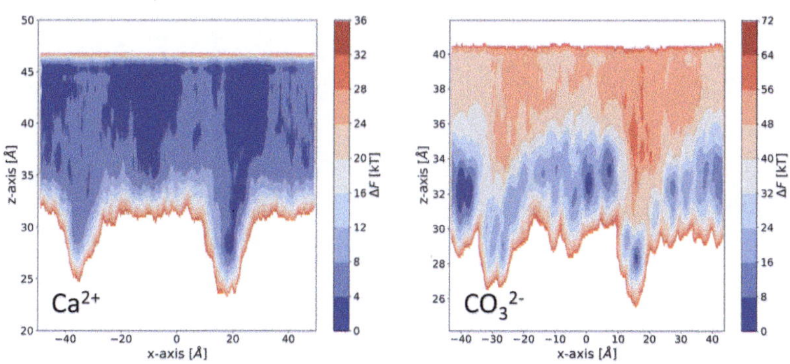

Fig. 9 Free energy profiles obtained from enhanced sampling metadynamics simulations of a single Ca^{2+} (left) and a single CO_3^{2-} (right) in water in the presence for surface vacancy type B.

this defect in our simulations – there is a barrier associated with the movement of ions. The effect can be seen stretching as far as 10 Å away from the surface.

By combining metadynamics with MD simulations, we have been able to gain further insight into the ion–surface interactions which dictate the observed ion cluster behaviour. We find that, in response to the altered surface topography, high and low energy regions emerge which stretch over a relatively large distance into the solution. These regions will control ionic diffusion at and close to the surface.

3.4 Large-scale MD simulations at vacancy type B

To further test our hypothesis, we performed two additional sets of simulations at a SAM surface of size 170 × 80 Å with a vacancy type B surface feature. Unlike the set of simulations previously discussed, in this case the initial configurations contain ions of Ca^{2+} and $CO_3{}^{2-}$ randomly dispersed in solution at two concentrations: 0.41 M and 0.69 M. Our objective is to probe whether the surface feature will have a noticeable effect on the ionic system behaviour using an equilibrium starting configuration rather than the non-equilibrium configurations with clusters placed close to defects as discussed above. The simulations were performed for 35 ns for the 0.41 M system and for 70 ns for the 0.69 M system.

In Fig. 10, we report the radial distribution function (RDF) between each unique –COOH functional group position along the *x*-axis and Ca^{2+} and $CO_3{}^{2-}$ ions for the last 5 ns of each simulation. Individual 1-dimensional RDFs, calculated as a function of the distance from the –COOH group to the ions, follow a dashed line. The RDFs have been plotted side by side, creating an overall 2-dimensional representation of the ionic distribution across and away from the surface. For clarity, we also provide a schematic of the corresponding system configuration by representing C-atoms on the functional group as black dots and ions as blue.

From the visual representation of the ions, we can see that there is clustering of ions close to and within the defect, running parallel to the *y*-axis and around 70–

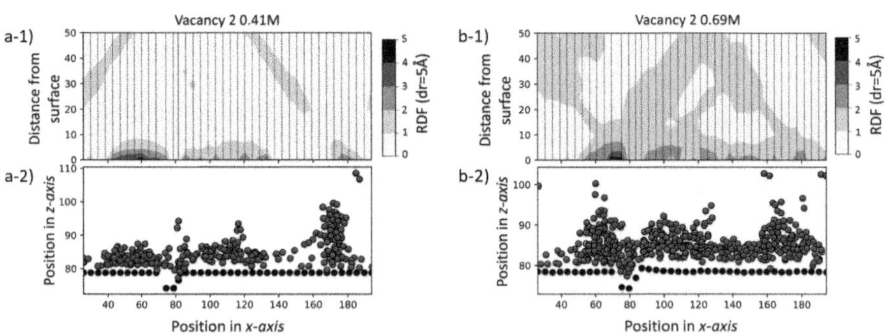

Fig. 10 (a-1) RDF of the large-scale simulation in a solution of 0.41 M $CaCO_3$. (b-1) RDF of the large-scale simulation in a solution of 0.69 M $CaCO_3$. Each dashed line corresponds to a 1D RDF profile as a function of the –COOH to ion distance. (a-2) Black dots show –COOH functional group positions in the *xz*-system cross-section, while blue ones show ion positions for 0.41 M. (b-2) Black dots show –COOH functional group positions in the *xz*-system cross-section, while blue ones show ion positions for 0.69 M.

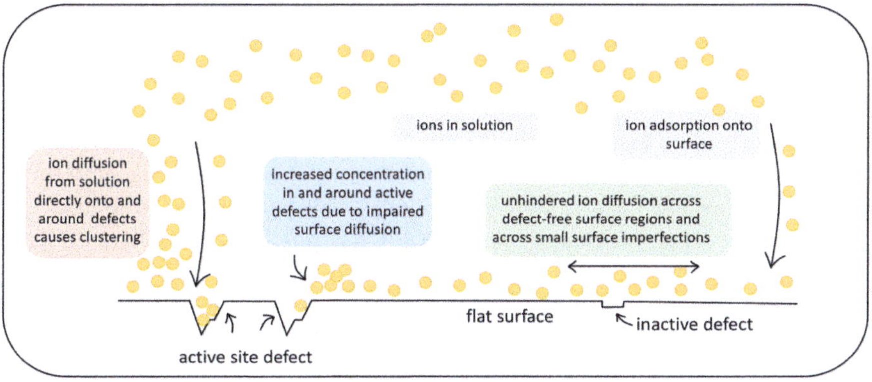

Fig. 11 Schematic summary of the observed effects that atomic-scale surface defects can exert on ionic substrates.

80 Å along the *x*-axis. Our RDF analysis reveals that for both cases, the surface generally attracts ions, as we see 2–3 times higher probability of ionic concentration in most surface areas compared to a random distribution. However, we can see a further distinct increase in and around the defect site of 3–5 times higher ionic distribution. More details on the calculation of the RDF distributions, and details of how the RDF evolves in time for the duration of each simulation, are provided in the ESI.†

The observations made here indicate that, even in a simulation environment where ions are initially distributed randomly, the presence of an active surface defect can affect the behaviour of ion clustering at the surface, which in turn can affect the rate at which ionic clusters form and grow.

4 Conclusions

This work investigates the effect of atomic-scale defects in a flexible SAM on the behaviour of the solvent, as well as on that of ionic $CaCO_3$ clusters. Our analysis shows that the presence of features in the surface topography alters the local behaviour of the solvent by providing sites where the adsorbed state of a solvent molecule is longer lived in comparison to a defect-free surface. Furthermore, the presence of surface defects affects the ion cluster behaviour at the SAM surfaces. Certain surface features in our investigation are clearly able to reduce the mobility of ions across the surface. When placed immediately next door to steps or vacancies, ions preferentially diffuse away from these features, showing that there is a barrier associated with diffusion across them. This is further supported by the WTmetaD simulations on the ion mobility across vacancy type B, where we see long-ranged disruption in the free energy. In simulations with significantly higher ionic concentrations, where ions are randomly dispersed in solution to start with, we see the same effects, where ions notably cluster around defect sites.

Our findings have important consequences for understanding heterogeneous nucleation. Firstly, only certain defect sites will act as active nucleation sites and therefore we can expect that not all surface features will aid nucleation, as indicated in Fig. 11 by the label inactive defects.

Secondly, the presence of active defects affects the solvent and solute behaviour, limiting the mobility of ions around the defect. We speculate that there are two mechanisms involved. Ions will cluster either when descending close to an active site from solution or when diffusing around on the surface before reaching an active site (as indicated in Fig. 11). Both mechanisms will result in higher concentration of ions in the solution around the defect. We recall that classical nucleation theory (like many rare event theories) involves an assumption of quasi-equilibrium between species in the initial state. It is nevertheless a theory about kinetics, not thermodynamics. The system passes along a single coordinate (traditionally defined as cluster size but modern computational methods have greatly widened this choice).[46] The enhanced local concentrations are consistent with an assumption of quasi-equilibrium at the surface but, as the local initial state for the nucleation process, it will also increase the probability of forming a critical nucleus at these sites.

In the context of heterogeneous nucleation, the implications of our observations perhaps help explain why and how few surfaces operate as effective nucleating agents. Surfaces are generally understood to stabilise a nucleus by lowering its interfacial energy and hence, aid nucleation (as would be predicted by the Joule–Thompson relation), but these interfacial energies are always assumed to be those of the bulk interface, ignoring local effects caused by defects. Reliable nucleating agents are in reality hard to find. If only certain surface topological features stimulate a significant increase in local supersaturation, as indicated by our work, then a defect-free region, which would be the majority of the surface area, will not be particularly good at inducing clustering in itself. Nuclei that form in the absence of surface active sites may be stabilised by the surface, however the probability of their formation will not be enhanced. In the presence of an active defect, the increase in local supersaturation, under the mechanisms we propose, will enhance the probability of nuclei formation. The concentration of active defects, however, will always be limited and only mildly dependent on the surface chemistry. This means that the number of nuclei that can form at active sites at any one time will also be limited. In the context of thermodynamics, these nuclei will therefore be stabilised by the fact that their number is self-limited and system solute concentration is not depleted upon their formation, until they reach critical size and crystal formation is underway.

The results presented in this work demonstrate that even small features in flexible surfaces can affect solvent and solute behaviour at and a few Ångstroms away from the surface. The presence and density of these atomic-scale defects can alter the course of nucleation in a heterogeneous environment. Understanding the role of these active sites is essential for the design of effective nucleating agents.

Conflicts of interest

There are no conflicts to declare.

Acknowledgements

The authors gratefully acknowledge the financial support of the Engineering and Physical Sciences Research Council (EPSRC), grant no. EP/R018820/1 which funds the Crystallization in the Real World consortium. *Via* our membership of the UK's HEC Materials Chemistry Consortium, which is funded by the EPSRC (EP/R029431), this work used the UK Materials and Molecular Modelling Hub for computational resources, MMM Hub, which is partially funded by the EPSRC (EP/T022213) and the ARCHER2 UK National Supercomputing Service (http://www.archer2.ac.uk).

Notes and references

1 L. Addadi and S. Weiner, *Angew. Chem., Int. Ed. Engl.*, 1992, **31**, 153–169.

2 S. Weiner and P. M. Dove, *Rev. Mineral. Geochem.*, 2003, **54**, 1–29.

3 S. Weiner and L. Addadi, *Annu. Rev. Mater. Res.*, 2011, **41**, 21–40.

4 S. Kim, C. Wei and S. Kiang, *Org. Process Res. Dev.*, 2003, **7**, 997–1001.

5 N. Variankaval, A. S. Cote and M. F. Doherty, *AIChE J.*, 2008, **54**, 1682–1688.

6 S. J. Dyer and G. M. Graham, *J. Pet. Sci. Eng.*, 2002, **35**, 95–107.

7 J. MacAdam and S. A. Parsons, *Rev. Environ. Sci. Bio/Technol.*, 2004, **3**, 159–169.

8 A. Antony, J. H. Low, S. Gray, A. E. Childress, P. Le-Clech and G. Leslie, *J. Membr. Sci.*, 2011, **383**, 1–16.

9 M. A. Levenstein, C. Anduix-Canto, Y.-Y. Kim, M. A. Holden, C. G. Niño, D. C. Green, S. E. Foster, A. N. Kulak, L. Govada, N. E. Chayen, S. J. Day, C. C. Tang, B. Weinhausen, M. Burghammer, N. Kapur and F. C. Meldrum, *Adv. Funct. Mater.*, 2019, **29**, 1808172.

10 L. M. Hamm, A. J. Giuffre, N. Han, J. Tao, D. Wang, J. J. D. Yoreo and P. M. Dove, *Proc. Natl. Acad. Sci. U. S. A.*, 2014, **111**, 1304–1309.

11 M. L. Whittaker, P. M. Dove and D. Joester, *MRS Bull.*, 2016, **41**, 388–392.

12 A. Mcpherson and P. Shlichta, *Science*, 1988, **239**, 385–387.

13 A. Markus Travaille, L. Kaptijn, V. Paul, B. Hulsken, J. A. A. W. Elemans, R. J. M. Nolte and H. van Kempen, *J. Am. Chem. Soc.*, 2003, **125**, 11571–11577.

14 E. Saridakis and N. E. Chayen, *Trends Biotechnol.*, 2009, **27**, 99–106.

15 J. P. Mithen and R. P. Sear, *J. Chem. Phys.*, 2014, **140**, 084504.

16 D. M. Duffy and J. H. Harding, *Surf. Sci.*, 2005, **595**, 151–156.

17 T. F. Whale, M. A. Holden, A. N. Kulak, Y.-Y. Kim, F. C. Meldrum, H. K. Christenson and B. J. Murray, *Phys. Chem. Chem. Phys.*, 2017, **19**, 31186–31193.

18 M. A. Holden, T. F. Whale, M. D. Tarn, D. O'Sullivan, R. D. Walshaw, B. J. Murray, F. C. Meldrum and H. K. Christenson, *Sci. Adv.*, 2019, **5**, eaav4316.

19 J. Aizenberg, A. J. Black and G. M. Whitesides, *J. Am. Chem. Soc.*, 1999, **121**, 4500–4509.

20 C. L. Freeman, J. H. Harding and D. M. Duffy, *Langmuir*, 2008, **24**, 9607–9615.

21 J. Aizenberg, A. J. Black and G. M. Whitesides, *Nature*, 1998, **394**, 868–871.

22 C. L. Freeman, Q. Hu, M. H. Nielsen, J. Tao, J. J. De Yoreo and J. H. Harding, *J. Phys. Chem. C*, 2013, **117**, 5154–5163.

23 K. Sprenger, Y. He and J. Pfaendtner, Probing How Defects in Self-assembled Monolayers Affect Peptide Adsorption with Molecular Simulation, in *Foundations of Molecular Modeling and Simulation*, ed. R. Snurr, C. Adjiman and D. Kofke, Springer, Singapore, 2016, pp. 21–35.

24 G. Chong, I. U. Foreman-Ortiz, M. Wu, A. Bautista, C. J. Murphy, J. A. Pedersen and R. Hernandez, *J. Phys. Chem. C*, 2019, **123**, 27951–27958.

25 C. Vericat, M. E. Vela and R. C. Salvarezza, *Phys. Chem. Chem. Phys.*, 2005, **7**, 3258–3268.

26 A. Ulman, *Chem. Rev.*, 1996, **96**, 1533–1554.

27 J. Wang, R. M. Wolf, J. W. Caldwell, P. A. Kollman and D. A. Case, *J. Comput. Chem.*, 2004, **25**, 1157–1174.

28 J. Wang, W. Wang, P. A. Kollman and D. A. Case, *J. Mol. Graphics Modell.*, 2006, **25**, 247–260.

29 P. Raiteri and J. D. Gale, *J. Am. Chem. Soc.*, 2010, **132**, 17623–17634.

30 P. K. Ghorai and S. C. Glotzer, *J. Phys. Chem. C*, 2007, **111**, 15857–15862.

31 H. A. Lorentz, *Ann. Phys. Chem.*, 1881, **248**, 127–136.

32 D. Berthelot, *C. R. Hebd. Seances Acad. Sci.*, 1898, 1703–1855.

33 C. L. Freeman, J. H. Harding, D. J. Cooke, J. A. Elliott, J. S. Lardge and D. M. Duffy, *J. Phys. Chem. C*, 2007, **111**, 11943–11951.

34 S. Plimpton, *J. Comput. Phys.*, 1995, **117**, 1–19.

35 A. P. Thompson, H. M. Aktulga, R. Berger, D. S. Bolintineanu, W. M. Brown, P. S. Crozier, P. J. Veld, A. Kohlmeyer, S. G. Moore, T. D. Nguyen, R. Shan, M. J Stevens, J. Tranchida, C. Trott and S. J. Plimpton, *Comput. Phys. Commun.*, 2022, **271**, 108171.

36 R. W. Hockney and J. W. Eastwood, *Computer Simulation Using Particles*, Hilger, Bristol, 1988.

37 S. Nosé, *J. Chem. Phys.*, 1984, **81**, 511.

38 W. G. Hoover, *Phys. Rev. A: At., Mol., Opt. Phys.*, 1985, **31**, 1695.

39 L. Martinez, R. Andrade, E. G. Birgin and J. M. Martinez, *J. Comput. Chem.*, 2009, **30**, 2157–2164.

40 A. Barducci, G. Bussi and M. Parrinello, *Phys. Rev. Lett.*, 2008, **100**, 020603.

41 A. Barducci, M. Bonomi and M. Parrinello, *WIREs Computational Molecular Science*, 2011, **1**, 826–843.

42 M. Bonomi, D. Branduardi, G. Bussi, C. Camilloni, D. Provasi, P. Raiteri, D. Donadio, F. Marinelli, F. Pietrucci, R. A. Broglia and M. Parrinello, *Comput. Phys. Commun.*, 2009, **180**, 1961–1972.

43 G. A. Tribello, M. Bonomi, D. Branduardi, C. Camilloni and G. Bussi, *Comput. Phys. Commun.*, 2014, **185**, 604–613.

44 W. Humphrey, A. Dalke and K. Schulten, *J. Mol. Graphics*, 1996, **14**, 33–38.

45 G. Van Rossum and F. L. Drake, *Python 3 Reference Manual, CreateSpace*, Scotts Valley, CA, 2009, p. 242.

46 B. Peters, *J. Phys. Chem. B*, 2015, **119**, 6349–6356.

Faraday Discussions

PAPER

Precrystallization solute assemblies and crystal symmetry

Monika Warzecha, †[a] Lakshmanji Verma, †[b] Rajshree Chakrabarti,†[b] Viktor G. Hadjiev, [c] Alastair J. Florence,[a] Jeremy C. Palmer[b] and Peter G. Vekilov *[bd]

Received 1st November 2021, Accepted 22nd December 2021
DOI: 10.1039/d1fd00080b

Solution crystallization is a part of the synthesis of materials ranging from geological and biological minerals to pharmaceuticals, fine chemicals, and advanced electronic components. Attempts to predict the structure, growth rates and properties of emerging crystals have been frustrated, in part, by the poor understanding of the correlations between the oligomeric state of the solute, the growth unit, and the crystal symmetry. To explore how a solute monomer or oligomer is selected as the unit that incorporates into kinks and how crystal symmetry impacts this selection, we combine scanning probe microscopy, optical spectroscopy, and all-atom molecular simulations using as examples two organic materials, olanzapine (OZPN) and etioporphyrin I (EtpI). The dominance of dimeric structures in OZPN crystals has spurred speculation that the dimers preform in the solution, where they capture the majority of the solute, and then assemble into crystals. By contrast, EtpI in crystals aligns in parallel stacks of flat EtpI monomers unrelated by point symmetry. Raman and absorption spectroscopies show that solute monomers are the majority solute species in solutions of both compounds. Surprisingly, the kinetics of incorporation of OZPN into kinks is bimolecular, indicating that the growth unit is a solute dimer, a minority solution component. The disconnection between the dominant solute species, the growth unit, and the crystal symmetry is even stronger with EtpI, for which the (010) face grows by incorporating monomers, whereas the growth unit of the (001) face is a dimer. Collectively, the crystallization kinetics results with OZPN and EtpI establish that the structures of the dominant solute species and of the incorporating solute complex do not correlate with the symmetry of the crystal lattice. In a broader context, these findings illuminate the immense complexity of crystallization

[a]*EPSRC, CMAC, Future Manufacturing Research Hub, c/o Strathclyde Institute of Pharmacy and Biomedical Sciences, Technology and Innovation Centre, 99 George Street, Glasgow, G1 1RD, UK*
[b]*Department of Chemical and Biomolecular Engineering, University of Houston, 4726 Calhoun Rd., Houston, TX 77204-4004, USA. E-mail: vekiov@uh.edu*
[c]*Texas Center for Superconductivity, University of Houston, 3369 Cullen Blvd., Suite 202, Houston, Texas 77004-50024, USA*
[d]*Department of Chemistry, University of Houston, 3585 Cullen Blvd., Houston, TX 77204-5003, USA*
† These authors contributed equally to this work.

scenarios that need to be explored on the road to the understanding and control of crystallization.

Introduction

Crystals are defined by their symmetry.[1] Crystal symmetry combines point symmetry elements, such as inversion centers, mirror planes, and rotation axes, typical of finite-size objects, with translation.[2] The translational symmetry of silicon, germanium, and other semiconductor crystals guides their electronic band structure and the behaviors of charge carriers that define their utility.[3] The point symmetries of diamond and other precious stones are lower than that of a sphere and govern their direction-dependent properties, including the substantial anisotropy of their refractive indexes, their most appealing property.[4] Alternatively, the lack of an inversion center in quartz and other polar crystals preconditions their piezoelectricity.[5] The hexagonal molecular arrangement in normal ice institutes a density lower than that of water and enforces ice accumulation on the surfaces of lakes and seas, which, in turn, facilitated the rise of life.[6]

Translational symmetry dictates exact rules for how a molecule in a crystal orients and positions with respect to any other crystal molecule. It arises during crystallization. The origin of point symmetry elements, however, is far less understood. Crystal point symmetry is certainly uncorrelated to molecular symmetry.[7,8] Water molecules carry two-fold symmetry, yet at standard conditions they crystalize as hexagonal ice. At elevated pressures and low temperatures, water arranges in several other distinct crystal lattices. The variability of crystal symmetries that a single molecule can adopt, often called polymorphism, is another example of the disconnection between molecular and crystal symmetries.[9]

The classical theories of crystal nucleation and growth assume that molecules incorporate into a crystal individually. This perspective inevitably commands that, similarly to translational symmetry, point symmetry arises during crystallization. Recent observations in complex biological, geological, and synthetic environments have exposed cases of nonclassical crystallization employing liquid, amorphous, and crystalline precursors.[10–15] The agency of the mesoscopic precursors only marginally modifies our understanding of the correlation between solute and crystal symmetries since the ordered structures of nanocrystals originate during their own nucleation;[11] amorphous, liquid, or partially ordered[16,17] particles attain symmetry upon conversion to crystals or association to a host crystal.

A more intriguing question is posed by microscopic crystallization precursors that present as solute dimers and higher order oligomers, which form owing to the unavoidable interactions between solute monomers[18–21] and may, in some cases, mirror the structural units of the crystal.[22–26] It is easy to presume that the common structural motifs in the solution and in the crystal represent the solute species that associate to the kinks, which would imply that crystal symmetry elements arise in the solution prior to crystallization. The elementary acts of incorporation into kinks, however, have only been directly visualized for relatively few crystals.[27–43] Given the constraints to directly identify the growth unit, the promotion of the solute oligomers to the rank of the incorporating species has remained largely unchallenged and the origin of crystal symmetry is obscure.

To test how solute oligomers transition into symmetry-related crystal blocks, here we put forth the kinetics of step growth as an indicator of the crystal growth unit. We then explore the correlation of the growth unit with the state of the solute and the crystal symmetry. We use as examples two molecular crystals, olanzapine (OZPN) and etioporphyrin I (EtpI). The results demonstrate that the crystal growth units may diverge from both the dominant solute species and the crystal structural units and inform us that the correlation between the symmetry of a solute oligomer and that of the crystal structure is far from a general law.

OZPN: growth by incorporation of dimers that capture a minor fraction of the solute

Olanzapine (OZPN, 2-methyl-4-(4-methyl-1-piperazinyl)-10*H*-thieno-[2,3-*b*] [1,5] benzodiazepine, Fig. 1a) is a benzodiazepine derivative used to treat the positive (hallucinations and delusions) and negative (social withdrawal) symptoms of schizophrenia, bipolar disorder, and other psychoses.[44] OZPN which precipitates in more than 60 individual crystal forms.[45] In most of these forms, OZPN molecules arrange in a centrosymmetric dimer, comprised of two conformational enantiomers (Fig. 1c).[45] The dominance of dimeric structures has encouraged the supposition that the dimers form in the solution, where they are expected to be the dominant species, and then incorporate in the crystal as a whole.[46,47]

To test the third part of the "dimer is the growth unit" hypothesis for OZPN, that OZPN crystals grow by the incorporation of dimers, we monitor the growth of $2OZPN \cdot EtOH \cdot 2H_2O$ crystals from a 1/1 (v/v) ethanol/water mixture using time-resolved *in situ* AFM.[47] Crystals prepared by cooling crystallization are light yellow in color and show a rhombohedral morphology with large {002} faces parallel to the substrate (Fig. 1b), suitable for observation by atomic force microscopy (AFM); for details of the experimental procedures, see ref. 49. Time-resolved *in situ* AFM monitoring reveals that the {002} faces grow by the incorporation of the solute into steps produced by screw dislocations (Fig. 2b).[47] The

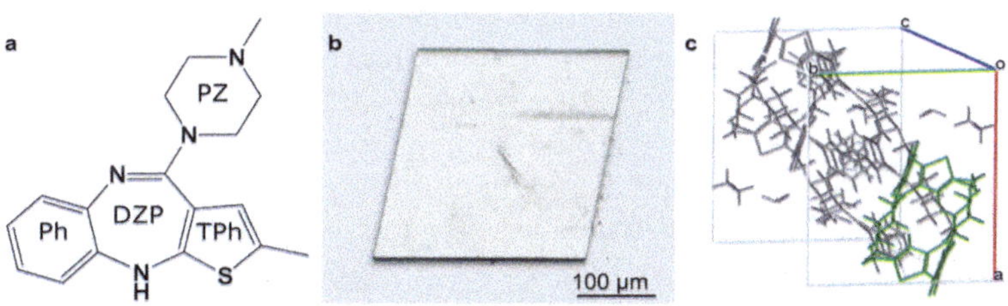

Fig. 1 The structure of OZPN and its crystals. (a) The structure of the OZPN molecule: Ph, benzene; DZP, diazepine; TPh, thiophene; PZ, piperazine rings. (b) Optical micrograph of an OZPN crystal, in which the (002) face faces upwards. (c) The crystal structure of the dihydrate ethanoate mixed solvate 2OZPN·EtOH·2H2O in space group $P2_1/c$ (Cambridge Structural Database REFCODE WEXQEW[48]). One centrosymmetric OZPN dimer is highlighted in green.

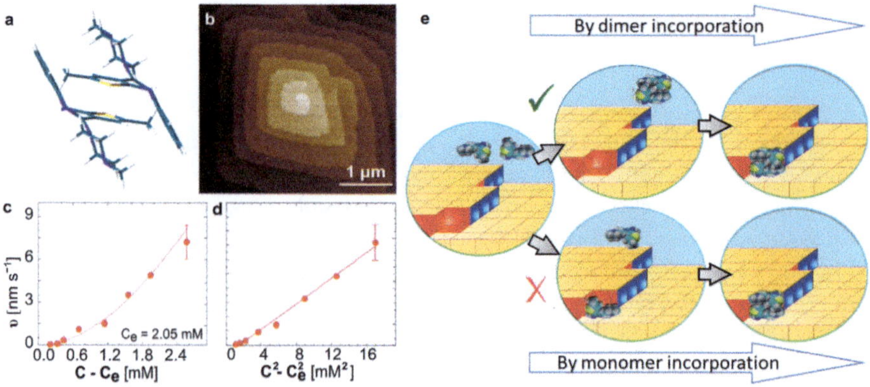

Fig. 2 The growth of OZPN crystals. (a) The structure of the OZPN crystal dimer. (b) *In situ* AFM image of the surface of a (002) face of an OZPN crystal at $C = 3.87$ mM. New crystal layers are generated by a screw dislocation. (c) The velocity ν of steps in the [110] direction as a function of OZPN concentration C in 1/1 (v/v) EtOH/H_2O. (d) The linear correlation between ν and $(C^2 - C_e{}^2)$ has $R = 0.95$. The dotted lines in (c and d) depict the relation $v = \beta_D K_D \Omega_D (C^2 - C_e{}^2)$, where β_D is the kinetic coefficient for growth by dimer incorporation, $\Omega_D = 2\ \Omega_M = 0.94\ \text{nm}^3$ is the volume occupied by a dimer in the crystal, and K_D is the dimerization equilibrium constant. Error bars in (c and d) indicate the standard deviation of v, determined as the slope of the displacement–time correlations. Data in (c and d) are from ref. 49. (e) Schematic of two alternative growth mechanisms. Upper path: two OZPN monomers form a dimer in the solution, which incorporates into the crystal as a whole. Lower path: the two monomers incorporate sequentially, forming a dimer in the crystal.

step velocity ν was determined from the slope of step displacement against a reference point as a function of time as OZPN steps grew at steady rates over extended periods. Assuming that steps grow by the incorporation of solute monomers[50] suggests a monomolecular reaction and generates a linear correlation between ν and the solute concentration C_{OZPN}, $\nu = \beta\Omega(C_{\text{OZPN}} - C_e)$, where subtracting the solubility C_e accounts for the reversibility of molecular attachment. Here, Ω is the molecular volume in the crystal, and β is an effective kinetic coefficient, which includes the kinetic parameters for the selected growth mechanism, direct incorporation or *via* adsorption on the terraces.[51,52] Linear $\nu(C_{\text{OZPN}})$ correlations have been observed for numerous solution grown crystals.[10,27,29,53–66] Surprisingly, OZPN displays a superlinear $\nu(C_{\text{OZPN}})$ dependence, which extends to concentrations more than twice the solubility C_e (Fig. 2c and d).

We excluded four potential causes of the apparent growth acceleration at high supersaturation:[49] mesoscopic OZPN-rich clusters,[47,67] an additional source of OZPN molecules to the steps, inaccurate solubility, increasing kink density at higher supersaturation[29,50,68] and step pinning by impurities.[53] We propose that the superlinear $\nu(C_{\text{OZPN}})$ is a consequence of crystal growth by the incorporation of dimers present in equilibrium with OZPN monomers in the growth solution (Fig. 2e). Elevated OZPN concentrations shift the dimerization equilibrium towards dimers and nonlinearly enhance the dimer concentration. Kinetic analysis[49] demonstrates that if monomers dominate in the solution, but growth occurs by the attachment dimers, then the step velocity ν correlates with the total solute concentration C_{OZPN}, as $\nu = \beta_D\Omega_D K_D(C_{\text{OZPN}}{}^2 - C_e{}^2)$, where β_D is the kinetic coefficient for growth by dimer incorporation, Ω_D is the volume that a dimer occupies in the crystal, and K_D is the dimerization equilibrium constant. By contrast, if the crystals grow by the incorporation of the majority monomers, then we recover the expected linear correlation $\nu = \beta_M\Omega_M(C_{OZPN} - C_e)$, where β_M is the kinetic coefficient for growth by monomer incorporation and Ω_M is the volume that a monomer occupies in the crystal. Importantly, a linear rate law is also obtained if the crystal grows by the incorporation of dimers in a solution dominated by dimers, $\nu = \beta_D\Omega_M(C_{OZPN} - C_e)$; notably, in the latter case ν scales with the monomer volume Ω_M.

This analysis reveals that if the dominant solute species, whether it be monomer or dimer, is the one that incorporates in the kinks, the kinetics of layer growth will be linear. If the crystals grow by the incorporation of solute dimers that exist in equilibrium with a majority of monomers, a quadratic correlation between the step velocity and the analytical concentration of the solute ensues. The $\nu(C_{OZPN})$ data for OZPN are consistent with this latter functional relation (Fig. 2c and d).

The discrepancy between the linear kinetic law expected for growth by the incorporation of dimers, which capture a majority of the solute, and the superlinear $\nu(C_{OZPN})$ dependence measured by AFM is the first indication that dimers do not dominate in OZPN solutions. To further explore the monomer–dimer solute dynamics, we employ Raman spectroscopy; for details of the experimental procedures, see ref. 49. We compare OZPN spectra at a low concentration, 0.005 M, to spectra at relatively high concentrations, up to 0.043 M, and to spectra of the corresponding solid form, $2OZPN \cdot EtOH \cdot 2H_2O$, in which OZPN is arranged as dimers (Fig. 3a). To assign the origins of the observed peaks, we model spectra for an OZPN monomer and an OZPN dimer using density functional theory (DFT)[49] (Fig. 3a). The model spectra for the OZPN monomer show multiple Raman peaks between 1200 and 1500 cm^{-1} (Fig. 3a), whereas the model spectra for the dimer show strong peaks around 1050 cm^{-1} and in the range 1500–1600 cm^{-1}.[49]

The Raman spectra of OZPN water ethanoate solvate crystals show noticeable dimer peaks similar to those of the spectra obtained for solution samples at high concentrations (Fig. 3a). The low-concentration solution spectra reveal strong monomer peaks and no dimer peaks. The Raman spectra of increasing intermediate concentrations in both solvents show the gradual contraction of the monomer peaks and a growth in the intensity of the dimer peaks (Fig. 3a). The concentration dependencies of the monomer and dimer Raman peak intensities provides the opportunity to assess the OZPN dimerization constant K_D.[49] We obtain $K_D = 2.7 \pm 0.1\ M^{-1}$. Mass balance calculations inform us that the dimer concentration C_D in the growth solution is 0.01 mM at the solubility of 2.05 mM; C_D increases to 0.05 mM at the highest tested total OZPN concentration in the growth studies, $C = 4.45$ mM (Fig. 2c). The superlinear increase of C_D dictates the quadratic $\nu(C)$ correlation measured by AFM (Fig. 2c and d).

All-atom umbrella sampling molecular dynamics (for details, see ref. 49) assessment of the potentials of mean force F between two OZPN monomers in $EtOH/H_2O$ reveal a sharp minimum at the center-of-mass separation of 0.43 nm (Fig. 3b). The conformations of the two monomers occupying the minimum oscillate around the structure of the dimer present in most OZPN crystal forms (Fig. 3b, inset). At larger separations, the dimer dissociates, however, F exhibits a shallow secondary minimum that links to loose dimers with variable

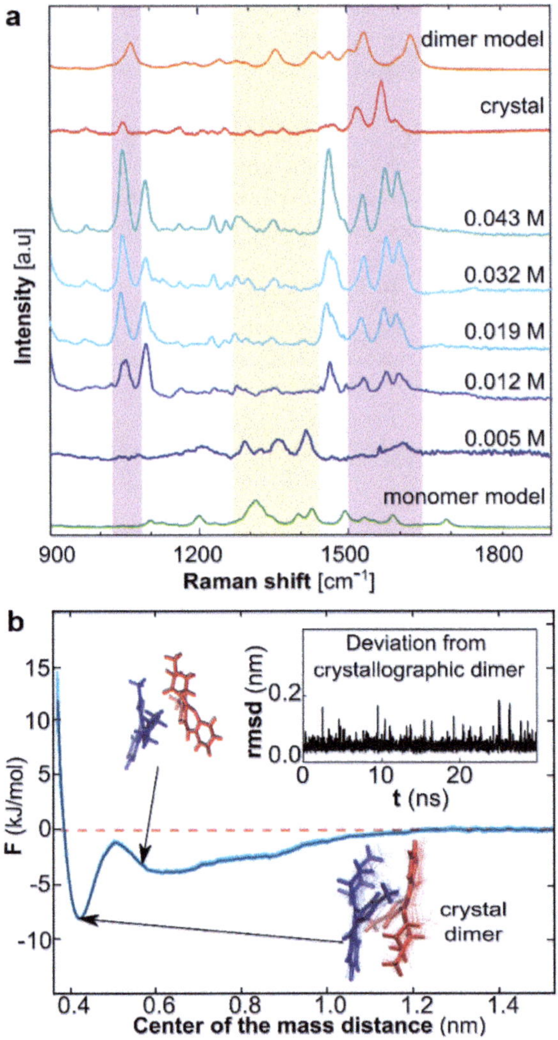

Fig. 3 OZPN dimers in solution. (a) Raman spectra of OZPN dissolved in 1/1 (v/v) EtOH/H_2O at the listed concentrations and of the solid crystalline solvate are compared to spectra for the OZPN monomer and dimer calculated using DFT. The yellow stripe highlights characteristic monomer peaks, whereas the pink stripes highlight those typical of dimers. (b) The potential of mean force *F* between two OZPN monomers in 1/1 (v/v) EtOH/H_2O computed using all-atom umbrella sampling molecular dynamics. The line width represents the computational uncertainty. Insets: top left, a representative configuration of the two monomers at separations longer than the deep minima. Top right, the root-mean-squared deviations of the positions of all OZPN atoms in the dimers occupying the deep minima from the dimer structure in the 2OZPN·EtOH·$2H_2O$ crystals as functions of time are small, indicating that the dimers assembled in the solution are nearly identical to the dimer found in the crystal structure. Lower right, representative snapshots of the configuration of the dimer (the constituent monomers are shown in red and blue, respectively) occupying the deepest *F* minimum, which is nearly identical to that of the dimer found in the crystal structure (one of the ten conformations is highlighted for clarity). The data for some of the spectra in (a) and the mean force potential in (b) are from ref. 49.

conformations (Fig. 3b). We evaluated the free energy of OZPN dimerization in EtOH/H_2O from the potential value at the deep minimum and obtained -8.1 kJ mol^{-1}.[49] The dimerization constant K_D was evaluated as the ratio of the integrals of F over the closest range minimum, *i.e.*, the bound state and the unbound state, and is 2.6 ± 0.8 M^{-1}. The similarity of the computed K_Ds to the values determined from the concentration responses of the monomer and dimer Raman peaks presents an independent validation of the simulations. Further MD calculations identify the reason for faster growth by dimers as their stronger adsorption on the crystal surface, supplemented by additional dimerization on the surface, which creates a readily available pool of dimers that can incorporate as a whole into the steps.[49]

The results with OZPN establish a deviation from the classical mechanisms of crystallization, which assume that crystals grow by the successive association of solute molecules as monomers. We show that a preformed centrosymmetric solute dimer is the preferred growth unit for OZPN crystals, despite the fact that dimers comprise a minority of the solute population in the solution bulk. Importantly, this study highlights that the growth unit can be identified by correlating solute oligomerization to the solute incorporation rate law. Furthermore, the kinetic, structural, and spectroscopic analyses of OZPN solutions and the growth of OZPN crystals emphasize the disparity between the majority species in the bulk solution, the microenvironment adjacent to the growth surface and both the crystal structure and the structure of the incorporating solute unit.

EtpI: distinct growth mechanisms of the anisotropic crystal faces

Etioporphyrin I (EtpI, Fig. 4a) represents a class of compounds whose solid-state structure (Fig. 4b) offers promising optical and electronic properties for use as semiconductors, solar cells, and field-effect transistors.[69–71] In contrast to OZPN, EtpI has a unique unsolvated triclinic crystal structure built of single molecules stacked in parallel pillars (Fig. 4c).[69] Currently, the Cambridge Structural Database (CSD) search of etioporphyrin I results in a single solid-state form belonging to the non-centrosymmetric P1 crystallographic symmetry group (Fig. 4c).

Owing to their nearly square cross-sections, EtpI crystals expose two faces to AFM observation: (010) and (001). The presence of a (101) face in the crystal

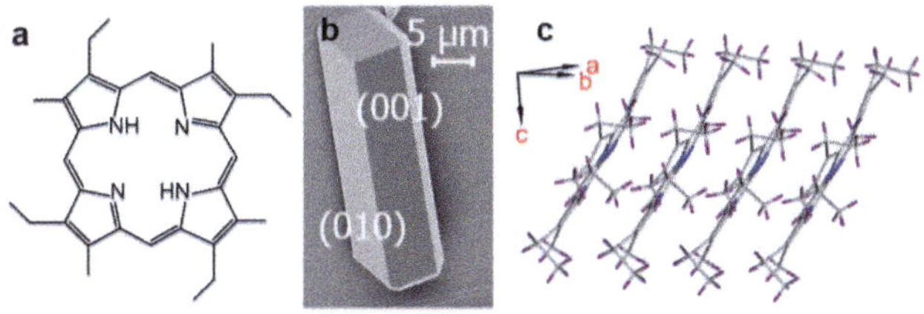

Fig. 4 EtpI and its crystals. (a) The EtpI molecule. (b) Scanning electron micrograph of an EtpI crystal. The (010) and (001) faces are labeled. (c) Molecular arrangement of EtpI in the crystal; Cambridge Structural Database REFCODE WOBVUF.[69] Carbon is shown in grey, nitrogen in blue, and hydrogen in purple.

morphology enforces distinct shapes, parallelogram for the (010) face and trapeze for the (001) face. *In situ* AFM monitoring exposes that, similar to the (002) OZPN face, both the (010) and (001) EtpI faces readily grow in 1-octanol by the association of molecules into steps generated by screw dislocations (Fig. 5a and b). Time resolved AFM measurements of the step velocity v reveal that on the (010) face, v scales linearly with the solute concentration C_{EtpI}, whereas on the (001) face, v shows a superlinear dependence of the step velocity v and C_{EtpI} (Fig. 5c and d).

Eliminating the trivial reasons that may drive a superlinear $v(C)$ dependence (inaccurate solubility value, paucity of kinks, and step inhibition by uncontrolled foreign substances), the two distinct correlations manifest monomolecular kinetics of solute incorporation on the (010) face and bimolecular reaction on the (001) face. A feasible mechanism that guides the divergent kinetics laws on the two crystal faces relies on monomers that dominate the solute speciation. Step growth on the (010) face, which exhibits monomolecular incorporation rate law, selects the monomers as incorporating species, since this incorporation mode motivates faster step growth than the incorporation of potential oligomers. By contrast, the bimolecular kinetics of step growth on the (010) face manifests two monomers combining into dimers, which remain a minority solute species, but still propel growth by dimers on the (010) face faster than by the incorporation of the majority solute monomers.

We characterized the EtpI oligomerization in octanol solutions by vibrational spectroscopy—the concentration of EtpI in octanol, controlled by the low solubility $C_e = 0.17$ mM, is below the sensitivity of Raman detection on the background of the strong octanol spectra—and augmented the spectroscopy by all-atom MD simulations of the dimerization equilibrium of EtpI dissolved in octanol. Similarly to OZPN, the two methods concertedly expose that the monomers represent the majority of the total C_{EtpI} and the monomer concentration increases proportionally to C_{EtpI}. This proportionality supports the proposal that the (010) face, on which the step velocity increases linearly with C_{EtpI}, grows by the incorporation of monomers.

The concentration of dimers in a solution with a total EtpI concentration equal to the solubility, 0.17 mM, is *ca.* 0.0002 mM, and it increases to about 0.002 mM at

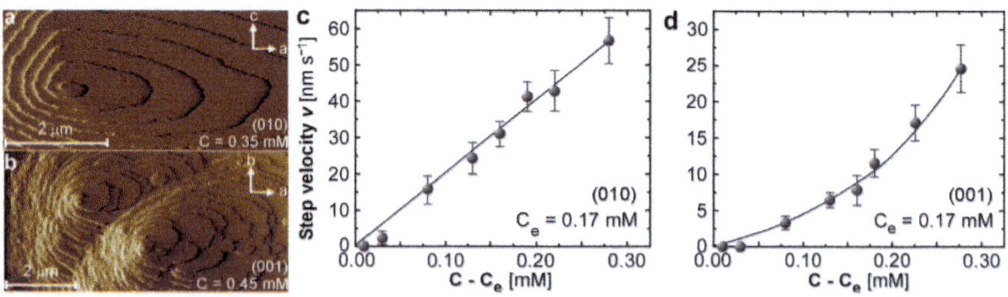

Fig. 5 The growth of the (010) and (001) faces of EtpI crystals. (a and b) Generation of new crystal layers by a screw dislocation outcropping on the respective faces. (c and d) The velocity v of steps in the [100] direction as a function of EtpI concentration C in octanol. Error bars denote standard deviation from the average of about 30 measurements for each data point. Solid lines correspond to best fits to linear, in *C*, and quadratic, in *D*, kinetics laws.

the highest $C_{\mathrm{EtpI}} = 0.45$ mM, at which step velocities were measured (Fig. 5).[72] The lower dimer content in EtpI solutions than in OZPN solutions is due to the lower total EtpI concentration, dictated by the lower solubility of this compound. The superlinear increase of the dimer concentration with higher C_{EtpI} manifests as a quadratic increase of the step velocity on the (001) face, concurrently with the proposed mechanism of growth by dimer incorporation on that face. Notably, the good correspondence of the $v(C_{\mathrm{EtpI}})$ correlation to a quadratic rate law minimizes the potential contributions of trimers and higher oligomers—expected to enforce an even steeper $v(C_{\mathrm{EtpI}})$ increase—to growth on the (001) EtpI face.

The solute pathway into kinks and the selection of growth units

A solute molecule may arrive at a kink either directly from the solution[51,73,74] (Fig. 6a) or after adsorption on the terraces between steps, followed by diffusion towards the kinks[50,52] (Fig. 6b). For olanzapine, detailed measurements of the velocities of closely spaced steps revealed that they grow substantially slower than steps separated by more than *ca.* 250 nm.[49] Furthermore, the dynamics of steps with uneven front and rear terraces indicated that the solute incorporates into steps asymmetrically and the lower terrace is preferred as a solute conduit.[49] Both observations are incompatible with direct solute incorporation.[59,60] Importantly, they comply with predictions of the surface diffusion models[32,75–80] and argue that the solute species that incorporate into kinks on OZPN (002) crystal surfaces reach the steps after diffusion along the crystal surface, similarly to numerous solution grown crystals.[49,59,77,79] All-atom MD simulations of the adsorption of OZPN monomers and dimers established that faster step growth by dimers is due to their stronger adsorption on the terraces between steps and the supplementary dimerization of monomers occurring in the adsorbed state.[49] This latter result may be interpreted to imply that the crystal surface is crucial for the selection of growth by dimer incorporation as it serves as a reservoir where dimers accumulate.

Mechanically transferring the OZPN insights to the two faces of EtpI would suggest that the solute reaches the steps on the (001) face, which grows by associating dimers, *via* surface diffusion, whereas steps on the (010) face, which preferentially incorporate monomers, recruit solute directly from the solution, where the monomers dominate. AFM measurements of the growth of steps of two distinct heights h, however, refute this conjecture. We compare the growth of steps as high as two lattice parameters to the growth of single-height steps. If the solute reaches the steps *via* the crystal surface, the step supply field is constrained

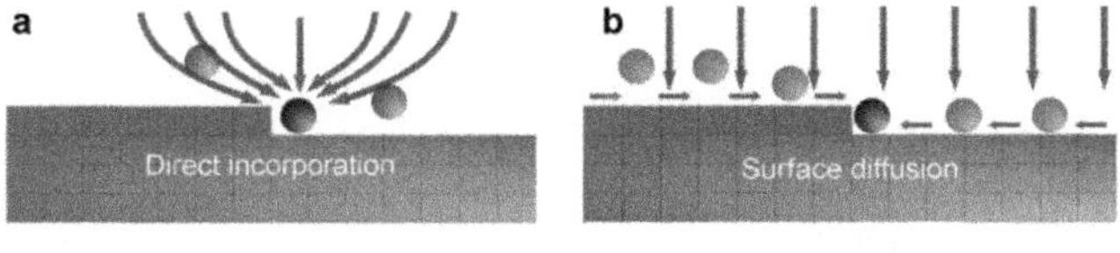

Fig. 6 Schematic of two pathways from solution to kinks: direct incorporation in (a), and *via* adsorption on the terraces followed by diffusion towards the steps in (b). Arrows indicate supply fluxes of the solute, depicted as spheres, towards the steps.

to two dimensions, which stunts the growth of steps of double height. Concurrently, analytical models of step growth mediated by surface diffusion predict that ν scales with h^{-1}.[52,77,81] By contrast, if the steps feed directly from the solution, the supply field is three-dimensional and abundant for twinned steps. Closed-form expressions for this growth mode predict negligible $\nu(h)$ correlation.[51,73] Time resolved *in situ* AFM measurements reveal that on both the {001} and {010} faces the velocities of steps as high as the lattice parameter are close to those of steps of single height (Fig. 7 and 8). The comparable rates of growth of twinned and single-height steps affirm that both faces of etioporphyrin I uniquely prefer the direct incorporation pathway.

Short of full scale MD simulations of the incorporation of monomers and dimers into kinks on the (001) and (010) faces of EtpI, we speculate that the selection of distinct solute species to incorporate into the kinks on the (010) and (001) faces of EtpI crystals is driven by the divergent structures of these two faces. The (010) face is relatively smooth (Fig. 9a) and the simple structure of the kink grants easy access for the incorporation of the majority monomers. By contrast,

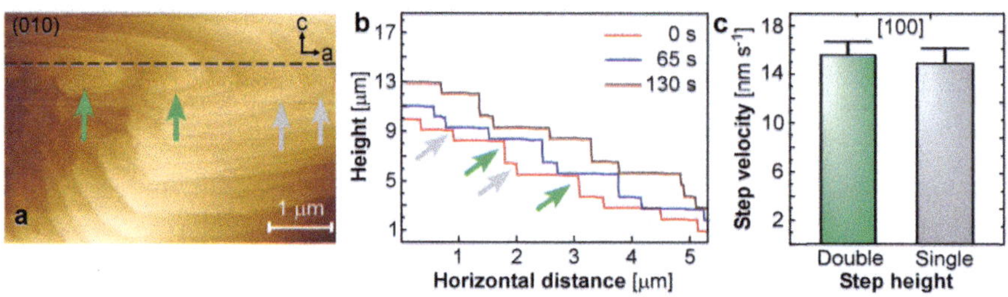

Fig. 7 Growth of single and double height steps on a (010) face of an EtpI crystal at C = 0.25 mM. (a) An AFM image of the growth of single (silver arrows) and double (green arrows) height steps of a (010) face. (b) The evolution of the surface profile along the dotted line in (a). Double-height steps (green arrows) advance over lengths similar to those of single-height steps (silver arrows). (c) Comparison of the velocities in the [100] direction of single and double height steps. The averages of 10 double and 10 single height steps are shown. Error bars represent the respective standard deviations.

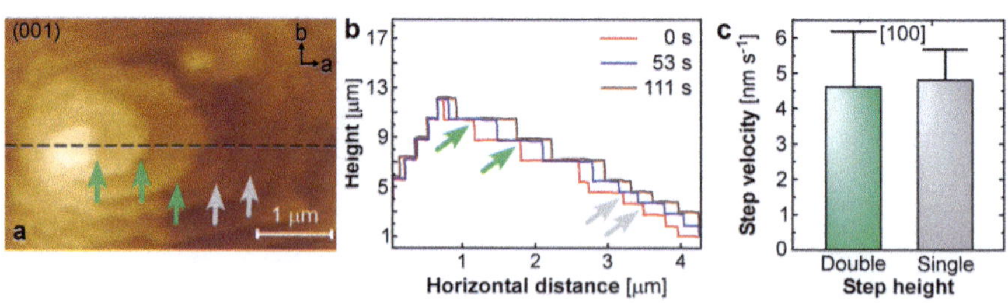

Fig. 8 Growth of single and double height steps on a (001) face of an EtpI crystal at C = 0.25 mM. (a) An AFM image of the growth of single (silver arrows) and double (green arrows) height steps of a (001) face. (b) The evolution of the surface profile along the dotted line in (a). (c) Comparison of the average velocities in the [100] direction of single and double height steps. Error bars represent the standard deviations from the averages over measurements of 10 double and 10 single height steps.

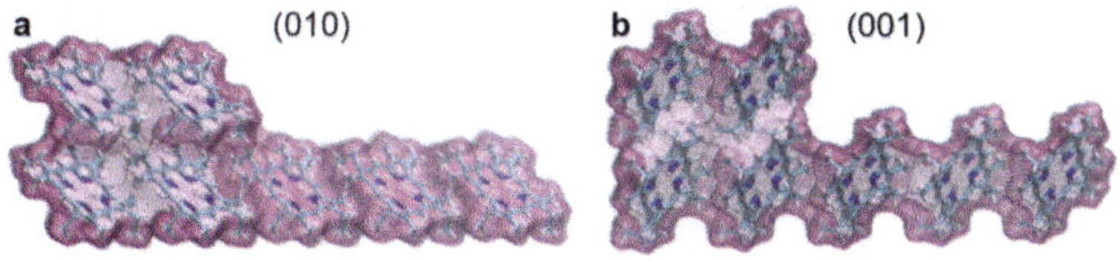

Fig. 9 Models of the molecular structures of steps on the (010), in (a), and (001), in (b), faces of EtpI crystals.

the (001) face is pierced by dimples and grooves. We surmise that monomers may be trapped in the dimples at the step edge and erect growth-incompetent configurations, whereas dimers, which fit seamlessly into the dimples, provide continuous growth.

Importantly for the correlations between crystal symmetry and the growth unit, discussed here, the finding that the (001) EtpI face grows by the association of dimers, which reach the kinks directly from the solution, demonstrates that solute adsorption on crystal surfaces that prefer dimers and promote further dimerization, as with OZPN, is not a necessary condition for this growth mode.

Conclusions

Olanzapine and etioporphyrin I reveal the lack of general rules to correlate the majority solute species and crystal structural blocks that carry the crystal symmetry. Olanzapine is an example where the crystal growth unit diverges from the majority solute monomers and matches a centrosymmetric crystal structural element. With etioporphyrin I, the monomeric crystal growth unit selected by one of the faces is identical to both the crystal structural element and the majority solute species, whereas the dimeric growth unit on the other face represents just one of the elements of a vast variety of dimers that cohabit in the solution and diverge from the crystal lattice motif.

Crystal symmetry, unfortunately, does not open a shortcut to identifying the crystal growth unit and cannot reliably simplify theories and models that target molecular-level understanding of crystal growth.

Conflicts of interest

There are no conflicts to declare.

Acknowledgements

This work was supported by the National Science Foundation (award no. DMR-2128121), the Welch Foundation (Grant E-1882), and the State of Texas through the Texas Center for Superconductivity at the University of Houston. Computational resources were provided by the Hewlett Packard Enterprise Data Science Institute at the University of Houston and the Texas Advanced Computing Center at the University of Texas at Austin. The authors acknowledge the CMAC National Facility, housed within the University of Strathclyde's Technology and Innovation Centre, and funded with a UKRPIF (UK Research Partnership Investment Fund) capital award, SFC ref H13054, from the Higher Education Funding Council for

England (HEFCE) and the EPSRC Future Continuous Manufacturing and Advanced Crystallisation Research Hub (grant ref: EP/P006965/1).

References

1 B. K. Vainshtein, *Fundamentals of Crystals: Fundamentals of Crystals. Symmetry, and Methods of Structural Crystallography*, Springer, Berlin, 2nd edn, 1994.
2 C. Kittel, *Introduction to solid state physics*, John Wiley and Sons, New York, 6th edn, 1986.
3 N. N. Greenwood and A. Earnshaw, *Chemistry of the Elements*, Butterworth-Heinemann, 2nd edn, 1997.
4 V. M. A. Rajendran, *Materials science*, Tata McGraw-Hill Pub., New Delhi, 2004.
5 Y. Saigusa, in *Advanced Piezoelectric Materials*, ed. K. Uchino, Woodhead Publishing, 2nd edn, 2017, pp. 197–233, DOI: 10.1016/B978-0-08-102135-4.00005-9.
6 V. F. Petrenko and R. W. Whitworth, *Physics of Ice*, Oxford University Press, Oxford, 2002.
7 T. V. Barker, *Nature*, 1923, **111**, 632.
8 A. Gavezzotti, *Acc. Chem. Res.*, 1994, **27**, 309–314.
9 J. Bernstein, *Polymorphism in Molecular Crystals*, Oxford University Press, Oxford, 2007.
10 O. Gliko, N. Neumaier, W. Pan, I. Haase, M. Fischer, A. Bacher, S. Weinkauf and P. G. Vekilov, *J. Am. Chem. Soc.*, 2005, **127**, 3433–3438.
11 D. Li, M. H. Nielsen, J. R. I. Lee, C. Frandsen, J. F. Banfield and J. J. De Yoreo, *Science*, 2012, **336**, 1014–1018.
12 J. Baumgartner, A. Dey, P. H. H. Bomans, C. Le Coadou, P. Fratzl, N. A. J. M. Sommerdijk and D. Faivre, *Nat. Mater.*, 2013, **12**, 310.
13 H. Coelfen, M. Antionetti, *Mesocrystals and Nonclassical Crystallization*, Wiley and Sons, 2008, ISBN: 978-0-470-02981-7.
14 J. J. De Yoreo, P. U. P. A. Gilbert, N. A. J. M. Sommerdijk, R. L. Penn, S. Whitelam, D. Joester, H. Zhang, J. D. Rimer, A. Navrotsky, J. F. Banfield, A. F. Wallace, F. M. Michel, F. C. Meldrum, H. Cölfen and P. M. Dove, *Science*, 2015, **349**, aaa6760.
15 S. Lee, H. S. Wi, W. Jo, Y. C. Cho, H. H. Lee, S.-Y. Jeong, Y.-I. Kim and G. W. Lee, *Proc. Natl. Acad. Sci. U. S. A.*, 2016, **113**, 13618–13623.
16 Y. Politi, R. A. Metzler, M. Abrecht, B. Gilbert, F. H. Wilt, I. Sagi, L. Addadi, S. Weiner and P. U. P. A. Gilbert, *Proc. Natl. Acad. Sci. U. S. A.*, 2008, **105**, 17362–17366.
17 A. I. Lupulescu and J. D. Rimer, *Science*, 2014, **344**, 729–732.
18 M. J. Thomason, C. R. Seabourne, B. M. Sattelle, G. A. Hembury, J. S. Stevens, A. J. Scott, E. F. Aziz and S. L. M. Schroeder, *Faraday Discuss.*, 2015, **179**, 269–289.
19 W. Tang, H. Mo, M. Zhang, S. Parkin, J. Gong, J. Wang and T. Li, *J. Phys. Chem. B*, 2017, **121**, 10118–10124.
20 V. Cheynier, R. Schneider, J.-M. Salmon and H. Fulcrand, in *Comprehensive Natural Products II*, ed. H.-W. Liu and L. Mander, Elsevier, Oxford, 2010, pp. 1119–1172, DOI: 10.1016/B978-008045382-8.00088-5.
21 A. Idrissi, P. Damay, K. Yukichi and P. Jedlovszky, *J. Chem. Phys.*, 2008, **129**, 164512.

22 C. A. Hunter, J. F. McCabe and A. Spitaleri, *CrystEngComm*, 2012, **14**, 7115–7117.
23 K. R. Back, R. J. Davey, T. Grecu, C. A. Hunter and L. S. Taylor, *Cryst. Growth Des.*, 2012, **12**, 6110–6117.
24 R. Davey, G. Dent, R. Mughal and S. Parveen, *Cryst. Growth Des.*, 2006, **6**(8), 1788–1796.
25 S. Parveen, R. J. Davey, G. Dent and R. G. Pritchard, *Chem. Commun.*, 2005, 1531–1533, DOI: 10.1039/B418603F.
26 M. Gardner, A. J. Guerin, C. A. Hunter, U. Michelsen and C. Rotger, *New J. Chem.*, 1999, **23**, 309–316.
27 S.-T. Yau, D. N. Petsev, B. R. Thomas and P. G. Vekilov, *J. Mol. Biol.*, 2000, **303**, 667–678.
28 S.-T. Yau, B. R. Thomas, O. Galkin, O. Gliko and P. G. Vekilov, *Proteins: Struct., Funct., Genet.*, 2001, **43**, 343–352.
29 S.-T. Yau, B. R. Thomas and P. G. Vekilov, *Phys. Rev. Lett.*, 2000, **85**, 353–356.
30 S.-T. Yau, B. R. Thomas and P. G. Vekilov, *J. Cryst. Growth*, 2001, **232**, 188–194.
31 M. Sleutel, D. Maes, L. Wyns and R. Willaert, *Cryst. Growth Des.*, 2008, **8**, 4409–4414.
32 M. Sleutel, C. Vanhee, C. V. de Weerdt, K. Decanniere, D. Maes and L. W. R. Willaert, *Cryst. Growth Des.*, 2008, **8**, 1173–1180.
33 M. Sleutel, R. Willaert, C. Gillespie, C. Evrard, L. Wyns and D. Maes, *Cryst. Growth Des.*, 2009, **9**, 497–504.
34 M. Sleutel, J. Lutsko, A. E. S. Van Driessche, M. A. Durán-Olivencia and D. Maes, *Nat. Commun.*, 2014, **5**, 5598.
35 A. J. Malkin, Y. G. Kuznetsov, R. W. Lucas and A. McPherson, *J. Struct. Biol.*, 1999, **127**, 35–43.
36 A. J. Malkin, Y. G. Kuznetsov and A. McPherson, *J. Cryst. Growth*, 1999, **196**, 471–488.
37 Y. G. Kuznetsov, A. J. Malkin, R. W. Lucas, M. Plomp and A. McPherson, *J. Gen. Virol.*, 2001, **82**, 2025–2034.
38 A. McPherson, A. J. Malkin, Y. G. Kuznetsov and M. Plomp, *Acta Crystallogr., Sect. D: Biol. Crystallogr.*, 2001, **57**, 1053–1060.
39 A. J. Malkin and A. McPherson, in From fluid-solid interfaces to nanostructural engineering, *Assembly in hybrid and biological systems*, ed. J. J. De Yoreo and X. Y. Lui, Plenum/Kluwer Academic, New York, 2004, vol. 2, pp. 201–238.
40 A. J. Malkin and R. E. Thorne, *Methods*, 2004, **34**, 273–299.
41 D. K. Georgiou and P. G. Vekilov, *Proc. Natl. Acad. Sci. U. S. A.*, 2006, **103**, 1681–1686.
42 S.-T. Yau and P. G. Vekilov, *Nature*, 2000, **406**, 494–497.
43 S.-T. Yau and P. G. Vekilov, *J. Am. Chem. Soc.*, 2001, **123**, 1080–1089.
44 B. Fulton and K. L. Goa, *Drugs*, 1997, **53**, 281–298.
45 R. M. Bhardwaj, L. S. Price, S. L. Price, S. M. Reutzel-Edens, G. J. Miller, I. D. H. Oswald, B. F. Johnston and A. J. Florence, *Cryst. Growth Des.*, 2013, **13**, 1602–1617.
46 Y. Sun, C. J. Tilbury, S. M. Reutzel-Edens, R. M. Bhardwaj, J. Li and M. F. Doherty, *Cryst. Growth Des.*, 2018, **18**, 905–911.
47 M. Warzecha, R. Guo, R. M. Bhardwaj, S. M. Reutzel-Edens, S. L. Price, D. A. Lamprou and A. J. Florence, *Cryst. Growth Des.*, 2017, **17**, 6382–6393.

48 I. Wawrzycka-Gorczyca, P. Borowski, J. Osypiuk-Tomasik, L. Mazur and A. Koziol, *J. Mol. Struct.*, 2007, **830**, 188–197.
49 M. Warzecha, L. Verma, B. F. Johnston, J. C. Palmer, A. J. Florence and P. G. Vekilov, *Nat. Chem.*, 2020, **12**, 914–920.
50 W. K. Burton, N. Cabrera and F. C. Frank, *Philos. Trans. R. Soc., A*, 1951, **243**, 299–360.
51 A. A. Chernov, *Modern Crystallography III, Crystal Growth*, Springer, Berlin, 1984.
52 G. H. Gilmer, R. Ghez and N. Cabrera, *J. Cryst. Growth*, 1971, **8**, 79–93.
53 J. J. de Yoreo and P. G. Vekilov, in *Biomineralization*, ed. P. M. Dove, J. J. de Yoreo and S. Weiner, De Gruyter, Berlin, 2003, 57–93.
54 J. J. De Yoreo, T. A. Land and B. Dair, *Phys. Rev. Lett.*, 1994, **73**, 838–841.
55 T. A. Land, A. J. Malkin, Y. G. Kuznetsov, A. McPherson and J. J. Deyoreo, *Phys. Rev. Lett.*, 1995, **75**, 2774–2777.
56 H. H. Teng, P. M. Dove, C. A. Orme and J. J. De Yoreo, *Science*, 1998, **282**, 724–727.
57 T. A. Land, T. L. Martin, S. Potapenko, G. T. Palmore and J. J. De Yoreo, *Nature*, 1999, **399**, 442–445.
58 Y. G. Kuznetsov, A. A. Chernov, P. G. Vekilov and I. L. Smol'skii, *Crystallogr. Rep.*, 1987, **32**, 584–587.
59 P. G. Vekilov, Y. G. Kuznetsov and A. A. Chernov, *J. Cryst. Growth*, 1992, **121**, 643–655.
60 P. G. Vekilov, Y. G. Kuznetsov and A. A. Chernov, *J. Cryst. Growth*, 1992, **121**, 44–52.
61 P. G. Vekilov, in *Studies and Concepts in Crystal Growth*, ed. H. Komatsu, Pergamon, Oxford, 1993, pp. 25–49.
62 P. G. Vekilov and F. Rosenberger, *J. Cryst. Growth*, 1996, **158**, 540–551.
63 P. G. Vekilov and F. Rosenberger, *J. Cryst. Growth*, 1998, **186**, 251–261.
64 O. Gliko, I. Reviakine and P. G. Vekilov, *Phys. Rev. Lett.*, 2003, **90**, 225503.
65 D. N. Petsev, K. Chen, O. Gliko and P. G. Vekilov, *Proc. Natl. Acad. Sci. U. S. A.*, 2003, **100**, 792–796.
66 Y. Qutub, I. Reviakine, C. Maxwell, J. Navarro, E. M. Landau and P. G. Vekilov, *J. Mol. Biol.*, 2004, **343**, 1243–1254.
67 M. Warzecha, M. S. Safari, A. J. Florence and P. G. Vekilov, *Cryst. Growth Des.*, 2017, **17**, 6668–6676.
68 M. A. Lovette and M. F. Doherty, *Phys. Rev. E*, 2012, **85**, 021604.
69 C.-M. Che, H.-F. Xiang, S. S.-Y. Chui, Z.-X. Xu, V. A. L. Roy, J. J. Yan, W.-F. Fu, P. T. Lai and I. D. Williams, *Chem.–Asian J.*, 2008, **3**, 1092–1103.
70 M. H. Hoang, Y. Kim, M. Kim, K. H. Kim, T. W. Lee, D. N. Nguyen, S.-J. Kim, K. Lee, S. J. Lee and D. H. Choi, *Adv. Mater.*, 2012, **24**, 5363–5367.
71 M. H. Hoang, Y. Kim, S.-J. Kim, D. H. Choi and S. J. Lee, *Chem.–Asian J.*, 2011, **17**, 7772–7776.
72 L. Verma, M. Warzecha, R. Chakrabarti, V. G. Hadjiev, J. C. Palmer and P. G. Vekilov, *Isr. J. Chem.*, 2021, **61**, 818–827.
73 A. A. Chernov, *Phys.-Usp.*, 1961, **4**, 116–148.
74 A. A. Chernov, *Contemp. Phys.*, 1989, **30**, 251–276.
75 M. Maruyama, K. Tsukamoto, G. Sazaki, Y. Nishimura and P. G. Vekilov, *Cryst. Growth Des.*, 2009, **9**, 127–135.

76 G. Sazaki, M. Okada, T. Matsui, T. Watanabe, H. Higuchi, K. Tsukamoto and K. Nakajima, *Cryst. Growth Des.*, 2008, **8**, 2024–2031.
77 T. A. Land, J. J. DeYoreo and J. D. Lee, *Surf. Sci.*, 1997, **384**, 136–155.
78 K. N. Olafson, M. A. Ketchum, J. D. Rimer and P. G. Vekilov, *Cryst. Growth Des.*, 2015, **15**, 5535–5542.
79 K. N. Olafson, J. D. Rimer and P. G. Vekilov, *Phys. Rev. Lett.*, 2017, **119**, 198101.
80 G. Ehrlich and F. G. Hudda, *J. Chem. Phys.*, 1966, **44**, 1039–1052.
81 K. N. Olafson, M. A. Ketchum, J. D. Rimer and P. G. Vekilov, *Proc. Natl. Acad. Sci. U. S. A.*, 2015, **112**, 4946–4951.

araday Discussions

ROYAL SOCIETY OF CHEMISTRY

PAPER

Manipulation of amorphous precursors to enhance zeolite nucleation†

Deependra Parmar, Zhiyin Niu, Yu Liang, Heng Dai and Jeffrey D. Rimer *

Received 23rd November 2021, Accepted 14th December 2021
DOI: 10.1039/d1fd00096a

Crystallization in media comprised of amorphous precursors is becoming a more common phenomenon for numerous synthetic, biological, and natural materials that grow by a combination of classical and nonclassical pathways. Amorphous phases can exhibit a wide range of physicochemical properties that may evolve during the course of nucleation and crystal growth. This creates challenges for establishing causal relationships between amorphous precursor properties and their effect(s) on the selection of mechanistic pathways of crystallization and ultimately the properties of the crystalline product. In this study, we examine ways to manipulate the composition and colloidal stability of amorphous (alumino)silicate precursors that are prevalent in nanoporous zeolite syntheses. Changes in the amorphous precursor properties are evaluated on the basis of their ability to enhance rates of crystal formation. Here, we use fumed silica as the primary silicon source and examine the effects of infusing the source or growth medium with additional alkali metal, which serves as an inorganic structure-directing agent to facilitate the formation of porous crystal structures. We also assess the impact of adding a polymer additive, which reduces the colloidal stability of precursors, wherein we posit that the confined pockets of solution within the interstitial spaces of the precursor aggregates play an important role in regulating the rate of zeolite crystallization. Three commercially relevant zeolites (mordenite, SSZ-13, and ZSM-5) were selected for this study based on their diverse frameworks and methods of preparation. Our findings reveal that alkali infusion significantly reduces the crystallization times for mordenite and SSZ-13, but has little impact on ZSM-5 synthesis. Conversely, we find that polymer addition markedly enhanced the rates of crystallization among all three zeolites, suggesting that this method may be a general approach to reduce zeolite synthesis times. Given the relatively high costs associated with commercial zeolite production, identifying new methods to improve the efficiency of hydrothermal syntheses can have significant practical implications beyond the fundamental benefits of developing new routes to tailor nonclassical crystallization.

Department of Chemical and Biomolecular Engineering, University of Houston, Houston, TX, 77204, USA. E-mail: jrimer@central.uh.edu

† Electronic supplementary information (ESI) available. See DOI: 10.1039/d1fd00096a

1. Introduction

The ubiquitous presence and important role(s) of amorphous precursors in crystallization spans numerous inorganic and organic systems.[1–6] Precursors can be diverse species that include (but are not limited to) oligomeric networks, clusters, and nanoparticles. They can function as sites for heterogeneous nucleation in two-step processes,[1,7–9] serve as growth units in crystallization by particle attachment (CPA),[10,11] or may be bystanders that regulate the supply of solute to/from the growth solution. One method of influencing crystal formation to achieve products with tailored physicochemical properties is through the manipulation of the precursor structure and/or composition. Precursors are generally amorphous but can possess local structure, as well as variations in composition, such as the degree of solvation or the occlusion of foreign species (*e.g.* ions, organics, *etc.*), over the course of crystallization. It is also possible that precursors evolve these properties with concomitant changes in their roles in nucleation and/or growth. Understanding causal relationships between precursor properties and crystallization mechanisms is nontrivial; however, it is recognized that developing improved knowledge of precursor assembly and their impact on pathways of crystal growth holds tremendous potential for crystal engineering.

One of the most complex systems of crystallization is that of zeolites, which are (alumino)silicates with multidimensional networks of nanopores.[12,13] Zeolites are natural clays but are also produced synthetically for widespread industrial applications as catalysts, sieves, and sorbents for separations or ion exchange.[14–19] Zeolites crystallize with a broad range of Si/Al ratios and require the use of a structure-directing agent to facilitate the formation of cages/channels and/or compensate the negative charge of the Al sites within the zeolite framework (*i.e.* function as extra-framework cations). Organics used for this purpose commonly possess quaternary ammonium moieties, whereas organic-free syntheses often employ alkali metals.[20–24] The high concentrations of silicon and aluminum sources necessary to achieve a sufficiently large zeolite yield generate sol–gel media comprised of soluble species (monomers and oligomers) and a substantial quantity of amorphous species. It is generally believed that nucleation occurs by two-step processes, although this area of research for zeolites is not well understood. The multitude of (alumino)silicate species makes it challenging to identify a single growth unit, thereby negating the use of supersaturation as a reliable metric for predicting crystallization.[25] Moreover, growth units can potentially evolve in their speciation and concentration throughout zeolite synthesis,[26–29] which allows for multiple crystallization pathways to occur simultaneously (*e.g.* both classical and nonclassical mechanisms).[4,30]

In this study, we begin by reviewing recent literature from our group and others examining the assembly and role of precursors in zeolite crystallization, with an emphasis on the diverse properties of amorphous species. Recent work from our group has shown that the manipulation of amorphous particle properties, such as the distribution or concentration of alkali metals, or the ability to control the colloidal stability of precursors are two approaches to markedly impact the rate of crystallization. In order to assess the broader applicability of these two approaches, we conduct a series of experiments for zeolites with diverse properties (structure and composition) prepared by different methods involving inorganic and organic structure-directing agents.

2. Zeolite crystallization mechanisms

There is ample evidence for zeolite growth by classical monomer addition (Fig. 1A).[31,32] *Ex situ* atomic force microscopy (AFM) images by Anderson and coworkers of numerous zeolite crystals extracted at the end of syntheses reveal layers with step heights equivalent to the unit cell dimension(s),[33,34] suggestive of a layer-by-layer growth mechanism. *In situ* AFM imaging of zeolite A (LTA type) at low supersaturation has provided direct evidence of layer generation and

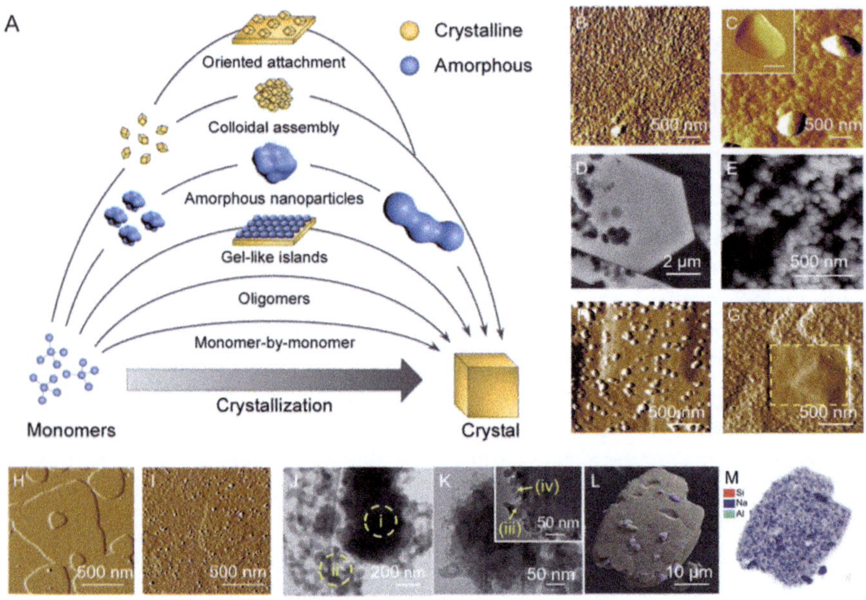

Fig. 1 (A) Illustration of crystallization pathways observed in zeolite synthesis involving amorphous (blue) and crystalline (gold) precursors. (B and C) *In situ* AFM images of zeolite A after (B) 60 min and (C) 312 min of growth at 45 °C by (nearly) oriented attachment.[35] Inset: nanocrystal marked with the arrow. Scale bar equals 500 nm. (D) SEM image and (E) AFM image of a partially dissolved ZSM-5 crystal in NH_4F revealing evidence of colloidal assembly.[37] (F) *In situ* AFM images of a zeolite A surface growth at 35 °C showing gel like islands.[35] (G) Enlarged scan size of panel F showing an area (dashed line) after continuous imaging in tapping mode. (H–M) Evidence of growth by amorphous nanoparticle attachment. (H and I) *In situ* AFM images of silicalite-1 surface growth after (H) 0 min and (I) 60 min of growth at 60 °C, showing the attachment of amorphous precursors.[26] (J) TEM image of SSZ-13 solids extracted once Bragg peaks are detected in the powder XRD patterns, showing populations of (i) amorphous precursors and (ii) SSZ-13 crystals.[30] (K) High resolution image of a crystal in panel (J) showing (iii) precursor attachment and (iv) smaller (unknown) particulates surrounding crystals that are putative intermediates of the disorder-to-order transition. (L and M) FE-SEM-EDX mapping of a mordenite crystal extracted near the end of crystallization where a high density of Na^+ ions (blue) is associated with residual amorphous particles.[40] Reproduced with permissions from Springer Nature for images (B), (C), (F) and (G) (copyright 2018), from the American Association for the Advancement of Science for images (D) and (E) (copyright 2021), from John Wiley & Sons, Inc. for images (H) and (I) (copyright 2019), from the American Chemical Society for images (J) and (K) (Copyright 2015), and from the American Chemical Society for images (L) and (M) (copyright 2020).

spreading.[31] These collective examples indicate that the classical pathway is predominant at later stages of zeolite synthesis where supersaturation is low and amorphous precursors have largely been consumed. At higher supersaturation, but still in the absence of amorphous precursors, *in situ* AFM measurements of zeolite A surface growth have captured evidence of two distinct nonclassical growth pathways.[35] The first is the (nearly) oriented attachment of nanocrystallites (Fig. 1B and C), where the faceted crystal deposit (Fig. 1C, inset) continues to grow post-attachment. A second pathway is the formation of gel-like islands (Fig. 1F and G) in growth media devoid of any particulates (*i.e.* so-called clear solutions). Recently, dissolution experiments have been used to extract information of zeolite crystallization at early stages using NH_4F as an etchant to indiscriminately dissolve Al–O and Si–O bonds in zeolite crystals.[36] Analysis of silicalite-1 (MFI type) by this approach showed that dissolution preferentially occurs at the center of coffin-shaped crystals where nonclassical pathways presumably lead to more defects (Fig. 1D) compared to the exterior rim of crystals, which grow by a classical pathway, leading to fewer defects. Examination of partially dissolved zeolite MFI crystals by AFM revealed clusters of spheroidal crystals (Fig. 1E), which were confirmed by transmission electron microscopy (TEM) to be arranged in crystallographic registry.[37] These images provide evidence of colloidal assembly at the early stages of growth (*i.e.* the interior of the crystal) where ripening at later synthesis times obscures evidence of this nonclassical pathway – analogous to phenomena observed for other minerals, such as magnetite.[38,39]

At higher (alumino)silicate concentrations representative of most zeolite syntheses, growth mixtures are sol–gels comprised of both amorphous precursors and soluble species (monomers and oligomers).[41] Several studies of different zeolite crystal structures have shown evidence of crystallization by particle attachment (CPA). *In situ* AFM measurements of silicalite-1 (MFI) growth reveal complete coverage of surfaces with amorphous precursors (Fig. 1H and I) where layers advance by simultaneous processes of CPA and monomer/oligomer addition.[26] TEM images of solids extracted from a SSZ-13 (CHA type) synthesis at the point where the first Bragg peaks were detected in powder X-ray diffraction (XRD) patterns reveal a mixture of amorphous precursors (Fig. 1J(i)) and rough spheroidal crystals (Fig. 1J(ii)).[30] High resolution images of these crystals show amorphous precursors attached to the surface (Fig. 1K(iii)), as well as smaller particulates of unknown structure (Fig. 1K(iv)) that are putative intermediates during the disorder-to-order transition of CPA. Time-resolved studies of mordenite (MOR type) crystallization by field emission scanning electron microscopy with energy dispersive X-ray spectroscopy (FE-SEM-EDX) show remnants of undissolved amorphous precursors (regions of higher Na^+ density in Fig. 1L and M) at the end of crystallization when powder XRD seemingly indicates synthesis is complete.[40] In all of these examples (Fig. 1H–M), nonclassical CPA is a prominent growth mechanism.

The complex sol–gels used to prepare zeolites are derived from mixtures of silicon and aluminum sources. Most silicon sources lead to the formation of particles (Fig. 2A) with a core–shell structure comprised of siliceous cores and aluminous shells. Reagents are generally added in such high quantities that the solids do not fully dissolve, and thus amorphous precursor particles persist through much of the crystallization period. The presence of Al-rich shells can impact the rate of silica dissolution (*i.e.* exchange between the solid and solution species), whereas the selection of the silicon source can lead to particles with

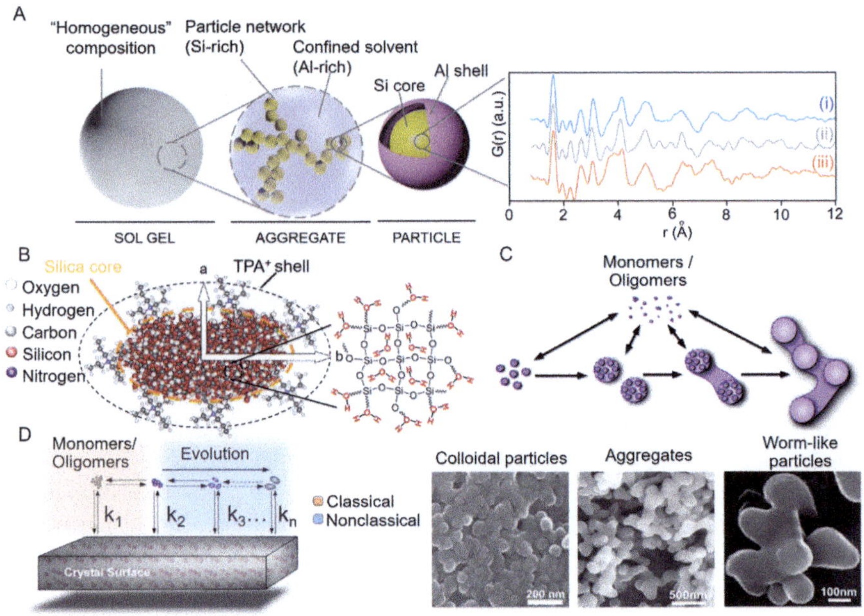

Fig. 2 (A) Complexity of sol–gels from the macroscopic to atomic scale (left to right). Gels are comprised of amorphous silica particles where aluminum coats the particles, which can impact their rate of dissolution. The local microstructure of the silicon sources also varies, as indicated by X-ray PDF analysis (far right)[52] of three common reagents: fumed silica (blue, (i)), potassium silicate (grey, (ii)), and colloidal silica (orange, (iii)). (B) One of the most extensively characterized precursors are the nanoparticles of silicalite-1 synthesis formed by the hydrolysis of TEOS. The particles are comprised of hydrated amorphous silica (callout) with tetrapropylammonium cations used as the organic structure-directing agent forming a shell.[26] (C) Amorphous precursors can evolve in size and shape from colloidal particles to aggregates where ripening leads to the fusing of precursors and potential formation of worm-like particles. Representative scanning electron micrographs of each stage are shown below the scheme for solids extracted from a zeolite L synthesis at periodic times during the induction period.[50] (D) Amorphous precursors can also evolve in microstructure, as illustrated for the nanoparticles in panel B where hydrothermal treatment leads to more ordered precursors.[26] We previously showed that more ordered precursors have a greater propensity to be involved in nonclassical CPA, whereas precursors that preserve their initial, more disordered state, are bystanders, thereby allowing growth to predominantly occur by monomer/oligomer addition. Reproduced with permission from the American Chemical Society for image (A) (copyright 2018), from John Wiley & Sons, Inc. for images (B) and (D) (copyright 2019), and from the American Chemical Society for image (C) (copyright 2016).

distinctly different microstructures. The latter was revealed from X-ray pair distribution function (PDF) analysis of three as-received reagents (Fig. 2A), which showed a histogram of atom–atom distances. All sources exhibit a common SiO_4 local structure and the lack of long-range order; however, sources such as fumed silica possess intermediate-scale structure (*i.e.* more ordered peaks). PDF profiles of the other two sources (colloidal silica and potassium silicate) contain broad, diffuse features that do not overlap beyond 4.5 Å, suggesting differences in Si–O–Si bond connectivity. These subtle differences in local microstructure can potentially impact zeolite nucleation and growth in ways that are not fully understood.

One of the most widely studied zeolites is silicalite-1 (MFI type) where syntheses commonly employ a fourth silicon source, tetraethyl orthosilicate (TEOS), which leads to the assembly of nanoparticles (1–6 nm)[26] with a hydrated amorphous silica core (Fig. 2B) and a shell of the organic structure-directing agent, tetrapropylammonium. These particles contain water, analogous to other systems such as amorphous calcium carbonate (ACC),[42–46] phosphate (ACP),[47,48] and oxalate (ACO)[49] that are the precursors to diverse biominerals. In zeolite syntheses, the silicon sources analyzed by PDF (Fig. 2A) are more common and lead to the formation of larger particles (Fig. 2C). Over the course of hydrothermal treatment, zeolite precursors can evolve in both size and shape. The scheme in Fig. 2C and the corresponding SEM images document the changes observed to precursors of zeolite L (LTL type) during the induction period where aggregation, fusing, and ripening can lead to worm-like particles.[50,51] Prior studies have also shown that precursors can evolve in microstructure. For example, the nanoparticles in silicalite-1 syntheses (Fig. 2B) can undergo a transition to more ordered particles at a local scale, but still lack the long-range order of a zeolite crystal. We recently reported[26] that the judicious selection of organic additives can facilitate structural reordering, which in turn promotes nonclassical CPA; however, systems where the precursors preserve a more disordered state are more likely to grow by monomer/oligomer addition (Fig. 2D). Thus, it is possible to tune the pathway of crystallization by controlling the microstructural evolution of amorphous precursors. It remains to be determined if similar phenomena govern crystallization pathways in other minerals. Indeed, precursor "evolution" has been reported for ACC and ACP systems where temporal changes in precursor hydration can impact nucleation and growth, with concomitant effects on the polymorph selection or crystal habit.[2,3,44]

3. Methods to enhance the rate of zeolite crystallization

Prior studies of zeolite L (LTL) crystallization have shown that the selection of the silicon source can have a notable impact on the induction time.[50] In a recent study,[53] we observed zeolite L nucleation within 8–10 h (Fig. 3A) using either colloidal or fumed silica; however, substitution of these sources with potassium silicate reduced the induction time nearly five-fold. Our hypothesis for more rapid nucleation was the spatial distribution of potassium cations within the interior of amorphous silicate particles. Elemental mapping of samples prepared from as-received reagents (Fig. 3B) revealed K^+ ions distributed throughout the amorphous silica particles. The presence of the alkali metal in the synthesis medium is critical since K^+ ions function as inorganic structure-directing agents to facilitate the formation of pores in the LTL crystal structure. The infusion of precursors with the alkali (Fig. 3C) was deemed significant to promote zeolite nucleation, which putatively occurs either on the surface or within the interior of amorphous precursors. The potential pathways of zeolite nucleation in complex sol–gel media are diverse and generally not well understood, but the ability to alter precursor composition with respect to the inorganic structure-directing agent is seemingly influential. To test this effect, a modified fumed silica was prepared by incubating the silicon source in $KBr_{(aq)}$ solution for one week to infuse K^+ within the

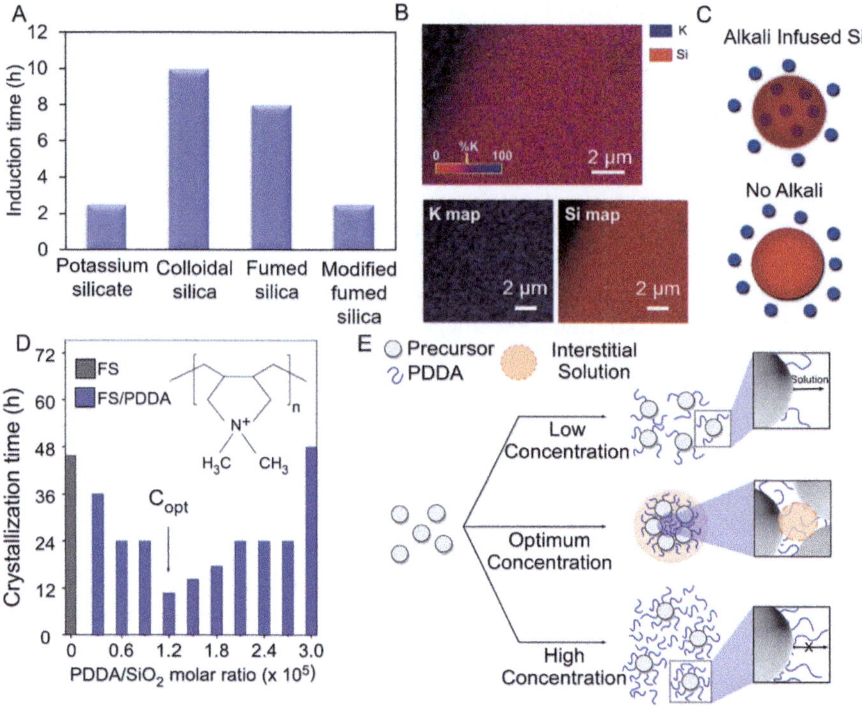

Fig. 3 (A) Induction time of zeolite L (LTL) syntheses with three as-received silicon sources (potassium silicate, fumed silica, and colloidal silica) and a modified fumed silica incubated in 2 M $KBr_{(aq)}$ solution at room temperature for 7 days.[53] Times are estimated from the first appearance of Bragg peaks in the powder XRD patterns. (B) FE-SEM-EDX elemental mapping of modified fumed silica showing K^+ ions (blue) evenly distributed throughout the particle.[53] (C) Idealized scheme of alkali-infused amorphous silica particles (*e.g.* potassium silicate and modified fumed silica) and particles without alkali infusion (*e.g.* fumed and colloidal silica). (D) Crystallization time of SSZ-13 in the absence (grey) and in the presence (blue) of PDDA at increasing polymer concentration (or PDDA/SiO_2 molar ratio). Data is taken from Dai *et al.*[60] with the shortest crystallization time at an optimal PDDA concentration (C_{opt}) around 0.2 wt% polymer (labelled with an arrow). (E) Cartoons illustrating how increased coverage of polymer on amorphous precursors with increasing polymer concentration putatively impacts the precursor–solution exchange, precursor aggregation, and the generation of confined interstitial pockets of solution between aggregated precursors at C_{opt}.[60] Reproduced with permission from John Wiley and Sons, Inc. for images (A), (B) and (C) (copyright 2018).

amorphous particles, which was confirmed by elemental analysis.[53] Zeolite L syntheses using the modified fumed silica resulted in similar induction time as potassium silicate (Fig. 3A). In a broader context, the infusion of amorphous precursors with metals has implications for many crystalline materials. For example, nearly all natural zeolites are comprised of alkali and/or alkaline earth metals.[13,54,55] In biomineralization, metals are also prevalent and can function as crystal growth modifiers.[56–58] This was recently demonstrated in calcium carbonate by De Yoreo and coworkers,[59] who showed that ACC infused with Mg^{2+} ions modifies precursor hydration and inhibits its evolution, which impacts the kinetics of $CaCO_3$ crystallization and crystal size.

Among the examples discussed, it is evident that the physicochemical properties of the amorphous precursors, such as microstructure or composition, can impact growth pathways and the kinetics of zeolite crystallization; however, we recently observed that the colloidal stability of precursor particles is another factor that can be manipulated in ways that markedly accelerate the rate of zeolite crystallization, analogous to the effects of alkali-infused precursors. The impact of colloidal stability was tested on syntheses of zeolite SSZ-13 (CHA type) using sol–gels prepared with fumed silica and *N*,*N*,*N*-trimethyl-1-1-adamantammonium (TMAda) as an organic structure-directing agent.[61,62] The addition of the polymer polydiallyldimethylammonium (PDDA) to the synthesis medium resulted in a non-monotonic trend (Fig. 3D), where low quantities of PDDA lead to a reduction in the SSZ-13 crystallization time with increased polymer concentration until reaching an optimal concentration (C_{opt}), above which a further increase in the polymer concentration results in longer crystallization times. Interestingly, similar trends were not observed with other polyamines, which may be attributed to the fact that PDDA is comprised of quaternary amines, similar to the structure-directing agent TMAda. Our hypothesis for this trend in crystallization kinetics is outlined in Fig. 3E, where low and high PDDA concentrations result in low and high coverage of the polymer on the surfaces of the amorphous precursors, respectively, that do not induce colloidal aggregation. We posit that low PDDA coverage does not significantly obstruct the exchange of (alumino)silicate species between precursors and solution, but does inhibit Ostwald ripening. The latter effect preserves the smaller size of precursors (*i.e.* higher surface area), which putatively enhances zeolite nucleation. Small particle size is also preserved at high PDDA coverage; however, it is likely that these conditions adversely impact the exchange of species between solid and solution states, thus leading to longer crystallization times. At C_{opt}, it was observed by multiple light scattering techniques[60] that PDDA promotes the aggregation of precursors. We proposed a hypothesis that the solution within the interstitial spaces of the aggregated precursors promotes nucleation. It is likely that higher local supersaturation can be generated in these confined pockets of solution *via* the dissolution of (alumino)silicate species. The presence of PDDA within these interstitial regions may also be critical. For instance, we found that the optimal concentration of PDDA can facilitate the condensation of free silicate monomer (or oligomers) in solution, which could lead to the formation of clusters to promote nucleation by a two-step mechanism.

Herein, we examine the generalizability of the two phenomena outlined in Fig. 3. We first go beyond zeolite L to determine if alkali-infused precursors or media can accelerate the rate of crystallization for other zeolites. We then examine if PDDA addition to diverse zeolite synthesis mixtures can have the same impact on crystallization time as was observed for zeolite SSZ-13. To perform these experiments, we selected three commercially relevant zeolites with different crystal structures: SSZ-13 (CHA, small pores), ZSM-5 (MFI, medium pores), and mordenite (MOR, large pores). The choice of structures targeted materials formed under synthesis conditions in the presence or absence of organics and disparate alkali metals (*e.g.* potassium and sodium). We also assessed whether the combination of both approaches, polymer addition to mixtures with alkali infused precursors, could lead to synergistic outcomes in zeolite crystallization.

4. Experimental methods

4.1 Materials

The following chemicals were purchased from Sigma Aldrich: sodium aluminate (technical grade), sodium hydroxide (NaOH, 98%), fumed silica (SiO_2, 99.8%), colloidal silica (SiO_2, 40% in water), aluminum hydroxide (76.5 wt% $Al(OH)_3$, reagent grade), and polydiallyldimethylammonium chloride (PDDAC, MW = 150k, 20 wt% in H_2O). Potassium bromide (>99%), tetrapropylammonium hydroxide (TPAOH, 40% in H_2O) and *N*,*N*,*N*-trimethyl-1-1-adamantammonium hydroxide (TMAdaOH, 25 wt% in H_2O) were purchased from Alfa Aesar and SACHEM Inc., respectively. Sodium hydroxide (NaOH, 1 M) was procured from Supelco, while sodium bromide (NaBr, 99.7%) was purchased from J.T. Baker. Deionized (DI) water was produced with an Aqua Solutions RODI-C-12A purification system (18.2 MΩ). All reagents were used as received without further purification.

4.2 Preparation of zeolites

The generation of modified fumed silica using sodium as the alkali metal was performed according to a previously reported procedure,[53] wherein as-received fumed silica was incubated in 2 M NaBr (or KBr) solution (10 g fumed silica in 200 g of 2 M salt solution) at room temperature for 7 days. Modified (or alkali-infused) fumed silica was recovered by two centrifuge and wash cycles with DI water and then dried in an oven at 50 °C and ambient pressure. The same procedure was repeated using pure DI water (10 g fumed silica in 200 g of water) in place of 2 M salt for an identical incubation time. The resulting water-treated fumed silica was isolated by the same centrifuge/washing procedure.

The synthesis of zeolite SSZ-13 (CHA) was performed according to a reported protocol[30] using TMAda as the organic structure-directing agent. In a typical synthesis, sodium hydroxide solution (50 wt%, 0.18 g, 0.0023 mol) and TMAdaOH (0.95 g, 0.0011 mol) were first mixed with DI water (8.13 g, 0.4960 mol), and stirred for 15 min at room temperature. Aluminum hydroxide (0.06 g, 0.0006 mol) was then added to the mixture, which was stirred for an additional 15 min. To this mixture was added fumed silica (0.68 g. 0.0113 mol) to produce a final molar composition of 0.052 $Al(OH)_3$: 1.0 SiO_2 : 0.2 NaOH : 44 H_2O : 0.1 TMAda. The growth mixture was aged at room temperature for 4 h under continuous stirring and then transferred to a 23 ml Teflon-lined stainless steel acid digestion bomb (Parr Instruments). The sealed autoclave was heated in a Thermo Scientific Precision oven under rotation (30 rpm) and autogenous pressure at 180 °C for different synthesis times.

The synthesis of zeolite mordenite (MOR) was performed according to a reported protocol[63] using an organic-free recipe with sodium as the inorganic structure-directing agent. In brief, sodium hydroxide (98%, 0.54 g, 0.019 mol) was dissolved in DI water (25.92 g, 1.428 mol), followed by the addition of sodium aluminate (0.50 g, 0.003 mol) with continuous stirring for 15 min. To this mixture was added fumed silica (3.03 g, 0.050 mol) and the resulting sol–gel was aged overnight at room temperature under continuous stirring. The aged solution was then transferred to a Teflon-lined acid digestion bomb and placed in an oven at 165 °C under rotation for different synthesis times.

The synthesis of zeolite ZSM-5 (MFI) was performed according to a reported protocol[64] using tetrapropylammonium (TPA) as an organic structure-directing agent. First, NaOH (1 M, 2.39 g, 0.0023 mol) and TPAOH (0.8624 g, 0.0017 mol) were dissolved in DI water (10.34 g, 0.574 mol). To this mixture was added sodium aluminate (0.0827 g, 0.0011 mol) with continuous stirring for 15 min; and then the silicon source (1.02 g, 0.017 mol) was added to the solution. The growth solution was aged for 2 h at room temperature and transferred into an acid digestion bomb. Unless otherwise stated, ZSM-5 was synthesized at 180 °C under static conditions for different times. For all zeolite syntheses, solids were extracted by the removal of the acid digestion bombs from the oven and cooling in a water bath to room temperature. The samples were subjected to the same centrifugation and washing procedure until neutral pH was achieved.

For studies of modified fumed silica, the silicon source in the above recipes was replaced by either alkali-infused or water treated sources (86 wt% SiO_2). For the former, additional quantities of Na (or K) and Br in the modified silicon source were not accounted in the molar composition of the recipe. A recipe of SSZ-13 with potassium as the alkali was adopted from Liang *et al.*[65] with fumed silica as the silica source to achieve a synthesis gel with molar composition 5.4 SiO_2 : 1.0 Al_2O_3 : 13.6 K_2O : 1093 H_2O. In a typical synthesis, potassium hydroxide solution was added in DI water until well mixed. To this solution was added aluminum sulfate hydrate, followed by continuous stirring until the mixture was homogenous prior to the addition of the silica source, which included fumed silica (FS), K-infused fumed silica (FS-a), and fumed silica with 15 wt% KBr (equivalent to KBr present in K-infused fumed silica). Growth solutions were aged for 24 h at room temperature prior to being transferred to a 60 mL polypropylene laboratory bottle (VWR) for syntheses at 100 °C in a ThermoFisher Precision oven under static conditions.

For syntheses involving the addition of PDDA, the above recipes were altered by adding polymer prior to the addition of aluminum and silicon sources. The PDDA/SiO_2 molar ratio used for each zeolite growth mixture was selected by screening syntheses for the optimal composition leading to the fastest crystallization time. Syntheses of mordenite and SSZ-13 with colloidal silica followed the same protocol as mentioned above where fumed silica was replaced with colloidal silica while keeping the molar compositions the same.

4.3 Materials characterization

The as-synthesized materials were characterized for their crystallinity and phase purity by powder X-ray diffraction (XRD) on a Rigaku diffractometer (Cu Kα radiation). Analysis of the bulk elemental composition was performed using energy dispersive X-ray spectroscopy (EDX) on a JEOL SM-31010/METEK EDAX system. The morphology of the zeolite crystals was determined by scanning electron microscopy (SEM) using a FEI 235 dual-beam (focused ion-beam) system. The aggregation of silica particles was studied by dynamic light scattering using a Brookhaven Instruments BI-200SM machine equipped with a TurboCorr Digital Correlator, a red HeNe laser diode (35 mW, 637 nm), and a decalin bath that was filtered to remove dust. The compositions of the solutions used for the DLS experiments were based on a reported protocol[66] using a dilute, aluminum-free solution to mimic the zeolite synthesis conditions (molar composition of 1

SiO_2 : 1 NaOH : 350 H_2O). DLS experiments were conducted at a fixed scattering angle of 90°, and the autocorrelation function was collected over a 1 min time-frame with at least 5 measurements for each sample.

Oblique illumination microscopy (OIM) measurements were performed to assess the PDDA-induced aggregation of silica precursors using a Nanosight LM10-HS microscope (Nanosight Ltd) equipped with a green laser ($\lambda = 532$ nm) operating at 25 ± 0.1 °C. OIM measures the Brownian motion of individual particles (or clusters). The composition of solutions prepared for OIM experiments was based on a reported protocol[60] using a molar composition of 1 SiO_2 : 1 NaOH : 1760 H_2O : 0.1 TPAOH. Textural analysis of the different silicon sources was performed by N_2 adsorption/desorption using a Micromeritics ASAP 2020 instrument. For all zeolite syntheses, the pH of the growth medium (or extracted supernatant solution) was measured using a Thermo Scientific Orion 3 Star meter.

5. Results and discussion

5.1 Alkali-infused precursors as crystallization accelerants

Here, we assess the broader applicability of alkali-infused sol gel media to reduce the time of zeolite crystallization for the synthesis of the zeolite mordenite (MOR). Crystallization was carried out in an organic-free medium using Na^+ as the structure-directing agent,[63] and an alkaline sol–gel mixture (pH 12–13) prepared

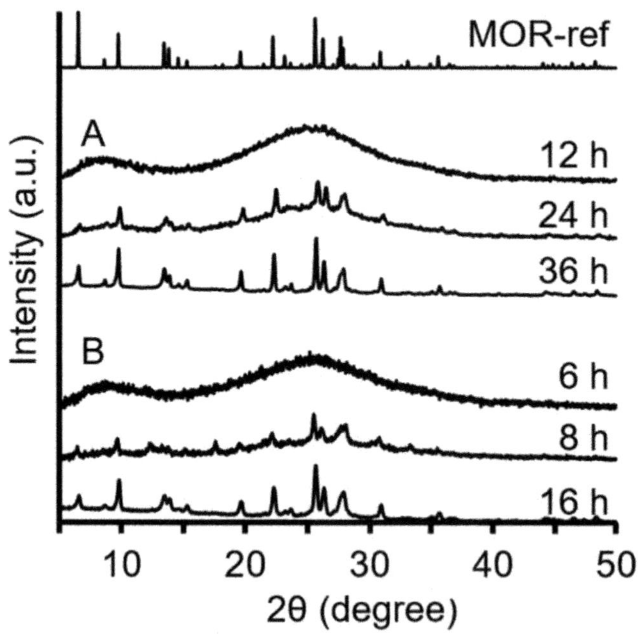

Fig. 4 Time-elapsed powder XRD patterns of the solids extracted from mordenite synthesis at various times of hydrothermal treatment at 165 °C using a molar composition of 18 SiO_2 : 1 Al_2O_3 : 6.8 NaOH : 509.3 H_2O. Comparisons are made between syntheses employing two different silicon sources: (A) fumed silica (sample FS) and (B) fumed silica infused with sodium ions (sample FS-a). Syntheses were performed using growth mixtures of identical molar composition, with the exception of excess NaBr in the modified fumed silica (15% NaBr by weight).

with fumed silica (sample FS) as the silicon source. The time-resolved powder XRD patterns of solids extracted from growth mixtures at various heating times reveal nucleation within 24 h and a fully crystalline product after *ca.* 36 h (Fig. 4A). The same experiments were then performed using a modified fumed silica that was pre-treated in a sodium bromide solution for one week. Elemental analysis of the extracted solid powder was performed by energy dispersive X-ray spectroscopy (EDS) to confirm the occlusion of the alkali in the modified amorphous silica (Table S1†). The resulting sodium-infused fumed silica (sample FS-a) was used to prepare a mordenite growth mixture with the same composition as the control. The time-resolved powder XRD patterns of the solids extracted from syntheses with FS-a (Fig. 4B) resulted in a marked reduction in both the induction time (<8 h) and crystallization time (*ca.* 16 h). We also conducted an experiment where additional NaBr was added to the growth mixture to assess the effects of excess Na^+ in FS-a and whether pre-treatment was necessary for alkali infusion. Our findings showed that the synthesis using FS as the source and the addition of NaBr (equivalent to the quantity in FS-a) resulted in a crystallization time (Fig.-S1A†) comparable to that of FS-a, suggesting that the effect of the alkali occurs *in situ* without the need for pre-treatment. To assess whether the alkali in solution impacts mordenite crystallization, we replaced FS with colloidal silica (a nonporous precursor). Additional NaBr in the synthesis mixtures reduces the crystallization time (Fig. S1B†), indicating that a higher concentration of Na^+ ions in the growth medium, present either in the solid or solution, impacts the rates of mordenite crystallization.

The evolution of amorphous precursors from both FS and FS-a sources was evaluated using DLS to measure changes in the hydrodynamic diameter of silicate particles with heating time. In order to prepare solutions with the appropriate volume fraction of dispersed silica particles for scattering experiments, we adapted a previously established protocol,[66] wherein a dilute zeolite synthesis mixture is prepared without the addition of alumina to avoid coagulation. Over a 24 h period of hydrothermal treatment, we observe a monotonic increase in amorphous particle size for sample FS (Fig. 5A and S2A†). Conversely, the particle size of the alkali-treated sample FS-a exhibits a step change increase with initial heating, followed by a more pronounced monotonic increase in size with time (Fig. 5A and S2B†). The exact cause for the modified fumed silica reducing the mordenite synthesis time is not fully understood. In addition to the effect of alkali infusion in sol gel media, similar to those observed for K^+ in zeolite L (Fig. 3A), there could be contributing factors due to particle aggregation, akin to the effects observed for PDDA in SSZ-13 synthesis (Fig. 3D). To differentiate these two potential effects, we altered the pre-treatment of fumed silica by removing NaBr and using pure DI water for an equal amount of time. When the extracted powder (sample FS-w) was used as the silicon source for a mordenite synthesis (Fig. S3†), we observed a crystallization time (*ca.* 16 h) similar to that of FS-a, but with a slightly longer induction time (<16 h). This seems to suggest that changes to the properties of the fumed silica during pre-treatment, irrespective of alkali infusion, are responsible for the altered kinetics of mordenite crystallization.

One property that is affected by pre-treatment is the porosity of the fumed silica. Textural analysis of the as-received fumed silica (FS) confirms the presence of micropores (Fig. 5B). If we consider amorphous silica spheres with an average diameter of 150 nm (equivalent to FS in Fig. 5A), the estimated external surface

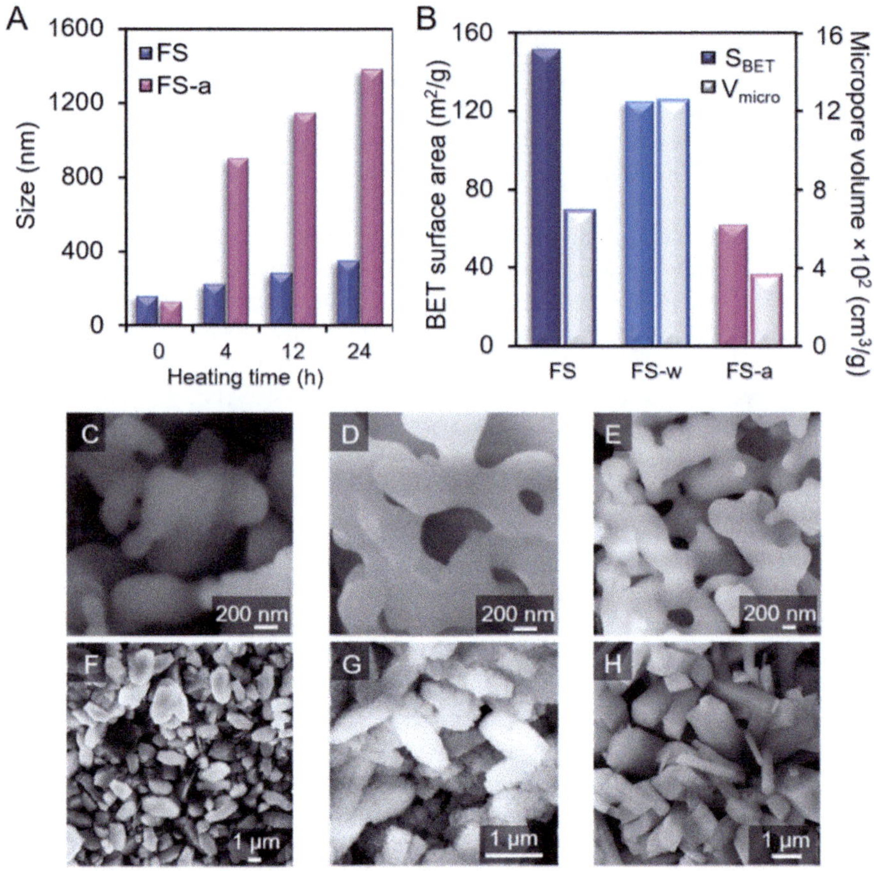

Fig. 5 (A) Dynamic light scattering measurements of the amorphous precursor size as a function of the hydrothermal treatment time. The heating temperature is 180 °C and the molar composition of the solution prepared for DLS is 1 SiO_2 : 1 NaOH : 350 H_2O (*i.e.* a diluted zeolite synthesis composition without the alumina source). Comparisons are made between two different silicon sources: fumed silica (sample FS) and fumed silica infused with sodium ions (sample FS-a). (B) Textural analysis of silicon sources used for all zeolite syntheses prior to hydrothermal treatment. Nitrogen adsorption/desorption measurements were used to obtain the BET surface area (left *y*-axis) and micropore volume (right *y*-axis) of amorphous precursors. A *t*-plot method was used to calculate the micropore volume. (C–E) Scanning electron micrographs of amorphous precursors extracted from mordenite syntheses. The growth mixtures were prepared with three different conditions and hydrothermal treatment times: (C) FS (6 h), (D) FS-w (8 h), and (E) FS-a (6 h). (F–H) Scanning electron micrographs of fully crystalline mordenite prepared with three different silicon sources and synthesis times: (F) FS (36 h), (G) FS-w (16 h), and (H) FS-a (16 h).

area is 10–20 m^2 g^{-1}. This accounts for 10% of the total BET surface area of FS (Fig. 5A), which indicates that the majority of the surface area is attributed to the micropores. After pre-treatment with water (FS-w), there is a two-fold increase in the micropore volume without a significant loss of the BET surface area. Pre-treatment with NaBr (FS-a) leads to a reduction in the micropore volume and

BET surface area, presumably due to the occupancy of Na^+ and Br^- ions within the micropores of amorphous silica particles. The exact impact of micropores on mordenite crystallization remains elusive; however, there are clear differences among FS, FS-w, and FS-a in terms of precursor evolution with concomitant impact on the size and morphology of the mordenite crystals. Scanning electron microscopy (SEM) images of fumed silica after a short period of hydrothermal treatment (Fig. 5C–E) reveal differences that cannot be discerned from DLS measurements. For example, sample FS-a (Fig. 5E) and sample FS-w (Fig. 5D) both resemble worm-like particles (Fig. 2C), whereas sample FS (Fig. 5C) is comprised of nanosized domains similar to the original sample prior to hydrothermal treatment. These differences among the amorphous precursors translate into disparate morphologies of the final mordenite crystals, as verified by SEM (Fig. 5F–H). Notably, the crystals obtained from FS (Fig. 5F) are somewhat faceted crystals with non-uniform morphology, FS-w produces rough crystals (Fig. 5G and S3†), analogous to the finned zeolites reported by Dai *et al.*,[67] and FS-a results in more faceted crystals (Fig. 5H).

We further assessed the impact of fumed silica pretreatment on two different zeolites prepared with organics: SSZ-13 (CHA)[30] and ZSM-5 (MFI)[64] using TMAda and TPA as the structure-directing agents, respectively. Syntheses using water-treated fumed silica (FS-w, Fig. S4 and S5†) did not significantly alter the crystallization time compared to growth conditions using fumed silica (FS) as the silicon source (Fig. 6 and S6–S8†). This indicates that the enhanced rate of crystallization observed for mordenite using FS-w is not universal. A more extensive parametric study is needed to relate the effects of FS-w to factors such as the growth medium composition (*e.g.* alkali metals, organics, *etc.*) and the final properties of zeolite crystals. Syntheses using alkali-infused fumed silica (FS-a) resulted in consistent results for mordenite and SSZ-13, similar to those observed for zeolite L (Fig. 2A). The preparation of mordenite and SSZ-13 with the

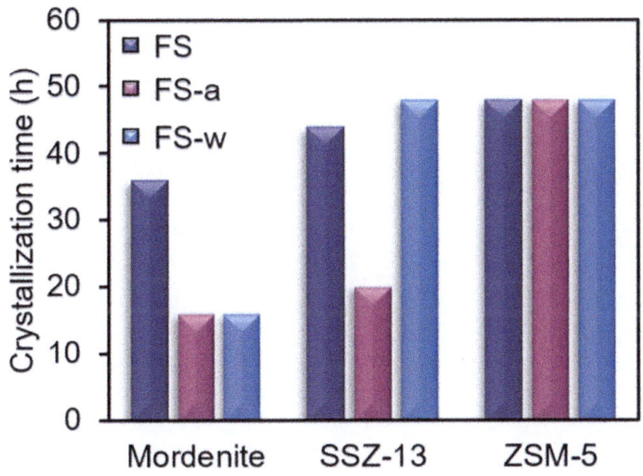

Fig. 6 Times to reach full crystallinity for the syntheses of mordenite (MOR), SSZ-13 (CHA), and ZSM-5 (MFI) using FS, FS-a, and FS-w as the silicon sources. Syntheses were performed under hydrothermal conditions at temperatures of 165, 180, and 170 °C, respectively.

sodium-infused silicon source led to more than a 50% reduction in the crystallization time compared to syntheses using the as received fumed silica (Fig. 6 and S6†). A similar effect was also observed using colloidal silica with additional NaBr (Fig. S1D†). This is unexpected for SSZ-13 given that K^+ ions (Fig. S9†) are the structure-directing agent used in organic-free preparation of the CHA crystal structure.[65] Conversely, Na^+ ions are used in organic-free syntheses of ZSM-5;[68] however, the substitution of FS with FS-a did not have an appreciable effect on ZSM-5 crystallization time (Fig. 6), suggesting that the organic structure-directing agent plays a more important role in ZSM-5 crystallization under the conditions selected for this study.

5.2 PDDA as a crystallization accelerant

Syntheses of mordenite and ZSM-5 using fumed silica (FS) as a silicon source were tested with the addition of PDDA polymer. The powder XRD patterns of solids

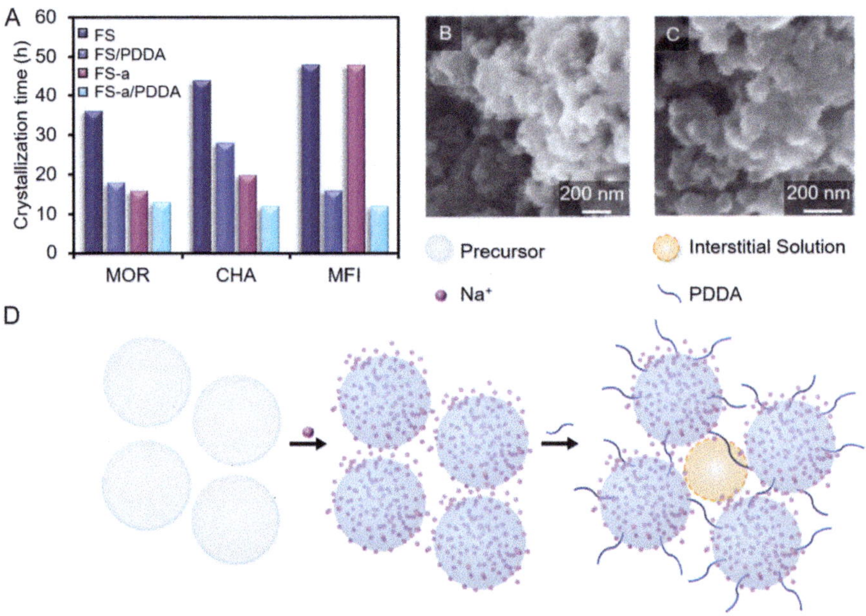

Fig. 7 (A) Times to reach full crystallinity for the syntheses of mordenite (MOR), SSZ-13 (CHA), and ZSM-5 (MFI) using four conditions: FS, FS/PDDA, FS-a, and FS-a/PDDA. Data for FS and FS-a are taken from Fig. 6. The PDDA : SiO_2 molar ratios for MOR (6.0×10^{-6}) and MFI (2.4×10^{-5}) were determined by pre-screening for the optimal PDDA concentration (Fig. S9†). The molar ratio used for CHA (1.2×10^{-5}) was set equal to C_{opt} in Fig. 3D; however, experiments by Dai *et al.*[60] used fumed silica from a different vendor than the source used in this study, which accounts for differences in crystallization times. (B and C) High magnification scanning electron micrographs of the amorphous precursors extracted from a mordenite synthesis after 8 h of hydrothermal treatment using (B) FS/PDDA and (C) FS-a/PDDA combinations. (D) Cartoon illustrating the infusion of the silicon source (fumed silica) with alkali ions during pre-treatment in 2 M $NaBr_{(aq)}$, followed by the PDDA-mediated assembly of precursors into clusters at an optimal PDDA : SiO_2 molar ratio.

extracted from syntheses at a fixed period of hydrothermal treatment and varying polymer contents were compared to approximate the PDDA concentration resulting in the fastest zeolite crystallization time (Fig. S10†). Syntheses in these optimal PDDA-containing growth mixtures (Fig. 7A) resulted in 2- and 3-fold reductions in the crystallization time for mordenite (Fig. S11†) and ZSM-5 (Fig. S12†), respectively, compared to their nominal (polymer-free) syntheses. This finding is consistent with the effect of PDDA on the SSZ-13 crystallization time (Fig. 3D, 7A, and S13†). Likewise, the presence of PDDA inhibits the evolution of amorphous precursors into worm-like particles observed in the absence of polymer (Fig. 5C–E). SEM images of solids extracted from mordenite syntheses using FS (Fig. 7B) and FS-a (Fig. 7C) confirm that PDDA preserves the original size of the amorphous precursors. Similar observations were made for growth solutions of ZSM-5 (Fig. S12 and S14†). As previously discussed (Fig. 3E), the smaller precursor size leads to a higher specific surface area that enables the facile exchange of (alumino)silicate species between the solid and solution. In the interstitial regions of PDDA-induced aggregates, the confined volume has the potential to concentrate these species in the presence of PDDA, which has been shown to facilitate silica condensation.

We previously performed scattering experiments to assess whether the presence of organic structure-directing agents influenced precursor aggregation.[60] For instance, it was previously shown for SSZ-13 that the organic structure-directing agent TMAda did not impact the ability of PDDA to induce precursor aggregation at the optimal polymer concentration. In this study, we conducted oblique illumination microscopy (OIM) measurements of solutions mimicking ZSM-5 synthesis and confirmed that its organic structure-directing agent, TPA, did not inhibit precursor aggregation in PDDA-containing solutions (Fig. S15†). These studies were not performed for mordenite since its synthesis is carried out in organic-free media.

5.3 Synergistic effects of precursor modification

To assess the potential synergy between PDDA and alkali-infused fumed silica (FS-a), we also performed experiments where FS was replaced with FS-a using the same PDDA concentrations. Our findings revealed that the combination of both modifications (Fig. 7A) resulted in further reductions of the crystallization time for mordenite and SSZ-13, beyond what was achieved with each individual precursor modification (Fig. S16 and S17†). This confirms the synergetic effect of altering both the composition and colloidal stability of precursors in zeolite crystallization. This is roughly illustrated in Fig. 7D, where we posit that the interstitial confined region (shaded yellow area) is a potential site for zeolite nucleation in syntheses involving the combined use of FS-a as the silicon source and PDDA as a polymer additive. It was shown in Fig. 4A that the enhanced concentration of alkali metal in sol gel media (FS-a) promotes particle aggregation, similar to the effect of PDDA. When FS-a and PDDA are combined, it is likely that the presence of both Na^+ ions and PDDA in the confined interstitial solution among precursor aggregates leads to the observed synergistic enhancement of zeolite nucleation for mordenite and SSZ-13; however, this effect was not observed for ZSM-5, wherein PDDA-containing solutions with and without FS-a exhibited similar crystallization times (Fig. 7A and S14†).

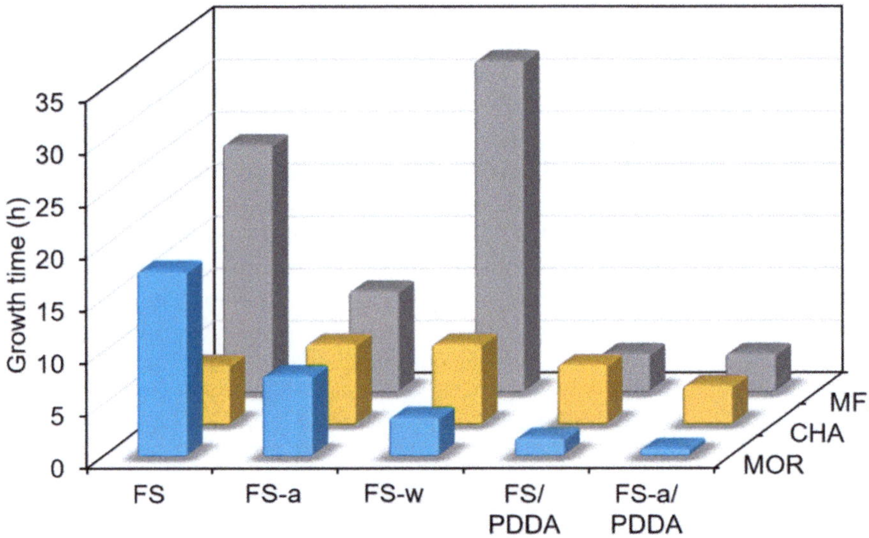

Fig. 8 Three dimensional plot of zeolite crystal growth times for all three structures (MOR, CHA, MFI) for the five synthesis conditions tested: FS, FS-a, FS-w, FS/PDDA, and FS-a/PDDA. The growth times listed here are obtained from powder XRD patterns of solids extracted at periodic synthesis times, wherein the estimated induction time (*i.e.* first appearance of Bragg peaks in the powder XRD) is subtracted from the approximate crystallization time (*i.e.* fully crystalline powder XRD).

We analyzed the effect that each precursor modification has on the times of nucleation and growth for the three zeolite structures tested in this study. These times were approximated from powder XRD patterns extracted at periodic synthesis times. The induction time was estimated as the first appearance of Bragg peaks, with values for each synthesis condition reported in Fig. S18.† The difference between the total crystallization time (Fig. 6 and 7A) and the induction time (Fig. S18†) was used to determine the time of growth (Fig. 8). For SSZ-13, the effects of alkali infusion and PDDA both had larger impacts on the induction time. The same trend was observed for ZSM-5 in the presence of PDDA. For mordenite, the opposite trend was observed, wherein alkali infusion and PDDA had more pronounced effects on the time of growth (Fig. 8). For mordenite syntheses, each modification to the synthesis mixture resulted in a reduced growth time, with the synergistic combination FS-a/PDDA resulting in an 18-fold reduction relative to that of the control (FS). For SSZ-13 syntheses, the growth time is not significantly impacted by any of the precursor modifications, indicating that the reduced crystallization time is attributed to accelerated nucleation. For ZSM-5 syntheses, the only reductions in the total crystallization time were observed for PDDA-containing solutions where reductions in the times of nucleation (Fig. S18†) and growth (Fig. 8) relative to those of the control were observed. Collectively, our findings reveal that precursor modifications can have notable impacts on nucleation and/or growth times with trends among zeolites that are not predictable.

6. Conclusions

The amorphous-to-crystalline transformations observed in this study for zeolite syntheses are common to many other systems where the impact of the physicochemical properties of amorphous precursors on the processes of crystal nucleation and growth is often elusive. In this study, we tested a small sampling of zeolites among the approximately 250 known structures; thus, it is uncertain if the trends observed in this study are broadly applicable to structures beyond the three investigated (MOR, CHA, and MFI). Here, we examined the effects that two modifications to amorphous precursors had on zeolite crystallization, focusing on syntheses using fumed silica as the primary silicon source. Our findings revealed that colloidal stability impacts all three zeolites, wherein the addition of a polymer additive, PDDA, can induce precursor aggregation with concomitant reductions in crystallization time. A working hypothesis was presented where we posit that the interstitial (confined) regions between aggregated precursors lead to enhanced rates of crystallization, coupled with the presence of PDDA that can facilitate silicate condensation (potentially enhancing the rate of nucleation).

Alkali infusion had a pronounced impact on SSZ-13 and mordenite, while ZSM-5 was the only zeolite tested that was unaffected by the addition of Na^+ ions. It remains to be determined if the reduced crystallization time for mordenite using the FS-a modified silicon source is a general outcome for organic-free syntheses. Likewise, it is uncertain if ZSM-5 crystallization being unresponsive to the substitution of FS with FS-a represents an anomaly among zeolite syntheses using organic structure-directing agents. Future investigations that expand this work to other zeolites could potentially shed more light on the impact of precursor modifications. It is evident, though, that these facile changes to amorphous precursors could be easily implemented in industrial syntheses. The ability to reduce crystallization times by the extent demonstrated in this study could have a significant impact on the economics of commercial production. Moreover, the concept of alkali infusion could have potential parallels with heteroatom insertion in zeotypes. It is too premature to make these comparisons; however, similar concepts have been validated for interzeolite transformations where the ion exchange of a parent zeolite (seed) with metals has been used to generate daughter zeolites (product) with occluded metals.[69,70] It remains to be determined if similar outcomes can be emulated in amorphous precursors to impact the crystallization pathways and resultant physicochemical properties of zeolites, zeotypes, and related materials that occur by nonclassical pathways.

Conflicts of interest

There are no conflicts to declare.

Acknowledgements

This work was primarily supported by the Department of Energy, Office of Basic Energy Sciences, Materials Science Division, Grant DE-SC0021384. Additional support was provided by The Welch Foundation (E-1794).

References

1 J. Baumgartner, A. Dey, P. H. Bomans, C. Le Coadou, P. Fratzl, N. A. Sommerdijk and D. Faivre, *Nat. Mater.*, 2013, **12**, 310–314.

2 E. Beniash, R. A. Metzler, R. S. Lam and P. Gilbert, *J. Struct. Biol.*, 2009, **166**, 133–143.

3 Y. U. Gong, C. E. Killian, I. C. Olson, N. P. Appathurai, A. L. Amasino, M. C. Martin, L. J. Holt, F. H. Wilt and P. Gilbert, *Proc. Natl. Acad. Sci. U. S. A.*, 2012, **109**, 6088–6093.

4 A. I. Lupulescu and J. D. Rimer, *Science*, 2014, **344**, 729–732.

5 Y. Politi, T. Arad, E. Klein, S. Weiner and L. Addadi, *Science*, 2004, **306**, 1161–1164.

6 P. R. ten Wolde and D. Frenkel, *Science*, 1997, **277**, 1975–1978.

7 L. F. Filobelo, O. Galkin and P. G. Vekilov, *J. Chem. Phys.*, 2005, **123**, 014904.

8 P. G. Vekilov, *Cryst. Growth Des.*, 2010, **10**, 5007–5019.

9 T. Yamazaki, Y. Kimura, P. G. Vekilov, E. Furukawa, M. Shirai, H. Matsumoto, A. E. Van Driessche and K. Tsukamoto, *Proc. Natl. Acad. Sci. U. S. A.*, 2017, **114**, 2154–2159.

10 J. J. De Yoreo, P. U. Gilbert, N. A. Sommerdijk, R. L. Penn, S. Whitelam, D. Joester, H. Zhang, J. D. Rimer, A. Navrotsky and J. F. Banfield, *Science*, 2015, 349.

11 Z. Sheng, H. Li, K. Du, L. Gao, J. Ju, Y. Zhang and Y. Tang, *Angew. Chem., Int. Ed.*, 2021, **60**, 13444–13451.

12 J. Cejka, H. van Bekkum, A. Corma and F. Schueth, *Introduction to Zeolite Molecular Sieves*, Elsevier, 2007.

13 M. E. Davis, *Ind. Eng. Chem. Res.*, 1991, **30**, 1675–1683.

14 A. Chawla, R. Li, R. Jain, R. J. Clark, J. G. Sutjianto, J. C. Palmer and J. D. Rimer, *Mol. Syst. Des. Eng.*, 2018, **3**, 159–170.

15 U. Olsbye, S. Svelle, M. Bjørgen, P. Beato, T. V. Janssens, F. Joensen, S. Bordiga and K. P. Lillerud, *Angew. Chem., Int. Ed.*, 2012, **51**, 5810–5831.

16 A. Primo and H. Garcia, *Chem. Soc. Rev.*, 2014, **43**, 7548–7561.

17 F. Su and C. Lu, *Energy Environ. Sci.*, 2012, **5**, 9021–9027.

18 S. Wang and Y. Peng, *Chem. Eng. J.*, 2010, **156**, 11–24.

19 B. M. Weckhuysen and J. Yu, *Chem. Soc. Rev.*, 2015, **44**, 7022–7024.

20 M. Dusselier and M. E. Davis, *Chem. Rev.*, 2018, **118**, 5265–5329.

21 S.-H. Lee, C.-H. Shin, D.-K. Yang, S.-D. Ahn, I.-S. Nam and S. B. Hong, *Micropor. Mesopor. Mat.*, 2004, **68**, 97–104.

22 K. Na, M. Choi and R. Ryoo, *J. Mater. Chem.*, 2009, **19**, 6713–6719.

23 W. Park, D. Yu, K. Na, K. E. Jelfs, B. Slater, Y. Sakamoto and R. Ryoo, *Chem. Mater.*, 2011, **23**, 5131–5137.

24 S. I. Zones, K. Jayanthi, J. Pascual, D. Xie and A. Navrotsky, *Chem. Mater.*, 2021, **33**, 2126–2138.

25 R. Jain, A. J. Mallette and J. D. Rimer, *J. Am. Chem. Soc.*, 2021, **143**, 21446–21460.

26 M. K. Choudhary, M. Kumar and J. D. Rimer, *Angew. Chem.*, 2019, **131**, 15859–15863.

27 J. D. Rimer, J. M. Fedeyko, D. G. Vlachos and R. F. Lobo, *Chem. –Eur. J.*, 2006, **12**, 2926–2934.

28 J. D. Rimer, O. Trofymluk, R. F. Lobo, A. Navrotsky and D. G. Vlachos, *J. Phys. Chem. C*, 2008, **112**, 14754–14761.
29 J. D. Rimer, D. G. Vlachos and R. F. Lobo, *J. Phys. Chem. B*, 2005, **109**, 12762–12771.
30 M. Kumar, H. Luo, Y. Román-Leshkov and J. D. Rimer, *J. Am. Chem. Soc.*, 2015, **137**, 13007–13017.
31 M. K. Choudhary, R. Jain and J. D. Rimer, *Proc. Natl. Acad. Sci. U. S. A.*, 2020, **117**, 28632–28639.
32 A. I. Lupulescu and J. D. Rimer, *Angew. Chem.*, 2012, **124**, 3401–3405.
33 J. R. Agger, N. Pervaiz, A. K. Cheetham and M. W. Anderson, *J. Am. Chem. Soc.*, 1998, **120**, 10754–10759.
34 P. Cubillas and M. W. Anderson, Synthesis mechanism: Crystal growth and nucleation, in *Zeolites and Catalysis: Synthesis, Reactions and Applications*, ed. j. Cejka, A. Corma and S. Zones, Wiley-VCH, 2010, pp. 1–55.
35 M. Kumar, M. K. Choudhary and J. D. Rimer, *Nat. Commun.*, 2018, **9**, 1.
36 K. N. Bozhilov and V. Valtchev, *Microsc. Microanal.*, 2019, **25**, 816–817.
37 K. N. Bozhilov, T. T. Le, Z. Qin, T. Terlier, A. Palčić, J. D. Rimer and V. Valtchev, *Sci. Adv.*, 2021, **7**, eabg0454.
38 G. Mirabello, A. Ianiro, P. H. Bomans, T. Yoda, A. Arakaki, H. Friedrich, G. de With and N. A. Sommerdijk, *Nat. Mater.*, 2020, **19**, 391–396.
39 J. D. Rimer, *Nat. Mater.*, 2020, **19**, 375–376.
40 A. Chawla, N. Linares, R. Li, J. García-Martínez and J. D. Rimer, *Chem. Mater.*, 2020, **32**, 3278–3287.
41 K. N. Olafson, R. Li, B. G. Alamani and J. D. Rimer, *Chem. Mater.*, 2016, **28**, 8453–8465.
42 M. A. Bewernitz, D. Gebauer, J. Long, H. Cölfen and L. B. Gower, *Faraday Discuss.*, 2012, **159**, 291–312.
43 L. Dai, X. Cheng and L. B. Gower, *Chem. Mater.*, 2008, **20**, 6917–6928.
44 J. Ihli, W. C. Wong, E. H. Noel, Y.-Y. Kim, A. N. Kulak, H. K. Christenson, M. J. Duer and F. C. Meldrum, *Nat. Commun.*, 2014, **5**, 3169.
45 A. S. Schenk, H. Zope, Y.-Y. Kim, A. Kros, N. A. Sommerdijk and F. C. Meldrum, *Faraday Discuss.*, 2012, **159**, 327–344.
46 P. J. Smeets, K. R. Cho, R. G. Kempen, N. A. Sommerdijk and J. J. De Yoreo, *Nat. Mater.*, 2015, **14**, 394–399.
47 L. B. Gower, *Chem. Rev.*, 2008, **108**, 4551–4627.
48 O. A. Tertuliano and J. R. Greer, *Nat. Mater.*, 2016, **15**, 1195–1202.
49 J. Ihli, Y.-W. Wang, B. Cantaert, Y.-Y. Kim, D. C. Green, P. H. Bomans, N. A. Sommerdijk and F. C. Meldrum, *Chem. Mater.*, 2015, **27**, 3999–4007.
50 M. Kumar, R. Li and J. D. Rimer, *Chem. Mater.*, 2016, **28**, 1714–1727.
51 N. Ren, B. Subotić, J. Bronić, Y. Tang, M. Dutour Sikirić, T. Mišić, V. Svetličić, S. Bosnar and T. Antonić Jelić, *Chem. Mater.*, 2012, **24**, 1726–1737.
52 R. Li, A. Chawla, N. Linares, J. G. Sutjianto, K. W. Chapman, J. G. Martínez and J. D. Rimer, *Ind. Eng. Chem. Res.*, 2018, **57**, 8460–8471.
53 R. Li, N. Linares, J. G. Sutjianto, A. Chawla, J. Garcia-Martinez and J. D. Rimer, *Angew. Chem., Int. Ed.*, 2018, **57**, 11283–11288.
54 D. L. Bish and D. W. Ming, *Natural Zeolites: Occurrence, Properties, Applications*, Walter de Gruyter GmbH & Co KG, 2018.
55 G. Gottardi and E. Galli, *Natural Zeolites*, Springer Science & Business Media, 2012.

56 B. G. Alamani, J. D. Gale and J. D. Rimer, *Cryst. Growth Des.*, 2021, **21**, 3375–3383.

57 S. Farmanesh, B. G. Alamani and J. D. Rimer, *Chem. Commun.*, 2015, **51**, 13964–13967.

58 W.-C. Huang, L.-M. Lyu, Y.-C. Yang and M. H. Huang, *J. Am. Chem. Soc.*, 2012, **134**, 1261–1267.

59 D. Wang, A. F. Wallace, J. J. De Yoreo and P. M. Dove, *Proc. Natl. Acad. Sci. U. S. A.*, 2009, **106**, 21511–21516.

60 H. Dai, J. Claret, E. L. Kunkes, V. Vattipalli, N. Linares, C. Huang, M. Fiji, J. García-Martinez, A. Moini and J. D. Rimer, *Angew. Chem., Int. Ed.*, 2022, **61**, e202117742.

61 D. W. Fickel, E. D'Addio, J. A. Lauterbach and R. F. Lobo, *Appl. Catal., B*, 2011, **102**, 441–448.

62 C. Paolucci, I. Khurana, A. A. Parekh, S. Li, A. J. Shih, H. Li, J. R. Di Iorio, J. D. Albarracin-Caballero, A. Yezerets and J. T. Miller, *Science*, 2017, **357**, 898–903.

63 F. J. Machado, C. M. López, M. a. A. Centeno and C. Urbina, *Appl. Catal., A*, 1999, **181**, 29–38.

64 Y. Shen, T. T. Le, D. Fu, J. E. Schmidt, M. Filez, B. M. Weckhuysen and J. D. Rimer, *ACS Catal.*, 2018, **8**, 11042–11053.

65 Y. Liang, A. J. Jacobson and J. D. Rimer, *ACS Mater. Lett.*, 2021, **3**, 187–192.

66 A. I. Lupulescu, W. Qin and J. D. Rimer, *Langmuir*, 2016, **32**, 11888–11898.

67 H. Dai, Y. Shen, T. Yang, C. Lee, D. Fu, A. Agarwal, T. T. Le, M. Tsapatsis, J. C. Palmer and B. M. Weckhuysen, *Nat. Mater.*, 2020, **19**, 1074–1080.

68 S. D. Kim, S. H. Noh, K. H. Seong and W. J. Kim, *Microporous Mesoporous Mater.*, 2004, **72**, 185–192.

69 S. Goel, Z. Wu, S. I. Zones and E. Iglesia, *J. Am. Chem. Soc.*, 2012, **134**, 17688–17695.

70 S. Goel, S. I. Zones and E. Iglesia, *J. Am. Chem. Soc.*, 2014, **136**, 15280–15290.

Faraday Discussions

PAPER

Simulating intergrowth formation in zeolite crystals: impact on habit and functionality†

Mollie Trueman, *[a] Duncan Akporiaye[b] and Michael W. Anderson [a]

Received 23rd November 2021, Accepted 20th January 2022
DOI: 10.1039/d1fd00097g

A kinetic Monte-Carlo methodology is presented for simulating crystal growth in materials which contain stacking faults. By simulating a large number of potential growth and dissolution events, a representation of the crystal is generated at various stages throughout the crystallisation, allowing the effects of disorder on the evolution of crystal habit and nanoscale surface topography to be explored. As examples, simulations were performed on two intergrown zeolite materials – zeolite T and zeolite beta. In both zeolite T and zeolite beta, simulations demonstrate how an intergrown structure leads to a characteristic roughening of certain crystal facets. In zeolite beta, this is accompanied by the development of internal defects which shows a non-homogeneous distribution. Results of simulations are validated by direct comparison to experimental scanning electron microscopy, atomic force microscopy and X-ray diffraction data. All simulations are performed using the CrystalGrower software package with modifications to account for disorder and should be generally applicable to all classes of crystals.

Introduction

The presence of defects in crystalline materials can lead to major deviation from the expected chemical and physical properties of the bulk materials, making understanding defect formation an important aim. A major class of defects is stacking faults, which occur when the periodicity of stacked layers in a crystal structure is interrupted, giving rise to two-dimensional, planar defects.[1,2] Stacking faults can arise as a result of stresses applied to a crystal structure, or can form as a consequence of 'mistakes' in the crystallisation process. Such mistakes can often be rationalised through comparison to related crystal structures which share a common crystallographic plane. For certain materials it is also possible to

[a]*Centre for Nanoporous Materials, School of Chemistry, The University of Manchester, Oxford Road, Manchester M13 9PL, UK. E-mail: mike.anderson@manchester.ac.uk*

[b]*SINTEF Industry, PO Box 124, Blindern, 0314 Oslo, Norway*

† Electronic supplementary information (ESI) available. See DOI: 10.1039/d1fd00097g

observe extended, three dimensional domains of related but different crystal structures within a single crystallite. These are commonly referred to as intergrowths.

Understanding stacking faults and intergrowths holds particular significance in the field of porous materials. Defects have the potential to disrupt pore structure, thereby introducing diffusional barriers and causing myriad changes to both catalytic and sorption properties of materials. These changes include variation in the rate of catalyst deactivation and changes to the product selectivity.[3–5] An interesting consequence of stacking faults and intergrowths is the impact to the external appearance of the crystal, both at the nanoscale and at larger length scales. Materials containing stacking faults frequently exhibit features such as roughening of certain facets, terrace stacking and high likelihood of twin formation.[2,6–8] These external changes correlate strongly to internal disorder and therefore contain a wealth of information. In spite of this, external features are often poorly understood and are rarely discussed in depth within literature.

Atomic force microscopy (AFM) and high-resolution scanning electron microscopy (HRSEM) have been used to great effect in studying surface structure in porous materials.[9–17] AFM has proved to be a particularly powerful tool for studying intergrown and epitaxial growth of materials, revealing information on how different crystallographic phases interface with one-another at the nanoscale in both zeolites[6,8,18] and metal–organic frameworks (MOFs).[19–21] Whilst microscopy techniques provide valuable insight, interpretation of images is often complex, and the link between crystal structure and the appearance of crystal surfaces is often non-intuitive. A strategy to aid interpretation of this data is to use computer simulations of crystal growth and attempt to replicate both crystal morphology and surface topology. This allows our understanding of crystal growth processes to be tested, and offers the potential to extract additional information from data. In light of this, the CrystalGrower simulation package was developed as a means to rapidly simulate 3D maps of crystal surfaces using a kinetic Monte-Carlo method, allowing detailed comparison to images obtained using scanning probe microscopy.[22,23] CrystalGrower achieves this aim by partitioning crystal structures into units, then reassembling these units *in silico*, thereby simulating the surface topology at each stage of the crystallisation. This is realised using a model which divides crystal surfaces into site types for growth and dissolution based on the number and type of intracrystalline bonds which could be formed if a growth unit attached at a certain site. For each of these identified site types, probabilities of growth or dissolution are calculated using estimates of crystallisation free energies. By simulating a large number of possible growth or dissolution events, a three-dimensional representation of the crystal surface at every stage of the crystallisation is generated, which can be compared directly to microscopy data.

In earlier reported versions of CrystalGrower (v1.4. and earlier), it was not possible to incorporate stacking faults or intergrowths into the model.[22,23] This prevented the application of the software to materials which contained a high level of disorder. In this study, a modified version of the CrystalGrower Monte Carlo approach has been developed and applied to create models of crystal growth and dissolution in two disordered zeolite frameworks which contain high levels of stacking faults, namely zeolite T and zeolite beta. An emphasis will be placed on exploring the link between internal disorder in these materials and the external appearance of crystals. The techniques outlined are general and the approach taken could be easily adapted to fit a wide range of problems, paving the way for wider applications of coarse grained Monte-Carlo simulations to aid in understanding the subtleties of crystal growth.

Crystal growth model

Simulations were run using the CrystalGrower simulation package.[22,23] The scheme used is adapted from that proposed for use in the program MONTY, as reported in 2004 by Meekes *et al.*[24] Crystallisation was modelled as a series of reversible transitions between a growth solution and a crystal surface, with the mother phase parameterised by bulk values and the crystal surface broken down into a limited number of site types, where site type is determined by the number of neighbouring units with which an attaching growth unit can form an interaction.[25,26] Monte Carlo 'trial moves' corresponded to attachment or dissolution of single growth units at surface sites. Probabilities of growth and dissolution at different site types were evaluated as functions of changes in Gibbs free energy. To estimate energy changes of growth or dissolution, the approximation that growth units at different surface site types have equal entropy and volume was invoked. This simplification is justified by the reasoning that differences in these values between growth sites will be negligible relative to the differences between solution and crystal phases. By making this approximation, the change in Gibbs free energy for attachment at a given growth site were found simply by considering the Gibbs free energy change in forming the bulk crystal, and applying a site-dependent offset for all the bonding interactions which cannot form if a unit attaches at a given position at the surface. This allowed the change in Gibbs free energy change to be simplified to:

$$\Delta G_{\mathrm{i}} = \Delta\mu - \Delta U_{\mathrm{i}} \tag{1}$$

where ΔG_{i} equals the Gibbs free energy change for a transition from the solution phase to a crystal surface site i, and $\Delta\mu$ represents the difference in chemical potential per growth unit between the bulk mother phase and the bulk crystal phase. The parameter ΔU_{i} was found by identifying the interactions which remain broken when a unit attaches at a given surface site i and summing their strengths. To achieve this, bonding interactions were identified within the crystal structure and assigned a crystallisation free energy by the user. As might be expected, the energy associated with such interactions strongly correlates to the strength of bonds in the crystal structure, but this relationship is not an exact one. Formation of intracrystalline bonds necessitates desolvation, which offsets the strength of these energies.[27]

If the attachment and dissolution of growth units is assumed to be thermodynamically reversible, it follows that the relative rates of attachment and dissolution obey the relationship:

$$\frac{P_{\mathrm{i}}^{\mathrm{grow}}}{P_{\mathrm{i}}^{\mathrm{etch}}} = \exp\left(\frac{\Delta G_{\mathrm{i}}}{k_{\mathrm{B}}T}\right) = \exp\left(\frac{\Delta\mu - \Delta U_{\mathrm{i}}}{k_{\mathrm{B}}T}\right) \tag{2}$$

where $P_{\mathrm{i}}^{\mathrm{grow}}$ is the probability of growth at site i, $P_{\mathrm{i}}^{\mathrm{etch}}$ is the probability of dissolution at site i and k_{B} is the Boltzmann constant. On its own eqn (2) is insufficient to determine the probabilities of attachment and detachment. The final form of the probabilities used took the form of the most computationally efficient equations which remain consistent with eqn (2):[24]

$$P_{\mathrm{i}}^{\mathrm{grow}} = \exp\left(\frac{-0.5\Delta U_{\mathrm{i}} + 0.5\Delta\mu}{k_{\mathrm{B}}T}\right) \tag{3}$$

$$P_{\mathrm{i}}^{\mathrm{etch}} = \exp\left(\frac{0.5\Delta U_{\mathrm{i}} - 0.5\Delta\mu}{k_{\mathrm{B}}T}\right) \tag{4}$$

When the programme is executed, at each iteration the number of sites for each different site type will be summed, and the probability of growth or dissolution occurring at that site will be calculated. These probabilities will be summed, and a growth or dissolution event will be selected randomly, with the likelihood of a given event occurring proportional to the values found using eqn (3) or (4). The programme then iterates over a large number of growth and dissolution events, building up an image of the crystal habit and nanoscale surface structure at each stage.

Partitioning of crystal structures

To incorporate stacking faults into the growth model, careful consideration had to be put into the partitioning of zeolite structures into growth units. Evidence from AFM and HRTEM studies show that in zeolite structures, surfaces terminate such that each tetrahedral atom loses a minimal number of interactions. These surfaces correspond to those where cages are complete, where possible, at the crystal surface. Since such studies indicate that cages persist temporally at crystal surfaces, it follows that cages are a sensible unit of growth for simulations. CrystalGrower interfaces with a software named ToposPro to generate a partitioning of a crystal structure into cage-like units through the "natural tiling" algorithm, developed by Blatov *et al.*[28–30] Whilst the natural tiling method is the preferred choice of partitioning for systems which contain low levels of defects, it is intrinsically difficult to adapt to disordered systems. In this study, the focus was on replicating stacking faults which form when two crystal structures have closely related structures. The use of complete cages, would have resulted in cages which were not present in both structures, and did not express the structural similarity between the pairs of closely related crystal structures. As such, for simulations containing stacking faults, crystal structures were instead divided into units which were simple and common to both crystal structures. It is possible that this change in approach will lead to discrepancies in the nanoscale surface structure, however at the length scales much greater than a unit cell, such an approach should be effective.

Although the natural tiling partitioning method was not used for the majority of simulations reported in this work, two simulations were run of the pure polymorphs of zeolite beta. These simulations were run to allow for comparison of crystal habit from models with and without stacking faults. The partitioning of crystal structures in these examples is as reported in our previous work.[22,23]

Supersaturation profile and crystallisation free energies

The approach outlined above requires knowledge of values for supersaturation ($\Delta\mu$) and the crystallisation free energies associated with different bonding interactions (ΔU_i). Determination of these values is a matter of particular difficulty for zeolite crystals. Synthetic zeolites are typically grown using hydrothermal methods, which involve conditions of elevated temperature and pressure. This is normally achieved at a lab scale by heating precursors in sealed vessels, and this type of synthesis does not readily lend itself to observation of crystallisation processes as they occur.[31] To further complicate matters, zeolites are typically grown from gels containing aluminosilicate species of varying degrees of polymerisation, which are known from NMR studies to undergo complex speciation.[32–34] Additionally, synthesis typically occurs from multicomponent systems and often includes a so called templating agent, which may be either organic or inorganic cations. Such templates are known to play a strong role in determining both the crystal structure of the product and the rate of crystallisation, suggesting they are influential in controlling supersaturation. Each of these factors combine to make it difficult to identify values for supersaturation and crystallisation free energies in zeolite synthesis. Calculation of these values using an *ab initio* method is beyond the scope of this work, and values were instead estimated through an empirical approach.

An initial estimate of the range of $\Delta\mu$ and ΔU_i value ranges was made by considering the crystallisation of zeolites in terms of the degree of condensation of each tetrahedral site. This is commonly denoted by the Q^n notation, where Q^4 describes a tetrahedral site with four Si–O–Si or Si–O–Al linkages and Q^1 describes a tetrahedral site with one Si–O–Si or Si–O–Al, and three Si–O–H linkages. In this context, a zeolite synthesis gel is a mixture of Q^0, Q^1, Q^2, Q^3 and Q^4 sites, deriving from precursors which are most often amorphous in nature. In contrast to this, bulk zeolite structures will contain mostly Q^4 sites, with Q^3 sites present in some structures. Two attempts have been made to measure the heat evolved from the below reaction:

With a value of −4.2 kcal mol^{-1} measured by Pokrovski *et al.* using an NMR approach[35] and a value of −0.57 kcal mol^{-1} measured by Tagirov *et al.* using a calorimetric approach.[36] This suggests both $\Delta\mu$ value and ΔU_i values should also be on this order of magnitude. To select values of ΔU_i suitable for use in simulations, an empirical method was used whereby a large number of simulations were run, with each ΔU_i value varied stepwise within the range of 1–10 kcal mol^{-1}. Aspect ratios of simulated crystals were compared to as-synthesized materials, and the combination of values which gave the best agreement was chosen for future simulations. The value for $\Delta\mu$ was not held constant, but rather changed through the course of the simulation to mimic changes in supersaturation that occur throughout crystallisation. X-ray diffraction studies have demonstrated that, in a typical hydrothermal synthesis of a zeolite sample, the growth rate follows a sigmoidal profile where an induction period is followed by a period of crystal growth in which the rate of crystallisation holds constant.[31] After a period of time, this crystal growth drops off and slows to nothing towards the end of the crystallisation period, at which point crystals exist in equilibrium with the solution phase. This suggests the supersaturation profile of a zeolite crystallisation should feature a period of high supersaturation followed by a drop off to equilibrium. Simulations began with a period of high supersaturation, with an initial supersaturation value far in excess of what would be naturally observed (100 kcal mol^{-1}), which allowed crystals to nucleate and reach a reasonable size within a reasonable timescale. In order to mimic synthetic conditions, supersaturation was then reduced to 5 kcal mol^{-1} for the remainder of the growth period. The value of $\Delta\mu$ was then dropped slowly until a point was reached where the rate of dissolution processes was exactly equal to the rate of growth processes. At this point the system was taken to be in equilibrium and $\Delta\mu$ was held at this value for the remainder of the simulation.

Two simulations were run using the natural tiling method of structure partitioning, as examples of structures free from stacking faults. $\Delta\mu$ was assigned the same profile for these simulations as all other simulations. Simulations run using the natural tiling method are run such that every silicate linkage is assigned the same crystallisation free energy. ΔU_i values are calculated by the software based on this value. A value of 3 kcal mol^{-1} per silicate linkage was chosen for these simulations, on the basis that this value was found to be appropriate in similar materials, in earlier publications.[22,23]

Stacking faults

To simulate the formation of stacking faults and intergrowths, layers had to be able to grow on more than one set of crystallographic positions, relative to the layer below. This meant it was necessary to introduce the idea of 'blocked sites', such that when a layer was growing, no growth at that layer's alternative positions could occur. This avoided a situation in which growth units would occupy positions impossibly close to each other. To achieve this, the code was modified such that an interaction in the crystal structure could be marked as a 'clash interaction' between growth units. If growth at a given site would result in a clash, then this site was a blocked site and the probability of growth is set to zero.

To illustrate the nature of the changes required to model disorder, consider Fig. 1a which shows a simple 2d hexagonal arrangement in a crystallising layer. In this system, the current layer occupies the purple sites and a layer crystallising above the layer depicted would occupy either the red or green interstitial sites. To allow the model freedom to form stacking faults, crystallisation must be allowed to occur at the red, green or purple sites within any layer, but once a layer begins forming, different sites must be treated differently. The figure illustrates three cases of how growth could occur within the existing layer. These must each be handled differently by the code. In the first case, labelled 1 in the figure,

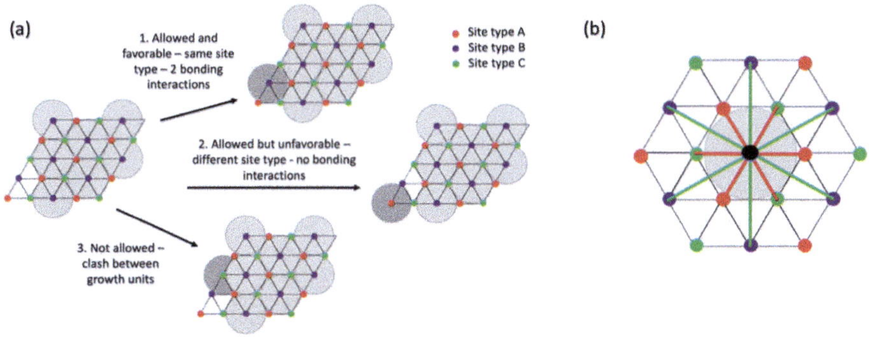

Fig. 1 (a) Three scenarios of attachment of a growth unit at the edge of a hexagonal layer. (b) Interactions to a single growth unit (black) which would be included in a CrystalGrower model representing outwards growth of a layer of a hexagonal crystal. Green lines represent location of bonding interactions, whilst red lines represent location of 'clash interactions'.

attachment occurs at the edge of a growing layer, in such a way that bonding interactions can form between the current layer and the attaching unit. Scenario 2 in the diagram shows a unit attaching at a site near to the current layer, but too far away to form any bonding interactions. Both of these two attachments are allowed, but due to the dependence of growth rate on the number and strength of interactions formed, given by eqn (3), these will occur at different rates in accordance with the outlined approach. The final case, labelled as 3 in Fig. 1a, shows growth occurring at a position which would result in a clash with regions of the crystal which are already grown. This is clearly something which cannot occur in a sensible model of crystal growth and a mechanism must therefore exist to ensure this site is blocked. To implement this, 'clash interactions' were added between growth sites when attachment at both sites is impossible. When the programme ran, in addition to detecting the number of bonding interactions which are present at each growth site, the modified version of the code also checks for 'clash interactions'. If one of these is detected, the site is blocked and the probability of growth is set to zero. Fig. 1b shows how these interactions would be assigned for the hexagonal model in Fig. 1a. 'Clash interactions' are shown in red whilst bonding interactions are shown in green.

Processing of simulation data

CrystalGrower outputs simplified *.xyz* files in which each growth unit is represented by a single point, and all lengths are scaled such that 1 Å equals the unit cell length in the a direction. These co-ordinates were converted to a full set of atomic positions using an in-house python script, using data obtained from the relevant CIF files. PXRD patterns were simulated for the complete set of atomic coordinates using the software Debyer. Debyer calculates diffraction patterns using Debye's scattering equation from a set of atomic co-ordinates.[37] Patterns were calculated for the 2θ range 5–50° with a step size of 0.01° and a wavelength of 1.54 Å. Small simulations were run to generate crystals at a size where calculation of patterns was feasible. These were run at 100 000 iterations, under conditions which were otherwise identical to longer simulations. Pore size distribution was calculated using the PorosityPlus software.[38] A probe radius of 2 Å was used with a box size of 126 × 126 × 126 Å^3. Visualisation of 3D output was done using Ovito.[39]

Results and discussion

Zeolite beta

The structure of zeolite beta can be broken down into a stacked series of identical layers as shown in Fig. 2a.[7,40] Consecutive layers are related to one another by a 90° rotation about the axis running perpendicular to the layer, and a translation relative to the layer below. Most reported synthetic procedures of zeolite beta result in a highly faulted intergrowth of two polymorphs, which are commonly referred to as polymorph A (BEA) and polymorph B (BEB), shown in Fig. 2b. These polymorphs differ in the positioning of the stacked layers. Fig. 2c shows how the positioning of consecutive layers varies in the two polymorphs when viewed down the [001] axis. It can be seen that the location of layers in polymorph A spirals upwards in a helical arrangement imparting chirality to the crystal structure. This chirality makes pure polymorph A a prized synthetic target,[41–44] but pure end-members of either polymorph continue to prove challenging to realise synthetically. The differing stacking positions result in different arrangements of pores for each of the two polymorphs of zeolite beta. This is clearly observable when the crystal structures of the two polymorphs are viewed down the [100] direction. In polymorph A these pores stack in an ABABAB sequence, whereas in polymorph B an ABCABC sequence is observed.[7,40] This is depicted in Fig. 2b.

A zeolite beta sample was prepared for comparison with the simulation results. The powder XRD pattern for this material, as shown in Fig. S1,† can be compared to simulated patterns for intergrown zeolite beta polymorph A (BEA) and zeolite beta polymorph B (BEB) in varying ratios, produced using DIFFaX software and shown in Fig. S2.†[45] This comparison suggests a composition of 50% BEA 50% BEB, which is a typical value for a standard zeolite beta synthesis.[1,7]

To model the crystallisation of zeolite beta in CrystalGrower, the structure was first broken down into small structural units common to both polymorph A and

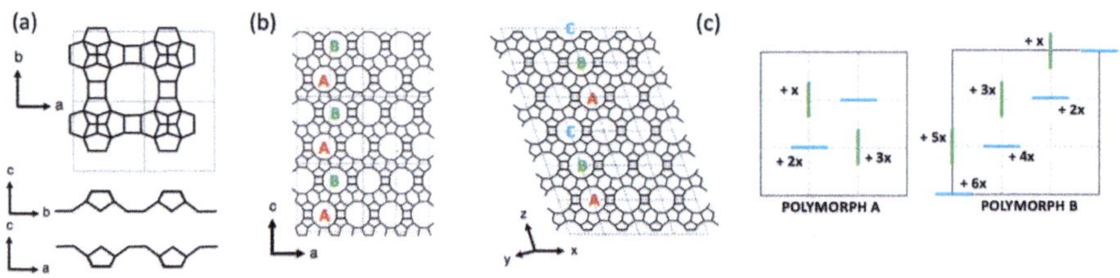

Fig. 2 (a) The layer unit of zeolite beta viewed down three different directions. (b) Crystal structures of zeolite beta polymorph A (left) and zeolite beta polymorph B (right). Letters indicate the ABAB and ABCABC sequences of pore positioning in polymorph A and B, respectively. (c) The relative positioning of consecutive layers in pure zeolite beta polymorphs, both viewed down the [001] axis, with the square outline representing the dimensions of a polymorph A unit cell. The $+x$ value indicates a shift of 6.55 Å coming out of the page. The blue and green colours, along with the differing orientation of lines, represent the rotation of 90° about the [001] axis between consecutive layers.

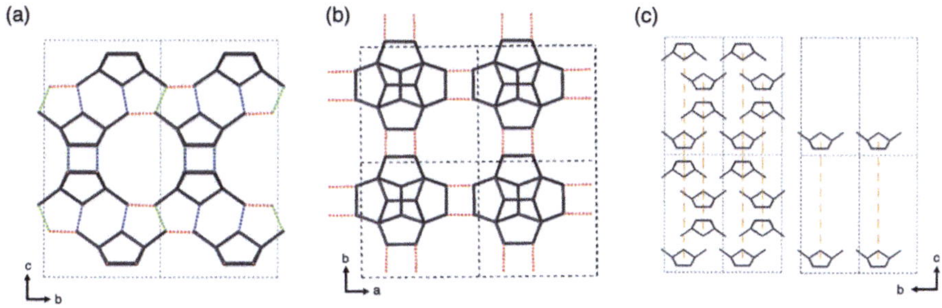

Fig. 3 (a) Zeolite beta polymorph A divided into growth units, with bonding interactions added. Viewed along the [100] direction. Three types of interaction are depicted. Bidentate interactions within layers are represented in red, bidentate interactions between layers are shown in green and tetradentate interactions between layers are shown in blue. (b) Bidentate interactions between growth units within each structural layer. (c) Long range interactions between growth units which were added to give control between polymorph A and polymorph B. Where these are assigned to be favourable interactions, a preference for polymorph A results.

polymorph B. These are shown in Fig. 3a and b. As a simple unit from which the whole structure could be assembled, it was reasoned that it should be possible to develop a model for crystallisation using this unit. It was noted that, since there is no evidence for the persistence of this unit at the surface or in solution, it is likely that nanoscale, sub-unit cell features are not replicated using this partitioning. After partitioning the structure, bonding was added to the model as illustrated by Fig. 3. Bonds were added along the ⟨100⟩ and ⟨010⟩ directions in the {00*l*} planes such that growth units with equivalent positioning to that seen in the BEA or BEB crystal structure form favourable interactions with one another. These bonding interactions correspond to the bidentate bonding interactions depicted in Fig. 3b. Bonds were also added between {00*l*} planes. Despite the different positioning of growth units, in both the BEA and BEB crystal structures each growth unit binds to two growth units in the adjacent layer. A tetradentate interaction is formed with one growth unit and a bidentate interaction is formed to another. Fig. S3† explores the impact of varying the relative strengths of the crystallisation free energies associated with these three different kinds of interaction.

As an initial approximation, it was assumed that the number of silicate linkages formed is the sole determinant of crystallisation free energy, with each linkage assigned a strength of 2 kcal mol^{-1}. Fig. 4 shows the simulated morphology of a zeolite beta crystal grown with this approximation, alongside simulations run using CrystalGrower's natural tiling method, assuming a crystal structure of either pure BEA or pure BEB. Expression of {001} facets is associated with incorporation of disorder, and a roughening of these facets is seen. SEM images of as synthesised zeolite beta are shown for comparison. It can be seen that the as-synthesised material adopts a crystal habit bound by eight {*h*0*l*} faces and two {00*l*} faces. {00*l*} faces are shown to possess a much rougher texture than that observed on other facets, with many squared features with edges running parallel to the *a* and *b* directions. This texture is replicated well in simulations, however the {*h*0*l*} faces are perhaps flatter and more clearly defined in SEM images, compared to the simulation. Comparison of SEM to Fig. S3† shows that strengthening of the between-layer bidentate interaction results in improved

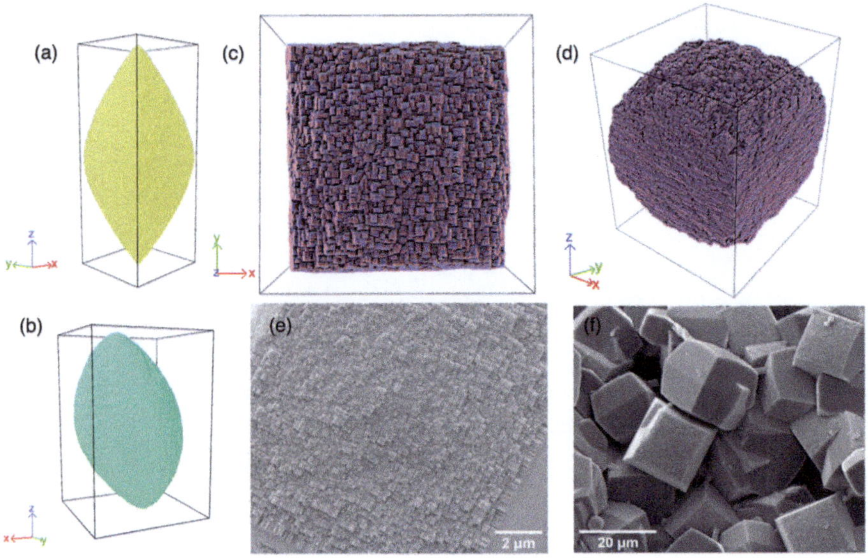

Fig. 4 (a and b) Simulated crystal of pure polymorphs of zeolite beta (no disorder allowed) grown at 200 °C, using CrystalGrower's natural tiling method, with each silicate linkage assigned an interaction strength of 3 kcal mol^{-1}. (a) Simulated crystal of beta polymorph A, (b) simulated crystal of beta polymorph B. (c and d) Simulated crystal of zeolite beta, grown at 200 °C allowing a disordered structure with each silicate linkage assigned an interaction strength of 2 kcal mol^{-1}. (e and f) Scanning electron micrographs of the as synthesised zeolite beta crystals.

definition of these facets. It is clear from Fig. 4 that incorporation of disorder into the simulation results in a morphology, which much more closely mimics that seen in reality.

The results of simulations were explored further by taking an internal cross section of the simulated crystals. Doing so revealed an interesting non-

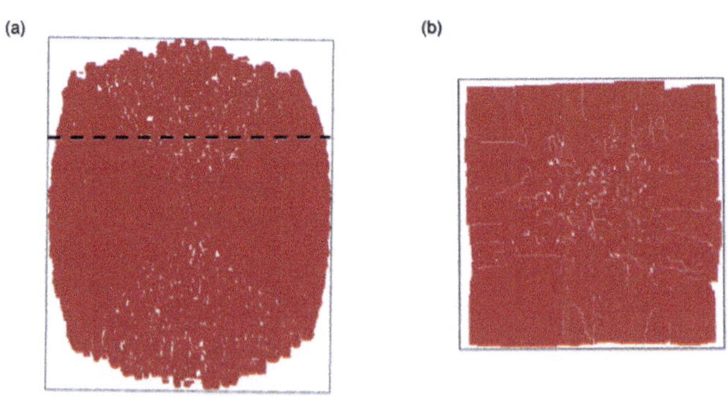

Fig. 5 (a) A cross section of a simulated zeolite beta crystal, through the (100) plane at the centre of the crystal, showing an 'hourglass' distribution of defects. This is thought to be generated through different levels of defect generation at each crystal facet. (b) A (001) cross section running along the dashed line shown in (a). A higher density of defects is seen towards the centre of the cross section.

homogeneous distribution of defects within zeolite beta crystals. It can be seen in Fig. 5 that the crystal can be divided into zones radiating out from the centre point to the edges of each crystal facet, which roughly correspond to the region of crystal grown from attachment at each facet. It is clear that the volume corresponding to {00*l*} facets has a high density of defects, and that voids develop in the crystal structure in the vicinity of these defects. Pore size distributions (shown in Fig. S4†) were calculated for each of the distinct zones in zeolite beta, confirming an increase of 10–20 Å pores in the high defect zone. The reasoning behind the occurrence of this non-homogeneous distribution is thought to be due to differences in growth *via* the two types of facet ({*h*0*l*} and {00*l*}). Growth at the {00*l*} faces requires 2d nucleation of a layer of zeolite beta but at this point there in no preference for either a polymorph A type layer or a polymorph B type layer. At high supersaturation {00*l*} faces will have multiple layers nucleating and expanding at the same time and where a polymorph A layers meets a polymorph B layer, a line defect will be formed. Growth at {*h*0*l*} faces also requires 2d nucleation, but in contrast to {00*l*} faces, there will now be a preference for one polymorph over another. When each {001} layer expands through growth at {*h*0*l*} faces, it will be much more difficult for a region of polymorph A to nucleate on a layer of polymorph B, since doing so would sacrifice bonding interactions to the current layer of growth, thereby raising the 2d nucleation barrier. An interesting piece of evidence that supports our assertion that defects should generate non homogeneously in zeolite beta is the presence of the so called ‘optical hourglass’ phenomenon which has been previously reported in zeolite beta.[46] When large crystals of zeolite beta are viewed through cross polars, it is possible to observe a faint hourglass inside the crystal, which could potentially be caused by slight differences in refractive index in different sections of the crystal. A similar hourglass pattern has been observed in the work of Weckhuysen and co-workers, who found that when crystals of zeolite beta were stained with fluorescent molecules they were seen to preferentially occupy an hourglass shaped region of the crystal.[47] The authors speculated this could be due to a high level of faulting in this region of the crystal, which is consistent with the results of our simulations.

Within the scientific community, there is sustained effort to synthesise pure polymorph A and polymorph B zeolite beta end members. Strategies for synthesizing novel zeolite materials commonly involve the addition of cationic ‘templating agents’ which direct the synthesis towards a certain crystal structure.[31] Although much remains uncertain as to the exact role played by these additives, it is often postulated that the fit of these ions within the pores results in favourable interactions between the templating ions which favour certain frameworks, whilst sterically inhibiting the formation of others.[48,49] To empirically mimic the action of a potential templating agent, the model for zeolite beta was expanded through addition of an interaction between growth units separated by three layers along the *c* direction, with the same relative positioning seen in polymorph A, as shown in Fig. 3c. By making this interaction favourable, the growth of polymorph A of zeolite beta was made more probable, whilst assigning a negative value to this interaction allowed the formation of polymorph A to be discouraged, favouring the formation of polymorph B by default. The effects of varying this interaction strength are illustrated by the simulated crystals in Fig. 6a. This allowed the shape of zeolite beta crystals to be predicted along the spectrum between the two pure polymorphs. It is interesting to note that even when one of the polymorphs is

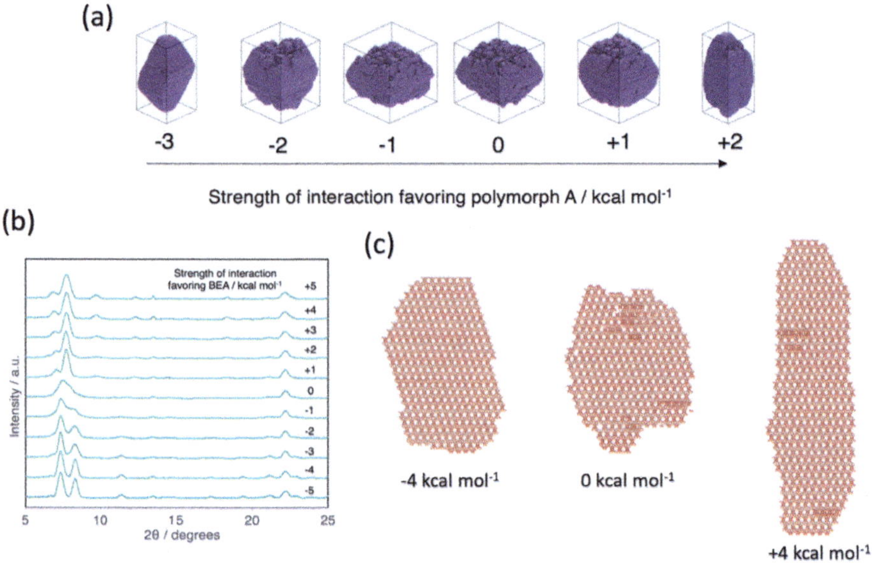

Fig. 6 (a) The effect on crystal habit of changing the strength of the interaction favouring beta polymorph A. At negative interaction strengths polymorph B is favoured. As interaction strength is increased, formation of polymorph A becomes more favourable, with crystals adopting a square prismatic habit. (b) Simulated powder X-ray diffraction patterns showing a gradual change in crystal structure as the strength of the interaction favouring polymorph A is increased. (c) Internal (100) cross sections showing variation in crystal structure with modification of strength of interaction favouring polymorph A. Note that the size of crystals in (c) is much smaller than in (a) to allow pore structure to be seen.

strongly favoured, a high degree of defect formation is still observed in zeolite beta crystals. The output of simulations was used to generate simulated powder X-ray diffraction patterns as shown in Fig. 6b. These patterns showed strong similarity to those generated using DIFFAX software (as shown in Fig. S2†), and demonstrated how simulations can be used to obtain data which can be directly compared with measurements from samples. Simulated diffraction patterns confirmed a gradual change in crystal structure, which can be seen by taking cross sections of the simulated crystals, as depicted in Fig. 6c.

Zeolite T

Zeolite T is an intergrowth of two similar but distinct zeolite frameworks – offretite and erionite. Both erionite and offretite belong to the ABC-6 zeolite family which is constructed from planar layers featuring six rings of tetrahedral sites arranged in a hexagonal motif as depicted in Fig. 7a. Adjacent layers can occupy one of three positions relative to the layer below and each member of the ABC-6 zeolite family is described by a unique sequence of stacking positions (commonly termed A, B and C positions). Erionite and offretite can be described by stacking sequences of AABAAC and AABAAB, respectively, as illustrated by Fig. 7b.

For the purpose of this study, two batches of zeolite T crystals were synthesised. Sample T-OFF contains a higher level of offretite, whilst the second sample, labelled T-ERI contains higher levels of erionite. SEM micrographs,

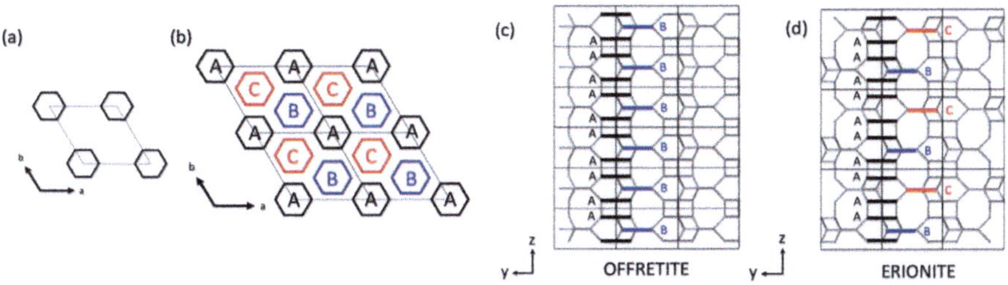

Fig. 7 (a) The arrangement of tetrahedral sites which makes up the building unit of layers of ABC-6 zeolites. (b) Possible layer positions for stacking in ABC-6 zeolites. (c) Offretite crystal structure viewed along the [100] direction. Coloured lines emphasise the AABAAB stacking sequence in offretite. (d) Erionite crystal structure viewed along the [100] direction, with coloured lines emphasising the AABAAC stacking sequence of erionite.

depicted in Fig. 8, reveal both T-OFF and T-ERI to crystallise as hexagonal prisms, displaying {10−10} and {001} facets. Fig. S5† shows the powder X-ray diffraction patterns for as-synthesised samples. Through comparison of these patterns to simulated diffraction patterns, generated using the DIFFaX program and shown in Fig. S6,†[45] the relative quantities of erionite were determined to be between 70% and 80% for T-ERI. None of the peaks unique to erionite were detected in the X-ray pattern of T-OFF, but simulated patterns indicate that when the quantity of erionite drops below 60%, the powder pattern will be virtually indistinguishable from that of pure offretite.

As a first stage in developing a model, the crystal structures of offretite and erionite were partitioned into closed cages. The cancrinite cage was selected as a growth unit for use in CrystalGrower, since both erionite and offretite crystal structures could be built solely from cancrinite cages, and because AFM results revealed surfaces terminating with terraces with an average height of 1.1 ± 0.1 nm (across 100 terrace measurements), consistent with complete cancrinite cages, as shown in Fig. 9. Whilst this does not imply that cancrinite cages are persistent in solution, it does suggest that transitions between complete cancrinite cages represents a rate determining step of crystallisation, suggesting these cages are

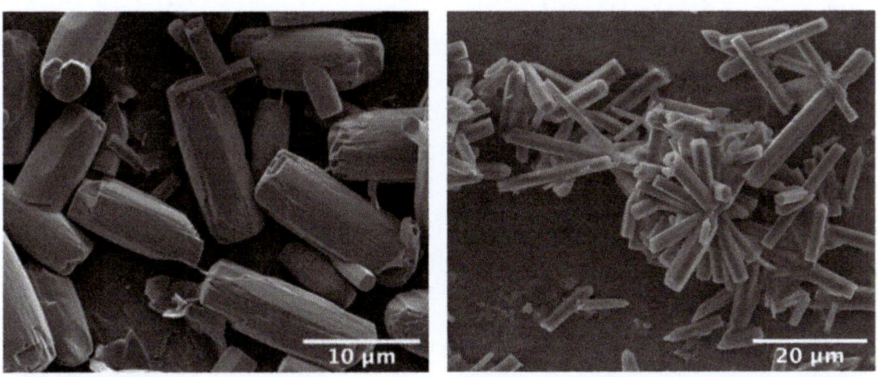

Fig. 8 SEM images of T-OFF (left) and T-ERI samples, showing both materials crystallise as hexagonal prisms.

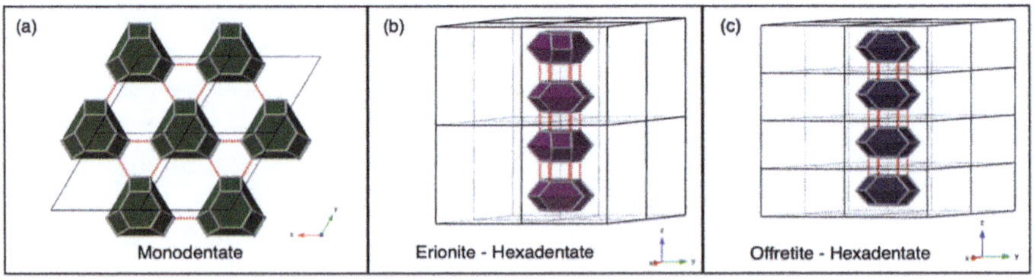

Fig. 9 Modes of connectivity in zeolite T – shown as red dashed lines (a) monodentate connectivity between cancrinite cages in the ab plane. (b) and (c) hexadentate connectivity between cancrinite cages in erionite and offretite. Note that whilst adjacent cages have the same orientation in offretite, in erionite, neighbouring cages are related through a 180° rotation about the *c* axis.

a suitable unit of growth for coarse-grained simulations of crystallisation. This approach parallels previous work on zeolite L, in which AFM results revealed formation of cancrinite cages to be rate determining for crystal growth.[11]

In both offretite and erionite, cancrinite cages connect in two distinct ways: cages connect in the (001) plane through monodentate linkages, and stack along the *c* axis connecting through hexadentate linkages. Whilst hexadentate linkages are observed in both crystal structures, these interactions are not identical in the offretite and erionite crystal structures. In offretite, cancrinite cages are in the same orientation either side of the hexadentate linkage; whilst in erionite, cages are related through a 180° rotation about the *c* axis. These modes of connectivity are shown in Fig. 10b and c.

The crystallisation free energies associated with formation of hexadentate linkages in erionite and offretite were initially approximated to be equal. Fig. S7† shows how the shape of simulated crystals relates to the relative crystallisation free energies of the hexadentate and monodentate linkages. It can be seen that increasing the energy associated with formation of the hexadentate linkage relative to the monodentate value results in an elongation of the crystal along the

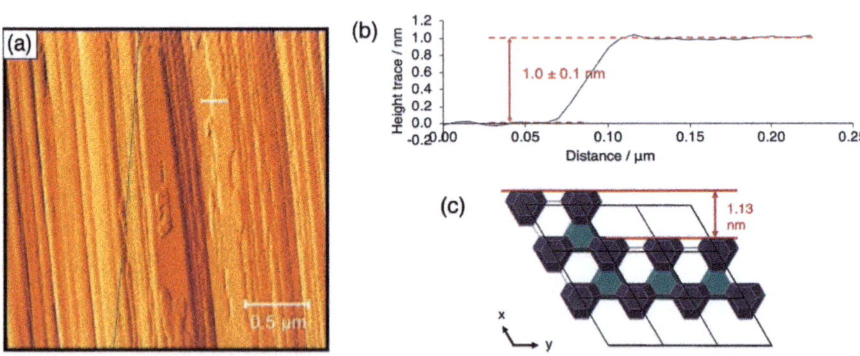

Fig. 10 (a) Atomic force micrograph (error signal) of T-OFF {10–10} facet. White line towards the top of micrograph corresponds to the height trace in (b). (c) Depicts possible surface termination in offretite crystals.

c axis. The average crystallisation free energy of a monodentate interaction in zeolite structures is widely accepted to be in the order of 1–4.5 kcal mol^{-1}.[35,36] With this in mind, values of 7.0 kcal mol^{-1} and 2.0 kcal mol^{-1} for hexadentate and monodentate linkages respectively, were found to give the best fit to experimentally observed interaction strengths in T-OFF, whilst in T-ERI the best fit was found to lie in the region between 10.0 kcal mol^{-1} for hexadentate linkages and 3.0 kcal mol^{-1} for monodentate linkages. This is in line with previously reported results from our group – in simulations of zeolite L a value of 2.0 kcal mol^{-1} was determined to be a good fit for monodentate interactions.[11]

Assigning equivalent crystallisation free energies to hexadentate linkages in offretite and erionite results in a scenario where the orientation of cancrinite cages is entirely random. If different interaction strengths are assigned based on whether adjacent cancrinite cages are in the same or different orientations, then it is possible to favour the formation of one of the two crystal structures. The former case mimics the crystallisation of zeolite T, whilst the latter resembles

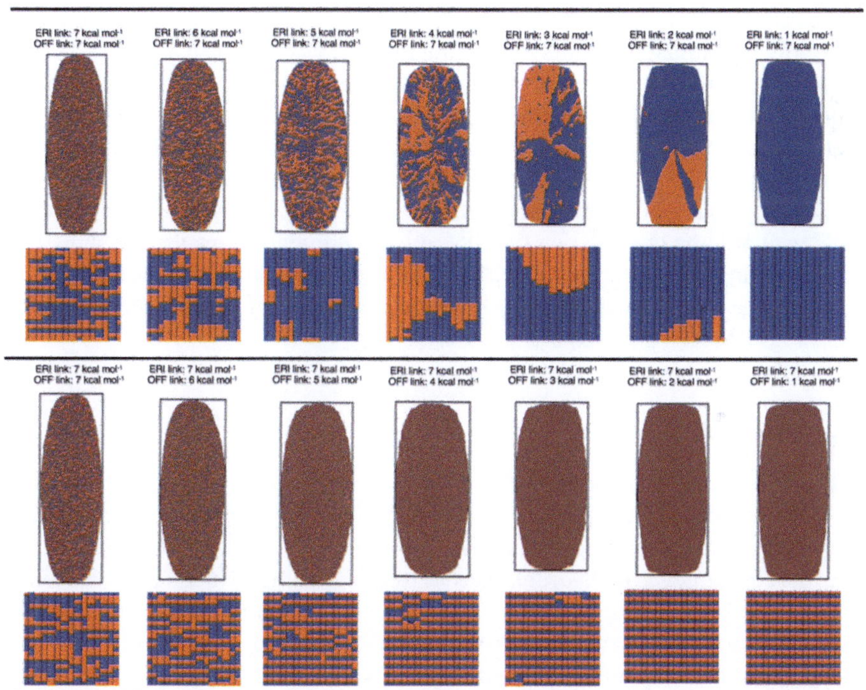

Fig. 11 Internal cross sections of simulated crystals showing how varying the relative strengths of hexadentate linkages effects change in the crystal structure. The upper panel shows the effect of weakening erionite type hexadentate linkages whilst holding offretite hexadentate linkages at a constant value of 7 kcal mol^{-1}. The lower panel shows the effect of holding erionite-type hexadentate linkages at 7 kcal mol^{-1} whilst offretite linkages are weakened. In each diagram, blue spheres represent cancrinite cages in one orientation, whilst orange spheres represent cancrinite cages related by a 180° rotation about the c axis, hence blocks of solid colour correspond to regions of pure offretite whilst regions with strict alternation of orange and blue layers represents the domains of erionite. Square images depict an enlarged subsection of the simulated crystal above them. All simulations were run at 140 °C.

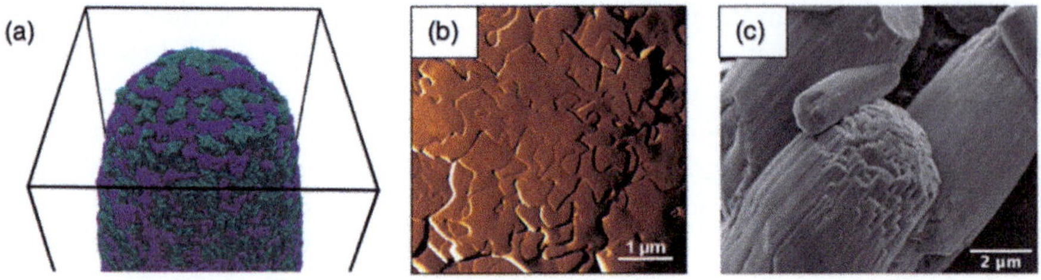

Fig. 12 (a) Simulated offretite crystal grown at 140 °C, with monodentate linkages at 2 kcal mol^{-1}, offretite hexadentate linkages at 7 kcal mol^{-1} and erionite hexadentate linkages at 5 kcal mol^{-1}, (b) atomic force micrograph of the {001} face of T-OFF crystals (c) scanning electron micrograph of T-OFF crystals showing the rough texture of the {001} face.

offretite containing stacking faults, such as that formed from mixtures containing tetramethyl ammonium cations as the templating agent. Fig. 11 illustrates the relationship between crystal structure and the relative energies of the two hexadentate interactions. It can be seen that weakening erionite type linkages relative to offretite type linkages results in large regions with cancrinite cages in the same orientation. Within these regions the crystal structure is that of pure offretite, and incorporation of erionite-type stacking faults results in twinning, giving a region where the crystal structure and cancrinite cages are rotated 180° relative to the other section.

When simulations favour either offretite or erionite, results suggest terrace stacking should be observed on the {001} facets. Simulation results are shown in Fig. 12. A similar texture is revealed in AFM and SEM imaging of {001} facets of high offretite. AFM results suggest edge heights between 20 and 100 nm, with larger edges observed *via* SEM.

Conclusions

This study has demonstrated how Monte-Carlo techniques provide a powerful tool for simulating crystal growth in microporous frameworks. It has been demonstrated that a simple, coarse-grained model can replicate features which arise as a consequence of disorder in crystals. Roughening of {00*l*} facets in zeolite T and zeolite beta crystals is successfully replicated in simulations, generating an interesting non-homogenous distribution of defects in zeolite beta crystals. Whilst this methodology proved effective in replicating many features in the systems studied, it is important to consider the limitations of this methodology. Several factors are known to play a role in determining the rates and outcomes of crystallisation, yet are not considered with the CrystalGrower methodology. Such factors include the roles of mass transfer (both at the crystal surface and in solution), growth unit conformation and non-classical growth mechanisms. The incorporation of these factors into CrystalGrower is left to future work. Additionally, the approach taken here estimates both $\Delta\mu$ and ΔU_i values using an empirical approach. Although this method is sufficient for replicating surface fine structure, if an *ab initio* method could be identified for calculating input parameters, this would allow the software to be used in a fully predictive manner.

In the works of Gale and coworkers, atomistic molecular-dynamics simulations were used to determine attachment rates of molecules at the surface of urea crystals,[50,51] these values were then used in kinetic Monte-Carlo simulations of crystal growth, allowing simulation of crystal growth on a micron scale. This work demonstrates the feasibility of a fully computational approach to simulating crystallisation down to the micron scale in molecular crystals, however the complex solution speciation in zeolite crystallisation will likely complicate determination of input parameters for this class of material. In spite of the limitations outlined, this work demonstrated that good agreement to experimental results can be achieved with this method, and that kinetic Monte-Carlo simulations can be used to rationalise stacking fault formation in crystalline materials.

Conflicts of interest

Michael Anderson and Mollie Trueman are shareholders and directors of CrystalGrower Ltd. which distributes the software used in this publication

Acknowledgements

The authors thank ESPRC and SINTEF for funding.

Notes and references

1 T. Willhammar and X. Zou, *Z. Kristallogr. Cryst. Mater.*, 2013, **228**, 11–27.

2 C. Hammond, in *The Basics of Crystallography and Diffraction*, 2015, pp. 20–27.

3 K. A. Łukaszuk, D. Rojo-Gama, S. Øien-Ødegaard, A. Lazzarini, G. Berlier, S. Bordiga, K. P. Lillerud, U. Olsbye, P. Beato, L. F. Lundegaard and S. Svelle, *Catal. Sci. Technol.*, 2017, **7**, 5435–5447.

4 M. W. Anderson, M. L. Occelli and J. Klinowski, *J. Phys. Chem.*, 1992, **96**, 388–392.

5 R. L. Smith, S. Svelle, P. del Campo, T. Fuglerud, B. Arstad, A. Lind, S. Chavan, M. P. Attfield, D. Akporiaye and M. W. Anderson, *Appl. Catal., A*, 2015, **505**, 1–7.

6 N. S. John, S. M. Stevens, O. Terasaki and M. W. Anderson, *Chem.–Eur. J.*, 2010, **16**, 2220–2230.

7 J. M. Newsam, M. M. J. Treacy, W. T. Koetsier and C. B. D. Gruyter, *Proc. R. Soc. London, Ser. A*, 1988, **420**, 375–405.

8 R. L. Smith, W. A. Sławiński, A. Lind, D. S. Wragg, J. H. Cavka, B. Arstad, H. Fjellvåg, M. P. Attfield, D. Akporiaye and M. W. Anderson, *Chem. Mater.*, 2015, **27**, 4205–4215.

9 P. Cubillas and M. W. Anderson, in *Zeolites and Catalysis*, Wiley-VCH Verlag GmbH & Co. KGaA, Weinheim, Germany, 2010, pp. 1–55.

10 M. A. Holden, P. Cubillas, M. P. Attfield, J. T. Gebbie and M. W. Anderson, *J. Am. Chem. Soc.*, 2012, **134**, 13066–13073.

11 R. Brent, P. Cubillas, S. M. Stevens, K. E. Jelfs, A. Umemura, J. T. Gebbie, B. Slater, O. Terasaki, M. A. Holden and M. W. Anderson, *J. Am. Chem. Soc.*, 2010, **132**, 13858–13868.

12 M. K. Choudhary, M. Kumar and J. D. Rimer, *Angew. Chem., Int. Ed.*, 2019, **58**, 15712–15716.

13 S. M. Stevens, P. Cubillas, K. Jansson, O. Terasaki, M. W. Anderson, P. A. Wright and M. Castro, *Chem. Commun.*, 2008, 3894.

14 P. Cubillas, S. M. Stevens, N. Blake, A. Umemura, C. B. Chong, O. Terasaki and M. W. Anderson, *J. Phys. Chem. C*, 2011, **115**, 12567–12574.

15 M. K. Choudhary, R. Jain and J. D. Rimer, *Proc. Natl. Acad. Sci. U. S. A.*, 2020, **117**, 28632–28639.

16 M. Kumar, M. K. Choudhary and J. D. Rimer, *Nat. Commun.*, 2018, **9**, 2129.

17 A. I. Lupulescu and J. D. Rimer, *Science*, 2014, **344**, 729–732.

18 J. R. Agger, N. Hanif, C. S. Cundy, A. P. Wade, S. Dennison, P. A. Rawlinson and M. W. Anderson, *J. Am. Chem. Soc.*, 2003, **125**, 830–839.

19 F. I. Pambudi, M. W. Anderson and M. P. Attfield, *Chem. Sci.*, 2019, **10**, 9571–9575.

20 F. I. Pambudi, M. W. Anderson and M. P. Attfield, *Faraday Discuss.*, 2021, **231**, 112–126.

21 R. Wagia, I. Strashnov, M. W. Anderson and M. P. Attfield, *Cryst. Growth Des.*, 2018, **18**, 695–700.

22 M. W. Anderson, J. T. Gebbie-Rayet, A. R. Hill, N. Farida, M. P. Attfield, P. Cubillas, V. A. Blatov, D. M. Proserpio, D. Akporiaye, B. Arstad and J. D. Gale, *Nature*, 2017, **544**, 456–459.

23 A. R. Hill, P. Cubillas, J. T. Gebbie-Rayet, M. Trueman, N. de Bruyn, Z. al Harthi, R. J. S. Pooley, M. P. Attfield, V. A. Blatov, D. M. Proserpio, J. D. Gale, D. Akporiaye, B. Arstad and M. W. Anderson, *Chem. Sci.*, 2021, **12**, 1126–1146.

24 S. X. M. Boerrigter, G. P. H. Josten, J. van de Streek, F. F. A. Hollander, J. Los, H. M. Cuppen, P. Bennema and H. Meekes, *J. Phys. Chem. A*, 2004, **108**, 5894–5902.

25 W. Kossel, *Ann. Phys.*, 1934, **413**, 457–480.

26 J. W. Mullin, *Crystallization*, Butterworth-Heinemann, 4th edn, 2001.

27 S. Piana and J. D. Gale, *J. Am. Chem. Soc.*, 2005, **127**, 1975–1982.

28 O. Delgado Friedrichs, M. O'Keeffe and O. M. Yaghi, *Acta Crystallogr., Sect. A: Found. Crystallogr.*, 2003, **59**, 22–27.

29 V. A. Blatov, O. Delgado-Friedrichs, M. O'Keeffe and D. M. Proserpio, *Acta Crystallogr., Sect. A: Found. Crystallogr.*, 2007, **63**, 418–425.

30 N. A. Anurova, V. A. Blatov, G. D. Ilyushin and D. M. Proserpio, *J. Phys. Chem. C*, 2010, **114**, 10160–10170.

31 C. S. Cundy and P. A. Cox, *Microporous Mesoporous Mater.*, 2005, **82**, 1–78.

32 S. L. Burkett and M. E. Davis, *J. Phys. Chem.*, 1994, **98**, 4647–4653.

33 M. Haouas, D. P. Petry, M. W. Anderson and F. Taulelle, *J. Phys. Chem. C*, 2009, **113**, 10838–10841.

34 S. D. Kinrade and T. W. Swaddle, *Inorg. Chem.*, 1988, **27**, 4253–4259.

35 G. S. Pokrovski, *Mineral. Mag.*, 1998, **62**, 1194–1195.

36 B. Tagirov, J. Schott, J. C. Harrichoury and J. Escalier, *Geochim. Cosmochim. Acta*, 2004, **68**, 1333–1345.

37 M. Wojdyr, 2009, http://www.unipress.waw.pl/debyer/.

38 G. Opletal, T. C. Petersen, S. P. Russo and A. S. Barnard, *J. Phys. Mater.*, 2018, **1**, 016002.

39 A. Stukowski, *Modell. Simul. Mater. Sci. Eng.*, 2010, **18**, 015012.

40 J. B. Higgins, R. B. LaPierre, J. L. Schlenker, A. C. Rohrman, J. D. Wood, G. T. Kerr and W. J. Rohrbaugh, *Zeolites*, 1988, **8**, 446–452.

41 A. Corma, M. Moliner, Á. Cantín, M. J. Díaz-Cabañas, J. L. Jordá, D. Zhang, J. Sun, K. Jansson, S. Hovmöller and X. Zou, *Chem. Mater.*, 2008, **20**, 3218–3223.
42 G. Zhang, B. Wang, W. Zhang, M. Li and Z. Tian, *Dalton Trans.*, 2016, **45**, 6634–6640.
43 T. Lu, L. Zhu, X. Wang, W. Yan, W. Shi and R. Xu, *Inorg. Chem. Front.*, 2018, **5**, 1640–1645.
44 M. Tong, D. Zhang, W. Fan, J. Xu, L. Zhu and W. Guo, *Nat. Publ. Gr.*, 2015, 1–10.
45 M. M. J. Treacy, 2010, http://www.public.asu.edu/~mtreacy/DIFFaX.html.
46 J. Sun, G. Zhu, Y. Chen, J. Li, L. Wang, Y. Peng, H. Li and S. Qiu, *Microporous Mesoporous Mater.*, 2007, **102**, 242–248.
47 F. C. Hendriks, J. E. Schmidt, J. A. Rombouts, K. Lammertsma, P. C. A. Bruijnincx and B. M. Weckhuysen, *Chem.–Eur. J.*, 2017, **23**, 6224.
48 D. W. Lewis, D. J. Willock, C. R. A. Catlow, J. M. Thomas and G. J. Hutchings, *Nature*, 1996, **382**, 604–606.
49 R. Brent, A. J. W. Lobo, D. W. Lewis and M. W. Anderson, *J. Phys. Chem. C*, 2010, **114**, 18240–18246.
50 S. Piana and J. D. Gale, *J. Cryst. Growth*, 2006, **294**, 46–52.
51 S. Piana, M. Reyhani and J. D. Gale, *Nature*, 2005, **438**, 70–73.

Faraday Discussions

PAPER

Theoretical aspects of the growth of a non-Kossel crystal from vapours: the role of advacancies†‡

Alexey Redkov * and Sergey Kukushkin

Received 7th November 2021, Accepted 20th December 2021
DOI: 10.1039/d1fd00083g

The growth of an arbitrary multicomponent non-Kossel crystal *via* the Burton–Cabrera–Frank mechanism is studied, considering the effect of advacancies and their recombination with adatoms on the surface. An analysis is carried out for two cases: growth due to vapours and growth due to chemical reactions. The analytical expressions are found for the rate of advancement of a group of equidistant steps and the crystal growth rate considering the properties of all the species involved in the growth process. Gallium nitride is used as an example to show that the effect is stronger at higher temperatures and in the presence of dopants that may increase the vacancy concentration. The impact of applied mechanical stress on the growth rate and mechanisms of its influence on the growth kinetics are discussed. It is demonstrated that the contribution of the vacancies to the total mass transfer depends on the type of applied stress. Tensile stresses increase the concentration of advacancies and total recombination rate, whereas compressive stresses lead to the opposite result. Some effects inherent to the multicomponent systems being considered are discussed.

1. Introduction

A major trend in the high-tech industry in recent decades is a transition to more and more complex crystalline materials with new properties that allow the creation of devices with better characteristics. Instead of simple, well-studied single-component materials, like silicon wide-bandgap semiconductors, such as gallium and aluminium nitrides and silicon carbides (GaN, AlN, SiC), are now often used in the production of LEDs and high electron-mobility transistors.[1–3] These materials ensure high operating temperatures, radiation stability, high breakdown voltage and much more. Efficient solar cells are made from perovskites – complex chemical compounds made up of many different atoms.[4–6] Many

Institute for Problems in Mechanical Engineering, Saint-Petersburg, 199178, Russia. E-mail: avredkov@gmail.com

† This paper was not presented at the meeting.

‡ Electronic supplementary information (ESI) available. See DOI: 10.1039/d1fd00083g

prospective optical crystals, superconductors and biomimetic materials are also multicomponent.[7–10] Much attention is also paid today to metal–organic-frameworks[11] (MOFs) and other types of frameworks[12] that are promising for storing hydrogen or methane for energy,[13] CO_2 capture,[14] catalysis or gas separation[15] and many other applications.[16] MOFs usually have a crystalline-like structure, and their growth can be treated as the growth of a very complex crystal with many features inherent to crystals, for example, screw dislocations (Fig. 1) and spiral growth.[17]

Besides the reasons above, many promising future materials are likely to have a multicomponent nature because almost all single-component materials are already well studied and have found their applications in technology. It is obvious that, for the growth of high-quality crystals and MOFs, an understanding is needed of the basic processes taking place during growth, both on the atomic scale (for example, the incorporation of building units into the lattice) and the macroscale (mass transfer or diffusion of components toward the surface). The impact of different types of defects in the crystal on the growth process, different kinds of fluctuations (temperature, component fluxes) and many other factors should be studied to optimise growth conditions. The impact of mechanical stresses is also of great importance because many of the devices are based on heterostructures; some stress usually arises during the growth of these crystalline thin films that affects the growth process. MOF[18] composite materials and even MOF-on-MOF heterostructures[19,20] are gradually finding new applications. During the synthesis of the latter, the same problems arise as are observed in the heteroepitaxy of classical semiconductors, which are already solved for single-component systems. Many of these problems are well-described and analysed, including spiral growth,[21,22] step-bunching,[23] nucleation,[24] growth of nanowires,[25] step–step interaction,[26] impact of stress[27–29] and impurities.[30] However, although the single-component analytical models describe multicomponent systems well in certain cases,[31] it is obvious that the latter demonstrate some peculiarities and effects caused by the presence of several components.[32,33] Thus, the properties of each component (diffusion coefficients, equilibrium concentration, *etc.*) should be considered to accurately describe the general growth process. The experimental techniques nowadays allow direct analysis of the growth process of multicomponent crystals at the sub-unit-cell level.[34–36] Modelling software has also seen development in recent years. Crystal growth, considering surface phenomena, such as surface diffusion, reactions, and defects (dislocations), is easy to model using molecular dynamics,[37] Monte Carlo,[38] cellular automata[23] and other methods, even on a personal computer, though with

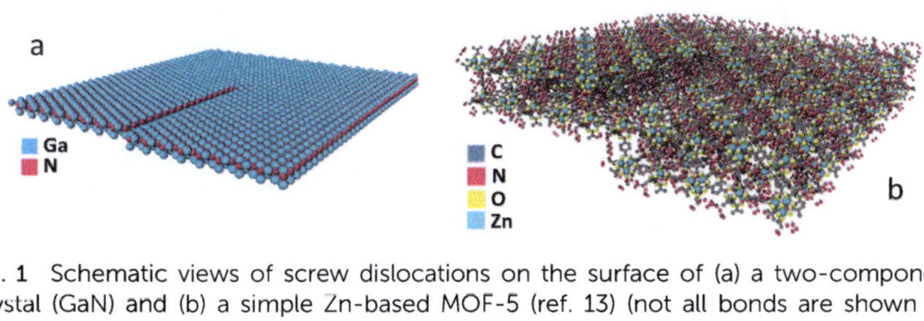

Fig. 1 Schematic views of screw dislocations on the surface of (a) a two-component crystal (GaN) and (b) a simple Zn-based MOF-5 (ref. 13) (not all bonds are shown for simplicity).

some limitations.[39] Thus, we are gradually obtaining new instruments to analyse atomistic and kinetic phenomena in multicomponent systems and compare them with the theory.

All of this, as was mentioned also by Chernov in his classical papers,[40] makes it of utmost importance to modify the existing theory of crystal growth to describe the growth of multicomponent non-Kossel crystals. Several valuable studies and models on the growth in multicomponent systems[41–44] mostly describe the growth of particular crystals, growth regimes and methods in demand in modern technology and thus cannot be used directly to describe the growth of an arbitrary crystal. There are also important theoretical papers on the growth of arbitrary multicomponent crystals[45–48] or due to chemical reactions,[49] but there are still many effects and dependencies that may greatly affect the growth rate and have not yet been analysed for multicomponent cases.

One of these issues is the impact of advacancies on the growth of the multicomponent crystal *via* the terrace-step-kink mechanism. Pimpinelli *et al.*[50] showed that advacancies might play an important role during single-component crystal growth/evaporation because they ensure additional mass transfer mechanisms at elevated temperatures. This effect was also studied both theoretically[51] and experimentally on silicon by Latyshev and colleagues.[52–54] The nucleation and surface recombination of advacancies may both facilitate the step-flow growth of a crystal or lead to its evaporation. The authors of ref. 55 and 56 have also shown how the vacancies themselves, by incorporation into the steps, may affect the advancement of the step and cause growth or dissolution of the crystal depending on the applied stress. This effect can be used for the formation of porous crystals. We suppose that the same phenomena should occur in multicomponent systems, and thus it is important to theoretically study multicomponent crystal growth considering surface recombination and advacancies.

This work continues the series of papers devoted to the growth and evolution of crystals and pores in single- and multicomponent systems by various mechanisms,[57–59] and its primary goal is to study the growth of a multicomponent crystal *via* the classical Burton–Cabrera–Frank (BCF) mechanism, while considering the effect of advacancies on the surface, their recombination with adatoms and applied mechanical stress.

2. Formulation of the problem

Here, we consider the growth of a multicomponent non-Kossel crystal *via* a terrace-step-kink mechanism, proposed in the classical paper of Burton, Cabrera and Frank.[21] The problem is close to the one studied in ref. 32, where equations were derived for the growth rate in different cases. The key difference is that here, we focus on the effect of advacancies,[50,52] their recombination and stress on the growth rate of the crystal, whereas advacancies were not considered in ref. 32. In the first part of the paper, we study the growth of the multicomponent crystal from the vapours or fluxes of its components, as in molecular-beam epitaxy (MBE). In the second part, we analyse the difference if the crystal grows due to chemical reactions, which usually happens in most cases of the epitaxy of industrial heterostructures, such as during the metal–organic (MOCVD) or halide–hydride epitaxy (HVPE) of gallium nitride.[60] The last section is devoted to an analysis of applied stress on the multicomponent crystal growth rate. Note that

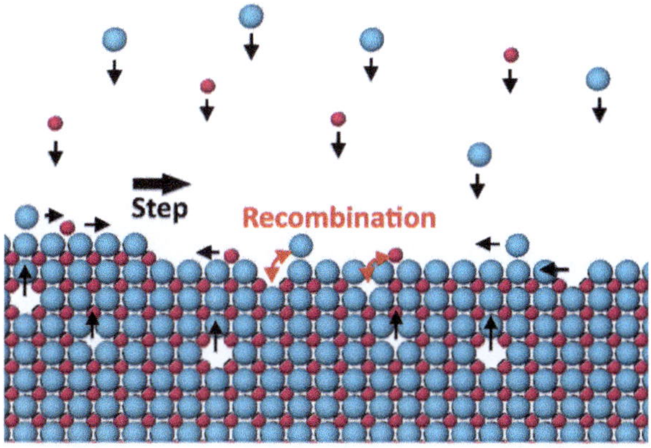

Fig. 2 Schematic representation of the processes being considered for the example of a two-component crystal.

the stress also may lead to a change in advacancy concentration and thus affects the mechanism considered. Schematically, the processes are presented in Fig. 2.

In the first case (Fig. 2), all the components of the non-Kossel crystal (or building units in the case of MOFs) are deposited from the gas phase to the terraces on the crystal surface. Then, they may diffuse along the terrace, evaporate, recombine with the advacancies, or incorporate into the kinks on the steps. The same occurs for vacancies. Being dissolved in the crystal as a gas of 'void',[55,56] they may diffuse toward the surface, become advacancies, diffuse over the surface, recombine with adatoms of the same type or simply 'evaporate' back to the bulk crystal.[56] In this paper, we consider that interaction on the terrace is only possible for advacancies and adatoms of the same type (*i.e.*, located in the same sublattice); no cross-interaction occurs between different components (adatom–adatom, advacancy–advacancy, adatom–advacancy). Such cross-interaction is possible only at the kink sites, and it leads to the formation of the next crystalline cell.

The increased component flux leads to an increased adatom concentration on the surface, the appearance of the surface flux toward the step and thus to the crystal growth. However, at increased surface concentrations of adatoms, the recombination rates at the terrace are higher. This causes the concentration of the advacancies on the terrace to be lower than its equilibrium value (which is maintained near the step), and a backward surface flux of advacancies arises that affects the step rate of advancement and total crystal growth rate.

The main difference of the second case is that we consider the growth of a crystal due to a chemical reaction when adatoms of different components are being deposited, not from the intrinsic vapours, but due to chemical reactions near the surface. We treat thermal evaporation of the adatoms in this case as negligible compared to the rate of the reverse chemical reaction,[32] which carries the adatoms back to the gas phase much faster.

In cases 1 and 2, we consider the vacancy concentration in the bulk as constant and equal to its equilibrium value. In the third case, the problem formulation is almost the same as in case 1, but we will analyse how the stress applied to the crystal affects the advacancy concentration on the surface. The main goal is to derive analytical expressions that describe the rate of advancement of the steps on a crystal surface considering all these phenomena and then analyse additional effects that may arise due to the presence of several components.

3. Multicomponent crystal growth rate in the presence of advacancies

3.1 Growth from vapours

Let us consider the first case, which is the growth of a multicomponent crystal from the vapours of its components (or a MOF from its building units). We suppose that components are being deposited from the gas phase to the terrace, then independently diffuse toward the steps and incorporate into the kinks according to the reaction:[32]

$$\nu_1 A_{1(g)} + \nu_2 A_{2(g)} + \ldots + \nu_L A_{L(g)} \leftrightarrow C_{(s)} \tag{1}$$

where L is the number of types of atoms that build up the crystal, A_i is the type of component (*e.g.*, Ga, N, Si) or MOF building unit, $C_{(s)}$ is the multicomponent crystalline cell and ν_i is the stoichiometric coefficient of the i-th atom type in the crystal. We note that we treat the adatoms as not interacting with each other on the terrace; that is, reaction (1) takes place only at the kinks. At the same time, each type of adatom may interact with advacancies of the same type on the surface. The system of equations describing the total flux of adatoms of the i-th type from the gas phase (subscript '↑') $j_{ai\uparrow}$ and advacancies from the bulk crystal toward the terrace $j_{vi\uparrow}$ can be written in the general case as follows (see Fig. 3):

$$\begin{cases} j_{vi\uparrow}(x) = \dfrac{n^0_{vi}}{\tau_{vi}}\alpha_{vi} - \dfrac{n_{vi}(x)}{\tau_{vi}} - K_i n_{ai}(x) n_{vi}(x) + K_i n^0_{ai} n^0_{vi} \\ j_{ai\uparrow}(x) = \dfrac{n^0_{ai}}{\tau_{ai}}\alpha_{ai} - \dfrac{n_{ai}(x)}{\tau_{ai}} - K_i n_{ai}(x) n_{vi}(x) + K_i n^0_{ai} n^0_{vi} \end{cases} \tag{2}$$

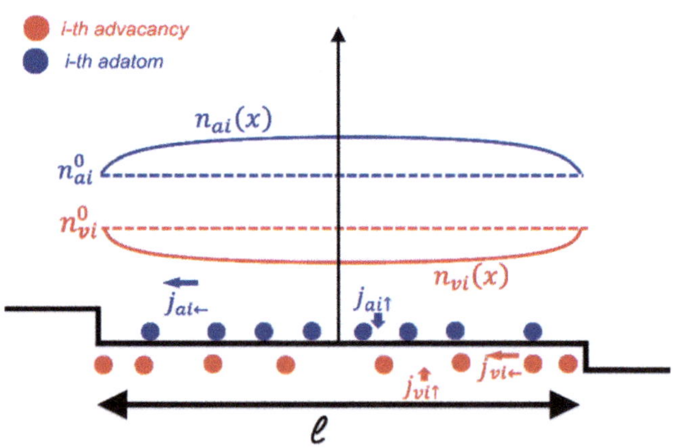

Fig. 3 Schematic representation of the mathematical problem being solved for the adatoms and advacancies of the i-th component. Surface fluxes $j_{vi\leftarrow}$, $j_{ai\leftarrow}$, as well as the fluxes from the gas phase $j_{ai\uparrow}(x)$ and bulk crystal $j_{vi\uparrow}(x)$, are shown.

Here, $n_{ai}(x)$ $n_{vi}(x)$ are the surface concentrations of adatoms (building units) and advacancies of the i-th type, respectively; n^0_{ai} and n^0_{vi} are their equilibrium values; τ_{ai} and τ_{vi} are the evaporation times of the adatom of the i-th type to the gas phase and corresponding advacancy to the bulk crystal. Let us also introduce an additional parameter $\xi_{ai} = P_i/P^0_i - 1$, where P_i is the partial pressure of the i-th component, P_{i0} is the equilibrium pressure, and ξ_{ai} is the supersaturation of the i-th component. Following BCF,[21] we will also introduce $\alpha_{ai} = P_i/P^0_i$ and $\alpha_{vi} = C_{vi}/C^0_{vi}$, where C_{vi} is the concentration of vacancies of the i-th type in the bulk crystal, and C^0_{vi} is its equilibrium value. We note that in the current case (case 1), we treat $\alpha_{vi} = 1$ because there is no excess of vacancies, but in case 3, we will use it and thus solve the problem in general. Furthermore, we will discuss in detail the origin and effect of α_{vi} and vacancy supersaturation ($\xi_{vi} = \alpha_{vi} - 1$) on the growth process. As in previous papers[32,55] on BCF growth, we consider that P_i and C_{vi} are constant over the crystal surface, whereas the surface concentrations $n_{ai}(x)$ and $n_{vi}(x)$ depend on the distance from the step.

The first term on the right side of the equations describes the deposition of the adatoms and advacancies of the i-th component on the terrace, the second is their evaporation to the gas phase and to the bulk crystal correspondingly, and K_i is the coefficient[50] that describes the recombination of adatoms and advacancies of the i-th type on the surface. The last term describes the generation of advacancies and adatoms of the i-th type on the terrace at equilibrium that, according to the principle of detailed balance, is equal to the recombination rate. Using the approach proposed in ref. 21, we also introduce functions $\varphi_i(x) = P_i/P^0_i - n_{ai}(x)/n^0_{ai}$ and $\psi_i(x) = C_{vi}/C^0_{vi} - n_{vi}(x)/n^0_{vi}$. Using them, one can write simple equations for the surface (subscript '←') fluxes of advacancies $j_{vi\leftarrow}(x)$ and adatoms $j_{ai\leftarrow}(x)$ in terms of ψ_i and φ_i:

$$\begin{cases} j_{vi\leftarrow}(x) = -D_{vi}\,\mathrm{grad}\,n_{vi}(x) = D_{vi}n^0_{vi}\,\mathrm{grad}\,\psi_i \\ j_{ai\leftarrow}(x) = -D_{ai}\,\mathrm{grad}\,n_{ai}(x) = D_{vi}n^0_{ai}\,\mathrm{grad}\,\varphi_i \end{cases} \quad (3)$$

where D_{ai} and D_{vi} are the surface diffusion coefficients of the adatoms and vacancies of i-th type, respectively. Then, considering that the step velocity is small and the distribution of adatoms and advacancies can be considered stationary and using the equation of continuity ($j_\uparrow = \mathrm{div}\,j_\leftarrow$), we may rewrite (2) and (3) as follows:

$$\begin{cases} D_{vi}\dfrac{\partial^2\psi_i}{\partial x^2} = \dfrac{\psi_i}{\tau_{vi}} - K_i n^0_{ai}(\alpha_{ai}\alpha_{vi} - \psi_i\alpha_{ai} - \alpha_{vi}\varphi_i - 1) \\ D_{ai}\dfrac{\partial^2\varphi_i}{\partial x^2} = \dfrac{\varphi_i}{\tau_{ai}} - K_i n^0_{vi}(\alpha_{ai}\alpha_{vi} - \psi_i\alpha_{ai} - \alpha_{vi}\varphi_i - 1) \end{cases} \quad (4)$$

We note that in deriving (4), we neglected the terms of the second order of magnitude ($\psi_i\varphi_i$). To solve the system for a set of equidistant steps with interstep distance l, one should set boundary conditions (see Fig. 3); right at the step ($x = l/2$), the concentrations of adatoms have their equilibrium values ($n_{vi}(l/2) = n^0_{vi}$, $n_{ai}(l/2) = n^0_{ai}$), which means that $\psi_i(l/2) = \xi_{vi}$, $\varphi_i(l/2) = \xi_{ai}$. Note, that we consider the simplest case and do not take into account the Schwoebel barrier and other effects near the step. At the centre of the terrace between the steps, the surface flux should be equal to zero because the distribution must be symmetric $\psi'_i(0) = 0$, $\varphi'_i(0) = 0$.

The solution of the system (see ESI‡) allows us to find the distribution functions of the advacancies and adatoms of each component:

$$\begin{cases} n_{vi}(x) = n_{vi}^{0}\left(\alpha_{vi} - \xi_{vi}\dfrac{\mathrm{ch}(x/\lambda_i)}{\mathrm{ch}\left(\dfrac{l}{2\lambda_i}\right)} - I_i n_{ai}^{0}\tau_{v}\left(\dfrac{\mathrm{ch}(x/\lambda_i)}{\mathrm{ch}\left(\dfrac{l}{2\lambda_i}\right)} - 1\right)\right) \\ n_{ai}(x) = n_{ai}^{0}\left(\alpha_{ai} - \xi_{ai}\dfrac{\mathrm{ch}(x/\lambda_i)}{\mathrm{ch}\left(\dfrac{l}{2\lambda_i}\right)} - I_i n_{vi}^{0}\tau_{a}\left(\dfrac{\mathrm{ch}(x/\lambda_i)}{\mathrm{ch}\left(\dfrac{l}{2\lambda_i}\right)} - 1\right)\right) \end{cases} \tag{5}$$

where

$$I_i = \frac{K_i(\alpha_{ai}\alpha_{vi} - 1)}{1 + K_i n_{vi}^{0}\alpha_{vi}\tau_{ai} + K_i n_{ai}^{0}\alpha_{ai}\tau_{vi}}, \lambda_i = \left(\frac{1}{\lambda_{ai}^{e\,2}} + \frac{1}{\lambda_{vi}^{e\,2}} + \frac{1}{\lambda_{ai}^{r\,2}} + \frac{1}{\lambda_{vi}^{r\,2}}\right)^{-1}$$

and λ_i is the characteristic length, which is determined by both the diffusion lengths of the vacancies and adatoms of the i-th component in all the processes considered (recombination and evaporation, see Fig. 4) and by definition is smaller than the smallest diffusion length of the corresponding component (see ESI‡).

The last term in the brackets of the right part of the equations is the interchange term. It describes the dependence of stationary concentrations of the adatoms and advacancies on each other due to recombination if their surface concentrations are slightly altered from equilibrium.

Knowing the distribution functions, one can calculate the diffusion fluxes of the vacancies and adatoms toward the step.

$$\begin{cases} j_{ai\leftarrow}\Big|_{x=\frac{l}{2}} = \dfrac{D_{ai}n_{ai}^{0}}{\lambda_i}\left(\xi_{ai} - I_i n_{vi}^{0}\tau_{ai}\right)\mathrm{th}\left(\dfrac{l}{2\lambda_i}\right) \\ j_{vi\leftarrow}\Big|_{x=\frac{l}{2}} = \dfrac{D_{vi}n_{vi}^{0}}{\lambda_i}\left(\xi_{vi} - I_i n_{ai}^{0}\tau_{vi}\right)\mathrm{th}\left(\dfrac{l}{2\lambda_i}\right) \end{cases}.$$

Now, we should consider that a stoichiometric relationship must be preserved between the fluxes right at the step in the stationary process of growth. Using the approach proposed in ref. 32 and a chain of transformations (see ESI‡), we

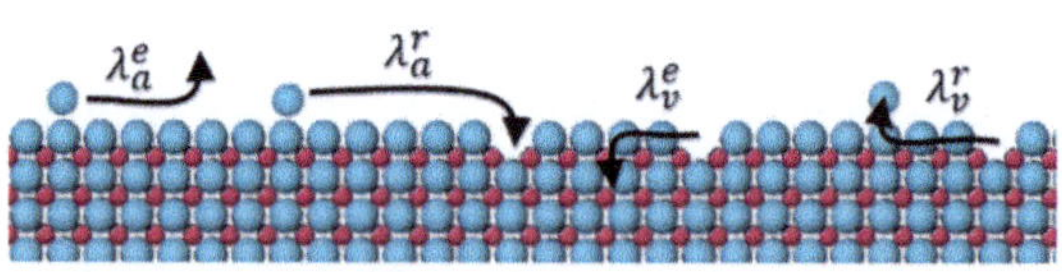

Fig. 4 Schematic representation of the diffusion lengths of the adatoms of the i-th component due to evaporation λ_a^e and recombination λ_a^r, and the same for the advacancies.

develop simple equations for the total fluxes of the advacancies $j_{v\leftarrow}$ and adatoms $j_{a\leftarrow}$ toward the step:

$$\begin{cases} j_{a\leftarrow} = D_a^{os}(\xi_a - I_a) \\ j_{v\leftarrow} = D_v^{os}(\xi_v - I_v) \end{cases}$$

where the following notation is used:

$$D_a^{os} = \left(\sum_{i=1}^{L} \frac{\lambda_i \nu_i^2}{D_{ai} n_{ai}^0 \text{th}\left(\frac{l}{2\lambda_i}\right)}\right)^{-1}, D_v^{os} = \left(\sum_{i=1}^{L} \frac{\lambda_i \nu_i^2}{D_{vi} n_{vi}^0 \text{ th}\left(\frac{l}{2\lambda_i}\right)}\right)^{-1}.$$

D_a^{os} and D_v^{os} are the generalised surface diffusion coefficients of the adatoms and advacancies, and $\xi_a = \sum_{i=1}^{L} \nu_i \xi_{ai}$ and $\xi_v = \sum_{i=1}^{L} \nu_i \xi_{vi}$ are the generalised supersaturations of the adatoms and advacancies, correspondingly. Generalised interchange terms are also introduced:

$$I_v = \sum_{i=1}^{L} I_i \nu_i n_{ai}^0 \tau_{vi}, I_a = \sum_{i=1}^{L} I_i \nu_i n_{vi}^0 \tau_{ai} \tag{6}$$

We note that the generalised coefficients D_a^{os} and D_v^{os} are, by definition, determined by the components for which the values of $D_{ai} n_{ai}^0 \text{th}(l/2\lambda_i)/\lambda_i \nu_i^2$ and $D_{vi} n_{vi}^0 \text{th}(l/2\lambda_i)/\lambda_i \nu_i^2$ are minimal – that is, by the slowest components (low diffusion coefficients), or components that are sparse on the surface (low concentration).

Knowing the fluxes, calculating the rate of advancement ϑ^{gr} of a group of equidistant steps at interstep distance l, which depends both on the influx of adatoms and outflux of vacancies, is straightforward:[32]

$$\vartheta^{gr} = \frac{2(j_{a\leftarrow} - j_{v\leftarrow})}{n_s} = \frac{2}{n_s}\left(D_a^{os}(\xi_a - I_a) - D_v^{os}(\xi_v - I_v)\right). \tag{7}$$

Here, $\frac{1}{n_s}$ is the surface area of the single crystalline cell in the step. We note that eqn (7) is derived for the general case when both supersaturations of adatoms and advacancies are present. In this case (case 1), we treat $\xi_v = 0$, and thus the step rate of advancement depends only on the supersaturation in the gas phase (ξ_a) and interchange terms I_a and I_v:

$$\vartheta^{gr} = \frac{2}{n_s}\left(D_a^{os}(\xi_a - I_a) + D_v^{os} I_v\right). \tag{8}$$

Now, knowing the rate of advancement of a group of equidistant steps in such a system, it is easy to calculate the spiral growth rate R:

$$R = \frac{n_s \omega \vartheta^{gr}}{19\rho_c} = \frac{2\omega}{19\rho_c}\left(D_a^{os}(\xi_a - I_a) + D_v^{os} I_v\right) \tag{9}$$

where ω is the volume of the crystalline cell and the interstep distance is $l = 19\rho_c$. $\rho_c = \gamma\omega/k_B T \xi_a$ is the critical radius of the 2D multicomponent nuclei on the surface,[61] and γ is the surface energy of the multicomponent crystal. Note that coefficients D_a^{os} and D_v^{os} also depend on the interstep distance, and thus growth rate R (eqn (9)) has a rather complex dependence on the supersaturation.

3.2 Growth due to chemical reactions

The second important case is the growth of a crystal due to a chemical reaction. The general model for such growth was considered in ref. 32. The reaction can be written as follows:

$$\nu_1 A_{1(g)} + \nu_2 A_{2(g)} + \ldots + \nu_N A_{N(g)} \leftrightarrow C_{(s)} + \nu_{N+1} B_{1(g)} + \ldots + \nu_{N+M} B_{M(g)} \quad (10)$$

It is very similar to (1), but $A_{i(g)}$ here are initial chemical reagents instead of crystal components (building units), $B_{i(g)}$ are reaction products, and ν_i are now the stoichiometric coefficients of the chemical reaction (10). N is the number of initial reagents and M is the number of reaction products. The solid crystal $C_{(s)}$ still consists of L different types of atoms. An example is the MOCVD or HVPE of gallium nitride.[60] In this case, as was shown in ref. 32, we may consider the process of crystal growth as follows. Each of the L separate components (building units) is being delivered onto the crystal surface by some chemical subreaction of (10) (see Fig. 5).

$$\begin{cases} A_{11} + \ldots + A_{1N_1} \leftrightarrow C_{1(ads)} + B_{11} + \ldots + B_{1M_1} \\ A_{21} + \ldots + A_{2N_2} \leftrightarrow C_{2(ads)} + B_{21} + \ldots + B_{2M_2} \\ \ldots \\ A_{L1} + \ldots + A_{LN_L} \leftrightarrow C_{L(ads)} + B_{L1} + \ldots + B_{LM_L} \end{cases} \quad (11)$$

where $C_{i(ads)}$ denotes the adatom (building unit) of the crystal of the i-th type. N_i and M_i are the total number of reagents and products of the i-th subreaction, respectively. As in ref. 32, we omitted the stoichiometric coefficients of these subreactions for simplicity; however, in the general case, they may differ from unity. An example of such decomposition for GaN halide vapour phase epitaxy is presented in the ESI.‡ A_{ij} and B_{ij} may be intermediate reagents and may differ from any of the initial reagents $A_{i(g)}$ or products $B_{i(g)}$ of the reaction (10). Each subreaction may involve an arbitrary number of subreagents. Products of some subreactions may be initial reagents for other reactions.

Note that decomposition of the main reaction into subreactions is needed mainly to find the kinetic parameters of adatom deposition and desorption at the

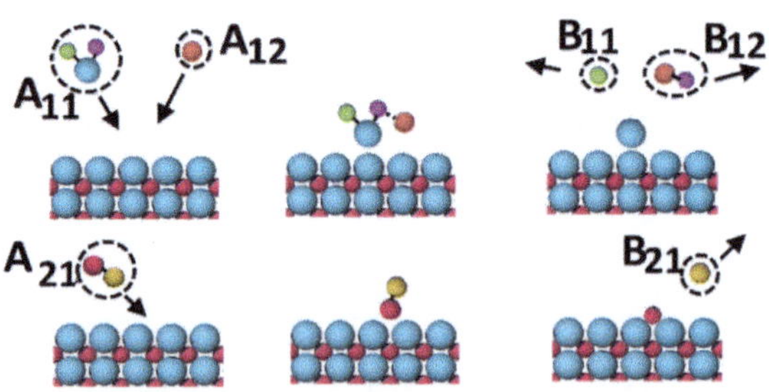

Fig. 5 Schematic representation of the growth process of a two-component crystal at which the adatoms arrive due to chemical subreactions (eqn (11)).

surface. Such decomposition will formally allow us to find the supersaturation of adatoms of each component, their lifetimes before evaporation (determined by the reverse chemical reaction) and their diffusion lengths (see details in ref. 32 and an example in the ESI‡). In the general case, these parameters can be written as follows:

$$\begin{cases} \xi_{\mathrm{a}1} = \dfrac{K_1}{K_1^0} - 1; \tau_{\mathrm{a}1} = \dfrac{1}{k_{\mathrm{r}1}\prod\limits_{k=1}^{M_1} \mathrm{B}_{1k}}; \lambda_{\mathrm{a}2}^{\mathrm{e}\ 2} = \dfrac{D_{\mathrm{a}1}}{k_{\mathrm{r}1}\prod\limits_{k=1}^{M_1} \mathrm{B}_{1k}} \\ \xi_{\mathrm{a}2} = \dfrac{K_2}{K_2^0} - 1; \tau_{\mathrm{a}2} = \dfrac{1}{k_{\mathrm{r}2}\prod\limits_{k=1}^{M_2} \mathrm{B}_{2k}}; \lambda_{\mathrm{a}2}^{\mathrm{e}\ 2} = \dfrac{D_{\mathrm{a}2}}{k_{\mathrm{r}2}\prod\limits_{k=1}^{M_2} \mathrm{B}_{2k}} \\ \cdots\cdots\cdots \\ \xi_{\mathrm{a}L} = \dfrac{K_L}{K_L^0} - 1; \tau_{\mathrm{a}L} = \dfrac{1}{k_{\mathrm{r}L}\prod\limits_{k=1}^{M_L} \mathrm{B}_{Lk}}; \lambda_{\mathrm{a}2}^{\mathrm{e}\ 2} = \dfrac{D_{\mathrm{a}L}}{k_{\mathrm{r}L}\prod\limits_{k=1}^{M_L} \mathrm{B}_{Lk}} \end{cases} \tag{12}$$

where $K_i = \prod\limits_{k=1}^{N_i} \mathrm{A}_{ik} \Big/ \prod\limits_{k=1}^{M_i} \mathrm{B}_{ik}$, K_i^0 is the equilibrium constant of the i-th subreaction from (11) and $k_{\mathrm{r}i} n_{\mathrm{a}i}^0$ is the rate constant of the i-th reverse chemical subreaction at equilibrium. Knowing the lifetimes and supersaturations of the adatoms (12) and keeping in mind that nothing has changed for the advacancies with respect to Section 3.1, we may perform the same chain of transformations as in the previous section using the adatom parameters from (12). The final expression for the rate of advancement of the steps and growth rate of the crystal will be exactly the same as in (8) and (9), but with redefined values of $D_{\mathrm{a}}^{\mathrm{os}}$, ξ_{a}, I_{v} and I_{a}, based on parameters (12).

Let us analyse what changes for I_{v} and I_{a} in the case of growth due to a chemical reaction. The main impact, as can be seen from eqn (6) and (12), is that the chemical reaction determines the lifetimes of the adatoms. Even at fixed supersaturations, the higher pressure of the chemical reagents leads to faster chemical reactions and thus lower residence times. This lowers the effect of recombination because the adatoms will probably be evaporated due to the chemical reactions rather than recombination. In the opposite case, during growth at lower pressures, the chemical reaction rate is also low, and thus recombination with the advacancies is more probable and terms I_{v} and I_{a} are larger. We note that at the lowest pressures, when the thermal evaporation is comparable with the chemical reaction rates, the proposed model is not correct.

3.3 Effect of mechanical stresses

Let us now consider how the presence of elastic stresses affects the growth rate of a multicomponent crystal *via* the considered BCF mechanism accounting for the advacancies. In ref. 27 and 29, it was shown that, in the single-component case, the rate of advancement of the straight step (ϑ^{gr}) depends on the local stress as follows:

$$\vartheta^{\mathrm{gr}}(\sigma) = \vartheta^{\mathrm{gr}}(0)\left(1 - \frac{\sigma\omega}{k_{\mathrm{B}}T\xi}\right)$$

where σ is the isotropic elastic stress, ω is the atomic volume, and ξ is the single-component supersaturation. From the thermodynamic point of view, the chemical potential of the atom in an elastically strained crystal increases, and as a consequence, a larger concentration of adatoms is required on the terrace near the step to balance the flux of adatoms evaporating from the step to the terrace per unit of time. The surface equilibrium concentration of adatoms near the step is thus increased by a factor of $(1 + \sigma\omega/k_BT)$, whereas the adatom concentration far from the step remains equal to its initial value without stress n_0. From the kinetic point of view, this causes the appearance of a diffusion flux of adatoms that is directed from the step toward the terrace. An excess of adatoms evaporates from the terrace into the gas phase. Thus, if the crystal that was initially at equilibrium conditions with the gas phase is subjected to mechanical stresses σ, it will begin to dissolve as if it were at undersaturation of $\xi = -\sigma\omega/k_BT$. The step will acquire a negative rate of advancement proportional to the difference in chemical potentials. The supersaturation in the gas phase must be increased up to $\xi = \sigma\omega/k_BT$ to balance the system. Then, the new concentration of adatoms on the terrace will correspond to the equilibrium concentration at the step, and the diffusion outflux of adatoms from the step will stop.

Note that the system in ref. 29 was the step under the influence of the local stress created by a dislocation, and thus the authors considered the advancement of the single step, not their group. The latter was caused by the fact that a dislocation creates a non-uniform stress field, and thus consideration of the group of steps becomes a complicated task. In our case, we consider the uniform stress σ and for simplicity did not take into account the stresses caused by the dislocation itself. Carrying out a similar chain of reasoning as in ref. 29 and using approaches from ref. 32 for a multicomponent system, one can obtain a completely analogous equation for the rate of advancement of a group of equidistant steps:

$$\vartheta^{gr}(\xi_a, \sigma) = \frac{2}{n_s}\left(D_a^{os}\left(\xi_a - \frac{\sigma\omega}{k_BT}\right)\right) \tag{13}$$

where $\omega = \sum_{i=1}^{L} \nu_i\omega_i$ is, as before, the total volume of the crystalline cell of the multicomponent crystal, $\xi_a = \sum_{i=1}^{L} \nu_i\xi_{ai}$ is the multicomponent supersaturation in the absence of stress and D_a^{os} is the generalised diffusion coefficient. We note that this equation does not consider the effect of advacancies. For this, we may use the approach from Section 3.1 and eqn (7), which considers vacancy supersaturation ξ_v. As shown in ref. 56, the applied mechanical stress increases the bulk concentration of vacancies of i-th type C_{vi} relative to its equilibrium value C_{vi}^0 according to the equation:

$$\frac{C_{vi}}{C_{vi}^0} = \exp\left(\frac{\sigma\omega_i}{k_BT}\right) \approx 1 + \frac{\sigma\omega_i}{k_BT}$$

where σ is positive when tensile. We note that the source of the vacancies might be dislocations or other defects in the crystal and we treat them as effective enough to provide increased concentration in the bulk. Thus, the vacancy supersaturation arises, which equals $\xi_{vi} = C_{vi}/C_{vi}^0 - 1 = \sigma\omega_i/k_BT$. The total vacancy supersaturation is then $\xi_v = \sum_{i=1}^{L} \nu_i\xi_i \approx \sigma\omega/k_BT$. Substituting this value into eqn (7),

Table 1 Comparison of expressions for multicomponent crystal spiral growth rate for different cases

Growth process	R – rate of spiral growth	Parameters
(1) Growth from vapours in classical BCF (diffusion-limited case) from ref. 32	$\frac{2\omega}{19\rho_c} D_a^{os} \xi_a$	$D_a^{os} = \left(\sum_{i=1}^{L} \frac{\nu_i^2 \lambda_i}{D_i n_{ai}^0 \mathrm{th}\left(\frac{19\rho_c}{2\lambda_i}\right)} \right)^{-1}$ $\xi_a = \sum_{i=1}^{L} \nu_i \xi_{ai}, \xi_{ai} = \frac{P_i}{P_{i0}} - 1, \lambda_i^2 = D_i \tau_i$
(2) Growth from vapours, with advacancies	$R = \frac{2\omega}{19\rho_c} (D_a^{os}(\xi_a - I_a) + D_v^{os} I_v)$	$D_v^{os} = \left(\sum_{i=1}^{L} \frac{\lambda_i \nu_i^2}{D_{vi} n_{vi}^0 \ \mathrm{th}\left(\frac{19\rho_c}{2\lambda_i}\right)} \right)^{-1}$ $I_v = \sum_{i=1}^{L} I_i \nu_i n_{ai}^0 \tau_{vi}, I_a = \sum_{i=1}^{L} I_i \nu_i n_{vi}^0 \tau_{ai}$ $\lambda_i^2 = \left(\frac{1}{\lambda_{ai}^{e\,2}} + \frac{1}{\lambda_{vi}^{e\,2}} + \frac{1}{\lambda_{ai}^{r\,2}} + \frac{1}{\lambda_{vi}^{r\,2}} \right)^{-1}$
(3) Growth due to chemical reaction, with advacancies	$R = \frac{2\omega}{19\rho_c} (D_a^{os}(\xi_a - I_a) + D_v^{os} I_v)$	$\xi_{ai} = \frac{\prod_{k=1}^{N_i} A_{ik}}{\prod_{k=1}^{M_i} B_{ik}} - 1, \lambda_i^2$–see text
(4) Growth from vapours, with advacancies and applied stress σ	$\frac{2\omega}{19\rho_c} (D_a^{os}(\xi_a(\sigma) - I_a) - D_v^{os}(\xi_v(\sigma) - I_v))$	$\xi_a(\sigma) = \xi_a - \frac{\sigma\omega}{k_B T}, \xi_v(\sigma) = \frac{\sigma\omega}{k_B T}$

obtained in the previous part of the paper, and combining it with eqn (13), we develop an equation for the rate of advancement of the multicomponent step in the presence of mechanical stress, which considers the recombination of vacancies and adatoms:

$$\vartheta^{gr}(\sigma) = \frac{2}{n_s}\left(D_a^{os}\left(\xi_a - \frac{\sigma\omega}{k_B T} - I_a\right) - D_v^{os}\left(\frac{\sigma\omega}{k_B T} - I_v\right)\right).$$

We note that the interchange terms I_a and I_v (see eqn (6)) are also dependent on supersaturations of the vacancies and adatoms and thus on the presence of stress. However, in the first order of magnitude of $\sigma\omega/k_B T$, this dependence can be neglected.

Finally, we may write the equation for the spiral growth rate of the crystal in the case considered:

$$R(\sigma) = \frac{n_s \omega \vartheta^{gr}(\sigma)}{19\rho_c} = \frac{2\omega}{19\rho_c}\left(D_a^{os}\left(\xi_a - \frac{\sigma\omega}{k_B T} - I_a\right) - D_v^{os}\left(\frac{\sigma\omega}{k_B T} - I_v\right)\right). \tag{14}$$

This expression is also valid for the case of growth due to chemical reaction, but D_a^{os} and ξ_a depend on the pressures of the chemical reagents (see Section 3.2) in this case, and the reverse reaction rate governs diffusion lengths.

4. Discussion

In the previous sections, we found the growth rates of the multicomponent crystal considering the presence of advacancies and their recombination with adatoms in the following cases: growth from vapours, growth due to chemical reactions and growth in the presence of elastic stress. The resulting growth rates and parameters used are summarised in Table 1. The equation for the growth rate of the multicomponent crystal *via* classical BCF (without advacancies and recombination) from ref. 32 is also shown.

Let us analyse these cases in more detail. The first case (Table 1, row 2) is the growth from vapours when adatoms are being deposited onto the surface right from the gas phase. The expression for the growth rate contains the interchange terms I_a and I_v. These terms are dependent on the recombination rate and surface concentrations of the adatoms and advacancies. If recombination is very low ($n_{vi}^0 \to 0$ or $K_i \to 0$), and it can be neglected, the resulting equation translates into the equation obtained in ref. 32 for the description of multicomponent crystal growth from vapours without advacancies (row 1). Then, if only one component is present in the system ($L = 1$), the resulting equation coincides with the one obtained in the classical papers[21] of BCF for the single-component case and its modifications.[59] We note that the interchange term for adatoms I_a is dependent on their lifetime and the equilibrium concentrations of the opposite species – advacancies, which is logical. The larger the concentration of advacancies, the higher the recombination rate and the lower the concentration of adatoms at the terrace far from the step. The larger the lifetime of the adatom, the higher the probability that it will recombine with the advacancy. This reduces the diffusive flux of the adatoms toward the step and its rate of advancement (and thus I_a comes with the ‘minus’ sign in eqn (8)). However, the situation is the same for the advacancies. A high recombination rate also decreases their concentration at the terrace, and thus the advacancy flux arises from the step toward the terrace, which positively affects the step rate of advancement (I_v comes with the 'plus' sign in eqn (8)). At zero supersaturations, the interchange terms also turn to zero, as expected. We note that the recombination in the presence of excess adatoms, as was shown earlier, leads to a decrease in the concentration of advacancies on the surface relative to the equilibrium. Therefore, a constant flux of vacancies from the bulk toward the surface (and thus of adatoms to the bulk crystal) should also appear proportional to I_v. This flux additionally leads to the growth of a crystal, even in the absence of the steps or screw dislocations. This is in an additional effect caused by the presence of advacancies not related to spiral or step-flow growth that will be studied and compared with spiral growth in future reports.

We also note that multicomponent crystals and MOFs may consist of many types of atoms ($L \gg 1$) or building units and have very complicated structures. Some of these atoms or units are buried in the 'depths' of the crystalline cell, covered by other types of atoms, and thus do not come out to the crystal surface (see Fig. 1b). Recombination of such atoms is much less probable due to additional barriers because the adatoms must diffuse through the upper layers of the crystalline cell toward the vacancy.

The second case is the growth of the multicomponent crystal due to a chemical reaction (row 3). The main difference from the previous case, as was shown in ref. 32 and in Section 3.2, is that the diffusion lengths of the adatoms become dependent on the absolute pressures of the reaction products. This gives an additional control for the growth process. One can selectively change the diffusion lengths of the adatoms of certain components by adjusting the partial pressures of the corresponding chemical reagents, even at fixed supersaturations. This allows the controlling of D_a^{os} and the kinetics of the growth. The higher pressures weaken the effects of recombination because the adatoms evaporate

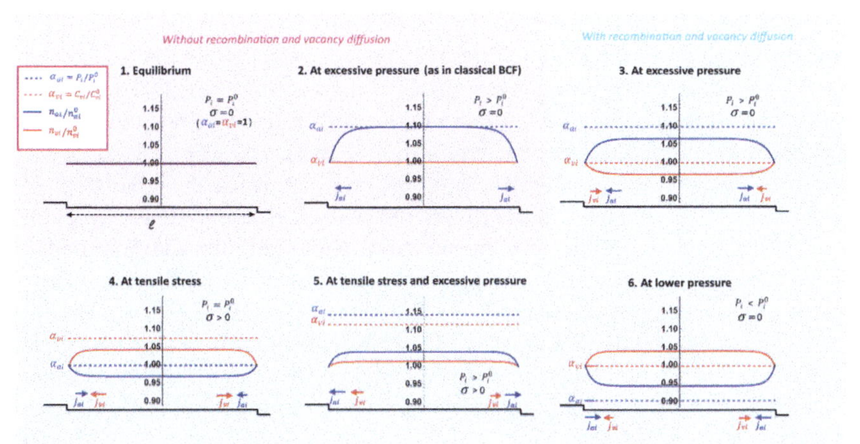

Fig. 6 Qualitative comparison of the distribution functions of the adatoms n_{ai} (blue) and advacancies n_{vi} (red) of the i-th component normalized to the equilibrium concentrations n_{ai}^0 and n_{vi}^0 at different growth conditions. The directions of the corresponding surface fluxes j_{ai} and j_{vi} near the steps are also shown.

due to a chemical reaction rather than recombine. The lower pressures lead to an increased residence time of adatoms, a larger interchange term I_a, and greater recombination effects.

In the third case, when the multicomponent crystal grows in the presence of applied stress, the latter plays a double role. First, it increases the chemical potential of atoms in the crystal, thereby reducing effective multicomponent supersaturation in the gas phase: $\xi_a(\sigma) = \xi_a(0) - \sigma\omega/k_BT$. Second, the applied stress leads to a change in the vacancy and advacancy concentrations. In the case of tensile stresses, the advacancy concentration increases and, therefore, recombination also increases. In the case of compressive stress, in contrast, the concentration of advacancies reduces and recombination decreases. Let us note that the expression for the growth rate in the presence of stress is not symmetric with respect to vacancies and adatoms. This is because we neglected that the increased pressure of the components also affects the vacancy concentration. Indeed, the increased gas pressure is equivalent to the applied compressive stress and thus, in the formula (row 4), the vacancy supersaturation ξ_v also should be reduced proportionally due to the increase in adatom supersaturation ξ_a. But in the real growth processes of heterostructures, the stress arising due to the lattice and thermal expansion coefficient mismatch is usually much higher (tens or hundreds of MPa) than the gas pressure in growth chamber (up to 100 kPa at HVPE), and thus we neglect this effect.

Fig. 6 qualitatively demonstrates the distribution of the adatoms and advacancies of one of the components (for simplicity) and the surface fluxes j_{ai} and j_{vi} on the terrace in several different cases to make it easier for the reader to grasp the process.

Although the importance and significance of the impact of vacancies has been already shown in ref. 52 and 53, let us also estimate it on the gallium nitride example using the approach of Pimpinelli *et al.*[50] Earlier, we have shown (see Section 3.1) that the process itself, fluxes and distribution of adatoms are dependent on λ_i. λ_i is the characteristic length determined by all the diffusion lengths of the vacancies and adatoms on the surface (see Fig. 4), and, by definition, it is less than the smallest length in the system.

For definiteness, we will analyse the situation when GaN grows in an excess of nitrogen, and thus the growth is governed by the adatoms of gallium. Let us compare the diffusion lengths of gallium due to recombination and due to evaporation. We should know the equilibrium surface concentrations of Ga adatoms n^0_{aGa} and vacancies n^0_{vGa} and their lifetimes. Let us use the conditions of the experiment,[62] at which the MBE growth of GaN was performed at a temperature of 730 °C. The equilibrium concentration of the gallium adatoms at the surface thus can be estimated[33] as $n^0_{aGa} = 4 \times 10^7$ at cm^{-2}, the lifetime of the adatoms[63] equals $\tau_{aGa} \sim 10^{-7}$ s, and their diffusion coefficient is $D_{Ga} \sim 2 \times 10^{-8}$ m^2 s^{-1}. The diffusion length due to evaporation can then be estimated as $\lambda^e_{Ga} = \sqrt{D_{Ga}\tau_{aGa}} \sim 45$ nm. To estimate the vacancy concentration, we will use data from ref. 64, where the authors, using the positron annihilation method, have shown that the bulk concentration of the Ga vacancies depends on the growth conditions and doping and may substantially vary in a range of $C^0_{vGa} = 10^{16}$–10^{19} at cm^{-3}. We should note that such a bulk concentration of vacancies in gallium nitride is quite high, and C_{v0} is usually less in the growth of non-doped GaN. However, for the case of such doping, the surface concentration of vacancies can then be lower-estimated as $n^0_{\mathrm{vGa}} = {C^0_{\mathrm{vGa}}}^{2/3} \sim 10^{11}$ at cm^{-2}. We note that the real concentration of advacancies should be even higher because an advacancy has a lower energy state than a bulk vacancy (fewer bonds are broken in the case of an advacancy). Comparing n^0_{aGa} and n^0_{vGa}, one can note that the surface concentration of gallium advacancies is higher than the surface concentration of adatoms. Next, the adatom diffusion length due to recombination $\lambda^{\mathrm{r}}_{\mathrm{aGa}}$ can be estimated as the square root of the reciprocal of the respective surface densities of the adatoms and advacancies, $\lambda^{\mathrm{r}}_{\mathrm{aGa}} \sim 20$ nm. For the recombination time, one can estimate $\tau^{\mathrm{r}}_{\mathrm{aGa}} \sim {\lambda^{\mathrm{r}}_{\mathrm{aGa}}}^2/D_{\mathrm{Ga}} \sim 3 \times 10^{-8}$ s. Finally, estimates have shown that the gallium diffusion length due to recombination in the example considered is at least of the order of the diffusion length due to evaporation (or even shorter), and thus recombination may affect the growth rate in the manner described in this paper.

We note that the flux of the adatoms is inversely proportional (see Table 1) to the shortest diffusion length, and thus if $\lambda^{\mathrm{r}}_{\mathrm{aGa}}$ is twice less than $\lambda^{\mathrm{e}}_{\mathrm{aGa}}$ (as in our estimates), then the growth rate will be at least twice that given by the classical BCF theory. Thus, the effect cannot be neglected during consideration of the crystal growth at higher temperatures when the concentration of advacancies is high or when some dopants are present, which also may increase the vacancy concentration.

Let us also discuss the applicability of the results obtained and the effect of advacancies in different systems. Above, we considered the simplest case of crystal growth from vapours or molecular fluxes, which are mainly used in the epitaxial growth of semiconductor thin films and heterostructures (MBE, MOCVD, *etc.*). To describe the growth and the effect of recombination we used the classical BCF model of surface diffusion of adatoms towards the steps and constant pressures over the whole surface. However, in other growth processes, such as growth from solutions, other mechanisms can play an important role:

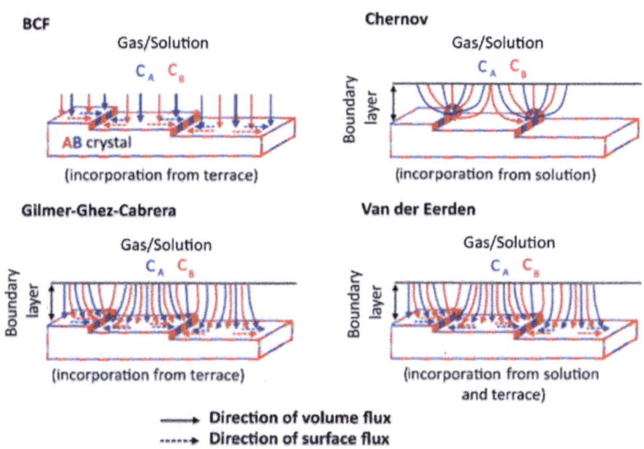

Fig. 7 Schematic representation of the growth mechanisms of a two-component crystal AB. In the BCF model, only surface diffusion and incorporation of adatoms into the steps are taken into account (there is no depletion and no direct incorporation of atoms in the gas phase), whereas other mechanisms consider volume diffusion in the boundary layer and incorporation of atoms into the step only from the gas/solution (Chernov[22]), or only from the terrace (GGC[65]) or from both of them (VE[66]). C_{A} and C_{B} are the concentrations of the components in the gas/solution.

Chernov,[22] Gilmer–Ghez–Cabrera[65] (GGC) or Van der Eerden[66] (VE). These mechanisms consider the boundary layer in which volume diffusion of the species takes place and limits the growth. The main difference between these mechanisms is summarized in Fig. 7. In the Chernov mechanism, only the volume diffusion of atoms proceeds in the boundary layer, and their incorporation proceeds directly from the solution into the steps. The GGC mechanism considers both surface and volume diffusion but incorporation into the steps proceeds indirectly, only through the adsorbed state on the terrace (not from the gas or solution). Finally, the most complex and general VE mechanism describes both types of diffusion (bulk and surface), as well as the direct incorporation of atoms from the gas or solution into the step and indirect incorporation of adatoms from the terrace.

Considering these mechanisms, it can be concluded that the mass transfer due to surface fluxes plays the biggest role in the BCF mechanism, whereas in the Chernov mechanism there is no adatom surface flux at all, and therefore there is no advacancy back-flux. In this sense, the mechanisms can be sorted by the expected impact of recombination and advacancy diffusion on the step advancement and crystal growth rate in the following descending order: BCF, GGC, VE, Chernov. This corresponds to their gradual transition from complete surface diffusion (BCF) to complete volume diffusion (Chernov). Considering the above mechanisms, the proposed effect of the advacancies should be the most pronounced with the growth of crystals at high temperature from neat undiluted vapours (which usually proceeds in BCF mode at low supersaturations) in the case of crystals having a large adatom surface diffusion length and low vacancy formation energy: doped semiconductors, some metals[67,68] *etc.* On the contrary, advacancies and recombination should have a smaller effect when the crystal grows from a dilute solution (which proceeds in Chernov mode).

We also note that in an arbitrary multicomponent system, different components may incorporate *via* different mechanisms. For example, a component which is in excess in the gas phase may incorporate *via* the BCF mechanism (*via* surface diffusion), whereas others, which are in short supply, may incorporate *via* the Chernov, VE or GGC mechanisms. In this case, the effect of the advacancies will be greater for the component incorporating *via* surface diffusion and lower for the components incorporating into the step directly from the gas phase or solution.

We also considered the generation of vacancies in the bulk crystal effective enough to ensure their constant flux toward the surface. This can take place if there is a significant number of defects in the volume of the crystal. This assumption allowed us to consider the vacancies incorporating into the steps *via* the BCF mechanism. However, if there are only a few defects in the crystal bulk and the vacancy generation rate is low, a depletion of vacancies should be observed in the near-surface region. In this case, the incorporation of vacancies into the steps will proceed *via* one of the other mechanisms shown in Fig. 7 (for example, the GGC mechanism which takes into account volume diffusion) and thus additional calculations should be performed to find the final effect of the advacancies.

Finally, in the multicomponent growth process both from gas or solution, several combinations of incorporation mechanisms for different components and vacancies are possible, and to find the total growth rate the approach from ref. 55

and 32 may be used. First, one should derive the expressions for the fluxes of atoms and vacancies of each component toward the step *via* the respective mechanism (BCF, Chernov, GGC, VE, *etc.*), and second, one should impose the stoichiometric condition on the resulting fluxes right at the step. In Section 3.1, this was done for the case when all of the components and vacancies incorporate *via* the same BCF mechanism, but the approach itself allows one to find the expressions for an arbitrary combination of growth mechanisms and estimate the impact of the advacancies in each particular case of growth. Additional studies have to be performed in order to understand the role of advacancies in the growth from melts, which was not considered in this paper.

5. Conclusions

The role of advacancies and their recombination with adatoms in the growth of a multicomponent crystal *via* the Burton–Cabrera–Frank mechanism is studied in the cases of growth from a vapour of its components and growth due to chemical reactions. It is shown that the recombination of advacancies and adatoms may, on the one hand, reduce the total surface concentration of adatoms and their flux toward the step and, on the other hand, cause the flux of advacancies from the step toward the terrace. The dependence of the rate of advancement of the steps on the supersaturation of each component and their properties, diffusion coefficients, stoichiometric coefficient and equilibrium concentration is found analytically, as well as the spiral growth rate of the crystal. It is shown that the process is governed by the component that has the lowest product of equilibrium concentration and diffusion coefficient. The diffusion length of the components in such a system may be dependent both on the recombination and evaporation rates or due to the reverse chemical reaction rate (in the case of growth due to chemical reactions). Using an example of the growth of gallium nitride, it is shown that the effect is especially strong in cases where the crystal is grown under conditions that ensure a high concentration of vacancies of certain components (*i.e.*, at high doping levels). The impact of recombination on the growth process in the absence of steps is also discussed, which may cause crystal growth, even without nucleation or step-flow.

The effect of mechanical stresses on the growth process is studied and shown to both increase the equilibrium vapour pressure over the surface and also to affect the recombination, adatom and advacancy surface fluxes. Applied tensile stress increases the concentration of vacancies in the bulk crystal and the concentration of advacancies. This increases the impact of the advacancies on the total mass transfer toward/away from the step and may even cause dissolution of the crystal due to a higher influx of vacancies toward the step than the influx of adatoms. The compressive stress, in contrast, decreases the concentration of advacancies and lowers the impact of recombination on the total growth rate. The results obtained allow describing the growth of certain types of multicomponent crystals or MOFs due to chemical reactions in the presence of elastic stresses considering the effect of the advacancies. The theory proposed in this paper can be further developed taking into account the Ehrlich–Schwoebel barrier of both the adatoms and advacancies, backstress effect of the growth spiral, surface reconstruction and other effects known for single-component systems. The results can be used in the analysis and explanation of the extensive existing experimental data, selection of optimal growth conditions and for further development of crystal growth theory.

Author contributions

AR: conceptualization, formal analysis, project administration, funding acquisition, investigation, writing – review & editing, methodology. SA: supervision, investigation.

Conflicts of interest

There are no conflicts to declare.

Acknowledgements

A. V. Redkov thanks the Russian Science Foundation (grant no. 21-72-00032) for the support of this work. The authors thank the developers of the OVITO[69] program, which was used to prepare graphics.

References

1 J. W. Orton and C. T. Foxon, *Rep. Prog. Phys.*, 1998, **61**, 1.
2 S. N. Mohammad, A. A. Salvador and H. Morkoc, *Proc. IEEE*, 1995, **83**, 1306–1355.
3 *Process technology for silicon carbide devices*, ed. C. M. Zetterling, 2002.
4 M. T. Hörantner, T. Leijtens, M. E. Ziffer, G. E. Eperon, M. G. Christoforo, M. D. McGehee and H. J. Snaith, *ACS Energy Lett.*, 2017, **2**, 2506–2513.
5 D. Gets, D. Saranin, A. Ishteev, R. Haroldson, E. Danilovskiy, S. Makarov and A. Zakhidov, *Appl. Surf. Sci.*, 2019, **476**, 486–492.
6 Y. Zhao, A. M. Nardes and K. Zhu, *Faraday Discuss*, 2015, **176**, 301–312.
7 S. E. Alexandrov, K. S. Tyurikov, D. A. Kirilenko, A. V. Redkov and A. A. Lipovskii, *Adv. Mater. Interfaces*, 2017, **4**, 1700241.
8 A. Voloshin, E. Rudneva, V. Manomenova, N. Vasilyeva, S. Kovalev, G. Emelchenko and A. Zhokhov, *Crystals*, 2019, **9**, 390.
9 D. V. Semenok, I. A. Troyan, A. G. Ivanova, A. G. Kvashnin, I. A. Kruglov, M. Hanfland and et al, ., *Mater. Today*, 2021, **48**, 18–28.
10 Y. Y. Kim, K. Ganesan, P. Yang, A. N. Kulak, S. Borukhin, S. Pechook, L. Ribeiro, R. Kröger, S. Eichhorn, S. Armes, B. Pokroy and F. C. Meldrum, *Nat. Mater.*, 2011, **10**, 890–896.
11 M. S. Yao, K. I. Otake, Z. Q. Xue and S. Kitagawa, *Faraday Discuss.*, 2021, **231**, 397–417.
12 Y. Li, S. Tang, A. Yusov, J. Rose, A. N. Borrfors, C. T. Hu and M. D. Ward, *Nat. Commun.*, 2019, **10**, 1–7.
13 N. L. Rosi, J. Eckert, M. Eddaoudi, D. T. Vodak, J. Kim, M. O'Keeffe and O. M. Yaghi, *Science*, 2003, **300**, 1127–1129.
14 H. H. Wang, L. Hou, Y. Z. Li, C. Y. Jiang, Y. Y. Wang and Z. Zhu, *ACS Appl. Mater. Interfaces*, 2017, **9**, 17969–17976.
15 Y. Zhang, X. Feng, S. Yuan, J. Zhou and B. Wang, *Inorg. Chem. Front.*, 2016, **3**, 896–909.

16 J. Casaban, Y. Zhang, R. Pacheco, C. Coney, C. Holmes, E. Sutherland and et al, ., *Faraday Discuss.*, 2021, **231**, 312–325.
17 P. Cubillas, M. W. Anderson and M. P. Attfield, *Chem.–Eur. J.*, 2012, **18**, 15406–15415.
18 C. W. Zhao, Y. A. Li, X. R. Wang, G. J. Chen, Q. K. Liu, J. P. Ma and Y. B. Dong, *Chem. Commun.*, 2015, **51**, 15906–15909.
19 L. Chai, J. Pan, Y. Hu, J. Qian and M. Hong, *Small*, 2021, **17**, 2100607.
20 C. Liu, L. Lin, Q. Sun, J. Wang, R. Huang, W. Chen and et al, ., *Chem. Sci.*, 2020, **11**, 3680–3686.
21 W. K. Burton, N. Cabrera and F. C. Frank, *Philos. Trans. R. Soc., A*, 1951, **243**, 299–358.
22 A. A. Chernov, *Phys.-Usp.*, 1961, **4**, 116.
23 M. Załuska-Kotur, H. Popova and V. Tonchev, *Crystals*, 2021, **11**, 1135.
24 D. Kashchiev, *Nucleation*, Elsevier, 2000.
25 V. G. Dubrovskii and J. Grecenkov, *Cryst. Growth Des.*, 2015, **15**, 340–347.
26 N. Akutsu, *Cryst. Growth Des.*, 2019, **19**, 2970–2978.
27 B. Van der Hoek, J. P. Van Der Eerden, P. Bennema and I. Sunagawa, *J. Cryst. Growth*, 1982, **58**, 365–380.
28 I. Sunagawa and P. Bennema, *J. Cryst. Growth*, 1981, **53**, 490–504.
29 N. Cabrera and M. M. Levine, *Philos. Mag.*, 1956, **1**, 450–458.
30 K. Sangwal, *Prog. Cryst. Growth Charact. Mater.*, 1996, **32**, 3–43.
31 T. Akasaka, Y. Kobayashi and M. Kasu, *Appl. Phys. Lett.*, 2010, **97**, 141902.
32 A. V. Redkov and S. A. Kukushkin, *Cryst. Growth Des.*, 2020, **20**, 2590–2601.
33 A. V. Redkov and S. A. Kukushkin, *Cryst. Growth Des.*, 2021, **21**, 4914–4926.
34 Z. Dong, L. Zhang, S. Wang and L. Luo, *Nano Energy*, 2020, **70**, 104527.
35 F. Panciera, Z. Baraissov, G. Patriarche, V. G. Dubrovskii, F. Glas, L. Travers, U. Mirsaidov and J-C. Harmand, ., *Nano Lett.*, 2020, **20**, 1669–1675.
36 J. Ihli, D. C. Green, C. Lynch, M. A. Holden, P. A. Lee, S. Zhang, I. Robinson, S. Webb and F. C. Meldrum, *Angew. Chem., Int. Ed.*, 2019, **58**, 17328–17334.
37 M. Salvalaglio, C. Perego, F. Giberti, M. Mazzotti and M. Parrinello, *Proc. Natl. Acad. Sci.*, 2015, **112**, E6–E14.
38 M. W. Anderson, J. T. Gebbie-Rayet, A. R. Hill, N. Farida, M. P. Attfield, P. Cubillas, V. A. Blatov, D. M. Proserpio, D. Akporiaye, B. Arstad and J. D. Gale, *Nature*, 2017, **544**, 456–459.
39 K. E. Blow, D. Quigley and G. C. Sosso, arXiv preprint arXiv:2104.13104., 2021.
40 A. A. Chernov, *J. Cryst. Growth*, 2004, **264**, 499–518.
41 I. D. Matukov, D. S. Kalinin, M. V. Bogdanov, S. Y. Karpov, D. K. Ofengeim, M. S. Ramm, J. S. Barash, E. N. Mokhov, A. D. Roenkov, Y. A. Vodakov, M. G. Ramm, H. Helava and Y. N. Makarov, *J. Cryst. Growth*, 2004, **266**, 313–319.
42 T. Hanada, *Phys. Rev. Mater.*, 2019, **3**, 103404.
43 R. Cadoret, *J. Cryst. Growth*, 1999, **205**, 123–135.
44 S. A. Kukushkin, S. S. Sharofidinov, A. V. Osipov, A. V. Redkov, V. V. Kidalov, A. S. Grashchenko, I. P. Soshnikov and A. F. Dydenchuk, *ECS J. Solid State Sci. Technol.*, 2018, 7(9), P480.
45 M. Takata and A. Ookawa, *J. Cryst. Growth*, 1977, **42**, 35–40.
46 H. Pfeiffe and W. Haubenreiss, *Phys. Status Solidi B*, 1979, **96**, 287.
47 M. Gerald, *J. Chem. Phys.*, 1964, **40**, 683.
48 T. A. Cherepanova, *J. Cryst. Growth*, 1981, **52**, 319–326.

49 P. Cermelli and M. Jabbour, *Proc. R. Soc. London, Ser. A*, 2005, **461**, 3483–3504.
50 A. Pimpinelli and J. Villain, *Phys. A*, 1994, **204**, 521–542.
51 M. J. Rost, L. Jacobse and M. T. Koper, *Nat. Commun.*, 2019, **10**, 1–7.
52 S. V. Sitnikov, A. V. Latyshev and S. S. Kosolobov, *J. Cryst. Growth*, 2017, **457**, 196–201.
53 A. S. Petrov, S. V. Sitnikov, S. S. Kosolobov and A. V. Latyshev, *Semiconductors*, 2019, **53**, 434–438.
54 S. Kosolobov, *Sci. Rep.*, 2019, **9**, 1–6.
55 A. V. Redkov, S. A. Kukushkin and A. V. Osipov, *CrystEngComm*, 2020, **22**, 5280–5288.
56 A. V. Redkov, *Phys. Solid State*, 2019, **61**, 2385–2389.
57 A. V. Redkov, A. V. Osipov and S. A. Kukushkin, *Phys. Solid State*, 2015, **57**, 2524–2531.
58 A. V. Redkov, S. A. Kukushkin and A. V. Osipov, *J. Phys. C: Solid State Phys.*, 2018, **1124**, 022006.
59 A. V. Redkov, S. A. Kukushkin and A. V. Osipov, *J. Cryst. Growth*, 2020, **548**, 125845.
60 M. Bockowski, M. Iwinska, M. Amilusik, M. Fijalkowski, B. Lucznik and T. Sochacki, *Semicond. Sci. Technol.*, 2016, **31**, 093002.
61 S. A. Kukushkin and A. V. Osipov,*Phys.-Usp* 1998, **41**, 983.
62 C. Adelmann, J. Brault, D. Jalabert, P. Gentile, H. Mariette, G. Mula and B. Daudin, *J. Appl. Phys.*, 2002, **91**, 9638–9645.
63 S. A. Kukushkin, V. N. Bessolov, A. V. Osipov and A. V. Luk'yanov, *Phys. Solid State*, 2001, **43**, 2229–2233.
64 K. Saarinen, P. Seppälä, J. Oila, P. Hautojärvi, C. Corbel, O. Briot and R. L. Aulombard, *Appl. Phys. Lett.*, 1998, **73**, 3253–3255.
65 G. H. Gilmer, R. Ghez and N. Cabrera, *J. Cryst. Growth*, 1971, **8**, 79–93.
66 J. P. Van der Eerden, *J. Cryst. Growth*, 1982, **56**, 174–188.
67 Q. S. Mei and K. Lu, *Philos. Mag. Lett.*, 2008, **88**, 203–211.
68 A. Forty, *London, Edinburgh Dublin Philos. Mag. J. Sci.*, 1952, **43**(344), 949–957.
69 A. Stukowski, *Modell. Simul. Mater. Sci. Eng.*, 2010, **18**, 015012.

Faraday Discussions

ROYAL SOCIETY OF CHEMISTRY

DISCUSSIONS

Growing crystals by design: general discussion

Michael Anderson, Matthew Bennett, Ruel Cedeno, Marta K. Dudek, Kristen Fichthorn, Aaron R. Finney, Ian Ford, Colin Freeman, Alan Hare, Connor Hewson, Adam Hill, Joonsoo Kim, Christine Kirschhock, Christian Kuttner, Fiona Meldrum, Sten O. Nilsson Lill, Rachel Pooley, Ivo B. Rietveld, Jeffrey Rimer, Kevin Roberts, Jutta Rogal, Matteo Salvalaglio, Jan Sefcik, Wenhao Sun, Damien Thompson, Jincheng Tong, Mollie Trueman and Peter Vekilov

DOI: 10.1039/d2fd90022j

Wenhao Sun opened the discussion of the paper by Kristen Fichthorn: Kristen, I really enjoyed reading your paper (https://doi.org/10.1039/d1fd00091h). To me, it suggests a theoretical pathway to computing the kinetic Wulff shape, which often is not the same as the thermodynamic Wulff shape. Unfortunately, as you show, your calculation is very expensive. I'm wondering if there are any heuristics or faster ways to reach a kinetic Wulff shape, perhaps from DFT-computable parameters?

Kristen Fichthorn responded: Kinetics is so system-dependent. I'm not certain there is an easy universal approach to the kinetic Wulff shape. In this work, we used DFT-computable parameters to get at linear-facet growth rates (though we did not explicitly mention this in the current paper, we did mention it in the past[1,2]). This could be done with considerable computational expense. Is there a faster way? In a past study, we could show (with the support of experiments) that it was the case that one facet was completely covered by a self-assembled monolayer of capping agent, while the other major facet present during growth did not have surfactant for the same solution-phase chemical potential. From there, we could conclude that solution-phase ion adsorption occurred only on the one facet, but not on the other, leading to certain experimentally observed kinetic shapes.[3,4] These DFT calculations were relatively easy, but the result was not "automatic".

1 J. Kim, J. Cui and K. A. Fichthorn, *ACS Nano*, 2021, **15**, 18279–18288.
2 X. Qi, Z. Chen, T. Yan and K. A. Fichthorn, *ACS Nano*, 2019, **13**, 4647–4656.
3 M. J. Kim, S. Alvarez, Z. Chen, K. A. Fichthorn and B. J. Wiley, *J. Am. Chem. Soc.*, 2018, **140**, 14740–14746.
4 K. A. Fichthorn and Z. Chen, *J. Vac. Sci. Technol. A*, 2020, **38**, 023210.

Wenhao Sun added: I just had a thought about the kinetic Wulff shape; different facets ought to have different solubility, we should be able to compute this. The chemical potential is $\mu = dG/dN$, and the surface energy is $\gamma = dG/dA$. So we can do a chain rule kind of thing: $\mu = dG/dA \times dA/dN$. In other words, the chemical potential (and thereby the solubility of the surface) is $\mu = \gamma \times$ [surface area of a monomer]. Perhaps using this, we could find an atomistic description based on the crystal structure for conditions where the flux of atoms from one surface to another (based on their different solubilities) results in a nanoparticle morphology that diverges from the thermodynamic Wulff shape... but I have to think about this more.

Kristen Fichthorn responded: So, it is true that differences in chemical potential provide a thermodynamic driving force for certain kinetic crystal shapes. However, I believe a formalism such as the one you propose would need more – a kinetic coefficient or knowledge of a mechanism. It would be great if we could totally predict kinetics from thermodynamics – catalysis might not be so much of a challenge in this case, for example. However, away from equilibrium, kinetics rules and there are so many different mechanisms with so many different rates. Uncovering these mechanisms and their ramifications will lead to much research for many, many years to come.

Christine Kirschhock commented: Your model suggests an intergrowth between ccp and hcp is necessary to explain the observed plate growth. This can be analysed by diffraction, determining the degree of stacking faults, for example using DIFFaX. Has this been attempted? Your model also should allow to deduce the faulting parameters, which then could be compared with the experimental observations.

Kristen Fichthorn answered: We could predict the stacking fault density based on our calculations and it would be interesting to compare this to experimental results. However, to our knowledge, the experiments have not been attempted.

Ruel Cedeno asked: What specific property of the iodine ion could rationalize its ability to favor the anisotropic growth of thin Cu plates (ionic radius, polarizability)? Do you observe chemisorption or orbital overlap between Cu and I or is it purely physisorption? Finally, do these findings extend to other halogen ions (*e.g.* bromide, chloride)?

Kristen Fichthorn replied: Iodine chemisorbs to the Cu surface – it binds quite strongly. From our analysis, probably the main factor for how iodine affects diffusion is its size. It restricts Cu binding on tightly packed Cu(111) to a greater degree than on more loosely packed Cu(100). As for Br and Cl, we studied the interaction of Cl with the Cu surfaces quite extensively: though the chemical interaction of Cl with the Cu surfaces is stronger than I, it turns out that the hydration free energy drives stronger I binding on Cu.[1] Also, interestingly, Cl binding drives faster diffusion on Cu(100) than on Cu(111) – opposite to what we find here.[2]

1 M. J. Kim, M. A. Cruz, Z. Chen, M. Brown, K. A. Fichthorn and B. J. Wiley, *Chem. Mater.*, 2021, **33**, 881–891.
2 J. Kim, J. Cui and K. A. Fichthorn, *ACS Nano*, 2021, **15**, 18279–18288.

Jutta Rogal queried: What happens to the iodine as the system grows? Does the coverage remain the same and if so where do the additional iodine atoms come from? And if the system grows another Cu layer, is it grown underneath the iodine adlayer?

Kristen Fichthorn answered: Cu atoms bind underneath iodine on the {100} facets, which leads to iodine naturally floating above the surface as the layer grows. On the {111} facets, Cu atoms bind in the same layer with I at low coverages. While the exact growth mechanism is less clear on the {111} facets, our calculations show that an intermixed Cu–I layer is less favored energetically than a layer in which I resides on top of Cu. These results (which we have not published) also suggest that an I layer floats on top of Cu as Cu atoms accumulate on the {111} facets.

Christian Kuttner commented: Your findings appear to be similar to the growth of gold nanoplates, nanotriangles, and nanoprisms. There are two types of mechanisms in the literature, namely, "plate-like" and "face-blocking" growth.[1] You showed the first one which relies on stacking faults. However, the alternative "face-blocking" relies on additional surfactant molecules and their selective adsorption at the most favourable facets. How do you see the competition between the two mechanisms? But I guess including anisotropy in the surface chemistry, *e.g.*, due to different surfactant densities at specific facets, would make the simulations too complex, right?

1 C. Kuttner, M. Mayer, M. Dulle, A. Moscoso, J. M. López-Romero, S. Förster, A. Fery, J. Pérez-Juste and R. Contreras-Cáceres, *ACS Appl. Mater. Interfaces*, 2018, **10**(13), 11152–11163.

Kristen Fichthorn responded: So, in our first paper on this topic,[1] we determined that under the conditions most relevant to experiments, the Cu surfaces are devoid of HDA capping agent. In that work, we did consider the co-adsorption of HDA and I – that calculation is highly do-able. In fact, we were able to demonstrate that the "face-blocking" (or probably more correctly "side-blocking") mechanism likely underlies the growth of Cu nanowires, as our first-principles calculations showed the side facets of penta-twinned Cu nanowires were covered with HDA and Cl, but the ends contained only Cl.[2]

1 M. J. Kim, M. A. Cruz, Z. Chen, M. Brown, K. A. Fichthorn and B. J. Wiley, *Chem. Mater.*, 2021, **33**, 881–891.
2 M. J. Kim, S. Alvarez, Z. Chen, K. A. Fichthorn and B. J. Wiley, *J. Am. Chem. Soc.*, 2018, **140**, 14740–14746.

Alan Hare asked: I guess you can exclude chloride from the iodide system in theory; but are you considering a possible chloride impurity? If so, and if this impurity occurs experimentally in "the real world", how might this affect the design outcome?

Kristen Fichthorn responded: In a study preceding the one presented here,[1] we considered possible co-adsorption of Cl and I on Cu(100) and Cu(111). Those calculations indicated there was no co-adsorption – the system either contained adsorbed Cl or adsorbed I, both with possible HDA coverage. However, first-principles calculations are limited by the size of the unit cell we can consider, so we cannot rule out the possibility of dilute impurities of any kind: Cl, defects, *etc.* Actually, experiments did indicate that dilute Cl was needed for plates to form, but the exact role of Cl was unclear. It is possible that Cl affected nucleation of the nanoplate seeds. Another possibility is this: our "ball stacking analog experiments" indicated higher index facets occur at the nanoplate sides during growth. We did not probe I adsorption and diffusion or Cl co-adsorption on those surfaces – we just assumed I was the main player and that adsorption was as strong and as slow as on the {100} facets. At any rate, the role of Cl in plate nucleation and/or growth is unclear.

1 M. J. Kim, M. A. Cruz, Z. Chen, M. Brown, K. A. Fichthorn and B. J. Wiley, *Chem. Mater.*, 2021, **33**, 881–891.

Matteo Salvalaglio asked: How did you enumerate the states used to define your Markov chain model? Do you have recommendations on strategies for the construction of kinetic models like your Markov chain from atomistic simulation data?

Kristen Fichthorn replied: One of the reasons we can successfully apply our model to this system (as well as to similar systems, where we applied it in the past[1,2]) is because the Cu plate structure is highly anisotropic, suggesting little Cu accumulation on the {111} facets. We are, in essence, using these calculations to predict how wide the plates will grow before adding another layer of height. In this case, the number of states is equal to the number of sites for a Cu atom to bind on the basal facet. For our initial configurations, we simply assume that a Cu atom is equally likely to bind to each of the N sites available to it.

If we accounted for accumulation on the {111} facets, the Markov chain calculations would rapidly become untenable, as there are too many permutations, which would lead to an intractable number of states. In this more difficult case, we have advocated combining absorbing Markov chain calculations with kinetic Monte Carlo, in an approach we call Local Superbasin Kinetic Monte Carlo.[3] In this approach, one can identify and treat groups of states connected by small barriers into superbasins – one can calculate the net exit time from such groups, as well as where the exits are likely to occur. This approach can greatly accelerate kinetic Monte Carlo simulations.

As far as building models, one needs to have a map of all relevant states of the system, as well as the rates of all important kinetic processes that take the system between various states. From there, one constructs the Markov matrix, as well as the transient and recurrent matrices. The procedure has been discussed in the literature.[3,4]

1 X. Qi, Z. Chen, T. Yan and K. A. Fichthorn, *ACS Nano*, 2019, **13**, 4647–4656.
2 J. Kim, J. Cui and K. A. Fichthorn, *ACS Nano*, 2021, **15**, 18279–18288.
3 K. A. Fichthorn and Y. Lin, *J. Chem. Phys.*, 2013, **138**, 164104.
4 B. Puchala, M. L. Falk and K. Garikipati, *J. Chem. Phys.*, 2010, **132**, 134104.

Wenhao Sun further commented: I had another idea for the kinetic Wulff shape. Different surfaces should have different solubility. If you have different solubilities, then the local concentration in the immediate vicinity of the two surfaces would be different. By Fick's first law, this would result in a flux of atoms from one surface to another. Perhaps this could be the origin of a kinetic Wulff shape…

Kristen Fichthorn replied: I believe your proposed mechanism could occur in certain systems. However here, the metal atoms bind so strongly that they stick where they hit. It could be that certain surfactants or capping agents can alter the rates at which the atoms reach certain surfaces, but it seems unlikely the solubility mechanism would apply to these metal systems.

Sten O. Nilsson Lill opened a general discussion of the paper by Colin Freeman: Based on your results, would you suggest to always include defects as a control experiment and study the effects when investigating heterogeneous nucleation with MD-simulations? Surface defects could for example, be randomly generated to see any major impacts.

Colin Freeman responded: This depends on the effect being examined. Certainly if you are trying to understand the rate of nucleation, the site of nucleation or the very early stages of nucleation, then including defects could be very important as these will govern ion concentrations and potential early clustering.

Jincheng Tong asked: The examples you present show the cracks that can induce nucleation, however they are very big (even in the micrometre scale), it is more likely these cracks are like edges of the substrate that can change nucleation dynamics. Do you have any idea if there is any case study that shows the atomic scale defects can change crystallization dynamics? Or do you think using a mixture of SAMs with different chain lengths that is able to form a non-flat surface, may have a similar effect on nucleation?

Colin Freeman answered: The experimental examples shown do involve larger defects. These are long but will not be micrometres in width, but are still larger than the atomic scale. Observing atomic sized defects in a surface is very difficult and therefore it is hard to identify examples where these scale defects have been shown to be significant. The results we present which highlight an enhanced concentration gradient at a defect may be applicable to larger defects as well, depending on their specific features. There has been a lot of work in the literature on building different patterned SAM systems using microcontact printing and examining crystallisation on these (see ref. 1 for example). These can produce patterns of crystal formation on the scale of tens of microns or less.

1 J. Aizenberg, *Adv. Mater.*, 2004, **16**, 1295–1302.

Ruel Cedeno asked: Since both the solvent and the solute ions can selectively cluster around the active defect sites, what is the ratio of the effective local concentration with respect to the bulk concentration? What are its implications on nucleation modelling where the bulk concentration is used for the driving force term?

Colin Freeman answered: Where we have used larger numbers of ions (bulk 0.41 M and 0.69 M) we can see significantly increased numbers of ions close to the defect (Fig. 10; 17 and 18 in the ESI of our article, https://doi.org/10.1039/d1fd00082a) – 3–5 times as many as would be expected for an equilibrium distribution of ions through the whole solution. Converting this to an effective concentration would, however, require a definition of a volume around the defect, as these simulations have a fixed number of ions present in the simulation box rather than a bulk fixed concentration (as in a constant chemical potential simulation). So the ratio obtained from the simulation is not the true local/bulk ratio. However, the difference between local and bulk concentration is certainly large and would affect both the surface free energy of the cluster and the kinetics of attachment of solute ions.

Christian Kuttner queried: If defects can promote nucleation, would it be reasonable to assume that defects can also suppress nucleation in special cases?

Colin Freeman replied: In the work we have performed we would suggest that defects can be either inactive (in which case they have little to no effect on the localised concentration of the nucleating species) or active (where they help concentrate the nucleating species). It is possible that a defect could actively lower the local concentration around it, but this would not suppress nucleation as this would cause very little effect on the overall concentration across the whole surface or at the active defect sites. It would also be possible for a defect to destabilise a growing nucleus by increasing its interfacial energy.

Perhaps the most important factor would be the concentration of defects. If the concentration of defects is low, then it is possible to grow clusters large enough to exceed the critical size without depleting the solute concentration of the local solution too much. If the concentration of defects is very high, then large numbers of sub-critical clusters may be produced, greatly depleting the solution concentration near the interface and producing diffusion-limited control.

Aaron R. Finney commented: Equilibrium molecular dynamics simulations are performed and, as already mentioned, the supersaturation at the surface and in the bulk (far from the surface in the simulations) is the same and the thermodynamic driving force for nucleation is the same too. An increased concentration of ions locally in the active defects could lead to a reduction of the kinetic barriers for nucleation, but then the very slow diffusion of ions could mean that translation and rotation of ions to adopt a crystalline lattice arrangement is potentially hindered. How do the authors reconcile these two effects? Furthermore, could the clustering of ions in defects facilitate nucleation in other ways *e.g.*, by leading to increased short-range ordering of ions and removal of water?

Colin Freeman responded: We are suggesting that the increased concentration of ions around the defect could lead to an increased number of “attempts” to cross the nucleation barrier which makes nucleation more probable at these points. The long range diffusion of these ions in these regions is lower but we have not seen evidence that their basic mobility is reduced (*i.e.*, they still rotate, vibrate and librate quickly) but we have not examined this in detail. We think we are seeing a more general reduction in their configurational freedom (as has been observed for other molecules/ions at surfaces, *e.g.* ref. 1). We have not seen any evidence to suggest that nucleation is being enhanced with increased short range order or a removal of water but these effects are certainly possibilities and would merit further investigation.

1 C. L. Freeman and J. H. Harding, *J. Phys. Chem. C*, 2014, **118**, 1506–1514.

Joonsoo Kim opened a general discussion of the paper by Peter Vekilov: Are there any other hypotheses trying to find a relationship between attaching a unit's symmetry and overall crystal symmetry?

Peter Vekilov responded: A common assumption in a large segment of the crystallization literature has been that the structural motifs of the crystal represent the solute species that associate to the kinks,[1–5] which would imply that crystal symmetry elements arise in the solution prior to crystallization. The inconsistencies of this assumption have been pointed out before our work (https://doi.org/10.1039/d1fd00080b), and it cannot explain how crystal forms with distinct structural motifs arise in the same solution. The elementary acts of incorporation into kinks, however, have only been directly visualized for relatively few crystals, mostly of large protein molecules.[6–22] Given these constraints to directly identify the growth unit, the promotion of the solute oligomers to the rank of the incorporating species had remained largely unchallenged until our recent results.

1 S. A. Kulkarni, E. S. McGarrity, H. Meekes and J. H. ter Horst, *Chem. Commun.*, 2012, **48**, 4983–4985.
2 A. Spitaleri, C. A. Hunter, J. F. McCabe, M. J. Packer and S. L. Cockroft, *CrystEngComm*, 2004, **6**, 490–493.
3 R. Davey, G. Dent, R. Mughal and S. Parveen, *Cryst. Growth Des.*, 2006, **8**, 1788–1796.
4 S. Parveen, R. J. Davey, G. Dent and R. G. Pritchard, *Chem. Commun.*, 2005, 1531–1533.
5 J. Chen and B. L. Trout, *J. Phys. Chem. B*, 2008, **112**, 7794–7802.
6 S.-T. Yau, D. N. Petsev, B. R. Thomas and P. G. Vekilov, *J. Mol. Biol.*, 2000, **303**, 667–678.
7 S.-T. Yau, B. R. Thomas, O. Galkin, O. Gliko and P. G. Vekilov, *Proteins: Struct., Funct., Genet.*, 2001, **43**, 343–352.
8 S.-T. Yau, B. R. Thomas and P. G. Vekilov, *Phys. Rev. Lett.*, 2000, **85**, 353–356.
9 S.-T. Yau, B. R. Thomas and P. G. Vekilov, *J. Crystal Growth*, 2001, **232**, 188–194.
10 M. Sleutel, D. Maes, L. Wyns and R. Willaert, *Cryst. Growth Des.*, 2008, **8**, 4409–4414.
11 M. Sleutel, C. Vanhee, C. V. de Weerdt, K. Decanniere, D. Maes and L. W. R. Willaert, *Cryst. Growth Des.*, 2008, **8**, 1173–1180.
12 M. Sleutel, R. Willaert, C. Gillespie, C. Evrard, L. Wyns and D. Maes, *Cryst. Growth Des.*, 2009, **9**, 497–504.
13 M. Sleutel, J. Lutsko, A. E. S. Van Driessche, M. A. Durán-Olivencia and D. Maes, *Nat. Commun.*, 2014, **5**, 5598.
14 A. J. Malkin, Y. G. Kuznetsov, R. W. Lucas and A. McPherson, *J. Struct. Biol.*, 1999, **127**, 35–43.
15 A. J. Malkin, Y. G. Kuznetsov and A. McPherson, *J. Crystal Growth*, 1999, **196**, 471–488.
16 Y. G. Kuznetsov, A. J. Malkin, R. W. Lucas, M. Plomp and A. McPherson, *J. Gen. Virol.*, 2001, **82**, 2025–2034.
17 A. McPherson, A. J. Malkin, Y. G. Kuznetsov and M. Plomp, *Acta Crystallogr., Sect. D: Biol. Crystallogr.*, 2001, **57**, 1053–1060.
18 A. J. Malkin and A. McPherson, in *From fluid-solid interfaces to nanostructural engineering*, ed. J. J. De Yoreo and X. Y. Lui, Plenum, Kluwer Academic, New York, 2004, vol. 2, Assembly in hybrid and biological systems, pp. 201–238.
19 A. J. Malkin and R. E. Thorne, *Methods*, 2004, **34**, 273–299.
20 D. K. Georgiou and P. G. Vekilov, *Proc. Natl. Acad. Sci. U. S. A.*, 2006, **103**, 1681–1686.
21 S.-T. Yau and P. G. Vekilov, *Nature*, 2000, **406**, 494–497.
22 S.-T. Yau and P. G. Vekilov, *J. Am. Chem. Soc.*, 2001, **123**, 1080–1089.

Ian Ford commented: The non-linear dependence of the step velocity on bulk concentration might indicate that the surface concentration of the depositing species is not equal to the bulk concentration. We have shown that this can emerge in a recent analysis of data on calcite growth: the observed critical step length of a screw dislocation source is consistent with a reduction in supersaturation due to diffusion through the boundary layer. Please refer to poster P45 at this meeting, or ref. 1. Do you think that this might explain your non-linear kinetics?

1 R. Darkins, I. J. McPherson, I. J. Ford, D. M. Duffy and P. R. Unwin, *Cryst. Growth Des.*, 2022, **22**, 982–986.

Peter Vekilov replied: The deviations from linearity due to depletion of the surface supersaturation at low solution flow rates depend on the surface kinetics, as highlighted in my previous answer above to Professor Ian Ford. The kinetics of step growth for both olanzapine and etioporphyrin I are sufficiently slow that diffusion supplies a sufficient feed of solute even in the absence of forced convection (our measurements were carried out in unstirred solutions). More importantly, insufficient solute supply to the steps enforces a sublinear correlation between the step velocity and the solute concentrations as the surface depletion is more severe at high supersaturations, at which the steps grow faster and consume more solute. The observed correlation with both olanzapine and etioporphyrin I are superlinear and I argue that they have nothing to do with insufficient solute supply to the steps.

Matthew Bennett said: Crystal growth is not easy. Do you think that studies like this, that do not provide the full answer/theory, close the gaps of the unknown, helping general understanding of the basics?

Peter Vekilov responded: I do not think that anyone would disagree with the statement that crystal growth is very complex. This is also the main message of our paper, that efforts to identify the solute unit that incorporates into the crystal cannot use shortcuts that only consider the crystal structure or the solution speciation. We present a method to identify the crystal growth unit that starts with measurements of the chemical kinetics at the kinks. We then demonstrate that a reliable identification of the molecular species that associates to the kinks requires a combined study of all three elements of crystal growth, the state of the solute in the solution, the kinetics of incorporation, and the crystal structure, by synergistic application of both experiment and high-power computer modeling. I strongly disagree that the denunciation of shortcuts retrogresses our understanding of crystal growth. On the contrary, I would argue that the proposed method to identify the crystal growth unit is a major step towards closing the knowledge gaps and promoting the understanding of crystal growth basics. Besides its fundamental implications, results due to our method may find immediate use in computer models of anisotropic crystal growth rates aimed to predict and control crystal sizes and morphologies for numerous technological applications.

Wenhao Sun commented: I really enjoyed your paper, showing that one surface grows due to monomer attachment, and the other results from dimer attachment. If you can control the ratio of dimers in solution, can the ratio of the growth rates of different surfaces be tailored?

Peter Vekilov answered: What an excellent idea! If an additive is found that binds preferentially to either dimers or monomers, it would shift the dimerization equilibrium, the ratio of monomers to dimers, and the relative rates of growth of the faces, which incorporate exclusively either monomers or dimers. This approach appears to be a potentially powerful tool for crystal habit control by modifying the solution properties of the solute, rather than by selective interactions with the anisotropic crystal faces, as pursued in most current strategies.

Wenhao Sun asked: So, Peter, you made a comment that different surfaces cannot have different solubility. I just wanted to address that with a thought-experiment. Suppose you have some material with very anisotropic surface energies. (001) is high energy and (100) is low energy. If you make two fake, non-equilibrium, very flat nanoparticles with the same number of atoms in each; one nanoparticle that exposes (001) and the other that exposes (100), you would have two nanoparticles with different solubility, right? This is based on Gibbs–Thomson or Ostwald–Freundlich arguments on the solubility.

So if you divide through both nanoparticles by the number of atoms (which is the same between the two nanoparticles), you would arrive at the conclusion that the differences in solubility arise from the differences in solubility of the surfaces. If you have different solubilities, then the local concentration in the immediate vicinity of the two surfaces would be different. This would result in a flux of atoms from one surface to another. Perhaps this could be the origin of a kinetic Wulff shape…

Peter Vekilov replied: The solubility is the concentration of a solution in equilibrium with a crystal. For large crystals, the exchange of molecules with the adjacent solution occurs exclusively at the kinks. The crystal edges at any temperature higher than 0 K are rounded and represent rough crystal surfaces rich in kinks. The kinks are defined as surface sites where a molecule has exactly half of the bonds that it would have in the crystal bulk. Thus, the free energy of a molecule binding to a kink is the same for kinks on all crystal faces; this is true for both centrosymmetric and polar crystal structures. Accordingly, solubility measurements based on observations of the step dynamics (steps neither grow nor dissolve at the solubility) yield solubilities which are identical across all facets and equal to the value measured in bulk crystallization experiments (https://doi.org/10.1039/d1fd00080b).[1–5] Thus, for large crystals the solubility is fully independent of the surface free energies of the facets which are clearly distinct.

Going back to your question, if {111} facets have higher surface free energy than {100} facets (I slightly modified your argument since you could not fully facet crystals with either {100} or {001} facets) crystals terminated exclusively by {111} facets would have higher total free energy and dissolve in a nearly equilibrated solution in the presence of crystals fully faceted by {100} faces, if they are sufficiently small. If both facets are present on the same small crystal at near-equilibrium, the sizes of the {111} and {100} facets will adjust by growth or dissolution until their ratios comply with the Wulff rule.

How small is sufficiently small? The answer is hidden in the Gibbs–Thomson relation, which for 3D crystals attains the form

$$\frac{c_{eR}}{c_e} = \exp\frac{2\Omega\gamma}{Rk_BT}$$

Here c_{eR} is the solubility of a crystal of size R, c_e is the solubility of a very large crystal, Ω is the molecular volume, k_BT is the thermal energy, and γ is the surface free energy. The Gibbs–Thomson relation argues that for crystals with small R, facets with distinct γ will equilibrate with solutions of different c_{eR}. The crucial governing parameter here is the capillary length

$$\frac{2\Omega\gamma}{k_BT}$$

which is typically a few nanometers. Thus for R greater than the capillary length, the factor in the exponential function on the right-hand side will be vanishingly small and $c_{eR} \cong c_e$. The small solubility differences between facets of moderate sizes, about 1 μm, is the reason why equilibrium shapes dictated by the Wulff construction, take several years to reach.[6,7]

1 M. Warzecha, L. Verma, B. F. Johnston, J. C. Palmer, A. J. Florence and P. G. Vekilov, *Nat. Chem.*, 2020, **12**, 914–920.
2 K. N. Olafson, M. A. Ketchum, J. D. Rimer and P. G. Vekilov, *Proc. Natl. Acad. Sci. U. S. A.*, 2015, **112**, 4946–4951.
3 O. Gliko, N. Neumaier, W. Pan, I. Haase, M. Fischer, A. Bacher, S. Weinkauf and P. G. Vekilov, *J. Am. Chem. Soc.*, 2005, **127**, 3433–3438.
4 P. G. Vekilov, Y. G. Kuznetsov and A. A. Chernov, *J. Crystal Growth*, 1992, **121**, 643–655.
5 P. G. Vekilov, Y. G. Kuznetsov and A. A. Chernov., *J. Crystal Growth*, 1992, **121**, 44–52.
6 D. Iwanov and D. Nenow, *J. Cryst. Growth*, 1992, **123**, 195–202.
7 D. Iwanov and D. Nenow, *J. Cryst. Growth*, 1999, **198–199**, 96–100.

Ivo B. Rietveld remarked: It is very interesting that you can show the different growth rate dependencies for two different surfaces (010 and 001) in the case of etioporphyrin crystals, as the 010 surface growth rate appears to depend on the molecular concentration, while the 001 surface depends on the dimer concentration.

How have you been able to experimentally distinguish between the growth dynamics of the two surfaces?

Peter Vekilov responded: The cross section of the etioporphyrin I crystals is nearly square. As a result, some of the crystals deposited on a substrate exhibit (010) facets, whereas others show (001) facets. This is in contrast to olanzapine crystals, which are shaped as plates along the (001) plane and always land on substrates with the (002) face facing upwards and exposed to AFM observation. The presence of a (101) face in the crystal morphology of etioporphyrin I enforces distinct shapes, a parallelogram for the (010) face and trapeze for the (001) face, which we use to identify the face monitored by AFM.

Jan Sefcik commented: There are two reasons for differences between solute concentration in the bulk solution and at the solution–solid interface: the first is due to diffusion limitations, which would lead to solute being attached to the interface of a growing crystal faster than it can be replenished by diffusion from the bulk solution, so the interfacial concentration would be lower than that in the bulk; the second is due to a local dispersion force field of the interface, where the interfacial concentration would be higher than that in the bulk.[1] In the second case, the interfacial concentration enhancement would be in thermodynamic equilibrium with the bulk solution, so the solute activity (and its chemical potential) would be the same as in the bulk. However, local concentration and ordering could be very different, which could facilitate non-classical/two-step nucleation pathways in the interfacial region. Also, dimer concentrations may be very different at the interface compared to the bulk solution.

1 D. McKechnie, S. Anker, S. Zahid, P. A. Mulheran, J. Sefcik and K. Johnston, *J. Phys. Chem. Lett.*, 2020, **11**, 2263–2271.

Peter Vekilov replied: This paper by McKechnie *et al.* suggests that glycine accumulates at the interface between an aqueous solution and mineral oil and this accumulation may boost heterogeneous nucleation of glycine crystals at the oil–water interface. I have no problem with that finding, even though I would prefer some experimental evidence in support. My own personal experience suggests that any such interfacial effects would be strongly dependent on the solute and the substance on the other side of the interface. Not all solutes will accumulate at a particular interface and if a solute accumulates at one interface, it will not necessarily adsorb on all others. Now, the suggestion appears to be that a similar accumulation of solute would occur at the interface of a solute with its own crystal and support nucleation of new crystallites at the interface with a growing larger crystal; the nucleated crystallites then incorporate into the large crystal, converting the growth mode to non-classical. The first part of this scenario appears trivial: solute adsorbed in the surface of a growing crystal diffuses along the surface and feeds the crystal growth, this represents the surface diffusion pathway, one of the two classical mechanisms of crystal growth. The adsorbed solute may form two-dimensional nuclei that initiate new crystal layers, another part of the classical crystal growth mechanisms put forth by Stranski and Kaischew in the 1930s.[1–3] Can the absorbed solute form 3D nuclei of new crystals? Highly unlikely in the absence of additional factors. The critical radius of 3D nuclei is two-fold larger than that of 2D nuclei, which contributes to a nucleation barrier for 3D nucleation several-fold higher than that for 2D nucleation. The nucleation barrier enters the nucleation rate law in the exponent, so that a several-fold higher barrier would suggest that 3D nucleation pretty much never happens. There may be scenarios that help 3D nucleation. They include elastic strain of the top crystal layers strong enough to shift the applicable rules from those of homoepitaxy (*i.e.*, crystal growth) to those of heteroepitaxy. Another type of event that may support 3D nucleation is the adsorption of larger foreign molecules at the crystal surface that favorably interact with a 3D nucleus of the solute. Recent liquid-phase TEM observations reveal the nucleation of crystallites not at the interface of a larger growing crystal, but at a certain distance from it. A lot is still not clear about the mechanism of this nucleation, current models appear to focus on solvent structuring that may stretch up to 1 nm from the crystal surface.

1 I. N. Stranski and R. Kaischew, *Z. Phys. Chem.*, 1934, **B26**, 114–116.
2 R. Kaischew, *Z. Phys.*, 1936, **102**, 684–690.
3 R. Kaischew and I. N. Stranski, *Z. Phys. Chem.*, 1937, **B35**, 427–432.

Ian Ford said: Finite element modelling of the deposition of calcite onto a surface in a typical flow cell has shown that the growth rate can be reasonably independent of flow rate and yet the supersaturation at the surface can still differ substantially from that in the bulk. Please see poster P45 at this meeting or ref. 1. Increasing the flow rate is not an entirely reliable way to exercise control over the surface conditions.

1 R. Darkins, I. J. McPherson, I. J. Ford, D. M. Duffy and P. R. Unwin, *Cryst. Growth Des.*, 2022, **22**, 982–986.

Peter Vekilov responded: This cited paper presents an ingenious way to evaluate the supersaturation on the surface of a calcite crystal growing in a typical Bruker Nasoscope AFM fluid cell.[1] The proposed method relies on the critical step length that can be measured by AFM. Indeed, the paper shows that the surface supersaturation is lower than that in the bulk. Notably, the flow rates analysed in the paper were low – up to 100 ml per hour. For a cell with 0.5 ml volume and about 1 cm length, this flow rate corresponds to a linear rate of about 0.5 mm s^{-1}. Other studies of the coupling of solute convective-diffusive transport and crystal interfacial kinetics employ flow rates from several cm s^{-1} (ref. 2) to m s^{-1}.[3] No wonder at such slow flow rates the boundary layer thickness δ is significant and the surface supersaturation, low. Classical analyses by Chernov[4,5] demonstrate that the ratio of the surface to the bulk supersaturations is

$$\frac{\sigma_s}{\sigma_b} = \left(1 + \frac{\delta h \beta}{Dl}\right)^{-1}$$

where h and l are the step height and step separation, respectively, β is the step kinetic coefficient, and D is the solute diffusivity. The presence of β in this relation highlights that the ratio

$$\frac{\sigma_s}{\sigma_b}$$

reflects the coupling of the solute supply to the incorporation kinetics and is not fully governed by the solution flow; faster β require lower δ, reached at higher flow rate, for σ_s to approach σ_b. Furthermore, for β of order 1 mm s^{-1}, D of order 10^{-5} cm^2 s^{-1}, and $h/l \approx 10^{-2}$ (roughly representative of calcite step kinetics 6), $\sigma_s \approx \sigma_b$ if δ <10 μm. The lowest δ analysed in Darkins *et al.*[1] is about 120 μm. It appears that there is no contradiction between the results of the beautiful study by Darkins *et al.*[1] and classical understating of how solute transport couples with the dynamics at the kinks.

1 R. Darkins, I. J. McPherson, I. J. Ford, D. M. Duffy and P. R. Unwin, *Cryst. Growth Des.*, 2022, **22**, 982–986.
2 M. Sleutel, A. E. S. Van Driessche, W. Pan, E. K. Reichel, D. Maes and P. G. Vekilov, *J. Phys. Chem. Lett.*, 2012, 1258–1263.
3 N. A. Booth, A. A. Chernov and P. G. Vekilov, *J. Crystal Growth*, 2002, **237–239**, 1818–1824.
4 A. A. Chernov, *Sov. Phys. Uspekhi*, 1961, **4**, 116–148.
5 A. A. Chernov, *Modern Crystallography III, Crystal Growth*, Springer, Berlin, 1984.
6 H. H. Teng, P. M. Dove, C. A. Orme and J. J. De Yoreo, *Science*, 1998, **282**, 724–727.

Marta K. Dudek asked: In MD simulations, were conformations in monomers different to those of dimers and in crystals, and do you know how they relate to gas phase (or solution phase) global minimum conformations?

Peter Vekilov answered: The constituent monomers in a dimer were allowed all translational, rotational, and molecular degrees of freedom on the pathway from fully separated monomers to monomers in a stable dimer. So, yes, the simulations did consider all possible dimer configurations.[1] We never considered etioporphyrin I in a gas phase, I am not sure if it can melt and evaporate without decaying first.

1 L. Verma, M. Warzecha, R. Chakrabarti, V. G. Hadjiev, J. C. Palmer and P. G. Vekilov, *Isr. J. Chem.*, 2021, **61**, 818–827.

Kevin Roberts commented: The transformation from the creation of intermolecular structures with short range order within the solution state and at the solution-mediated crystal growth interface, to that of the 3D periodic structures needed for the development of the bulk crystallographic structures, have a rate-limiting growth processes defined by the creation of stoichiometric periodic bonding chains (PBCs). As an example, the crystal structure of the dicarboxylic acid adipic acid $HOOC(CH_2)_4COOH$, has infinite chains of hydrogen-bonded carboxylic acid dimers which are PBCs and hence provide strong stoichiometric growth-promoting interactions. In contrast, the closely related monocarboxylic acid caproic acid ($H_3C(CH_2)_4COOH$) has a single ended molecular structure resulting in a PBC comprising of a combination of carboxylic acid hydrogen-bonded dimers (strong) and van der Waals interactions (weaker) which when combined have lower growth-promoting capability due to the weak link between the alkyl chain's terminal methyl groups. The latter are much weaker than the former and therefore are rate limiting to growth. I was wondering as to which of these two cases fit the compound you presented in your paper?

Peter Vekilov answered: As stated in our paper in this issue (https://doi.org/10.1039/d1fd00080b), crystal symmetry combines point symmetry elements, such as an inversion center, mirror planes, and rotation axes, typical of finite-size objects, with translation. Translational symmetry dictates exact rules for how a molecule in a crystal orients and positions with respect to any other crystal molecule. It arises during crystallization. The origin of point symmetry elements, however, is far less understood. The results that we present elucidate the origin of point symmetry, a subject of intense controversy in the literature. We show that in some cases molecular dimers that carry the structure of a unit in the crystal lattice may form in the solution and incorporate in the crystal as a whole. Clearly, the periodic bond chains that you allude to would form as the preformed dimer incorporates in the crystal lattice. Our results also argue that the formation of the PBCs is not the rate limiting step, which determines whether the crystal grows by incorporation of monomers, which are much more abundant in the solution, or of the minority monomers. We show that the preference for growth by dimers is dictated not by the faster (than that of monomers) incorporation of dimers at kinks, but rather by the greater energy of adsorption of the dimers on the terraces between steps. Owing to the energetically-favorable adsorption of dimers, the surface layer is enriched with dimers, and growth by incorporation of dimers is much faster than that by incorporation of monomers.[1]

1 M. Warzecha, L. Verma, B. F. Johnston, J. C. Palmer, A. J. Florence and P. G. Vekilov, *Nat. Chem.*, 2020, **12**, 914–920.

Rachel Pooley asked: In the case of etioporphyrin I, you stated that from the measured kinetics of terrace growth, that the incorporating species on the (010) face are monomeric and on the (001) face the kinetics are consistent with a bimolecular reaction, corresponding to dimeric growth units. However, AFM scans as presented in Fig. 8 in your paper show step heights on the (001) face of EtpI of both double *and* single height, corresponding to dimeric and monomeric growth, respectively. There is also no significant difference between the growth rates of the double and single height steps. This suggests that monomeric growth must have some role to play in the growth of the (001) face. Conversely, in Fig. 9 of your article you present an analysis of the shape of the (001) surface, and present a hypothesis that " …monomers may be trapped in the dimples at the step edge and erect growth-incompetent configurations, whereas dimers, which fit seamlessly into the dimples, provide continuous growth." How can the observation of single height steps on the (001) face, growing at a comparable rate to the double height steps, be consistent with this assertion that the face is only growing by dimer inclusion and that monomeric species form a "growth incompetent configuration"?

Peter Vekilov responded: The structure of etioporphyrin I crystals (Fig. 4c in our article, https://doi.org/10.1039/d1fd00080b) suggests that a scenario, by which the step heights on the (001) and the (010) faces are equal to a dimension of solute dimers, would require a dimer structured as two monomers positioned next to one another in a single plane and with a minimal number of contacts. This is not the structure of the dimer predicted by all-atom MD simulations.[1] As expected, the etioporphyrin I dimer comprises two monomers positioned in two roughly parallel planes, an orientation, which maximizes the number and the strength of contacts between the two monomers. Furthermore, double height steps arise due to the dynamics of adjacent steps, some of which move slightly faster than others and catch up with them. In turn, the double height steps decay into single height steps. These observed behaviors imply that there are no fundamental differences between the mechanism of incorporation of solute molecules into steps of distinct height. Then the equal velocities of double and single height steps are evidence for step growth by direct incorporation of solute molecules and not by surface diffusion, as argued in the text of our article (https://doi.org/10.1039/d1fd00080b). Lastly, if we had a mixed mechanism of growth by incorporation of both monomers and dimers, the correlation between the step velocity and the solute concentration would be a power between one and two. It is, in fact, strictly quadratic on the (001) face and strictly linear on the (010) face.

1 L. Verma, M. Warzecha, R. Chakrabarti, V. G. Hadjiev, J. C. Palmer and P. G. Vekilov, *Israel J. Chem.*, 2021, **61**, 818–827.

Damien Thompson queried: Can you explain more about how your model predicts oligomer geometries in solution? Does solvent structuring around olanzapine play a role in stabilising the dimer? What solvent effect and/or olanzapine conformational effects block formation of higher-order structures, such as trimer, tetramer *etc.*?

Peter Vekilov responded: In the MD simulations, the two monomers on the way to forming a dimer were allowed all translational, rotational, and molecular degrees of freedom. We evaluated the free energies of all possible dimer conformations. We identified the conformation that is characterized with the lowest free energy. For olanzapine, the dimer configuration that occupies the minimum of the mean free force potential between the two monomers is identical to that in the crystals.[1] We never considered higher order oligomers. The similarity of the computer Raman spectra to those measured at higher olanzapine concentrations suggests that higher order oligomers are poorly represented in the solution. The simulations explicitly accounted for all atoms of two solvents, 1 : 1 water : ethanol mixture and chloroform. The similarity between the dimer structures predicted for these two relatively different solvents suggests that at least for olanzapine, the solvent properties exert small effects on the dimer structure.

1 M. Warzecha, L. Verma, B. F. Johnston, J. C. Palmer, A. J. Florence and P. G. Vekilov, *Nat. Chem.*, 2020, **12**, 914–920.

Sten O. Nilsson Lill said: In your Raman DFT-models of olanzapine, it seems you didn't include water or solvent which are present in the experiment and also in the final crystal. Do you expect any impact on the results for the models by solvent inclusion or does it only affect higher wavenumbers?

Peter Vekilov replied: The models did consider the effects of the solvents.[1] We generated Raman spectra of olanzapine dissolved in ethanol and water and chloroform using COSMO (COnductor-like Screening MOdel)2 and the DMol3 module in Materials Studio. COSMO treats each solvent as a continuum with a permittivity ε surrounding the solute molecules outside of a molecular cavity.[2] COSMO derives the polarization charges of the continuum, caused by the polarity of the solute, from a scaled-conductor approximation. The computed and measured spectra of olanzapine in a water–ethanol mixture and in chloroform are similar, suggesting that the solvent effects are not very strong.

1 M. Warzecha, L. Verma, B. F. Johnston, J. C. Palmer, A. J. Florence and P. G. Vekilov, *Nat. Chem.*, 2020, **12**, 914–920.
2 A. Klamt and G. Schüürmann, *J. Chem. Soc., Perkin Trans. 2*, 1993, 799–805.

Adam Hill asked: Would you expect different results if you selected a solvent that specifically breaks up the observed dimers (*e.g.*, by forming strong aromatic interactions or strong hydrogen bonds with the monomers)? If so, could the differences (or similarities) in crystal growth behaviour in different solvents be used to further prove the reliance of the growth mechanism on preformed solute dimers or monomers?

Peter Vekilov answered: Absolutely. We in fact tried this experiment with olanzapine. An anhydrous form IV of olanzapine crystals, grown in a polymer dispersion, was found, in which olanzapine molecules are not arranged in dimers.[1] We wanted to study the growth mechanism of this form and the solute properties in the solution from which they grow, as a crucial benchmark for our studies. Unfortunately, our extensive efforts failed to produce stable crystals of olanzapine form IV. Differential calorimetry scans of the generated solids indicate the presence of crystals with a melting temperature equal to that of form IV. These crystals converted to the stable anhydrous form I, before we could record the powder X-ray diffraction pattern or the Raman spectra.

1 S. Askin, J. K. Cockcroft, L. S. Price, A. D. Gonçalves, M. Zhao, D. A. Tocher, G. R. Williams, S. Gaisford and D. Q. M. Craig, *Cryst. Growth Des.*, 2019, **19**, 2751–2757.

Kristen Fichthorn opened the discussion of the paper by Jeffrey Rimer: Polymers added to nanoparticle suspensions can cause depletion forces that induce aggregation of the nanoparticles. Do you think this could play a role in the systems you are studying?

Jeffrey Rimer replied: This is an excellent point. We do not believe that depletion forces are playing a large role in the observed aggregation of silica nanoparticles. This argument is based on the fact that we surveyed a number of different polymers and only observed this effect with one (PDDA). If depletion forces were acting upon silica nanoparticles, we would anticipate a more general phenomenon; however, we cannot rule out the potential that depletion forces have a small contribution. According to our hypothesis, we posit the concentration of (alumino)silicate nutrient species in the interstitial spaces of aggregated particles is higher than that of the bulk solution, which could facilitate faster rates of zeolite nucleation and crystal growth. Further experiments are required to identify what role (if any) depletion forces play in the observed aggregation of precursors with the addition of PDDA.

Christine Kirschhock said: Many authors suggest zeolite nucleation and growth is mediated by the liquid phase. Do you suspect the concentrations and speciation within the confined liquid in the pockets you create is different compared to the bulk liquid? May such an assumption explain your observations?

Jeffrey Rimer answered: This is an excellent question. The short answer is that we are not certain if there will be differences in the speciation of soluble (alumino) silicates between the bulk and pockets. Our expectation is that the confined volume could have a different level of supersaturation, which could, with the aid of polymers encapsulated within these spaces, direct the formation of distinct (alumino)silicate species. It is possible that these species could be a driving force that accelerates zeolite nucleation. This hypothesis is also qualitatively consistent with our observations for the effects of alkali infusion in amorphous precursors, which we discuss in a previous publication.[1]

1 H. Dai, J. Claret, E. L. Kunkes, V. Vattipalli, N. Linares, C. Huang, M. Fiji, J. García-Martinez, A. Moini and J. D. Rimer, *Angew. Chem., Int. Ed.*, 2022, **61**, e202117742.

Michael Anderson commented: The driving force in zeolite crystallisation is driven by the difference in free energy between the gel and the zeolite (for a gel preparation). This will in large part be given by the Qn species in the gel which can be measured by ^{29}Si solid-state NMR? Is this something that you did?

Jeffrey Rimer replied: This is an excellent question. In our previous publication,[1] we performed ^{29}Si solid-state NMR on the fully crystalline samples instead of partially-crystalline materials to reveal if there is any pronounced effect of PDDA concentration on the final product. Characterization of partially crystalline solids extracted from syntheses using Si NMR is non-trivial and often leads to significant peak broadening. To this end, we used UV-Raman spectra to examine partially crystalline solids removed at different times of hydrothermal treatment. These measurements revealed a much faster (and progressive) development of peaks associated with 4- and 6-membered rings of the CHA framework structure when PDDA is present. The use of UV-Raman was a surrogate for powder XRD with the latter providing more accurate assessment of structure development to estimate the time of nucleation. We unfortunately do not have Si NMR data to gain more insight into the progressive changes in Qn species, but this information would indeed be very useful to determine how PDDA influences crystallization in ways that cannot be easily discerned by other techniques.

1 H. Dai, J. Claret, E. L. Kunkes, V. Vattipalli, N. Linares, C. Huang, M. Fiji, J. García-Martinez, A. Moini and J. D. Rimer, *Angew. Chem., Int. Ed.*, 2022, **61**, e202117742.

Marta K. Dudek asked: What do you think is the molecular mechanism of polymer-related enhancement in crystallization of zeolites? Or more specifically, why do you think PDDA polymer works and others not so much?

Jeffrey Rimer replied: We appreciate this question. The unique behavior of PDDA relative to the other polymers examined in this study is not well understood. PDDA induces bridging flocculation of precursors within a narrow range of polymer concentration, which accelerates nucleation. We also observe that PDDA can promote the precipitation of soluble silicates from light-scattering experiments. In our previous publication,[1] identical DLS studies were performed in solutions with pseudo monomer EMP and polymers PB, PDDAM, and PEIM. These studies were conducted over a range of modifier concentrations. The results showed that only one other additive (*i.e.* PDDAM) induced moderate particle aggregation (to a lesser extent than PDDA) at a concentration above the critical concentration of PDDA (please see Fig. S19, showing the comparison of PDDA and PDDAM reproduced from the supporting information of our previous work[1]). The other additives showed no appreciable change in the hydrodynamic diameter of dispersed colloidal silica particles. This indicates a uniqueness for PDDA; however, this cannot be solely associated with the quaternary ammonium functional group since other polymers possess the same functionality. We surmise that the structure of PDDA enables configurational degrees of freedom that facilitate the bridging of two or more colloidal particles. The mechanism for this mode of action is elusive, and would likely require more detailed investigation (*e.g.* molecular modelling) to elucidate the unique behavior of PDDA compared to other polyquats.

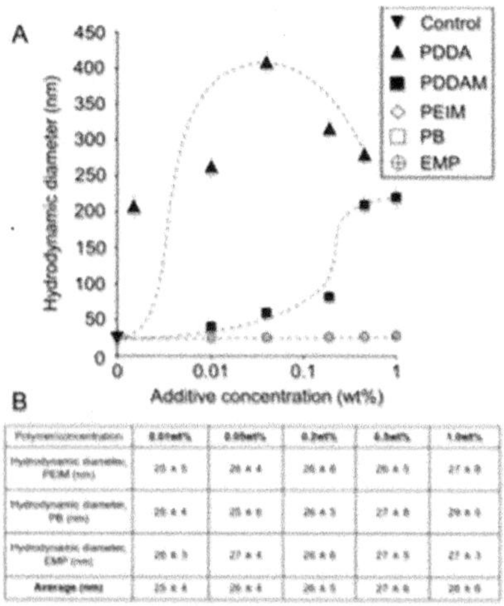

Polymer/concentration	0.01wt%	0.05wt%	0.2wt%	0.5wt%	1.0wt%
Hydrodynamic diameter, PEIM (nm)	25 ± 5	26 ± 4	26 ± 6	26 ± 5	27 ± 9
Hydrodynamic diameter, PB (nm)	26 ± 4	25 ± 6	26 ± 3	27 ± 8	29 ± 6
Hydrodynamic diameter, EMP (nm)	26 ± 3	27 ± 4	26 ± 6	27 ± 5	27 ± 3
Average (nm)	25 ± 4	26 ± 4	26 ± 5	27 ± 6	28 ± 6

Figure S19. (A) DLS measurements of colloidal silica particle size with different concentrations of polymer (PDDAM, PB, and PEIM) or monomer (EMP). Solutions were prepared with a diluted SSZ-13 synthesis gel composition in the absence of the Al source using the molar composition 1 SiO_2: 1 NaOH: 1760 H_2O: 0.1TMAda. Error bars span two standard deviations and dashed lines are interpolations to guide the eye. (B) Table for the overlapping data (PEIM, PB, and EMP) in panel (A) that exhibit no appreciable change in hydrodynamic diameter with increasing modifier concentration.

1 H. Dai, J. Claret, E. L. Kunkes, V. Vattipalli, N. Linares, C. Huang, M. Fiji, J. García-Martinez, A. Moini and J. D. Rimer, *Angew. Chem., Int. Ed.*, 2022, **61**, e202117742.

Christian Kuttner commented: I wonder if the molecular weight of PDDA plays a role in the formation and volume of the interstitial volume. The molecular weight of 150 kDa is quite high. What do you expect to happen when you reduce the molecular weight of PDDA chains?

Jeffrey Rimer responded: This is an excellent question. In our previous paper,[1] we tested the effect of PDDA molecular weight on the SSZ-13 crystallization time; however, we did not focus on smaller molecular weight, but rather compared three cases: monomer, 150 kDa PDDA, and 450 kDa PDDA. For the monomer and higher molecular weight polymer, the crystallization time was longer than that of the nominal case (150 kDa PDDA). As shown in our current manuscript (https://doi.org/10.1039/d1fd00096a), the monomer also results in a longer crystallization time than the optimal concentration of PDDA (150 kDa). It is uncertain if the crystallization time can be reduced even further by selecting a polymer in the range 50–100 kDa molecular weight. To our knowledge, these materials are not commercially available, but this question does pose an interesting topic for future investigation.

1 H. Dai, J. Claret, E. L. Kunkes, V. Vattipalli, N. Linares, C. Huang, M. Fiji, J. García-Martinez, A. Moini and J. D. Rimer, *Angew. Chem., Int. Ed.*, 2022, **61**, e202117742.

Jan Sefcik asked: Aluminium has to be present for the synthesis of zeolites; what would be a way to find out where aluminium is and what it does? It is well known that aluminium also plays an important role in aluminosilicate speciation in the liquid phase.[1,2] Are there models available that can capture the effects of alkali cations on aluminosilicate speciation to provide insight on how zeolite solubility and liquid phase supersaturation depends on liquid phase composition and specific cations?

1 J. Šefčík and A. V. McCormick, *Microporous Mater.*, 1997, **10**, 173–179.
2 J. Šefčík and A. V. McCormick, *Chem. Eng. Sci.*, 1999, **54**, 3513–3519.

Jeffrey Rimer responded: Experimentally it is possible to track the evolution of Al and aluminosilicate species using solution Al NMR. Elemental analysis by EDX or ICP of samples extracted at periodic times can also be used to track changes in Al concentration between solution and solid phases during zeolite synthesis. In the solid state, the scenario is more complicated and techniques such as EXAFS and pair distribution function (PDF) analysis can be used to investigate the local environment of Al species in the solid phase. Theoretically, models such as the one described in ref. 1 can also be used to assess changes in Al distribution. We were able to find a DFT[1] study and a mathematical model[2] that examined the effects of alkali cations on Al speciation in synthesis solutions.

1 J. D. Gale, A. L. Rohl, H. R. Watling and G. M. Parkinson, *J. Phys. Chem. B*, 1998, **102**(50), 10372–10382.
2 J. Provis and J. Van Deventer, *Chem. Eng. Sci.*, 2007, **62** (9), 2318–2329.

Ivo B. Rietveld opened the discussion of the paper by Mollie Trueman: The software that you have presented (https://doi.org/10.1039/d1fd00097g) is very interesting and it seems to me that it could also be used for teaching students crystallisation, or at least to help students understand the connection between a molecule and its crystal formation.

Could you tell me on which platforms this software can run? How many resources are needed to calculate the crystal growth (is it possible to run it on a laptop for example). What is your opinion about using the software within a teaching program for university students on crystallisation?

And my last question is because the presentation was mainly on modelling the crystallisation of zeolites, to what level is the software suitable to model crystallisation of molecular species (drug molecules for example)?

Mollie Trueman answered: The software is cross platform and has been tested on Linux, Windows and MacOS systems. It can operated either through the command line or through a graphical user interface (which is also cross platform). Simple simulations can be easily run using laptops. I think teaching would be a great application of CrystalGrower. I think it could be useful in explaining concepts such as the role of screw dislocations in crystal growth and the link between inter-molecular bonds and crystal shape. We would be happy to hear from you if you have any ideas on how this could be used in teaching! In terms of modelling crystallisation of molecular species, this is something we handle using individual molecules as growth units. When we look at zeolites, we assume each of the covalent interactions formed is of the same strength, which provides us a starting point for our interaction strength. Finding the interaction strength in molecular crystals is more challenging, since it usually involves a larger number of interactions with variable strength. We attempt to fit these to experimental data, but the larger the number of interactions, the more difficult this process becomes. We're working with theoretical groups to develop methods which we hope will allow for prediction of interaction strengths in these systems.

Adam Hill queried: Could you comment on what systems this approach to disorder could be extended? Could you model the morphological effect of substitutional defects in systems such as solid solutions?

Mollie Trueman answered: This approach for modelling disorder could be adapted to substitutional defects fairly easily. In order to handle stacking faults we had to encode a system which allowed multiple growth units to compete for the same growth site. This could easily be adapted to a system containing substitutional defects.

Michael Anderson asked: Are you able to use the CrystalGrower software to look at dendritic growth by incorporating diffusion limitations?

Mollie Trueman responded: Currently this is not possible in the software since we simplify the solution phase to be parameterised by a single value (the $\Delta\mu$ parameter). This simplification allows us to run simulations to micron sizes but restricts us from handling situations where mass transfer plays a role in determining crystal shape. In order to handle dendritic growth we would need to reconsider how we handle the liquid phase, and this is something we would be keen to explore in future work.

Jutta Rogal asked: In KMC simulations, if you have the correct rate constants for the processes, you recover the appropriate timescale; considering the approximations that you make for the attachment/detachment probabilities, is the timescale in your simulations sensible and how does it compare to experimental timescales?

Follow up: if you are fitting the probabilities to, for example, match experimental growth rates, how confident are you that you capture all important processes?

Mollie Trueman said: Timescales recovered agree fairly well under most growth conditions, but under conditions of high supersaturation agreement is poor. We do not believe we fully capture all important processes, for example we neglect mass transfer in the solution phase and do not account for non-classical processes. Whilst we do not believe these factors are likely to have a major impact on the systems studied in this paper we do hope to incorporate these factors into our model in later work.

Ruel Cedeno queried: In the CrystalGrower program, what experimental data are required as input to the model? Do you fix which facets appear in the final form? How does the model account for the different growth rates of each facet (anisotropy)?

Mollie Trueman responded: We do not fix which facets appear in the final form. The input consists of a connected net of growth units and interactions, which are then used to simulate crystal growth using a classical Kossel-style model. The difficulty lies in identifying sensible values to use as interaction strengths in the net, since these contain components from bonds formed within the crystal structure as well as an offset which corresponds to desolvation of the growth unit. Here we use experimental data to help us arrive at values which agree with what we see experimentally and can be used in further simulations.

Christine Kirschhock asked: You indicated that the normal approach to CrystalGrower depends on a cif file which extracts the topology. However, zeolites can create intergrowth, which cannot be described by a single periodic crystal structure. Examples are zeolite beta and zeolite W. Especially in the case of zeolite W, the MER/PHI intergrowth also leads to combination of tiles of the two parent topologies, using half tiles of the two frameworks. How do you implement this difficulty in your modelling?

Mollie Trueman replied: In previous work from our research group, we have used AFM to study the surfaces of zeolites and other porous materials and found that step heights at surfaces correlate to complete cages.[1,2] This suggests these cage-like units represent sensible growth units for growth simulations. In our methodology, partitioning of zeolite crystal structures into these closed cages can be achieved through an interface with the software ToposPro,[3] which applies a natural tiling algorithm to identify closed cage structures. In the case of intergrown phases, this methodology falls apart since each structure has its own set of cages and these often do not tesselate with one another. In the structures discussed in this work, and in the case of the PHI/MER system, growth units must be selected carefully such that they are common to both of the intergrowing crystal phases, such as the half tiles you describe for zeolite W. Currently we do not have an automated way to identify such units, and it is necessary to manually produce input files for CrystalGrower in order to simulate growth in these cases.

1 P. Cubillas and M. W. Anderson, in *Zeolites and Catalysis*, Wiley-VCH Verlag GmbH and Co. KGaA, Weinheim, Germany, 2010, pp. 1–55.
2 R. Brent, P. Cubillas, S. M. Stevens, K. E. Jelfs, A. Umemura, J. T. Gebbie, B. Slater, O. Terasaki, M. A. Holden and M. W. Anderson, *J. Am. Chem. Soc.*, 2010, **132**, 13858–13868.
3 V. A. Blatov, A. P. Shevchenko and D. M. Proserpio, *Cryst. Growth Des.*, 2014, **14**, 3576–3586.

Jeffrey Rimer commented: Given that many zeolite mechanisms involve multiple growth units, particularly the dominance of nonclassical pathways where oligomers or particles can play a role in synergy with monomers, is the CrystalGrower software capable of modelling different growth units or multiple pathways simultaneously? Is it possible to speculate on how the results would potentially be altered by inclusion of more complex growth units?

Mollie Trueman commented in reply: I think this is a really important question and currently this is not something we can consider easily in our growth model. Given the growing evidence that particle attachment plays a role in the growth of zeolites (and other materials) it's clearly an important omission and something we need to consider incorporating in future work. Incorporating non-classical pathways into simulations would likely require reassessing how we handle the solution phase in simulations. Currently we use the simplification that the growth solution has a single, homogeneous driving force for attachment at all points on the surface. Although this simplification lowers the computational load and allows us to explore how the surface might look at larger length scales than typical computational methods, it compromises accuracy. Attempts to incorporate non-classical pathways, could potentially involve adding local regions of high supersaturation at surfaces to mimic particle attachment (although whether this would accurately describe the growth pathway is unclear). To speculate how this might influence results, these local regions of high supersaturation could lead to higher rates of 2d nucleation which would in turn result in a higher level of defects formed at the points where growth fronts meet one another. If these pathways were active on some facets, but not others, this may enhance the 'zoning' effect we see, where growth off certain facets gives rise to regions of the crystal with higher levels of defects. Whilst I think this would be an interesting avenue to explore, I feel we need greater understanding of how crystal growth by particle attachment occurs to accurately include this pathway in our model, and to fully understand how non-classical growth modes influence the resulting crystal.

Connor Hewson asked: Do you think this modelling software could be capable of accurately determining the BET surface area of large macromolecules like zeolites? Especially when large surface area materials are hard to define experimentally.

Mollie Trueman answered: The simulations we run can predict the presence of defects and mesoporosity – both of which will have consequences for the BET surface area in the materials. Whilst this may allow us to predict trends in how we would expect the BET surface areas to vary, we do not believe we are at a point where this data can be used to predict surface areas accurately. This is due in part to difficulties in defining these values experimentally, but also to the high sensitivity of these values to experimental conditions.

Fiona Meldrum suggested: Have you considered analysing your crystals using a high resolution tomography technique such as X-ray ptychography? Resolutions of 10 nm can be achieved on optimal samples as large at 10 microns.

Mollie Trueman replied: This is something we are looking into and hope to do in the future! We're hoping to use either X-ray or electron tomography to map the mesoporosity in zeolite beta crystals.

Alan Hare commented: It may be foolhardy of me to take silica out of a discussion about aluminosilicate crystals(!) … but as the rôle of aluminal species did not seem crystal clear, I thought I would comment as to some that might be active. Delegates at the *Faraday Discussions* meeting at UCL in 2007 will recall Akihiro Wakisaka's presentation[1] of the mass spectrometric experiment, electrospray ionisation (or ESI), which shed light on alkali metal chloride nucleation.

Later, inspired by this paper, I was able to reinterpret mass-to-charge ratio values (m/z) arising out of aluminium salt solutions. Admittedly with silica absent, the experiment in nitrate and chloride solutions[2] revealed a variety of aluminium(III) cations, including a monomer, dimer (several dimers, in fact), trimer, tetramer and pentamer.[3] While ambiguity of the m/z value mitigated

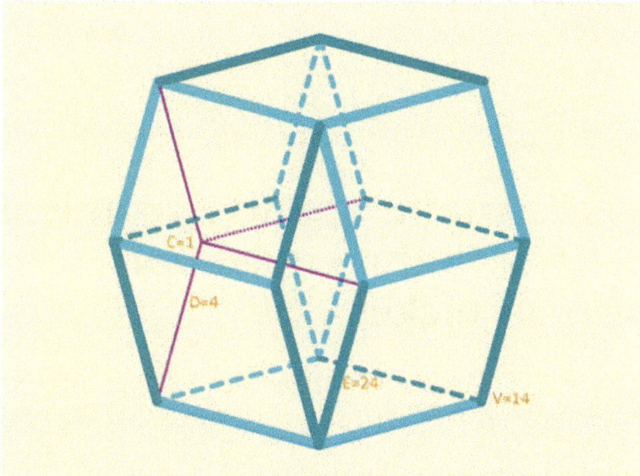

Fig. 1 A rhombic dodecahedron. A dodecahedron of the Biliński kind comprises four golden rhombohedra, two prolate and two oblate. With cations at its vertices (one internal), the Keggin structure may be formulated as a special case that can be identified with an Al_{13}^{3+} tridecamer in which the central aluminate tetrahedron $[AlO_4]^{5-}$ is caged. Like the cube – and unlike the sphere – the Biliński is a 3-d space-filler. When Aristotle allegedly mistook the tetrahedron (which like the cube is a Platonic solid) for a space-filler, could he have foreseen what Stanko Biliński first recognised only as recently as 1960?

against identification of a cyclic hexamer, its existence was not disproved. But neither was a linear hexamer prominent, nor indeed any likely $Al_nO_p(OH)_q(OH_2)_r^{(3n-2p-q)+}$ oligomer with $n > 6$ below $n = 13$. The pentamer and one particular dimer may be of interest: $Al_5O_4(OH)_6(OH_2)_2^+$ and $Al_2(OH)_4(OH_2)_2^{2+}$, respectively (each of these species was identified unambiguously, and both appeared to be tetrahedral). The former species may attach four of the latter – the OH^- or $[Al(OH)_4]^-$ ion seemingly playing an important role in the process – giving rise to a tridecameric polycation $[AlO_p(OH)_{4-p}][Al_3(OH)_{6+x}(OH_2)_{3-x}]_4^{(11-p-4x)+}$ such as the tripositive $[Al_3(OH)_8(OH_2)]_4[Al(OH)_4]^{3+}$ ($p = 0$, $x = 2$) or $AlO_4[Al_3(\text{-}OH)_7(OH_2)_2]_4^{3+}$ ($p = 4$, $x = 1$). The tetrahedral structure of the tridecamer (see Fig. 1 attached, below) could stabilise octahedral trimer chains, offering plenty of space for choride or nitrate anions to surround it. This would explain its long-term persistence reported in quasi-amorphous alumina[2] and its inhibition in the sulfate case. Of course, this is not to say that these species will necessarily populate an aluminosilicate system. But it is not beyond all credibility to suggest that, prior to formation of the zeolite framework, one or two of the smaller cations – monomer, dimer or trimer, perhaps – could play a part in its nucleation.

1 A. Wakisaka, *Faraday Discuss.*, 2007, **136**, 299–308.
2 T. Urabe and M. Tanaka, *Rapid Commun. Mass Spectrom.*, 2011, **25**, 2933–2942.
3 A. S. Hare, *RSC Adv.*, 2016, **6**, 86540–86559; Correction, 2017, **7**, 7694.

Conflicts of interest

There are no conflicts to declare.

Faraday Discussions

ROYAL SOCIETY OF CHEMISTRY

PAPER

Interplay of structural and dynamical heterogeneity in the nucleation mechanism in nickel

Grisell Díaz Leines, *[a] Angelos Michaelides [a] and Jutta Rogal [bc]

Received 24th November 2021, Accepted 10th December 2021
DOI: 10.1039/d1fd00099c

Gaining a fundamental understanding of crystal nucleation processes in metal alloys is crucial for the development and design of high-performance materials with targeted properties. Yet, crystallization is a complex non-equilibrium process and, despite having been studied for decades, the microscopic aspects that govern the crystallization mechanism of a material remain elusive to date. Recent evidence shows that the spatial heterogeneity in the supercooled liquid, characterised by extended regions with distinctive mobility and order, may be a key microscopic factor that determines the mechanism of crystal nucleation. These findings have advanced our view of the fundamental nature of crystallization, as most research has assumed that crystal clusters nucleate from random fluctuations in a 'homogeneous' liquid. Here, by analysing transition path sampling trajectories, we show that dynamical heterogeneity plays a key role in the mechanism of crystal nucleation in an elemental metal, nickel. Our results demonstrate that crystallization occurs preferentially in regions of low mobility in the supercooled liquid, evidencing the collective dynamical nature of crystal nucleation in Ni. In addition, our results show that low mobility regions form before and spatially overlap with pre-ordered domains that act as precursors to the crystal phase that subsequently emerges. Our results show a clear link between dynamical and structural heterogeneity in the supercooled liquid and its impact on the nucleation mechanism, revealing microscopic descriptors that could pave a novel way to control crystallization processes in metals.

1. Introduction

Crystal nucleation is one of the most fundamental physical processes, relevant to a large range of everyday phenomena, from cloud formation to drug development. Yet, despite being studied for decades, it is still not well understood at the

[a]*Yusuf Hamied Department of Chemistry, University of Cambridge, Lensfield Road, Cambridge CB2 1EW, UK. E-mail: gd466@cam.ac.uk*
[b]*Department of Chemistry, New York University, New York, NY 10003, USA. E-mail: jutta.rogal@nyu.edu*
[c]*Fachbereich Physik, Freie Universität Berlin, 14195 Berlin, Germany*

microscopic level.[1,2] During homogeneous nucleation, the formation of a crystal cluster is induced by fluctuations of competing translational and orientational orders in the supercooled liquid.[3] It has been assumed so far that a nucleus emerges from random fluctuations of order in a 'homogeneous' liquid state. However, several experimental and computational studies have revealed that liquids can exhibit much more complex behaviour, such as structural and dynamical heterogeneity,[3–10] where spatial regions with different structural features and mobility coexist in the liquid. Recently, it has been shown that the heterogeneities in the liquid are highly correlated with the nucleation ability of a material and the polymorph formed, revealing the fascinating collective behaviour of supercooled liquids and its connection to crystallization mechanisms.[3,9–18] However, our fundamental understanding of the nature of supercooled liquids is far from complete. Further studies are required to establish the general connections between liquid characteristics and crystal nucleation mechanisms in order to derive novel fundamental rules and microscopic descriptors. Not only are such insights important and interesting in their own right, but they could also guide future screenings in the rational control of crystal structures and the prediction of materials' properties.

Here, we focus on nucleation mechanisms during solidification in Ni. Elemental Ni has been widely used as a model system, for example, for nickel-based superalloys that are used as high temperature materials relevant in the transportation and energy sectors.[19] Consequently, understanding the crystallization processes in these metals is of great technological importance. Despite its relevance, even for simple metals like Ni, the microscopic mechanism of crystallization is not completely understood. Previously, we found that structural heterogeneity in the liquid plays an important role during solidification in Ni.[11,13] Pre-ordered liquid domains with distinctive bond-orientational order and fcc–hcp like structural features precede the formation of crystallites and act as precursors of the nucleation process, predetermining the polymorph that forms. We have also shown that pre-ordered liquid regions play a key role in the description of the reaction coordinate and the interfacial free energy[11] and, therefore, are key physical descriptors of the cluster structure and the mechanism. In a first attempt to derive predictive rules for the control of crystallization mechanisms, we have recently demonstrated that the ability of templates to modify the structural features of the liquid and, thus, its heterogeneities is directly linked to the nucleating ability of a template and the promoted polymorphs that emerge.[20] Precursor-mediated mechanisms, namely structural heterogeneity, have been shown to play a key role during the crystallization mechanism of several systems, such as ice, hard spheres,[17,18] and colloidal models.[15,16] Recently, a study of ice nucleation has revealed that dynamical heterogeneity is another key microscopic factor during crystal nucleation, where extended regions of reduced mobility in the melt precede the emergence of ice crystals, challenging the classical picture of diffusive single particle attachment to a growing crystal cluster.[10] It was also shown that relatively immobile regions are composed of structural hallmarks that resemble the final polymorph selected during ice nucleation, showing the clear correlation between structural and dynamical precursors during crystallization. In metallic liquids, it was shown that mobile atoms near the surface of nanowires promote crystallization by allowing a much more effective collective formation of crystal clusters.[21] Dynamical arrest and heterogeneity have also been found to be the hallmarks of glass transitions[22,23] where spatial correlations in the liquid emerge in terms of motion but not in structure. However, it remains to be understood how pre-ordered clusters in the liquid are correlated with regions of different mobility, in connection with the nucleation mechanism, and under which conditions higher or lower mobility yields crystallization or amorphisation. Given the significant role of structural heterogeneities in the crystallization and glass formation mechanisms in metals, the dynamical behaviour of the supercooled liquid in nickel alloys is bound to be a relevant factor in the nucleation mechanism. How the mobility of the liquid impacts the nucleation mechanism in Ni is, however, unknown.

In this work, by analysing a statistical ensemble of crystallization trajectories, obtained from transition path sampling simulations, we show that crystal nucleation in Ni occurs in extended domains of lowest mobility in the supercooled liquid. The regions of low mobility are spatially correlated with the emergence of structural precursors in the liquid that successfully yield critical fluctuations. Furthermore, we illustrate that a collective drop in mobility in the supercooled liquid precedes the formation of pre-ordered regions, indicating that dynamical arrest in the liquid plays a key role in facilitating pre-ordering and subsequent crystallization. Our results demonstrate that dynamical and structural heterogeneity in the liquid are key microscopic factors that promote crystal nucleation events. We expect that future investigations of the decoupling between dynamical arrest and structural pre-ordering in the supercooled liquid will be crucial to understand and predict the nucleating and glass forming ability of other metallic systems.

2. Simulation details and calculation of dynamical and structural properties

2.1 Computational setup

All simulations were performed for 3D periodic systems containing $N = 8788$ Ni atoms. The interactions between Ni atoms were modelled by an embedded atom model (EAM) potential.[24] The melting temperature of Ni for this potential is $T_\mathrm{m} = 1710$ K (ref. [25]) and we investigate nucleation at 20% undercooling, corresponding to $T = 1370$ K. All molecular dynamics (MD) simulations were performed with the programme package LAMMPS[26] using a time step of $\Delta t = 2$ fs, and the temperature and pressure were controlled by a Nosé–Hoover thermostat and barostat.

2.2 Dynamical heterogeneity

The relaxation dynamics in the supercooled liquid can be investigated by determining the self-intermediate scattering function[27]

$$F(\mathbf{q},t) = \langle \Phi(\mathbf{q},t) \rangle \tag{1}$$

with

$$\Phi(\mathbf{q},t) = \frac{1}{N}\sum_{j=1}^{N} \exp\left(i\mathbf{q}\left[\mathbf{r}_j(t) - \mathbf{r}_j(0)\right]\right), \tag{2}$$

where N is the number of particles, $\mathbf{r}(t)$ is the position of a particle at time t, and $\mathbf{q}$ is a reciprocal space vector. The dynamical heterogeneity (DH) in the supercooled liquid can be further characterised by the dynamical susceptibility $\chi_4(\mathbf{q},t)$[28,29]

$$\chi_4(\mathbf{q},t) = N[\langle|\Phi(\mathbf{q},t)|^2\rangle - \langle\Phi(\mathbf{q},t)\rangle^2]. \tag{3}$$

In an isotropic system, F and χ_4 can be evaluated for a given value of q_0 as the average over independent directions, with $F(q_0,t) = \langle F(\mathbf{q},t)\rangle_{\|\mathbf{q}\|=q_0}$ and $\chi_4(q_0,t) = \langle\chi_4(\mathbf{q},t)\rangle_{\|\mathbf{q}\|=q_0}$, respectively. The choice of q_0 controls the length scale at which the dynamical processes are observed. Here, we set q_0 to the first peak of the isotropic structure factor

$$S(q) = 1 + \frac{4\pi\rho}{q}\int_0^\infty \mathrm{d}r\, r\sin(rq)[g(r) - 1], \tag{4}$$

where ρ is the density of the liquid and $g(r)$ the radial distribution function. In liquid Ni at 20% undercooling, our simulations result in a value of $q_0 = 3.075$, correspondingly. We find that the supercooled liquid exhibits only weak DH with an almost exponential relaxation of $F(q_0,t)$. The time of maximum heterogeneity t_0, where the dynamics in the range of nearest neighbours are most heterogeneous, corresponds to the time where $\chi_4(q_0,t)$ has a maximum. For the investigated system, the corresponding value is $t_0 = 2.64$ ps at the level of supercooling examined.

Spatially resolved information about the DH is obtained by evaluating the dynamical propensity (DP)[10,30] of each particle,

$$\mathrm{DP}_i(t_0) = \left\langle\frac{\|r_i(t_0) - r_i(0)\|^2}{\mathrm{MSD}}\right\rangle_{\mathrm{ISO}}. \tag{5}$$

The average is taken over the so-called isoconfigurational ensemble,[10,31,32] where a number of MD trajectories are initiated from the same configuration with random velocities drawn from the Maxwell–Boltzmann distribution for a given temperature. MSD denotes the mean squared displacement for time t_0. DP values of <1.0 indicate particles that are less mobile than the average, and DP >1.0 for particles that are more mobile.

2.3 Structural characterisation

To identify solid- and liquid-like particles in the simulations, the criterion introduced by ten Wolde *et al.*[33] based on Steinhardt bond order parameters[34,35] was used. A bond between two particles i and j is characterised as solid if $s_{\mathrm{ij}} = \sum_{m=-6}^{6} q_{6m}(\mathrm{i})q^*_{6m}(\mathrm{j}) > 0.5$, where q_{6m} are the complex vectors given by the sum over spherical harmonics with $l = 6$. In addition, the average correlation over neighbouring atoms, $\langle s_{\mathrm{i}}\rangle = 1/N_{\mathrm{nn}}\sum s_{\mathrm{ij}}$, was considered to improve the definition of solid particles at the interfaces between solid clusters and the liquid. A particle i with at least 7 solid bonds and $\langle s_{\mathrm{i}}\rangle > 0.6$ is identified as solid. The size of the largest cluster composed of solid particles is denoted by n_{s}.

The local structure around each particle was characterised using the averaged local bond-order parameters[36] $\bar{q}_4$ and $\bar{q}_6$. Details concerning the computation of the corresponding $\bar{q}_4$, $\bar{q}_6$ reference map for face-centred cubic (fcc), body-centred cubic (bcc), hexagonal closed-packed (hcp), and liquid Ni at 20% undercooling can be found in our previous publication.[13]

3. Results

3.1 Supercooled liquid Ni exhibits mild dynamical heterogeneity

A long MD trajectory of 10 ns was run in the supercooled liquid at 20% undercooling to equilibrate the systems. From this trajectory, 200 configurations were randomly selected to compute the dynamical propensity of particles in the liquid. For each configuration, 100 MD simulations with $t_0 = 2.64$ ps were performed to calculate the ensemble average using eqn (5).

The probability distribution of DP values in the supercooled liquid is shown in Fig. 1. The distribution peaks at around DP = 1 and is fairly narrow, reflecting the rather small degree of dynamical heterogeneity in the liquid at 20% undercooling. Similar to previous studies,[10,30] the top and bottom 5% of the distribution are identified as the most mobile (MM, DP > 1.236) and most immobile (MI, DP < 0.728) particles, respectively. Together with the probability distribution, a representative configuration of the supercooled liquid is shown in Fig. 1, where the atoms are coloured according to their DP values. The spatial distribution of fast (red) and slow (blue) atoms appears to be mostly random and there is no strong indication of low mobility domains, as expected for a liquid with little dynamical heterogeneity. Our analysis of larger undercoolings, near the nucleation temperature ($\Delta T \sim 30\%$), showed similar behaviour with a slight broadening of the distribution of DP values, indicating that the magnitude of dynamical heterogeneity is not remarkably increased with supercooling. The mild dynamical heterogeneity in supercooled liquid Ni is probably associated with the poor glass forming ability of elemental metals. In contrast, binary alloys and other glass forming materials have been shown to exhibit significant dynamical heterogeneity in the liquid, where MI regions yield dynamical arrest and suppress the emergence of structural precursors of crystallization.[3,30] Nevertheless, regions of varying mobility (MI and MM) can be distinguished from random fluctuations in the liquid at 20% undercooling (Fig. 1b). In the next section, we further analyse

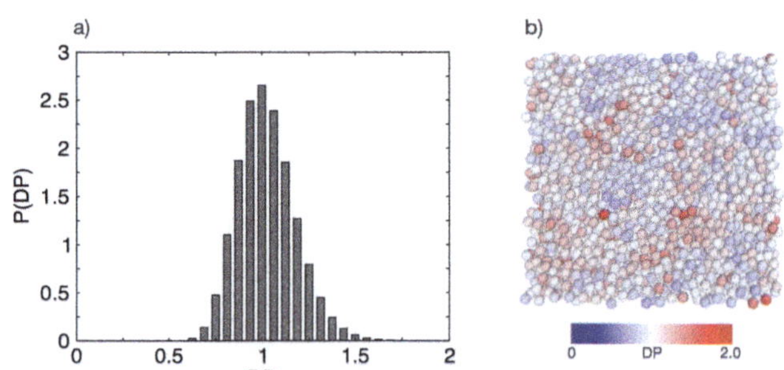

Fig. 1 (a) Probability distribution of DP values in liquid Ni at 20% undercooling; the distribution is centred around DP = 1 and is fairly narrow. (b) Snapshot of a representative configuration in the supercooled liquid; atoms are coloured according to their DP value, indicating high (red) and low (blue) mobility.

the correlation of MM and MI regions to structural precursors in the liquid, in connection to the nucleation mechanism during solidification in Ni.

3.2 Structural pre-ordering without a decrease in mobility does not promote nucleation

Using transition path sampling (TPS)[37–40] simulations, we established in our previous work[11,13] that homogeneous nucleation in Ni proceeds *via* the initial formation of precursors from which the crystalline phase emerges. These precursors are composed of solid-like particles that have higher bond-orientational order than the liquid but less than any of the crystalline phases. This pre-ordering in the liquid is associated with a significant contribution to the nucleation barrier ($\approx$1 eV at 20% undercooling with a barrier of $\Delta G = 4.21$ eV) and plays a crucial role in the reaction coordinate describing the nucleation mechanism.[11] In addition, we have recently shown that the efficiency of precursors to facilitate the formation of a crystalline bulk phase strongly depends on their structural characteristics.[20] Here, we focus on pre-critical clusters with $n_s = 50$ that are predominantly (>90%) composed of pre-structured liquid particles. The distribution of $\bar{q}_4$, $\bar{q}_6$ values of the pre-critical clusters is shown in the left graph of Fig. 2 (purple and pink), together with the distributions of bulk fcc (red), hcp (green), bcc (blue), and liquid (grey). The clusters have been extracted from 400 trajectories of the transition path ensemble (TPE) of homogeneous nucleation in Ni (details concerning the TPS simulations are given in ref. [11] and [13]). As indicated above, pre-ordered liquid particles exhibit a bond-orientational order that differs from both those of the liquid and the crystalline phases. The purple distribution represents pre-critical clusters from trajectories that successfully nucleate a crystalline phase and continue to grow, whereas the pink distribution corresponds to clusters that eventually dissolve. Effective precursors (purple) that successfully initiate the emergence of the crystalline phase and grow beyond the critical nucleus size show a clear increase in bond-orientational order compared to unsuccessful pre-critical clusters (pink) that shrink and dissolve.

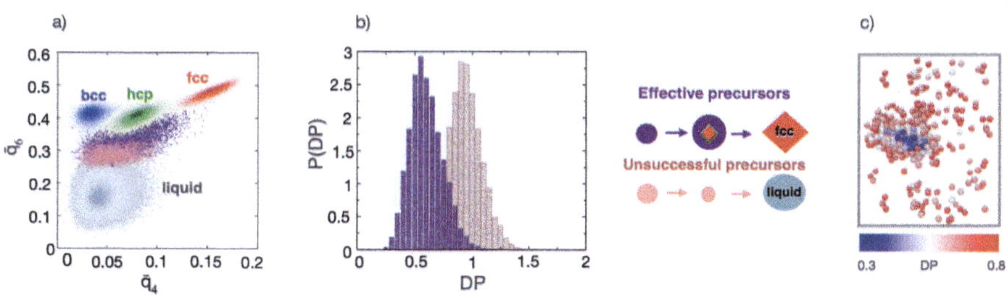

Fig. 2 (a) Map showing the distribution of averaged local bond-order parameters $\bar{q}_4$ and $\bar{q}_6$ for the crystalline bulk phases fcc (red), hcp (green), and bcc (blue), as well as the liquid (grey), together with the values for pre-critical clusters that successfully continue to grow (purple) or dissolve again (pink). Figure adapted from ref. [20]. (b) Probability distribution of DP values for effective (purple) and unsuccessful (pink) precursors; the distribution of the unsuccessful precursor is comparable to the one of the liquid shown in Fig. 1a, while the one of the effective precursors is clearly shifted to lower DP values. (c) Representative configuration of an effective precursor; only atoms with DP < 0.728 (most immobile) are shown; atoms are coloured according to their DP value.

The dynamical propensity of atoms in the pre-critical clusters was obtained by selecting 200 configurations with $n_s = 50$ from the TPE of our previous study[11,13] and running 100 MD simulations with $t_0 = 2.64$ ps for each configuration to calculate the ensemble average using eqn (5). The corresponding probability distributions of DP values are shown in the middle graph of Fig. 2. Atoms comprising effective precursors show a strong decrease in DP compared to the liquid, whereas the DP distribution of atoms in unsuccessful precursors is very similar to the one of the liquid shown in Fig. 1a. This is rather interesting, since these particles are characterised as solid according to the criterion outlined in Section 2.3 and, correspondingly, exhibit a higher bond-orientational order than the liquid, evidenced by their $\bar{q}_4$, $\bar{q}_6$ distribution. Still, their dynamical properties are comparable to those of liquid particles.

By comparing the structural and dynamical properties of effective and unsuccessful precursors, it appears that the structural and dynamical heterogeneity affect the nucleation process differently. While both effective and unsuccessful precursors show an increase in bond-orientational order compared to the liquid, the DP is reduced only for particles in effective structural precursors that yield critical fluctuations. Structural pre-ordering in the liquid without a decrease in mobility, therefore, does not promote the nucleation of the crystalline bulk phase. A snapshot of a configuration with an effective precursor is shown on the right in Fig. 2. The atoms are coloured according to their DP value and only atoms with DP < 0.728 (most immobile) are shown. These particles clearly form a region of lowest mobility in which the largest pre-structured cluster that succeeds in nucleating is embedded.

3.3 Structural pre-ordering emerges in low mobility regions

To follow the evolution of low mobility regions during nucleation, that is regions comprised of the MI particles, we analyse 100 AB-paths (trajectories that successfully crystallise from the liquid state A to the solid state B) from the TPE of our previous study.[11,13] In addition to the largest solid cluster n_s, the largest cluster

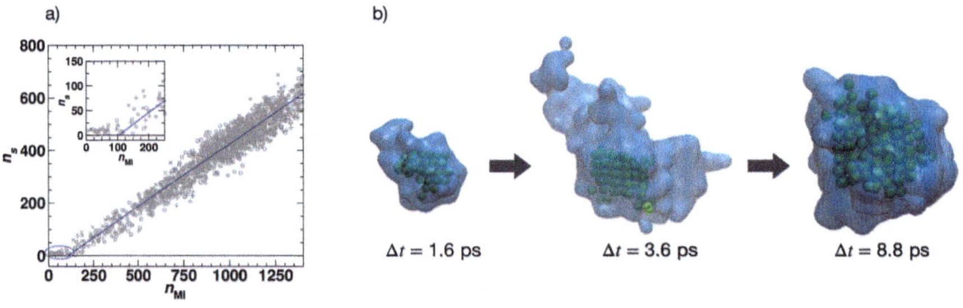

Fig. 3 (a) Correlation between the largest solid cluster, n_s, and the largest cluster of most immobile particles, n_{MI}. Up to $n_{MI} \approx 100$, n_s is approximately constant (marked by the blue ellipse) before it increases linearly with n_{MI}. The data are extracted from five representative AB-trajectories of the TPE (marked by various symbols). The blue line shows a linear fit to the data. (b) Three configurations along a representative AB-trajectory; the blue, transparent region denotes the largest MI cluster, n_{MI}, the green spheres show the largest solid cluster, n_s; the time below each configuration corresponds to the transition time, that is the time after leaving the stable liquid state.

composed of MI particles n_{MI} is determined for each time slice in the AB-trajectories. In Fig. 3a, the number of particles in the largest solid cluster is plotted as a function of the number of particles in the largest MI cluster for five representative trajectories. Up to $n_{MI} \approx 100$, the size of the n_s is small and remains roughly constant, indicating that the structural pre-ordering in the supercooled liquid is preceded by a distinctive dynamical arrest of spatial regions in the supercooled liquid and a decoupling of the dynamical and structural properties of the melt during induction time. For larger sizes of n_{MI}, n_s increases linearly with n_{MI}, showing a strong correlation of the MI regions with structural fluctuations that yield crystallization. The regions of low mobility are always larger than the largest solid clusters, and for stable clusters that continue to grow and crystallize, the overlap between particles that belong to both n_s and n_{MI} is >90%, showing that crystallites nucleate in regions of lowest mobility in the supercooled liquid. Consequently, the low mobility regions encompass the regions of increased structural ordering. The linear correlation between n_s and n_{MI} with a slope of $n_{MI}/n_s \approx 2$ suggests that, within a spherical approximation, the radius of the low mobility region is ≈25% larger than the radius of the largest solid cluster.

The largest MI cluster is visualised, together with the largest solid cluster for three configurations, along a representative AB-trajectory in Fig. 3b. The times in Fig. 3b correspond to the transition time without the residence time in the stable liquid state, since trajectories in the TPE correspond only to the actual transition itself, while the stable states are sampled separately. It is clearly visible that the largest solid cluster (green spheres) is located within the regions of low mobility (blue transparent) at different times along the nucleation process. The behaviour depicted in Fig. 3b is typical of all trajectories in the ensemble.

4. Conclusions

We have analysed the structural and dynamical properties of atoms during homogeneous nucleation in nickel. Our results reveal a clear correlation between the formation of low mobility regions in the liquid and the formation of structural precursors that precede the emergence of the crystalline bulk phase. Our findings are similar to the results reported for the nucleation of ice,[10] even though Ni, as an elemental metal, exhibits much less dynamical heterogeneity in the supercooled liquid. Still, we find that the dynamical propensity of the atoms plays a key role in the initial formation of structurally pre-ordered regions in the liquid.

Most notably, the distribution of DP values for successful precursors that continue to grow and crystallise into the bulk phase is significantly shifted compared to the distribution of DP values for liquid particles. In contrast, solid clusters that dissolve have DP values similar to the average value of the liquid. The dynamical properties are thus crucial in the description of the nucleation mechanism. As the solid clusters continue to grow, they are always embedded in a region of low mobility and atoms in the vicinity of the solid cluster exhibit clearly decreased DP values.

Understanding the role of dynamical heterogeneity in nucleation mechanisms has implications that go beyond the description of homogeneous nucleation. In particular, our findings further support the idea that in heterogeneous nucleation, structural templating is not necessarily the only decisive factor, but the efficiency of nucleating agents might also be determined by their ability to modify the dynamical properties of the liquid. Including dynamical heterogeneity in the study of nucleation mechanisms, for both homogeneous and heterogeneous nucleation, provides an additional perspective on the underlying atomistic processes that are key to controlling the central parameters during crystallisation, including nucleation rates and polymorph selectivity.

Author contributions

All authors contributed to the conceptualisation of the study. GDL and JR performed the investigation, the formal analysis and visualisation of the research data, and wrote the original draft. All authors reviewed and edited the final draft.

Conflicts of interest

There are no conflicts of interest to declare.

Acknowledgements

The authors would like to thank Martin Fitzner for fruitful discussions and support in preparing the simulations for the dynamical heterogeneity of the liquid. JR acknowledges financial support from the Deutsche Forschungsgemeinschaft (DFG) through the Heisenberg Programme project 428315600. GDL acknowledges support from Conacyt-Mexico through fellowship Ref. No. 220644. GDL and AM acknowledge support from the Isaac Newton Trust Grant Ref No: 20.40(h). The authors acknowledge computing time from UKCP (EP/P022561/1) and the UK Materials and Molecular Modelling Hub, which is partially funded by EPSRC (EP/P020194/1 and EP/T022213/1).

Notes and references

1 G. C. Sosso, J. Chen, S. J. Cox, M. Fitzner, P. Pedevilla, A. Zen and A. Michaelides, *Chem. Rev.*, 2016, **116**, 7078–7116.
2 K. E. Blow, D. Quigley and G. C. Sosso, *J. Chem. Phys.*, 2021, **155**, 040901.
3 J. Russo, F. Romano and H. Tanaka, *Phys. Rev. X*, 2018, **8**, 021040.
4 D. Gebauer, M. Kellermeier, J. D. Gale, L. Bergström and H. Cölfen, *Chem. Soc. Rev.*, 2014, **43**, 2348–2371.
5 P. R. Ten Wolde and D. Frenkel, *Science*, 1997, **277**, 1975–1978.
6 P. R. Ten Wolde and D. Frenkel, *Phys. Chem. Chem. Phys.*, 1999, **1**, 2191–2196.
7 T. Zhang and X. Y. Liu, *J. Am. Chem. Soc.*, 2007, **129**, 13520.
8 S. Prestipino, A. Laio and E. Tosatti, *Phys. Rev. Lett.*, 2012, **108**, 225701.
9 J. Russo and H. Tanaka, *Sci. Rep.*, 2012, **2**, 505–512.
10 M. Fitzner, G. C. Sosso, S. J. Cox and A. Michaelides, *Proc. Natl. Acad. Sci. U. S. A.*, 2019, **116**, 2009–2014.
11 G. Díaz Leines and J. Rogal, *J. Phys. Chem. B*, 2018, **122**, 10934–10942.
12 H. Tanaka, *Eur. Phys. J. E: Soft Matter Biol. Phys.*, 2012, **35**, 113.
13 G. Díaz Leines, R. Drautz and J. Rogal, *J. Chem. Phys.*, 2017, **146**, 154702.
14 Q. Zhang, J. Wang, S. Tang, Y. Wang, J. Li, W. Zhou and Z. Wang, *Phys. Chem. Chem. Phys.*, 2019, **21**, 4122–4135.

15 W. Lechner, C. Dellago and P. G. Bolhuis, *Phys. Rev. Lett.*, 2011, **106**, 085701.
16 P. Tan, N. Xu and L. Xu, *Nat. Phys.*, 2014, **10**, 73.
17 T. Schilling, H. J. Schöpe, M. Oettel, G. Opletal and I. Snook, *Phys. Rev. Lett.*, 2010, **105**, 025701.
18 H. J. Schöpe, G. Bryant and W. van Megen, *Phys. Rev. Lett.*, 2006, **96**, 175701.
19 R. C. Reed, *The Superalloys: Fundamentals and Applications*, Cambridge University Press, Cambridge, 2006.
20 G. Díaz Leines and J. Rogal, 2021, arXiv:2106.00147.
21 P. Zhang, J. J. Maldonis, Z. Liu, J. Schroers and P. M. Voyles, *Nat. Commun.*, 2018, **9**, 1129.
22 M. D. Ediger, *Annu. Rev. Phys. Chem.*, 2000, **51**, 99.
23 J. P. Garrahan, *Proc. Natl. Acad. Sci. U. S. A.*, 2011, **108**, 4701–4702.
24 S. M. Foiles, M. I. Baskes and M. S. Daw, *Phys. Rev. B: Condens. Matter Mater. Phys.*, 1986, **33**, 7983–7991.
25 J. J. Hoyt, D. Olmsted, S. Jindal, M. Asta and A. Karma, *Phys. Rev. E: Stat., Nonlinear, Soft Matter Phys.*, 2009, **79**, 020601.
26 S. Plimpton, *J. Comput. Phys.*, 1995, **117**, 1–19.
27 L. Berthier and G. Biroli, *Rev. Mod. Phys.*, 2011, **83**, 587–645.
28 C. Donati, S. Franz, S. C. Glotzer and G. Parisi, *J. Non-Cryst. Solids*, 2002, **307–310**, 215–224.
29 C. Toninelli, M. Wyart, L. Berthier, G. Biroli and J.-P. Bouchaud, *Phys. Rev. E: Stat., Nonlinear, Soft Matter Phys.*, 2005, **71**, 041505.
30 G. C. Sosso, J. Colombo, J. Behler, E. Del Gado and M. Bernasconi, *J. Phys. Chem. B*, 2014, **118**, 13621–13628.
31 A. Widmer-Cooper and P. Harrowell, *J. Chem. Phys.*, 2007, **126**, 154503.
32 J. Colombo, A. Widmer-Cooper and E. Del Gado, *Phys. Rev. Lett.*, 2013, **110**, 198301.
33 P. R. ten Wolde, M. J. Ruiz-Montero and D. Frenkel, *Phys. Rev. Lett.*, 1995, **75**, 2714–2717.
34 P. J. Steinhardt, D. R. Nelson and M. Ronchetti, *Phys. Rev. B: Condens. Matter Mater. Phys.*, 1983, **28**, 784–805.
35 S. Auer and D. Frenkel, *Adv. Polym. Sci.*, 2005, **173**, 149–207.
36 W. Lechner and C. Dellago, *J. Chem. Phys.*, 2008, **129**, 114707.
37 C. Dellago, P. Bolhuis and P. L. Geissler, *Adv. Chem. Phys.*, 2002, **123**, 1–78.
38 T. S. van Erp and P. G. Bolhuis, *J. Comput. Phys.*, 2005, **205**, 157–181.
39 T. S. van Erp, *Phys. Rev. Lett.*, 2007, **98**, 268301.
40 P. G. Bolhuis, *J. Chem. Phys.*, 2008, **129**, 114108.

Faraday Discussions

ROYAL SOCIETY OF CHEMISTRY

PAPER

Superlattice ordering transitions driven by short-range structure in barium calcium carbonates

Michael. L. Whittaker, †*[a] Efrat Pri-gal,[b] Asher Schmidt [b] and Derk Joester [a]

Received 9th November 2021, Accepted 2nd February 2022
DOI: 10.1039/d1fd00086a

Balcite ($Ba_xCa_{1-x}CO_3$) is a synthetic analog of rhombohedral carbonate minerals like calcite and dolomite that is disordered on both the cation and anion sublattices. Here, we show that multiple exotic superlattice structures, including a dolomite analog that we call balcomite, can form from balcite at elevated temperatures. The second-order balcite-to-balcomite conversion at temperatures between 150–600 °C is driven by the preference of barium and calcium for different oxygen coordination numbers and facilitated by local carbonate reorientation. At elevated pressure, further superlattice order arises from cation segregation in all three dimensions, producing a supercell with the same $R\bar{3}m$ symmetry as balcite but 6× larger. This highly ordered structure relaxes back to the balcomite structure upon returning to ambient conditions. None of the three naturally occurring polymorphs of $Ba_{0.5}Ca_{0.5}CO_3$ (alstonite, paralstonite, barytocalcite) formed from balcite despite being putatively energetically favored. Instead, alstonite transformed to a balcomite-like structure *via* a first-order process after transiently converting to a paralstonite-like structure *via* a second-order process. Together, these results show that high temperature transformation pathways between structures in the barium calcium carbonate system can be driven by coarsening and are facilitated by similarity in short-range order, conceptually analogous to previously described low-temperature transformations. Many of the exotic high temperature carbonate structures are unstable, but may participate in transformation pathways between naturally observed metastable mineral phases, suggesting important roles for ephemeral phases in shaping past and current mineral distributions.

[a]*Department of Materials Science and Engineering, Northwestern University, 2220 Campus Dr. Evanston, IL, 60208, USA. E-mail: mwhittaker@lbl.gov*

[b]*Schulich Faculty of Chemistry and Russell Berrie Nanotechnology Institute, Technion – Israel Institute of Technology, Haifa 32000, Israel*

† Current address: Energy Geosciences Division, Lawrence Berkeley National Laboratory, 1 Cyclotron Rd, Berkeley, 94720, CA, USA.

Introduction

Mixed-cation $CaCO_3$–$BaCO_3$ phases play a unique biogeochemical role as alkaline carbonate sediments,[1,2] paleoclimate proxies,[3–5] and biologically induced precipitates[6,7] including the first known cyanobacterial intracellular biominerals.[8,9] All natural and experimentally synthesized structures in this system lie within the metastability window[10] approximately 20 kJ mol^{-1} above ground-state, phase-separated calcite ($CaCO_3$) and witherite ($BaCO_3$), and below amorphous calcium barium carbonate (ACBC, $Ba_xCa_{1-x}CO_3$).[11,12] However, simulations indicate that only a few of many possible structures occur naturally in the lithosphere despite relatively small energy differences between myriad metastable phases.[13] If differences in stability are small relative to the barriers to interconversion between phases, as they appear to be in the $CaCO_3$–$BaCO_3$ system,[14] a better understanding of mineral transformation pathways may offer important insight into how these minerals are distributed and their implications for reconstructing past conditions.[15]

Despite its persistence for months in aqueous solution,[12] balcite ($Ba_{0.5}Ca_{0.5}CO_3$ with $R\bar{3}m$ symmetry) is unstable relative to calcite and witherite[11,12] and, putatively, to three naturally occurring $Ba_{0.5}Ca_{0.5}CO_3$ polymorphs (alstonite, paralstonite, barytocalcite).[2,16,17] Both the carbonate anion orientation and cation distribution are highly disordered in balcite,[18,19] while the naturally occurring polymorphs are thought to have both cation and carbonate order.[17,20] Here, we investigate the recrystallization of $Ba_{0.5}Ca_{0.5}CO_3$ polymorphs at elevated temperatures and find evidence for complex crystallization pathways among many exotic carbonate structures. Importantly, we show that long-range ordered structures form through continuous, second-order transformation from precursors with similar short-range order, but that interconversion between structures with distinct short-range order occurs *via* discontinuous, first-order transitions.

Results

hr-pXRD, Rietveld refinement, and *in situ* MAXS/WAXS

Balcite was synthesized under ambient conditions and subsequently annealed at a controlled rate in order to observe the products and pathways by which it transforms. As-synthesized balcite with $x = 0.50$ (Fig. 1a) was annealed to 600 °C

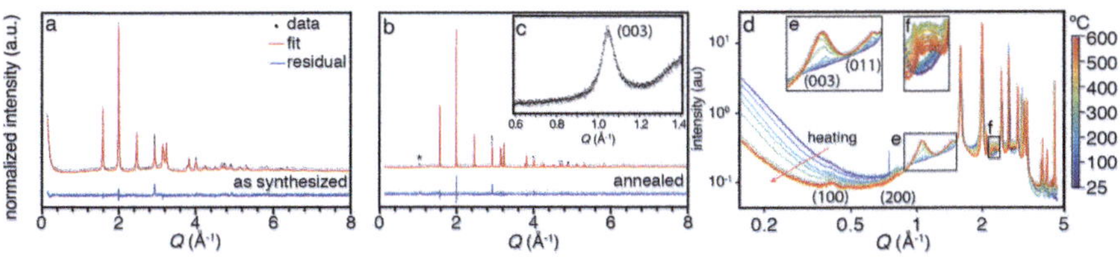

Fig. 1 (a) hr-pXRD of as-synthesized balcite and (b) balcite annealed to 600 °C. (c) A small, broad peak emerged at $Q = 1.05$ Å^{-1}, consistent with a superlattice reflection along the c-axis. (d) *In situ* WAXS/MAXS of balcite during heating from 25–600 °C. (e) Gradual emergence of superlattice reflections. (f) Abrupt formation of peaks associated with carbonate ordering.

(Fig. 1b and c) at a rate of 10 °C minute^{-1} and held for 10 minutes, which led to a significant decrease in the width of Bragg peaks that were initially very broad. As-synthesized balcite had an average grain size of 34 ± 4 nm, consistent with previous measurements,[18] while annealed balcite grains exceeded 250 nm, as determined by Rietveld refinement. Loss of diffraction intensity at scattering vectors $Q < 0.5$ Å^{-1} in the annealed sample is consistent with a reduction in small-angle scattering from nanoscale grains (Fig. 1b and c). Scattering intensity at $Q <$ 0.5 Å^{-1} in as-synthesized balcite shifted to lower Q in medium- and wide-angle X-ray scattering (MAXS/WAXS) collected simultaneously while heating *in situ*, and diffraction peaks also grew narrower as the temperature increased (Fig. 1d–f). Together, these observations indicate that initially nanoscale grains coarsened at elevated temperature. Coarsening began around 150 °C, with grains appearing to reach a constant size by 500 °C, at which point the peak widths and small-angle scattering intensity did not change appreciably.

Coarsened grains also adopted crystalline order that was distinct from the as-synthesized nanocrystalline material. Most notably, a peak at $Q = 1.05$ Å^{-1} emerges around 150 °C and grows until around 500 °C, concurrent with coarsening. The gradual appearance of this reflection during *in situ* heating is indicative of compositional evolution on alternating cation lattice planes, and is consistent with the self-segregation of barium and calcium that were initially randomly mixed. This peak is analogous to the (003) superlattice ordering peak observed in dolomite[21,22] that is associated with the segregation of Ca and Mg to alternating lattice planes, doubling the unit cell along the c axis. Cation superlattice order that is consistent with $R\bar{3}c$ symmetry in a unit cell with $c = 2c_{\mathrm{balcite}}$ persisted after the sample was cooled, an indication of quenched crystallographic reorganization that was irreversible at ambient conditions.

Additional peaks at $Q = 0.405$ Å^{-1}, 0.810 Å^{-1}, 1.43 Å^{-1}, and 2.20 Å^{-1} emerged over this temperature range during *in situ* heating (Fig. 1d), which occurred within a hermetically sealed aluminum crucible, but were not observed after the sample was returned to ambient temperature and pressure (Fig. 1b and c). This suggests that additional structural evolution occurs at elevated pressure inside the sealed crucible, which is likely due to trapped structural water in the as-synthesized material (see section on NMR). These peaks are consistent with a superlattice structure that forms along both the a and c axes. In addition to the superlattice peak at $Q = 1.05$ Å^{-1}, peaks $Q = 0.405$ Å^{-1} and 0.810 Å^{-1} correspond to (100) and (200) reflections from a cell with $a = 3a_{\mathrm{balcite}} \cong 17.95$ Å, and the peak at $Q = 1.433$ Å^{-1} is a superlattice reflection of the balcite (022) at $Q = 2.866$ Å^{-1} (Fig. 1e). Thus, under pressure the system evolves towards an $R\bar{3}m$ supercell with $a/c \cong 1$ in which cations are ordered in three dimensions. However, this structure is unstable, and only the cation superlattice ordering along the c axis remains upon returning to ambient conditions.

Carbonate anion orientation, which is a key differentiator between calcite and balcite structures, is tightly linked to the cation coordination and sensitive to different ratios of barium and calcium.[18] It is expected that the formation and (de)stabilization of both the $R\bar{3}c$ superlattice and $R\bar{3}m$ supercell structures are mediated by carbonate anion orientation on a sublattice for which the carbon remains essentially fixed. While the most prominent peaks that gradually emerge upon heating under pressure are predominantly associated with cation ordering (Fig. 1e), a set of peaks associated with carbonate order appeared abruptly around

500 °C above a dramatically increased background (Fig. 1f). These structures appear to relax back to the dolomite-like superlattice structure upon returning to ambient conditions, suggesting that short-range restructuring occurs on both cation and anion lattices in response to changing pressure. To understand the mechanisms behind these transformations, the local structure of both carbonates and cations were investigated by solid-state NMR and XANES/EXAFS spectroscopy.

Solid-state NMR

Short-range carbonate order in balcites with different barium content was compared to natural $Ba_{0.5}Ca_{0.5}CO_3$ polymorphs alstonite and barytocalcite using direct excitation (DE) magic angle spinning (MAS) ^{13}C NMR to determine whether the cation composition affects the carbonate environments. Peak position and width (isotropic chemical shift, δ_{iso} ($\delta\nu$), in ppm) were used to determine the uniqueness and degree of disorder, respectively, of each carbonate environment. The barium content of balcite was shown by the ^{13}C DE-MAS NMR spectra to have a negligible effect on carbonates in the as-synthesized material, yet a significant impact on its structure after heating (Fig. 2a and b).

Balcites precipitated from aqueous solutions with barium fractions of x_{feed} = 0.24, 0.31, 0.41, 0.51 exhibited one major peak centered at 167.6 ppm with a peak-

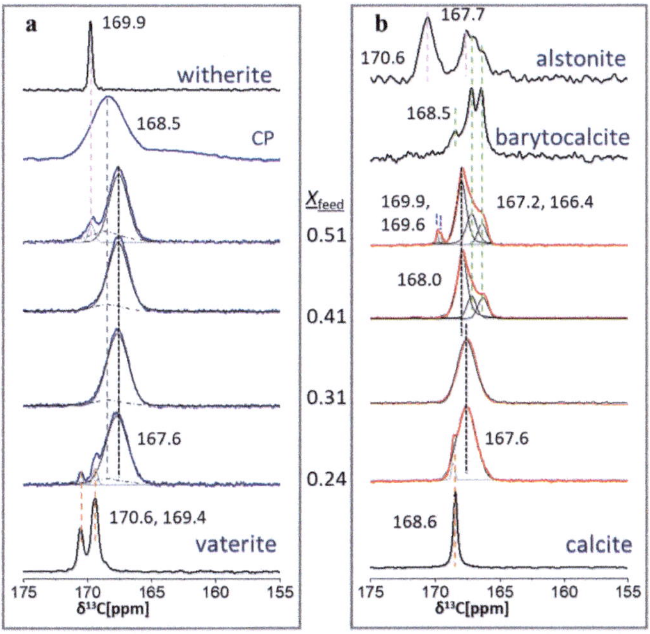

Fig. 2 ^{13}C DE MAS NMR spectra of balcite, barytocalcite, and alstonite; the spectra of vaterite, calcite and witherite are added for comparison. (a) As-synthesized (blue) balcite samples with x_{feed} 0.24–0.51 and (b) after heating to 600 °C and cooling back to room temperature (red). A representative ^{13}C CP MAS spectrum of the balcite samples ((a) second from top, x_{feed} = 0.41; denoted CP) demonstrates the occurrence of disordered hydrated carbonates in all balcite samples. Vertical dashed lines are provided to guide the eye to the different isotropic chemical shifts.

width of 2.0 ppm, ten-fold broader than that of calcite (168.6; <0.2 ppm) and witherite (169.9; <0.3 ppm).[23] The same chemical shifts and large peak widths across the various balcite compositions are indicative of disordered carbonate environments that do not depend on the barium content. Additionally ^{13}C CP MAS spectra identify that as-synthesized balcite contains disordered, hydrated carbonates.[23] Cross polarization (CP) NMR from hydrogen to carbon atoms is a complementary technique that selectively enhances the peaks of carbonate species in the vicinity of hydrogen. A notable hydrated carbonate peak (168.5; 3.0 ppm) in the CP MAS spectra (Fig. 2a; CP) of as-synthesized balcite is also present in DE MAS spectra (Fig. 2a; rectangle) and shows a hydrated carbonate content of ~15% in the three lower x balcites, and ~25% in the higher x balcite.

Balcite was annealed to 600 °C at 10 °C minute^{-1} and cooled back to 25 °C at the same rate and re-examined by DE-MAS NMR (Fig. 2b). The balcite peak at 167.6 (1.6) ppm in $x = 0.24$ and 0.31 samples did not change appreciably after heating, suggesting that short-range carbonate heterogeneity is retained. Vaterite present in the $x = 0.24$ balcite sample (Fig. 2a) transformed to calcite upon heating (Fig. 2b), as evidenced by the replacement of the two vaterite peaks at 169.4 and 170.6 ppm with a single narrow calcite peak at 168.6 ppm. Absence of the peaks in the CP MAS spectra of heated balcites confirmed that heating dehydrated all carbonates, indicating that the disordered carbonate environments in the annealed materials are due to intrinsic structural disorder and not to defects induced by residual water molecules.

Upon annealing the $x = 0.41$ and 0.51 balcite samples the balcite peak splits into at least three distinct components at 168.0, 167.2, and 166.4 ppm, each with ~1.0 ppm linewidth. Higher barium content resulted in higher intensity for the 167.2 ppm peak, changing the intensity ratio from approximately 4 : 1 : 1 at $x = 0.41$ to 4 : 2 : 1 at $x = 0.51$. Thus, three unique carbonate sites are established upon heating, of which at least one changes with differing amounts of calcium and barium. Witherite remained in the annealed $x = 0.51$ sample as a secondary phase.

Alstonite and barytocalcite also contain multiple carbonate environments (Fig. 2b; top), some of which coincide with the chemical shift of those identified in annealed balcite. However, the presence of unique peaks and varying ratios between the peaks distinguished the three phases. Barytocalcite exhibits three peaks, 168.5 (0.7), 167.2 (1.3) and 166.4 (0.5) ppm. At least two are found at the same chemical shift as in annealed balcite, and have an intensity ratio similar to balcite with $x = 0.51$ (and same linewidths of 1.0 ppm). A third peak at 168.5 ppm is far lower in intensity relative to the other two peaks. Alstonite exhibited four peaks with an intensity ratio of approximately 8 : 6 : 1 : 1. A broad, intense peak at 170.6 (1.2) ppm clearly distinguishes it from all other structures examined. However, alstonite also had three peaks at 167.7 (0.5), 167.1 (1.3) and 166.2 (0.5) ppm that appear consistent with those in balcite and barytocalcite.

Mineral samples with natural ^{13}C abundance had much lower signal-to-noise compared to isotopically enriched synthetic samples. Nonetheless, in all cases, multiple carbonate environments were present, at least one of which (167.2 ppm) appears to be present in each and is sensitive to the barium content. X-ray absorption spectroscopy at both the calcium and barium K-edge was used to obtain a complementary picture of the local structure and better understand the carbonate chemical shift sensitivity from the perspective of the cations.

XANES/EXAFS

Cation coordination numbers (CN) and symmetry of carbonate oxygens in the first coordination shell of the cations are reflected in the near-edge structure (XANES, Fig. 3a and d), while bond lengths and distances were extracted from fits to EXAFS data (Fig. 3b, c, e, f and Table 1) for calcite, witherite, and balcite with $x = 0.5$ and the ACBC from which it formed.

The position of the calcium K-edge edge (defined here as the maximum in the first derivative of the intensity) is very sensitive to the coordination number (CN), with a shift of 0.7 ± 0.1 eV towards higher energy (relative to the calcite edge) for each additional oxygen beyond the sixth.[24] The edge energy of ACBC and balcite were both shifted 0.5 eV relative to calcite, an indication that the average CN in both materials was between six and seven. A weak pre-edge feature corresponds to the 1s → 3d transition, ≈10 eV below the edge. This feature is a singlet and is markedly more intense in ACBC, indicating that the configuration of the first shell lacks an inversion center and is more disordered than calcite or barytocalcite.[25]

Data in the calcium K-edge EXAFS region were fit in *R*-space with a structural model based on calcite (Fig. 3b, c and Table 1). Two separate distances were needed for each Ca–O_1 (first shell), Ca–C_1 (second shell), and Ca–O_2 (third shell) for both ACBC and balcite. ACBC had shorter Ca–O_1 distances (2.19, 2.39 Å) than balcite (2.37, 2.55 Å). The CN-weighted average Ca–O distance was 2.32 Å in ACBC compared to 2.42 Å in balcite, indicating that first-shell oxygen was more tightly

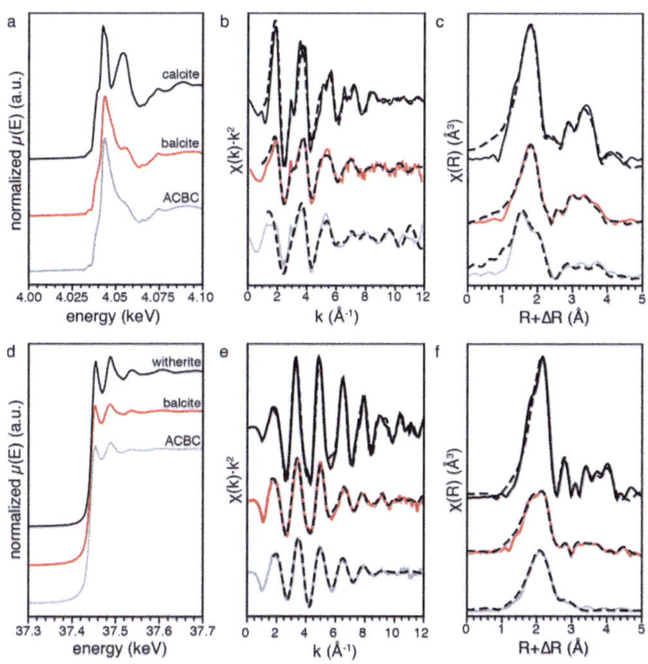

Fig. 3 X-ray absorption spectroscopy at the calcium K-edge (a–c) and barium K-edge (d–f) of calcite, balcite with $x = 0.5$, and ACBC with $x = 0.55$. (a and d) XANES, (b and e) *k*-space EXAFS, and (c and f) *R*-space EXAFS.

Table 1 XAS fitting parameters

	Ca K-edge			Ba K-edge		
	CN	R Ca–O (Å)	σ^2 (Å^2)	CN	R Ba–O (Å)	σ^2 (Å^2)
Calcite	6	2.39 (0.01)	0.010 (0.001)	—	—	—
ACBC	2.28, 4.22	2.19, 2.39 (0.01)	0.001, 0.004 (0.001)	3.17, 4.83	2.68, 2.83 (0.01)	0.005, 0.007 (0.002)
Balcite	4.27, 2.23	2.37, 2.55 (0.02)	0.007, 0.006 (0.004)	1.74, 6.26	2.55, 2.73 (0.01)	0.006, 0.007 (0.001)
Witherite	—	—	—	3, 4, 2	2.73, 2.85, 2.91 (0.02)	0.004, 0.008, 0.008 (0.004)

bound in the amorphous structure. The Debye–Waller factor, σ^2, was appreciably higher in balcite (0.007, 0.006 Å^2) than ACBC (0.001, 0.004 Å^2). In addition, balcite required the inclusion of Ca–Ba (fourth shell) for the fit to converge.

Barium K-edge spectra of ACBC and balcite were qualitatively very similar, and differed only slightly from witherite (Fig. 3d). The edge energies of ACBC and balcite were both shifted −3.7 eV relative to witherite, an indication that the CN in these two materials was similar and less than nine. However, no correlation between oxygen CN and edge energy shift has been determined at the barium K-edge to our knowledge. In the EXAFS region, data were fit in *k*-space with a model based on witherite, but with a variable coordination number to account for the shift in edge energy in XANES. The CN of barium is eight in both ACBC and balcite, with CN weighted oxygen distances of 2.77 Å and 2.69 Å. This trend is opposite that of calcium, suggesting that crystallization of ACBC to balcite redistributes oxygen from barium to calcium while maintaining the same overall CN. Thus, transformation between ACBC and balcite leads to long-range crystalline order through subtle changes in short-range interactions.

Natural $Ba_{0.5}Ca_{0.5}CO_3$ polymorphs

To determine whether balcite could be synthesized from naturally occurring minerals by heating, alstonite from Alston mine, Cumberland, England was annealed under the same heating schedule as balcite (600 °C at a rate of 10 °C minute^{-1} and held for 10 minutes, and cooling to 25 °C at the same rate, Fig. 4).

Alstonite undergoes a transformation to a structure resembling paralstonite starting at ~150 °C (Fig. 4). Paralstonite then converts back to alstonite (with larger lattice parameters due to the elevated temperature) starting around 530 °C. Upon holding this re-crystallized alstonite at 600 °C for 30 minutes, a transient conversion to another unknown intermediate is observed, which then rapidly converts to a balcite/balcomite-like structure.

Discussion

Heating balcite did not produce $Ba_{0.5}Ca_{0.5}CO_3$ polymorphs alstonite, paralstonite, or barytocalcite. Alstonite and paralstonite have similar aqueous solubilities to barytocalcite[26] and similar calculated cohesive energies,[14] suggesting that they too should be more stable than balcite, which is unstable and nearly ~20 kJ mol^{-1}

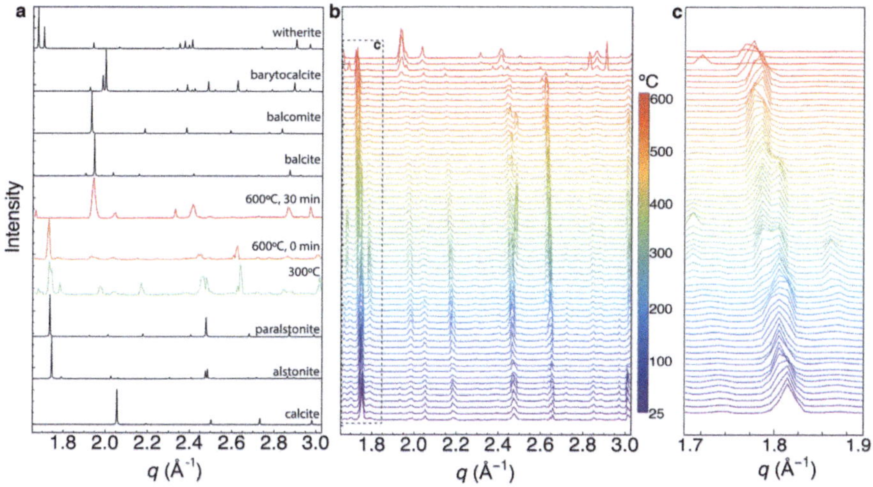

Fig. 4 (a) *In situ* WAXS of alstonite and paralstonite compared to XRD reference patterns for balcite, barytocalcite, calcite and witherite. (b) Alstonite transforms to paralstonite at 600 °C through a number of structural intermediates at lower temperatures.

above the energetic ground state.[12] The sequence of transformations alstonite → paralstonite → alstonite → balcomite suggests that alstonite is metastable relative to paralstonite at room temperature, but becomes more stable as the temperature increases. Surprisingly, barytocalcite did not form despite its reported stability on phase diagrams.[1,2,27] Together, these observations suggest that relative stability is not the primary driving force for transformations among metastable phases in this system.

We observed the formation of balcomite (Fig. 5) through continuous transformation upon annealing balcite, and a similar balcomite-like structure upon heating alstonite. This confirms that balcomite is the stable phase of $Ba_{0.5}Ca_{0.5}CO_3$ at elevated temperatures, and that perhaps a supercell balcomite-like phase is more stable under pressure given that the *in situ* experiments were conducted in a hermetically sealed crucible that was likely subjected to elevated pressure. These experiments also establish that paralstonite and alstonite interconvert *via* continuous, second-order transformations. Together, these observations indicate that continuous transformations are facilitated by similar short range order, but that transitions between structures with dissimilar local symmetry (*i.e.*, alstonite and balcomite) occur through discontinuous first-order transitions. We discuss the supporting evidence for both cation and anion ordering in the superlattice and supercell structures, culminating in a hypothetical structure for metastable end-member balcomite.

Cation ordering

Disordered cation distributions in as-synthesized balcite,[18,19] which originate from the ACBC precursor,[12,18] evolve towards superlattice and supercell ordering during heating. The observation of coarsening crystallites indicates that the ionic mobility is sufficient to undergo displacive transformation, but the structure nonetheless evolves continually. Specifically, cation ordering was observed in hr-

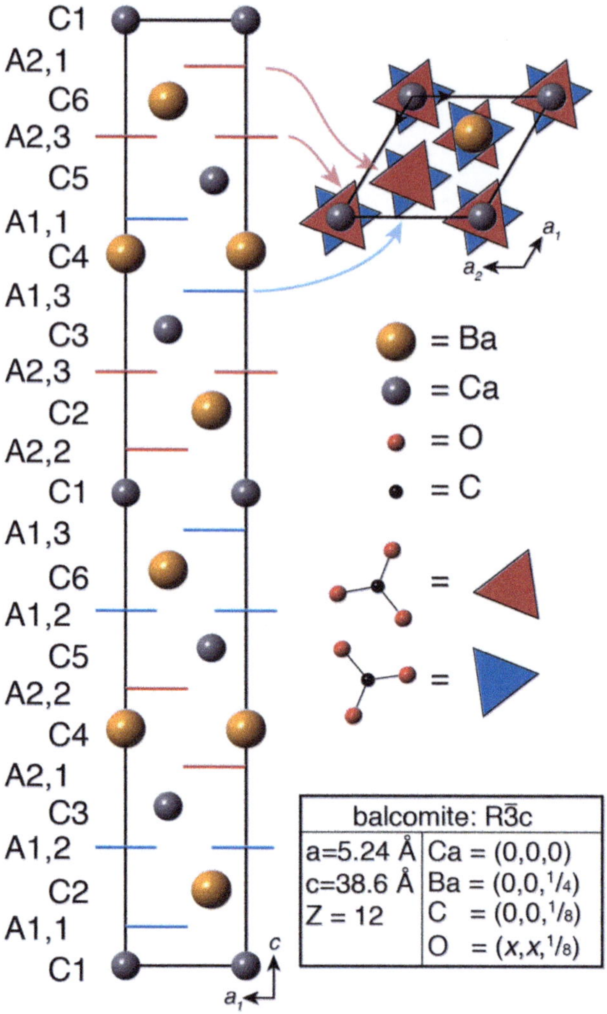

Fig. 5 Proposed structure of balcomite, towards which balcite transformed (but did not fully convert), at elevated temperature.

pXRD and *in situ* WAXS as the emergence of superlattice reflections (Fig. 1b and c). This is very similar to cation superlattice ordering of calcium and magnesium in dolomite[21,28] ($Ca_{0.5}Mg_{0.5}CO_3$, spacegroup $R\bar{3}$) and barium and magnesium in norsethite[29] ($Ba_{0.5}Mg_{0.5}CO_3$ spacegroup $R\bar{3}m$).

Despite the symmetrical similarities between dolomite, norsethite, and balcomite, there are important differences, both structurally and energetically, between these phases. While dolomite and norsethite impart stability with respect to phase-separation,[30,31] balcite is unstable under ambient conditions.[12] And while both dolomite and norsethite have calcite-like carbonate lattice sites that provide uniform 6-fold oxygen coordination number to each cation, annealed balcite exhibits at least three distinct carbonate environments (Fig. 3 and Table 1) with coordination numbers that differ between calcium (6–7) and barium (8–9).

Carbonate ordering

Cation ordering was accompanied by carbonate reorganization (Fig. 1d–f). Carbonates in as-synthesized balcite are highly disordered and are largely indistinguishable from those in ACBC (Fig. 2a and b). The cation CN does not change appreciably between ACBC and balcite, and the carbonates are only subtly redistributed such that they are slightly more closely bound to barium and slightly less so for calcium (Fig. 3 and Table 1). The balcite ^{13}C NMR peak is five to ten times broader than that of calcite and witherite and more closely resembles mixed-cation lattices such as magnesian calcites.[32] Surprisingly, neither the peak position nor linewidth is affected by the barium content. The redistribution of this peak intensity into three distinct carbonate environments after heating confirms that the structural transformation observed in hr-pXRD is partially a result of carbonate reordering that occurs alongside cation segregation.

Carbonate rotational disorder at high temperatures is common. Calcite ($R\bar{3}c$) undergoes a second-order phase transition to 'carbonate disordered' calcite IV approaching a critical temperature of $T_c = 970$ °C.[33,34] This transition is characterized by the rotation of carbonates by 60° about their threefold axes, effectively inverting about the carbon center. Above 970 °C calcite IV becomes further disordered into a 'carbonate-melted' calcite V with $R\bar{3}m$ symmetry.[35] The 'anomalous' increase in the stability of calcite relative to aragonite at higher temperatures arises from this carbonate rotational disorder due to the accompanying increase in configurational entropy.[36] Of particular relevance to this work, the temperature at which this transition occurs is dramatically reduced as the substitution of barium for calcium increases.[37]

There are no reports in the Inorganic Crystal Structure Database of barium with less than 8-fold coordination, and low barium coordination numbers in the (6-fold coordinated) calcite lattice have been shown to be highly unfavorable.[38,39] Witherite ($BaCO_3$ spacegroup *Pmcn*) undergoes a first-order transition to a calcite-type phase with $R3m$ symmetry at 811 °C,[40] both with 9-fold coordination. This $R3m$ phase has the same cation sublattice as calcite, but the apices of the carbonates are rotated 30° relative to those in the calcite lattice, and point the same direction in all carbonate planes. This allows one plane of carbonates to participate in bidentate coordination to the cation, while the other is in monodentate coordination, increasing the cation coordination number through a subtle change in anion orientation.

Balcomite structure

Based on evidence from hr-pXRD and *in situ* MAXS/WAXS, supported by NMR and XANES/EXAFS of precursor and reference materials, and using analogous carbonate systems to infer expected behavior of carbonate-cation local order, the balcomite structure is best understood as alternating cation and anion planes stacked along the *c*-axis in a calcite or dolomite-like unit cell.

Cations and anions each exhibit a different stacking sequence in the balcomite structure. Balcomite has dolomite-like cation ordering in which calcium-rich and barium-rich planes alternate along the *c*-axis. There are three unique cation planes in the balcite structure (C1, C2, and C3), but the introduction of superlattice ordering in balcomite doubles that number (*i.e.*, C4, C5, C6). The carbonate anions also lie on three unique planes (A1, A2, A3), but they can be oriented in one of two ways in each lattice site (1 or 2), giving a total of six different anion planes (A1,1; A1,2; A1,3; A2,1; A2,2; A2,3). Unlike other related carbonate structures, the orientation of carbonates follows a double-repeating pattern, whereby two A1 planes are followed by two A2 planes in order to simultaneously satisfy the preferred coordination number of both calcium (6) and barium (9).

Although the complete transformation to pure balcomite was not observed, the balcomite appears to be a metastable end-member structure towards which both balcite and alstonite transform. The result is an exotic superlattice structure with $R\bar{3}c$ symmetry in which 9-fold coordination around barium sites and 6-fold coordination around calcium sites can be achieved through continuous transformation from $R\bar{3}m$ balcite.

Implications for mineralogy

Second-order transitions were observed between both balcite and balcomite, and alstonite and paralstonite. Both systems appear to interconvert through some combination of local cation segregation and carbonate reorientation. These observations highlight the importance of local order on the pathways by which minerals interconvert. Metastable minerals, such as alstonite and paralstonite, may exist only under conditions in which the appropriate precursor was also favored. In the case of balcite and balcomite, these conditions are associated with the crystallite size of the intermediate phase. A reduction in the system energy through coarsening may reduce the driving force for the transformation to a more stable phase, but it appears that the more important consideration is whether any compositional changes that occur during coarsening give rise to increasing structural similarity. In addition, requirement to nucleate distinct local order may increase the kinetic barrier to transformation, allowing metastable polymorphs to persist for extended periods.

Experimental

Consumables

Unless otherwise noted, all aqueous solutions were prepared using ultra-pure water (ρ = 18.2 MΩ cm) prepared with a Barnstead NanoDiamond UF + UV purification unit. Reagent grade (>99%) $BaCl_2 \cdot 2H_2O$, $CaCl_2 \cdot 2H_2O$, $Na_2CO_3 \cdot 2H_2O$, $CaCO_3$, $BaCO_3$, (Sigma-Aldrich), were used without further purification. Iceland spar calcite was obtained from Dave's Down to Earth Rock Shop in Evanston, IL, USA. Witherite (cat #M21424), barytocalcite, alstonite, and paralstonite reference material was provided by the Field Museum of Natural History in Chicago, IL USA.

ACBC/balcite synthesis

Amorphous calcium barium carbonate (ACBC) was synthesized by rapid mixing of a cold aqueous solution of the alkaline earth chlorides ($[Ca^{2+}] + [Ba^{2+}] = 1$ M, pH 5.6–5.7, solution A) with a cold aqueous solution of Na_2CO_3 (1 M, pH 11.7, solution B). The mole fraction of Ba x_{feed} was varied between 0 and 1. For instance, to prepare solution A with $x_{feed} = 0.50$, 3.053 g $BaCl_2 \cdot 2H_2O$ (125 mmol) and 1.838 g $CaCl_2 \cdot 2H_2O$ (125 mmol) were dissolved in water to a final volume of 25 mL. For solution B, 3.100 g Na_2CO_3 (250 mmol) was dissolved in water to a final volume of 25 mL. Both solutions were then stored on ice for 2 h. Ice cold solution

B was rapidly added to solution A, mixed rapidly and thoroughly by vigorous shaking for 30 s.

To isolate ACBC, the resulting white precipitate was immediately collected by vacuum filtration through a 100 mm diameter Whatman no. 4 cellulose filter paper, washed 3 times with 5 mL cool water ($T = 10$ °C), and dried under vacuum (R.T., $p \approx 10$ Pa $= 10^{-1}$ Torr, 24 h). ACBC crystallized within 24 hours when kept in its mother liquor at ambient conditions. Balcite was then collected by vacuum filtration using the same procedure as for ACBC, with an additional drying step at 110 °C for 1 hour.

Samples for NMR were prepared through the same procedure, but with 20% of the Na_2CO_3 enriched with ^{13}C to enhance the NMR sensitivity. Solutions with nominal $x_{feed} = 0.05$, 0.10, 0.15, 0.30, 0.40, 0.50 and 0.60 were used for NMR experiments.

High resolution synchrotron powder diffraction and Rietveld refinement

High resolution powder X-ray diffraction was acquired at sector 11-ID of the Advanced Photon Source at Argonne National Lab. A wavelength of $\lambda = 0.414529$ Å (29.9097 keV) was set with a Si(111) double crystal (bounce-up geometry) and focused with a sagittally bent Si(111) crystal in the horizontal and a 1 meter Si/Pt mirror in the vertical, enabling a resolution of $2\theta = 0.001°$. Rietveld refinement was performed with GSAS using the EXPGUI graphical user interface.[41] The scattering vector, Q, is defined as $Q = 4\pi \sin \theta/\lambda$ with $Q = 2\pi/d$ where d is the real-space distance between scatterers.

In situ synchrotron wide angle X-ray scattering

Low mass aluminum pans (TZero) were filled with approximately 20 mg of powder and hermetically sealed by crimping an aluminum lid to the top of the pan. The pans were annealed from 25 °C to 600 °C and cooled back to 25 °C at a rate of 10 °C min^{-1} in a Linkam DSC600 temperature stage controlled with Linksys32 software. WAXS patterns were collected every minute at intervals of 10 °C. Diffuse rings consistent with a polycrystalline material were observed in two-dimensional CCD images. These images were radially integrated using GSAS-II.

Solid-state NMR

^1H and ^{13}C NMR measurements were carried out on 500 and 300 MHz AVANCE III (Bruker) spectrometers using triple-resonance magic angle spinning (MAS) NMR probes with 4 mm zirconia rotors. Samples were spun at 5000 and 10 000 $\pm$ 2 Hz. Direct ^{13}C excitation (DE) echo experiments were carried out with a 5.0 μs $\pi/2$, 10.0 μs π pulse widths, an echo interval τ equal to the rotor period T_R (200 and 100 μs), and TPPI ^1H decoupling at a level of 100 kHz. Relaxation delays of 1000 and 2400 s were employed to obtain fully relaxed spectra of the crystalline polymorphs, respectively. Cross polarization (CP) MAS echo experiments (indirect excitation) were carried out with a 5.0 μs $\pi/2$, 10.0 μs π pulse widths, an echo interval t (200 and 100 μs) identical to the rotor period T_R, a ^1H TPPI decoupling level of 100 kHz, Hartmann–Hahn RF levels were matched at 50 kHz, Hartmann–Hahn rf levels set at 50 kHz for ^{13}C, and ramped ^1H level (70–100 kHz), with contact times of 0.5, 1, and 2 ms and relaxation delays of 3 s for ^{13}C. Up to 16k transients were acquired. The chemical shifts of ^{13}C and ^1H are reported relative to tetramethylsilane (TMS) and adamantane, respectively, with respective accuracies of ±0.05 and ±0.1 ppm; linewidths ($\delta\nu$; full width at half maximum, FWHM) were determined with ±0.1 ppm accuracy. Peak areas were calculated by deconvolution using DMFIT[42] and Topspin.

X-ray absorption spectroscopy

ACBC ($x = 0.50$) was synthesized as described above, and approximately half was left in the mother liquor to crystallize to balcite. Measurements were performed at the Dow–Northwestern–Dupont collaborative access team (DND-CAT), Sector 5-BM-D of the Advanced Photon Source at Argonne National Laboratory. Samples for Ca K-edge measurements were spread on 8 μm Kapton tape and kept dry under a stream of He gas. For measurements performed at the Ba K-edge, powders were loaded into 1 mm OD quartz capillaries with 100 μm thickness (Charles Supper). Calibration at the Ca K-edge was performed with a geological Iceland spar (calcite) standard, for which the edge energy was set at 4038 eV. ACBC samples were measured relative to this value. Molybdenum foil (Mo K-edge at 20.000 eV) was used to calibrate the Ba K-edge to an absolute scale.

Both the X-ray absorption near edge spectra (XANES) and extended X-ray absorption fine structure (EXAFS) spectra were analyzed using the Demeter package.[43] Normalization and background subtraction were carried out in Athena using Autobk. χ-Data were weighted by k, k^2, and k^3 between $1.6 < k < 11.5$ Å^{-1} for the Ca K-edge and between $2 < k < 14$ Å^{-1} for the Ba K-edge. Fitting of the data to the crystal structures was performed using Artemis. The theoretical photoelectron scattering amplitudes and phase shifts were calculated using FEFF1. To generate the feff.inp file, which contains a list of the atomic coordinates centered at the absorbing atom, Atoms[44] was used to calculate the atomic coordinates centered at Ca. The total theoretical $\chi(k)$ for each model was then constructed from the most important scattering paths and fit in R-space (Ca K-edge) or k-space (Ba K-edge). Multiple scattering was not considered. The coordination number, bond distance, Debye–Waller factor, amplitude reduction factor, S_o^2, and shift in energy origin, ΔE_o, were determined previously.[45] Coordination numbers and Debye–Waller factors are highly correlated and were fit separately, but in each case with all other parameters floating. The validity of the fits was determined by the R-factor, which was minimized to ≤0.01.

Data in the calcium K-edge EXAFS region were fit in R-space with a structural model based on calcite (Fig. 3b, c and Table 1). Two separate distances were needed for each Ca–O_1 (first shell), Ca–C_1 (second shell), and Ca–O_2 (third shell) for both ACBC and balcite. ACBC had shorter Ca–O_1 distances (2.19, 2.39 Å) than balcite (2.37, 2.55 Å). The CN-weighted average Ca–O distance was 2.32 Å in ACBC compared to 2.42 Å in balcite, indicating that first-shell oxygen was more tightly bound in the amorphous structure. The Debye-Waller factors, σ^2, were appreciably higher in balcite (0.007, 0.006 Å^2) than ACBC (0.001, 0.004 Å^2). In addition, balcite required the inclusion of Ca–Ba (fourth shell) for the fit to converge.

Five separate Ba–O distances in the witherite reference material were reduced to three to limit the number of free parameters given the resolution of the measurements (~0.02 Å). Two separate oxygen distances and a CN of eight produced the best fit for both ACBC and balcite. Ba–O distances were longer in ACBC (2.68, 2.83) than balcite (2.55, 2.73 Å), with the CN-weighted averages of 2.77

Å and 2.69 Å, respectively. Only a single Ba–C_1 (second shell) was needed for a convergent fit to ACBC, while balcite required two Ba–C_1 distances, two Ba–O_2 (third shell) distances, Ba–Ca (fourth shell), and a Ba–O_3 (fifth shell).

Author contributions

All authors have given approval to the final version of the manuscript.

Conflicts of interest

There are no conflicts of interest.

Acknowledgements

AS and EP acknowledge the Israel Science Foundation grant 2001/17 for their financial support. AS and EP acknowledge the valuable assistance of Dr Shifi Kababya (Schulich Faculty of Chemistry, Technion, Israel).

References

1 A. F. Shatskiy, K. D. Litasov and Y. N. Palyanov, Phase relations in carbonate systems at pressures and temperatures of lithospheric mantle: review of experimental data, *Russ. Geol. Geophys.*, 2015, **56**(1–2), 113–142.

2 L. L. Chang, Subsolidus phase relations in the systems $BaCO_3$-$SrCO_3$, $SrCO_3$-$CaCO_3$, $BaCO_3$-$CaCO_3$, *J. Geol.*, 1965, **73**(2), 346–368.

3 M. Dietzel, N. Gussone and A. Eisenhauer, Co-precipitation of Sr^{2+} and Ba^{2+} with aragonite by membrane diffusion of CO_2 between 10 and 50 degrees C, *Chem. Geol.*, 2004, **203**(1–2), 139–151.

4 A. J. Tesoriero and J. F. Pankow, Solid Solution Partitioning of Sr^{2+}, Ba^{2+}, and Cd^{2+} to calcite, *Geochim. Cosmochim. Acta*, 1996, **60**(6), 1053–1063.

5 V. Mavromatis, K. E. Goetschl, C. Grengg, F. Konrad, B. Purgstaller and M. Dietzel, Barium partitioning in calcite and aragonite as a function of growth rate, *Geochim. Cosmochim. Acta*, 2018, **237**, 65–78.

6 S. D. Škapin and I. Sondi, Homogeneous Precipitation of Mixed Anhydrous Ca-Mg-Ba-Sr carbonates by Enzyme-Catalyzed Reaction, *Cryst. Growth Des.*, 2005, **5**(5), 1933–1938.

7 I. Sondi and E. Matijević, Homogeneous Precipitation by Enzyme-Catalyzed Reactions. 2. Strontium and Barium Carbonates, *Chem. Mater.*, 2003, **15**, 1322–1326.

8 K. Benzerara, F. Skouri-Panet, J. Li, C. Ferard, M. Gugger, T. Laurent, E. Couradeau, M. Ragon, J. Cosmidis, N. Menguy, I. Margaret-Oliver, R. Tavera, P. Lopez-Garcia and D. Moreira, Intracellular Ca-carbonate biomineralization is widespread in cyanobacteria, *Proc. Natl. Acad. Sci. U. S. A.*, 2014, **111**(30), 10933–10938.

9 E. Couradeau, *et al.*, An early-branching microbialite cyanobacterium forms intracellular carbonates, *Science*, 2012, **336**(6080), 459–462.

10 M. Aykol, S. S. Dwaraknath, W. Sun and K. A. Persson, Thermodynamic limit for synthesis of metastable inorganic materials, *Sci. Adv.*, 2018, **4**, eaaq0148.

11 M. L. Whittaker, W. Sun, K. DeRocher, J. Saivenkataraman, G. Ceder and D. Joester, Structural basis for metastability in amorphous barium calcium carbonate, *Adv. Funct. Mater.*, 2018, **28**, 1704202.

12 M. L. Whittaker, W. Sun, D. O. Duggins, G. Ceder and D. Joester, Dynamic Barriers to Crystallization of Calcium Barium Carbonates, *Cryst. Growth Des.*, 2021, **21**(8), 4556–4563.

13 W. Sun, S. T. Dacek, S. P. Ong, G. Hautier, A. Jain, W. D. Richards, A. C. Gamst, K. A. Persson and G. Ceder, The thermodynamic scale of inorganic crystalline metastability, *Sci. Adv.*, 2016, **2**, e1600225.

14 R. Chuliá-Jordán, D. Santamaria-Perez, J. Ruiz-Fuertes, A. Otero-de-la-Roza and C. Popescu, Crystal Structure of $BaCa(CO_3)_2$ Alstonite Carbonate and Its Phase Stability upon Compression, *ACS Earth Space Chem.*, 2021, **5**(5), 1130–1139.

15 R. M. Hazen, D. Papineau, W. Bleeker, R. T. Downs, J. M. Ferry, T. J. McCoy, D. A. Sverjensky and H. Yang, Mineral evolution, *Am. Mineral.*, 2008, **93**(11–12), 1693–1720.

16 D. Spahr, L. Bayarjargal, V. Vinograd, R. Luchitskaia, V. Milman and B. Winkler, A new $BaCa(CO_3)_2$ polymorph, *Acta Crystallogr., Sect. B: Struct. Sci., Cryst. Eng. Mater.*, 2019, **75**(Pt 3), 291–300.

17 P. N. Gavryushkin, A. B. Belonoshko, N. Sagatov, D. Sagatova, E. Zhitova, M. G. Krzhizhanovskaya, A. Rečnik, E. V. Alexandrov, I. V. Medrish, Z. I. Popov and K. D. Litasov, Metastable structures of $CaCO_3$ and their role in transformation of calcite to aragonite and postaragonite, *Cryst. Growth Des.*, 2021, **21**(1), 65–74.

18 M. L. Whittaker and D. Joester, ACBC to Balcite: Bioinspired Synthesis of a Highly Substituted High-Temperature Phase from an Amorphous Precursor, *Adv. Mater.*, 2017, **29**(26), 1606730.

19 E. Seknazi, D. Levy, I. Polishchuk, A. Katsman and B. Pokroy, Experimental and Theoretical Insights into the Bioinspired Formation of Disordered Ba-Calcite, *Adv. Funct. Mater.*, 2020, **30**, 1805028.

20 L. Bindi, A. C. Roberts and C. Biagioni, The crystal structure of alstonite, $BaCa(CO_3)_2$: an extraordinary example of 'hidden' complex twinning in large single crystals, *Mineral. Mag.*, 2020, **84**(5), 699–704.

21 R. J. Reeder, Constraints on Cation Order in Calcium-rich Sedimentary Dolomite, *Aquat. Geochem.*, 2000, **6**, 213–226.

22 V. A. Drits, D. K. McCarty, B. A. Sakharov and K. L. Milliken, New insight into structural and compositional variability in some ancient excess-Ca dolomite, *Can. Mineral.*, 2005, **43**, 1255–1290.

23 S. Kababya, A. Gal, K. Kahil, S. Weiner, L. Addadi and A. Schmidt, Phosphate-water interplay tunes amorphous calcium carbonate metastability: spontaneous phase separation and crystallization vs stabilization viewed by solid state NMR, *J. Am. Chem. Soc.*, 2015, **137**(2), 990–998.

24 F. E. Sowrey, L. J. Skipper, D. M. Pickup, K. O. Drake, Z. Lin, M. E. Smith and R. J. Newport, Systematic empirical analysis of calcium-oxygen coordination environment by calcium K-edge XANES, *Phys. Chem. Chem. Phys.*, 2004, **6**(1), 188–192.

25 J. L. Fulton, *et al.*, Understanding the effects of concentration on the solvation structure of Ca^{2+} in aqueous solution. I: The perspective on local structure from EXAFS and XANES, *J. Phys. Chem. A*, 2003, **107**(23), 4688–4696.

26 R. M. Garrels, M. E. Thompson and R. Siever, Stability of some carbonates at 25C and one atmosphere total pressure, *Am. J. Sci.*, 1960, **258**, 402–418.

27 L. L. Chang, Subsolidus phase relations in the systems $BaCO_3$-$SrCO_3$, $SrCO_3$-$CaCO_3$ and $BaCO_3$-$CaCO_3$, *J. Geol.*, 1965, **73**(2), 346–368.

28 R. J. Reeder and Y. Nakamura, The Nature of Ordering and Ordering Defects in Dolomite, *Phys. Chem. Miner.*, 1982, **8**, 29–35.

29 H. Effenberger and J. Zemann, Single crystal X-ray investigation of norsethite, $BaMg(CO_3)_2$: one more mineral with an aplanar carbonate group, *Z. Kristallogr. Cryst. Mater.*, 1985, **171**(3–4), 275–280.

30 C. Pimentel and C. M. Pina, The formation of the dolomite-analogue norsethite: Reaction pathway and cation ordering, *Geochim. Cosmochim. Acta*, 2014, **142**, 217–223.

31 A. V. Radha and A. Navrotsky, Thermodynamics of Carbonates, *Rev. Mineral. Geochem.*, 2013, **77**(1), 73–121.

32 J. Wang and U. Becker, Structure and carbonate orientation of vaterite ($CaCO_3$), *Am. Mineral.*, 2009, **94**(2–3), 380–386.

33 N. Ishizawa, Calcite V: a hundred-year-old mystery has been solved, *Powder Diffr.*, 2014, **29**(S1), S19–S23.

34 N. Ishizawa, H. Setoguchi and K. Yanagisawa, Structural evolution of calcite at high temperatures: phase V unveiled, *Sci. Rep.*, 2013, **3**, 2832.

35 M. Farhadi Khouzani, D. M. Chevrier, P. Güttlein, K. Hauser, P. Zhang, N. Hedin and D. Gebauer, Disordered amorphous calcium carbonate from direct precipitation, *CrystEngComm*, 2015, **17**(26), 4842–4849.

36 M. T. Dove and B. M. Powell, Neutron Diffraction Study of the tricritical orientational order/disorder phase transition in calcite at 1260K, *Phys. Chem. Miner.*, 1989, **16**, 503–507.

37 J. J. Lander, Polymorphism and Anion Rotational Disorder in the Alkaline Earth Carbonates, *J. Chem. Phys.*, 1949, **17**(10), 892–901.

38 J. M. Astilleros, C. M. Pina, L. Fernández-Díaz and A. Putnis, The effect of barium on calcite {1014} surfaces during growth, *Geochim. Cosmochim. Acta*, 2000, **64**(17), 2965–2972.

39 M. Menadakis, G. Maroulis and P. G. Koutsoukos, Incorporation of Mg^{2+}, Sr^{2+}, Ba^{2+} and Zn^{2+} into aragonite and comparison with calcite, *J. Math. Chem.*, 2009, **46**(2), 484–491.

40 S. M. Antao and I. Hassan, $BaCO_3$: high-temperature crystal structures and the Pmcn -> R3m phase transition at 811 degrees C, *Phys. Chem. Miner.*, 2007, **34**(8), 573–580.

41 B. H. Toby and R. B. Von Dreele, GSAS-II: the genesis of a modern open-source all purpose crystallography software package, *J. Appl. Crystallogr.*, 2013, **46**(2), 544–549.

42 D. Massiot, F. Fayon, M. Capron, I. King, S. L. Calvé, B. Alonso, J.-O. Durand, B. Bujoli, Z. Gan and G. Hoatson, Modelling one- and two-dimensional solid-state NMR spectra, *Magn. Reson. Chem.*, 2002, **40**, 70–76.

43 B. Ravel and M. Newville, ATHENA, ARTEMIS, HEPHAESTUS: data analysis for X-ray absorption spectroscopy using IFEFFIT, *J. Synchrotron Radiat.*, 2005, **12**(4), 537–541.

44 B. Ravel, ATOMS: crystallography for the X-ray absorption spectroscopist, *J. Synchrotron Radiat.*, 2001, **8**, 314–316.

45 M. L. Whittaker, W. Sun, K. A. DeRocher, S. Jayaraman, G. Ceder and D. Joester, Structural Basis for Metastability in Amorphous Calcium Barium Carbonate (ACBC), *Adv. Funct. Mater.*, 2018, **28**(2), 1704202.

Faraday Discussions

ROYAL SOCIETY OF CHEMISTRY

PAPER

On the mechanism of calcium carbonate polymorph selection *via* confinement

Alexander Katsman, [a] Iryna Polishchuk[a] and Boaz Pokroy *[ab]

Received 30th November 2021, Accepted 5th January 2022
DOI: 10.1039/d1fd00111f

Organisms deposit various biominerals in the course of their biomineralisation. The most abundant of these is calcium carbonate, which manifests itself in several polymorphs. While organisms possess the ability to control the specific polymorph deposited, the exact mechanism by which polymorph selection takes place is not yet fully understood. Because biominerals often grow within confined spaces, one of the suggested possibilities was that polymorph selection might be an outcome of confinement. Confining conditions have indeed been extensively shown to have a strong impact on the nucleation and crystal growth of calcium carbonate and, in particular, on its polymorph selection. However, despite numerous studies on the crystal growth of calcium carbonate in confined spaces, the mechanism of polymorph selection under confinement has not been elucidated. Herein, we discuss previously reported results and suggest a mechanistic explanation of the observed selective formation of calcite or aragonite or vaterite. We consider the possible effects of charged confining inner surfaces and of the sizes of the confining pores, and discuss whether the predominantly precipitating phase is amorphous calcium carbonate. We also discuss two possible scenarios of crystallization from amorphous calcium carbonate under conditions of confinement: *via* solid-state transformation or *via* a mechanism of dissolution–reprecipitation.

Introduction

During the biomineralisation process, organisms deposit various biominerals that serve multiple functions. The most abundant biomineral is calcium carbonate, which manifests itself in several polymorphs. Organisms possess the ability to control the specific polymorph deposited.[1,2] The exact mechanism by which polymorph selection takes place is not yet fully understood. Among the several proposed mechanisms are crystal growth templated by insoluble organic templates,[3] effects of soluble organic macromolecules,[4–7] a combination of both of those,[8] or stabilisation of a more soluble polymorph followed by transformation

[a]Department of Materials Science and Engineering, Technion-Israel Institute of Technology, 32000, Haifa, Israel. E-mail: bpokroy@technion.ac.il
[b]Russell Berrie Nanotechnology Institute, Technion-Israel Institute of Technology, 32000, Haifa, Israel

to a more stable phase *via* the Ostwald Lussac rule of stages. It was proposed that since biomineralisation often takes place in small volumes within pores and tubules, confinement might be an additional mechanism controlling calcium carbonate polymorphism.[9–11] To test that hypothesis various experiments were carried out in which calcium carbonate was grown in confined volumes.[12–16] It was shown that when calcium carbonate is grown under confining conditions, such as within track-etched (TE) membranes, vesicles or picolitre droplets, it was possible to stabilise various polymorphs such as vaterite, aragonite or calcite.[12,14,17,18] It was also found that among other factors, the type of membrane (manufacturer, polymer type, *etc.*), surface chemistry and pore diameter strongly controlled the specific polymorph formed.[17]

Even after acquiring all the above-mentioned experimental data, we still lack clear proof that polymorph selectivity is affected by confinement. Therefore, the mechanism governing polymorph selection is still not obvious. Nevertheless, it was clear that the nucleation of calcium carbonate crystalline phases from supersaturated solutions in confined environments is dependent on thermodynamics and kinetics, and hence on variables such as temperature, pressure, supersaturation and interphase energies, and that these can be influenced in turn by the confinement of matter and by the presence of additives. A summary of several studies that we found to show calcium carbonate polymorph control by confinement is presented in Table 1.

It is apparent from the table that no single parameter can explain the specific polymorph selection and that it probably derives from a combination of factors, and possibly even from several mechanisms, depending on the experimental conditions.

Revisiting the effect of confinement on polymorph selection

Crystallisation of biominerals, other than calcium carbonate, under confinement have been studied by several groups.[19–22] It was found that metastable or transient

Table 1 Calcium carbonate polymorphs formed in confined environments

Type of confinement	Pore diameter	Temperature, °C	Pressure, bar	First phase precipitated	Final polymorph
TE-membrane polycarbonate[13]	0.2–3.0 μm	3–4 °C, RT	1	ACC	Calcite
	10 μm	3–4 °C, RT	1	Calcite	Calcite
TE-membrane polycarbonate with a coating of PVP (polylactam) which increases hydrophilicity[14]	200 nm	RT	1	ACC	Vaterite
Droplets[15]	100 μm	RT	1	ACC	Vaterite, calcite
TE-membrane polycarbonate[12]	25 nm	RT	1	ACC	Aragonite
	50–800 nm	RT		ACC	Aragonite + calcite
	1200 nm				Calcite
Fungal hyphae tubular cells[16]	1.5–5.0 μm	134 °C	2.5	ACC	Aragonite

phases appear under nanoscopic confinement. In particular, early stages of polymer crystallization under nanoscopic confinement are characterized by a mesomorphic precursor phase packed differently from the final crystal.[20] The stabilization of various metastable crystalline forms has been explained based on surface energy contributions compensating the energetic advantage of a more stable crystalline polymorph. It was also assumed that polymorph selectivity is influenced by critical size constraints: the preference for a specific metastable phase within smaller pores is attributed to its smaller critical nucleus size as compared to that of other forms.[21] The latter assumption, however, seems to be irrelevant in the case of nucleation in pores larger than 100 nm. At the same time, nucleation of transient phases (amorphous calcium carbonate (ACC) in the case of calcium carbonate) and the metastable crystalline polymorphs of calcium carbonate (aragonite and vaterite) can indeed be a result of smaller interface energies of these phases.

It was clearly established by Meldrum *et al.* that micro-confinement stabilises ACC.[17] Therefore, it is reasonable to assume that in all cases in which confinement exists (below a certain pore threshold), hydrated ACC is the first phase to precipitate. Its transformation to crystalline $CaCO_3$ polymorphs can then take place *via* two possible routes: *via* the dissolution–reprecipitation mechanism[23,24] or *via* a solid-state transformation (SST).[25–28]

If confinement exists, our first assumption is that nucleation most probably takes place on the inner wall of the confining template. The apparent activation energy of nucleation is a complex parameter that integrates a thermodynamic barrier (work done in formation of a critical nucleus) and an effective kinetic barrier (originating from desolvation) of solute ions, as well as activation energy for the attachment of ions to the critical nucleus, and structural rearrangements. The thermodynamic barrier to the nucleation of a new phase is determined by the corresponding difference between the free energy and the energy required for the formation of a new interface. It is reasonable to assume that the nucleation of $CaCO_3$ is heterogeneous and is controlled by the electrostatic interaction between the hydrated calcium ions and the substrate (the space-confining wall in the present case).

Charged surfaces in confinement

In the case of calcite the $(10\bar{1}4)$ facets exhibit minimal surface energy, thus yielding the rhombohedral shape of calcite crystals. However, when highly hydrated ACC is formed in a confined volume with negatively charged inner surfaces, the direct crystallisation of calcite crystals with non-charged $(10\bar{1}4)$ surfaces can possibly be suppressed kinetically. In such a case, electrostatic interaction of calcium and carbonate ions with charged surfaces may result in the nucleation of $CaCO_3$ that exhibits charged rather than neutral surfaces. As calculated by Leeuw and Parker,[29] the corresponding energies of the hydrated charged surfaces of aragonite and vaterite appear to be slightly smaller than those of calcite.

The thermodynamic barrier for nucleation of a given phase depends on the interface energies involved. For heterogeneous nucleation of a phase P on the inner surface of a confined volume (the nucleus size is assumed to be much smaller than the confining pore radius) within a second phase A, which can be either a liquid solution (LS) or solid ACC (Fig. 1) the thermodynamic barrier for a disk-shaped nucleus can be written as follows:

$$\Delta G_c = \frac{4}{27}\frac{(\Delta\gamma)^3}{(\eta\Delta G_v)^2} \tag{1}$$

where ΔG_v is the driving force for transformation; $\Delta\gamma = 3(\gamma_{1A}^P + \gamma_{1s}^P) - \gamma_s^A$; where γ_{1A}^P and γ_{1s}^P are the top P/A and base P/(inner surface) interface energies of the facets parallel to the substrate, γ_s^A is the surface energy between phase A and the inner surface; where $\eta = d/r = (\gamma_{1A}^s + \gamma_{1A}^P)/\gamma_{2A}^P$; r and d are the radius and the thickness of the disk, respectively; γ_{2A}^P is the P/A lateral cylindrical interface energy.

By simplifying the case of nucleation in a liquid solution, such as $\gamma_{1LS}^P \approx \gamma_{1LS}^s \approx \gamma_s^A \equiv \gamma_1^P$; $\gamma_{2LS}^P \equiv \gamma_2^P$; $\Delta\gamma \approx 5\gamma_1^P$; $\eta \approx 2\gamma_1^P/\gamma_2^P$, we can obtain the ratio of thermodynamic barriers for re-nucleation of aragonite and calcite on the inner surface of a confining volume in the LS:[16]

$$\frac{\Delta G_c^{arag}}{\Delta G_c^{cal}} \approx \frac{\gamma_1^{arag}(\gamma_2^{arag})^2}{\gamma_1^{cal}(\gamma_2^{cal})^2}\frac{(\Delta G_v^{cal})^2}{(\Delta G_v^{arag})^2} \tag{2}$$

According to Leeuw and Parker,[29] in the case of a calcite nucleus revealing (0001) charged facets, the interfacial energies of the calcite/LS top surfaces are $\gamma_1^P \simeq \gamma_1^{cal} \simeq 0, 38\ \mathrm{J\ m^{-2}}$, and for the side facets $(10\bar{1}0)$, $(10\bar{1}1)$ the lateral energies are $\gamma_2^P \simeq \gamma_2^{cal} \simeq 0.75\ \mathrm{J\ m^{-2}}$ and $0.63\ \mathrm{J\ m^{-2}}$, respectively.

In the case of an aragonite nucleus, interfacial energies of the aragonite/LS top surfaces of the {010}Ca- and {110}Ca-hydrated surfaces are, respectively, $\gamma_1^P \simeq \gamma_1^{arag} \simeq 0.24\ \mathrm{J\ m^{-2}}$, and for the side facets $\gamma_2^P \simeq \gamma_2^{arag} \simeq 0.56\ \mathrm{J\ m^{-2}}$, respectively (for estimations of the lateral interface energy we used the averaged value of lateral interface energies).

In the case of a vaterite nucleus, the vaterite/LS base and side interfacial energies of the hydrated surfaces {010}Ca and {110}Ca are $0.91\ \mathrm{J\ m^{-2}}$ and $0.52\ \mathrm{J\ m^{-2}}$, respectively, which are higher than those of aragonite or calcite. On the other hand, the energies of the {010}CO_3 and {011}CO_3 surfaces are $0.22\ \mathrm{J\ m^{-2}}$ and $0.26\ \mathrm{J\ m^{-2}}$, respectively, which are smaller than those of aragonite or calcite, and therefore, under certain conditions, such as when the confining inner surfaces are positively charged, the nucleation of vaterite *via* dissolution–reprecipitation (with negatively charged surfaces terminated by CO_3^-) can be

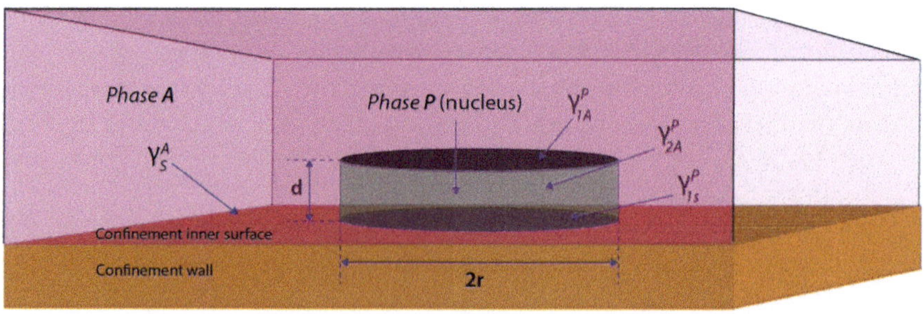

Fig. 1 Schematic representation of heterogeneous nucleation of a phase P on the inner surface of a confined volume within a second phase A, which can be either a LS or solid ACC.

favourable. This scenario is supported by the study of Schenk *et al.*,[14] who grew calcium carbonate within 200 nm Millipore TE membranes coated with polyvinyl pyrrolidone (PVP) and showed that single crystals of vaterite are formed, and PVP-functionalised surfaces indeed possessed positive charges in aqueous media.

According to Kawano *et al.*,[30] the driving force for calcite nucleation from an initial solution is greater than that for aragonite nucleation, $|\Delta G_v^{cal}| > |\Delta G_v^{arag}|$, while the ratio of the driving forces $(\Delta G_v^{cal}/\Delta G_v^{arag})$ do not exceed 1.04. Use of the above-mentioned values of negatively charged interface energies yields the ratio $(\Delta G_c^{arag}/\Delta G_c^{cal}) \approx 0.45$, implying that despite the higher thermodynamic stability of calcite, the nucleation of aragonite is more favourable in confined volumes with negatively charged inner surfaces.

According to the nucleation mechanism described by Lippmann,[31] heterogeneous nucleation of aragonite on negatively charged surfaces should result in tangential orientation of the *c*-axis of aragonite and radial orientation of its *b*-axis. Such orientation was indeed observed in our recent study, where aragonite needles grown within tubular organic cells were found to be oriented along the [001] direction.[16] Those results provided additional evidence supporting the feasibility of aragonite nucleation on negatively charged inner surfaces in confined volumes.

When considering aragonite *vs.* calcite, Lippman[31] showed that although calcite is more stable thermodynamically than aragonite, under certain given conditions the crystallisation rate of aragonite exceeds that of calcite. This higher crystallisation rate of aragonite may be related to the larger number of oxygens (9 in total) surrounding the calcium ion, in contrast to that of calcite (only 6). We therefore suggest that the flux of ions to the nucleus is faster in the case of aragonite, thereby facilitating its more favourable growth. The same would apply to incoming CO_3^{2-} groups, given that their thermal vibrations are reportedly more restrained in the aragonite structure than in that of calcite. This last factor would entail a more efficient capture and retention of both CO_3^{2-} and Ca^{2+} ions by the active sites of aragonite nuclei.[31] It is reasonable to assume that in the ACC atomic structure, which forms more easily than that of calcite, there are also 9 oxygens surrounding the calcium ions.

In the case of SSTs of ACC to aragonite or calcite, the interface energies γ_{1ACC}^{P} and γ_{1s}^{P} can be substantially different. Given that the length of the Ca–O bonds in ACC reportedly fits better to the length of the aragonite structure (see below), it is reasonable to expect the aragonite/ACC interface energies to be smaller than those of calcite/ACC or calcite/LS interface energies.

Assuming, for example, that $\gamma_{1ACC}^{arag} = \alpha\gamma_1^{arag}$ and $\gamma_{2ACC}^{arag} = \alpha\gamma_2^{arag}$ (where $\alpha < 1$ and $\gamma_{1s}^{arag} \approx \gamma_s^{ACC} \approx \gamma_1^{arag}$), then $\Delta\gamma_{ACC\text{-}arag} = 3(\alpha + 1)\gamma_1^{arag} - \gamma_1^{arag} = (3\alpha + 2)\gamma_1^{arag}$ and $\eta^{arag} = (1 + \alpha)\gamma_1^{arag}/\alpha\gamma_2^{arag}$. Retaining the above assumption for calcite (*i.e.*, $\gamma_{1ACC}^{cal} \approx \gamma_1^{s} \approx \gamma_s^{ACC} \equiv \gamma_1^{cal}$, $\gamma_{2ACC}^{cal} = \gamma_2^{cal}$, $\Delta\gamma \approx 5\gamma_1^{cal}$, $\eta \approx 2\gamma_1^{cal}/\gamma_2^{cal}$), then the ratio of thermodynamic barriers for solid-state nucleation of aragonite and calcite in ACC will be:

$$\frac{\Delta G_c^{arag}}{\Delta G_c^{cal}} \approx F(\alpha)\frac{\gamma_1^{arag}(\gamma_2^{arag})^2}{\gamma_1^{cal}(\gamma_2^{cal})^2}\frac{(\Delta G_v^{cal})^2}{(\Delta G_v^{arag})^2}, \tag{3}$$

where $F(\alpha) \equiv \left(\frac{3\alpha + 2}{5}\right)^2\left(\frac{2\alpha}{1+\alpha}\right)^3$. For $\alpha = 1$, the factor $F(\alpha) = 1$ and it decreases with decreasing α (Fig. 2). For example, for $\alpha = 0.8$, this factor and the ratio (3) are approximately halved. Apparently, therefore, the nucleation of aragonite inside ACC is kinetically much more favourable than the nucleation of calcite, thus providing a solid-state ACC-to-aragonite transformation.

According to estimations by Koishi *et al.*,[32] the interface energies γ_s^A for an LS/substrate and for an ACC/substrate ($\gamma_s^{LS} \approx \gamma_s^{ACC} \sim 100$ mJ m^{-2}) are similar to that of the calcite/LS interface energy, while the calcite/substrate interface energy $\gamma_{1s}^{cal} \sim 50$ mJ m^{-2} is smaller than that of ACC, $\gamma_{1s}^{cal} < \gamma_s^{ACC}$, which is an additional reason for the crystallization of calcite from ACC. It should be noted, that according to the results of Koishi *et al.*,[32] the ACC/LS interface energy $\gamma_{ACC\text{-}LS} \sim 50$ mJ m^{-2} is smaller than that of calcite/LS $\gamma_{cal\text{-}LS} \sim 100$ mJ m^{-2}, which can yield predominant nucleation of ACC from supersaturated LS. Unfortunately, the aragonite/ACC interface energies, as well as the aragonite/LS and aragonite/substrate energies, are unknown.

To summarise this section on the effect of confinement with charged inner surfaces on polymorph selection, we can state that the first phase to precipitate is ACC. This is followed by nucleation and growth of a crystalline calcium carbonate phase *via* two possible routes (Fig. 3):

• Dissolution–reprecipitation mechanism when dissolution of ACC feeds the growth of crystal nuclei at the confined inner surfaces within LS. In the case of charged surfaces, the lower interfaces energies of those charged interfaces in aragonite (or vaterite) as compared to calcite can provide a lower energy barrier to aragonite (or vaterite) nucleation. Based on the calculated energies of hydrated interfaces, negatively charged surfaces can be assumed to favour nucleation of aragonite, whereas positively charged surfaces favour nucleation of vaterite.

• SST of ACC to a crystalline phase *via* prior dehydration of ACC. Transformation of ACC to aragonite by SST occurs faster than its transformation to calcite, owing to the faster nucleation of aragonite than of calcite at the confinement inner surface/ACC interface. This is attributable in turn to the lower ACC/aragonite interface energies than those of calcite. In both mechanisms, nucleation of the less thermodynamically stable polymorph is predominant, because of its lower energy barriers resulting from smaller interface energies of the nuclei.

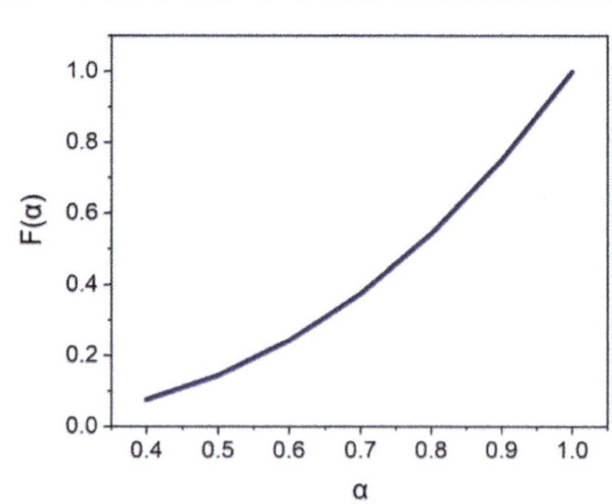

Fig. 2 Calculated function $F(\alpha)$ showing a decrease in the thermodynamic barrier for aragonite nucleation in ACC as a function of the decrease in aragonite/ACC interface energy ratio, $\alpha = \gamma_{1ACC}^{arag}/\gamma_{1s}^{arag}$.

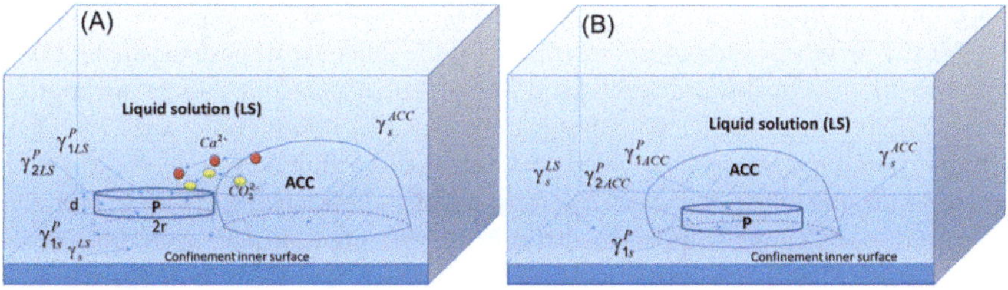

Fig. 3 Schematic representation of the heterogeneous nucleation of a crystalline calcium carbonate phase P on the inner surface of a confined volume *via* dissolution–reprecipitation mechanism (A) or *via* SST of ACC to a crystalline calcium carbonate phase P (B).

Size effect of confinement

Now that we understand how confinement combined with charged inner surfaces leads to polymorph selection, we can attempt to explain the fascinating results of Zeng *et al.*[12] In their study of the quantitative relationship between the aragonite/calcite percentage and the membrane pore diameter in the range of 25–1200 nm, those authors found a linear dependence of the aragonite percentage on the inverse pore diameter.

Zeng *et al.* suggested that the number of aragonite nucleation sites is proportional to the surface area, that the amount of precipitated $CaCO_3$ is proportional to the volume, and that the crystal growth rate is constant. For cylindrical tubes the surface area is proportional to the diameter of the confining tube d, and the volume is proportional to d^2; the ratio of the surface to the volume, however, remains proportional to $1/d$. This explanation, although persuasive, nevertheless seems insufficient to explain the results, since it would assume that nucleation of calcite occurs within the entire confined volume (proportional to d^3 for spherical pores), whereas it appears to take place mainly on the inner membrane surface. As such, the relative increase in surface relative to the bulk is identical for calcite and aragonite.

Similar reasoning can be applied to considering the effect of size on the volume fraction of ACC. Assuming the subunits of ACC such as $CaCO_3$ molecules (and possibly their small clusters) form at the inner confinement surface, and then can return to the confined volume, one can estimate the size effect on nucleation rate of ACC. Concentration of such ACC subunits, C_{su}, should be proportional to the confinement surface, S_c, and inversely proportional to the confined volume, V_c, and thus $C_{su} \sim S/V \sim 1/d$. Nucleation of ACC in the confined volume occurs *via* formation of subunit clusters, and is proportional to the concentration of clusters with some critical size, which, in turn, is proportional to the concentration of elementary units, C_{su}. Therefore, nucleation rate, and hence the volume fraction of ACC would increase with decreasing pore radius as $1/d$. Actually, this increase of ACC volume fraction with decreasing confined volume has been recorded by many researchers. One can assume that the volume fraction of aragonite in the final crystalline structure formed in confinement is proportional to the transient volume fraction of ACC.

A number of other factors are considered by Zeng *et al.*[12] as possibly responsible for the increase in aragonite fraction with the decrease in pore diameter: (1) confinement affects ion transport and leads to an evolving supersaturation profile, which may influence polymorph formation. (2) Confinement reduces the probability of nucleation (*i.e.*, it increases nucleation induction times) owing to a reduction in the number of available ions and the elimination of advection and convection. Such elimination is most significant for the thermodynamically favoured phase (calcite) and leads to the formation and stabilisation of the metastable phases (aragonite). (3) For both of these polymorphs, nucleation and growth processes are obstructed by water molecules solvating the cations, which may be more significant for the denser aragonite structure; thus, confinement may lead to reduced levels of hydration or facilitate dehydration, thereby favouring aragonite. However, none of the above-mentioned factors was able to explain the observed near-linear relationship between aragonite fracture and the inverse pore diameter, $1/d$.[12]

Crystallisation of ACC within cylindrical pores that contain liquid solutions can be considered as nucleation and growth under external pressure. This pressure develops owing to the capillary effect of a liquid confined within cylindrical pores:

$$p = \frac{\gamma}{R}, \tag{4}$$

where γ is the surface tension of the liquid solution and R is the radius of a cylindrical pore. In the case of aqueous solutions ($\gamma_{\text{water}} \approx 72$ mJ m^{-2}), the pressure within cylindrical pores of diameters ranging from 200 nm to 25 nm (as in Zeng *et al.*'s work) may reach values as high as $(7 \div 29)$ bars, which in principle can influence the nucleation rate of different polymorphs of calcium carbonate.

Assuming that the first phase precipitated is ACC, how would this pressure affect the amorphous-to-crystalline transformation?

Aragonite is denser than calcite, 2.93 g cm^{-3} as compared to 2.71 g cm^{-3}.[33] The difference in density corresponds to a difference in the molar volume: $\Delta V_{\text{mol}} = V_{\text{cal}} - V_{\text{arag}} = M_{\text{mol}} \cdot \frac{\rho_{\text{arag}} - \rho_{\text{cal}}}{\rho_{\text{cal}}\rho_{\text{arag}}} \approx 2.8\ \text{cm}^3$, where $M_{\text{mol}} \simeq 100.1$ g mol^{-1} is the molar mass of calcium carbonate. This difference in the molar volume results in an additional difference in the free energies $P\Delta V \approx (2 \div 8)$ J mol^{-1} in favour of aragonite. At the same time, the difference in free energy between calcite and aragonite, $\Delta G_{\text{cal-arag}}$, in relation to the dissolved $[Ca^{2+}][CO_3^{\ 2-}]$ under standard conditions, was found to be 1.25 kJ mol^{-1} in favour of calcite.[33] Thus, the difference of $(2 \div 8)$ J mol^{-1}, as compared to $\Delta G_{\text{cal-arag}}$, seems insufficient to explain a change in the thermodynamic preference for calcite or in the kinetics of the polymorph nucleation.

On the other hand, SST of a hydrated ACC to a crystalline polymorph should be preceded by its dehydration. Small pressures can influence the microstructure of hydrated ACC formed from a liquid solution, and in this manner influence the SST.

As mentioned above, in order to convert the ACC phase into calcium carbonate crystals *via* SST, dehydration must take place prior to the crystallisation of anhydrous crystalline nuclei.[25,34] The process of ACC dehydration was initially proposed to be thermodynamically favoured because the enthalpy of hydrated

ACC particles (measured by isothermal acid solution calorimetry) is higher than that of anhydrous ones (measured by differential scanning calorimetry).[35] This thermodynamic argument has often been used to describe the SST of biogenic ACC under ambient conditions.[35–37] However, more recent experiments, in which the same technique was used to assess the enthalpy of ACC before and after its dehydration, revealed that the enthalpy of the dehydrated ACC particles is higher.[38,39] The thermodynamic stability of ACC is described by the molar free energy ΔG. When assessing the entropic contribution due to water incorporation, Alberic *et al.*[38] obtained a constant molar free energy for the transformation of ACC to calcite, regardless of its water content. The counterbalancing effect of the enthalpy and the entropy contributions resulted in no thermodynamic preference for any hydration level from 0.01 to 1.3 H_2O at 25 °C.[38] Given these circumstances, even a small pressure difference can play a decisive role.

The activation energy of such dehydration varies in the range 50–250 kJ mol^{-1},[25,39] and can be decreased by a pressure σ for about $\sigma\Delta V$, where ΔV is the activation volume of the process. Assuming that the dehydration results in an increase in density of about 0.5 g cm^{-3}, say from 1.5 g cm^{-3} to 2.0 g cm^{-3}, the activation volume can be estimated as ~17 cm^3 per mole (according to calculations by Bushuev *et al.*,[40] the excess volume of the anhydrous ACC–water mixture $CaCO_3\cdot nH_2O$ is 6 cm^3 per mole for $n = 0.6$.). For pressures ~30 bars (as estimated for maximum capillary pressures) this would result in a decrease in activation energy to ~50 J mol^{-1}, which seems at first glance too low to influence the kinetics of the dehydration process. Assuming the dehydration rate is proportional to $\exp\left(\frac{\sigma\Delta V}{R_g T}\right) \approx 1 + \frac{\sigma\Delta V}{R_g T}$, where R_g is the gas constant, it can be seen that the second term is about 0.02 for room temperature and cannot influence the dehydration rate significantly.

We can assume that the capillary pressure can stimulate initial dehydration. In this way some densification of ACC can be achieved, thereby providing the following SST to one of the anhydrous calcium carbonate polymorphs.

The atomic structure of hydrated ACC was analysed by Bushuev *et al.*[40] using molecular dynamics simulations. According to their results, the peak of the pair distribution function in the hydrated ACC corresponds to a Ca–O distance very close to 2.5 Å. This distance is also close to the Ca–O distances in aragonite, where the Ca–O_9 polyhedrons are formed by nine Ca–O pairs consisting of two pairs with distances 2.45, 2.53, 2.54, 2.66 Å (each two) and one pair with the shortest distance of 2.41 Å. All these distances are substantially larger than the Ca–O distance (2.36 Å) in calcite[41] (the greater density of aragonite than of calcite derives from the O–O distances between oxygen molecules belonging to different CO_3 groups: in aragonite these O–O distances are smaller than the corresponding O–O distances in calcite[41]).

Apparently, nucleation of aragonite inside ACC is easier than nucleation of calcite, because aragonite/ACC interface energies are smaller than those of calcite/ACC. This supposition is validated by the closer Ca–O distances in ACC and aragonite than in calcite, so the aragonite/ACC interface can be considered as semi-coherent whereas the calcite/ACC interface should be considered as mainly incoherent. Further support comes from the study of Fernandez-Martinez *et al.*[42] showing that pressure further alters the short-range order of ACC to fit that of aragonite rather than that of calcite. It was also shown that polymorphism of the ACC structure can be influenced by pH values of the solution from which the ACC precipitated or by the presence of some cations such as Mg^{2+}.[42] More specifically, calcite-like ACC is obtained at pH ≈ 8.75, vaterite-like ACC at pH ≈ 9.8 and higher, while an aragonite-like ACC structure was identified in ACC from shells of freshwater snails and in synthetic samples of ACC doped with Mg^{2+}.[42] We can assume that confinement, in some cases, may alter the pH of the solution or the concentrations of cation in the pores, thereby affecting a short-range order in ACC, that further determines the predominant crystalline polymorph formed *via* solid ACC/crystalline calcium carbonate transformation.

To summarise the effect of pore size in confinement, we can state that the size effect of confinement can be attributed to the effect on formation of ACC which then transforms to aragonite; it can be also attributed to capillary pressures, which accelerate the initial (dehydrating) stage of hydrated ACC particles and further push the ACC short-range order closer to that of aragonite than to that of calcite. The confinement size could also influence the short-range order of ACC by controlling the local pH. In addition, an effect of cation concentrations in the solution within confined volumes cannot be ruled out.

Concluding remarks

Preceding the formation of crystalline $CaCO_3$ biominerals, the amorphous precursor phase ACC often appears in confined environments. Competitive transformation of ACC to aragonite, vaterite or calcite is determined not only by thermodynamic driving forces but also by nucleation kinetics, which are highly sensitive to energies of new interfaces of the crystalline nuclei. The smaller interfacial energies of aragonite/organic negatively charged inner confining surfaces and of aragonite/ACC interfaces as compared to those of calcite can provide the predominant formation of aragonite in micro- and nano-confined conditions, *via* possible mechanisms such as dissolution–reprecipitation and SST. On the other hand, confinement within spaces whose inner surfaces are positively charged may yield nucleation and growth of vaterite *via* a dissolution–reprecipitation mechanism. The confinement size effect can be attributed to the increased nucleation of ACC; it is also sensitive to capillary pressures, which increase with a decrease in the size of confined volume and accelerate the initial stage of ACC dehydration. Dehydrated ACC particles can further transform to different crystalline polymorphs depending on the predominant ACC short-range order, while the latter can also be influenced by capillary pressures or by the pH of the local solution with the confinement.

Conflicts of interest

There are no conflicts to declare.

References

1 H. A. Lowenstam, Minerals Formed by Organisms, *Science*, 1981, **211**(4487), 1126–1131.

2 S. Weiner and P. M. Dove, An overview of biomineralization processes and the problem of the vital effect, *Rev. Mineral. Geochem.*, 2003, **54**, 1–29.

3 L. B. Gower, Biomimetic Model Systems for Investigating the Amorphous Precursor Pathway and Its Role in Biomineralization, *Chem. Rev.*, 2008, **108**(11), 4551–4627.

4 M. F. Butler, N. Glaser, A. C. Weaver, M. Kirkland and M. Heppenstall-Butler, Calcium carbonate crystallization in the presence of biopolymers, *Cryst. Growth Des.*, 2006, **6**(3), 781–794.

5 S. Kirboga and M. Oner, Investigation of calcium carbonate precipitation in the presence of carboxymethyl inulin, *CrystEngComm*, 2013, **15**(18), 3678–3686.

6 A. W. Xu, M. Antonietti, H. Colfen and Y. P. Fang, Uniform hexagonal plates of vaterite CaCO3 mesocrystals formed by biomimetic mineralization, *Adv. Funct. Mater.*, 2006, **16**(7), 903–908.

7 G. Falini, S. Albeck, S. Weiner and L. Addadi, Control of aragonite or calcite polymorphism by mollusk shell macromolecules, *Science*, 1996, **271**(5245), 67–69.

8 A. Rao, P. Vasquez-Quitral, M. S. Fernandez, J. K. Berg, M. Sanchez, M. Drechsler, A. Neira-Carrillo, J. L. Arias, D. Gebauer and H. Colfen, pH-Dependent Schemes of Calcium Carbonate Formation in the Presence of Alginates, *Cryst. Growth Des.*, 2016, **16**(3), 1349–1359.

9 L. M. Diez, F. M. Villa, A. H. Gimenez and F. T. Garcia, Streaming Potential of Some Polycarbonate Microporous Membranes When Bathed by Licl, Nacl, Mgcl2, and Cacl2 Aqueous-Solutions, *J. Colloid Interface Sci.*, 1989, **132**(1), 27–33.

10 R. K. Given and B. H. Wilkinson, Kinetic Control of Morphology, Composition, and Mineralogy of Abiotic Sedimentary Carbonates, *J. Sediment. Petrol.*, 1985, **55**(1), 109–119.

11 G. Falini, S. Fermani, M. Gazzano and A. Ripamonti, Polymorphism and architectural crystal assembly of calcium carbonate in biologically inspired polymeric matrices, *J. Chem. Soc., Dalton Trans.*, 2000, **21**, 3983–3987.

12 M. L. Zeng, Y. Y. Kim, C. Anduix-Canto, C. Frontera, D. Laundy, N. Kapur, H. K. Christenson and F. C. Meldrum, Confinement generates single-crystal aragonite rods at room temperature, *Proc. Natl. Acad. Sci. U. S. A.*, 2018, **115**(30), 7670–7675.

13 E. Loste, R. J. Park, J. Warren and F. C. Meldrum, Precipitation of calcium carbonate in confinement, *Adv. Funct. Mater.*, 2004, **14**(12), 1211–1220.

14 A. S. Schenk, E. J. Albarracin, Y. Y. Kim, J. Ihli and F. C. Meldrum, Confinement stabilises single crystal vaterite rods, *Chem. Commun.*, 2014, **50**(36), 4729–4732.

15 J. Cavanaugh, M. L. Whittaker and D. Joester, Crystallization kinetics of amorphous calcium carbonate in confinement, *Chem. Sci.*, 2019, **10**(19), 5039–5043.

16 A. Livne, S. C. Mijowska, I. Polishchuk, W. Mashikoane, A. Katsman and B. Pokroy, A fungal mycelium templates the growth of aragonite needles, *J. Mater. Chem. B*, 2019, **7**(37), 5725–5731.

17 F. C. Meldrum and C. O'Shaughnessy, Crystallization in Confinement, *Adv. Mater.*, 2020, **32**(31), 2001068.

18 C. C. Tester, R. E. Brock, C. H. Wu, M. R. Krejci, S. Weigand and D. Joester, In vitro synthesis and stabilization of amorphous calcium carbonate (ACC) nanoparticles within liposomes, *CrystEngComm*, 2011, **13**(12), 3975–3978.

19 C. L. Jackson and G. B. Mckenna, The Glass-Transition of Organic Liquids Confined to Small Pores, *J. Non-Cryst. Solids*, 1991, **131**, 221–224.

20 M. Beiner, Nanoconfinement as a tool to study early stages of polymer crystallization, *J. Polym. Sci., Part B: Polym. Phys.*, 2008, **46**(15), 1556–1561.

21 J. M. Ha, J. H. Wolf, M. A. Hillmyer and M. D. Ward, Polymorph selectivity under nanoscopic confinement, *J. Am. Chem. Soc.*, 2004, **126**(11), 3382–3383.

22 Y. C. Zuo, Y. Z. Zhang, R. D. Huang and Y. J. Min, The effect of nanoconfinement on the glass transition temperature of ionic liquids, *Phys. Chem. Chem. Phys.*, 2019, **21**(1), 22–25.

23 P. Bots, L. G. Benning, J. D. Rodriguez-Blanco, T. Roncal-Herrero and S. Shaw, Mechanistic Insights into the Crystallization of Amorphous Calcium Carbonate (ACC), *Cryst. Growth Des.*, 2012, **12**(7), 3806–3814.

24 J. D. Rodriguez-Blanco, S. Shaw and L. G. Benning, The kinetics and mechanisms of amorphous calcium carbonate (ACC) crystallization to calcite, *via* vaterite, *Nanoscale*, 2011, **3**(1), 265–271.

25 J. Ihli, W. C. Wong, E. H. Noel, Y. Y. Kim, A. N. Kulak, H. K. Christenson, M. J. Duer and F. C. Meldrum, Dehydration and crystallization of amorphous calcium carbonate in solution and in air, *Nat. Commun.*, 2014, **5**, 3169–3179.

26 E. M. Pouget, P. H. H. Bomans, J. A. C. M. Goos, P. M. Frederik, G. de With and N. A. J. M. Sommerdijk, The Initial Stages of Template-Controlled CaCO3 Formation Revealed by Cryo-TEM, *Science*, 2009, **323**(5920), 1455–1458.

27 E. M. Pouget, P. H. H. Bomans, A. Dey, P. M. Frederik, G. de With and N. A. J. M. Sommerdijk, The Development of Morphology and Structure in Hexagonal Vaterite, *J. Am. Chem. Soc.*, 2010, **132**(33), 11560–11565.

28 D. Gebauer, A. Volkel and H. Colfen, Stable Prenucleation Calcium Carbonate Clusters, *Science*, 2008, **322**(5909), 1819–1822.

29 N. H. de Leeuw and S. C. Parker, Surface structure and morphology of calcium carbonate polymorphs calcite, aragonite, and vaterite: An atomistic approach, *J. Phys. Chem. B*, 1998, **102**(16), 2914–2922.

30 J. Kawano, N. Shimobayashi, A. Miyake and M. Kitamura, Precipitation diagram of calcium carbonate polymorphs: its construction and significance, *J. Phys.: Condens. Matter*, 2009, **21**(42), 425102–425112.

31 F. Lippmann, *Sedimentary Carbonate Minerals*, Springer-Verlag: Berlin, 1973.

32 A. Koishi, A. Fernandez-Martinez, A. E. S. Van Driessche, L. J. Michot, C. M. Pina, C. Pimentel, B. Lee and G. Montes-Hernandez, Surface Wetting Controls Calcium Carbonate Crystallization Kinetics, *Chem. Mater.*, 2019, **31**(9), 3340–3348.

33 S. A. T. Redfern, Structural variations in carbonates, *High-Temperature and High-Pressure Crystal Chemistry*, 2000, vol. 41, pp. 289–308.

34 M. Saharay, A. O. Yazaydin and R. J. Kirkpatrick, Dehydration-Induced Amorphous Phases of Calcium Carbonate, *J. Phys. Chem. B*, 2013, **117**(12), 3328–3336.

35 A. V. Radha, T. Z. Forbes, C. E. Killian, P. U. P. A. Gilbert and A. Navrotsky, Transformation and crystallization energetics of synthetic and biogenic

amorphous calcium carbonate, *Proc. Natl. Acad. Sci. U.S.A.*, 2010, **107**(38), 16438–16443.

36 R. T. DeVol, C. Y. Sun, M. A. Marcus, S. N. Coppersmith, S. C. B. Myneni and P. U. P. A. Gilbert, Nanoscale Transforming Mineral Phases in Fresh Nacre, *J. Am. Chem. Soc.*, 2015, **137**(41), 13325–13333.

37 A. Gal, K. Kahil, N. Vidavsky, R. T. DeVol, P. U. P. A. Gilbert, P. Fratzl, S. Weiner and L. Addadi, Particle Accretion Mechanism Underlies Biological Crystal Growth from an Amorphous Precursor Phase, *Adv. Funct. Mater.*, 2014, **24**(34), 5420–5426.

38 M. Alberic, L. Bertinetti, Z. Y. Zou, P. Fratzl, W. Habraken and Y. Politi, The Crystallization of Amorphous Calcium Carbonate is Kinetically Governed by Ion Impurities and Water, *Adv. Sci.*, 2018, **5**(5), 1701000.

39 M. P. Schmidt, A. J. Ilott, B. L. Phillips and R. J. Reeder, Structural Changes upon Dehydration of Amorphous Calcium Carbonate, *Cryst. Growth Des.*, 2014, **14**(3), 938–951.

40 Y. G. Bushuev, A. R. Finney and P. M. Rodger, Stability and Structure of Hydrated Amorphous Calcium Carbonate, *Cryst. Growth Des.*, 2015, **15**(11), 5269–5279.

41 E. Macias-Sanchez, M. G. Willinger, C. M. Pina and A. G. Checa, Transformation of ACC into aragonite and the origin of the nanogranular structure of nacre, *Sci. Rep.*, 2017, **7**, 12728–12739.

42 A. Fernandez-Martinez, B. Kalkan, S. M. Clark and G. A. Waychunas, Pressure-Induced Polyamorphism and Formation of 'Aragonitic' Amorphous Calcium Carbonate, *Angew. Chem. Int. Ed.*, 2013, **52**(32), 8354–8357.

Faraday Discussions

PAPER

The role of solvation in proton transfer reactions: implications for predicting salt/co-crystal formation using the ΔpK_a rule†

Aurora J. Cruz-Cabeza, [*ab] Matteo Lusi, [c] Helen P. Wheatcroft[b] and Andrew D. Bond [d]

Received 4th November 2021, Accepted 12th January 2022
DOI: 10.1039/d1fd00081k

The ΔpK_a rule is commonly applied by chemists and crystal engineers as a guideline for the rational design of molecular salts and co-crystals. For multi-component crystals containing acid and base constituents, empirical evidence has shown that $\Delta pK_a > 4$ almost always leads to salts, $\Delta pK_a < -1$ almost always leads to co-crystals and ΔpK_a between −1 and 4 can be either. This paper reviews the theoretical background of the ΔpK_a rule and highlights the crucial role of solvation in determining the outcome of the potential proton transfer from acid to base. New data on the frequency of the occurrence of co-crystals and salts in multi-component crystal structures containing acid and base constituents show that the relationship between ΔpK_a and the frequency of salt/co-crystal formation is influenced by the composition of the crystal. For unsolvated co-crystals/salts, containing only the principal acid and base components, the point of 50% probability for salt/co-crystal formation occurs at $\Delta pK_a \approx 1.4$, while for hydrates of co-crystals and salts, this point is shifted to $\Delta pK_a \approx -0.5$. For acid–base crystals with the possibility for two proton transfers, the overall frequency of occurrence of any salt (monovalent or divalent) *versus* a co-crystal is comparable to that of the whole data set, but the point of 50% probability for observing a monovalent salt *vs.* a divalent salt lies at $\Delta pK_{a,II} \approx -4.5$. Hence, where two proton transfers are possible, the balance is between co-crystals and divalent salts, with monovalent salts being far less common. Finally, the overall role played by the "crystal" solvation is illustrated by the fact that acid–base complexes in the intermediate region of ΔpK_a tip towards salt formation if ancillary hydrogen bonds can exist. Thus, the solvation strength of the lattice plays a key role in the stabilisation of the ions.

[a]*Department of Chemical Engineering, School of Engineering, University of Manchester, UK. E-mail: aurora.cruzcabeza@manchester.ac.uk*
[b]*Chemical Development, Pharmaceutical Technology & Development, AstraZeneca, Macclesfield, UK*
[c]*Department of Chemical Sciences, Bernal Institute, University of Limerick, Limerick, Ireland*
[d]*Yusuf Hamied Department of Chemistry, University of Cambridge, Cambridge, UK*
† Electronic supplementary information (ESI) available. See DOI: 10.1039/d1fd00081k

Introduction

Proton transfer reactions between an acid and a base are responsible for many natural phenomena, from the existence of salts to the zwitterionic nature of common amino acids in environments amenable to life, or the interactions between proteins and substrates. From a chemical point of view, acid–base equilibria can easily be studied. In particular, the activity (or concentration) of ionised species can be measured in solution and acid dissociation constants (K_a) calculated and reported, typically in the form of pK_a values. Moreover, robust computational models have been developed to predict aqueous pK_a values from the molecular structure.[1–3]

Proton transfer is also common in the solid state, as demonstrated by the large number of molecular salts[4] and zwitterionic crystal structures reported in the Cambridge Structural Database (CSD).[5] In molecular crystals, proton transfer reactions generally lie fully to one side of the chemical equation, yielding either a co-crystal where no proton transfer occurs or a salt where a proton is transferred from an acid to a base. Exceptions include dynamic systems, such as solid-state proton conductors,[6] or cases in which protons appear to be located part way between the acid and the base.[7] The characterisation of such systems can be complicated by the inherent difficulties associated with locating H atoms using X-ray diffraction data, and in some cases the extent of proton transfer may be temperature dependent.[8,9] Because of this, complementary characterisation techniques, such as XPS,[10–12] solid-state NMR[13] or neutron diffraction[14,15] (amongst others[16]), often need to be used to locate precisely the position of the acid H atoms. There are also cases where both ionised and non-ionised species of an acid or base are present in the same crystal.[17]

The ionisation of active pharmaceutical ingredients (APIs) into salts plays a crucial role in the formulation of medicines.[4,18] Salts are often more soluble than their corresponding non-ionised forms, thereby affording drugs with improved pharmacokinetics.[19] In recent years, co-crystals have been shown to be a viable alternative to salts since they can also improve the solubility of APIs[20] whilst offering other advantages, such as being less prone to hydration.[21] Co-crystals are often designed using crystal engineering concepts,[22–25] whereby suitable co-crystal formers are selected based on their potential to form strong molecule-to-molecule interactions (synthons), typically based on hydrogen bonds.[26–28] Very often, the targeted interactions involve an acid and a base. Indeed, hydrogen bonds are particularly strong, and therefore most attractive for crystal engineering strategies, for acid–base pairs with $\Delta pK_a \approx 0$, where $\Delta pK_a = pK_a(\text{protonated base}) - pK_a(\text{acid})$.[29,30] However, such strategies may be compromised by proton transfer in the solid state, thereby yielding a salt rather than the intended co-crystal product.

As a rule of thumb, salt formation in a molecular crystal is typically considered to be likely if the ΔpK_a for the constituent acid and base is larger than 3. In 2005, Bhogala *et al.*[31] noted that negative values of ΔpK_a always lead to the formation of co-crystals. This so-called "ΔpK_a rule" has been commonly used for the rational design of salts and co-crystals, supported by the free availability of experimental and calculated pK_a data. In 2012, Cruz-Cabeza revised the rule on the basis of a survey of *ca.* 6500 crystal structures and pK_a values calculated for the molecular components.[32] It was found that $\Delta pK_a > 4$ almost always leads to salts, $\Delta pK_a < -1$

almost always leads to co-crystals and ΔpK_a between −1 and 4 can be either. The intermediate region of ΔpK_a accommodates what Childs has referred to as the salt/co-crystal continuum.[33]

Given the enduring popularity of the ΔpK_a rule amongst chemists and crystal engineers,[34–36] this paper sets out to review the theoretical basis of the rule, highlighting (in particular) the role of solvation for determining the outcome of proton transfer reactions in the solid state. When applying the ΔpK_a rule to predict salt/co-crystal formation, a key question is how the crystal lattice influences proton transfer, particularly in the intermediate region close to $\Delta pK_a \approx 0$. Some new data extracted from an extensive set of crystal structures demonstrates how the ΔpK_a rule might be fine-tuned for specific classes of crystals, such as crystalline hydrates. Analysis of those structures also reveals the important role played by ancillary hydrogen-bonds in stabilising ionisation in the solid-state.

Methods

pK_a and ΔpK_a calculations

pK_a values were calculated using ChemAxon's Marvin software suite (version 21.13),[37] either through the visualiser or *via* the command line executable for the bulk processing of many compounds. The calculator returns pK_a values for both acids and protonated bases (depicted in the visualiser in red and blue, respectively). The ΔpK_a for an acid–base pair was then calculated by subtracting the pK_a values of the strongest acid and strongest base, where $\Delta pK_a = pK_a(BH^+) - pK_a(A)$. For systems where a second proton transfer is possible, a second value of ΔpK_a, denoted $\Delta pK_{a,II}$, was also calculated. For this calculation, the stoichiometry of the system was accounted for when considering the second-strongest acid and base. For example, the first ΔpK_a uses the lowest $pK_a(A)$ and the highest $pK_a(BH^+)$ in the crystal structure (*i.e.* the strongest acid and strongest base) and the second $\Delta pK_{a,II}$ would use the second lowest $pK_a(A)$ and the second highest $pK_a(BH^+)$. Depending on the crystal composition and stoichiometry, the second base could be the same as the first for a $H_2A:B$ system with a 1 : 2 stoichiometry (resulting in $\{A^{2-}\}\{BH^+\}_2$) or it could be the second strongest basic functional group in the base for a system with a 1 : 1 stoichiometry (resulting in $\{A^{2-}\}\{BH_2{}^{2+}\}$). The ΔpK_a values were obtained from the calculated pK_a values using a simple Python script. Finally, it was recorded whether the acid–base pair for each ΔpK_a calculation was located within the same molecule or in different molecules. This differentiates ΔpK_a values for intermolecular proton transfer (salt formation) from $\Delta pK_{a,self}$ values for intramolecular proton transfer (zwitterion formation).

Energy calculations

Energy calculations were performed with Gaussian 16 (ref. 38) using the B97d level of theory and the 6-31+G(d,p) basis sets.[39] For the calculation of $\Delta E^{\circ HA\cdot B}_{ion\text{-}gas}$, ions, as well as neutral species, were optimised separately and the term $\Delta E^{\circ HA\cdot B}_{ion\text{-}gas}$ was calculated as the difference between the electronic energy of the ions minus the electronic energy of the neutral species. Implicit solvation models (SMD)[40] for toluene, 2-heptanone and water were also used for some calculations. For the calculation of the potential energy for proton transfer in acid–base pairs, molecular models were generated manually using Avogadro[41,42] and initially optimised using the MMFF94s forcefield.[43] The resulting model for the acid–base dimer was then optimised in the gas phase using Gaussian. In all cases, the acid and base were oriented to form an A–H···B hydrogen bond. The resulting geometry (containing neutral species in all cases) was then used to run constrained optimisations in various media (gas phase, toluene, 2-heptanone and water). Here, the distance between the acid proton and the acceptor atom of the base was constrained progressively in the range of −0.8 to +2.4 Å in steps of 0.1 Å. Energies are reported relative to the best non-ionised gas-phase geometry in each case.

Datasets of crystal structures

Crystal structure datasets were built following the steps depicted in Fig. 1. The CSD version 5.42 (ref. 44) was first filtered for structures containing any combination of the atom types H, C, O, S and halogens, and with determined 3D coordinates. Only organic structures were thus retrieved. This yielded over 215k structures of single-component crystals and over 61k structures of multi-component crystals.

The CSD Python API interface was then used to generate SMILES strings for all components in the crystal structures, and the stoichiometry and nature of the components were established. Components with a positive charge were designated as cations, those with a negative charge as anions, and those with separation of charges within the same compound as zwitterions (which could also be anions or cations depending on the overall charge). Cations and zwitterions were further classified as quaternary (Q4) if the positive charge was not due to proton transfer (*e.g.* N^+ bonded to four other C atoms). The SMILES of neutral components were checked and flagged as water, solvent (if within the 18 most common solvent types in the CSD) or main molecular component. Most of the total 276k retrieved crystal structures passed successfully through the pK_a calculator, but some failed due to errors, disorder or uncommon SMILES strings. The final dataset-1 contained nearly 259k crystal structures with calculated pK_a values.

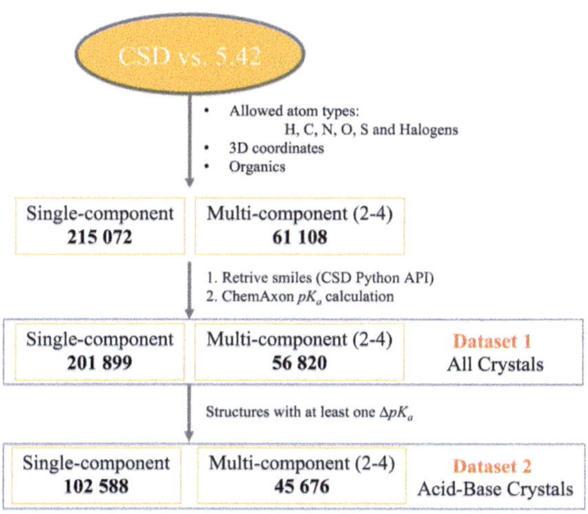

Fig. 1 Process for the retrieval of dataset-1 and dataset-2.

Dataset-1 was then filtered for structures containing at least one acid–base pair, and thus one ΔpK_a value, resulting in dataset-2, containing over 148k crystal structures. A Python script was written to analyse all data. The number of protons transferred in each crystal structure was recorded, as well as the location of the strongest acid and base. If the strongest acid and base were contained within the same molecule, ΔpK_a was classified as $\Delta pK_{a,self}$ and these data were used for the analysis of trends in zwitterions and neutral compounds. If the strongest acid and base were contained within different molecules in the crystal, this was classified as a standard ΔpK_a. Second strongest acids and bases were also considered and $\Delta pK_{a,II}$ was recorded for relevant structures.

Full interaction maps

Full interaction maps (FIMs)[45] were calculated for pairs of acids and bases using a water oxygen as a probe and displaying only the third contour with the standard setting of 6.0 (higher localised special probability). This was done using Mercury version 2021.2.0.[46]

Hydrogen bond analysis in salts and co-crystals with $\Delta pK_a \approx 0$

In total, 818 structures of co-crystals and monovalent salts with $\Delta pK_a \approx 0$ (within an absolute tolerance of 1 unit, ΔpK_a values between −0.5 and 0.5) were identified in dataset-2. These crystal structures were loaded into Conquest and the hydrogen-bonding environments of the acid–base pairs were analysed. First, the set was filtered to structures containing an acid–base contact involving a carboxylic acid/carboxylate. This resulted in a subset of 531 structures. The ancillary hydrogen bonds (HBs) around the carboxylic acid/carboxylate were then analysed further and structures grouped into three subsets: structures with 0, 1 or 2 ancillary HBs. For this purpose, Conquest queries were built for the different HB environments, with HBs identified as follows:

• The HB donor is either an N or an O atom (no C atoms).

• The HB acceptor is either O or the O^- in the carboxylic acid or carboxylate, respectively.

• The distance between the non-H atoms involved in the HB is less than the sum of their van der Waals radii.

• The angle of the HB ($N/O–H\cdots O/O^-$) is between 120 and 180°.

• Both intramolecular and intermolecular HBs were recorded.

Theoretical background

Gibbs free energy and ΔpK_a

For the dissociation of a monoprotic acid (HA) in water, the equilibrium is quantified by the acid dissociation constant, K_a^{HA}:

$$\mathrm{HA(aq)} + \mathrm{H_2O(l)} \xrightleftharpoons{K_a^{HA}} \mathrm{A^-(aq)} + \mathrm{H_3O^+(aq)}$$

For a monoprotic base (B) accepting a proton in water, it is convenient also to quantify the equilibrium using the acid dissociation constant of the protonated base, $K_a^{BH^+}$:

$$BH^+(aq) + H_2O(l) \xrightleftharpoons{K_a^{BH^+}} B(aq) + H_3O^+(aq)$$

Since an equilibrium between HA and B in aqueous solution is a combination of the acid and base equilibria, the equilibrium constant can be expressed as a quotient of the individual acid dissociation constants:

$$HA(aq) + B(aq) \rightleftharpoons A^-(aq) + BH^+(aq)$$

$$K_{\text{ion-water}}^{HA\cdot B} = \frac{K_a^{HA}}{K_a^{BH^+}}$$

Taking logarithms equates the log of the equilibrium constant for the proton transfer reaction in water to ΔpK_a:

$$\log K_{\text{ion-water}}^{HA\cdot B} = \log K_a^{HA} - \log K_a^{BH^+} = pK_a^{BH^+} - pK_a^{HA} = \Delta pK_a$$

Using the thermodynamic relation between the standard Gibbs free energy of a reaction and its equilibrium constant, the standard free energy change for proton transfer between the acid and base in water is related to ΔpK_a as follows:

$$\Delta G_{\text{ion-water}}^{\circ HA\cdot B} = -RT \ln K_{\text{ion-water}}^{HA\cdot B} = -2.3RT \log K_{\text{ion-water}}^{HA\cdot B} = -2.3RT\Delta pK_a$$

At 298 K, $\Delta G_{\text{ion-water}}^{\circ HA\cdot B} = -5.71\ \Delta pK_a$. Thus, positive values of ΔpK_a correspond to negative $\Delta G_{\text{ion-water}}^{\circ HA\cdot B}$ and therefore favour proton transfer from acid to base. Negative values of ΔpK_a correspond to positive $\Delta G_{\text{ion-water}}^{\circ HA\cdot B}$ and favour non-ionised acid and base. Each additional unit of ΔpK_a changes $\Delta G_{\text{ion-water}}^{\circ HA\cdot B}$ by ± 5.71 kJ mol^{-1}. For $\Delta pK_a = 0$, $\Delta G_{\text{ion-water}}^{\circ HA\cdot B} = 0$, giving equal quantities of ionised and non-ionised species in water. Most pK_a data refer to an aqueous solution at 298 K. However, the values may change drastically in other solvents and media.[47]

Extension to molecular crystals

The formation of a multi-component molecular crystal comprising an acid and a base may yield either a salt if proton transfer occurs or a co-crystal if there is no proton transfer. A thermochemical cycle summarising the associated free energy changes is shown in Fig. 2. For the co-crystal, the free energy of formation is simply the lattice free energy relative to the non-interacting molecules in the gas phase, $\Delta G_{\text{latt}}^{\circ \text{cocrystal}}$ (assuming no significant conformational change for the molecules upon crystallisation). For the salt, the free energy of formation is the sum of the free energy for proton transfer in the gas phase, $\Delta G_{\text{ion-gas}}^{\circ HA\cdot B}$, plus the lattice free energy of the salt relative to the non-interacting ions in the gas phase, $\Delta G_{\text{latt}}^{\circ \text{salt}}$. Hence, the formation of a salt or co-crystal (under thermodynamic control) depends on the balance between the free energy for proton transfer in the gas phase and the difference between the lattice free energies of the salt and co-crystal:

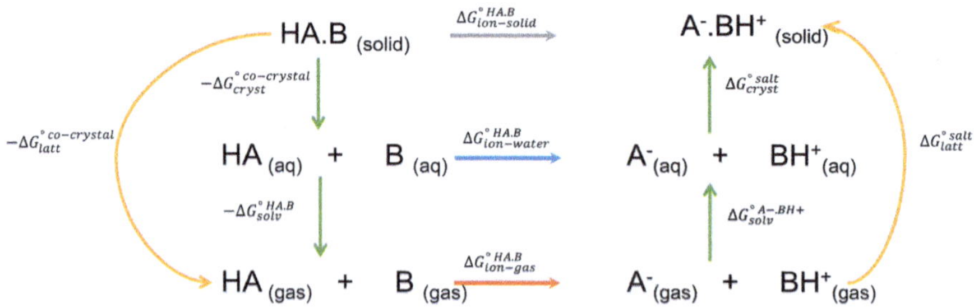

Fig. 2 Thermochemical cycle summarising the free energy changes associated with the formation of a molecular salt or co-crystal.

$$\Delta G^{\circ \mathrm{HA \cdot B}}_{\text{ion-solid}} = \Delta G^{\circ \mathrm{HA \cdot B}}_{\text{ion-gas}} + \left(\Delta G^{\circ \text{salt}}_{\text{latt}} - \Delta G^{\circ \text{cocrystal}}_{\text{latt}}\right) \tag{1}$$

If $\Delta G^{\circ \mathrm{HA \cdot B}}_{\text{ion-solid}} < 0$, a salt will be favoured over a co-crystal. Conversely, if $\Delta G^{\circ \mathrm{HA \cdot B}}_{\text{ion-solid}} > 0$, a co-crystal will be preferred.

Evaluating $\Delta G^{\circ \mathrm{HA \cdot B}}_{\text{ion-solid}}$: the gas-phase route

Since temperature and entropic effects remain challenging to compute and their relative changes are typically small,[48] the standard free energy differences can be approximated to potential energy changes at 0 K, so:

$$\Delta G^{\circ \mathrm{HA \cdot B}}_{\text{ion-solid}} \approx \Delta E^{\mathrm{HA \cdot B}}_{\text{ion-solid}} = \Delta E^{\mathrm{HA \cdot B}}_{\text{ion-gas}} + \left(\Delta E^{\text{salt}}_{\text{latt}} - \Delta E^{\text{cocrystal}}_{\text{latt}}\right) \tag{2}$$

where the first term, $\Delta E^{\mathrm{HA \cdot B}}_{\text{ion-gas}}$, is the energy for proton transfer in the gas phase, and the second term is the difference in lattice energies between the resulting salt and co-crystal.

Although less commonly reported than pK_a values, proton affinities (PA) in the gas phase can be measured or computed, which enables $\Delta E^{\circ \mathrm{HA \cdot B}}_{\text{ion-gas}}$ to be determined as follows:

$$\mathrm{PA(acid)} = \Delta E^{\circ}\{\mathrm{A^-(g) + H^+(g) \rightarrow HA(g)}\}$$

$$\mathrm{PA(base)} = \Delta E^{\circ}\{\mathrm{B(g) + H^+(g) \rightarrow BH^+(g)}\}$$

$$\Delta E^{\circ \mathrm{HA \cdot B}}_{\text{ion-gas}} = \mathrm{PA(base)} - \mathrm{PA(acid)}.$$

The second term in eqn (2) is usually highly negative because lattice energies of salts are significantly more negative than those of co-crystals due to the coulombic contributions in the salt, and the first term in eqn (2) is usually highly positive (energy is required for proton transfer). It is the balance between $\Delta E^{\circ \mathrm{HA \cdot B}}_{\text{ion-gas}}$ and $(\Delta E^{\text{salt}}_{\text{latt}} - \Delta E^{\text{co-crystal}}_{\text{latt}})$ that determines whether a co-crystal would form.

In this context, ΔpK_a can be viewed as a proxy estimate for $\Delta E^{\circ \mathrm{HA \cdot B}}_{\text{ion-gas}}$. In general, the strengths of acids and bases in aqueous solution do not necessarily correspond to the proton affinities in the gas phase, but a good linear dependence between the gas phase and aqueous acidity has been shown for organic acids of the type typically found in molecular crystals.[49] If HA is a strong acid and B is a strong base (leading to a large positive ΔpK_a), $\Delta E^{\circ HA\cdot B}_{ion\text{-}gas}$ is likely to be less positive and amply compensated by the lattice energy gain of the salt. If HA and B are a weak acid/base (leading to a negative ΔpK_a), $\Delta E^{\circ HA\cdot B}_{ion\text{-}gas}$ will have a larger positive value which could outweigh the energy gain of the salt lattice and therefore favour the co-crystal. For intermediate cases, the balance is delicate.

If a crystal structure (or structures) is known, the difference in lattice energy between the salt and co-crystal can be explicitly calculated for a given system.‡ For example, Mohamed *et al.* have shown the energetic balance between a salt and co-crystal for a series of pyridine/carboxylic acid crystals.[50] In this series, the acid and base interact directly in the crystal structure so the salt/co-crystal balance can be explored by shifting the acid proton across an O–H···N hydrogen bond. In some cases, a single-well potential was found with a clearly favoured position (either co-crystal or salt), but other cases showed a very shallow potential with very little energetic difference between either situation.[50] Similar studies on a series of hydrazone/dicarboxylic acid crystals produced comparable results.[51] Further to this, commonly applied DFT-D methods have been shown to often result in incorrect proton transfer in acid–base cocrystals.[53] Hence, the second term in eqn (2) may need careful consideration on a crystal-by-crystal basis. The biggest limitation here, of course, is the requirement to determine the crystal structure(s) in order to make the calculations. To predict salt/co-crystal formation from the molecular structure alone, the entire process would require crystal structure prediction (CSP) of both the co-crystal and salt systems followed by accurate computation of the lattice energies.[52] Hence, the full evaluation of eqn (2) is generally impractical as a predictive tool.

Evaluating $\Delta G^{\circ HA\cdot B}_{ion\text{-}solid}$: the aqueous route

Following the thermochemical cycle in Fig. 2, $\Delta G^{\circ HA\cdot B}_{ion\text{-}solid}$ can be calculated from aqueous solution data as follows:

$$\Delta G^{\circ HA\cdot B}_{ion\text{-}solid} = \Delta G^{\circ HA\cdot B}_{ion\text{-}water} + \left(\Delta G^{\circ salt}_{cryst} - \Delta G^{\circ cocrystal}_{cryst}\right) \tag{3}$$

where ΔG°_{cryst} is the standard free energy change upon crystallisation of the salt or co-crystal from the constituent components in aqueous solution. This is the opposite process to that defining the solubility product, and hence $\Delta G^{\circ}_{cryst} = RT\ \ln K_{sp}$, where K_{sp} is the product of the molar concentrations of the molecular constituents in solution. The first term of eqn (3) has been shown earlier to be $-2.3RT\Delta pK_a$, so eqn (3) can be re-written as:

$$\Delta G^{\circ HA\cdot B}_{ion\text{-}solid} = -2.3RT\left(\Delta pK_a - \log\left(\frac{K^{\circ salt}_{sp}}{K^{\circ cocrystal}_{sp}}\right)\right). \tag{4}$$

‡ If a classical modelling approach is used, the ionisation term and the lattice energy terms will be computed separately. However, if a quantum modelling approach is used, the lattice energy will be optimised together with the ionisation state. A single optimisation of a crystal structure with DFT-d will result in either a salt or a co-crystal for a single well potential scenario since it will optimise the total energy of the lattice simultaneously.

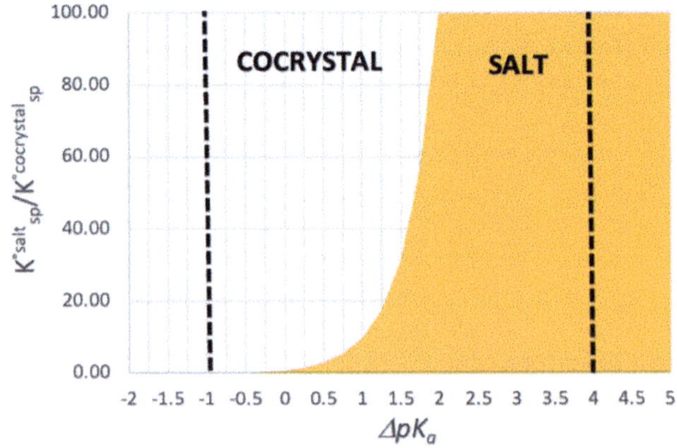

Fig. 3 Diagram showing the regions where salt (yellow) and co-crystal (no fill) are favoured as a function of the ΔpK_a of the acid–base pair and the relative solubility of the salt and the co-crystal.

Using eqn (4), the sign of $\Delta G^{\circ HA\cdot B}_{\text{ion-solid}}$ and, hence, the relative thermodynamic stability of the salt and co-crystal, can be visualised through a simple diagram (Fig. 3). For each ΔpK_a value, the value of $\log\left(\frac{K_{sp}^{\circ\,salt}}{K_{sp}^{\circ\,cocrystal}}\right)$ at which the free energy of proton transfer is 0 represents the boundary between the co-crystal and salt regions. This diagram illustrates neatly the ΔpK_a rule. Co-crystals dominate when $\Delta pK_a < 0$. When $\Delta pK_a > 4$, salts dominate in most solubility ratio ranges. In the intermediate region of ΔpK_a, the outcome is strongly sensitive to the relative solubility product of the salt *versus* the co-crystal.

While ΔpK_a can easily be measured or calculated, the second term in eqn (4) cannot be quantified unless both a co-crystal and a salt exist for the system, and the solubility products can be measured. This is almost never the case. Hence, eqn (4) is no more practical than eqn (2) as a general predictive tool.

Finally, it should be stressed that the balance between ΔpK_a and the solubility ratio may be complex. For example, in the case where an acid becomes significantly weaker than the expectation in aqueous solution due to a change to a poorly solvating solvent, this may be compensated by a change in the relative solubilities of the co-crystal and the salt. For example, upon changing from water to an organic solvent, the effective ΔpK_a may decrease, but this is likely to be compensated by a decrease in the solubility product of the salt relative to the co-crystal. An example of such behaviour is seen for cediranib maleate.[54] Hence, the applicability of expectations based on an aqueous environment must be carefully considered when transferring to other solvents.

Practical application of the ΔpK_a rule

Although the preceding sections add theoretical context to ΔpK_a as a predictor for salt/co-crystal formation, the practical application of the ΔpK_a rule is effectively based on empirical evidence. While it is unclear when the observation that salts would typically form for acid and bases with $\Delta pK_a > 3$ was made, this knowledge has been used for decades in the pharmaceutical industry. In 2012, Cruz-Cabeza probed the rule in the CSD and showed that $\Delta pK_a > 4$ almost always leads to salts and $\Delta pK_a < -1$ almost always leads to co-crystals. In the intermediate region, the probability of salt formation was found to follow the empirical trend:

$$P_{obs}(\text{salt, \%}) = 17\ \Delta pK_a + 28$$

In this context, with over 1 million crystal structures now available in the CSD, we set out in this work to probe further the experimental observations of salt and co-crystal formation, in order to shed light on the impact of crystal composition (particularly hydration), the location of the acid and base groups, the local hydrogen-bonding environment and the possibility of further proton transfers on the ΔpK_a rule.

Results

Impact of solvation on the potential energy of acid–base proton transfer

To demonstrate the effect of the acid–base relative strength, as well as solvation, on the ionisation of acid–base pairs, we selected acetic acid (AA) as a model acid together with three bases with increasing base strength: THF, pyridine (PYR) and trimethylamine (TMA). The computed ΔpK_a and $\Delta E^{\circ HA\cdot B}_{\text{ion-gas}}$ values in the gas phase are summarised in Fig. 4. As expected, as ΔpK_a increases, $\Delta E^{\circ HA\cdot B}_{\text{ion-gas}}$ decreases, from 605 kJ mol^{-1} for AA:THF to 330 kJ mol^{-1} for AA:TMA.

To expand on these values, and to illustrate the impact of solvation, the potential energy curve for proton transfer for each acid–base pair was calculated in different solvation environments: (a) gas-phase; (b) toluene; (c) 2-heptanone; (d) water. The solution environments were accounted for implicitly using SMD solvation models. Toluene and 2-heptanone were chosen because their dielectric constants (toluene = 2.3, 2-heptanone = 12) are similar to those anticipated for neutral molecular solids and molecular salts, respectively.[55] Hence, they are intended to give a coarse approximation of the situation that may be found in the crystalline state. They are not necessarily implied to be common laboratory

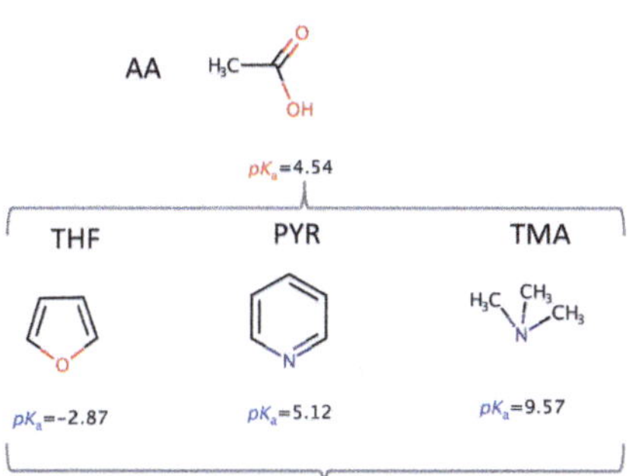

ΔpK_a	-7.41	0.58	5.03
$\Delta E^{\circ HA.B}_{ion-gas}$	605 kJ/mol	505 kJ/mol	330 kJ/mol

Fig. 4 Calculated ΔpK_a values and gas-phase proton transfer energies for acid–base pairs formed between AA with three bases (THF, PYR, TMA) of increasing strength.

solvents (indeed, 2-heptanone is rare in practice). For the calculations, the AA molecule was located next to the base so that an A–H···B hydrogen bond was formed, and the geometry was optimised. The potential energy was then calculated as a function of the position of H^+ within the A–H···B hydrogen bond, as outlined in the Methods section. The results are shown in Fig. 5.

For AA:THF ($\Delta pK_a = -7.4$), none of the environments favour proton transfer from acid to base because of the high $\Delta E^{\circ HA\cdot B}_{\text{ion-gas}}$ value for this pair. For AA:TMA ($\Delta pK_a = 5.0$), a similar situation is seen in the gas phase, but there is a progressive emergence of an energy minimum for proton transfer as the dielectric constant of the solvent increases. In aqueous solution, the minimum is pronounced and the ionised species are favoured over the neutral species by over 20 kJ mol^{-1}. For AA:PYR ($\Delta pK_a = 0.6$), an intermediate situation is seen, whereby the curve for aqueous solution resembles that for AA:TMA in 2-heptanone. Hence, the balance between $\Delta E^{\circ HA\cdot B}_{\text{ion-gas}}$ and the extent to which the ionised species are stabilised by solvation is clearly seen. To extend this picture to the solid state, the key question is how the crystal lattice impacts the outcome. The empirical evidence established for the ΔpK_a rule suggests that the effect of the lattice is never sufficient to influence the outcome for acid–base pairs with $\Delta pK_a < -1$ or $\Delta pK_a > 4$, but that it can have a significant effect in the intermediate region. The remainder of this paper considers whether existing crystallographic information can be used to enhance the predictive power of the ΔpK_a rule in the intermediate region.

Ionisation and solvation statistics in the CSD (dataset-1)

Dataset-1 contains 258 719 crystal structures of organic compounds containing between one and four different molecular components. Dataset-1 was analysed by crystal type and composition, as described in the Methods section. Crystals with only one main component were classified either as neutral (no proton transfer) or zwitterionic, with the further distinction that some zwitterions are not the result of proton transfer (referred to as zwitterion-Q4). Multi-component systems were classified as co-crystals (neutral main components), co-crystals containing zwitterions, salts, salts-Q4 (quaternary salts) and ionic co-crystals. Some more

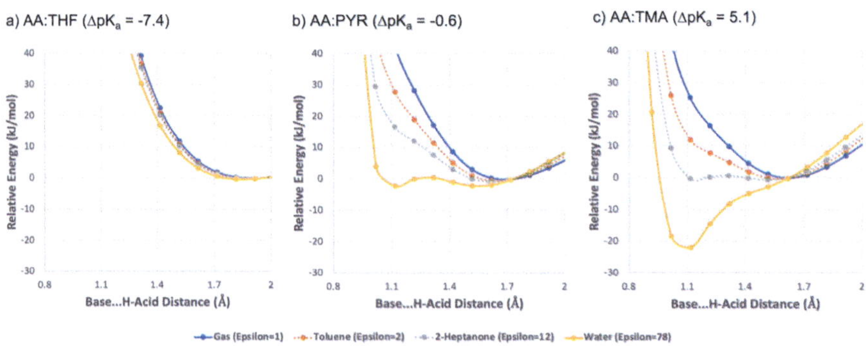

Fig. 5 Computed potential energy curves for the proton transfer reaction (as a function of the H···B distance) between the three acid–base pairs in four different solvation environments. The energy is given relative to the most stable neutral acid–base pair configuration. The implicit SMD solvation model is used at the #B97d/6-31+G(d,p) level of theory.

Table 1 Crystal types classified by their ionisation state and composition, number of crystal structures in dataset-1 (*N*) and percentage of those observed as unsolvated, hydrates or solvates

				Percentage of dataset (%)		
Crystal type	#MC[a]	Composition[b]	*N*	Unsolvated	Hydrate	Solvate
Neutral	1	N	215 491	92	4	5
Zwitterion	1	Z	2112	60	34	6
Zwitterion-Q4	1	Q4–Z	3677	86	10	4
Co-crystal	2	N N	9464	89	6	5
Co-crystal with Z	2	N Z	349	73	21	6
Co-crystal with ZQ4	2	N Q4–Z	197	85	7	8
Salt	2	C^+ A^-	16 610	72	23	5
Salt Q4	2	Q4–C^+ A^-	6479	65	27	8
Salt with Z	2	Z^+ A^-	181	66	33	1
Salt with Z	2	C^+ Z^-	138	69	28	3
Salt with Z	2	Q4–Z^+ A^-	123	64	33	3
Co-crystal	3	N N N	248	91	6	3
Ionic co-crystal	3	C^+ A^- N	1364	76	20	4
Ionic co-crystal	3	C^+ A^- Z	85	82	18	0
Ionic co-crystal	4	C^+ A^- N N	46	100	0	0

[a] #MC = number of main components. [b] N = neutral non solvent or zwitterion; Z = zwitterion; Z^+ = cationic zwitterion; Z^- = anionic zwitterion; C^+ = cation; A^- = anion; Q4 = quaternary.

complex systems were found, for example salts with multiple anion/cations, but these were not analysed further. A summary of the classification data is given in Table 1.

The most common types of crystals in the dataset are those containing only one neutral component, followed by salts, co-crystals and quaternary salts. Of these crystal structures, 10% contain ionised species (salts or (zwitter)ionic co-crystals), of which almost 73% involve proton transfer and 27% are quaternary salts. The transfer of a single proton is most common amongst all of the salts, being observed in 74% of the structures, followed by the transfer of two protons in 23% of the structures. Zwitterions are observed in only 3% of the dataset, so structures with zwitterions remain relatively rare in the CSD. Zwitterionic compounds that are not a result of proton transfer (zwitterion-Q4) are more common than zwitterionic forms that arise from self-proton transfer. This is likely to be a simple consequence of the chemical space represented in the CSD, which reflects to a large extent the common preferences of synthetic chemists. Whilst quaternary systems are analysed here for the purpose of general statistics, these are not relevant to the ΔpK_a rule and thus are removed for the subsequent sections and analyses.

Most interesting in the current context is the relationship between salt/co-crystal formation and the hydration/solvation of the crystals. Whilst solvates make up only a small percentage of all crystal types, hydrates are much more common. In part, this is to be expected due the ubiquitous nature of water and its frequent use as a solvent, and several authors have discussed reasons why hydration is particularly common in molecular crystals.[21,56–59] However, while hydrates constitute up to 6% of crystals containing only neutral (non-zwitterion) species, the percentage increases significantly to around 30% for crystals containing ions and/or zwitterions.

ΔpK_a in multi-component crystals

Dataset-2 is a subset of dataset-1 containing all structures with at least one acid–base pair, and thus one ΔpK_a value. This set was used to determine the relative occurrence of co-crystals, mono-ionised salts and multi-ionised salts as a function of ΔpK_a. The first comparison involves the relative distribution of co-crystals *versus* mono-ionised salts where ΔpK_a is calculated as the difference between the strongest acid and the strongest base (which must be in different molecules). The results are shown in Fig. 6, where the data have been separated into unsolvated (neat) co-crystals and salts (Fig. 6a) and hydrates of co-crystals and salts (Fig. 6b).

The result for neat co-crystals and salts is very similar to that presented by Cruz-Cabeza in 2012, with just a small shift to higher ΔpK_a values at which the percentage of the dataset is 50%:50% for co-crystals and salts. The percentage value in the plot can be considered as the probability for an acid/base system of a given ΔpK_a to be ionised when crystallised in an unsolvated form. In Fig. 6a, the co-crystal:salt equivalence point is observed at $\Delta pK_a \approx 1.4$, compared to $\Delta pK_a \approx 1$ in the 2012 study. Interestingly, the data for hydrates of co-crystals and salts is significantly shifted to the left, with the equivalence point in that distribution found at $\Delta pK_a \approx -0.5$. The intermediate region of the plot ($-3 < \Delta pK_a < 2$) deviates noticeably from a sigmoidal shape. This may be influenced by the relatively smaller sample size, or it may indicate a greater degree of uncertainty in the apparent trend when applied to crystalline hydrates.

ΔpK_a for second proton transfer ($\Delta pK_{a,II}$)

For crystals in dataset-2 with the possibility for two proton transfers (*i.e.* with calculated values for both ΔpK_a and $\Delta pK_{a,II}$), Fig. 7 shows the relative occurrence of mono-ionised salts *versus* di-ionised salts as a function of $\Delta pK_{a,II}$. The intent

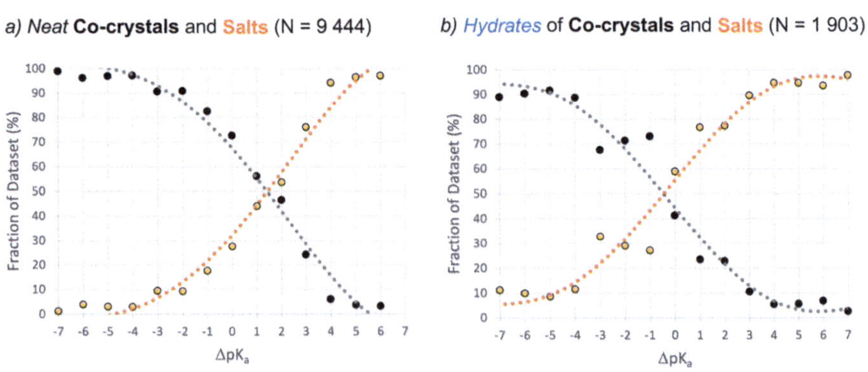

Fig. 6 Relative occurrence (%) of co-crystals (black) *versus* salts (orange) as a function of ΔpK_a (bins of one ΔpK_a unit) and hydration. Dataset-2 is filtered according to: no zwitterions, no Q4 ions, no other solvents, no or one proton transfer.

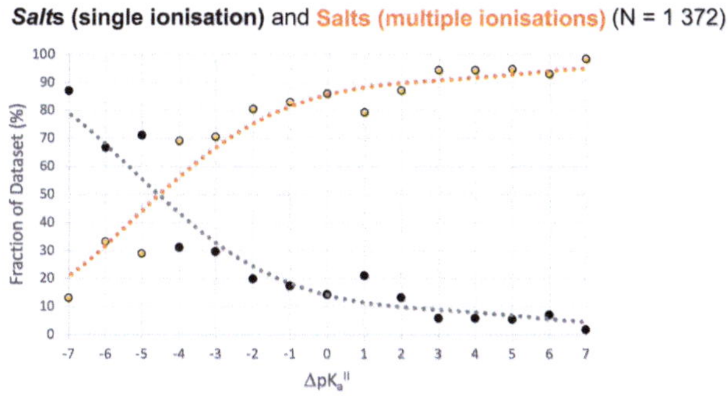

Fig. 7 Relative occurrence (%) of salts with a single ionisation (black) *versus* salts with multiple ionisations (orange) as a function of $\Delta pK_{a,II}$ (bins of one $\Delta pK_{a,II}$ unit). Dataset-2 is filtered according to: no zwitterions, no Q4 ions, no other solvents, one or more proton transfer.

here is to examine the effect of the crystal lattice of the mono-ionised salt on the second ionisation of an acid/base system. The number of observations in this case is reduced considerably compared to the co-crystal/salt dataset in Fig. 6, and thus more noise is seen on the apparent sigmoidal curves. Also, because of the smaller number of observations, the data are not separated into unsolvated *versus* hydrated forms. The data in Fig. 7 show a considerable shift to $\Delta pK_{a,II} = -4.5$ for the crossing point between the mono-ionised and di-ionised salts compared to that for the formation of any salt *versus* co-crystal as a function of ΔpK_a. Fig. 8 shows the number of proton transfers (0, 1 or 2) plotted against ΔpK_a and $\Delta pK_{a,II}$ in this dataset. Considering only ΔpK_a, the 50% probability of forming any salt (one or two proton transfers) falls between $\Delta pK_a = 0$ to 1, consistent with the expectations for co-crystals and salts seen in Fig. 6a. When looking at $\Delta pK_{a,II}$, however, the point of 50% probability for forming any salt (either one or two proton transfers) remains around $\Delta pK_{a,II} \approx 0$, but the probability of forming a salt with two proton transfers is clearly much greater than that of forming a salt with one proton transfer. Hence, where there is the possibility for two proton transfers, the choice is more frequently between a co-crystal and a divalent salt, with monovalent salts being far less common. This is consistent with the indications in Fig. 7, and highlights that the ΔpK_a rule does not apply within the same limits for ΔpK_a and $\Delta pK_{a,II}$ where there is a possibility for a second proton

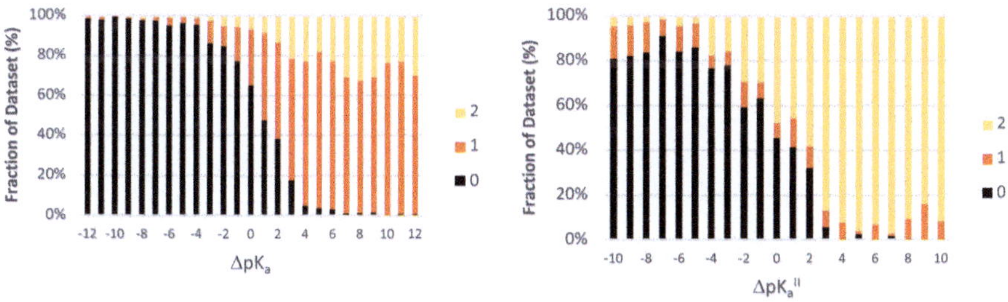

Fig. 8 Relative occurrence (%) of salts with 0, 1 or 2 proton transfers as a function of ΔpK_a and $\Delta pK_{a,II}$ (bins of one ΔpK_a value). The dataset is the same as that used to produce Fig. 7.

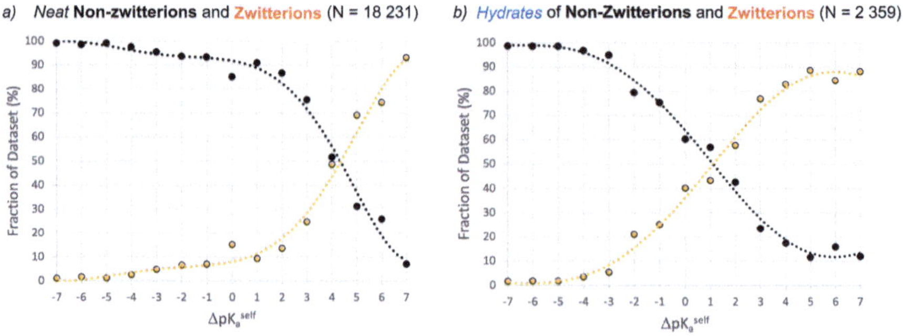

Fig. 9 Relative occurrence (%) of non-zwitterions (black) *versus* zwitterions (orange) as a function of $\Delta pK_{a,self}$ (bins of one ΔpK_a unit). Data for unsolvated forms (a) are compared to hydrates (b). Dataset-2 is filtered according to: only one main component, no Q4 zwitterions.

transfer. Here, of course, the stoichiometry of the acid : base could play a key role when the acid and/or base itself does not have two acid/basic sites but rather two molecules of the acid/base are required per molecule of base/acid. Practically, this emphasises the importance of exploring a larger range of acid–base ratios and stoichiometries during salt screening, especially for potential divalent salts.

ΔpK_a in zwitterions

Data were retrieved for unsolvated forms (one main component only) and hydrates (one main component plus water) of molecules potentially able to exist as a zwitterion (*i.e.* with a calculated $\Delta pK_{a,self}$ value). Fig. 9 shows that the equivalence point in the distribution of neutral and zwitterionic forms occurs at $\Delta pK_{a,self} \approx 4.1$, which is significantly higher than that found for multi-component crystals. Again, hydration has the same effect as in multi-component systems, lowering the equivalence point in the zwitterion set to $\Delta pK_{a,self} \approx 1$. It must be stressed that the data computed here are macro ionisation constants, which refer to the behaviour of the whole molecule to act as an acid or base, without any information on the microspecies that may be involved. Where the acid and base groups are sufficiently far apart within the molecule, it could be assumed that they would act independently, so $\Delta pK_{a,self}$ is a meaningful measure. Where the acid and base groups are not independent (*e.g.* in close proximity in the molecule or linked through resonance effects), this could have a significant influence on $\Delta pK_{a,self}$. Hence, while Fig. 9 serves as a guideline for potential self-proton transfer, the application of $\Delta pK_{a,self}$ to predict zwitterion formation should be viewed with a greater degree of caution since here the micro-ionisation constants play an important role.

Discussion

When transferring ΔpK_a values from the aqueous environment to make predictions about potential proton transfer in the solid state, the significant change in solvation environment can impact the overall expectations. Analysis of crystal structures in the CSD suggests in general that when ΔpK_a is sufficiently negative (<−1) or sufficiently positive (>+4), the influence of this change in environment is unlikely to overturn the indication of salt/co-crystal formation provided by ΔpK_a. When ΔpK_a lies in the intermediate range, however, the thermodynamic preference for proton transfer is strongly dependent on the solvation environment. The important new result in this paper is the demonstration that crystal composition has an important influence on the ΔpK_a rule. Most notably, the point of 50% probability for salt formation is shifted to lower ΔpK_a for crystalline hydrates compared to unsolvated forms. Indeed, the data show that salt formation in crystalline hydrates mirrors almost perfectly aqueous solution data: thus, for crystalline hydrates of acid/bases, $\Delta pK_a > 0$ should result in a salt and $\Delta pK_a < 0$ should result in a co-crystal. This finding begs the question as to which is the consequence of which: does hydration occur because of proton transfer? Or does proton transfer occur because of hydration?

In the regions where the non-ionised *versus* the ionised probabilities cross, a linear correlation can be found. The fitted equations in Table 2 may prove valuable for prediction of proton transfer in molecular crystals of various nature and compositions.

Concerning unsolvated co-crystals *versus* salts, the data suggest that unsolvated forms of acid–base pairs are usually worse at solvating the ionic species than water. To compensate for this, ΔpK_a must be positive for ionisation to occur.

Table 2 Linear fitting results for the intermediate ΔpK_a regions (ranges given) for the different ionisation systems. The fitted equations represent the probability of ionisation to occur at that given ΔpK_a. Above the intermediate ΔpK_a region, ionisation almost always occurs, whilst below the region, ionisation almost never occurs

System: neutral (a) *vs.* ionised (b)	Range	ΔpK_a cross	P_{obs}[a] (ion, %)	R^2
Monovalent *vs.* multivalent salt	[−7, 0]	−4.2	$11\Delta pK_a^{II} + 96$	0.89
Hydrated co-crystal *vs.* hydrated monovalent salt	[−5, 4]	−0.6	$10\Delta pK_a + 56$	0.94
Hydrated neutral *vs.* hydrated zwitterion	[−3, 5]	1.1	$11\Delta pK_a^{self} + 38$	0.98
Cocrystal *vs.* monovalent salt	[−2, 4]	1.3	$14\Delta pK_a + 32$	0.98
Neutral *vs.* zwitterion	[1, 7]	4.1	$15\Delta pK_a^{self} - 12$	0.97

[a] Ionisation, second ionisation or self-ionisation as appropriate.

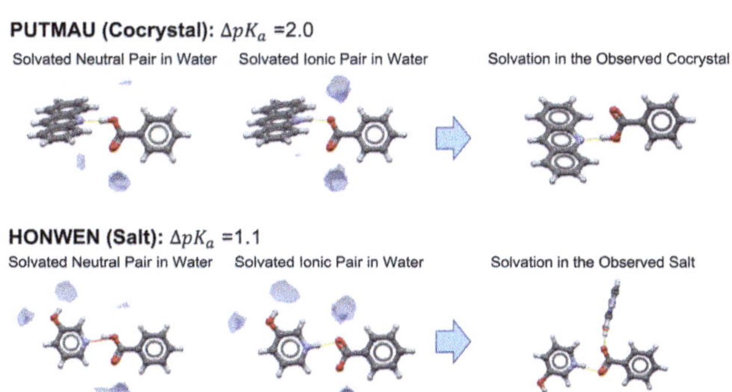

Fig. 10 Visualisation of water solvation of non-ionised *versus* ionised acid–base pairs with full interaction maps *versus* the resulting ionisation and interactions observed in their crystals.

Table 3 Salts and co-crystals with $-0.5 < \Delta pK_a < 0.5$ where the acid is an organic carboxylic acid. Statistics relate to the % of those carboxylic acids involved in 3, 2, 1 or 0 ancillary hydrogen bonds (aHBs)

	N total	Salt%	Co-crystal%
2 aHBs A/A$^-$	147	54%	46%
1 aHBs A/A$^-$	247	12%	88%
0 aHBs A/A$^-$	137	3%	96%

In order to illustrate this observation, we have selected two examples of benzoic acid forming a salt and a co-crystal with two different bases with ΔpK_a values between 1 and 2. Using the Mercury full interaction maps (FIMs) tool with a water oxygen as probe, we can visualise the ancillary hydrogen bonds (aHBs) where water is involved with the acid–base pair in aqueous solution in its non-ionised form *versus* its ionised form (Fig. 10). In aqueous solution, water forms one aHB with the carbonyl of the carboxylic acid in its neutral form whilst forming two strong aHBs with the carboxylate. For the crystal outcome to be predicted solely by ΔpK_a, the level of solvation of the acid–base pair in solution must be maintained in the solid state. In such a scenario, a positive ΔpK_a will imply salt formation, whilst a negative ΔpK_a will imply co-crystal formation. If solvation in the crystal is worse than in water (fewer aHBs), then the ΔpK_a value for the switch between salt and co-crystal must be shifted to higher values. In PUTMAU (Fig. 10), for example, there are no other HB donors in the acid or base, so no aHBs are possible and, thus, the outcome is a co-crystal. In HOMWEN (Fig. 10), however, there is a hydroxyl group available on the base able to form an aHB with the carboxylate, and thus a salt is observed in the solid state. For an acid and base with a ΔpK_a between 1 and 2, the experimentally derived probability for a salt to form would be between 50–60% (Table 2). In that scenario, a higher number of aHBs should result in a salt, whilst a lower number should result in a co-crystal.

To illustrate further the importance of crystalline aHBs to the ionisation outcome, we analyse the number of aHBs in acid–base co-crystals and salts with a $\Delta pK_a \approx 0$ (Table 3). 147 crystal structures were found where the carboxylic acid/carboxylate was involved in two aHBs, 247 in one aHB and 137 in no aHBs (Table 3).

Acid–base crystalline complexes able to afford two aHBs generate a solvation environment similar to water. Interestingly, nearly 50% of the 147 crystals with two aHBs are salts and the other 50% are co-crystals, mirroring the behaviour expected in aqueous solution where $\Delta pK_a \approx 0$. When only one or no aHBs are possible, co-crystals clearly dominate over salts. Here, when a reduction of aHBs occurs, the nature and strength of those aHBs would matter, with stronger HBs better able than weaker aHBs to stabilise the ionic species. This effect is clearly seen in the case of double ionisation where the $\Delta pK_{a,II}$ values for the switch between salt and co-crystal are shifted to negative values.

Conclusions

The ΔpK_a rule is widely used in crystal engineering for the design of salts and co-crystals, particularly for pharmaceutical compounds. Its origin is experimental, but associated with the thermodynamics of acid–base aqueous equilibria. There is indeed a simple thermodynamic relationship between ΔpK_a for an acid and a base and the corresponding free energy for proton transfer in aqueous solution ($\Delta G^{\circ\,\mathrm{HA\cdot B}}_{\mathrm{ion\text{-}water}} = -2.3RT\Delta pK_a$). Transferring this relationship to the solid state, although clearly convenient, comes with several challenges. Solvation plays a key role in the ionisation reaction and moving the acid–base species from aqueous solution (where the ΔpK_a defines their relative acid strength) into the crystals shifts the ΔpK_a expectations for ionisation to occur (as seen from experimentation). This experimental shift observation is what we refer to as the ΔpK_a rule. The rule works well for more extreme cases of ΔpK_a but loses its predictive power at ΔpK_a values between −1 and 4, where proton transfer in the solid state becomes a fine balance between the "inherent" tendency for proton transfer and the influence of the crystalline environment.

To illustrate the important role of the solvation environment on the ionisation outcome in the crystalline state, we have examined the applicability of the ΔpK_a rule in crystalline systems with different compositions. This reveals that the point of equal probability of salt/co-crystal formation lies at $\Delta pK_a \approx 1.1$ for neat (unsolvated) forms, but at around 0 for hydrates. In zwitterions, an even more positive ΔpK_a value of 4.1 is seen, whilst in multi-protic salts the point is negative, around $\Delta pK_a \approx -4.2$. New probability equations for ionisation (based on linear fittings to the CSD observations) are provided to enhance the predictive power of the ΔpK_a rule in the intermediate region for systems of various compositions.

We have also highlighted the importance of solvation in this intermediate region of ΔpK_a values for a few illustrative examples. Acid–base pairs that are stabilised by strong solvation in the crystalline state are more likely to be ionised than acid-base pairs that are not involved in any ancillary hydrogen bonds in the crystal. Finally stoichiometry and its impact to the hydrogen-bonding environment of the acid–base complex can also play a role.[60] We anticipate that this contribution will be useful for future applications of the ΔpK_a rule. The herein derived probability equations for crystals of various compositions, combined with an analysis of possible aHBs, can undoubtedly enhance the ΔpK_a rule's predictive power in the intermediate region.

Author contributions

AJCC initiated the project and performed the CSD searches and calculations. All authors contributed to the discussions, drafting of the paper and improvement of the data and made important contributions to the theoretical fundamentals. The authors thank Prof. Roger Davey for helpful discussions.

Conflicts of interest

There are no conflicts to declare.

Acknowledgements

AJCC thanks the Royal Society for an Industry Fellowship in AstraZeneca.

Notes and references

1 J. Manchester, G. Walkup, O. Rivin and Z. You, *J. Chem. Inf. Model.*, 2010, **50**, 565–571.

2 D. D. Perrin, B. Dempsey and E. P. Serjeant, *pK_a Prediction for Organic Acids and Bases*, Springer, 1981, vol. 1.

3 C. Lim, D. Bashford and M. Karplus, *J. Phys. Chem.*, 1991, **95**, 5610–5620.

4 D. A. Haynes, W. Jones and W. D. S. Motherwell, *J. Pharm. Sci.*, 2005, **94**, 2111–2120.

5 C. R. Groom, I. J. Bruno, M. P. Lightfoot and S. C. Ward, *Acta Crystallogr., Sect. B: Struct. Sci., Cryst. Eng. Mater.*, 2016, **72**, 171–179.

6 M. Sadakiyo, T. Yamada and H. Kitagawa, *J. Am. Chem. Soc.*, 2009, **131**, 9906–9907.

7 J. S. Stevens, S. Coultas, C. Jaye, D. A. Fischer and S. L. M. Schroeder, *Phys. Chem. Chem. Phys.*, 2020, **22**, 4916–4923.

8 J. S. Stevens, M. Walczak, C. Jaye and D. A. Fischer, *Chem.–Eur. J.*, 2016, **22**, 15600–15604.

9 C. C. Wilson, N. Shankland and A. J. Florence, *J. Chem. Soc., Faraday Trans.*, 1996, **92**, 5051–5057.

10 J. S. Stevens, S. J. Byard and S. L. M. Schroeder, *J. Pharm. Sci.*, 2010, **99**, 4453–4457.

11 S. Tothadi, T. R. Shaikh, S. Gupta, R. Dandela, C. P. Vinod and A. K. Nangia, *Cryst. Growth Des.*, 2021, **21**, 735–747.

12 P. T. Edwards, L. K. Saunders, A. R. Pallipurath, A. J. Britton, E. A. Willneff, E. J. Shotton and S. L. M. Schroeder, *Cryst. Growth Des.*, 2021, **21**(11), 6332–6340.

13 L. Rajput, M. Banik, J. R. Yarava, S. Joseph, M. K. Pandey, Y. Nishiyama and G. R. Desiraju, *IUCrJ*, 2017, **4**, 466–475.

14 L. H. Thomas, A. J. Florence and C. C. Wilson, *New J. Chem.*, 2009, **33**, 2486–2490.

15 M.-M. Blum, M. Mustyakimov, H. Rüterjans, K. Kehe, B. P. Schoenborn, P. Langan and J. C.-H. Chen, *Proc. Natl. Acad. Sci. U. S. A.*, 2009, **106**, 713–718.

16 S. M. Reutzel-Edens, *Pharmaceutical Salts and Co-crystals*, Royal Society of Chemistry, 2011, pp. 212–246.

17 T. Wang, J. S. Stevens, T. Vetter, G. F. S. Whitehead, I. J. Vitorica-Yrezabal, H. Hao and A. J. Cruz-Cabeza, *Cryst. Growth Des.*, 2018, **18**, 6973–6983.

18 S. M. Berge, L. D. Bighley and D. C. Monkhouse, *J. Pharm. Sci.*, 1977, **66**, 1–19.

19 A. T. Serajuddin, *Adv. Drug Delivery Rev.*, 2007, **59**, 603–616.

20 D. P. Elder, R. Holm and H. L. De Diego, *Int. J. Pharm.*, 2013, **453**, 88–100.

21 C. B. Aakeröy, M. E. Fasulo and J. Desper, *Mol. Pharmaceutics*, 2007, **4**, 317–322.

22 G. R. Desiraju, *Angew. Chem., Int. Ed.*, 2007, **46**, 8342–8356.

23 G. R. Desiraju, *J. Am. Chem. Soc.*, 2013, **135**, 9952–9967.

24 M. Karimi-Jafari, L. Padrela, G. M. Walker and D. M. Croker, *Cryst. Growth Des.*, 2018, **18**, 6370–6387.

25 M. K. Corpinot and D.-K. Bučar, *Cryst. Growth Des.*, 2019, **19**, 1426–1453.

26 A. Nangia and G. R. Desiraju, Supramolecular Synthons and Pattern Recognition, in *Topics in Current Chemistry*, E. Weber, Springer, Berlin, 1999, vol. 198.

27 G. R. Desiraju, *Angew. Chem., Int. Ed. Engl.*, 1995, **34**, 2311–2327.

28 J. A. Bis, P. Vishweshwar, D. Weyna and M. J. Zaworotko, *Mol. Pharmaceutics*, 2007, **4**, 401–416.

29 P. Gilli, L. Pretto and G. Gilli, *J. Mol. Struct.*, 2007, **844–845**, 328–339.

30 P. Gilli, L. Pretto, V. Bertolasi and G. Gilli, *Acc. Chem. Res.*, 2009, **42**, 33–44.

31 B. R. Bhogala, S. Basavoju and A. Nangia, *CrystEngComm*, 2005, **7**, 551–562.
32 A. J. Cruz-Cabeza, *CrystEngComm*, 2012, **14**, 6362–6365.
33 S. L. Childs, G. P. Stahly and A. Park, *Mol. Pharmaceutics*, 2007, **4**, 323–338.
34 G. Ramon, K. Davies and L. R. Nassimbeni, *CrystEngComm*, 2014, **16**, 5802–5810.
35 A. Lemmerer, S. Govindraju, M. Johnston, X. Motloung and K. L. Savig, *CrystEngComm*, 2015, **17**, 3591–3595.
36 A. Mukherjee and G. R. Desiraju, *Cryst. Growth Des.*, 2014, **14**, 1375–1385.
37 *ChemAxon version 21.13*, https://www.chemaxon.com.
38 M. J. Frisch, G. W. Trucks, H. B. Schlegel, G. E. Scuseria, M. A. Robb, J. R. Cheeseman, G. Scalmani, V. Barone, G. A. Petersson, H. Nakatsuji, X. Li, M. Caricato, A. V. Marenich, J. Bloino, B. G. Janesko, R. Gomperts, B. Mennucci, H. P. Hratchian, J. V. Ortiz, A. F. Izmaylov, J. L. Sonnenberg, D. Williams-Young, F. Ding, F. Lipparini, F. Egidi, J. Goings, B. Peng, A. Petrone, T. Henderson, D. Ranasinghe, V. G. Zakrzewski, J. Gao, N. Rega, G. Zheng, W. Liang, M. Hada, M. Ehara, K. Toyota, R. Fukuda, J. Hasegawa, M. Ishida, T. Nakajima, Y. Honda, O. Kitao, H. Nakai, T. Vreven, K. Throssell, J. A. Montgomery, Jr, J. E. Peralta, F. Ogliaro, M. J. Bearpark, J. J. Heyd, E. N. Brothers, K. N. Kudin, V. N. Staroverov, T. A. Keith, R. Kobayashi, J. Normand, K. Raghavachari, A. P. Rendell, J. C. Burant, S. S. Iyengar, J. Tomasi, M. Cossi, J. M. Millam, M. Klene, C. Adamo, R. Cammi, J. W. Ochterski, R. L. Martin, K. Morokuma, O. Farkas, J. B. Foresman, and D. J. Fox, *Gaussian 16*, Revision C.01, Gaussian, Inc., Wallingford CT, 2016.
39 S. Grimme, *J. Comput. Chem.*, 2006, **27**, 1787–1799.
40 A. V. Marenich, C. J. Cramer and D. G. Truhlar, *J. Phys. Chem. B*, 2009, **113**, 6378–6396.
41 *Avogadro version 1.1.1*, https://avogadro.cc.
42 M. D. Hanwell, D. E. Curtis, D. C. Lonie, T. Vandermeersch, E. Zurek and G. R. Hutchison, *J. Cheminf.*, 2012, **4**, 17.
43 T. A. Halgren, *J. Comput. Chem.*, 1996, **17**, 490–519.
44 C. R. Groom, I. J. Bruno, M. P. Lightfoot and S. C. Ward, *Acta Crystallogr., Sect. B: Struct. Sci., Cryst. Eng. Mater.*, 2016, **72**(2), 171–179.
45 P. A. Wood, T. S. G. Olsson, J. C. Cole, S. J. Cottrell, N. Feeder, P. T. A. Galek, C. R. Groom and E. Pidcock, *CrystEngComm*, 2013, **15**, 65–72.
46 C. F. Macrae, I. Sovago, S. J. Cottrell, P. T. A. Galek, P. McCabe, E. Pidcock, M. Platings, G. P. Shields, J. S. Stevens, M. Towler and P. A. Wood, *J. Appl. Crystallogr.*, 2020, **53**, 226–235.
47 B. G. Cox, *Acids and Bases: Solvent Effects on Acid-Base Strength*, Oxford University Press, Oxford, New York, 2013.
48 Y. Abramov, G. Sun, Y. Zhou, M. Yang, Q. Zeng and Z. Shen, *Cryst. Growth Des.*, 2019, **19**, 7132–7137.
49 I. E. Charif, S. M. Mekelleche, D. Villemin and N. Mora-Diez, *J. Mol. Struct.: THEOCHEM*, 2007, **818**, 1–6.
50 S. Mohamed, D. A. Tocher, M. Vickers, P. G. Karamertzanis and S. L. Price, *Cryst. Growth Des.*, 2009, **9**, 2881–2889.
51 L. Mazur, I. Materek, A. D. Bond and W. Jones, *Cryst. Growth Des.*, 2019, **19**, 2663–2678.

52 A. M. Reilly, R. I. Cooper, C. S. Adjiman, S. Bhattacharya, A. D. Boese, J. G. Brandenburg, P. J. Bygrave, R. Bylsma, J. E. Campbell, R. Car, D. H. Case, R. Chadha, J. C. Cole, K. Cosburn, H. M. Cuppen, F. Curtis, G. M. Day, R. A. DiStasio Jr, A. Dzyabchenko, B. P. van Eijck, D. M. Elking, J. A. van den Ende, J. C. Facelli, M. B. Ferraro, L. Fusti-Molnar, C.-A. Gatsiou, T. S. Gee, R. de Gelder, L. M. Ghiringhelli, H. Goto, S. Grimme, R. Guo, D. W. M. Hofmann, J. Hoja, R. K. Hylton, L. Iuzzolino, W. Jankiewicz, D. T. de Jong, J. Kendrick, N. J. J. de Klerk, H.-Y. Ko, L. N. Kuleshova, X. Li, S. Lohani, F. J. J. Leusen, A. M. Lund, J. Lv, Y. Ma, N. Marom, A. E. Masunov, P. McCabe, D. P. McMahon, H. Meekes, M. P. Metz, A. J. Misquitta, S. Mohamed, B. Monserrat, R. J. Needs, M. A. Neumann, J. Nyman, S. Obata, H. Oberhofer, A. R. Oganov, A. M. Orendt, G. I. Pagola, C. C. Pantelides, C. J. Pickard, R. Podeszwa, L. S. Price, S. L. Price, A. Pulido, M. G. Read, K. Reuter, E. Schneider, C. Schober, G. P. Shields, P. Singh, I. J. Sugden, K. Szalewicz, C. R. Taylor, A. Tkatchenko, M. E. Tuckerman, F. Vacarro, M. Vasileiadis, A. Vazquez-Mayagoitia, L. Vogt, Y. Wang, R. E. Watson, G. A. de Wijs, J. Yang, Q. Zhu and C. R. Groom, *Acta Crystallogr., Sect. B: Struct. Sci., Cryst. Eng. Mater.*, 2016, **72**, 439–459.

53 L. M. LeBlanc, S. G. Dale, C. R. Taylor, A. D. Becke, G. M. Day and E. R. Johnson, *Angew. Chem., Int. Ed.*, 2018, **57**, 14906–14910.

54 P. Praphanwittaya, P. Saokham, P. Jansook and T. Loftsson, *J. Drug Delivery Sci. Technol.*, 2021, **62**, 102359.

55 T. G. Cooper, K. E. Hejczyk, W. Jones and G. M. Day, *J. Chem. Theory Comput.*, 2008, **4**, 1795–1805.

56 G. R. Desiraju, *J. Chem. Soc., Chem. Commun.*, 1991, 426–428.

57 L. Infantes, L. Fábián and W. D. Sam Motherwell, *CrystEngComm*, 2007, **9**, 65–71.

58 A. Bajpai, H. S. Scott, T. Pham, K.-J. Chen, B. Space, M. Lusi, M. L. Perry and M. J. Zaworotko, *IUCrJ*, 2016, **3**, 430–439.

59 H. D. Clarke, K. K. Arora, H. Bass, P. Kavuru, T. T. Ong, T. Pujari, L. Wojtas and M. J. Zaworotko, *Cryst. Growth Des.*, 2010, **10**, 2152–2167.

60 P. Stainton, E. Nauha, T. Grecu, J. F. McCabe, T. Munshi, I. Scowen, H. C. Stephen Chan, S. Nilsson and N. Blagden, *Cryst. Growth Des.*, 2022, **22**(3), 1665–1679.

Faraday Discussions

PAPER

The structural pathway from its solvated molecular state to the solution crystallisation of the α- and β-polymorphic forms of *para* amino benzoic acid†

Ian Rosbottom, ‡[a] Thomas D. Turner, §[a] Cai Y. Ma, [a] Robert B. Hammond, [a] Kevin J. Roberts, *[a] Chin W. Yong [bc] and Ilian T. Todorov [b]

Received 30th November 2021, Accepted 27th January 2022

DOI: 10.1039/d1fd00112d

Para amino benzoic acid (PABA) has two well-characterised α- and β-polymorphic forms and, whilst both crystallise in the monoclinic space group $P2_1/n$, they have quite different crystal chemistry and crystallisability behaviour. Previous work has shown that the molecular conformation deformation energy in the crystalline state is higher for the β-form than for the α-form and that the lattice energy for the former converges more slowly than for the latter overall. This suggests that not only is there a higher barrier to crystallisation for the β-form but also that low solution supersaturations might be needed for it to preferentially nucleate. Additionally, solute cluster propensity and solute solvation energetic analysis highlight the importance of an aqueous solvation environment in inhibiting the α-form's strong OH⋯O carboxylic acid hydrogen bond (H-bond) dimer. Despite this, the detailed molecular-scale pathway from solvated molecules to 3D crystallographic structure still remains unclear, most notably regarding how the nucleation process is activated and how, as a result, this mediates the preferential formation of either of the two polymorphic forms. Molecular dynamics (MD) simulations coupled with FTIR studies and intermolecular synthon analysis address this issue through characterisation of the propensity of the incipient bulk synthons that are important in the crystallisation of the two polymorphic forms within the solution state. MD molecular trajectory analysis within crystallisation solutions reveals a greater propensity for OH⋯O synthons (both single H-bonds and homodimers) typical of the α-form and NH⋯O synthons found in both the α- and β-forms when compared to

[a]*Centre for the Digital Design of Drug Products, School of Chemical and Process Engineering, University of Leeds, Woodhouse Lane, Leeds, LS2 9JT, UK. E-mail: k.j.roberts@leeds.ac.uk*

[b]*Scientific Computing Department, Science and Technology Facilities Council, Daresbury Laboratory, Warrington, WA4 4AD, UK*

[c]*Division of Pharmacy and Optometry, University of Manchester, Manchester, UK*

† Electronic supplementary information (ESI) available. See DOI: 10.1039/d1fd00112d

‡ Current address: GlaxoSmithKline, Gunnels Wood Rd, Stevenage, SG1 2NY, UK.

§ Current address: School of Mechanical Engineering, University of Leeds, Leeds, LS2 9JT, UK.

aqueous solution but much lower propensities for the β-form's "fingerprinting" OH···N and π–π stacking synthons. In contrast, data from the aqueous solution environment reveals a much greater propensity for the β-form's π–π interaction synthons. IR dilution studies in acetonitrile in the carbonyl region reveal the presence of two C=O vibrational stretching bands, whose relative intensities vary as a function of solution dilution. These were assigned to the solvated PABA monomer and a COOH dimer of PABA. Similar data in ethanol shows a main C=O stretching band with a shoulder peak suggesting a similar monomer *vs.* dimer speciation may exist in this solvent. The IR data is consistent with the organic solvent MD data, albeit the corresponding analysis for the aqueous solution was precluded due to the latter's strong OH vibrational mode which restricted validation in aqueous solutions.

1. Introduction

In solution crystallisation processes, understanding and controlling the transition pathway associated with the assembly of molecules from their supersaturation solvated state into different polymorphic crystalline solids, represents a significant grand challenge for the physical–chemical sciences.[1]

Industrial utilisation of crystalline high-value materials is highly desirable, particularly so within the pharmaceutical and fine chemical industry, reflecting *e.g.* the high material purity enabled through the crystallisation process. To produce material in the crystalline state with the required yield and without compromising its stability, solution crystallisation still provided one of the most reliable and sustainable methods for manufacture. However, many organic materials display polymorphism, where such different crystal structures can have very different physico-chemical properties such as bioavailability, stability and morphology. However, crystallisation into the correct form is of vital importance when developing an effective and sustainable manufacturing process.

It has been hypothesised that the degree to which crystallisation conditions, such as solvent, temperature and concentration, can encourage the formation of intermolecular solute structuring in solution which can, in turn, influence crystallisation behaviour and, through this, result in different polymorphic form of the material. Knowledge of solution structural behaviour as a key part of the crystallisation design process could allow for a much more targeted process development which could, in principle, greatly reduce the cost and time needed.

Characterisation of the solid-state intermolecular interactions present in organic crystal structures has resulted in a good understanding of the crystal chemistry of these materials. Indeed, it has become possible to use molecular modelling to predict the physical properties of molecular crystals, such as their polymorphic stability,[2,3] thermal expansion,[4] morphology and surface chemistry,[5,6] habit modification,[7,8] surface wettability[9] and API/excipient compatibility.[10–12] In contrast, predicting the intermolecular chemistry of the solution phase can be much more demanding and one requires more sophisticated computational design tools, such as afforded by the application of molecular dynamics (MD) simulations. In this, extracting the details of the intermolecular interactions from successful MD simulations is far from straightforward, often relies on qualitative estimations from limited snapshots of the simulations.

Recent work by Yong *et al.*[13–15] has demonstrated a new and more efficient approach when setting up MD simulations notably by providing more flexibility within the user interface, *e.g.*, in interatomic forcefield selection and atom typing. The new approach also allows for more rational MD data analysis, notably in terms of searching for correlations between the relative positions of atoms within selected target molecules of interest through the use of distance and angular criteria to monitor the propensity of non-bonded intermolecular contacts.

This approach has been used to interrogate and characterise the propensity of incipient bulk synthons within the simulated solution-state through targeted analysis of the MD trajectory files. Through this holistic approach, intermolecular interactions between the solvated benzoic acid molecules present within hexane solutions were cross-correlated with the strongest intermolecular interactions found in the crystal structure of this material through studies which also revealed self-consistency with solution infra-red (IR) spectroscopy data.[15,16]

The above approach is extended through this work to examine the solution/solid-state pathways associated with *para*-amino benzoic acid crystallisation and with the specific aim to characterise the role played by solvent choice (ethanol, acetonitrile and water) in terms of mediating and controlling the resulting polymorphic form of the materials as identified through synthon analysis of the α- and β-polymorphic forms in the solid state. These MD simulations are coupled with experimental solution-state IR spectroscopy studies to support model validation.

2. *para*-Amino benzoic acid

This section provides a selective review of some key research related to the solution crystallisation of *para*-amino benzoic acid (PABA). A much longer version of Section 2 is provided in ESI.†

2.1 Molecular properties and crystallographic structures

para-Aminobenzoic acid (PABA), an organic compound, also known as vitamin B10, is a white crystalline substance with molecular formula $(NH_2)(C_6H_4)(COOH)$, consisting of three molecular moieties: an aromatic benzene ring substituted with carboxylic acid and amino groups which are *para* with respect to each other (Fig. 1).

The crystal structures of the α- and β-forms of *para*-aminobenzoic acid which are enantiotropically interrelated are discussed in Section S1.1, ESI.† Characteristic molecular descriptors and associated crystallographic structural data for PABA is summarised in Table S1.† Interestingly, the β-form is more close packed with a higher density and less void space than the α-form.[17] Analysis of the hydrogen bonds (H-bonds), as shown in Table S2,† reveals that they are stronger in the β-form structure than in the α-form, as evidenced by the shorter H-bond lengths in the β-form than in the α-form. The molecular descriptors for PABA (Table S1†) are compared with the representative pharmaceutical materials[3,18] in Section S1.1.†

2.2 Crystal chemistry and intermolecular interactions

The examinations of solid-state chemistry of α- and β-form crystal structures[19] found that the α-form has strong H-bonding carboxylic acid OH···O dimer

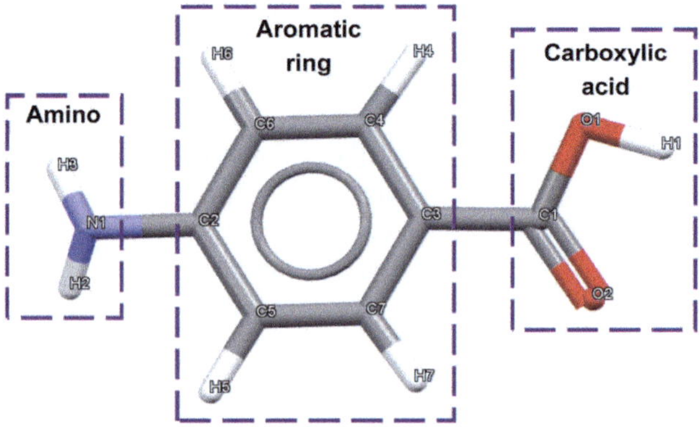

Fig. 1 The molecular structure of *para*-aminobenzoic acid with the atom notations and the three molecular moieties (functional groups): aromatic ring, carboxylic acid and amino group as indicated in the boxes.

interactions, π–π stacking and NH···O interactions (Fig. 2(a)) and that it crystallises with a needle- or lath-like morphology at higher temperatures. In contrast, the β-form has a characteristic 4-membered H-bonding ring comprising identical pairs of alternating OH···N and NH···O interactions and crystallises with a prismatic equant morphology (Fig. 2(b)).

Detailed molecular conformational analysis for these two polymorphic structures reveals some interesting differences (Fig. S3†). The rigid planar COOH dimers formed in the α-form crystal structure strongly hold the COOH group planar (Fig. 2(a)). The NH···O H-bond is also relatively planar, leading overall to the α-form's tetra-molecular core building block remaining planar. Therefore the carboxyl acid group was found to be very close to the planar conformational minima and also the slight pyramidal bend of amino group close to the pseudopyramidal conformational minima (Fig. S3†). On the contrast, the 4-membered H-bond ring in β-form structure distorts conformation of both functional groups as the OH···N H-bond pulls N atom towards OH group and the NH···O H-bond pulls the H atom towards the C=O group (Fig. 2(b)). Hence, the carboxyl acid group was found to be slightly rotated away from a planar conformation, whilst the amino group was found to be more pyramidal (Fig. S3†). This conformational distortion reflects the need to complete the β-form's tetra-molecular H-bonded ring structure. As the α-form PABA conformation is closer to the minimum

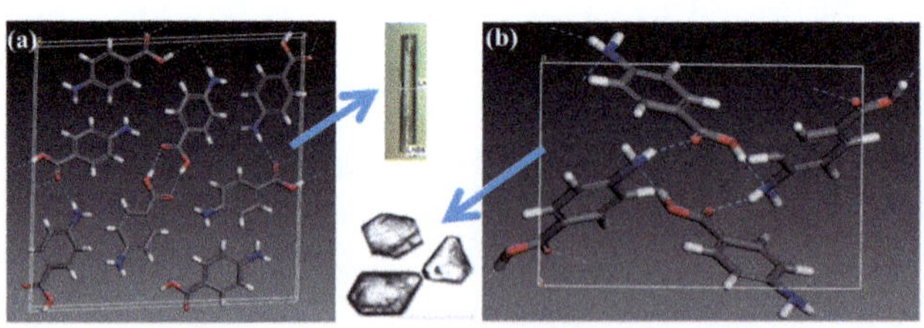

Fig. 2 The crystal packing structures of *para*-aminobenzoic acid: (a) α-form; (b) β-form with their H-bonding networks and also observed morphologies (data derived from ref. 19).

energy conformation (Fig. S3†), this would present a lower energy barrier to the crystallisation of α-form when compared to the β-form, consistent with the known challenges in crystallising the β-form. That said, whilst the conformation deformation penalty could imply lower crystallisability of the β-form, once formed the strength of its core 4 member H-bonding ring structure would suggest a strong and stable arrangement for its subsequent development and growth (Table S2†).

2.3 Solution properties, solvation and solubility

Synthon propensity studies,[19,20] using the COSMO-RS approach, was utilised to compare the relative populations of the energetically top-ranked synthons (Aα, Bα and Aβ, Bβ) in solution for a range of solvent systems (Fig. S5†). Synthon Aα associated with OH···O H-bonding interactions was found to be dominant for all solvents studied with the exception of water where the concentration of this synthon was found to be much lower and where π–π interactions were much more prevalent notably the π–π head-to-head (H2H) (synthon Bα) and head-to-tail (H2T) (synthon Aβ) interactions which were found to be the most stable. In this, calculation of the surface charge distributions for the synthons reveals that the π–π interaction synthons (Bα, Aβ) have greater polar surface area when compared to the OH···O H-bonding synthons (Aα, Bβ). Overall, whilst the data supports the preferential crystallisation of the α-form from most solvents (except water), it does suggest that the β-form would be more likely to be crystallised from aqueous solutions.

A more detailed analysis of the intermolecular chemistry associated with the solvation of PABA molecules has been provided using intermolecular grid-search methods[21–23] which reveals (Fig. S6†) the predicted solvation shell structures for 3 solvent systems, EtOH, AcN and water (further details can be found in Section S1.3, ESI†). The solvation energies confirm PABA to be well-solvated by the protic solvent EtOH which can interact strongly with both the aromatic ring through van der Waals interactions with the alkyl group and with the carboxylate through H-bonding. In contrast, the aprotic AcN solvates to a much lesser degree only forming only quite limited interactions with the COOH group compared to EtOH and water. Water, in contrast, fails to solvate the whole surface area of the PABA molecule apart from COOH group, particularly the amino group and the hydrophobic phenyl ring. This would also be consistent with its much lower solubility.

Solubility data,[24] obtained from solution–dissolution measurements (Fig. S7(a–c)†, upper plots) was analysed using van't Hoff plots for EtOH, AcN and water solvents (Fig. 3(a)). These reveal less than ideal solubility (activities −0.91, 0.43, 0.02, respectively) for all three solvents, consistent with a solution structure with an enhanced degree of solute/solute interactions and associated solute clustering.

2.4 Metastability, solute clustering and nucleation

Analysis of the slopes of the associated crystallisation on-set points (T_c) as a function of the solution cooling rate (q) probing the kinetic balance between the relative rates of supersaturation generation and nucleation, provides a helpful indication as to the relative crystallisabilities (Fig. S7(a–c),† lower plots) for the three solvents. The data shows distinctly different behaviour between the 3 solvents with water having a much flatter slope compared to EtOH and AcN, respectively, suggesting the rate-limiting parameters regarding crystallisability would have an order of (kinetics driven) EtOH > AcN > water (thermodynamics

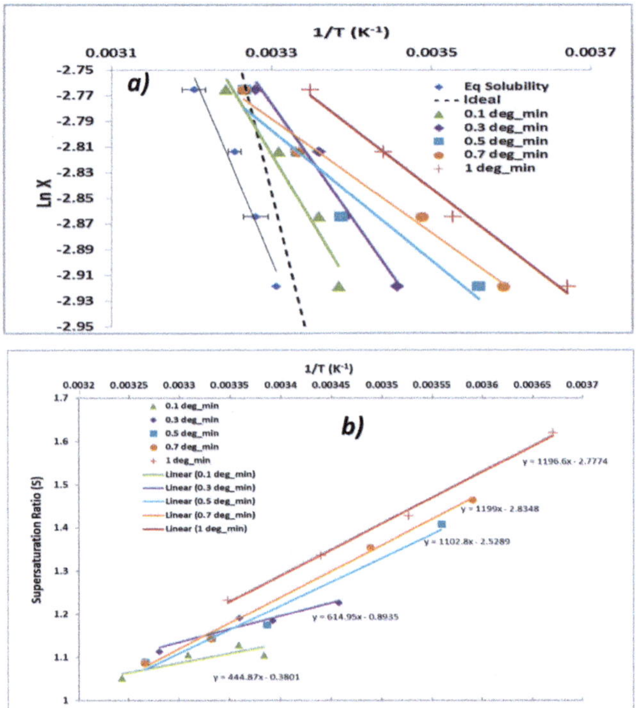

Fig. 3 (a) Crystallisation temperatures of PABA in EtOH with van't Hoff coordinates, showing marked differences in slope with respect to the ideal solubility line (- - -) and (b) supersaturation ratios in van't Hoff plotting (data derived from ref. 24).

driven). Also as shown in (Fig. 3(a)), the crystallisation on-sets in EtOH as a function of cooling rate show a marked difference in slope with respect to the ideal solubility line, moving away from the equilibrium solubility which is slightly less than ideal to greater than ideal with increasing cooling rate, under-cooling and hence supersaturation. Hence, qualitatively taking the slope as an indicator of the enthalpy of dissolution (ΔH_{diss}), ΔH_{diss} decreases from ideal solubility (2500 J mol^{-1}) to 488 J mol^{-1} (1.0 °C min^{-1}) as the solutions under-cool. This change in the apparent crystallisability with increased cooling rate can also be seen in the data when replotted in terms of critical supersaturation at T_c (Fig. 3(b)). This reveals that the highest cooling rates access a much wider range of supersaturations and nucleation cluster sizes across the solute concentration range (1.2–1.6) when compared to lowest cooling rates (1.05–1.10).

Analysis of the polythermal crystallisation behaviour as a function of solvent type and solution concentration indicated that the nucleation mechanism was mostly instantaneous and consistent with a two-step nucleation mechanism[24] except at low solute concentrations where it was progressive. Comparatively for the instantaneous cases, it could be said that EtOH was more instantaneous than AcN and water, this perhaps reflecting their relative solubilities.[24]

The strong evidence for solute clustering from the less than ideal equilibrium solubilities and instantaneous nucleation mechanism has been further examined using small angle X-ray scattering (SAXS).[25,26] The data confirmed the formation of

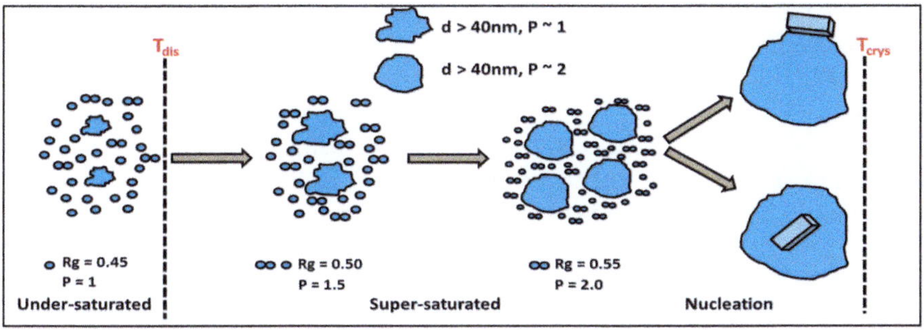

Fig. 4 The tentative structural schematic of α-PABA nucleation pathway from EtOH solutions, based on the modelling and experimental evidence obtained so far, indicating large liquid-like clusters surrounded by a population of monomer and dimerised PABA molecules. Note the pathway for β-PABA nucleated in water can have different parameters. R_g is the radius of gyration which is related to cluster size (nm) and P is the fractal dimensionality and related to interface structure where a higher value infers a more ordered interface (data derived from ref. 26).

solvent clusters within PABA in ethanolic solutions during cooling, highlighting at low Q an increase in cluster numbers and their size (>40 nm) during cooling. Such clustering behaviour is consistent with a number of previous studies on *e.g.* glycine,[27] citric acid,[28] and L-isoleucine.[29] Simultaneous analysis at high Q reveals solution de-supersaturation during cooling with evidence for formation of small cluster structures (*ca.* 1–2 nm) well matched to the overall dimensions of the Aα synthon.

A tentative nucleation pathway for PABA crystallisation from EtOH solutions is shown schematically in Fig. 4 which draws together the structural information obtained from the analysis of the modelling and experimental data to date. This indicates the progress of nano-scale assemblies from large disordered liquid-like PABA nano-clusters (dimers and monomers) in under-saturated condition, then the growth of these clusters in sizes under supersaturated conditions including the increasing degree of structural ordering at cluster/solution interface increasing the mass fractal dimension with their increasing interfacial smoothness and faceting at the interface and through to the formation of the crystalline phase (nuclei) by further growth. Overall, the data supports a two-step nucleation mechanism associated with the formation and development of nano-crystallite clusters together with the growth of smaller molecular clusters.

It is important to remark that the SAXS measurements require good contrast between the aggregated and continuum phases demanding, in turn, comparatively high solute concentrations. This aspect was limited to SAXS studies to EtOH solutions (only) and in this case only at the relatively high solute concentration typified by instantaneous nucleation. In effect, this restriction precluded SAXS studies over a wide range of concentrations in EtOH as well as on the AcN and aqueous solutions. Hence, at this stage we do not know how this model would apply more generally.

3. Materials and methods

3.1 Molecular dynamics simulations

Molecular modelling of intermolecular ordering in the solution state was performed using the integrated molecular dynamics (MD) software package developed by Daresbury Laboratory, namely DL_FIELD (MD input files, atom type standardisation and forcefield selection),[13] DL_POLY (MD simulations)[30] and DL_ANALYSER (MD trajectory file analysis)[14] which all form part of the DL_Software system.[31]

For this study, the Cartesian coordinates of a single molecule of PABA was input into DL_FIELD, where the concentration, forcefield and solvent was specified and the CONTROL, FIELD and CONFIG files needed for input into DL_POLY for MD simulations were constructed. The simulations were carried out using a 60 $Å^3$ periodic simulation box and with all intermolecular forces and their associated interaction energies being calculated using the Amber GAFF forcefield.[32]

Three crystallising solution systems were examined: notably AcN (polar, aprotic), EtOH (polar, protic) and water (polar, protic) consistent with previous studies regarding solvation energies,[23,33] solvent-dependant synthon propensities[20] and nucleation kinetics.[24,25] The simulations were carried out at a solute concentration of 0.1 g ml^{-1} and at a temperature of 20 °C which reflect the different PABA solubilities in these three solvents. This corresponded to solution supersaturations of *ca.* 1.2, 1.0 and 5.0, respectively.

To ensure effective equilibration, a constant energy simulation was run with constant scaling at 400 K for approximately 800 ps to remove any pre-existing solution chemistry from the solution box generation. The solution was then cooled to the sampling temperature of 293 K using a constant energy ensemble with constant scaling. The scaling was removed and, if the temperature was observed to remain around the sampling temperature, then the solution was deemed acceptably equilibrated. The solution was then run for a short time using the Berendsen NVT ensemble,[34] before being changed to the Berendsen NPT ensemble[34] until the box size no longer fluctuated. The sampling simulations were run for 10.0 ns each, with a time step of 0.002 ps using the Nosé–Hoover NPT ensemble with thermostat and barostat relaxation times of 4.0 ps and 5.0 ps, respectively.[35]

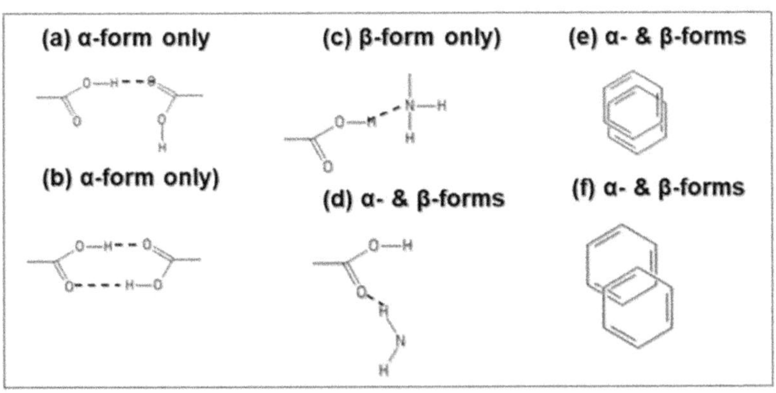

Fig. 5 Molecular diagrams highlighting the intermolecular interactions searched in the solution-state MD trajectory files for PABA using DL_ANALYSER in relation to the known polymorphic forms: carboxylic to carboxylic interactions associated with single OH···O H-bonded (a), and double OH···O (b) (α-form only) carboxylate group H-bonding interactions; amino to carboxylic group interactions associated with (c) NH···O (α- and β-forms) and (d) OH···N (β-form only) H-bonding; inter-phenyl ring π–π interactions associated with both occluded (e) and staggered (f) structures.

The standardised notation for the atom typing using the DANAI notation[14] within the DL_FIELD was exploited for searching within the HISTORY trajectory files resulting from the MD simulations using DL_ANALYSER. Drawing upon the approach developed through previous work on benzoic acid,[15] both the macro interactions that define the intermolecular interactions between functional groups of atoms as well as the micro interactions that define the exact atomic scale geometry of the selected interactions were characterised. The latter was based upon the key intermolecular synthons and their local inter-atomic geometry involved in the molecular assembly and crystallisation of the α- and β-forms of PABA.[6] Specifically, the trajectory files obtained from solution-state MD simulations were examined for the presence of the selected interacting groups, given in Fig. 5, in order to assess the extent to which the PABA molecules in their solution state undergo a degree of pre-ordering prior to crystallisation. In this analysis, any H-bonding interactions occurring at a distance of less than 2.5 Å and an interatomic angle of 120° or more were counted, whilst the close π–π parallel packing of the aromatic groups in PABA were only counted when they were closer than 4.5 Å and within a 20° angle or less.

Whilst the analysis here has focused on the above important H-bonded and non-bonded interactions, the more subtle "finger printing" differences between

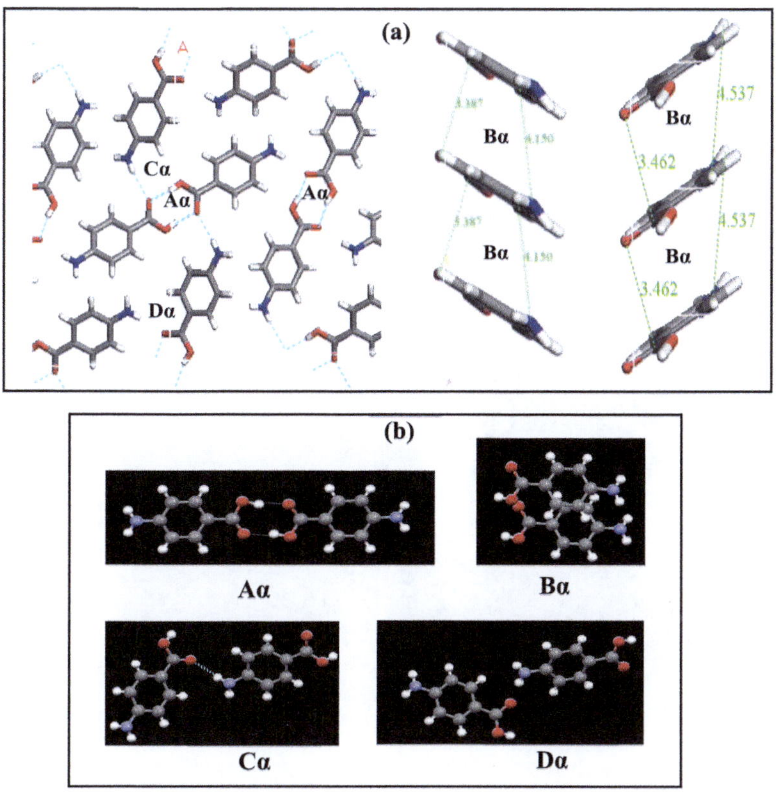

Fig. 6 Intermolecular diagrammatic representation of the key synthons present in the α-form of PABA: (a) packing diagrams; (b) molecular pair representation of the synthons highlighting the nature of their interactions (data derived from ref. 6 and 25).

the latter interactions between the α-form (H2H chains) and the β-form (H2T dimer pairs) were not probed. Overall, through this work, the extent to which the intermolecular synthons present in this pre-ordered solution state mirrors those present with the crystal structures following crystallisation was examined.

3.2 IR spectroscopy experiments

PABA (99.9%) in its α-form was purchased from Sigma Aldrich and used as supplied, and β-form PABA was obtained by slurry inter-conversion from the α-form at 5 °C using a 500 ml temperature-controlled jacketed crystallisation vessel. AcN (99%) was supplied by Alfa Aesar, EtOH (99.9%) by VWR and deionised water in-house.

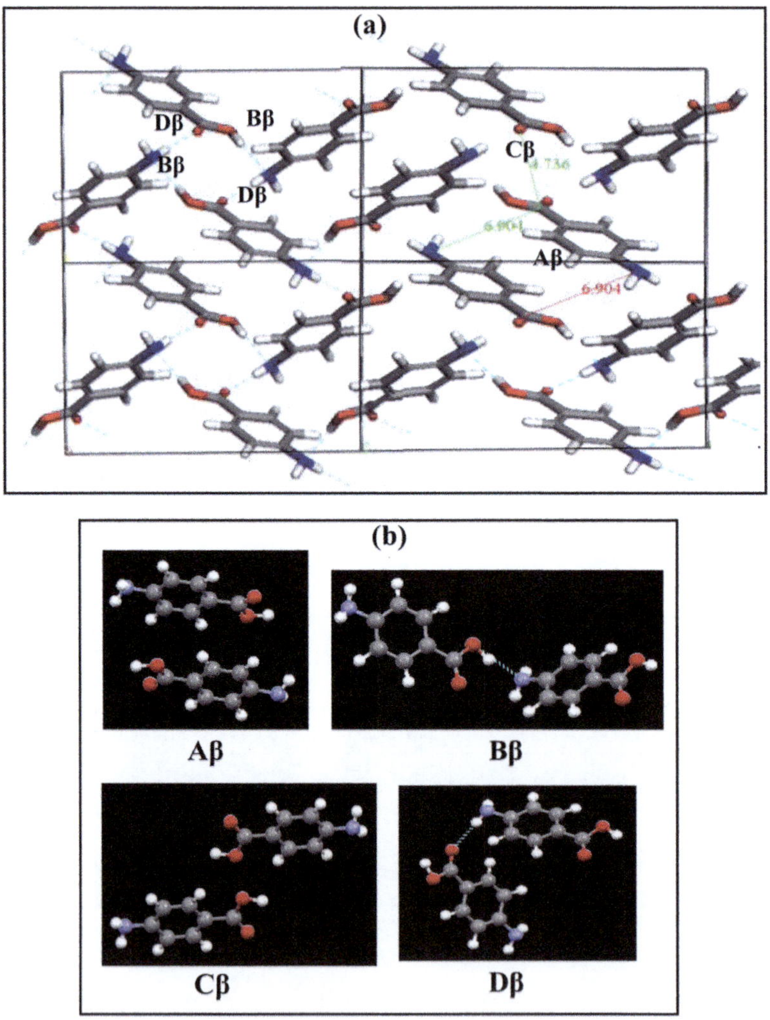

Fig. 7 Intermolecular diagrammatic representation of the key synthons present in the β-form of PABA: (a) packing diagrams; (b) molecular pair representation of the synthons highlighting the nature of their interactions (data derived from ref. 6 and 25).

Solution-state samples in EtOH and AcN were prepared using the α-form PABA dissolved in solvent in 10 ml screw top vials with magnetic stirring at 300 rpm which were left to equilibrate for 24 hours and passed through a 0.45 μm filter prior to use. The concentrations studied in EtOH were 147.7, 130, 110 and 80 g kg^{-1}, corresponding to saturation temperatures of 20, 3, −3 and −17 °C, respectively. The saturation concentrations of samples in AcN were 64.4, 60 and 55 g kg^{-1} corresponding to saturation temperatures of 20, 18 and 15 °C, respectively.

The IR spectra of PABA in solid and in solution forms were collected using a Nicolet Nexus FTIR spectrometer[36] in absorption mode with a zinc selanide ATR probe. The IR data was collected with an average of 32 scans at a spectral resolution of 0.482 cm^{-1} within a spectral range of 4000–500 cm^{-1}. Thermo-Galactic[36] software was used for the data collection and further analysis.

4. Results and discussion

4.1 Solid-state synthon analysis

A detailed analysis of the intermolecular synthons present in the solid state for the α- and β-forms of PABA has been previously published[6] and the data is summarised in Fig. 6, 7 and Table 1.

Analysis of the higher temperature α-form reveals the strongest synthon (Aα) to be associated with H-bonded carboxylic acid OH···O dimer interactions followed by, in decreasing strength, H2H π–π stacking interactions (Bα), NH···O (Cα) H-bonding interactions and displaced H2H van der Waals interactions (Dα).

In contrast, the strongest intermolecular synthons present within the low temperature β-form involve H2T π–π dimer pair interactions (Aβ), followed by, in decreasing strength, OH···N H-bonding interactions (Bβ), displaced H2T van der Waals interactions (Cβ) and NH···O H-bonding interactions (Dβ).

A noteworthy difference between the two structures lies in the presence of the strong carboxyl acid dimer (Aα) in the α-form which is not present in the β-form

Table 1 Synthon analysis for the α-form (a) and β-form (b) of PABA associated with identification of the strongest intermolecular interactions in the solid state representing *ca.* 70% of the total lattice energy and highlighting the differing physical chemical nature of the interactions between these two forms (data derived from ref. 6)

Bond	Multiplicity	Distance (Å)	Intermolecular energy (kcal mol^{-1})	% Contribution to lattice energy	Dominating interatomic interaction type
(a)					
Aα	1	8.23	−5.67	23.1	OH···O H-bonding
Bα	2	3.86	−2.68	21.8	H2H π–π stacking
Cα	1	7.90	−2.27	9.3	NH···O H-bonding
Dα	2	7.96	−2.26	18.4	Displaced H2H vdW
Total			−17.82	72.6	
(b)					
Aβ	1	4.17	−2.57	11.9	H2T π–π dimers
Bβ	2	8.11	−2.45	22.7	OH···N H-bonding
Cβ	2	5.73	−2.39	22.2	Displaced H2T vdW
Dβ	2	6.74	−1.46	13.6	NH···O H-bonding
Total			−15.17	70.4	

structure as well as strong H2T dimer pair interactions (Aβ) and the OH···N H-bonding interactions (Bβ) in the β-form, both of which are not present within the α-form structure.

4.2 MD simulations

Since the solubilities of PABA are very different, finding a single concentration that is realistic for an experimental crystallisation experiment was challenging. However, the 0.1 g ml^{-1} at 20 °C selected was found to be the closest concentration that was reasonably realistic for all three solvents. In these conditions, the AcN solution was supersaturated, the EtOH solution was saturated and the H_2O solution was extremely supersaturated.

4.2.1 Oxygen–oxygen H-bonding interactions. The propensity for the formation of singular OH···O H-bonding interactions among the carboxylic groups present within the PABA solutions are shown in Fig. 8(a) revealing that the greatest amount of singular OH···O interactions were found in the AcN solution, followed by the EtOH and then H_2O solutions. It could be argued that as the solubility of PABA is lower in AcN than EtOH, then it might be expected that there would be a greater degree of solute–solute interaction. However, the solubility of

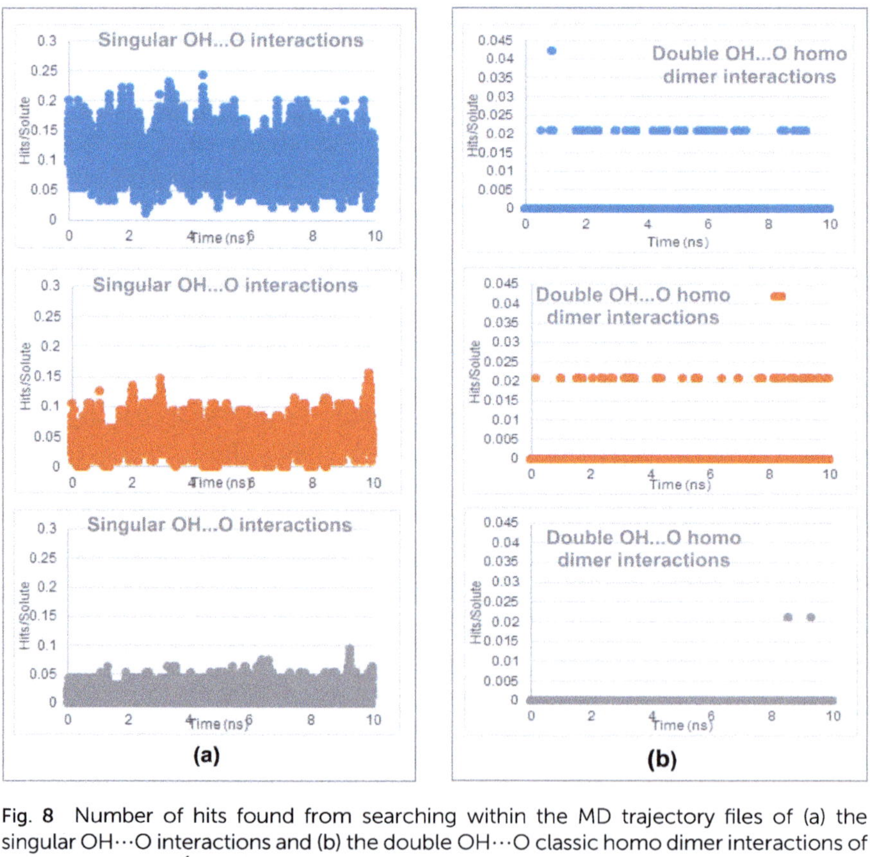

Fig. 8 Number of hits found from searching within the MD trajectory files of (a) the singular OH···O interactions and (b) the double OH···O classic homo dimer interactions of PABA in 0.1 g ml^{-1} solutions: AcN (blue), EtOH (orange) and H_2O (grey) over a 10 ns simulation. Hits are normalised relative to the amount of solute molecules per simulation.

PABA is the lowest in water, where it was found that the OH···O interactions between the solute were also found to be much lower than the same found in AcN and EtOH solutions. It would be, perhaps, more likely that the higher counts found for AcN (H-bond acceptor) reflects its lack of solvation ability with respect to the carboxyl acid group when compared to EtOH (H-bond donor and acceptor). In contrast, the low counts found in the aqueous solution simulations probably reflects the strong solvation ability of water molecules with respect to their strong (only) interactions with the carboxyl acid group.

Examination of the propensities of the OH···O homo dimers (Fig. 8(b)) associated with synthon Aα were also found in higher amounts in the AcN and EtOH solutions, in comparison to H_2O solutions where they were only observed twice in the 10 ns simulation. It was noteworthy also that these interactions were found in much lower amounts when compared to the singular OH···O interactions, and that the propensities were quite similar between AcN and EtOH.

4.2.2 Oxygen–nitrogen H-bonding interactions. Fig. 9 highlights the differences in the amounts of NH···O (α- and β-forms) and OH···N (β-form) H-bonding

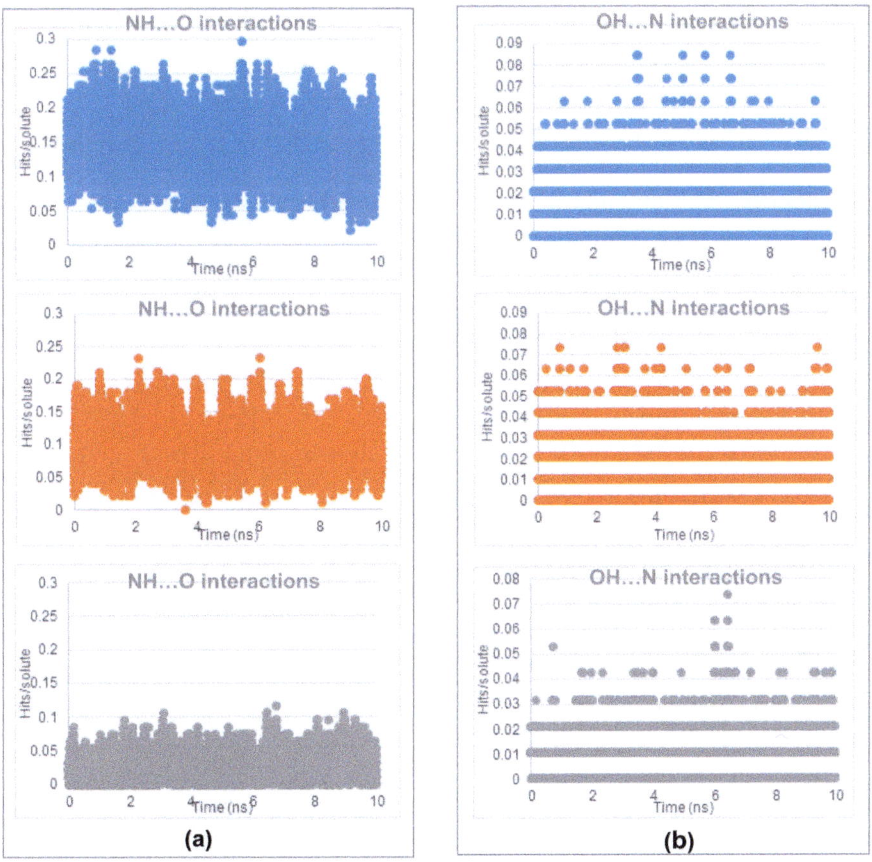

Fig. 9 Number of hits found from searching within the MD trajectory files of (a) the singular NH···O interactions and (b) the singular OH···N classic homo dimer interactions of PABA in 0.1 g ml^{-1} solutions: AcN (blue), EtOH (orange) and H_2O (grey) over a 10 ns simulation. Hits are normalised relative to the amount of solute molecules per simulation.

interactions (between amino and carboxylic groups) for the three different solvents with the latter found to be lower, when compared to the NH···O interactions. The NH···O interactions were found to be quite similar between AcN and EtOH but significantly lower in H_2O. The amount of OH···N interactions was found to be similar in AcN and EtOH, and, interestingly, only fractionally lower in H_2O solutions. The OH···N interactions, found only in the β-structure, were found to be lower in all solvents when compared to the NH···O interactions. However, it should also be noted that the amount of NH···O interactions was found to be higher than the OH···O interactions in all three solvents tested.

4.2.3 Strong van der Waals intermolecular bonding interactions. Fig. 10 compares the van der Waals interactions between the sandwich (a) and displaced (b) π–π synthons revealing the π displaced interactions are found in greater numbers when compared to the π sandwich interactions in both the AcN and EtOH solutions. This is consistent with current knowledge regarding π–π

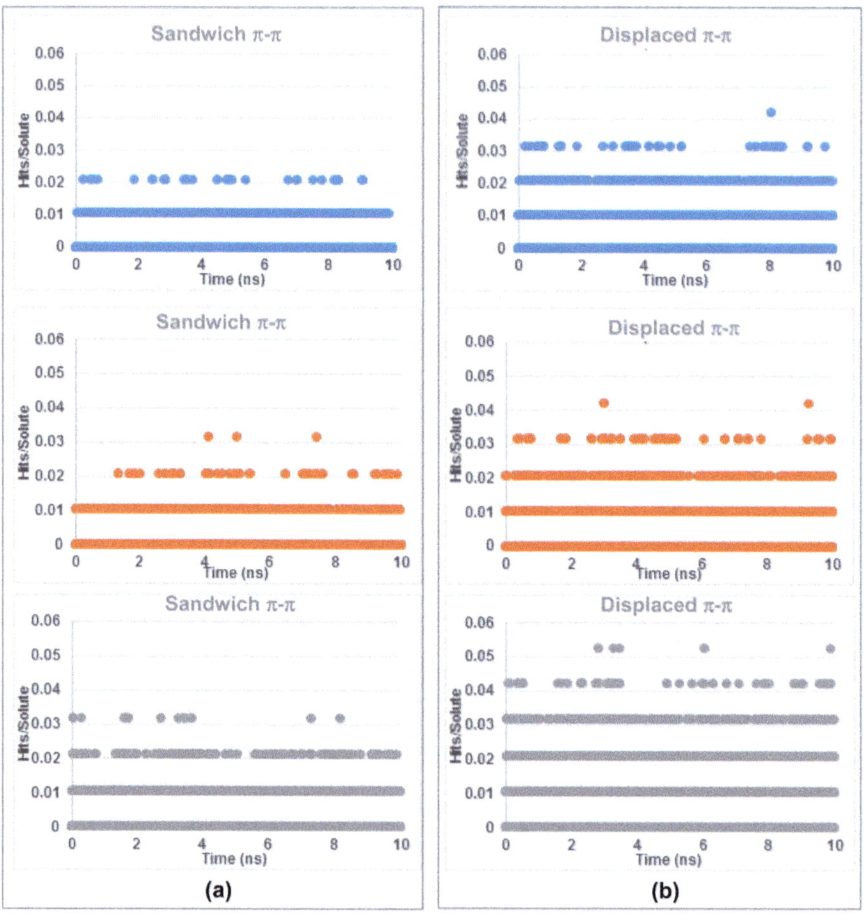

Fig. 10 Hits per PABA molecule found from searching within the MD trajectory files of (a) the sandwich π–π interactions and (b) the displaced π–π interactions of PABA in 0.1 g ml^{-1} solutions: AcN (blue), EtOH (orange) and H_2O (grey) over a 10 ns simulation. Hits are normalised relative to the amount of solute molecules per simulation.

interactions, which reflects that the electrostatic interactions between the negative ring π electrons and the positive ring hydrogens often results in a displaced orientation for the stacking interactions. In contrast to the H-bonding interactions, it is also interesting to note that the sandwich π–π stacking interactions were found to be slightly higher in water than EtOH. Instinctively, one might think that the hydrophobic tail of EtOH would be most likely to disrupt the π–π stacking interactions by forming apolar interactions with the phenyl ring. However, it is likely that the presence of the H-bonding donor proton attracts the EtOH molecules to the more polar COOH and NH_2 groups, resulting in the extra carbon having a negligible effect in decreasing the amount of π–π stacking interactions in the simulation. Overall, the data are consistent with the hydrophobic aromatic rings tending to close pack in a hydrophilic protic solvent such as water and the fact that EtOH in comparison with AcN tends to solvate PABA more equantly in 3D.

The absolute number of hits for each type of interaction throughout each 10 ns simulation are summarised in Table 2 to provide an improved quantification as to how the different intermolecular interactions change between the different solvents.

Table 2 shows that the absolute number of single OH···O interactions found in EtOH was found to be approximately half those found in AcN. However, in terms of the absolute number of complete dimers found, there was little difference between AcN and EtOH. As is clear in Fig. 8(a), the concentration of OH···O interactions found in the aqueous simulations were found to be very low.

4.2.4 Summary. In comparison to the OH···O interactions, the NH···O interactions were found to be only approximately 30% lower in EtOH than AcN, along with these interactions being approximately 80% lower in H_2O. This is a significant drop, but much less significant than the drop in OH···O interactions between AcN and H_2O, where the OH···O interactions found in H_2O were only a fraction of those found in the other solvents. The data shows only limited differences between the "finger printing" OH···N interactions in all three solvents, but importantly the π–π interactions do seem to be favoured in a β-form generating aqueous environment.

Table 2 Absolute number of hits for carboxylic–carboxylic, amino–carboxylic and π–π stacking interactions found in the three solvents at 0.1 g ml^{-1} at 20 °C

Atom–atom interactions	AcN	EtOH	H_2O	Associated synthons	Comments
Single OH···O	9.99	4.73	0.28	Aα	O both donor and acceptor, only in α-form
Double OH···O	0.068	0.066	0	Aα	O both donor and acceptor, only in α-form
NH···O	13.63	9.30	2.50	Cα, Dβ	N donor and O acceptor, in both forms
OH···N	1.28	1.44	1.12	Bβ	N acceptor, only in β-form
Sandwich π–π	0.0007	0.0010	0.0017	Bα, Aβ	H2H and H2T not differentiated, in both forms
Displaced π–π	0.0027	0.0031	0.0064	Dα, Cβ	H2H and H2T not differentiated, in both forms

Table 3 List of the major FTIR absorption bands (cm^{-1}) in the solid-state and solution state of PABA close to the carbonyl region. Note that the presence of a strong solvent absorption band precludes PABA characterisation in the other IR spectral regions (data derived from ref. 26)

Environment	ν carbonyl (cm^{-1})				ν aromatic ring (cm^{-1})
α-form solid		1658.56 OH···O (Aα)	1632.98 OH···O (Aα)	1622.20 NH···O (Cα)	1596.72 H2H (Bα)
β-form solid		1687.52 OH···N (Dβ)		Weak NH···O (Bβ)	1599.30 H2T (Aβ)
EtOH solution	1710.62 Solvated carboxylate	1684.89 Carboxylate dimer	1640.19 Carboxylate dimer		1606.26
AcN solution	1710.46 Solvated carboxylate	1678.50 Carboxylate dimer	1633.09 Carboxylate dimer		1605.54

4.3 Infrared spectroscopy measurements

Solid- and solution-state IR data are given in Fig. 11–13 and in Table 3.

4.3.1 Solid-state FTIR spectroscopy. Analysis of the solid state of PABA using IR spectroscopy is severely limited by strong solvent absorption which effectively masks many of the useful spectral features revealed for solution-state structural characterisation. Despite this, the carbonyl region is available. Fig. 11 shows the solid-state spectra in the carbonyl region revealing the characteristic absorption bands for the α- and β-forms. These data are also tabulated in Table 3.

4.3.2 Solution-state FTIR spectroscopy. Fig. 12 and 13 show the solution-state spectra recorded as a function of concentration in AcN and EtOH, respectively. For PABA in AcN solvent, the C=O region is well defined with three sharp

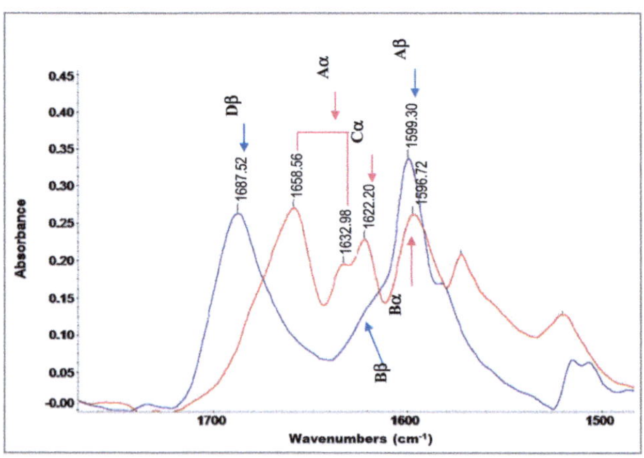

Fig. 11 Comparison of the carbonyl region of the α (red) and β (blue) IR spectra (data derived from ref. 26).

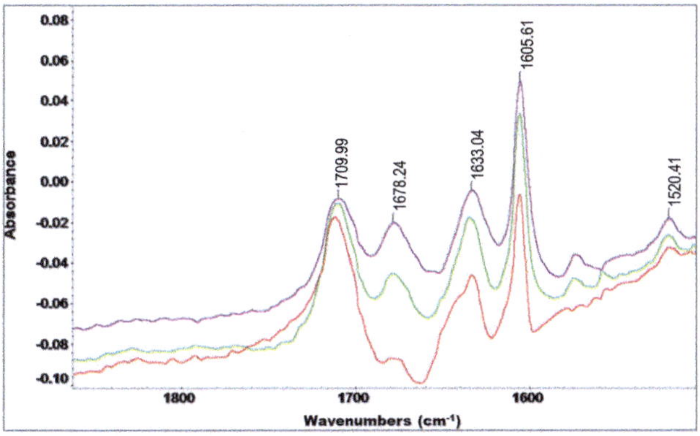

Fig. 12 IR spectra of PABA in AcN with three solution concentrations of 64.4 g kg^{-1} (purple), 60 g kg^{-1} (green) and 55 g kg^{-1} (red) at 20 °C, 18 °C and 15 °C (data derived from ref. 26).

absorption bands at 1709.99, 1678.24 and 1633.04 cm^{-1}, the two higher wavenumber bands could represent two carbonyl stretching environments due to their proximity to each other, within 40 cm^{-1} which is the expected range for related C=O stretching, monomer and dimer frequencies. This has been observed for a number of systems[37] and can be ascribed to the presence of more effective H-bonding from the solvent which increases the C=O stretching frequency and hence the C=O stretching frequency for a dimerised structure would be expected to lie at a lower wavenumber in comparison. This spectral assignment could thus indicate the presence of a solution-state dimer structure consistent with the analysis of the carbonyl peaks in the solid-state spectra of the α-form. The latter has been seen in a number of carboxylic acid systems including benzoic acid and tetrolic acid.[16]

In the dilution experiments for the AcN solution (Fig. 12), the C=O peak at 1678.24 cm^{-1} can be seen to decrease in intensity with respect to its neighbouring C=O peak at 1709.99 cm^{-1}. This seems to indicate that the peak at 1709.99 cm^{-1} is related to the solvated monomer C=O stretching vibration due to this intensity increase relative to the other C=O frequency as a function of increasing dilution. This will likely be caused by H-bonding between the C=O···H–Me resulting from the induced dipole on the AcN molecule. This dilution effect also seems to suggest that the peak at 1678.24 cm^{-1}, which decreases in intensity upon increasing solvent concentration, is the carbonyl stretching of the carboxylic acid dimer C=O. Upon dilution the two peaks shift in intensity with the peak relating to the solvated monomer intensity increasing, and the peak relating to the carboxylic acid dimer decreasing. This is a good indication that in relatively concentrated solutions of PABA in AcN there is a distribution of solvated monomers and carboxylic acid dimers of PABA.

For the PABA in EtOH solution, Fig. 13 shows that the C=O region is well resolved in EtOH with sharp absorption bands at 1685.72, 1643.13 and 1606.93 cm^{-1}, unlike the AcN carbonyl region, these two peaks are not within the expected 40 cm^{-1} wavenumber range and so might not be characterised as being representative of a distribution of monomers and dimers. However on closer

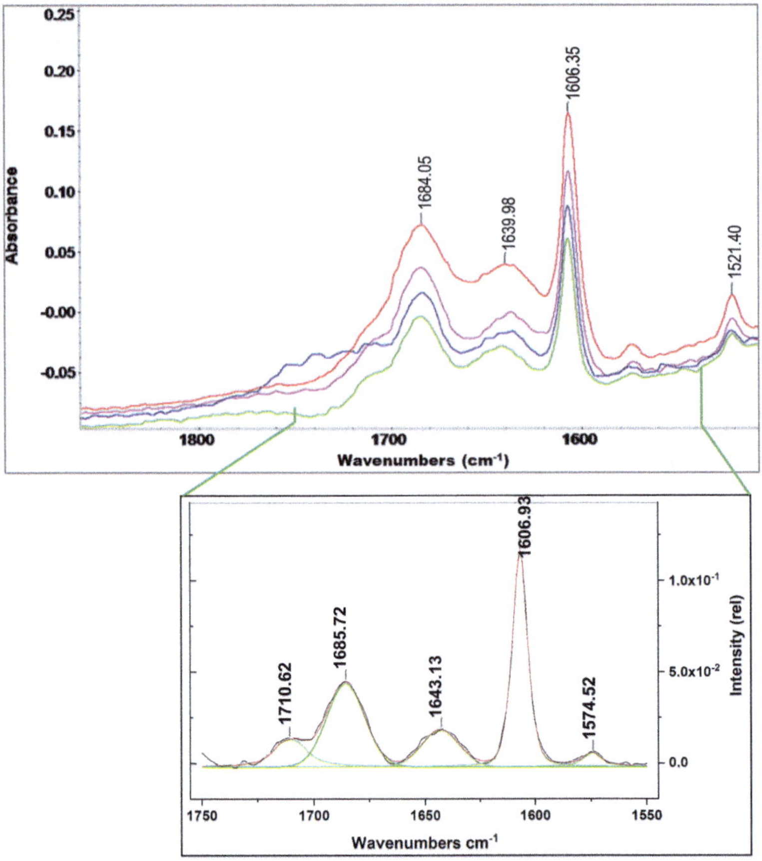

Fig. 13 IR spectra of PABA in EtOH with four solution concentrations of 147.7 g kg^{-1} (red), 130 g kg^{-1} (pink), 110 g kg^{-1} (blue) and 80 g kg^{-1} (green) at 20 °C, 3 °C, −3 °C and −17 °C. Note that the enlarged spectrum of PABA in EtOH at 80 g kg^{-1} was fitted to identify the peak position (1710.62 cm^{-1}) at the shoulder, the highlighted peak positions are fitted peak positions and hence differ from the above (data derived from ref. 26).

inspection of the data, a small shoulder peak is visible on the 1685.72 cm^{-1} carbonyl stretching band, which seems to increase in relative intensity with increasing dilution. Further baseline correction and peak fitting of this data set revealed a peak position at maximum intensity of 1710.62 cm^{-1} of the shoulder peak, which falls within the 40 cm^{-1} of $\Delta\nu$ expected for a dimer and monomer population in solution. As such, the data seems to mirror that for AcN solutions, with a major C=O band at ~1685.72 cm^{-1} and a less intense band at 1710.62 cm^{-1} for the two C=O stretches which would relate to the solvated monomer and dimer peaks in the AcN spectra at 1709.99 and 1678.24 cm^{-1}, respectively. This would suggest that in EtOH solutions PABA forms a dimerised molecular structure in saturated and under-saturated solutions as a majority species and which decreases at higher dilution ranges relative to AcN solutions.

4.3.3 Summary. The solid-state FTIR results provided an understanding of the two key structural synthons of the centro-symmetric carboxylic acid dimer (Fig. 2(a)) and the four membered H-bonded ring system (Fig. 2(b)) found in the crystal structures of α- and β-form PABA, respectively. The dilution experiments in the EtOH and AcN solutions revealed the presence of a distribution of solvated monomers and dimerised PABA molecules in these solutions.

Comparatively, it is noteworthy that the solution-state data show a more reasonable agreement with the IR absorption features which could be expected to template the α-form structure than with those expected to template the β-form structure, notably the strong feature at 1687.5 cm^{-1}.

Overall, the FTIR results of PABA in solid-state and in solutions have demonstrated strong evidence that PABA has the ability to self-associate in saturated solutions. This was confirmed using FTIR spectroscopy in both AcN and EtOH solutions, providing a strong link between the solution-state species present and the primary α-form structural synthon found in the crystal structure. This is a likely cause for the dominant crystallisation of the α-form crystal structure from solution and provides a basis of knowledge for further nucleation studies.

5. Concluding remarks

A multi-scale modelling approach has linked solid-state synthonic analysis, based upon the crystal structures of PABA, with solution-state characterisation using MD simulations. The latter, facilitated through a detailed description of the solution-state inter-molecular interactions, has been combined with molecular geometry snapshots taken from the MD trajectory files together with experimental infra-red spectroscopy data. This analysis has yielded details concerning how the functional groups in PABA aggregate within solvation environments, how this is mediated by solvent choice and how this relates to the polymorphic structures that are formed after crystallisation from the solution phase.

Drawing this new data together with a body of previous work has shed new light on the potential molecular pathway that directs solvated molecules within solution to preferentially crystallise in the α rather than β crystallographic form. Specifically, crystallisation of the β-form requires greater change in molecular conformation in the solid-state when compared to the α-form, hence providing a bigger barrier to crystallisation. Crystallisation of the β-form also requires the assembly of a larger cluster size to stabilise its structure when compared to the α-form consistent with low solution supersaturations needed to prepare this form. Solid-state characterisation linking to solution-state synthon propensity calculations reveal the α-form's carboxylic acid dimer (synthon Aα) to be a very strong interaction when compared to competing solute/solvent interactions and, thus, one that can be expected to dominate the polymorphism direction process. The exception to this seems to be in aqueous solutions where water, through its strong solvation of the carboxylate group, appears to disrupt the carboxylic acid dimer formation process and hence, at low aqueous supersaturations, the β-form can crystallise, albeit rather slowly. MD calculations were found to be consistent with the presence of intermolecular interactions between the carboxylic acid (synthon Aα) in solution as well as NH···O H-bonding (synthons Cα, Dβ), π–π (synthons Bα, Aβ) and aromatic displaced ring van der Waals interactions (synthons Dα, Cβ). The MD results are also consistent with the experimental crystallisability observations. In particular, OH···O (synthon A) interactions are stronger in AcN when compared to EtOH where the solvation is much stronger. In aqueous solutions, the hydrogen bonded interactions are much lower than for AcN and EtOH. However, in the case of the hydrophobic π–π interactions, their concentrations were found to be higher than for the cases of AcN and EtOH solutions. These is also only rather limited evidence for the presence of other key interactions, notably the β-form's "finger printing" OH··· N interactions (synthon Bβ) which is only present in the β-form and, at the current stage of analysis, it has not been possible to discriminate between the head-to-head (α-form) and head-to-tail (β-form) aromatic ring interactions.

Whilst the MD studies have provided important information concerning the solution-state structural environment within the solution state, MD simulation times have been far too short to enable the simulations to follow the pathway through to the nucleation event and this aspect is a clear restriction of the approach presented here.

Infra-red spectra in the solution phase (EtOH and AcN solvents) were found to be closely related to the α-form when compared to the β-form structure. The solution IR data was also found to be consistent with the presence of both solvated monomers and carboxylic acid dimers (synthon Aα) in solution, albeit strong solvent absorption precluded IR characterisation within aqueous solutions. The MD synthon propensity and PABA solvation studies all support this model notably through examining the relative populations of π–π dimer pair interactions (synthon Aβ) *versus* OH···O H-bonding dimers (synthon Aα) which suggests the latter templates the greater crystallisability of PABA in AcN with respect to EtOH associated with the latter's higher solvation energy.

In terms of how the crystallisation and polymorph selection process is mediated by solvent selection, van't Hoff analysis reveals less than ideal solubility, low solvation energy and, hence, favourable solute–solute interactions in the order of water > AcN > EtOH. This behaviour is matched by wider meta-stable zone widths and lower nucleation rates, respectively, encompassed within an instantaneous nucleation mechanism (except at low solute concentrations) which are consistent with the formation of solute pre-nucleation clusters in solution.

The above model is also supported through SAXS measurements in ethanolic solutions which illustrate the progress from poorly ordered pre-nucleation clusters in under-saturated solutions through to their development and the accompanying transformation of PABA molecular entities from monomers to dimers during the crystallisation process. These studies are, though quite limited in extent in terms of solvent type and solute concentration as the latter needs to be high enough to achieve good solute/solvent contrasting in the SAXS measurements. This has, in turn, restricted studies to ethanol solutions and, even then, only at concentrations known to be associated with an instantaneous nucleation mechanism hence precluding examining the mechanistic transformation to progressive nucleation at the lower concentrations. With reference to the schematic in Fig. 4, a further open question centres on whether the subsequent nucleation process progresses within the evolving solute clusters through an annealing recrystallisation mechanism of the solute clusters, or at the interface between the supersaturated solution and the cluster surfaces, for example by a secondary nucleation/templating mechanism. Further studies are clearly needed to elucidate this aspect.

Overall, this study further demonstrates the attractive potential for this comparatively user-friendly crystallographically-based multiscale modelling approach which integrates solid-state and MD simulations. Its strengths highlight the ease of input file construction, MD simulations and synthon analysis of the trajectories of organic molecular solutions. The synergy of these calculations with related high-level electronic structure DFT calculations of synthon polarisability and propensity has provided a further degree of granulation and quantification to the fine details of the solution chemistry. The study's further integration with a wide range of experimental data highlights the added-value of combining multi-scale modelling with multi-technique crystallisation characterisation using in-process analysis techniques. It also demonstrates the inherent strengths in exploring the molecular-scale transition pathway from solution through aggregation/nucleation to the direction of the polymorphic form of the resultant crystals which would be clearly inaccessible from a single modelling or experimental approach.

List of symbols and abbreviations

AcN	Acetonitrile
ATR	Attenuated total reflectance
Aα, Aβ	Synthon A of α-form, synthon A of β-form
DFT	Density functional theory
EtOH	Ethanol
FTIR	Fourier-transform infrared spectroscopy
H2H	Head-to-head
H2T	Head-to-tail
H-bond	Hydrogen bond
IR	Infra-red
MD	Molecular dynamics
PABA	*para*-Amino benzoic acid
q	Solution cooling rate (m s^{-1})
Q	Scattering vector (nm^{-1})
SAXS	Small angle X-ray scattering
T_c	Crystallisation on-set point (critical temperature) (°C)
vdW	van der Waals
ΔH_{diss}	Enthalpy of dissolution (kcal mol^{-1})
ν	IR spectrum wavenumber (cm^{-1})

Conflicts of interest

No potential conflict of interest was reported by the authors.

Acknowledgements

The co-authors are most grateful to all the various sponsors of the research herein this paper. Research into characterising the structural pathway associated with the nucleation of the PABA system was initiated through an EPSRC Critical Mass Programme (Molecules, Clusters and Crystals: Grants EP/I014446/1 and EP/I013563/1) in collaboration with the University of Manchester which involved the PhD programmes of two of the co-authors (IR[19] and TDT[26]). This work also builds upon software developments funded through the Synthonic Engineering Program supported by the EPSRC (Grant EP/I028293/1) in collaboration with Pfizer, Boeringer-Ingellheim, Novartis, and Syngenta. The MD modelling research presented here was supported through the Advanced Manufacturing Supply Chain Initiative associated with the 'Advanced Digital Design of Pharmaceutical Therapeutics' (Grant No. 14060) project. In the latter respect, some of the development of the DL_ANALYSER software was supported under the auspices of the EPSRC's Collaborative Computational Project No. 5 (CCP5) (Grant: EP/M022617/1).

References

1 *Nucleation – a Transition State to the Directed Assembly of Materials: Faraday Discussion*, 2015, vol. 179, pp. 1–560, ISSN: 1359-6640 (print); 1364-5498 (web).

2 R. B. Hammond, K. Pencheva and K. J. Roberts, *CrystEngComm*, 2012, **14**, 1069–1082.

3 C. Wang, I. Rosbottom, T. D. Turner, S. Laing, A. Maloney, A. Y. Sheikh, R. Docherty, Q. Yin and K. J. Roberts, *Pharm. Res.*, 2021, **38**, 971–990.

4 T. D. Turner, X. X. Lai and K. J. Roberts, *CrystEngComm*, 2018, **20**, 4099–4102.

5 A. A. Moldovan, I. Rosbottom, V. Ramachandran, C. M. Pask, O. Olomukhoro and K. J. Roberts, *J. Pharm. Sci.*, 2017, **106**, 882–891.

6 I. Rosbottom, K. J. Roberts and R. Docherty, *CrystEngComm*, 2015, **17**, 5768–5788.

7 N. Anuar, W. R. Wan Daud, K. J. Roberts, S. K. Kamarudin and S. M. Tasirin, *Cryst. Growth Des.*, 2012, **12**, 2195–2203.

8 T. D. Turner, L. E. Hatcher, C. C. Wilson and K. J. Roberts, *J. Pharm. Sci.*, 2019, **108**, 1779–1922.

9 I. Rosbottom, J. H. Pickering, B. Etbon, R. B. Hammond and K. J. Roberts, *Phys. Chem. Chem. Phys.*, 2018, **20**, 11622–11633.

10 T. T. H. Nguyen, R. B. Hammond, I. D. Styliari, D. Murnane and K. J. Roberts, *CrystEngComm*, 2020, **22**, 3347–3360.

11 V. Ramachandran, D. Murnane, R. B. Hammond, J. H. Pickering, K. J. Roberts, M. Soufian, B. Forbes, S. Jaffari, G. P. Martin, E. Collins and K. Pencheva, *Mol. Pharm.*, 2015, **12**, 18–33.

12 T. D. Turner, P. Gajjar, I. Fragkopoulos, J. Carr, T. T. H. Nguyen, D. Hooper, F. Clarke, N. Dawson, P. Withers and K. J. Roberts, *Cryst. Growth Des.*, 2020, **20**, 4252–4263.

13 C. W. Yong, *J. Chem. Inf. Model.*, 2016, **56**, 1405–1409.

14 C. W. Yong and I. T. Todorov, *Molecules*, 2017, **23**, 36.

15 I. Rosbottom, C. W. Yong, D. L. Geatches, R. B. Hammond, I. T. Todorov and K. J. Roberts, *Mol. Simul.*, 2021, **47**, 257–272.

16 R. J. Davey, G. Dent, R. K. Mughal and S. Parveen, *Cryst. Growth Des.*, 2006, **6**, 1788–1796.

17 M. Svard, F. L. Nordström, E.-M. Hoffmann, B. Aziz and A. C. Rasmuson, *CrystEngComm*, 2013, **15**, 5020–5031.

18 M. J. Bryant, S. N. Black, H. Blade, R. Docherty, A. G. P. Maloney and S. C. Taylor, *J. Pharm. Sci.*, 2019, **108**, 1655–1662.

19 I. Rosbottom, The Influence of the Intermolecular Synthons on the Molecular Aggregation, Polymorphism, Crystal Growth and Morphology of *p*-

Aminobenzoic Acid Crystals from Solution, PhD thesis, University of Leeds, UK, 2015.

20 I. Rosbottom, D. Toroz, R. B. Hammond and K. J. Roberts, *CrystEngComm*, 2018, **20**, 7543–7555.

21 R. B. Hammond, C. Y. Ma, K. J. Roberts, P. Y. Ghi and R. K. Harris, *J. Phys. Chem. B*, 2003, **107**, 11820–11826.

22 R. B. Hammond, R. S. R. S. Hashim, C. Y. Ma and K. J. Roberts, *J. Pharm. Sci.*, 2006, **95**, 2361–2372.

23 I. Rosbottom, J. H. Pickering, R. B. Hammond and K. J. Roberts, *Org. Process Res. Dev.*, 2020, **24**, 500–507.

24 T. D. Turner, D. M. C. Corzo, D. Toroz, A. Curtis, M. M. Dos Santos, R. B. Hammond, X. Lai and K. J. Roberts, *Phys. Chem. Chem. Phys.*, 2016, **18**, 27507–27520.

25 D. Toroz, I. Rosbottom, T. D. Turner, D. M. C. Corzo, R. B. Hammond, X. Lai and K. J. Roberts, *Faraday Discuss.*, 2015, **179**, 79–114.

26 T. D. Turner, Molecular Self-Assembly, Nucleation Kinetics and Cluster Formation Associated with Solution Crystallisation, PhD thesis, University of Leeds, UK, 2015.

27 A. S. Myerson and P. Y. Lo, *J. Cryst. Growth*, 1990, **99**, 1048–1052.

28 H. C. Groen and K. J. Roberts, *J. Phys. Chem. B*, 2001, **105**, 10723–10730.

29 N. Salwani Md Azmi, N. Anuar, N. Fitrah Abu Bakar, N. Fadhilah Kamalul Aripin and K. J. Roberts, *J. Cryst. Growth*, 2019, **519**, 91–99.

30 I. T. Todorov, W. Smith, K. Trachenko and M. T. Dove, *J. Mater. Chem.*, 2006, **16**, 1911–1918.

31 DL_Software is the collective term for a range of scientific software developed at the Daresbury Laboratory, spanning across multi-length and -time scales: https://www.scd.stfc.ac.uk/Pages/Materials-Modelling-Software.aspx.

32 W. D. Cornell, P. Cieplak, C. I. Bayly, I. R. Gould, K. M. Merz, D. M. Ferguson, D. C. Spellmeyer, T. Fox, J. W. Caldwell and P. A. Kollman, *J. Am. Chem. Soc.*, 1995, **117**, 5179–5197.

33 R. A. Sullivan, R. J. Davey, G. Sadiq, G. Dent, K. R. Back, J. H. Ter Horst, D. Toroz and R. B. Hammond, *Cryst. Growth Des.*, 2014, **14**, 2689–2696.

34 H. J. C. Berendsen, J. P. M. Postma, W. F. V. Gunsteren, A. DiNola and J. R. Haak, *J. Chem. Phys.*, 1984, **81**, 3684–3690.

35 W. G. Hoover, *Phys. Rev. A: At., Mol., Opt. Phys.*, 1985, **31**, 1695–1697.

36 Thermo, 2015, https://www.thermofisher.com/.

37 P. Novak, D. Vikić-Topić, Z. Meić, S. Sekušak and A. Sabljić, *J. Mol. Struct.*, 1995, **356**, 131–141.

Faraday Discussions

PAPER

On the crystal forms of NDI-C6: annealing and deposition procedures to access elusive polymorphs†

Inês de Oliveira Martins, [ab] Francesco Marin, [b] Enrico Modena [*a] and Lucia Maini [*b]

Received 25th November 2021, Accepted 19th January 2022

DOI: 10.1039/d1fd00100k

NDI-C6 has been extensively studied for its semiconducting properties and its processability. It is known to have several polymorphs and a high thermal expansion. Here we report the full thermal characterization of **NDI-C6** by combining differential scanning calorimetry, variable temperature X-ray powder diffraction, and hot stage microscopy, which revealed two different thermal behaviours depending on the annealing process. The ranking of stability was determined by the temperature and energy involved in the transitions: Form α is stable from RT up to 175 °C, Form β is metastable at all temperatures, Form γ is stable in the range 175–178 °C, and Form δ in the range 178–207 °C followed by the melt at 207 °C. We determined the crystal structure of Form γ at 54 °C from powder. The analysis of the thermal expansion principal axis shows that Form α and Form γ possess negative thermal expansion (*X*1) and massive positive thermal expansion (*X*3) which are correlated to the thermal behaviour observed. We were able to isolate pure Form α, Form β, and Form γ in thin films and we found a new metastable form, called Form ε, by spin coating deposition of a toluene solution of **NDI-C6** on Si/SiO_2 substrates.

Introduction

In the past few decades organic semiconductors (OSCs) have attracted a lot of attention due to their possible employment in solution-processed optoelectronic and electronic devices, such as organic field-effect transistors (OFETs).[1–3] One of the big advantages of solution-processing is the possibility to produce flexible substrates at low cost.[3–5] This allows various potential applications that cannot be achieved with wafer-based electronics, such as large-area flexible displays, RFID

[a]*PolyCrystalLine SPA, Via Della Cooperazione, Bologna, 29 40059, Medicina, Italy. E-mail: enrico.modena@polycrystalline.it*

[b]*Dipartimento di Chimica "G. Ciamician", Via Selmi 2 – Universitá di Bologna, I-40126, Bologna, Italy. E-mail: l.maini@unibo.it*

† Electronic supplementary information (ESI) available. CCDC 2124322. For ESI and crystallographic data in CIF or other electronic format see DOI: 10.1039/d1fd00100k

tags, wearable electronics, and biomedical devices.[6–8] Organic molecular materials tend to form polymorphs, which can exhibit very different conductive,[9] luminescent,[10] and mechanical properties.[11] In most cases, the control of the crystal structure is decisive to maximise the performances of the final devices.[6,7,12,13]

In the field of organic electronics, there are several properties of the active layer that have a key role in the performance of the devices, namely, charge carrier mobility, and stability.[12,14,15] Polymorphism has proven to be an important factor to obtain high-performance organic devices since in some cases metastable phases can present mobilities some orders of magnitude higher than the most stable phase.[6,7] In addition, it is known that some materials can show new phases that are not observed in bulk when thin films are deposited in specific substrates, this effect is described as surface induced polymorphism.[15–19] Therefore, it is of extreme importance to study the polymorphism of OSCs, both in bulk and in thin films since the charge transport in electronic devices occurs in the interface between the OSC and the dielectric, in just a few layers of molecules.[20]

Bis(naphthalene diimide) (NDI) derivatives are a particularly interesting family of organic materials. NDIs possess high electron affinity, good charge carrier mobility, and excellent thermal and oxidative stability, making them promising candidates for applications in organic electronics, photovoltaic devices, and flexible displays.[8,21] Recently, the structure–properties relationship and the polymorphism of these molecules has gained considerable attention,[4,22–25] in particular the influence of the alkyl side chain length on performance and crystal structures has been studied.[26,27]

In this work we present the polymorphism both in bulk and in thin films of *N*,*N*′-bis(*n*-hexyl)naphthalene-1,4,5,8-tetracarboxylicdiimide (**NDI-C6**), Fig. 1, a cheap small-molecule organic n-type semiconductor which has high solubility making it easy to process.

It is known that **NDI-C6** is prone to generating several polymorphs, and the number of crystal forms is higher compared to the analogous molecules with shorter or higher chain length, nonetheless only the crystal structure of the most stable polymorph (Form α) is available.[24] Up to now, four different polymorphs have been reported: Form α the stable form at RT, and Form β, Form γ and Form δ – which can be obtained by thermal treatement and all convert back to Form α at RT.[27] Several papers report the preparation of thin films based on **NDI-C6**,[4,26–29] but only recently have the thin films been characterized by X-ray diffraction and revealed the presence of a mixture of Form α and a metastable polymorph at room

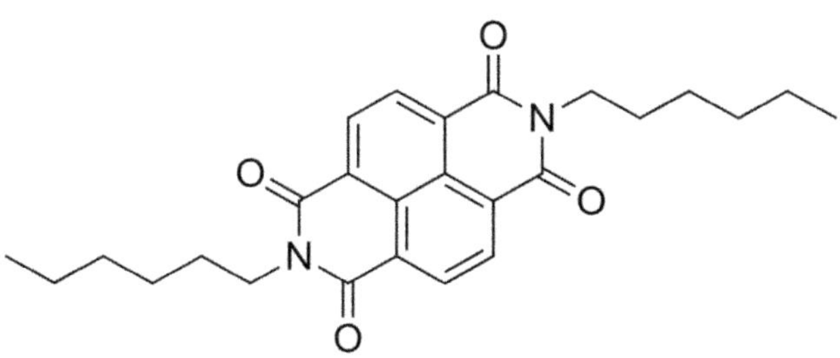

Fig. 1 Chemical structure of **NDI-C6**.

temperature.[26,27] Herein, the main goal of this work is to study and understand the polymorphic behaviour of **NDI-C6** in bulk and in thin films, with a special focus on the thermal behaviour. In fact, the annealing procedure is a common way to increase the performance of the devices, increasing the crystallinity,[30,31] but in the case of **NDI-C6** it can induce wanted or unwanted phase transitions. The synergic approach of the different thermal techniques such as hot stage optical microscopy (HSM), differential scanning calorimetry (DSC), and variable temperature X-ray powder diffraction (VT-XRPD), allowed us to determine the ranking in energy of the different polymorphs and to optimize different annealing procedures to obtain the desired phase in bulk and thin film. We were able to solve the structure of Form γ at high temperature by X-ray powder diffraction with synchrotron radiation. Furthermore, we optimized the thin film deposition to obtain pure phases and we discovered a new phase called ε.

Experimental

Materials

N,*N*′-Bis(*n*-hexyl)naphthalene-1,4,5,8-tetracarboxylic diimide (**NDI-C6**) is commercially available, it was purchased from Sigma-Aldrich and used as received.

Polymorph screening

The **NDI-C6** solubility was assessed in 12 different solvents at RT, 50 °C, 75 °C, and 100 °C when possible (see Table 1 in ESI†). Recrystallization by solvent evaporation was carried out in chloroform (CHF) and dichloromethane (DCM) at RT, in toluene (TOL) and *p*-xylene (PXY) at high temperature (60 °C) which presented the higher solubility.

For the remaining solvents and solvent mixtures of toluene and methanol (MET), with different ratios, slurry experiments at RT were performed. High temperature slurries were carried out for *N*,*N*-dimethylformamide (DMF) at 90 °C, 2-propanol (2PR), acetonitrile (ACN), and water (H_2O) at 50 °C. With the residual mother liquid obtained after filtration of the precipitate in the DMF solution, a high temperature (60 °C) evaporation experiment was set.

Crystallization by anti-solvent addition was performed testing TOL : MET (1 : 1.25 vol), PXY : MET (1 : 1.20 vol), CHF : *N*-heptane (HEP) (1 : 2 vol), DCM : acetone (ACT) (1 : 1.33 vol), and CHF : MET (1 : 1.5 vol).

Precipitation by gradient temperature was performed by crash cooling of DMF solutions and a solvent mixture of TOL : MET (1 : 1). Slow cooling crystallization was also tested with DMF solutions at different concentrations, using different cooling rates.

Solvothermal crystallization was carried out using TOL and PXY saturated solutions.

X-ray powder diffraction (XRPD)

All the precipitates obtained from the polymorph screening experiments, as well as the starting material, were characterized by XRPD to identify the crystalline phase. The XRD patterns were obtained using a Rigaku MiniFlex 600 diffractometer with Cu Kα ($\lambda = 1.54178$ Å) radiation from a copper sealed tube with 40 kV voltage and 15 mA current in Bragg–Brentano geometry, over the 2θ range of 3–40° with a step size of 1 0.01° (2θ) at a speed of 10.0° min^{-1} (2θ).

Differential scanning calorimetry (DSC)

The starting material was characterized by differential scanning calorimetry (DSC) using a Mettler-Toledo DSC-1 instrument. The samples were prepared by accurately weighing approx. 2–4 mg of the powder in aluminium closed pans (40 μL). The measurements were performed during two cycles of heating and cooling with different rates (2 °C min^{-1}, 5 °C min^{-1} and 10 °C min^{-1}) under a dry N_2 atmosphere (flow rate 80 mL min^{-1}). All the data were analysed using STARe software.

Hot stage microscopy (HSM)

The thermal behaviour of the material was studied by hot stage microscopy (HSM), the transitions were observed on a single crystal of Form α, grown by precipitation by gradient temperature with a slow cooling rate, using an OLYMPUS BX41 stereomicroscope equipped with a LINKAM LTS350 platinum plate for temperature control and VISICAM analyser. The single crystal is placed on a glass slide and covered with a coverslip; next, it is set inside of the sealed heating chamber.

During the experiment time-lapse images were taken using a NIKON DS FI3 high speed camera and analysed using Nikon NIS Elements software and Linksys32.

Variable temperature X-ray powder diffraction (VT-XRPD)

VT-XRPD in transmission mode was performed at the Paul Scherrer Institute (PSI) synchrotron radiation facility (Switzerland) at MS-X04SA beamline using a well filled and compacted capillary with **NDI-C6** powder. The MS powder diffractometer used works in Debye–Scherrer geometry and is equipped with a solid-state silicon microstrip detector, called MYTHEN (Microstrip sYstem for Time-rEsolved experimeNts).[32] The data were collected with a beam energy of 12.4 keV (1.0 Å). The characterization was executed in a temperature range from 20 °C to 195 °C, to observe the several transitions that occur during temperature variation.

VT-XRPD in reflection mode were performed on a PANalytical X'Pert Pro automated diffractometer with an X'Celerator detector in Bragg–Brentano geometry, using Cu Kα radiation ($\lambda = 1.5418$ Å) without a monochromator in the 2θ range of 3–30.0° under continuous scan mode, step size of 0.0167°, counting time of 24.765 s, Soller slit of 0.04 rad, 40 mA and 40 kV equipped with an Anton Paar TTK 450 system for measurements at a controlled temperature.

Thin films deposition

Various thin films were fabricated by spin-coating on 2 × 2 cm^2 SiO_2/Si substrates, previously cleaned by sonication in an ethanol bath, followed by sonication in an acetone bath and finally rinsed with deionised water and dried with compressed air. Deposition was carried out varying the process parameters: solvent, concentration, and speed. In addition, thermal annealing at 75 °C for 1

Table 1 Parameters used to deposit **NDI-C6** films by spin-coating on SiO_2/Si substrates

Solvent	Concentration	Speed	Number of layers	Thermal annealing
TOL	1 mg mL^{-1}	1000 rpm	1	No
				Yes
			2	No
			3	No
		750 rpm	1	No
	3 mg mL^{-1}	1000 rpm	1	No
DCM	1 mg mL^{-1}	1000 rpm	1	No
			2	No
			3	No
		750 rpm	1	No
				Yes
	3 mg mL^{-1}	1000 rpm	1	No
CHF	1 mg mL^{-1}	750 rpm	1	No
				Yes

hour was performed for some samples. The details of the process parameters are summarised in Table 1.

Structure determination from powder

Indexing, structure determination and refinement of Form γ was performed using TOPAS v5, a nonlinear least-squares optimization program written in the C++ programming language.[33,34] The X-ray powder diffraction pattern of Form γ at 54 °C was indexed with triclinic cell with parameters: $a = 4.855$ Å, $b = 6.471$ Å, $c = 19.946$ Å, $\alpha = 92.14°$, $\beta = 95.83°$, $\gamma = 104.22$ and volume 603 Å^3 and space group $P\bar{1}$. The simulated annealing was performed with a half molecule constrained on the inversion centre. The best solution was chosen for Rietveld refinements, which were performed with the software TOPAS. A shifted Chebyshev function with five parameters and a pseudo-Voigt function were used to fit background and peak shape, respectively. An overall thermal parameter was adopted for all atoms of non-hydrogen atoms. All the hydrogen atoms were fixed in calculated positions. The Rietveld refinement reached the following values: Rwp = 3.45%, R = 2.12%, GoF = 2.81.

PASCal

PASCal – Principal Axis Strain Calculator is a web-based tool used to determine the principal coefficients of thermal expansion from variable-temperature lattice parameters.[35] The input data needed by the tool is simply the value of the temperature and the unit cell parameters at each temperature, for that we used the diffractograms acquired at PSI and refined the cell parameters using Pawley refinement. The tool returns as output the principal axis of thermal expansion and the orientation of the principal axes relative to the axes of the unit cell.

Results and discussion

Polymorph screening

Recrystallization experiments in ACN, ACT, CHF, DCM, MET, TOL, H_2O, 2PR, ETA and PXY led to the isolation of pure Form α, either as a powder or single crystals.

The single crystals were obtained by temperature gradient precipitation with a slow cooling profile. Crash cooling experiments in DMF allowed the concomitant crystallization of Form α and Form β in powder mixture. Furthermore, a mixture of the two different forms was also attained by evaporating at 60 °C, the filtered mother liquor from high temperature slurry experiments in DMF.

Recrystallization gave pure Form α or a mix with Form β. However, we were able to obtain pure Form β by annealing Form α. Phases γ and δ were observed only at high temperature, and it was not possible to achieve them at room temperature in bulk. However, in thin films it was possible to isolate Form α, Form β and Form γ at room temperature.

Thermal properties

We performed DSC, HSM and VT-XRPD on **NDI-C6** to characterize its thermal behaviour and the results were compared to determine the different phases obtained during the thermal treatment.

DSC analysis was based on the first heating, cooling and second heating cycles.

During the first heating of the starting material, the DSC curve presents a solid–solid transition around 178 °C before the melting point at 207 °C, Fig. 2b. This coincides with what has been described before by Milita *et al.*[27] and is ascribable to a double transition: Form α → Form γ suddenly followed by the transition of Form γ → Form δ. The Form α → Form δ transition could be clearly observed by HSM at 180 °C (see Fig. S1 in ESI†).

The VT-XRPD data (Fig. 3) are consistent with the described behaviour, an extra peak belonging to Form γ, appeared during the transition (around 180 °C), and disappeared concomitantly with the peaks of Form α once the transition into Form δ was complete.

While the first heating cycle of the **NDI-C6** Form α was always reproducible, the cooling cycle showed two different behaviours depending on whether the sample was melted or not. When the DSC experiment was performed without reaching the melting point of **NDI-C6**, the cooling curve matched with the behaviour

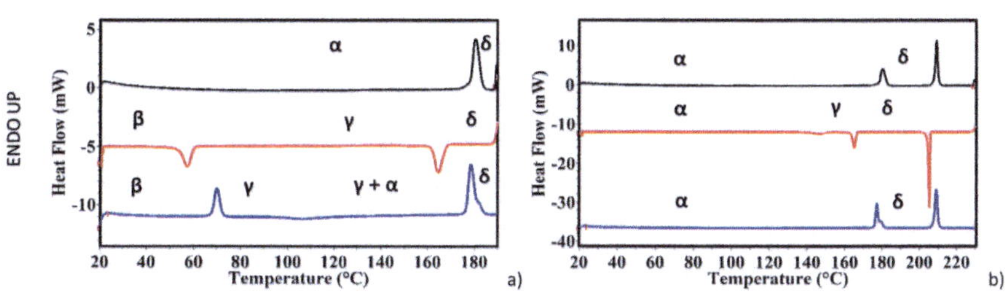

Fig. 2 DSC curves of **NDI-C6**, (a) heating cycle before melting, black line: first heating, up to the first transition Form α, γ → Form δ, red line: first cooling with two exothermic events, Form δ → Form γ at 163 °C and Form γ → Form β at 56 °C, blue line: second heating, with an endothermic peak at 69 °C Form β → Form γ, at 110 °C an exothermic peak related to the partial conversion of Form γ into Form α, and at 177 °C an endothermic peak Form γ + α → Form δ. (b) Heating cycle up to melting, black line: first heating, at 178 °C transition Form α, γ → Form δ, at 206 °C melting, red line: first cooling with three exothermic events recrystallization into Form δ at 205 °C, Form δ → Form γ at 165 °C, and Form γ → Form α at 143 °C, blue line: second heating, at 178 °C transition Form α, γ → Form δ, at 206 °C melting.

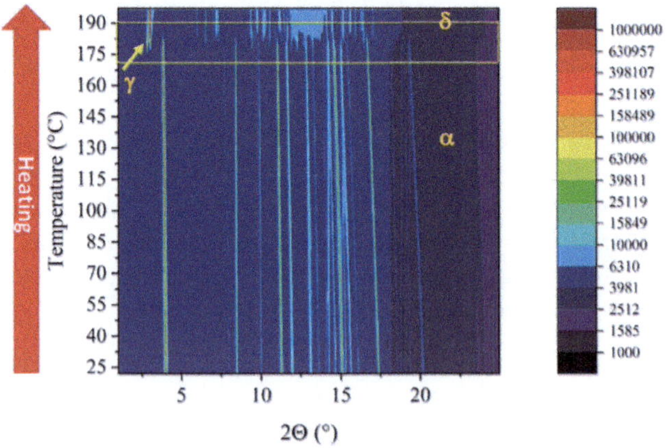

Fig. 3 VT-XRPD of **NDI-C6** at PSI synchrotron, 2D isolines of first heating until 195 °C with a transition of Form α → Form δ at 180 °C, and the presence of the Form γ peak is highlighted with an arrow.

already reported[27] with the Form δ → Form γ transition at 163 °C and followed the Form γ → Form β transition at 56 °C, ending up with Form β as a metastable phase at room temperature (Fig. 2a).

On the other hand, by heating up the sample above the melting point, we observed a different pathway. By recrystallization at 205 °C, Form δ was obtained which converted into Form γ at 165 °C; the latter, unexpectedly converted into Form α at 143 °C (Fig. 2b). The presence of Form α at RT was confirmed by the second heating which was characterized by the almost concomitant double transition at around 175 °C ending in Form δ. Furthermore, we confirmed this behaviour by VT-XRPD: the **NDI-C6** was melted and recrystallized on the sample stage. After melting, the sample is characterized by a strong preferential orientation, but the position of the first peak was used for the determination of the different phases and confirmed the presence of Form α at room temperature (see Fig. S3 in ESI†).

The thermal behaviour was studied by visual observation of the solid–solid transitions of **NDI-C6** by HSM. A crystal was heated above the melting point (214 °C), and then cooled at −10 °C min^{-1} as shown in Fig. 4. The liquid crystallized as Form δ at 203 °C, as observed also in DSC analysis. Decreasing the temperature, we observed a transition at 160 °C, that corresponded to the transition from Form δ to Form γ, then, at 145 °C, the transition Form γ → Form α occurred. No other transition was detected during the cooling, ending up with only Form α in the sample.

During the VT-XRPD collected at PSI, the sample was not melted and during the cooling, the transitions observed matched with the DSC analysis: Form δ converted into Form γ around 157 °C and then Form γ converted into Form β at 48 °C (see Fig. 2a and Fig. 5a). It is worth noting, that the remarkable thermal stability of Form γ was present in a temperature window of more than 100 °C and showed a high thermal contraction.

During the second heating, the DSC curve showed the conversion of Form β into Form γ at 69 °C then, an exothermic event occurred at 100 °C, which was

Fig. 4 Hot stage microscopy during the cooling after the melting of a single crystal **NDI-C6**, (a) recrystallization at 203 °C into Form δ, (b) Form δ → Form γ transition starting at 160 °C, (c) transition into Form γ completed, (d) transition Form γ → Form α occurring at 145 °C, (e) transition into Form α completed.

ascribed to the partial formation of Form α as confirmed by the VT-XRPD, see Fig. 5b. This event suggests that Form α is more stable than Form γ at temperatures lower than 175 °C, and that the conversion Form γ → Form α is promoted by the high temperature. In the temperature range 175–180 °C the two transitions – Form α → Form γ, and Form γ → Form δ – occurred.

The VT-XRPD data confirmed the behaviour observed in DSC, pure Form β is obtained by cooling the sample annealed to 195 °C (Fig. 5b) upon heating it converts into Form γ at 73 °C. It is worth noting that at around 108 °C the peaks associated with the interplanar distances 14.28 Å and 6.75 Å of Form α appeared, which are related to the exothermic transition observed in the DSC. Form α and

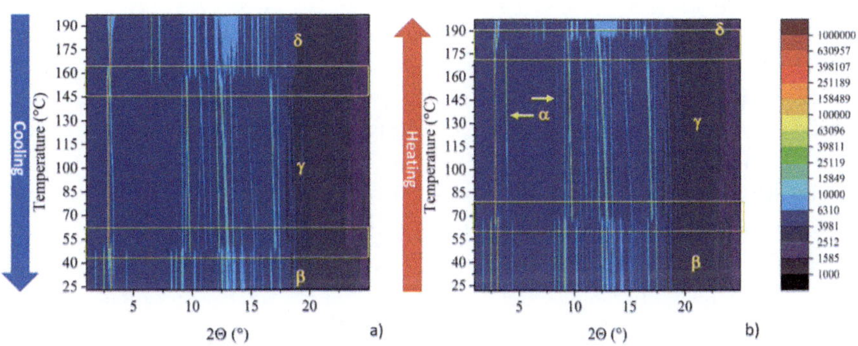

Fig. 5 VT-XRPD of **NDI-C6** at PSI synchrotron, (a) 2D isolines during first cooling from 195 °C to RT, starting at Form δ and ending at Form β. Two transitions are observable Form δ → Form γ at 157 °C and Form γ → Form β at 48 °C. (b) 2D isolines view of the second heating from RT to 195 °C, with two transitions: Form β → Form γ at 73 °C and Form γ → Form δ at 187 °C, it is also visible by the presence of two peaks of Form α, highlighted in the figure with arrows, the peak at lower 2*θ* is already present before the transition Form β → Form γ but the intensity increases considerably around 110 °C.

Form γ coexist up to 183 °C when Form α is fully converted into Form γ, then at 187 °C Form δ appears and the transition is completed at 190 °C.

The synergic approach of the different techniques allows us to rank the stability of the polymorphs: Form α is stable until 175 °C, then Form γ is stable until 178 °C when it converts into Form δ which is stable until the melting at 207 °C. Form β is always metastable but its energy should be between Form α and Form γ.

More difficult is rationalizing what causes the different thermal behaviour in cooling since they share common transitions: in fact, in both cases the high temperature Form δ converts into Form γ, only at this stage, the two pathways are observed. The different annealing process probably produces Form γ with different morphology or crystallinity which ends up in a different metastability of the crystals. When Form γ converts at high temperatures (145 °C) the thermal energy allows us to directly reach the stable Form α, instead, when the Form γ endures until 60 °C, it converts into Form β because the lower thermal energy is not enough to overcome the activation energy to reach Form α.

Thin films

We explored different combinations of solvent, deposition parameters and post-deposition treatment to obtain different films of different polymorphs. As expected, the crystal phases deposited are highly oriented and only the main peak is observed. The peak (100) of the Si at 33.00° (ref. 36) was used as the internal standard to determine the correct 2θ position of the peaks, and Table 2 summarizes the peak position used to identify the different phases.

By spin coating deposition of a low concentrated solution of **NDI-C6** in DCM, it was possible to obtain three different phases in the same film: Form α, Form β and a new form called Form ε, which was not observed before in the bulk analysis, Fig. 6. Form ε presents a strong peak at 5.2° that is not present in the other forms, and it clearly shows a new phase. Form ε was observed only in thin films crystallized on Si/SiO_2 substrates: it is worth noting that since Form ε could be obtained also by quenching the melted **NDI-C6** on Si/SiO_2 substrates, this suggests that the new form might be induced by the substrate.

By changing concentration, speed or the number of layers, we always obtained a mixture of phases with Form α always present.

In films prepared using CHF solutions, only the Form α and Form ε were present as a mixture. It was possible to obtain pure Form α by thermal annealing at 75 °C, Fig. 7.

By spin coating of toluene solutions with a speed of 1000 rpm, Form ε was obtained but it starts converting into Form β after a short time, Fig. 7.

Table 2 Peak position and related *hkl* plane observed in films of each phase

Form	*d* (Å)	*hkl*	2θ (°)
Form α	14.3	(001)	6.2
Form β	18.3	(001)	4.9
Form γ	19.8	(001)	4.4
Form δ	19.0	(002)	4.6
Form ε	17.0	—	5.2

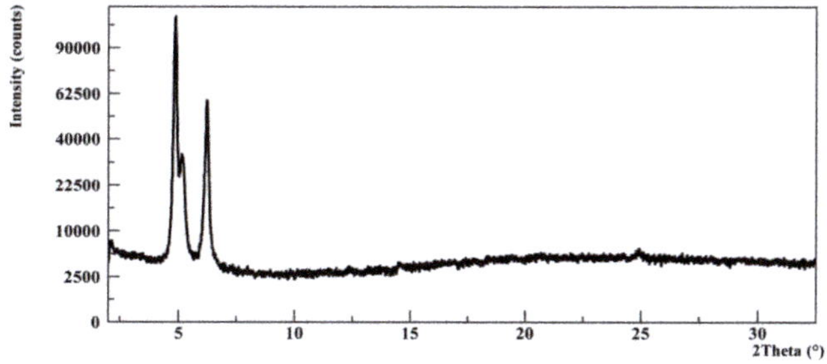

Fig. 6 Thin film prepared with a DCM solution showing a mixture of three different phases (Form α, Form β and Form ε).

A Form γ film is obtained by thermal treatment of Form β film at 75 °C (Fig. 7).

It is worth noting that the more polar and volatile solvents (DCM and CHF) favour the deposition of the stable Form α, although always concomitant with other metastable phases (Form β and Form ε) potentially promoted by the fast evaporation of the solvent. Toluene, on the other hand, favours the presence of ε, and despite being a high boiling solvent, the stable Form α is hardly observed. The main peaks observed in the thin films are ascribable to the alignment of the molecules on the substrate and it is worth noting that the β, γ and ε are characterized by peaks at lower angles than α, which indicates that **NDI-C6** molecules are less tilted in respect to the substrate as observed for the γ phase. The longer planar distance is also induced by different conformation of the chains as discussed in the next section.

Crystal structure

Structure solution of Form γ was obtained by simulated annealing with TOPAS 5. The X-ray powder pattern of Form γ at 54 °C was indexed with a triclinic cell with a volume comparable with the molecular volume, the simulated annealing was performed with a half molecule constrained on the inversion centre.

As observed for the NDI-C*n* molecules, the overall structures can be described as the stacking of layers of molecules, showing on the interface the alkyl chains. Conformation and length of the chains play a key role in the final structure tuning the aromatic core interactions. In Form γ, the alkyl chains of **NDI-C6** present

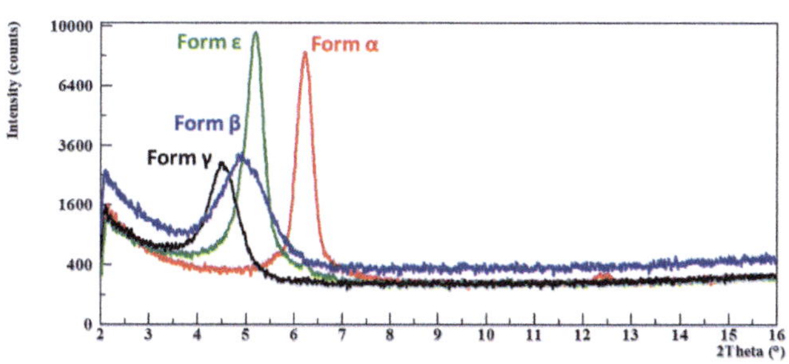

Fig. 7 Thin films of the different phases, identified by the characteristic peak of each phase, Form α in red, Form β in blue, Form γ in black and Form ε in green.

a quite different conformation in respect to Form α, although the carbon atoms do not adopt a full *anti* conformation, the chain is stretched, increasing the long axis of the molecule (see Fig. 8a and b). In both structures, the molecules form π–π interactions, with mainly columnar packing, the interplanar distance of Form γ is slightly longer than in Form α (3.5 Å instead of 3.4 Å), which allows the nearby columns to get tighter (see Fig. 8c and d). The different arrangement of the aromatic core can be better evaluated with the use of the stacking vector (SV) and the angles χ and ψ as described by Milita *et al.*[27] The SVs are similar in both Form α and Form γ (4.90 Å and 4.86 Å respectively), but they have quite a different orientation, hence different χ and ψ values (76.0° and 46.4° in Form α and 64.7° and 58.4° in Form γ). The clustering of the different χ/ψ values for the NDI-C*n* shows two different stacking modes, one for the short alkyl chains ($n < 7$) and the other for the longer ones. Interestingly the values obtained for Form γ are closer to the group of longer chains, where the energy contribution of the chains is predominant.

The thin films of Form α and Form γ are highly orientated and in both cases the highest peak is due to the crystallographic plane (001), which means that the planes (00*l*) are parallel to the substrate. Fig. 9 shows how the molecules of Form

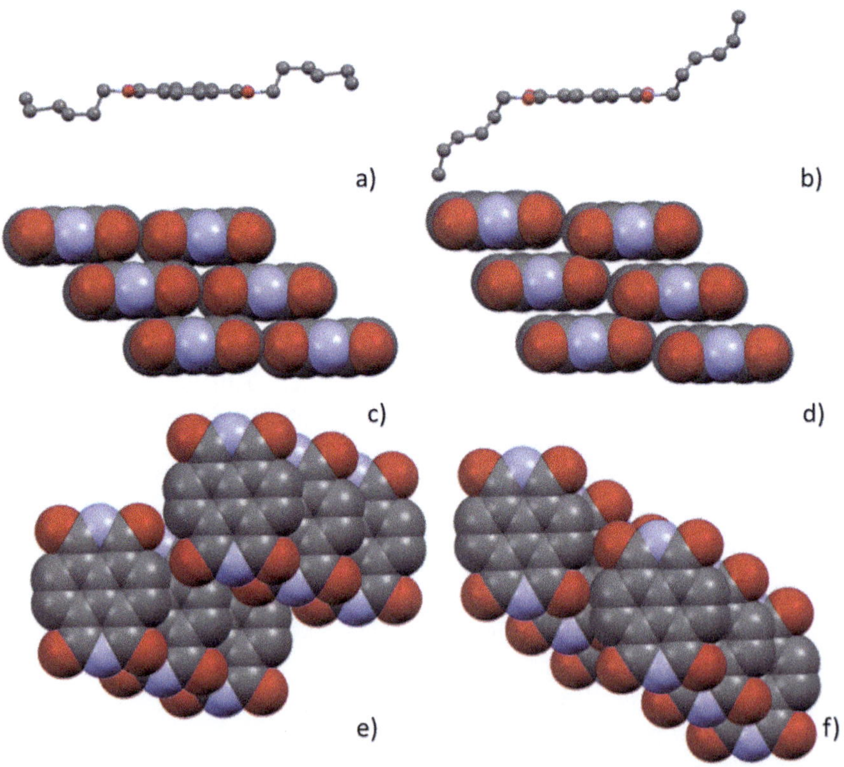

Fig. 8 Comparison of the structure of Form α (left) and Form γ (right): (a and b) different conformation of the chains in the **NDI-C6** structures (hydrogen omitted for clarity); (c and d) columnar arrangement of the aromatic core, view along the long direction of the aromatic core; (e and f) columnar arrangement of the aromatic core, view perpendicular to the aromatic core. In c–f hydrogen atoms and alkyl chains are omitted for clarity.

α and Form γ are organized on the substrate, and since in Form γ the **NDI-C6** molecules are less tilted in respect to the substrate than in Form α, the $d_{0\ 0\ 1}$ is longer and the first peak is observed at a lower 2θ angle than the peak of Form α.

Thermal expansion calculation

As mentioned previously, **NDI-C6** presents a big thermal expansion that could be noticed by the peak shift in the VT-XRPD. To have better insight into the thermal expansion mechanism, we used PASCal to analyse Form α and Form γ. The principal axis coefficients and the orientation of the tensors are summarized in Table 3.[35] It is worth noting that the tensors are a set of orthogonal axes (the principal axes) along which the material responds in a purely linear fashion, and in the case of the triclinic system they are not related to the unit-cell axes.

Looking at the results obtained, we can immediately see that both polymorphs have a negative thermal expansion (NTE) along one principal axis, this behaviour is uncommon but it has been described in different materials before, for example in BHH-BTBT.[37,38]

In Form α, the highest positive thermal expansion positive thermal expansion (PTE) of 308 MK^{-1} can be considered a huge expansion[39] and it affects the distances between the layers (see Fig. 10a), which involves mainly the alkyl chains, while the NTE occurs inside the layer and affects mainly the distances between the side-by-side aromatic core, these two effects promote the conversion into Form γ, where the alkyl chains are more stretched, and the columns are closer. As observed before, increasing the temperature allows to reach this new family of phases where the arrangement of the aromatic core is closer to the one observed for the NDIs with long chains and the molecules are less tilted with respect to the (0 0 1) plane, which can be detected by the first peak as low 2θ values.

It is worth noting that in Form γ the thermal expansion is highly anisotropic, with a colossal NTE ($X1 = -292\ MK^{-1}$) and a massive PTE ($X3 = 552\ MK^{-1}$) using the notation proposed by Goodwin *et al.*[39] and Henke *et al.*[40] In Fig. 10b the $X1$ and $X3$ are reported, and considering that the component of $X1$, $X2$ and $X3$ along the c

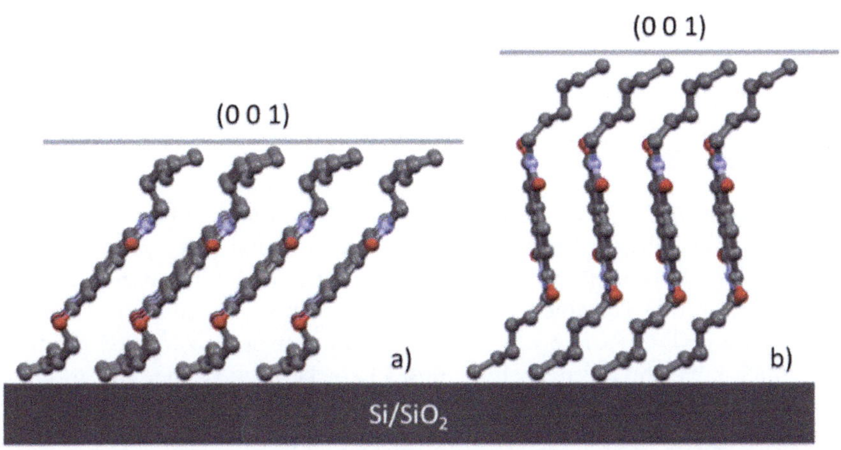

Fig. 9 Molecular arrangement of thin films oriented in the Si/SiO_2 substrates of (a) Form α and (b) Form γ.

Table 3 Values (α_X) of the principal axis of thermal expansion ($X1$, $X2$ and $X3$) and their orientation in regards to the cell axis, *a*, *b* and *c* for both Form α and Form γ

	Principal axis	α_X (MK^{-1})	*a*	*b*	*c*
Form α	*X*1	−87	−0.3966	−0.9180	0.0040
	*X*2	133	0.9847	−0.0715	0.1590
	*X*3	308	−0.6626	0.2685	0.6992
Form γ	*X*1	−292	0.1631	0.9478	−0.2740
	*X*2	193	0.9613	−0.2584	−0.0957
	*X*3	552	−0.7771	−0.5966	−0.2005

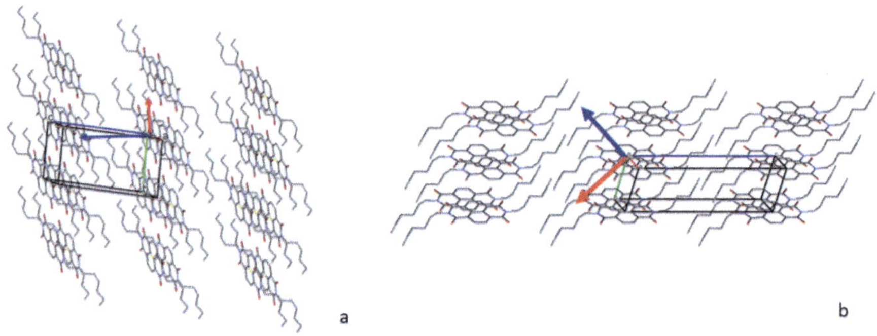

Fig. 10 Packing of the structures with *X*2 perpendicular to the figure, *X*1 in red and *X*3 in blue. The arrows indicate the direction of the tensors: (a) Form α and (b) Form γ.

axis are limited, it suggests that the main rearrangement occurs in the intralayer and the tilt of the molecules. This behaviour can explain the different thermal pathways, when the transition Form γ → Form α occurs at high temperature, it is possible to rearrange the chains and reach the stable Form α. On the other hand, if Form γ reaches a low temperature, the contraction that occurs during the cooling, which involves mainly the layer, compresses the alkyl chains preventing their folding, and instead of Form α, Form β is obtained, which is characterized by a longer axis.

Conclusion

We have studied in detail the thermal behaviour of **NDI-C6**, already known for the high number of polymorphs that can be detected by varying temperature. The careful study of the transition allowed us to determine the stability of the different polymorphs: Form α is stable up to 175 °C, then Form γ is stable until 178 °C when it converts into Form δ which is stable until it melts at 207 °C. Form β is always metastable and its energy is supposed to be between Form α and Form γ. This study revealed that depending on the thermal history, it is possible to achieve different phases at RT: Form α or Form β. The former is obtained by cooling of the melt while the latter is obtained by annealing until below the melting point (206 °C), this phase is metastable at RT, and it reconverts into Form α over time.

We were able to determine the crystal structure of Form γ, where the **NDI-C6** presents almost stretched alkyl chains, and an overall packing closer to the structure of NDI-C_n with $n > 7$. Form α and Form γ present exceptionally high PTE and NTE values, and the analysis of the principal coefficients of thermal expansion gave us insight into the solid–solid transition. In Form α, the highest PTE influences mainly the interlayer distance and upon heating, promotes the stretching of the alkyl chains and thus the conversions to Form γ. On the other hand, colossal NTE and massive PTE calculated on Form γ, are predominant in the *ab* plane, hence in the arrangement of the aromatic cores. Upon cooling, if the transition to Form α does not occur at high temperature, Form γ reaches low temperatures, and the contraction that occurs during the cooling prevents the folding of the alkyl chains and instead of Form α, Form β is obtained, which is characterized by a longer axis.

By different combinations of solvent, speed of deposition and thermal annealing, it was possible to obtain pure phases on the thin film. DCM and CHF promoted the formation of a mixture of Form α with different phases, while, by deposition of toluene solutions, a new polymorph labelled Form ε was observed. The new form is metastable and converts into Form β over time. Pure Form α and γ were obtained by thermal annealing at 75 °C for 1 h of the sample obtained with DCM and TOL solution, respectively. Form ε has been observed only on Si/SiO_2 substrates and further investigations are planned to confirm that it is a substrate-induced polymorph.

Author contributions

Inês de Oliveira Martins: investigation, experimental and computational analysis and writing of the manuscript. Francesco Marin: experimental and computational analysis. Enrico Modena: conceptualization and supervision. Lucia Maini: supervision, manuscript review and editing.

Conflicts of interest

The authors declare no conflicts of interest

Acknowledgements

We acknowledge the Paul Scherrer Institut, Villigen, Switzerland for provision of synchrotron radiation beamtime at beam-line MS-X04SA of the SLS (ID proposal 20201790) and would like to thank Nicola P. M. Casati for assistance. This project has received funding from the European Union's Horizon 2020 research and innovation programme under the Marie Skłodowska-Curie grant agreement no. 811284 (UHMob).

References

1 Y. Diao, L. Shaw, Z. Bao and S. C. B. Mannsfeld, Morphology Control Strategies for Solution-Processed Organic Semiconductor Thin Films, *Energy Environ. Sci.*, 2014, 7, 2145–2159, DOI: 10.1039/c4ee00688g.

2 J. Mei, Y. Diao, A. L. Appleton, L. Fang and Z. Bao, Integrated Materials Design of Organic Semiconductors for Field-Effect Transistors, *J. Am. Chem. Soc.*, 2013, **135**(18), 6724–6746, DOI: 10.1021/ja400881n.

3 X. Gao and Z. Zhao, High Mobility Organic Semiconductors for Field-Effect Transistors, *Sci. China: Chem.*, 2015, **58**(6), 947–968, DOI: 10.1007/s11426-015-5399-5.

4 C. Zhao, A. Li, X. Chen, M. U. Ali and H. Meng, Hysteresis Effect in Organic Thin Film Transistors Based on Naphthalene Tetracarboxylic Diimide Derivatives, *Appl. Phys. Lett.*, 2021, **118**, 193302, DOI: 10.1063/5.0045183.

5 K. Fukuda, Y. Takeda, M. Mizukami, D. Kumaki and S. Tokito, Fully Solution-Processed Flexible Organic Thin Film Transistor Arrays with High Mobility and Exceptional Uniformity, *Sci. Rep.*, 2014, **4**, 3947, DOI: 10.1038/srep03947.

6 S. Riera-Galindo, A. Tamayo and M. Mas-Torrent, Role of Polymorphism and Thin-Film Morphology in Organic Semiconductors Processed by Solution Shearing, *ACS Omega*, 2018, **3**, 2329–2339, DOI: 10.1021/acsomega.8b00043.

7 H. Chung and Y. Diao, Polymorphism as an Emerging Design Strategy for High Performance Organic Electronics, *J. Mater. Chem. C*, 2016, **4**(18), 3915–3933, DOI: 10.1039/c5tc04390e.

8 X. Gao and Y. Hu, Development of N-Type Organic Semiconductors for Thin Film Transistors : A Viewpoint of Molecular Design, *J. Mater. Chem. C*, 2014, **17**, 3099–3117, DOI: 10.1039/c3tc32046d.

9 E. Benvenuti, D. Gentili, F. Chiarella, A. Portone, M. Barra, M. Cecchini, C. Cappuccino, M. Zambianchi, S. G. Lopez, T. Salzillo, E. Venuti, A. Cassinese, D. Pisignano, L. Persano, M. Cavallini, L. Maini, M. Melucci, M. Muccini and S. Toffanin, Tuning Polymorphism in 2,3-Thienoimide Capped Oligothiophene Based Field-Effect Transistors by Implementing Vacuum and Solution Deposition Methods, *J. Mater. Chem. C*, 2018, **6**(21), 5601–5608, DOI: 10.1039/c8tc00544c.

10 C. Cappuccino, S. Canola, G. Montanari, S. G. Lopez, S. Toffanin, M. Melucci, F. Negri and L. Maini, One Molecule, Four Colors: Discovering the Polymorphs of a Thieno(Bis)Imide Oligomer, *Cryst. Growth Des.*, 2019, **19**, 2594–2603, DOI: 10.1021/acs.cgd.8b01712.

11 C. Cappuccino, L. Catalano, F. Marin, G. Dushaq, G. Raj, M. Rasras, R. Rezgui, M. Zambianchi, M. Melucci, P. Naumov and L. Maini, Structure–Mechanical Relationships in Polymorphs of an Organic Semiconductor (C4-NT3N), *Cryst. Growth Des.*, 2020, **20**, 884–891, DOI: 10.1021/acs.cgd.9b01281.

12 M. Mas-Torrent and C. Rovira, Role of Molecular Order and Solid-State Structure in Organic Field-Effect Transistors, *Chem. Rev.*, 2011, **111**(8), 4833–4856, DOI: 10.1021/cr100142w.

13 Y. Zhen, H. Dong, L. Jiang and W. Hu, Tailoring Crystal Polymorphs of Organic Semiconductors towards High-Performance Field-Effect Transistors, *Chin. Chem. Lett.*, 2016, **27**(8), 1330–1338, DOI: 10.1016/j.cclet.2016.06.023.

14 D. Liu, X. Xiao, Z. He, J. Tan, L. Wang, B. Shan and Q. Miao, Control of Polymorphism in Solution-Processed Organic Thin Film Transistors by Self-Assembled Monolayers, *Sci. China: Chem.*, 2020, **63**(9), 1221–1229, DOI: 10.1007/s11426-020-9793-2.

15 T. Salzillo, A. Campos, A. Babuji, R. Santiago, S. T. Bromley, C. Ocal, E. Barrena, R. Jouclas, C. Ruzie, G. Schweicher, Y. H. Geerts and M. Mas-torrent, Enhancing Long-Term Device Stability Using Thin Film Blends of

Small Molecule Semiconductors and Insulating Polymers to Trap Surface-Induced Polymorphs, *Adv. Funct. Mater.*, 2020, 2006115, DOI: 10.1002/adfm.202006115.

16 A. O. F. Jones, B. Chattopadhyay, Y. H. Geerts and R. Resel, Substrate-Induced and Thin-Film Phases : Polymorphism of Organic Materials on Surfaces, *Adv. Funct. Mater.*, 2016, **26**, 2233–2255, DOI: 10.1002/adfm.201503169.

17 L. Pithan, D. Nabok, C. Cocchi, P. Beyer, G. Duva, J. Simbrunner, C. Nicklin, P. Schäfer, C. Draxl, F. Schreiber, S. Kowarik, L. Pithan, D. Nabok, C. Cocchi, P. Beyer, G. Duva, F. Schreiber and S. Kowarik, Molecular Structure of the Substrate-Induced Thin-Film Phase of Tetracene, *J. Chem. Phys.*, 2018, **149**, 144701, DOI: 10.1063/1.5043379.

18 D. Reischl, C. Ro, P. Christian, E. Roblegg, H. M. A. Ehmann, I. Salzmann and O. Werzer, Surface-Induced Polymorphism as a Tool for Enhanced Dissolution: The Example of Phenytoin, *Cryst. Growth Des.*, 2015, **15**, 4687–4693, DOI: 10.1021/acs.cgd.5b01002.

19 A. O. F. Jones, Y. H. Geerts, J. Karpinska, A. R. Kennedy, R. Resel, C. Röthel, C. Ruzié, O. Werzer and M. Sferrazza, Substrate-Induced Phase of a [1] Benzothieno[3,2-*b*]Benzothiophene Derivative and Phase Evolution by Aging and Solvent Vapor Annealing, *ACS Appl. Mater. Interfaces*, 2015, **7**, 1868–1873, DOI: 10.1021/am5075908.

20 O. D. Jurchescu, Conductivity Measurements of Organic Materials Using Field-Effect Transistors (FETs) and Space-Charge-Limited Current (SCLC) Technique, in *Handbook of Organic Materials for Optical and (Opto)electronic Devices*, Woodhead Publishing Limited, 2013, pp. 377–383, DOI: 10.1533/9780857098764.2.377.

21 S. V. Bhosale, M. Al Kobaisi, R. W. Jadhav, P. P. Morajkar, L. A. Jones and S. George, Naphthalene Diimides : Perspectives and Promise, *Chem. Soc. Rev.*, 2021, **50**, 9845–9998, DOI: 10.1039/d0cs00239a.

22 Z. Ma, H. Geng, D. Wang and Z. Shuai, Influence of Alkyl Side-Chain Length on the Carrier Mobility in Organic Semiconductors : Herringbone *vs.* Pi–Pi Stacking, *J. Mater. Chem. C*, 2016, **4**(20), 22–24, DOI: 10.1039/c6tc00755d.

23 X. Guo, F. S. Kim, M. J. Seger, S. A. Jenekhe and M. D. Watson, Naphthalene Diimide-Based Polymer Semiconductors: Synthesis, Structure–Property Correlations, and n-Channel and Ambipolar Field-Effect Transistors, *Chem. Mater.*, 2012, **24**, 1434–1442, DOI: 10.1021/cm2034273.

24 D. Shukla, S. F. Nelson, D. C. Freeman, M. Rajeswaran, W. G. Ahearn, D. M. Meyer and J. T. Carey, Thin-Film Morphology Control in Naphthalene-Diimide-Based Semiconductors : High Mobility n-Type Semiconductor for Organic Thin-Film Transistors, *Chem. Mater.*, 2008, **20**(24), 7486–7491.

25 X. Chen, Y. He, M. U. Ali, Y. He, Y. Zhu, A. Li, C. Zhao, I. F. Perepichka and H. Meng, Isothianaphthene Diimide : An Air-Stable n-Type Semiconductor, *Sci. China: Chem.*, 2019, **62**, 1360–1364.

26 A. Welford, S. Maniam, E. Gann, X. Jiao, L. Thomsen, S. J. Langford and C. R. Mcneill, Influence of Alkyl Side-Chain Type and Length on the Thin Film Microstructure and OFET Performance of Naphthalene Diimide-Based Organic Semiconductors, *Org. Electron.*, 2019, **75**, 105378, DOI: 10.1016/j.orgel.2019.105378.

27 S. Milita, F. Liscio, L. Cowen, M. Cavallini, B. A. Drain, T. Degoussée, S. Luong, O. Fenwick, A. Guagliardi, B. C. Schroeder and N. Masciocchi, Polymorphism

in *N,N'*-Dialkyl-Naphthalene Diimides, *J. Mater. Chem. C*, 2020, **8**, 3097–3112, DOI: 10.1039/c9tc06967d.

28 D. Shukla, S. F. Nelson, D. C. Freeman, M. Rajeswaran, W. G. Ahearn, D. M. Meyer and J. T. Carey, Thin-Film Morphology Control in Naphthalene-Diimide-Based Semiconductors : High Mobility n-Type Semiconductor for Organic Thin-Film Transistors, *Chem. Mater.*, 2008, **20**(4), 7486–7491.

29 S. Ali, M. A. Jameel, A. Gupta, S. J. Langford and M. Shafiei, Capacitive Humidity Sensing Performance of Naphthalene Diimide Derivatives at Ambient Temperature, *Synth. Met.*, 2021, **275**, 116739, DOI: 10.1016/j.synthmet.2021.116739.

30 S. Cho, K. Lee, J. Yuen, G. Wang, D. Moses, A. J. Heeger, M. Surin and R. Lazzaroni, Thermal Annealing-Induced Enhancement of the Field-Effect Mobility of Regioregular Poly(3-Hexylthiophene) Films, *J. Appl. Phys.*, 2006, 114503, DOI: 10.1063/1.2400796.

31 Y. Yuan, G. Giri, A. L. Ayzner, A. P. Zoombelt, S. C. B. Mannsfeld, J. Chen, D. Nordlund, M. F. Toney, J. Huang and Z. Bao, Ultra-High Mobility Transparent Organic Thin Film Transistors Grown by an off-Centre Spin-Coating Method, *Nat. Commun.*, 2014, **5**, 3005, DOI: 10.1038/ncomms4005.

32 P. R. Willmott, D. Meister, S. J. Leake, M. Lange, A. Bergamaschi, M. Böge, M. Calvi, C. Cancellieri, N. Casati, A. Cervellino, Q. Chen, C. David, U. Flechsig, F. Gozzo, B. Henrich, S. Jäggi-Spielmann, B. Jakob, I. Kalichava, P. Karvinen, J. Krempasky, A. Lüdeke, R. Lüscher, S. Maag, C. Quitmann, M. L. Reinle-Schmitt, T. Schmidt, B. Schmitt, A. Streun, I. Vartiainen, M. Vitins, X. Wang and R. Wullschleger, The Materials Science Beamline Upgrade at the Swiss Light Source, *J. Synchrotron Radiat.*, 2013, **20**(5), 667–682, DOI: 10.1107/S0909049513018475.

33 A. A. Coelho, TOPAS and TOPAS-Academic: An Optimization Program Integrating Computer Algebra and Crystallographic Objects Written in C++: An, *J. Appl. Crystallogr.*, 2018, **51**(1), 210–218, DOI: 10.1107/S1600576718000183.

34 J. S. O. Evans, Advanced Input Files & Parametric Quantitative Analysis Using Topas, *Mater. Sci. Forum*, 2010, **651**, 1–9, DOI: 10.4028/www.scientific.net/MSF.651.1.

35 M. J. Cliffe and A. L. Goodwin, PASCal: A Principal Axis Strain Calculator for Thermal Expansion and Compressibility Determination, *J. Appl. Crystallogr.*, 2012, **45**(6), 1321–1329, DOI: 10.1107/S0021889812043026.

36 P. Zaumseil, High-Resolution Characterization of the Forbidden Si 200 and Si 222 Reflections, *J. Appl. Crystallogr.*, 2015, **48**, 528–532, DOI: 10.1107/S1600576715004732.

37 A. van Der Lee, G. H. Roche, G. Wantz, J. J. E. Moreau, O. J. Dautel and J. S. Filhol, Experimental and Theoretical Evidence of a Supercritical-like Transition in an Organic Semiconductor Presenting Colossal Uniaxial Negative Thermal Expansion, *Chem. Sci.*, 2018, **9**(16), 3948–3956, DOI: 10.1039/c8sc00159f.

38 Z. Liu, Q. Gao, J. Chen, J. Deng, K. Lin and X. Xing, Negative Thermal Expansion in Molecular Materials, *Chem. Commun.*, 2018, **54**, 5164–5176, DOI: 10.1039/c8cc01153b.

39 A. L. Goodwin, M. Calleja, M. J. Conterio, M. T. Dove, J. S. O. Evans, D. A. Keen, L. Peters and M. G. Tucker, Colossal Positive and Negative Thermal Expansion

in the Framework Material $Ag_3[Co(CN)_6]$, *Science*, 2008, **319**(5864), 794–797, DOI: 10.1126/science.1151442.

40 S. Henke, A. Schneemann and R. A. Fischer, Massive Anisotropic Thermal Expansion and Thermo-Responsive Breathing in Metal–Organic Frameworks Modulated by Linker Functionalization, *Adv. Funct. Mater.*, 2013, **23**(48), 5990–5996, DOI: 10.1002/adfm.201301256.

Faraday Discussions

ROYAL SOCIETY OF CHEMISTRY

DISCUSSIONS

Controlling polymorphism: general discussion

Ruel Cedeno, Aurora Cruz-Cabeza, Rik Drummond-Brydson, Marta K. Dudek, Katharina Edkins, Kristen Fichthorn, Aaron R. Finney, Ian Ford, Johanna Marie Galloway, Romain Grossier, Joonsoo Kim, Christian Kuttner, Lucia Maini, Fiona Meldrum, Mark Miller, Peter Morris, Sten O. Nilsson Lill, Boaz Pokroy, Sarah Price, Ivo B. Rietveld, Jeffrey Rimer, Kevin Roberts, Jutta Rogal, Matteo Salvalaglio, Jan Sefcik, Wenhao Sun, Stéphane Veesler, Peter Vekilov, Helen Wheatcroft, Michael Whittaker and Ran Zhao

DOI: 10.1039/d2fd90023h

Kristen Fichthorn opened the discussion of the paper by Jutta Rogal: I find the shapes of the nuclei that form in your studies interesting, as there is evidence that fivefold twinned nuclei and nuclei with stacking faults can occur[1,2] in nucleation of fcc systems.

1 J. C. E, L. Wang, Y. Cai, H. A. Wu and S. N. Luo, *J. Chem. Phys.*, 2015, **142**, 064704.
2 N. C. Karayiannis, R. Malshe, M. Kroger, J. J. de Pablo and M. Laso, *Soft Matter*, 2012, **8**, 844–858.

Wenhao Sun asked: I had a question about your DP parameter. It reminds me very much of the fluctuation–dissipation theorem, where a local parameter varies in a sort of normal distribution.

In the fluctuation dissipation theorem, we learn that the full-width-half-max of the fluctuations in energy corresponds to the heat capacity of the material. Since your displacement term in DP is sort of a measure of the kinetic energy, do you anticipate that if you change your system to another system that is not Ni, for example, Pb, which has a much higher heat capacity, that would show different phenomenology when it comes to nucleation from a supercooled solution?

Jutta Rogal answered: The DP is related to the dynamical heterogeneity in the liquid. It is not necessarily related to the kinetic energy (as also discussed in other questions). The heat capacity might, therefore, not be the decisive property. Other factors, such as, for example, the competition between an isotropic attraction and an anisotropic repulsion in the interaction between atoms, can greatly impact the glass-forming or crystallisation ability of a material[1] leading to different phenomenologies in the nucleation mechanism.

1 J. Russo, F. Romano and H. Tanaka, *Phys. Rev. X*, 2018, **8**, 021040.

Sarah Price said: Please describe the physical basis of the embedded atom model potential that you used for Ni. To what extent is this expected to be realistic in giving the anisotropy of the interactions between the Ni atoms in condensed phases? There is a huge difference in the anisotropy of the intermolecular interactions in water and in a metal like Ni, so I wondered whether the simulation model would be capturing the anisotropy that was present for Ni.

Jutta Rogal responded: The embedded atom method (EAM) potentials were developed to describe systems with metallic bonds where the energy is viewed as the energy of an atom embedded in the electron density of the other atoms (see ref. 1 for a review). The EAM potential that we have used in our study for Ni (https://doi.org/10.1039/d1fd00099c) has been fitted to reproduce experimentally determined properties of the solid phase (such as *e.g.*, equilibrium lattice constants, elastic constants, vacancy formation energies)[2] and it also shows very good agreement with properties of the liquid phase (structure factor, melting temperature, latent heat, volume change upon melting).[3,4] Furthermore, for homogeneous nucleation in Ni, the nucleation barriers have been shown to be in excellent agreement with experimental measurements.[5] Based on this we are confident that the employed EAM potential performs reasonably well in capturing the interactions in Ni.

1 M. S. Daw, S. M. Foiles and M. I. Baskes, *Mater. Sci. Rep.*, 1993, **9**, 251.
2 S. M. Foiles, M. I. Baskes and M. S. Daw, *Phys. Rev. B: Condens. Matter Mater. Phys.*, 1986, **33**, 7983.
3 S. M. Foiles, *Phys. Rev. B: Condens. Matter Mater. Phys.*, 1985, **32**, 3409.
4 D. Y. Sun, M. Asta and J. J. Hoyt, *Phys. Rev. B: Condens. Matter Mater. Phys.*, 2004, **69**, 024108.
5 J. Bokeloh, R. E. Rozas, J. Horbach and G. Wilde, *Phys. Rev. Lett.*, 2011, **107**, 145701.

Christian Kuttner asked: Do the size and number of the pre-critical clusters play an important role in the successfulness of nucleation and the dynamic properties? Mobility should scale with the cluster size, shouldn't it?

Jutta Rogal replied: As the crystalline cluster grows, the dynamical properties change, of course, and also the likelihood of a successful nucleation event changes. For our analysis we have chosen a size of solid clusters with $n_s = 50$ since up to this size the clusters are predominantly composed of pre-structured liquid, that is before the crystalline phase emerges. For these pre-structured regions we found that the distribution of dynamical propensity (DP) largely differs for clusters that continue to grow or shrink. Clusters that shrink have a distribution of DP values similar to the liquid while their structural environment is clearly distinct from the liquid. Clusters that continue to grow and from which the crystalline phase emerges show a clear shift to lower DP values.

We have also analysed the DP values along the transition path sampling trajectories of successful nucleation events. As the solid cluster grows, it is always embedded in a region of low mobility that, within a spherical approximation, has a radius that is about 25% larger than the radius of the largest solid cluster.

The change in mobility is therefore key in the initial stages of nucleation and continues to play a role during the growth of the solid cluster.

Mark Miller enquired: Is it possible to say what triggers a localised change in dynamic propensity that is large enough to lead to a crystallisation event? Is it simply an "equilibrium" dynamic fluctuation characteristic of the conditions (accepting that the parent phase is metastable)?

Rik Drummond-Brydson also asked: For the regions of low mobility, if there are no density differences then what causes these regions – is it just a fortuitous fluctuation of random walks?

Jutta Rogal responded: For homogeneous nucleation in this system where the supercooled liquid itself exhibits only little dynamical heterogeneity, it is not entirely clear what leads to the formation of low-mobility regions. The dynamical heterogeneity does depend on the supercooling and thus impacts the distribution of dynamical propensity. It is thus conceivable that these regions emerge from fluctuations depending on the environmental conditions. It also depends on the specific system, as supercooled liquids of, for example, water[1] or GeTe[2] already show significant dynamical heterogeneity.

Considering heterogeneous nucleation it would interesting to investigate how the dynamical properties of the liquid are modified in the presence of nucleating agents and how this impacts the nucleation ability.

1 M. Fitzner, G. C. Sosso, S. J. Cox and A. Michaelides, *Proc. Natl. Acad. Sci. U. S. A.*, 2019, **116**, 2009.
2 G. C. Sosso, J. Colombo, J. Behler, E. Del Gado and M. Bernasconi, *J. Phys. Chem. B*, 2014, **118**, 13621.

Ruel Cedeno asked: Can we consider this formation of low mobility regions as local density fluctuations (akin to classical nucleation theory)?

Jutta Rogal responded: We did not observe any changes in the density that were correlated with the changes in dynamical propensity in this system.

Mark Miller commented: Do you think dynamic propensity would be an important variable in even simpler, archetypal models than nickel, such as the Lennard-Jones fluid or even hard spheres (where there is no attraction between particles)?

Jutta Rogal responded: This might indeed be the case. Elemental Ni is already a rather simple system where the supercooled liquid exhibits only mild dynamical heterogeneity. Still, low mobility regions emerge that precede structural pre-ordering in the liquid. This might also occur in Lennard-Jones fluids or hard spheres.

Aaron R. Finney asked: How (computationally) expensive were the calculations of the dynamic propensity given that, in your paper, you indicated that 200 configurations were selected and each of these required 100 simulations? It seems like this property could be very valuable to interpret the results from nucleation simulations, in general.

Jutta Rogal responded: Apart from the system size and the interaction potential, the computational cost also depends on the time of maximum heterogeneity t_0. In Ni at 20% undercooling the time of maximum heterogeneity is rather short with $t_0 = 2.64$ ps (which determines the simulation length of each trajectory). Performing 100 simulations for each of the 200 configurations corresponds to an accumulated time of 53 ns which is entirely feasible.

For the Ni system with 8788 atoms using an EAM potential[1] and the software package Lammps,[2] one simulation took about 16 s on a single Intel(R) Xeon(R) CPU E5-2695 v3 @ 2.30GHz (about 27 min for 100 runs).

We also computed the dynamical propensity for every slice along our transition path sampling trajectories where each trajectory contains several hundred slices which was entirely possible considering the computational cost for each slice.

1 S. M. Foiles, M. I. Baskes and M. S. Daw, *Phys. Rev. B: Condens. Matter Mater. Phys.*, 1986, **33**, 7983.
2 S. Plimpton, *J. Comput. Phys.*, 1995, **117**, 1.

Joonsoo Kim asked: What is the qualitative difference between dynamical propensity (DP) and temperature? If measuring local temperature instead of DP, maybe particles with higher temperature have smaller undercooling, so the critical nucleus would be larger.

Jutta Rogal replied: The dynamical propensity is determined as an ensemble average over the so-called isoconfigurational ensemble. To compute this average for a given configuration, random Maxwell–Boltzmann distributed velocities are assigned to all particles and the squared displacement over the time of maximum heterogeneity is measured. This is normalised with respect to the mean squared displacement of all particles in the supercooled liquid. The DP therefore measures the relative mobility of each atom in a given configuration. The resulting DP value is not for any specific velocity of the particle but computed as an ensemble average over Maxwell–Boltzmann distributed velocities for each particle. The mean temperature of each particle in the isoconfigurational ensemble is the same and equal to the target temperature ($T = 1370$ K, 20% undercooling in our case).

Ian Ford enquired: Might there be a connection between dynamic propensity and local kinetic temperature? A low value of kinetic energy in the region might give rise to a spatial configuration that offered greater resistance to diffusive motion of the constituent particles in the subsequent simulations from which the dynamic propensity is derived. My thinking is that a patch of fluid that leads on towards a nucleation event is typically cooler than surrounding regions (in a kinetic temperature sense) because such a region is better able to accept the latent heat of crystallisation that it is about to receive. Without such a cool predisposition, the emerging nucleus would be more inclined to dissolve/melt due to its slightly elevated temperature.

Jutta Rogal responded: The dynamical propensity is determined as an ensemble average over the so-called isoconfigurational ensemble. To compute this average for a given configuration, random Boltzmann-distributed velocities are assigned to all particles and the squared displacement over the time of maximum heterogeneity is measured. This is normalised with respect to the mean squared displacement of all particles in the supercooled liquid. The DP therefore measures the relative mobility of each atom in a given configuration. The resulting DP value is not for any specific velocity of the particle but computed as an ensemble average over Boltzmann distributed velocities for each particle. The mean temperature of each particle in the isoconfigurational ensemble is the same and equal to the target temperature ($T = 1370$ K, 20% undercooling in our case).

We have also analysed the 'original' velocity distribution from the snapshots of the transition path sampling trajectories. Fig. 1(a) shows the velocity distribution of all atoms together with the corresponding Maxwell–Boltzmann distribution and Fig. 1(b) the velocity distribution of the most immobile particle (MI, DP < 0.728, bottom 5%) together with the Maxwell–Boltzmann distribution. The velocity distribution of the MI atoms is the same as for all other atoms and corresponds to the Maxwell–Boltzmann distribution at the simulation temperature of $T = 1370$ K. The MI particles do not show any difference in their velocity distribution.

In Fig. 2, a snapshot from a nucleation trajectory of the transition path ensemble is shown. In Fig. 2(a), only the MI atoms are shown forming a cluster in the supercooled liquid. In Fig. 2(b), only the slowest atoms ($v < 260$ m s^{-1}, bottom 5% of the speed distribution) are shown which are randomly distributed. There is no correlation between low dynamical propensity and low velocities.

Johanna Marie Galloway opened a general discussion of the paper by Michael Whittaker: In the droplets in the 'orchard', very high concentrations of calcium (1 M $CaCl_2$) were used in the original calcium carbonate paper.[1] Did you use these same high concentrations for your $(BaCa)_1CO_3$ system? As your barium system nucleates much faster in $\approx$1 hour rather than over many days, do you have plans

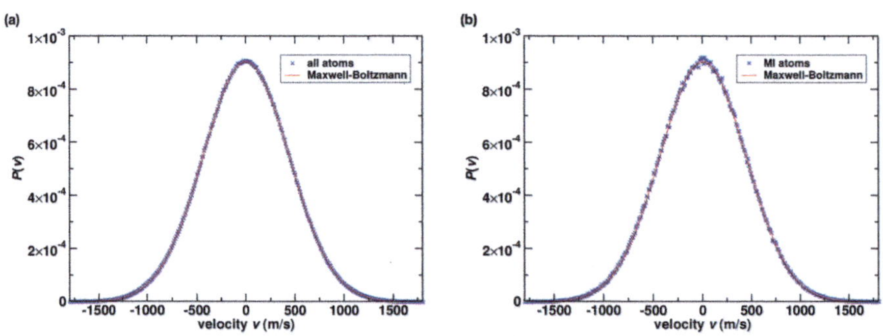

Fig. 1 (a) Velocity distribution of all Ni atoms along a nucleation trajectory from the transition path ensemble at $T = 1370$ K (blue crosses); (b) velocity distribution of the most immobile (MI, DP < 0.728) Ni atoms along a nucleation trajectory from the transition path ensemble at $T = 1370$ K (blue crosses); the red line is the corresponding Maxwell–Boltzmann distribution. The velocity distribution of the MI atoms does not deviate from the Maxwell–Boltzmann distribution.

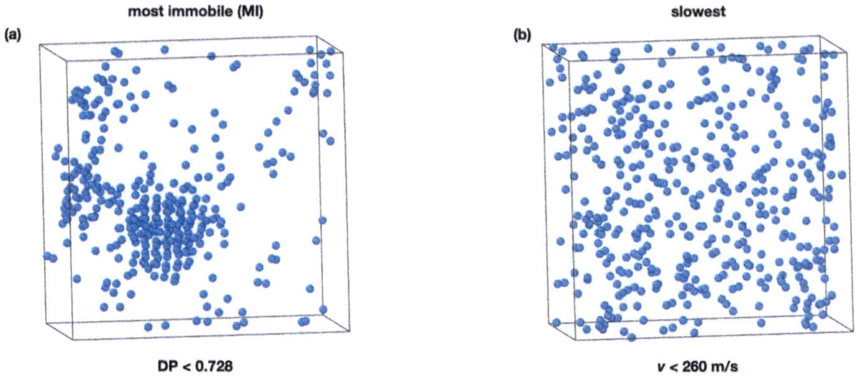

Fig. 2 Snapshot from a nucleation trajectory of the transition path ensemble at $T = 1370$ K; (a) only the most immobile (MI) atoms with DP < 0.728 (bottom 5% of the DP distribution) are shown; (b) only the slowest atoms with $\nu < 260$ m s^{-1} (bottom 5% of the speed distribution) are shown. There is no correlation between the MI and slowest atoms.

to investigate lower concentrations? These lower concentrations should provide accessible nucleation times for the barium calcium carbonate system, and it has been shown that lowering the concentration of reactants in calcium carbonate mineralisation in confinement leads to a polymorph change,[2] *i.e.* calcite at high concentrations, aragonite at low concentrations. You may also see polymorph control effects in your barium calcium carbonate system when crystallising in the confinement of the droplet system. You can see crystals from crystallisation in droplets at 100 mM $CaCl_2$ pretty easily, they just take a lot longer to turn up than your barium calcium carbonate crystals would.

1 J. Cavanaugh, M. L. Whittaker and D. Joester, *Chem. Sci.*, 2019, **10**, 5039–5043.
2 M. Zeng, Y.-Y. Kim, C. Andiux-Cantom and F. C. Meldrum, *Proc. Natl. Acad. Sci. U. S. A.*, 2018, **115**, 7670–7675.

Michael Whittaker answered: We used 1 M total salt concentration (Ca + Ba) in the experiments described here (https://doi.org/10.1039/d1fd00086a).

It is important to note that the driving force for precipitation from a supersaturated solution is established by the product of both the (fixed) initial cation activities within an emulsion drop as well as the carbonate activity, which is introduced *via* gas diffusion through the microfluidic chip. The supersaturation can therefore be varied by changing the carbonate concentration independently from the cation concentration, and reducing the carbonate concentration can produce arbitrarily low supersaturations even at relatively high cation concentration in the drops. The initial cation concentration is not necessarily a deterministic control on precipitation within the drops.

Furthermore, there is no reason to expect that the total initial cation concentration should control which polymorph forms from an amorphous precursor. Once the amorphous precursor forms, the solution within the drop re-equilibrates at the equilibrium solubility of the amorphous phase, and the significant (hours to days) lag between amorphous precipitation and crystallization makes it very likely that the cation concentration stabilizes at this lower value before crystal nucleation commences. Both calcite and aragonite remain supersaturated relative to the initially-formed amorphous phase (equally so within the precipitate and solution), and calcite always more so than aragonite, but at a much lower cation activity than the drop contained initially.

We believe that it is very likely to be the presence of a new precipitate interface, whose relative influence on the barrier to the nucleation is far stronger than that of the supersaturation, that is the primary control on crystallization rates to various polymorphs.

Kevin Roberts enquired: Are metastable polymorphs simply structures which are stable at the small domain sizes close to nucleation where in such cases the balance between surface and bulk free energies very much favours the surface component where the molecules lack the full intermolecular coordination inherent within the bulk structure. This is well known in organic systems. See *e.g.*, ref. 1 and also 2 and references therein for inorganic systems.

1 R. B. Hammond, K. Pencheva and K. J. Roberts, An examination of the polymorphic stability and molecular conformational flexibility as a function of crystal size associated with the nucleation and growth of benzophenone, *Faraday Discuss.*, 2007, **136**, 87–102.
2 R. B. Hammond, K. Pencheva and K. J. Roberts, Structural variability within, and polymorphic stability of, nano-crystalline molecular clusters of L-glutamic acid and D-mannitol, modelled with respect to their size, shape and 'crystallisability', *CrystEngComm*, 2012, **14**, 1069–1082.

Michael Whittaker answered: There are many reasons why a polymorph might be metastable. It is true that one mechanism for producing a metastable phase is for it to form under conditions in which it is thermodynamically stable and transition to distinct conditions for which it is not stable but is slow to transform. The interfacial energy may be an important contribution to the total free energy at small sizes, and growth may bring a formerly-stable phase into conditions for which it is metastable.

The case of balcite is different. Balcite forms from an even less stable amorphous precursor and is, in fact, unstable. Thus, balcite formation is possible because it lowers the system free energy relative to the amorphous precursor *via* crystallization. This does not necessarily have to do with the size of the particles involved in the transformation.

Wenhao Sun said: You should have a theorist calculate the stability of balcomite.

I think you have a very good point, there are probably many minerals that are indeed thermodynamically stable but require a difficult processing pathway for their formation. Balcomite is (possibly) a great example; of course we'd have to verify its stability *via* computation or calorimetry first. But in general, I agree that the pathway required, which involves the formation of a metastable, short-lived balcite phase that is then annealed at 600 °C into a balcomite phase, would not happen in Nature, but is a processing pathway to that phase.

This inspires me because much of our pursuit of functional materials is based on known minerals in nature. So if the diversity of known minerals could be expanded, that would also give a bigger play space for functional materials discovery and design.

Michael Whittaker responded: I share your interest in further investigating the stability and formation pathways of minerals. Determining the stability of balcite, including the configurational and vibrational entropies associated with the various types of disorder, would be an important next step in the barium–calcium–carbonate system.

Christian Kuttner opened a general discussion of the paper by Boaz Pokroy: You have explained very clearly that charge can control polymorphism – but what about weak charges? Where is the threshold when it comes to slightly non-neutral interfaces?

Boaz Pokroy responded: Excellent question which I honestly do not know the answer to.

Fiona Meldrum asked: You state that capillary pressure contributes to the formation of aragonite in the cylindrical pores. However, I contend that there is no capillary pressure as there is no liquid–vapour interface within the pore. In all experiments the membrane is immersed in bulk solution. Therefore, capillary pressure cannot contribute to the formation of aragonite in this system.

There are no bubbles in the pores as they are filled under vacuum. Further, the double diffusion set-up is such that no crystallisation would occur if there was a bubble.

Boaz Pokroy answered: I agree that in the absence of bubbles there is no capillary pressure. Yet, when 25 nm pores exist it is extremely difficult to extract small bubbles completely and it depends on the vacuum level and duration. There is no was to verify this. On the other hand, even in the absence of bubbles, curvature (nm-sized ACC particles) also induces pressure in the curved particles.

Stéphane Veesler asked: In your presentation you proposed 2 possible routes for phase transformation, my question concerns solid state transformation. Is there some experimental observation of solid–solid transformation of ACC or ikaite to a more stable phase of $CaCO_3$?

Boaz Pokroy replied: Indeed, there are several papers in which solid state transformation of ACC to crystalline $CacO_3$ have been observed. Usually this route is accompanied by dehydration.

Peter Morris said: As Meldrum *et al.*[1,2] observed, confinement can lead to the stabilisation of ACC, which is comparably short-lived under similar solution concentrations in bulk solution. Can you comment on the origins of this stabilising effect? You mentioned kinetic suppression in a particular case with negatively charged inner surfaces, is this applicable to other confinement conditions?

1 C. J. Stephens, S. F. Ladden, F. C. Meldrum and H. K. Christenson, *Adv. Funct. Mater.*, 2010, **20**, 2108–2115.
2 Y. Wang, M. Zeng, F. C. Meldrum and H. K. Christenson, *Cryst. Growth Des.*, 2017, **17**, 6787–6792.

Boaz Pokroy replied: I am not sure on the mechanism of stabilization but would assume it has to do with surface effects in confinement (ACC inner wall interface).

Romain Grossier asked: In Table 1 in your paper (https://doi.org/10.1039/d1fd00111f), when pore diameter is reported as a possible parameter on polymorphism selection through confinement, it's half the information: confinement is not "felt" with the same strength by the different possible polymorphs because their solubilities differ. In a fixed volume, or pore size, the most soluble phase will tend to be less "confined" (impact of system size on nucleation barrier height) than the less soluble phase.

This thermodynamical confinement, that shifts the nucleation barrier to higher values depending on both system size and solubility,[1] would need to be taken into account while comparing experiments, especially if not conducted in the exact same conditions (supersaturation relative to a fixed polymorph for example). Maybe pore size impact on polymorphism selection would then appear less confusing.

1 R. Grossier and S. Veesler, Reaching one single and stable critical cluster through finite-sized systems, *Cryst. Growth Des.*, 2009, **9**(4), 1917–1922.

Boaz Pokroy responded: I fully agree. The only purpose of presenting the table was to show that the published data on polymorph selection in $CaCO_3$ in confinement is not decisive and there are many open questions.

Michael Whittaker commented: I would like to ask a question about the effect of surface energy on nucleation *versus* growth. Nucleation often happens at defects, which necessarily have different charge and/or structure from average surface properties. Can you comment on how to distinguish between effects of nucleation on defects *versus* the effects of growth, and whether the average surface energy of an interface with a specific charge is indicative of conditions leading to preferential nucleation of a specific polymorph?

Boaz Pokroy responded: The critical step in polymorph selection is nucleation. This indeed usually takes place at defects or surfaces which are also defects by definition. As we showed by our calculations different charged surfaces can indeed influence the polymorph selection due to different crystal/surface inter-racial energies for different polymorphs.

Wenhao Sun said: Thank you Prof. Pokroy for the very interesting discussion on the role of charges and pores. My question is: if your argument is that surface charges is what affects nucleation (and presumably some form of heterogeneous nucleation), then why are pores needed at all? Also, most interfaces in solution are charged, in fact it is not so easy to get a charge neutral surface (you have to tune the pH to the point of zero charge (PZC)). So if this is the case, based on your arguments, shouldn't we very rarely see calcite? But this is not the case, since calcite is the equilibrium phase and we see it very frequently.

Boaz Pokroy answered: Firstly, I fully agree that it is almost impossible to really have a neutral surface.

Secondly, charged surfaces alone are not sufficient. The combination of charged surfaces and confinement in $CaCO_3$ is crucial. In confinement often the first phase to precipitate is ACC that immediately lowers the supersaturation. Then if the surface is negatively charged the prospects of aragonite nucleation is higher.

Fiona Meldrum commented: The effects of confinement on the calcium carbonate polymorph can be seen in relatively large pores. Given that the polymorph is defined at nucleation, then a nucleus forming on the pore wall would surely not feel the effect of the opposing wall? If surface chemistry is the dominating effect, then the same control over the polymorph should be observed in bulk solution. Your analysis has not considered any confinement effects. Notably, it is difficult to precipitate aragonite from solution under ambient conditions in the absence of magnesium ions. A large number of studies have addressed the influence of surfaces on calcium carbonate precipitation and they do not describe the effects on polymorph that you suggest.

Boaz Pokroy responded: You are correct, however, as stated if you do obtain ACC as the first phase in confinement then it is a different story than one small nucleolus in solution. Moreover, when you have pores of 25 nm diameter this might already be on the same order of magnitude of the stable nucleolus and then you cannot compare it to bulk.

Stéphane Veesler said: Concerning the discussion about the confinement effect on phase selection, I would add a comment.

Here, we are speaking of confinement by the volume, the first expected effect is 1 crystal per confined volume (can be a microfluidic droplet or inside a pore). What is known for a long time is because of this mononuclear nucleation, the nucleated crystal is a metastable phase because it being alone it cannot dissolve. Thus, small volumes are able to "freeze" the metastable phase.[1] This why it is interesting to use the confinement in polymorph or phase screening, because, contrary to in a large volume, the metastable phase will not be in contact with a more stable phase which will grow at its expense.

For the second effect of confinement discussed here, it is expected that by controlling the nature and the charge of the surface one can control the phase which is going to nucleate.

1 M. Ildefonso, E. Revalor, P. Punniam, J. B. Salmon, N. Candoni and S. Veesler, Nucleation and polymorphism explored *via* an easy-to-use microfluidic tool, *J. Cryst. Growth*, 2012, **342**(1), 9–12.

Sten O. Nilsson Lill opened a general discussion of the paper by Aurora Cruz-Cabeza: Considering your CSD search where you found hydrates are frequent for salts, *i.e.* part of the structure, what do you think the impact of explicit water solvation for your potential energy curves would be, or is implicit solvation a good model for your purpose?

Aurora Cruz-Cabeza responded: Solvation has a significant impact on the ionisation state of the acid–base pairs. We have used an implicit solvation model in conjunction with DFT because electronic methods are indeed needed to simulate the proton transfer. Explicit solvation would require simulations of large cells of solvent with molecular dynamics. These simulations would be very expensive with DFT methods so we feel that for the purpose of demonstrating the impact of solvent effects here, the implicit solvation models are a good and reasonably accurate approximation.

Marta K. Dudek asked: Because of the pronounced effect of solvent on the energy of the ionization *vs.* neutral state do you think it is possible to design specific crystallization conditions to obtain either salts or cocrystals (different polymorphs) for a given system? For example, in ref. 1 an ethionamide–salicylic acid system formed either cocrystals or salts depending on the conditions. Could we use calculations similar to yours to direct crystallization in one direction or another?

1. D. Bernasconi, S. Bordignon, F. Rossi, E. Priola, C. Nervi, R. Gobetto, D. Voinovich, D. Hasa, N. T. Duong, Y. Nishiyama and M. R. Chierotti, *Cryst. Growth Des.* 2020, **20**, 906–915.

Aurora Cruz-Cabeza responded: For the majority of cases, the protonation state of the final crystal form would ultimately be determined by the structure itself and not the crystallisation pathway. If the acid–base interaction remains the same, the conversion from a cocrystal to a salt through the shift of the proton would be very facile. It is the crystal structure, thus, that determines the protonation.

If polymorphs are possible for the specific system under study, like the ethionamide–salicyclic acid system you refer to, then the crystallisation conditions would direct the system towards one or another polymorph. It is the polymorph crystal structure which you select through the crystallisation conditions that would determine the ionisation state in the end.

Marta K. Dudek said: I was wondering whether you checked the quality of the crystal structures examined in the last part of the paper in terms of the method of localizing the position of a hydrogen atom between a HA–B pair? Do you have any estimation of how many of the analyzed crystals might actually be something in-between salts and cocrystals?

Aurora Cruz-Cabeza responded: Crystallographers determine the structures and submit them to the Cambridge Crystallographic Data Centre (CCDC). The CCDC stores them as described by the authors in the publication and thus they categorise them as ionised or not accordingly. We do our database work based on the identification of charges in the database. There may indeed be errors in structure determination and cases where X–H bonds are too long or too short. We cannot look at each individual structures but rather analyse the data of many (over 9 thousand) and plot the trends. The errors on hydrogen atom determinations will be there but hopefully few and thus just a bit of noise in the data. We are confident in the overall trends but notice that in some of the subsets the data numbers are small, so trends have more data noise.

Marta K. Dudek remarked: Maybe in the region of uncertain ΔpK_a values we should not differentiate between salts, cocrystals and anything in-between? Especially if we cannot experimentally establish the exact position of a proton.

Aurora Cruz-Cabeza answered: In the intermediate region of ΔpK_a there will be systems which are clearly a salt and systems which are clearly a cocrystal. There will also be other systems where the proton sits in the middle, shifts with temperature or it is hard to determine with a specific experimental technique, like in the work of Stevens cited in our paper (https://doi.org/10.1039/d1fd00081k). For such structures, indeed the nomenclature is also a grey area. Childs (cited in our paper) refers to these salt-cocrystal systems as a continuum.

Katharina Edkins commented: Since the pK_a is a thermodynamic characteristic, the crystallisation outcome (salt or co-crystal) predicted by the ΔpK_a rule will be the thermodynamically stable form. But there are cases where both materials exist and for example in the case of sulfamethazine–saccharine, the co-crystal form is metastable to the salt.[1] Can you please comment on the applicability or potential variation of the ΔpK_a rule to metastable co-crystals, which are still important to discover and investigate and may pose some advantages for the manufacturing process?

1 S. R. Perumalla, C. Wand, Y. Guo, L. Shi and C. C. Sun, Robust bulk preparation and characterization of sulfamethazine and saccharine salt and cocrystal polymorphs, *CrystEngComm*, 2019, **21**, 2089–2096.

Aurora Cruz-Cabeza responded: The pK_a rule cannot predict whether two components would crystallise together or not, but rather, if they were to crystallise together, what is the likelihood of them being a salt or a co-crystal. In the extreme regions, the ΔpK_a will tell you the answer with a $\Delta pK_a > 4$ almost always resulting in a salt and with a ΔpK_a.

Katharina Edkins stated: The pK_a of a compound is classically determined for aqueous solutions and can differ significantly in organic solvents. If the crystallisation could be conducted under exclusion of water and by selecting the correct solvent, would it be possible to enhance or supress co-crystal formation, even if only as metastable intermediates?

Aurora Cruz-Cabeza replied: I believe co-crystal and salt formation will ultimately be controlled by the resulting crystal structure. In the intermediate region of ΔpK_a, the same acid–base may lead to a salt or a co-crystal, as discussed in the previous question. Now, if we change solvent, the ΔpK_a rule still remains valid since it relates to the final outcome, but the solvent may indeed affect the outcome of crystallisation. Here, one would need to look at the ternary phase diagram and ensure that the work is being carried out in the region where the multicomponent crystal form is favoured (if such a region exists). The solvent/media indeed may impact initial aggregation states which then may impact the nucleation of a specific form. But once the form is formed and growing, the ionisation will be determined by the final crystal structure. If (as in many cases) this is the case, the acid is directly interacting with the base then the proton can easily move from the acid and base at any stage of the process and thus the final crystal structure will dictate the outcome whilst the solvent will dictate which structure forms.

Sarah Price remarked: The difference between a salt and a cocrystal is more than just an issue of proton location. When cocrystals first were being investigated as potentially advantageous ways of delivering drugs, there were a lot of discussions in the pharmaceutical industry about how the regulations on the manufacturing pharmaceutical products would be applied to cocrystals.[1] It now appears that locating the proton by computation is not an easy alternative to experiment. A recent paper, entitled "Pervasive delocalisation error causes spurious proton transfer in organic acid–base co-crystals"[2] shows the limitations of popular DFT functionals without sufficient exact exchange for studying this problem.

1 K. Izutsu, T. Koide, N. Takata, Y. Ikeda, M. Ono, M. Inoue, T. Fukami and E. Yonemochi, Characterization and quality control of pharmaceutical cocrystals, *Chem. Pharm. Bull.*, 2016, **64**(10), 1421–1430.
2 L. M. LeBlanc, S. G. Dale, C. R. Taylor, A. D. Becke, G. M. Day and E. R. Johnson, Pervasive delocalisation error causes spurious proton transfer in organic acid–base co-crystals, *Angew. Chem., Int. Ed.*, 2018, 57(45), 14906–14910.

Aurora Cruz-Cabeza replied: Prof. Price, thank you for sharing these references and for this important remark. I agree very much with your first point. Salts are known to be more hygroscopic than co-crystals and that, of course, would have implications to the way pharmaceuticals are manufactured and stored. I would refer to Christer Aakeröy's paper.[1]

With regards to the modelling, it is indeed very challenging. In these systems, just like in conformational polymorphism, modelling the balance of intramolecular and intermolecular terms correctly is very difficult. The LeBlanc *et al.* article you refer to is indeed revealing and highlights that in the intermediate zone of ΔpK_a, also the modelling may fail.

1 C. Aakeröy, M. E. Fasulo and J. Desper, Cocrystal or salt: does it really matter?, *Mol. Pharmaceutics*, 2007, **4**(3), 317–322.

Kevin Roberts commented: Multi-component organic systems are still at an early stage of discovery and the field is perhaps in its infancy when compared to inorganic systems where quaternary systems form a key component of semiconductor opto-electronic devices such as $Ga_xAs_yIn_{1-x}P_{1-y}$; this is similarly true in metallic alloys where solid solutions of mixed metal atoms are important. Due to their greater variety of bonding options and their anisotropic and "engineerable" molecular functionality, molecular solids have so much to offer in terms of structural diversity *e.g.* we now see quaternary systems in molecular crystals such as salt solvate co-crystal structures of drug molecules offering the tantalising thought of a future world of truly molecular formulations based upon mixtures of molecules rather than formulations based upon blends of powdered ingredients. Within such an innovation landscape there are the proponents of promoting the need definitions, rules and laws. My comment of "does it matter" regarding the definitions aspect reflects the vibrant field of solid-state organic chemistry coupled with the wistful feeling that, perhaps, we should let the innovation rip somewhat in the short/medium term and deal with definitions *etc.* when the field is more mature.

Helen Wheatcroft said: The thermodynamic cycle (Fig. 2) shows ionisation, solvation, crystallisation and lattice free energies. The aqueous ionisation free energy, and hence ΔpK_a, will be affected by a change in solvent system, while the solid and gaseous ionisation free energies and the lattice energies will be independent of the solvent. The solvation and crystallisation free energies will change to compensate for the change in the solution ionisation energy maintaining the thermodynamic cycle.

The pK_a of the acid and base will be shifted to different extents and often in opposite directions by the change in the solvation of the ions in a different solvent system. The shift in pK_a of acids in a solvent system is strongly influenced by the ability of the solvent to solvate the ions formed on dissociation and so this is similar for many Brønsted acids.[1] The shifts in the pK_a of bases in a given solvent system also tend to be consistent. Accordingly the ΔpK_a in this solvent will be different from that in water for a specific salt/co-crystal system and for each solvent system this will lead to a probability distribution for salt formation centred on a different ΔpK_a, but the change in solvent is not expected to affect the overall shape of the distribution.

However additional terms may be needed to account for processes such as ion pairing for solvent systems which solvate ions poorly, for example in solvents with dielectric constants <20.[2] It would be interesting to investigate whether this causes the ΔpK_a rule to break down for such solvents.

1 F. Rived, M. Roses and E. Bosch, *Anal. Chim. Acta*, 1998, **374**, 309–324.
2 J. R. Chipperfield, *Non-aqueous solvents*, OUP, 1999.

Aurora Cruz-Cabeza responded: Thank you for your very detailed comment, it is excellent. I agree with what you say and a thought related to this is that perhaps the pK_a of these acids and bases measured from solvents with a solvation ability similar to the crystal system may shift the ΔpK_a boundary towards a value of 0 – thus offering a better predictor of protonation in the solid state. However, since pK_a values in water are so common and can be predicted quite reliably, it is just more convenient to use the ΔpK_a rule and the newly derived equations in this paper with the pK_a values from water.

Helen Wheatcroft addressed Aurora Cruz-Cabeza and Kevin Roberts: I would like to respond to the earlier statement that the distinction between salts and co-crystals is mainly about semantics with a question. I would expect that the charge balance requirement in an organic salt crystal would reduce the likelihood that a solid solution (or deviations from 1 : 1 or 1 : 2 *etc.* stoichiometry) would form for organic salts in comparison with co-crystals. Is this the case?

Kevin Roberts replied: This is a most interesting question.

Firstly, I don't think that I was making the point about differences being about semantics but simply making a suggestion that in the new and evolving field of multicomponent organic crystal chemistry, it may be best not to rush towards definitions before the field is mature and scientific consensus emerges (see also my comment at 608).

In terms of salts *vs.* co-crystals *vs.* solid-solutions, it is known that long-chain hydrocarbons tend to form solid-solutions in both their neutral molecule[1] and salt states (soaps).[2] There are examples of acid soaps[3] whereby you have both salt and co-crystal interactions in the same structure but I am not sure there have been any rigorous studies as to whether solid-solutions of acid soap might form. Knowing how difficult it is to purify such systems, due to the presence of homologous impurities in solid solution, intuitively one would think the answer to such a question might be yes but clearly more work is needed in this interesting area. This area might be a particularly interesting area for pharmaceuticals as solid solutions tend to have lower crystallinity and be less stable and hence more soluble and bioavailable.

1 V. Chevallier, E. Provost, J. B. Bourdet, M. Bouroukba, D. Petitjean and M. Dirand, Mixtures of numerous different *n*-alkanes: 1. Structural studies by X-ray diffraction at room temperature—Correlation between the crystallographic long c parameter and the average composition of multi-alkane phases, *Polymer*, 1999, **40**, 2121–2128.
2 M. C. Costa, M. P. Rolemberg, A. J. A. Meirelles, J. A. P. Coutinho and M. A. Krähenbühl, The solid–liquid phase diagrams of binary mixtures of even saturated fatty acids differing by six carbon atoms, *Thermochim. Acta*, 2009, **496**(1–2), 30–37.
3 H. C. Kung and E. D. Goddard, Molecular association in fatty acid potassium soap systems. II, *J. Colloid Interface Sci.*, 1969, **29**(2), 242–249.

Aurora Cruz-Cabeza replied: Excellent question. I do not think that there is enough data on this to make a concluding statement. However, inorganic salts are known to form solid solutions, so I would expect organic ones can too if the right additives can be found.

Jan Sefcik asked: How important is the solvent of crystallisation for determining whether a co-crystal or salt forms? If solvent does not matter, the probability should not depend on crystallisation solvent. Is it consistent with experimental observations?

Aurora Cruz-Cabeza replied: So there are several aspects to this question. First, thermodynamics will determine whether cocrystal/salt formation is favoured. Second, if the multicomponent crystal is favoured, the ΔpK_a rule will determine whether it is a salt or a cocrystal. In the intermediate region of ΔpK_a, it is possible for salt/cocrystal polymorphism to exist. In this region, the solvent may impact which structure forms just in the same way solvent can impact polymorphism. The solvent will also impact whether the multicomponent crystal can form under a specific set of conditions. Here the relative solubilities of the components are important and ternary phase diagrams can be very helpful. Ultimately, however, the final protonation state for acid–base pairs in intermediate region of ΔpK_a should be determined by the crystal structure itself.

Jeffrey Rimer enquired: In the intermediate ΔpK_a range with co-crystals and salt, do these crystals form as separate polymorphs or do they form mixed phases?

Is it possible that one crystallizes as an intergrowth on the other if epitaxial relationships exist? Given that the acid and base retain molecular recognition for the co-crystal, one could envision scenarios where intergrowths can form. Is there any evidence of this?

Aurora Cruz-Cabeza answered: There are several layers to this question. A requirement here to describe your scenario is that two polymorphs of the acid–base pair exist and that one is a salt and the other is a co-crystal. There are very limited examples of this in the literature and one needs to then further consider the structural relationship between these two phases for the epitaxial growth.

First, let's consider the case where the salt and the co-crystal are iso-structural. In this case, most likely, the two phases cannot coexist. The crystal structure would determine the most stable protonation state and since energy barriers for acid–base proton transfer are very low, the transfer will just occur readily under the required conditions. For example, temperature has been shown to impact this (please see ref. 1). Once the proton transfer starts, the entire crystal will change into that structure under those conditions. Because of the structural similarity between the salt and the co-crystal, you would expect that epitaxial growth is possible but because the phases cannot coexist it will not be able to exist.

Second, let's consider the case where the salt and the co-crystal structures are different. Here epitaxial growth may be possible if there exists sufficient similarity between them; there is an epitaxial relationship. Perhaps one could produce the salt and then use this structure to crystallise the co-crystal onto. I am not aware of any experiments on this but I anticipate this will be a difficult experimental to undertake. In particular, finding a suitable system for this experiment will be difficult since the number of known salt co-crystal polymorphs is very small and then there is the requirement of the epitaxial relationship.

1 J. S. Stevens, M. Walczak, C. Jaye and D. A. Fischer, *Chem.–Eur. J.*, 2016, **22**, 15600–15604.

Sten O. Nilsson Lill remarked: A general comment on the difference of cocrystal *vs.* salt and the importance of locating the proton. If you have ion–ion interactions these are more strong and long-ranged compared to hydrogen bonding in a cocrystal. This may impact crystal growth and material properties (soft *vs.* hard). See a recent discussion in a paper by Nic Blagden.[1]

1 N. Blagden, *et al.*, *Cryst. Growth Des.*, 2022, **22**(3), 1665–1679.

Aurora Cruz-Cabeza responded: Thank you for your excellent comment. This is indeed why proton transfer is important. The same crystal structure as a salt may be more brittle or hydrate more easily than as a cocrystal. Despite the structures being the same, the position of the proton can heavily impact the properties of the structure. The article that you mention is a nice example of this.

Sarah Price opened a general discussion of the paper by Kevin Roberts: Please could you comment on the validation of the Amber GAFF force field for the PABA solution simulations? For example, how well did the relative lattice energies and structures calculated with GAFF reproduce both experiment and the results of periodic electronic structure calculations? This is a test of the model for the PABA interactions with itself. Was the solubility in the different solvents calculated and compared with experiment as a test of the force-field model for the solvent–PABA interactions?

Kevin Roberts responded: This is a generic and important question. Overall, we feel that Amber GAFF force field would perform reasonably well since it has been comprehensively developed as a general force field for organic molecules in condensed-phase through fitting to experimentally measurable data such as spectroscopic and thermodynamic measurements with electronic charges being normally obtained *ab initio* using DFT. Previous work has shown that the COMPASS II force field seems to fare best[1] for lattice energy calculations which is perhaps is not surprising given it was fitted to the largest set of test molecules. We have not so far done similar performance tests with Amber GAFF, but given the conventional wisdom that modern force fields fitted to large number and classes of molecules tend to perform better, we feel that it would also perform well. The lattice energies were not calculated with Amber GAFF as, at this stage, this force field has not been implemented in the HABIT98 code used for the synthon energy calculations.

In general, we use high-level DFT electronic structure calculations for geometric optimisation of molecular structures and electronic polarizability calculations but prefer to use force fields for modelling intermolecular interaction energies in organic structures where dispersive interactions are important. This reflects the fact, as highlighted above, that organic force fields tend to be very mature and reliable. In contrast developing good dispersion-corrected DFT is still an area of current research[2] and one that can be particularly challenging for higher molecular weight pharmaceutical compounds where the contribution of dispersion forces can become dominant.

We have not carried out any performance test through MD simulations in terms of predicting the PABA solubilities but we have calculated its solvation energies in ethanol, acetonitrile and water solutions using both MD[3] and force field[4] modelling and found them to be both consistent with each other and with the relative solubilities in these three solvents.

1 R. L. M. Robinson, D. Geatches, C. Morris, R. Mackenzie, A. G. P. Maloney, K. J. Roberts, A. Moldovan, E. Chow, K. Pencheva and C. Edge, Evaluation of force-field calculations of lattice energies on a large public dataset, assessment of pharmaceutical relevance, and comparison to density functional theory, *J. Chem. Inf. Model.*, 2019, **59** 4778–4792.
2 D. Geatches, I. Rosbottom, R. M. Robinson, P. Byrne, P. Hasnip, M. Probert, D. Jochym, A. Maloney and K. Roberts, Off-the-shelf DFT-D methods: are they now 'on-trend' for organic molecular crystals?, *J. Chem. Phys.*, 2019, **151**, 044106.
3 T. D. Turner, D. M. Camacho-Corzo, D. Toroz, A. Curtis, R. B. Hammond, X. Lai and K. J. Roberts, The influence of solution environment on the nucleation kinetics and crystalisability of *para*-aminobenzoic acid, *Phys. Chem. Chem. Phys.*, 2016, **18**, 27507–27520.
4 I. Rosbottom, J. Pickering, R. B. Hammond and K. J. Roberts, A digital workflow supporting the selection of solvents for optimizing the crystallizability of *p*-aminobenzoic acid, *Org. Process Res. Dev.*, 2020, **24**, 500–507.

Ruel Cedeno asked: It's interesting that the beta form has a higher density yet lower melting point compared to the alpha form. We have also observed this in a diastereomeric salt system.[1] I suppose this is a common feature of enantiotropic systems, *i.e.* the more closely packed structure is more stable at low temperatures while the less dense structure is entropically favored at higher temperatures leading to a higher melting point. Given the relatively small difference in free energy between PABA polymorphs, did you observe the inversion of stabilities in the MD simulations? In your paper, it was mentioned that the nucleation of the beta form has a higher energy barrier due to its higher molecular conformation deformation energy in the crystalline state. Thus, I would imagine it should require a higher driving force to overcome such barrier. Why does it preferentially nucleate at lower supersaturations (low driving force)?

1 T. Lerdwiriyanupap, G. Belletti, P. Tinnemans, R. Cedeno, H. Meekes, E. Vlieg and A. E. Flood, *Cryst. Growth Des.*, 2022, **22**, 1459–1466.

Kevin Roberts replied: Thank you for raising these most interesting points.

In the paper (https://doi.org/10.1039/d1fd00112d) I was reporting the literature but on reflection the lower MP for the beta form is probably simply the transformation from the beta to alpha form which seems also from our studies to take place over a range of temperatures. Strictly speaking it should not be feasible to measure the melting point for a low temperature enantiotropic form as it should simply transform without melting first. In our case the low temperature form has the higher density; I think that this is due to the close coupling of the beta form's 4-membered ring of alternating synthons Bβ and Dβ. One could speculate that as the temperature increases enhanced thermal vibration brings the carboxylates together through the strong Aα synthon. Maybe this could take place through a surface/surface melting interaction but we don't really know as of yet. We have though measured the thermal expansion of the two forms and this gives a bit of a clue regarding the solid phase transformation.[1] We have also studied the conversion in the solution phase with on line XRD where it is much faster than in the melt highlighting the solvent-mediated nature of the transformation process.[2] As stated in the paper the reverse transformation from alpha to beta can be very slow indeed, typically 2–3 weeks at 5 °C.

We do though see some polymorphs where the melting points do not match the expected stabilities, notably for *n*-methyl L-arginine HCl[3] and ritonavir.[4] In the latter case the stable form II has a stronger hydrogen bonding network than the metastable form I but it is in a structure which precludes the optimal close packing seen in the former case. We have not, as of yet, studied the free energy differences between the PABA polymorphs in the MD simulations having focussed our high performance computing resources on modelling the solution state as a function of solvent.

The beta form does appear to have a higher energy barrier to its formation due to its higher molecular conformation deformation energy in the crystalline state which is associated with the conformational adjustment needed to assemble the 4 membered ring of alternating synthons Bβ and Dβ. However, it needs a large cluster size to reach a stable structure compared to the alpha form which, in turn, we interpret as it needing a lower supersaturation with respect to the alpha form. The latter is mindful that in nucleation theory cluster size is inversely proportional to supersaturation. This effect we believe is generic to many polymorphic forms and maybe helpful regarding the development crystallisation R&D workflows.

1 T. D. Turner, X. Lai and K. J. Roberts, The thermal expansion coefficients of the alpha and beta polymorphic forms of *p*-aminobenzoic acid in relation to their bulk crystal chemistry, *CrystEngComm*, 2018, **20** 4099–4102.
2 T. D. Turner, S. Caddick, R. B. Hammond, K. J. Roberts and X. Lai, Kinetics of the aqueous–ethanol solution mediated transformation between the beta and alpha polymorphs of *para*-aminobenzoic acid, *Cryst. Growth Des.*, 2018, **18** 1117–1125.
3 S. Dharmayat, R. B. Hammond, C. Kilner, X. Lai, R. A. Palmer, B. S. Potter, C. M. Rayner and K. J. Roberts, Comparison of the crystal chemistry, the process conditions for crystallization and the relative structural stability of two polymorphic forms of NG-monomethyl-L-arginine hydrochloride, *Org. Process Res. Dev.*, 2008, **12**, 860–868.
4 C. Wang, I. Rosbottom, T. D. Turner, S. Laing, A. Maloney, A. Y. Sheikh, R. Docherty, Q. Yin and K. J. Roberts, Molecular, solid-state and surface structures of the conformational polymorphic forms of ritonavir in relation to their physical chemical properties, *Pharm. Res.*, 2021, **38**, 971–990.

Sten O. Nilsson Lill commented: In the alpha-form you have a strong dimer synthon and you say this will control crystal formation. What would be optimal for crystal growth, one synthon with a high stabilization, or several synthons with more similar stabilization energies? Could a strong synthon such as a closed COOH dimer be bad for overall crystal growth since you would need growth in 3D to build the crystal.

Kevin Roberts responded: The anisotropic molecular conformations for many organic materials can lead to anisotropic intermolecular packing within their 3D crystal structures. Hence, it's no surprise that organic crystals tend to crystallise in low symmetry crystal systems such as triclinic, monoclinic and orthorhombic and with greatly unequal unit cell dimensions.[1] The upshot of this is that intermolecular bonding in organic materials also tends to be anisotropic and that results in unequal crystal growth velocities on the different crystal habit planes leading mostly to non-equant needle-, lath- or plate-like crystal morphologies. In an ideal world, growth would be isotropic with strong bonding in all directions but often the chemistry does not permit this. In the case of the alpha form of PABA examination of the crystal chemistry[2] is helpful and this work reveals that the 3D nucleation process seems to be driven by the formation of the carboxylic acid dimer (synthon A). However, the subsequent growth process in 2D is a combination of several interactions: the strong pi–pi head-to-head van der Waals interactions (synthon B) along the *b*-axis coupled with 1D chains of hydrogen bonds formed from a combination of the strong OH···O dimer (synthon A) and the weaker NH···O (synthon C) hydrogen bonds. The hydrophobic van der Waals interactions are both strong and desolvate easier in polar solvents when compared to the other hydrophilic habit faces and thus the crystals tend to display needle- or lath-like crystal morphologies.[2–4]

As you point out in your question, a closed dimer such as synthon A is not by itself growth promoting and hence it needs another interaction in order to form a stoichiometric, *i.e.* crystallographically-repeating (referred to as a periodic bonding chains – PBC) measure of the strength of the growth process. In this case, it's the combination of synthons A and C that provides this stoichiometric link or PBC.

1 N. Anuar, S. N. Yusop and K. J. Roberts, Crystallisation of organic materials from solution: a molecular, synthonic and crystallographic perspective, *Crystallogr. Rev.*, 2022, accepted and in press.

2 I. Rosbottom, K. J. Roberts and R. Docherty, The solid state, surface and morphological properties of *p*-aminobenzoic acid in terms of the strength and directionality of its intermolecular synthons, *CrystEngComm*, 2015, 17 5768–5788.
3 D. Toroz, I. Rosbottom, T. D. Turner, D. M. C. Corzo, R. B. Hammond, X. Lai and K. J. Roberts, Towards an understanding of the nucleation of alpha-*para* amino benzoic acid from ethanolic solutions: a multi scale approach, *Faraday Discuss.*, 2015, **179**, 79–114.
4 T. D. Turner, D. M. Camacho-Corzo, D. Toroz, A. Curtis, R. B. Hammond, X. Lai and K. J. Roberts, The Influence of solution environment on the nucleation kinetics and crystalisability of *para*-aminobenzoic acid, *Phys. Chem. Chem. Phys.*, 2016, **18**, 27507–27520.

Matteo Salvalaglio remarked: In your work, you relate PABA polymorph selection with the stability of structural motifs of PABA associates in solution. A quantitative analysis of PABA intermolecular interaction motifs in solution obtained *via* NMR and confirmed with extensive MD simulations[1] suggests that, especially in polar solvents, the vast majority of PABA is present in solution in monomeric form.

In the same study, we find that associated PABA molecules can form a variety of highly dynamic adducts in solution, including a majority of pi-stacked structures. Such adducts display a very short persistence timescale, likely much shorter than the characteristic timescale for incorporation into the crystal phase. Even in apolar solvent hydrogen-bonded dimers, presenting the paradigmatic cyclic H-bond that emerges as the lowest energy dimer, represent a minority of the species in solution, likely due to the unfavourable configurational entropy associated with their required co-planar geometry. Due to these observations, we conclude that monomers are likely the growth unit of PABA crystals. The data presented in your paper on the abundance of intermolecular motifs and their frequency of formation in solution (Fig. 8–10) suggests a similar dynamic behaviour of PABA in your simulations. Did you investigate the finite-temperature stability and persistence of PABA associates in solution? Based on your results, do you think associates such as H-bonded dimers play a direct role as growth units in the nucleation and growth mechanisms leading to polymorph selection?

1 R. Bobvros, *et al.*, *Cryst. Growth Des.*, 2021, **21**, 436–448.

Kevin Roberts replied: Examining the wide panorama of crystallisation together with its underpinning crystal science teaches us quite clearly about the defining role that crystal lattice defects and their dimensionality play in determining the physical and chemical properties of crystals and their associated processing behaviour. Impurities underpin the operation of electronic and optoelectronic materials, non-stoichiometry and vacancies drive corrosion, milled materials tend to have lower chemical stability compared to when they are unmilled, tailor-made additives change crystal shape *etc.* We also know that the concentration of such defects does not have to be very high at all to have an effect, *e.g.* a few ppm of polymer additive can totally inhibit crystallisation in fuels under cold weather conditions [see *e.g.* ref. 1–3]. Essentially, defects play a catalytic role lowering the activation energy and promoting all kinds of effects. This backdrop is perhaps important in the analysis of the PABA data, *i.e.* concentration effects may impact on kinetics but it's the thermodynamics that drive the outcome.[4]

The key point of the PABA solution state MD analysis is that dimers are present in the solution state and their concentration relates to solvent selection. In the protic polar solvents dimer formation is inhibited but not eliminated by solvation of the carboxylate and hence it's no surprise that nucleation is slower and the metastable zone is wider. The aprotic solvents such as acetonitrile solvate less specifically and hence dimer formation at the point of nucleation seems to be easier and hence so is crystallisation. The carboxylate dimer is the strongest synthon and eventually it wins out against the competing interactions. This is particularly true for the case where the supersaturation gets high and where the solute concentration is high and metastable thermodynamically. At this point the nucleation cluster size gets smaller and then PABA desolvates and the alpha form undergoes primary nucleation as a dimer. Once formed the other dimers would be likely to assemble by secondary nucleation on the dimer template through pi–pi interactions between the dimer pairs. Morphological analysis[5] shows that the latter dominate the subsequent growth process through a needle-like morphology along the crystal structure's hydrophobic *b*-axis. Single crystal growth measurements[6] are consistent with a rough interface mechanism with linear growth kinetics. In that case, it can be quite hard to ascertain whether the growth is associated by addition of monomers or dimers. In contrast, the hydrophilic hydrogen-bonded habit surfaces along the *a*- and *c*-axes grow much slower.

This is the story so far which builds upon the work presented at Faraday 179,[2] see also the detailed review presented in the supplementary materials of this meeting's paper (https://doi.org/10.1039/d1fd00112d). The research on PABA including the further analysis of the MD data is ongoing but alas so far we have not had the opportunity to investigate the finite-temperature stability and persistence of PABA associates in solution.

1 B. P. Binks, P. D. I. Fletcher, N. A. Roberts, J. Dunkerley, H. Greenfield, A. Mastrangelo and K. Trickett, How polymer additives reduce the pour point of hydrocarbon solvents containing wax crystals, *Phys. Chem. Chem. Phys.*, 2015, **17**, 4107–4117.
2 K. Lewtas and R. D. Tack, Wax Crystallisation in Diesel Fuel: Habit Modification and the Growth of n-Alkane Crystals, in *Advances in Industrial Crystallization*, ed. J. Garside, R. J. Davey and A. G. Jones, Butterworth Heinemann, Oxford, 1991, pp. 166–179.
3 S. Anyckyj and C.B. Rupar, Ethylene-vinyl ester pour depressant for middle distillates, *U.S. Pat.*, 3,048,479, 1962.
4 T. D. Turner, D. M. Camacho-Corzo, D. Toroz, A. Curtis, R. B. Hammond, X. Lai and K. J. Roberts, The Influence of solution environment on the nucleation kinetics and crystalisability of para-aminobenzoic acid, *Phys. Chem. Chem. Phys.*, 2016, **18**, 27507–27520.
5 I. Rosbottom, K. J. Roberts and R. Docherty, The solid state, surface and morphological properties of *p*-aminobenzoic acid in terms of the strength and directionality of its intermolecular synthons, *CrystEngComm*, 2015, **17** 5768–5788.
6 D. Toroz, I. Rosbottom, T. D. Turner, D. M. Camacho Corzo, R. B. Hammond, X. Lai and K. J. Roberts, Towards an understanding of the nucleation of alpha-*para* amino benzoic acid from ethanolic solutions: a multi scale approach, *Faraday Discuss.*, 2015, **179**, 79–114.

Rik Drummond-Brydson opened the discussion of the paper by Lucia Maini: How dependent are the phase transitions on heating/cooling rate during heating and cooling?

Lucia Maini replied: In the NDI-C6 the solid state transitions were not really affected by the heating/cooling rate. We ran the DSC and HSM experiments with different rates but this did not affect the solid-state transition and their temperatures. We know that for other systems different events can be observed.[1]

1 D. Braga, F. Grepioni, L. Maini, D. Capucci, S. Nanna, J. Wouters, L. Aerts and L. Quere, *Chem. Commun.*, 2012, **48**, 8219–8221.

Peter Vekilov commented: Your results appear to indicate that upon changing temperature a crystal transforms from one crystal form to another. Such a transformation would require the molecules to move from their positions in the old form to their positions in the new stable form. On the other hand, the viscosity of a crystal is infinite, which would fully arrest all molecular motions. How do you think these molecular motions occur?

Lucia Maini responded: It is common the idea that crystals are the chemical cemetery, which evokes a static condition but on the contrary crystals hide a very dynamic nature.[1] In NDI-C6 the atoms of the alkyl chains are characterized by having a thermal parameter higher than the atoms of the core and presumably, when the temperature increases, the thermal parameters of chain increase and this probably promotes the transition. We cannot exclude the possibility that in the phases at high temperature (*i.e.* form δ) the chains could be characterized by dynamic disorder as observed also in other similar systems.[2]

1 P. Naumov, D. P. Karothu, E. Ahmed, L. Catalano, P. Commins, J. Mahmoud Halabi, M. B. Al-Handawi, L. Li, D. Prasad Karothu, E. Ahmed, L. Catalano, P. Commins, J. Mahmoud Halabi, M. B. Al-Handawi and L. Li, *J. Am. Chem. Soc.*, 2020, **142**, 13256–13272.
2 S. Milita, F. Liscio, L. Cowen, M. Cavallini, B. A. Drain, T. Degousée, S. Luong, O. Fenwick, A. Guagliardi, B. C. Schroeder and N. Masciocchi, *J. Mater. Chem. C*, 2020, **8**, 3097–3112.

Ran Zhao asked: What's the main difference for the phase transformation in bulk or powders and in thin films? Could you please comment on the reason that the new metastable form only appears in optimized spin coated films but not in other cases?

Lucia Maini replied: This a very good question. For us it was already surprising to observe a different behaviour depending on thermal treatment of the bulk. In both cases, upon heating, we obtain the form γ, but the size and defects of the crystals play a fundamental role in their kinetic stability and different pathways have been observed (see also ref. 1). The thermal study revealed the richness of the crystal structures of NDI-C6. The spin coating deposition seems to favour the formation of phases in which molecules are less tilted with respect to the substrate. In fact, form α, which is the thermodynamic stable form and is the only phase obtained by recrystallization, has never been obtained in a pure form in thin films but as a mixture with other phases. Form ε has been observed only in thin films and the planar distance of 17.0 Å suggests that the molecules are less tilted. Probably NDI-C6 is a surface-induced polymorph, but we would like to study better the role of the substrate, in particular if the polymorphism is induced by the smoothness of or by the atomic nature of the surface. Studies are in progress (see the review on SIP[2]).

1 G. Shi, S. Li, P. Shi, J. Gong, M. Zhang and W. Tang, *IUCrJ*, 2021, **8**, 584–594.
2 A. O. F. Jones, B. Chattopadhyay, Y. H. Geerts, R. Resel, A. O. F. Jones, R. Resel, B. Chattopadhyay and H. Geerts, *Adv. Funct. Mater.*, 2016, **26**, 2233–2255.

Ran Zhao enquired: In your system of NDI-C6, do you observe any differences in morphology for the same crystalline phase under different annealing processes? And when we talk about those obtained polymorphs, how do you quantify and rank their stability? Do you also consider how long they can stay stable besides their thermal behaviours (referring to the heating temperatures)?

Lucia Maini responded: Unfortunately we can study the morphology only of form α, and not of the forms obtained by the solid state transition; they are crystalline powders. By DSC, VT-XRPD and HSM analysis we ranked the different stabilities. Excepted for form β the solid state transitions are reversible within the timescale of the analysis, hence the temperature of transition as well as the energy involved were used to qualitatively rank the stability of the polymorphs. For form β, which converts into α in days, the stability was confirmed also by slurry experiment.

Rik Drummond-Brydson asked: Could you argue that the appearance of the epsilon polymorph in the thin film is effectively stabilisation of a particular polymorph in a confined structure? Would this polymorph disappear if the thin film was allowed to grow thicker do you think?

Lucia Maini replied: I do not think that thin film deposition and confined structures shares similarities, for example thin films are free to growth in two dimensions.

In our cases the thickness of the films is about 40 nm which corresponds roughly to 20–25 layers (depending on the polymorph). I do not exclude that by increasing the thickness the form ε would disappear, but it could be also hidden by the new phase on top.

Sarah Price commented: The argument that changing the conformation or packing of a molecule within a crystal is difficult is why some metastable polymorphs appear to be stable. Other solid state polymorphic phase transformations can be observed, as in the paper presented by Lucia Maini at this meeting (https://doi.org/10.1039/d1fd00100k). The difficulty in reorganizing the positions of molecules has led to the argument that all molecular solid state phase transitions are first-order, involving nucleation at a surface or defect.[1,2] Second-order phase transitions in which the transformation proceeds maintaining translational symmetry do occur in metals and some inorganic functional materials. Nonetheless there are reversible organic phase transitions that do occur with a phase boundary passing through the crystal, similar to a martensitic transformation, but are diffusionless first-order transformations, *e.g.* ref. 3. Cooperative motion, most commonly associated with second-order transitions, is compatible with nucleation and growth theory for the mechanism of solid-state transitions, but fitting some organic phase transitions into the classifications used for diffusionless atomic solid-state phase transformations is difficult.[3]

1 F. H. Herbstein, On the mechanism of some first-order enantiotropic solid-state phase transitions: from Simon through Ubbelohde to Mnyukh, *Acta Crystallogr., Sect. B: Struct. Sci.*, 2006, **62**(3), 341–383.
2 Y. Mnyukh, *Fundamentals of Solid-State Phase Transitions, Ferromagnetism and Ferroelectricity*, 1st Books Library, Bloomington, 2001.
3 V. K. Srirambhatla, R. Guo, D. M. Dawson, S. L. Price, A. J. Florence, Reversible, two-step single-crystal to single-crystal phase transitions between desloratadine forms I, II, and III, *Cryst. Growth Des.*, 2020, **20**(3), 1800–1810.

Kevin Roberts responded: It is probably true to say that whilst detailed studies of polymorphism in organic crystals is quite extensive, our knowledge of phase transformations between the polymorphs and the lattice defects such as twins and stacking faults formed by symmetry reduction[1] is much less well established. Enantiotropic phase transformations are consistent with solid–solid structural transformation but the mechanism of transformation between the phases is not always clear. This situation is often not helped by the inter-related phases having different settings for their unit cell dimensions. Even when the unit cells and crystal chemistry are quite similar, transformations may generate lattice strain leading to defect formation, plastic deformation, crystal damage and potentially fracture. From studies of inorganic materials we can see the value of looking into the group/subgroup relationships between the phases to understand the inter-relation between pre- and post-transformed structures. Pseudo symmetry can play a part and whilst two polymorphs may apparently have quite distinctly different unit cells and space groups, their crystal chemistry can be remarkably similar. An example of this is trinitrotoluence where the stable form I is monoclinic (space group $P2_1/c$) which has a centre of symmetry, a pseudo b-glide and an AABBAABB packing motif. In contrast, the metastable form II is orthorhombic (space group $Pb2_1a$) and has a real b-glide, a pseudo centre of symmetry and an ABABABAB packing motif. Both polymorphic structures can be related to a hypothetical "parent" phase (*Pbma*) containing both symmetry elements.[2] Such effects can also be seen through orientational disorder leading to the formation of plastic crystals,[3] *e.g.* CBr_4 which has a 3D disordered face-centred cubic crystal structure when crystallised from the melt but one which subsequently transforms to an ordered monoclinic phase with pseudo-cubic symmetry at room temperature. In a similar manner many paraffin structures crystallise into 2D disordered rotator phases which subsequently orders into lower symmetry structure upon cooling.[5] Organic crystals are also known to deform by mechanical deformation through the formation of ferroelastic (reversible) twin structures.[6,7] How such deformation behaviour would work with conformational flexible molecules and how the presence of highly directed hydrogen bonding would impede the same is perhaps an open question at this time. Overall, this whole domain area of organic crystal science: polymorphism, pseudo-symmetry, lattice defects and phase transformations, may benefit from having a dedicated *Faraday Discussions* in the future.

1 T. Hahn and H. Klapper, Twinning of crystals, *International Tables for Crystallography*, ed. A. Authier, Kluwer Academic Publishers, Dordrecht, 2003, vol. D, pp. 393–448.
2 H. G. Gallagher, K. J. Roberts, J. N. Sherwood and L. A. Smith, A theoretical examination of the molecular packing, intermolecular bonding and crystal morphology of 2,4,6-trinitrotoluene in relation to polymorphic structural stability, *J. Mater. Chem.*, 1997, 7 229–235.
3 J. N. Sherwood, *The Plastically Crystalline State: Orientationally Disordered Crystals*, Wiley, 1979, ISBN 0471997153.
4 M. More, J. Lefebvre and R. Fouret, Solid-state phase transition in carbon tetrabromide. II. The crystal structure of phase I, *Acta Crystallogr., Sect. B: Struct. Crystallogr. Cryst. Chem.*, 1977, *33*, 3862–3865.
5 P. K.Mukherjee, Phase transitions among the rotator phases of the normal alkanes: A review, *Phys. Rep.*, 2015, 588, 1–54.
6 T. Sasaki, Mechanical twinning in organic crystals, *CrystEngComm*, 2022, **24** 2527–2541.
7 E. R. Engel and S. Takamizawa, Versatile ferroelastic deformability in an organic single crystal by twinning about a molecular zone axis, *Angew. Chem., Int. Ed. Engl.*, 2018, 57 11888–11892.

Ivo B. Rietveld said: This is a remark in support of the remark made by Sally Price on solid–solid transitions. In particular for molecular crystals it is my experience that most solid–solid transitions that are observed in experiment are first order, that is, they are accompanied by a change in volume and in enthalpy within the system. Second order solid–solid transitions for molecular systems are in my experience rare, as I do not recall observing them experimentally. That of course, does not mean they do not exist, but they don't seem to be common. The transition dynamics are often a complicating factor, which often leads to erroneous conclusions about phase stability or even the existence of a phase equilibrium.

In inorganic systems, second order transitions are more common with, for example, transitions in magnetic behaviour. It may be possible to observe second order transitions in molecular plastic crystals, something I have not looked into yet, though.

Sarah Price responded: Thank you for this comment from a wealth of experimental observation. I agree that molecular plastic phases might prove an exception to the general rule that organic solid state phase transitions are first order. This could also apply to cases where part of the molecule, such as a hydrocarbon substituent, becomes dynamically disordered when the temperature increases.

Ivo B. Rietveld addressed Lucia Maini, Aurora Cruz-Cabeza and Kevin Roberts: It is clear that for certain specific molecules, we know how to obtain specific polymorphs, but to what level can we take a solvent from the shelf and crystallise a desired polymorph of any molecule? I understand that this is not an easy question, but what do we still need to reach the point of "polymorph/crystal on demand" (and let's stick to molecules for this question to give some semblance of setting some limits).

Lucia Maini replied: I would like to answer quoting my friend Joel Bernstein: "If I knew that, I wouldn't be here now but on my personal island in the Caribbean." We have made tremendous steps in crystal structure prediction, however how to crystallize the predicted polymorphs in the lab is still unknown. This is the missing step that the industries are asking continuously to avoid the nightmare of late appearing polymorphs.[1]

1 M. Mortazavi, J. Hoja, L. Aerts, L. Quéré, J. van de Streek, M. A. Neumann and A. Tkatchenko, *Commun. Chem.*, 2019, **2**(1), 1–7.

Kevin Roberts replied: As your question implies, there is not a comprehensive answer to your question but nonetheless there are a number of useful tools and associated workflows that have been developed and can be used. Here are 3 examples from the many available within the literature.

One can look at the solution environment using tools such quantum and statistical mechanical models such as COSMO-RS[1] which can examine the propensity for specified molecular or intermolecular structures as a function of solvent. This can be used to look at different molecular conformers or intermolecular interactions which are representative of the different polymorphic forms[2] and help enable the selection of a suitable solvent for a desired polymorphic form.

Polymorphs can also form as a function of the nucleation process and stabilities may be size dependant and so you can *e.g.* examine with molecular modelling the relative stabilities as a function of the size for optimised molecular clusters created based on their crystallographic structures.[3] Using this approach you can ascertain which cluster size stabilises which polymorph and use this to guide the selection of an appropriate supersaturation for the crystallisation process; in the latter case being mindful that the nucleation cluster size is roughly inversely proportional to the solution supersaturation.

The crystallisation of conformational polymorphs can involve understanding the often subtle balance between conformational stability and intermolecular packing which can be impacted by the different intermolecular interactions in the solid state, notably isotropic van der Waals interaction and directional hydrogen bonds. Molecular and crystallographic modelling can be used study this balance. Examination of the two polymorphic forms of the drug ritonavir demonstrate this quite nicely.[4] The metastable form has a close packed crystal structure that requires little distortion to the molecular conformation. As a result whilst it crystallises quite easily, its structure does not achieve its optimal hydrogen bonding. In contrast, crystallisation of the stable form requires distortion of the molecular conformation and hence its crystallisation can be quite slow. However, the distortion enables much more optimal hydrogen bonding, albeit in a less close packed structure.

1 See for example https://en.wikipedia.org/wiki/COSMO-RS, https://www.annualreviews.org/doi/10.1146/annurev-chembioeng-073009-100903.
2 I. Rosbottom, D. Toroz, R. B. Hammond and K. J. Roberts, Conformational and structural stability of the single molecule and hydrogen bonded clusters of para aminobenzoic acid in the gas and solution phases, *CrystEngComm*, 2018, **20**, 7543–7555.
3 See *e.g.* R. B. Hammond, K. Pencheva and K. J. Roberts, Structural variability within, and polymorphic stability of, nano-crystalline molecular clusters of L-glutamic acid and D-mannitol, modelled with respect to their size, shape and "crystallisability", *CrystEngComm*, 2012, **14**, 1069–1082.
4 C. Wang, I. Rosbottom, T. D. Turner, S. Laing, A. Maloney, A. Y. Sheikh, R. Docherty, Q. Yin and K. J. Roberts, Molecular, solid-state and surface structures of the conformational polymorphic forms of ritonavir in relation to their physical chemical properties, *Pharm. Res.*, 2021, **38**, 971–990.

Aurora Cruz-Cabeza replied: I agree this is indeed a very important question that we need to answer. Nucleation of different polymorphs and the impact of solvent, however, is something that is hard to predict. I hope with the development of modelling techniques, we will come closer to being able to answer these questions in the future.

Katharina Edkins addressed Ivo B. Rietveld, Lucia Maini, Aurora Cruz-Cabeza and Kevin Roberts: To pick up on the question on solvent influence on crystallisation/polymorph control, I think that this is a very important discussion to be had. To increase the complexity, we also have to take into account that in reality we are not crystallising from pure solvents. All solvents open to atmospheric conditions contain more or less water, which can have significant influence on the crystallisation outcome. In addition, organic solvents have been reported to form mesostructures or microheterogeneity in the presence of water, as is possibly best determined for acetonitrile/water mixtures.[1] For this particular binary solvent mixture, we could show that solute aggregation is influenced by the solvent–water interface of the microheterogeneity, which supports nucleation and growth of a hydrated crystal form.[2] Other relevant organic solvents have been reported to form mesostructures with water as well as with other organic solvents, which becomes relevant in anti-solvent crystallisation, but little is known about the structuring of these solvent mixtures, the lifetimes of the mesostructures and how these will influence crystallisation. Therefore, we have to not only investigate the influence of "pure" solvents on crystallisation but also account for potential structuring of the solvent and solvent mixtures.

1 Y. Marcus, The structure of and interactions in binary acetonitrile plus water mixtures, *J. Phys. Org. Chem.*, 2012, **25**(12), 1072–1085.
2 C. D. Jones, M. Walker, Y. Xiao, K. Edkins, *Chem. Commun.* 2019, **55**(33), 4864–4868.

Kevin Roberts answered: This is a really good point to make as in our work indeed "as received" ethanol which contains water was used. The water will also bind strongly to the carboxylate group and this might further explain the widened metastable zone for PABA that was observed in ethanol when compared to other solvents.[1] The formation of mesostructures in water containing solvent mixtures is a common problem in organic synthesis when organic and aqueous solvents are involved in different steps of the overall synthetic route and are not completely separated in downstream work-up. This is often referred to as "oiling out"[2,3] where a, say, water dispersion within the mixed solvent creates extra interfacial area within the solvent which can enhance crystallisation. We found no evidence for hydrate formation in the ethanol/water system but we did find evidence for solvate formation in the analogous ethanol/nitromethane mixed solvent system[4] but we did not investigate whether this behaviour was related to the creation of any mesostructures within the mixed solvent solutions.

1 T. D. Turner, D. M. Camacho-Corzo, D. Toroz, A. Curtis, R. B. Hammond, X. Lai and K. J. Roberts, The influence of solution environment on the nucleation kinetics and crystalisability of *para*-aminobenzoic acid, *Phys. Chem. Chem. Phys.*, 2016, **18**,27507–27520.
2 See *e.g.* J. Lu, Y.-P. Li, J. Wang, Z. Li, S. Rohani, and C.-B. Ching, Study on the oiling-out and crystallization for the purification of idebenone, *Org. Process Res. Dev.*, 2012, 16 442–446.
3 K. Tanaka and H. Takiyama, Effect of oiling-out during crystallization on purification of an intermediate compound, *Org. Process Res. Dev.*, 2019, 23, 2001–2008.
4 I. Rosbottom, C. Y. Ma, T. D. Turner, R. A. O'Connell, G. Sadiq, J. Loughrey, R. J. Davey and K. J. Roberts, Influence of solvent composition on the crystal morphology and structure of *p*-aminobenzoic acid crystallised from mixed ethanol and nitromethane solutions, *Cryst. Growth Des.*, 2017, 17 4151–4161.

Aurora Cruz-Cabeza responded: Yes, indeed Dr Edkins you make a very important comment here. Solvent mixtures may indeed result in completely new polymorphs and the mesostructure of those mixtures may indeed play a very important role in initiating the nucleation of novel forms. I look forward to reading more future work in this area.

Ivo B. Rietveld responded: Clearly, if solvents are structured around each other (and I also heard about water and ethanol structuring but I have no references ready) then they may also be structured around the solutes, which then also may affect crystallisation. Moreover, structuration itself can depend on how solvents or solutes and solvents are mixed together.

But in a sense there must be a lot of information on crystallisation from solvents and resulting polymorphs present in the literature (I remember Ghazala mentioning that crystallisation solvents are included in the CSD). Is there no way to analyse it with machine learning? It may even give some inroads into structuration even if it hasn't been reported explicitly. And then further, impurities and how this may alter nucleation. All worthwhile and difficult subjects to study.

Conflicts of interest

There are no conflicts to declare.

Faraday Discussions

PAPER

Micron-sized biogenic and synthetic hollow mineral spheres occlude additives within single crystals

Bartosz Marzec,[ab] Jessica Walker,[ac] Yasmeen Jhons,[a] Fiona C. Meldrum,[d] Michael Shaver[e] and Fabio Nudelman*[a]

Received 22nd November 2021, Accepted 7th December 2021
DOI: 10.1039/d1fd00095k

Incorporating additives within host single crystals is an effective strategy for producing composite materials with tunable mechanical, magnetic and optical properties. The type of guest materials that can be occluded can be limited, however, as incorporation is a complex process depending on many factors including binding of the additive to the crystal surface, the rate of crystal growth and the stability of the additives in the crystallisation solution. In particular, the size of occluded guests has been restricted to a few angstroms – as for single molecules – to a few hundred nanometers – as for polymer vesicles and particles. Here, we present a synthetic approach for occluding micrometer-scale objects, including high-complexity unicellular organisms and synthetic hollow calcite spheres within calcite single crystals. Both of these objects can transport functional additives, including organic molecules and nanoparticles that would not otherwise occlude within calcite. Therefore, this method constitutes a generic approach using calcite as a delivery system for active compounds, while providing them with effective protection against environmental factors that could cause degradation.

Introduction

The formation of mineralized tissues is a widespread phenomenon in nature. Organisms from all 5 kingdoms are known to precipitate more than 60 different types of mineral, each tailor-made for their biological functions.[1] Examples include vertebrate bone and teeth, mollusc shells, sea urchin spines, silica spicules in sponges and magnetite and greigite particles found in magnetotactic bacteria. The formation of mineralized tissues is a tightly regulated process,

[a]*EaStCHEM School of Chemistry, University of Edinburgh, Joseph Black Building, The King's Buildings, David Brewster Road, Edinburgh EH9 3FJ, UK. E-mail: Fabio.nudelman@ed.ac.uk*

[b]*JEOL UK Ltd, 1-2 Silver Court, Watchmead, Welwyn Garden City, AL7 1LT, UK*

[c]*Beamline I14, Diamond Light Source, Harwell Science and Innovation Campus, Didcot, Oxfordshire, OX11 0DE, UK*

[d]*School of Chemistry, University of Leeds, Woodhouse Lane, Leeds LS2 9JT, UK*

[e]*Department of Materials, School of Natural Sciences, The University of Manchester, UK*

where a water-insoluble organic matrix provides a three-dimensional framework and scaffold in which the mineral forms,[1] and water-soluble proteins and polysaccharides that participate in crystal nucleation and growth.[2–4] These additives can also become incorporated within the crystal, modifying mechanical properties.[5,6]

The ability of water-soluble macromolecules to be incorporated within crystalline hosts and to modify their morphologies and material properties has been the focus of significant attention over recent years, both in biomineralization and in bio-inspired crystallization studies.[7] It has led to a variety of synthetic approaches for the incorporation of different guest species within host crystals.[5,8–15] Calcium carbonate, and in particular the thermodynamically stable polymorph calcite, has been widely used as a model system in such studies. A range of substances, including small molecules,[8,15–17] nanoparticles,[11,18] carbon nanotubes,[19] gel networks,[20,21] polymer nano-objects,[5,22–24] and beads,[25] emulsions[26] and protein nanogels[27] have been incorporated within single crystals of calcite and other minerals.[11,28,29] This has led to the manufacture of single-crystal composites with unique optical, mechanical and magnetic properties.[11,13,30–33] Occlusion of micron-sized species within calcite in the form of *Escherichia coli*[34] and the algae *Chlorella* sp.[35] cells, on the other hand, has been met with limited success. In both cases, the cells were confined to the surface of the crystal and only partially occluded. This shows that there are still limitations in terms of the size of the guest species that can be incorporated within crystals. While additives ranging from single molecules to a few hundreds of nanometers in size can be occluded, the incorporation of larger, micron-sized guests is still challenging.

In order to be occluded within a single crystal, additives must bind to the crystal surface such that they are resident long enough to be incorporated, but do not strongly inhibit crystal growth. The primary strategy used to engineer incorporation has been functionalising particles with block copolymers rich in anionic groups including carboxylates, sulfates, sulfonates, phosphates and phosphonates.[9,10,14,36–38] Notably, nanoparticles functionalised with low charge hydroxyl-rich groups incorporate at very high levels due to their strong binding to the crystal surface and high colloidal stability in the crystallisation solution.[11] The latter enables high concentrations of particles to be employed. Control over parameters such as the nature of the functional groups, their density and chain length, therefore offers a basis for optimizing the incorporation of particles inside single crystals.

Incorporation within single crystals also offers an attractive means of protecting active compounds, where they are completely isolated from the environment within the impermeable crystal. This strategy has been demonstrated previously by occluding protein nanogels loaded with active compounds within calcite.[27] Here, we explore the potential of achieving much higher loading levels by occluding micron-sized structures within calcite single crystals. Two systems are investigated: in the first, 2 μm large cells of the calcifying unicellular marine alga *Emiliania huxleyi* ('Ehux') are entrapped. These cells produce coccoliths, which are disk-shaped assemblies of calcite single crystals that form an exoskeleton around the cells, termed the coccosphere (Fig. 1). In addition to the cytoskeleton and biomolecules normally present within any cell, Ehux can be pre-loaded with a variety of additives, including fluorescent molecules and inorganic nanoparticles, and can transport them into calcite hosts to produce single-crystalline functional

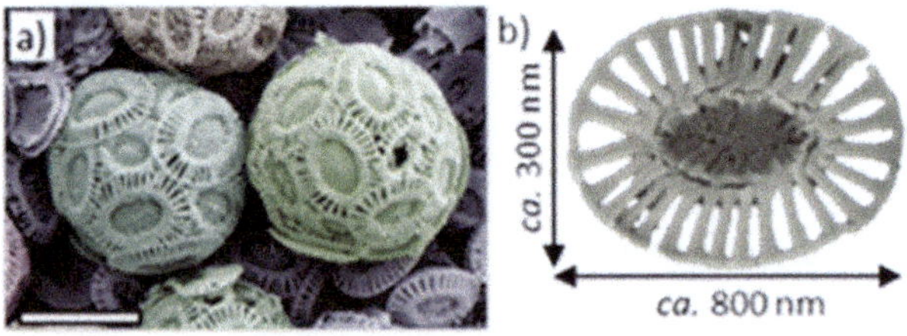

Fig. 1 (a) False-colored SEM image of *Emiliania huxleyi* coccospheres isolated from their growth medium. (b) An individual coccolith.

composites. In the second system, this methodology is extended to synthetic hollow calcite spheres, offering the ability to deliver large synthetic and biological structures occluded within single crystals. Notably, incorporation is achieved by functionalizing both types of additives with commercially-available poly(allylamine hydrochloride) (PAH). While additives used to direct calcium carbonate precipitation are usually anionic,[9,12,13,26,38] this work demonstrates the potential of using cationic additives to drive effective occlusion.

Results

Occlusion of *E. huxleyi* cells in calcite

Ehux cells were isolated from their growth medium (see Experimental) and re-dispersed in a supersaturated $CaCO_3$ growth mixture obtained by mixing equal volumes of 20 mM $CaCl_2$ and 20 mM $NaHCO_3$, and containing 10 μM of PAH. This polymer is positively-charged under these reaction conditions (≈pH 9) and has been previously shown to modify calcium carbonate precipitation, leading to the formation of thin films and fibers of calcite particles.[39] At specific time points (5 min, 20 min, 60 min, 24 h, 48 h and 96 h), samples were removed from the crystallization medium and imaged using scanning electron microscopy (SEM).

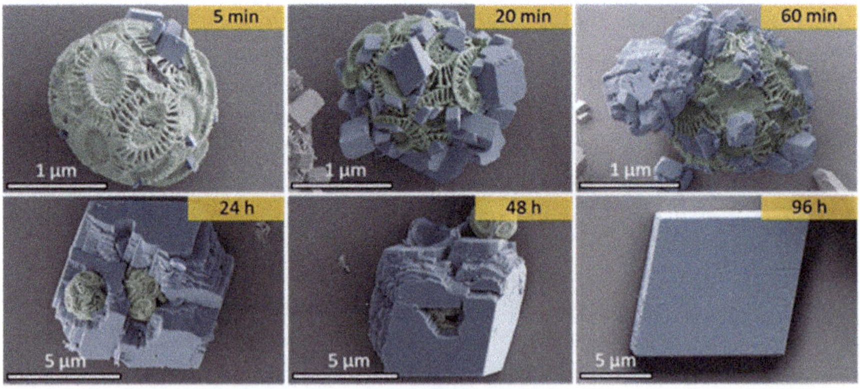

Fig. 2 Time-resolved false colored SEM images showing the occlusion of Ehux coccospheres within calcite single crystals. For clarity, the coccospheres have been highlighted green and the freshly deposited calcite units have been colored blue. Each image was taken from a different sample, collected at different time points.

As shown in Fig. 2, calcite crystals nucleated on the surface of the PAH-coated coccospheres. A few crystals were present after 5 min, and were more abundant after 20 and 60 min. The structure then developed such that a single crystal surrounded the cells after 24 h and 48 h, and the cells were completely encapsulated inside a single crystal after 96 h. The resulting particles were 10–20 μm in size and displayed well-defined rhombohedral morphologies. Raman spectroscopy confirmed that the crystals containing the Ehux cells were calcite (Fig. 3a), and polarized light microscopy showed that they were completely birefringent when viewed with crossed-polarizers (Fig. 3b and c).

To further confirm that the Ehux cells were entrapped within individual calcite crystals, occluded, the composite crystals were imaged using confocal laser scanning microscopy (CLSM) based on the detection of chlorophyll autofluorescence. Fluorescence was detected within the calcite (Fig. 4, left), in the focal planes that passed through the middle crystals (Fig. 4, right). This not only confirmed that the cells were indeed located within the calcite crystals, but also demonstrated that the aqueous interior of the cells were retained during crystallization.

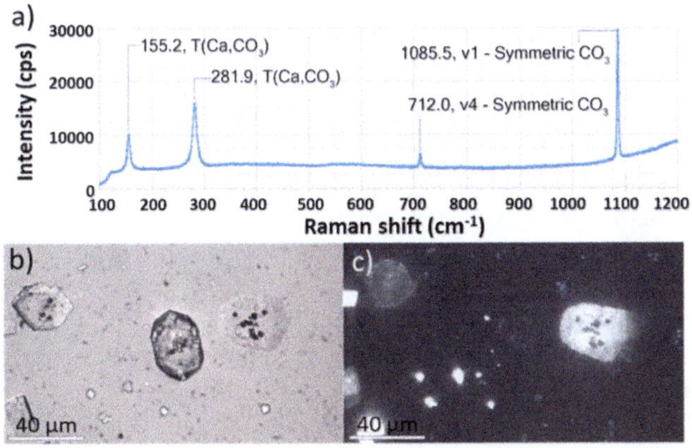

Fig. 3 (a) Raman spectroscopy of a crystal containing Ehux occlusion. (b) Bright field and (c) corresponding crossed-polarized image of calcite crystals containing Ehux occlusions recorded using a petrographic microscope. Scale bars: 40 μm.

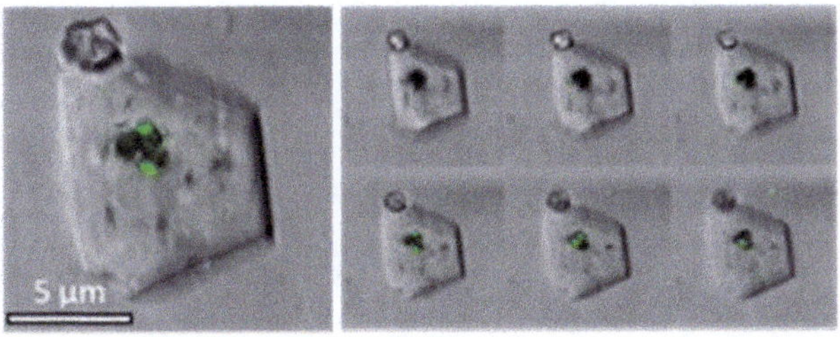

Fig. 4 Confocal microscopy image of a representative calcite crystal containing Ehux cells with the overlay of bright field and fluorescence. The right panel shows images of the crystal obtained at different focal planes. ($\lambda_{ex} = 458$ nm, $\lambda_{em} = 630–660$ nm).

Control experiments were also conducted in the absence of PAH and using different organic additives (Fig. 5a). With no PAH present, multiple well-defined calcite rhombohedra formed on the exterior of the coccosphere. These crystallites were randomly oriented. PAH concentrations above 15 μM resulted in the formation of a calcium carbonate film over the surface of the coccoliths, together with some rod- or fiber-shaped protrusions. Both thin films and fibers are characteristic of a polymer-induced liquid precursor ('PILP') phase,[40] which is known to form in the presence of PAH.[41] Experiments were carried out using polymers with different charges: neutral poly(ethylene oxide) (PEG) and negatively charged poly(acrylic acid) (PAA). As shown, 10 μM PEG had little effect on calcite formation, leading to the formation of *ca.* 1 μm calcite crystals on the coccospheres' surfaces. In contrast, addition of 10 μM PAA resulted in the deposition of rod-shaped mineral protrusions after 96 h.

Interaction of PAH with the coccospheres

To further probe the interaction between the polymers and the surface of the cells, we measured the ζ-potential of the coccospheres under different experimental conditions. As shown in Fig. 5b (left), coccoliths dispersed in $CaCl_2$ had a ζ-potential of −10 mV, which increased to + 12 mV upon addition of 5 μM of PAH, and + 16 mV on addition of 10 μM of PAH. This increase in the ζ-potential confirms that PAH binds to the surface of the cells, likely mediated *via* electrostatic interactions with the negatively charged coccospheres.

CLSM imaging of cells treated with PAH labelled with a fluorescent substituent further confirmed the adsorption of the PAH. Images taken at the focal plane corresponding to the surface of the algae (Fig. 6, top row) show fluorescent coccoliths. In images taken at the focal plane cutting through a cross-section of the cells (bottom row), the fluorescence signal was visible as a green ring surrounding dark cells, showing that the polymer was located on the coccosphere and absent from the interior of the algae.

Concentrations of PAH above 10 μM only resulted in small increases in the ζ-potential, reaching a plateau close to + 18 mV at concentrations of 13 μM and

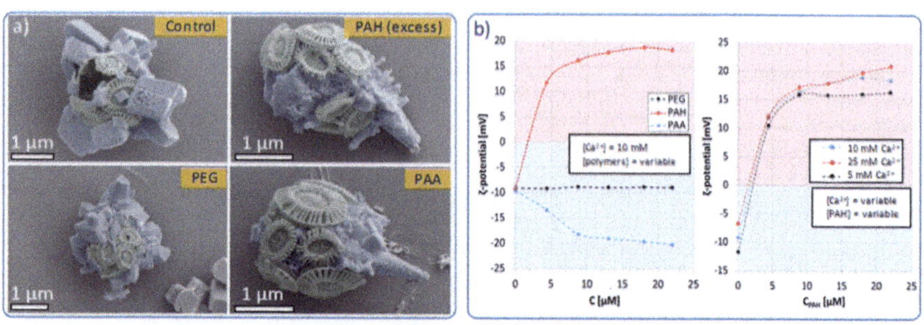

Fig. 5 Effects of polymeric habit modifiers on the morphology and ζ-potential of Ehux coccospheres. (a) False color SEM images of coccospheres placed in supersaturated $CaCO_3$ solutions containing no additives (control), or synthetic polymers, recorded after 96 hours. (b) ζ-Potential measurements recorded for Ehux coccospheres dispersed in 10 mM $CaCl_2$ containing variable concentrations of PAH, PEG and PAA (left), and coccospheres dispersed in mixtures containing various concentrations of $CaCl_2$ and different amounts of PAH (right).

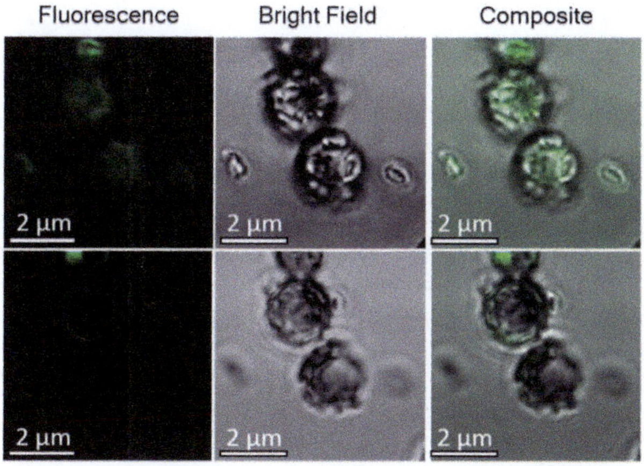

Fig. 6 Confocal microscopy images of Ehux cells dispersed in a growth mixture containing fluorescently-labelled PAH, recorded at two different focal planes: at 0 μm (top row), corresponding to the top surface of the coccospheres, and at −1 μm (bottom row), which corresponds to a cross-section through the coccosphere and the cells. Left column: fluorescence image. Middle column: bright field image. Right column: overlay of fluorescence and bright field images.

above (Fig. 5b). This shows that above [PAH] = 10 μM, the coccospheres became saturated with the polymer and the excess remains in the crystallization solution, where it modifies the formation of the calcite crystals, leading to the formation of fibrous structures (Fig. 5a). The interaction between PAH and the coccospheres was independent of the $CaCl_2$ concentration as the ζ-potential followed a similar trend and reached similar values when $[Ca^{2+}] = 5$, 10 and 25 mM (Fig. 5b, right).

As presented in Fig. 5b (left), neutral PEG did not affect the ζ-potential of the coccospheres, and the negatively charged PAA induced a change from −10 mV to −20 mV due to binding to the surface of the coccoliths. Given that both the polymer and the coccospheres are negatively charged, it is conceivable that such binding is mediated by calcium ions. In common with PAH, the negative surface charge reached a plateau at polymer concentrations above 10 μM, suggesting that the cells become saturated with PAA and that the excess remains in the solution.

Use of *E. huxleyi* cells to occlude fluorescent dyes within calcite

Based on these optimized occlusion conditions, the possibility of using the algae as delivery vehicles for non-biological materials was then explored. Ehux cells were cultured in the presence of a mixture of Nile Red (an organic fluorescent probe used to highlight cellular membranes)[42] and Triton-X100 (a surfactant disrupting cellular membranes and facilitating additives uptake), or CdTe quantum dots decorated with carboxylate functionalities, where these exhibit well-defined fluorescence.[43] The successful uptake of these materials was confirmed with CLSM, where both compounds were imaged inside the cells (Fig. 7a and b). Cells pre-loaded with the synthetic fluorescent probes were then used in a $CaCO_3$ crystallization experiment, resulting in calcite single crystals containing cellular occlusions. As shown in Fig. 7c and d, fluorescence images taken at different focal planes showed that fluorescence originating from the

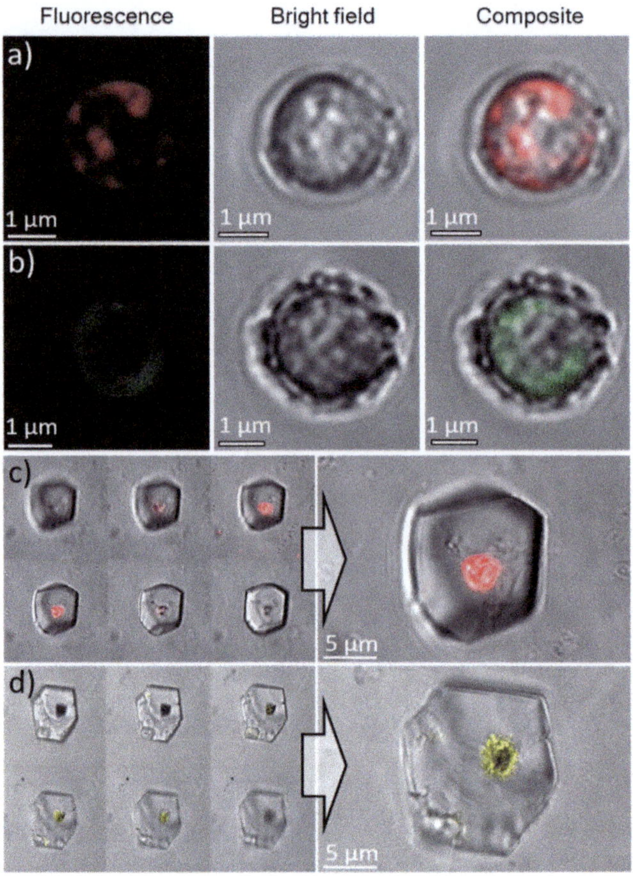

Fig. 7 Confocal microscopy images of Ehux cells pre-loaded with: (a) Nile Red, λ_{ex} = 514 nm, λ_{em} = 595–625 nm and (b) CdTe quantum dots, λ_{ex} = 458 nm, λ_{em} = 495–515 nm. Scale bars: 1 μm. Confocal microscopy images of calcite crystals containing Ehux cells pre-loaded with: (c) Nile Red and (d) CdTe quantum dots, inside calcite crystals. The images are the overlay of fluorescence and bright field images. Left panel: images at different focal planes.

synthetic probes was detected inside the crystals (Fig. 7c and d, left panel), demonstrating that the unicellular organisms can be used as carriers to occlude additives within single crystalline calcite.

Occlusion of polycrystalline hollow spheres of $CaCO_3$ in single crystalline calcite

Having demonstrated the successful occlusion of coccolithophores using PAH, we then extended our strategy to synthetic polycrystalline hollow calcite capsules. These were 2–4 μm in diameter (Fig. 8a) and were precipitated from a water/hexane mixture using the method of Fujiwara *et al.*[44] These behaved similarly to the coccospheres in seeds in $CaCO_3$ crystallization experiments in that they were only occluded within calcite crystals when PAH was present (Fig. 8b and c). Their occlusion was confirmed by fracturing the composite crystals, where SEM analysis revealed the presence of regular, spherical imprints of similar diameter

Fig. 8 (a) SEM image of synthetic hollow calcite spheres. (b) SEM image of a growing calcite crystal incorporating calcite spheres. (c) SEM image of a mature calcite crystal where occluding calcite spheres are no longer visible. (d) SEM image of calcite crystals crushed between two glass slides, revealing hollow imprints whose diameters correspond to the diameter of the occluded calcite capsules. Confocal microscopy images of a calcite single crystal containing occlusions of synthetic hollow spheres pre-loaded with an aqueous solution of HPTS ($\lambda_{ex} = 448$ nm, $\lambda_{em} = 490–510$ nm): (e) Z-stack of the crystal with images taken at different focal planes and (f) representative image of the crystal. The images are the overlay of fluorescence and bright field.

as the original calcite hollow spheres (Fig. 8d). These synthetic capsules were also explored as delivery vehicles, and were pre-loaded with an aqueous solution of a fluorescent probe 8-hydroxy-1,3,6-trisulfonic acid (HPTS). As shown in Fig. 8e and f, fluorescent calcite spheres were confirmed to be present with the calcite hosts. This system therefore provides a versatile and simple strategy for encapsulating large solution volumes – containing selected payloads – within single crystals.

Discussion

A wide range of guest species ranging from small molecules to nanoparticles, gels and submicron particles have been occluded within calcite single crystals. However, while it is straightforward to incorporate low concentrations of additives, achieving high loading levels has proven far more challenging. This has driven efforts to determine "design-rules" that govern successful incorporation, where this is clearly a complex process determined by multiple factors.

In order for additives in this size range to be incorporated it is essential that they are present on the crystal surface when the growth steps propagate. The coverage of additives on the surface is therefore determined by the concentration of additives present in the solution and their residence time on the surface, where higher concentrations and binding strengths will drive higher occlusion. However, it is not as simple as just increasing binding strength as strongly-binding additives also inhibit crystal growth, often leading to the formation of polycrystalline particles and sometimes a change in polymorph. Similarly, while it may appear straightforward to just increase the concentration of additive in the crystallisation solution, this can often lead to aggregation – particularly for the anionic additives that bind strongly to calcite. This ultimately reduces the concentration of free particles available to bind to the crystal surface.

A "sweet-spot" therefore exists in the binding strength of the additive to the crystal surface, and this has been seen in a number of systems. Gold nanoparticles functionalised with low charge, hydroxyl-rich proteins or synthetic polymers incorporated within calcite at exceptionally high levels, where these particles bind effectively and are extremely stable in the crystallisation solution.[11] A series of block copolymer micelles were also synthesized that exhibited different ratios of surface chains with either carboxylate (negatively charged) or hydroxyl (neutral) functional groups.[12] The occlusion efficiency did not directly scale with the carboxylate content of the steric stabilizers, where nanoparticles comprising 1 : 1 carboxylate/hydroxyl groups incorporated at higher levels than those that only possessed carboxylate chains. Adding further to the degree of complexity, occlusion also depends on the rate of crystal growth. Weakly-binding additives are predicted to incorporate better at high supersaturations, while the converse is expected for strongly-binding additives as there is less competition for their binding to the surface at low supersaturations.[45]

In the current work, we have shown that we can occlude particles that are an order of magnitude larger (≈ 2 μm) than any previously explored. This was achieved by functionalising the occlusion species with PAH, which is cationic under the reaction conditions. Our data also demonstrate that – due to the large size of the particles – the mechanism of incorporation within calcite single crystals is completely different to the processes responsible for the incorporation of small molecules and nanoparticles. Rather than binding to the crystal surface, the particles now act as "seeds" on which new crystals nucleate and then grow. In the vast majority of the systems explored, namely unmodified Ehux cells, and cells functionalized with PAA (anionic) and PEG (neutral) polymers, these crystallites grow to generate a polycrystalline shell. Only in the case of PAH were single crystals observed.

The mechanism by which this occurs is intriguing. Given that the calcite crystals that form on the PAH-functionalized coccospheres are oriented in multiple directions, their mutual continued growth will not lead to a single crystal product. A single crystal can therefore only develop if one crystal grows at the expense of the others, potentially *via* an Ostwald ripening process. That such a mechanism may operate is supported by the slow transformation, where perfect calcite single crystals are only observed after 4 days incubation in solution. Clearly, the PAH plays a significant role as single crystals were not observed in its

Table 1 Composition of the medium used to culture Ehux cells

Ingredient	Aldrich cat. no	Volume [mL]
Seawater	—	28.1
Guillard's (F/2) marine water enrichment solution (with silicate)	G9903	0.6
Antibiotics solution (penicillin–streptomycin–neomycin)	P4083	0.3
Parent culture	—	1.0
Total		30.0

absence. We postulate that the PAH offers judicious balance between promoting calcite nucleation on the surface of the PAH-functionalised particle, and retarding growth in solution, such that Ostwald ripening to a single large crystal can occur.

The incorporation of the micron-sized carrier particles within single crystals of calcite serves two main purposes: it prevents the leakage of the encapsulated active species to the external medium, and provides them with protection from environmental factors that can cause degradation, such as humidity, UV light and atmospheric oxygen. As calcite is biocompatible and dissolves readily at acidic pH values, it could serve for the oral delivery of different active pharmaceutical ingredients. That we can encapsulate both biological cells and synthetic hollow calcite spheres demonstrates the generality of our approach as a protection/delivery system, where commercially-available PAH can potentially be used to drive the occlusion of a wide range of additives that would not incorporate in its absence. This method can therefore be further optimized for targeted delivery applications, taking advantage of the host crystal lattice to provide occluded molecules with additional protection against environmental factors. In this respect, the use of synthetic hollow spheres of calcite would be preferable over the use of living organisms. The advantage of using micron-sized objects over nanoscale additives as carriers is that their internal volumes are several orders of magnitude larger ($\approx$14 μm^3) for a 3 μm-sized calcite hollow sphere, *versus* $\approx$ 0.0005 μm^3 for a 100 nm micelle. Thus, they enable the loading of much larger volumes of active species into calcite.

Conclusions

Building on previous work which has explored the occlusion of species ranging from organic molecules, to organic and inorganic nanoparticles, and sub-micron polymer particles within single crystals, we have here described a strategy that leads to the incorporation of micron-scale particles within calcite single crystals. Given the much larger volume occupied by these large particles within each calcite crystal as compared with the typical loadings of nanoparticles, this has great potential as an effective protection/delivery system for active compounds. Key to our approach was the functionalisation of the particles with poly(allylamine hydrochloride) (PAH) which is positively charged under the reaction conditions (pH 9). This strategy was demonstrated for two complementary particles – coccospheres and hollow calcium carbonate shells – and the encapsulation mechanism was shown to be completely distinct from that which leads to the occlusion of nanoparticles. Further work will address the mechanism by which PAH mediates the occlusion of micron-scale particles within the host crystal and will explore the use of these nanocomposites in delivery applications.

Experimental

Materials

All chemical reagents were purchased from Sigma-Aldrich and were used without further purification. Water used in the crystallization experiments was of HPLC grade and was purchased from Fisher Scientific. Seawater used to maintain algae cultures was provided by the Culture Collection of Algae and Protozoa (Scottish Association for Marine Science, Oban, United Kingdom).

Methods

***Emilainia huxleyi* ('Ehux') cultures.** Ehux (strain 982/1) culture was established using cells provided by the Culture Collection of Algae and Protozoa (Oban, UK). Parent Ehux cultures were used for subculturing once the concentration of cells stabilized at 3.5–106 cells per mL. The final composition of newly established culture media is presented in Table 1. The cells were incubated in an incubator (Panasonic MLR-352 Versatile Environmental Test Chamber) at 15 °C using the 12 h/12 h day/night cycle (photosynthetic photon flux density 200 μmol m^{-2} s^{-1}).

Calcite crystal growth – general experimental procedure. Calcium carbonate was precipitated by mixing equimolar aqueous solutions of calcium chloride and sodium bicarbonate. Unless stated otherwise, the initial concentrations of Ca^{2+} and HCO_3^- in the growth mixtures were adjusted to 20 mM. Crystallization was carried out in 10 mL glass vials equipped with *ca.* 3 mm × 5 mm glass substrates that had been cleaned with Piranha solution prior to calcite deposition.

Calcite containing *Emiliania huxleyi* ('Ehux') occlusions. When the concentration of Ehux cells in the growth medium reached 3 500 000 cells per mL, which was recorded after *ca.* 10 days since subculturing, 1 mL of the culture medium was placed in a centrifuge tube and was centrifuged at 2500 rpm for 10 minutes in order to separate the cells from seawater. The remaining liquid was decanted and the cells were re-dispersed in 1 mL of a 20 mM aqueous solution of $CaCl_2$. 4 μL of 0.25% aqueous solution of poly(allylamine hydrochloride) ('PAH', $M_w = 17.500$ g mol^{-1}, Sigma Aldrich cat. no. 283215) were added to the centrifuge tube and the mixture was gently shaken for about 1 minute. After that time, the Ehux containing PAH/$CaCl_2$ mixture was placed on the surface of a Piranha-cleaned glass substrate located at the bottom of 10 mL glass vial and 1 mL of freshly prepared 20 mM $NaHCO_3$ mixture was added to the Ehux/PAH/$CaCl_2$ solution. In order to achieve complete occlusion of coccospheres within single crystalline calcite host crystals, calcite was allowed to grow for at least 96 hours. After that time, the glass substrates were removed from the vials, washed with ethanol, dried using a stream of air, and imaged using optical and electron microscopes.

Partially mineralized samples used in time-resolved SEM experiments were obtained following the same procedure. However, in that case, the crystallization time was varied between 5 minutes and 48 hours.

Practical remarks: The occlusion of Ehux coccospheres was efficient, with above 85% (optical microscopy estimation) of coccospheres present in the sample becoming incorporated within calcite single crystals. In order to maximize the occlusion efficiency, the following steps were taken:

Coccospheres were treated gently, especially during the centrifugation. Any increase of the centrifugation speed above 2500 rpm led to increased numbers of loose coccoliths present in the sample, suggesting that the mechanically interlocked coccospheres were not able to withstand the increased centrifugal force.

The usage of clean water was avoided – Ehux cells burst when exposed to clean water due to the increased osmotic pressure, destroying the coccospheres. According to our observations, the coccospheres were preserved for the time required to carry out their occlusion when Ehux cells were dispersed in 20 mM solution of $CaCl_2$.

Using Ehux coccospheres as carriers for organic molecules. 10 μL of 1 mM solution of Nile Red (Sigma Aldrich cat. no. 19123) in DMSO was added to 30 mL of Ehux culture. The cells were incubated in the presence of the fluorescent dye for 12 hours (night cycle). After that time, they were separated from the growth medium using centrifugation, re-dispersed in an aqueous 20 mM solution of $CaCl_2$, and occluded within calcite single crystals using the procedure described above.

Practical remark: although the cells tolerated the presence of Nile Red (tested up to 10 μM) and DMSO (up to 1% v/v) in the medium, exposing Ehux cultures to these organic compounds for a longer time did not offer any substantial benefit. Also, we noted that when Nile Red was added to the Ehux culture medium during the day phase, it rapidly photodegraded to non-fluorescent derivatives.

Using Ehux coccospheres as carriers for inorganic nanoparticles. 25 μL of 1% aqueous solution of CdTe quantum dots ($\lambda_{em} = 570$ nm) coated with carboxylate functionalities (Sigma Aldrich cat. no. 777943) were added to 30 mL of Ehux culture. The cells were allowed to grow in the presence of the quantum dots for 12 hours (during the night phase). After that time, they were separated from the CdTe-containing growth medium using centrifugation, re-dispersed in 20 mM aqueous solution of $CaCl_2$ and occluded within single-crystalline calcite hosts using the method described above.

Synthesis of synthetic calcite hollow spheres ('SCHS'). SCHS were precipitated from water/hexane mixtures following a literature procedure reported by Fujiwara *et al.* Briefly: 32 mL of freshly prepared 3 M K_2CO_3 solution was rapidly mixed with 48 mL of hexane containing 0.67 g of Tween 80 and 0.33 g of Span 80. The obtained emulsion was then poured into 640 mL of 300 mM $CaCl_2$ solution in water and stirred at 400 rpm for 10 minutes. After that time, the precipitated solid particles were filtered from the remaining emulsion using a 0.22 μm Millipore membrane, washed with an excess of 10 mM aqueous solution of $CaCl_2$ and ethanol, and dried in an oven, at 110 °C. In order to precipitate SCHS pre-loaded with a fluorescent dye, 0.2 mg of 8-hydroxy-1,3,6-pyrenetrisulfonic acid ('HPTS') was added to the aqueous K_2CO_3 solution prior to emulsification.

Synthesis of poly(allylamine hydrochloride-*co*-(fluorescein)) ('PAH-FL'). Fluorescein-labelled poly(allylamine hydrochloride) was synthesised from 15 000 g mol^{-1} PAH (Sigma-Aldrich, cat. no. 283215) and fluorescein isothiocyanate (Sigma-Aldrich, cat. no. F4274) followed by recrystallisation and lyophilisation under vacuum, according to previously published procedures.[46]

Scanning electron microscopy ('SEM'). SEM imaging was performed using a Zeiss Evo scanning electron microscope equipped with a Gemini column and a field emission gun, operating at 3 kV. In order to investigate the dimensions and morphological details of the precipitated crystals, samples deposited on glass slides were fixed on aluminum stubs using double sided copper tape, coated with a thin layer (*ca.* 5 nm) of gold, and imaged using the secondary electron detector.

Confocal laser scanning microscopy. Crystals containing fluorescent occlusions were imaged using a Leica SP5 inverted confocal microscope equipped with a set of diode and gas lasers. Samples deposited on glass slides were coated with a thin layer of high refractive index oil and then were placed up-side down on a thin microscope cover glass. Focal plane merging acquisitions (*z*-stacks) were recorded at 4096 × 4096 resolution by averaging three frames recorded at the same focal length. Image analysis was carried out using the FiJi software package.

Raman microscopy. Raman spectra were acquired using the Renishaw inVia confocal Raman microscope equipped with a 785 nm IR laser. Spectra were recorded in the 1200 cm^{-1} to 100 cm^{-1} range and the total acquisition time was 60 seconds per spectrum.

Conflicts of interest

There are no conflicts to declare.

Acknowledgements

Financial support from the University of Edinburgh (J. W. and Y. J.) and the UK Biology and Biotechnology Research Council (BB/M029611/1) is gratefully acknowledged.

References

1 H. A. Lowenstam and S. Weiner, *On Biomineralization*, Oxford University Press, New York, 1989.

2 L. Addadi, J. Moradian, E. Shay, N. G. Maroudas and S. Weiner, *Proc. Natl. Acad. Sci. U. S. A.*, 1987, **84**, 2732–2736.

3 L. Addadi and S. Weiner, *Proc. Natl. Acad. Sci. U. S. A.*, 1985, **82**, 4110–4114.

4 S. Albeck, J. Aizenberg, L. Addadi and S. Weiner, *J. Am. Chem. Soc.*, 1993, **115**, 11691–11697.

5 Y.-Y. Kim, K. Ganesan, P.-C. Yang, A. N. Kulak, S. Borukhin, S. Pechook, L. Ribeiro, R. Kroeger, S. J. Eichhorn, S. P. Armes, B. Pokroy and F. C. Meldrum, *Nat. Mater.*, 2011, **10**, 890–896.

6 S. Weiner, L. Addadi and H. D. Wagner, *Mater. Sci. Eng., C*, 2000, **11**, 1–8.

7 F. C. Meldrum and H. Colfen, *Chem. Rev.*, 2008, **108**, 4332–4432.

8 D. C. Green, J. Ihli, P. D. Thornton, M. A. Holden, B. Marzec, Y.-Y. Kim, A. N. Kulak, M. A. Levenstein, C. Tang, C. Lynch, S. E. D. Webb, C. J. Tynan and F. C. Meldrum, *Nat. Commun.*, 2016, **7**, 13524.

9 A. Hanisch, P. C. Yang, A. N. Kulak, L. A. Fielding, F. C. Meldrum and S. P. Armes, *Macromolecules*, 2016, **49**, 192–204.

10 C. T. Hendley, L. A. Fielding, E. R. Jones, A. J. Ryan, S. P. Armes and L. A. Estroff, *J. Am. Chem. Soc.*, 2018, **140**, 7936–7945.

11 Y. Y. Kim, R. Darkins, A. Broad, A. N. Kulak, M. A. Holden, O. Nahi, S. P. Armes, C. C. Tang, R. F. Thompson, F. Marin, D. M. Duffy and F. C. Meldrum, *Nat. Commun.*, 2019, **10**, 5682.

12 Y. Y. Kim, L. A. Fielding, A. N. Kulak, O. Nahi, W. Mercer, E. R. Jones, S. P. Armes and F. C. Meldrum, *Chem. Mater.*, 2018, **30**, 7091–7099.

13 Y.-Y. Kim, J. D. Carloni, B. Demarchi, D. Sparks, D. G. Reid, M. E. Kunitake, C. C. Tang, M. J. Duer, C. L. Freeman, B. Pokroy, K. Penkman, J. H. Harding, L. A. Estroff, S. P. Baker and F. C. Meldrum, *Nat. Mater.*, 2016, **15**, 903–910.

14 A. N. Kulak, M. Semsarilar, Y. Y. Kim, J. Ihli, L. A. Fielding, O. Cespedes, S. P. Armes and F. C. Meldrum, *Chem. Sci.*, 2014, **5**, 738–743.

15 G. Magnabosco, M. Di Giosia, I. Polishchuk, E. Weber, S. Fermani, A. Bottoni, F. Zerbetto, P. G. Pelicci, B. Pokroy, S. Rapino, G. Falini and M. Calvaresi, *Adv. Healthcare Mater.*, 2015, **4**, 1510–1516.

16 S. Borukhin, L. Bloch, T. Radlauer, A. H. Hill, A. N. Fitch and B. Pokroy, *Adv. Funct. Mater.*, 2012, **22**, 4216–4224.

17 B. Marzec, D. C. Green, M. A. Holden, A. S. Cote, J. Ihli, S. Khalid, A. Kulak, D. Walker, C. Tang, D. M. Duffy, Y. Y. Kim and F. C. Meldrum, *Angew. Chem., Int. Ed.*, 2018, **57**, 8623–8628.

18 A. N. Kulak, P. Yang, Y.-Y. Kim, S. P. Armes and F. C. Meldrum, *Chem. Commun.*, 2014, **50**, 67–69.

19 M. Calvaresi, G. Falini, L. Pasquini, M. Reggi, S. Fermani, G. C. Gazzadi, S. Frabboni and F. Zerbetto, *Nanoscale*, 2013, **5**, 6944–6949.

20 H. Li, H. L. Xin, D. A. Muller and L. A. Estroff, *Science*, 2009, **326**, 1244–1247.

21 E. C. Keene, J. S. Evans and L. A. Estroff, *Cryst. Growth Des.*, 2010, **10**, 5169–5175.

22 L. A. Estroff and I. Cohen, *Nat. Mater.*, 2011, **10**, 810–811.

23 K. Rae Cho, Y.-Y. Kim, P. Yang, W. Cai, H. Pan, A. N. Kulak, J. L. Lau, P. Kulshreshtha, S. P. Armes, F. C. Meldrum and J. J. De Yoreo, *Nat. Commun.*, 2016, **7**, 10187pp.

24 Y.-Y. Kim, M. Semsarilar, J. D. Carloni, K. R. Cho, A. N. Kulak, I. Polishchuk, C. T. I. V. Hendley, P. J. M. Smeets, L. A. Fielding, B. Pokroy, C. C. Tang, L. A. Estroff, S. P. Baker, S. P. Armes and F. C. Meldrum, *Adv. Funct. Mater.*, 2016, **26**, 1382–1392.

25 C. Lu, L. Qi, H. Cong, X. Wang, J. Yang, L. Yang, D. Zhang, J. Ma and W. Cao, *Chem. Mater.*, 2005, **17**, 5218–5224.

26 Y. Ning, F. C. Meldrum and S. P. Armes, *Chem. Sci.*, 2019, **10**, 8964–8972.

27 O. Nahi, A. N. Kulak, T. Kress, Y.-Y. Kim, O. G. Grendal, M. J. Duer, O. J. Cayre and F. C. Meldrum, *Chem. Sci.*, 2021, **12**, 9839–9850.

28 Y. Ning, L. A. Fielding, J. Nutter, A. N. Kulak, F. C. Meldrum and S. P. Armes, *Angew. Chem., Int. Ed.*, 2019, **58**, 4302–4307.

29 A. Brif, G. Ankonina, C. Drathen and B. Pokroy, *Adv. Mater.*, 2014, **26**, 477–481.

30 I. Polishchuk, A. A. Bracha, L. Bloch, D. Levy, S. Kozachkevich, Y. Etinger-Geller, Y. Kauffmann, M. Burghammer, C. Giacobbe, J. Villanova, G. Hendler, C. Y. Sun, A. J. Giuffre, M. A. Marcus, L. Kundanati, P. Zaslansky, N. M. Pugno, P. U. P. A. Gilbert, A. Katsman and B. Pokroy, *Science*, 2017, **358**, 1294–1298.

31 S. C. Yang, F. Z. Liu, C. Wu and S. Yang, *Small*, 2016, **12**, 4028–4047.

32 S. J. Zhang, C. I. Pelligra, X. D. Feng and C. O. Osuji, *Adv. Mater.*, 2018, **30**, 1705794.

33 D. C. Green, M. A. Holden, M. A. Levenstein, S. H. Zhang, B. R. G. Johnson, J. Gala de Pablo, A. Ward, S. W. Botchway and F. C. Meldrum, *Nat. Commun.*, 2019, **10**, 206.

34 M. Zhang, H. Ping, W. Fang, F. Wan, H. Xie, Z. Zou and Z. Fu, *J. Mater. Chem. B*, 2020, **8**, 9269–9276.

35 M. Kim, M. G. Choi, H. W. Ra, S. B. Park, Y. J. Kim and K. Lee, *Materials*, 2018, **11**, 296.

36 Y. Ning, L. A. Fielding, L. P. D. Ratcliffe, Y. W. Wang, F. C. Meldrum and S. P. Armes, *J. Am. Chem. Soc.*, 2016, **138**, 11734–11742.

37 K. Rae Cho, Y. Y. Kim, P. Yang, W. Cai, H. Pan, A. N. Kulak, J. L. Lau, P. Kulshreshtha, S. P. Armes, F. C. Meldrum and J. J. De Yoreo, *Nat. Commun.*, 2016, **7**, 10187.

38 Y. Ning, L. Han, M. J. Derry, F. C. Meldrum and S. P. Armes, *J. Am. Chem. Soc.*, 2019, **141**, 2557–2567.

39 B. Cantaert, Y.-Y. Kim, H. Ludwig, F. Nudelman, N. A. J. M. Sommerdijk and F. C. Meldrum, *Adv. Funct. Mater.*, 2012, **22**, 907–915.

40 L. B. Gower, *Chem. Rev.*, 2008, **108**, 4551–4627.

41 L. Dai, E. P. Douglas and L. B. Gower, *J. Non-Cryst. Solids*, 2008, **354**, 1845–1854.

42 G. S. Aleman-Nava, S. P. Cuellar-Bermudez, M. Cuaresma, R. Bosma, K. Muylaert, B. E. Ritmann and R. Parra, *J. Microbiol. Methods*, 2016, **128**, 74–79.

43 K.-T. Yong, W.-C. Law, I. Roy, Z. Jing, H. Huang, M. T. Swihart and P. N. Prasad, *J. Biophotonics*, 2011, **4**, 9–20.

44 M. Fujiwara, K. Shiokawa, K. Morigaki, Y. Zhu and Y. Nakahara, *Chem. Eng. J.*, 2008, **137**, 14–22.

45 A. A. Chernov, *Modern Crystallography III*, Springer-Verlag, 1984.

46 G. Begum, S. Singh, N. Rangaraj, G. Srinivas and R. K. Rana, *J. Mater. Chem.*, 2010, **20**, 8563–8570.

Faraday Discussions

ROYAL SOCIETY OF CHEMISTRY

PAPER

Atomic-scale structures and dynamics at the growing calcite step edge investigated by high-speed frequency modulation atomic force microscopy

Kazuki Miyata,[abc] Yuta Kawagoe,[b] Naoyuki Miyashita,[b] Tomoki Nakagawa[c] and Takeshi Fukuma*[abc]

Received 8th November 2021, Accepted 15th November 2021

DOI: 10.1039/d1fd00084e

We have investigated the calcite growth mechanism by directly imaging atomic-scale structural changes at the growing step edges with high-speed frequency modulation atomic force microscopy (HS-FM-AFM). We compared the results with those previously obtained during dissolution, where a transition region (TR) consisting of a $Ca(OH)_2$ monolayer was found to be formed along the step edges as an intermediate state. We found that the TR is created not only during dissolution but also during the growth process. Steps with and without a TR coexist with a ratio of 7 : 3 in both dissolution and growth, implying that their primary reaction pathways should involve TR formation. While all the dissolving steps show a linear shape, the growing steps additionally present a complex non-linear shape with many kinks. The TRs formed along the linear steps present a fixed and uniform width, while those along the complex steps present a non-uniform and dynamically varying width. The acute and obtuse steps show similar TR formation probability, TR width, and step velocity during growth, while a TR is preferentially formed along an acute step during dissolution. For both step types, TRs during growth are wider than those during dissolution. Based on these findings, we present possible reaction pathways triggered by the adsorption of either CO_2 or HCO_3^- for the elementary steps in calcite growth.

1 Introduction

Calcite is the most abundant carbonate mineral on the Earth, and its growth and dissolution play critical roles in the global carbon cycle.[1–3] In addition, calcite has attracted much attention due to its importance in biomineralization,[4,5] carbon

[a]Nano Life Science Institute (WPI-NanoLSI), Kanazawa University, Kakuma-machi, Kanazawa 920-1192, Japan. E-mail: fukuma@staff.kanazawa-u.ac.jp; Fax: +81-76-234-4632; Tel: +81-76-234-4847

[b]Division of Electrical Engineering and Computer Science, Kanazawa University, Kakuma-machi, Kanazawa 920-1192, Japan

[c]Division of Frontier Engineering, Kanazawa University, Kakuma-machi, Kanazawa 920-1192, Japan

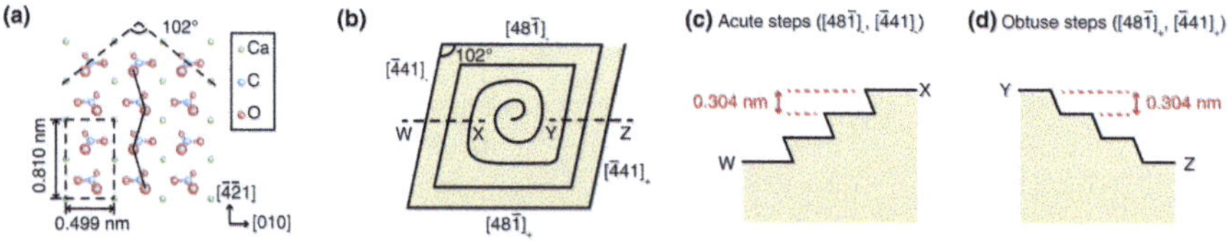

Fig. 1 Models of the calcite (10$\bar{1}$4) surface. (a) Atomic-scale surface structure consisting of Ca^{2+} and CO_3^{2-}. (b) Step arrangement around a screw dislocation during the calcite growth. (c) Acute steps. (d) Obtuse steps.

sequestration,[6,7] and optical engineering.[8,9] Thus, the mechanism of calcite growth and dissolution has intensively been studied by various methods.[10–20] In particular, atomic force microscopy (AFM)[21] has been a powerful tool for visualizing atomistic step flows, which has helped us to understand the nanoscale behaviors of the calcite growth and dissolution processes.[10,11,16,22–24]

In the growth process, screw dislocations serve as the source of the steps.[25] The created steps take a spiral shape near the center of the dislocation and are transformed into a linear shape as they propagate away from the center (Fig. 1b).[11] Meanwhile, the dissolution is initiated by forming small rectangular pits followed by propagation of the four linear steps surrounding the pits.[23,24] These basic behaviors have successfully been reproduced by kinetic Monte Carlo simulations, typically with empirically determined activation energy values for the different step edge structures (*i.e.*, acute and obtuse steps shown in Fig. 1c and d)[26] and solution conditions (*e.g.*, pH, degree of supersaturation, and various additives).[14,27,28] However, it is essential to understand the atomistic behaviors at the step edges to understand the origin of these different activation energies.

Compared with the nanoscale step flow behaviors, the atomistic events at the step edges have not been well understood due to the difficulties in their direct observation. So far, AFM observations of the step flow with varying pH and Ca^{2+}/CO_3^{2-} ratios have been performed to obtain insights into the chemical reactions associated with the adsorption or desorption of ions at the step edges.[15,17] Meanwhile, molecular dynamics (MD) simulations were performed to investigate the free energy landscape for Ca^{2+} and CO_3^{2-} and the stability of water near the step edges.[19,20,29] Based on these results, a possible model for the atomistic kink formation was proposed.[20] However, direct observations of atomistic events are strongly demanded to validate the mechanisms suggested by these previous experiments and simulations.

We have recently developed high-speed frequency modulation AFM (HS-FM-AFM) to solve such problems.[30] The developed system allows in-liquid atomic-resolution imaging at ~0.5 s per frame.[31] With this technique, we previously performed direct imaging of the atomistic events at the step edges during the calcite dissolution in water.[30] The obtained images revealed the existence of a layer-like structure, the "transition region (TR)", along the step edges. The TR typically has a width of a few nanometers and a height of 0.1–0.2 nm. This height is lower than the atomistic step (0.304 nm) at the calcite (10$\bar{1}$4) surface. With the help of MD simulations, we attributed the TRs to a $Ca(OH)_2$ monolayer formed as an intermediate state during the dissolution. Furthermore, we also reported their statistical properties,[31] behaviors in a small etch pit,[32] and possible roles in the step dissolution kinetics.[32] However, the method has not been applied to studies on the calcite growth process. Thus, the existence of the TR and its possible roles in the growth process remain unclear.

In this study, we have investigated the atomic-scale mechanism of the calcite growth process by directly imaging atomistic events at the growing step edges in a supersaturated solution. The obtained images confirm the existence of TRs in the growth process. Their structures and dynamics are clarified and compared with those observed during the dissolution. Based on these findings, we present possible models to explain the atomic-scale growth mechanism at the step edges.

2 Experimental

2.1 Sample preparation

In this study, we used a calcite single crystal (Crystal Base Co, Ltd.) with a size of 5 $\times$ 5 $\times$ 2 mm^3. This crystal was fixed onto a sample holder for our AFM system with epoxy glue. The fixed crystal was cleaved by a razor blade. Immediately after the cleavage, 50 μl solution was dropped onto the sample surface to perform AFM experiments in the deposited liquid.

The calcite supersaturated solution used for the AFM imaging of the growth process was prepared as follows. We prepared an 11.6 mM $CaCl_2$ solution by dissolving its powder form (10043-52-4, FUJIFILM Wako Pure Chemical) into Milli-Q water. In the same way, we prepared 11.6 mM $NaHCO_3$ solution from its powder form (31212-25, Nacalai Tesque). These two solutions and Milli-Q water were mixed with a ratio of 1 : 1 : 7 to prepare the supersaturated solution with a relative supersaturation (σ) of 21. σ is defined by $(C - C_{eq})/C_{eq}$, where C and C_{eq} denote the real and saturated calcite concentrations, respectively. For this solution, C (=1.3 mM) is 22 times higher than C_{eq} (=58 μM).[10]

1.3 mM HCO_3^-, produced by the ionization of the dissolved $NaHCO_3$, should be subject to the following equilibrium reactions.

$$CO_2 + H_2O \rightleftharpoons H^+ + HCO_3^-, \tag{1}$$

$$HCO_3^- \rightleftharpoons H^+ + CO_3^{2-}. \tag{2}$$

The ionization constants of reactions (1) and (2) are $K_{a1} = 4.25 \times 10^{-7}$ M and $K_{a2} = 5.6 \times 10^{-11}$ M, respectively. From these parameters, the pH can be calculated to be 8.3. This pH is slightly higher than the experimentally measured value (~7.8) due to the dissolution of CO_2 from the air. By taking into account this carbon increase, we expect that the solution should contain 2.6 mM Cl^-, 1.3 mM Ca^{2+}, Na^+, and HCO_3^-, 60 μM CO_2, and 3.6 μM CO_3^{2-}.

2.2 AFM experiments

We used a custom-built HS-FM-AFM with a low-noise cantilever deflection sensor,[33–35] highly stable photothermal cantilever excitation system,[35,36] high-speed scanner,[37,38] and low-latency wideband PLL and FM-AFM controller.[39–41] Details of the design and performance of this system were reported previously.[30]

All the HS-FM-AFM imaging was performed in the constant frequency shift (Δf) mode at room temperature. The growth process was imaged in the supersaturated solution ($\sigma = 21$), while the dissolution process was imaged in Milli-Q water. We used commercially available silicon cantilevers (AC55, Olympus) with a typical spring constant k, quality factor Q, and resonance frequency f_0 in an aqueous environment of 80 N m^{-1}, 10 and 1.5 MHz, respectively. The AFM tip was coated with a 15 nm silicon film by a dc sputter coater (K575XD, Emitech) for cleaning and stabilizing the tip surface.[42]

3 Results and discussion

3.1 HS-FM-AFM imaging

We performed HS-FM-AFM imaging of the growing steps on the calcite $(10\bar{1}4)$ surface in the supersaturated solution. The obtained images reveal the variations in the step edge structure and dynamics. Among the 84 steps observed in this study, 55 steps (65%) show a linear edge with a small number of kinks, as shown in Fig. 2a and c. Meanwhile, 29 steps (35%) show a complex non-linear edge with many kinks, as shown in Fig. 3a and c. Thus, the linear ones seem to be more common. Both linear and complex steps were imaged with and without a TR, which demonstrates that a TR can be formed even during the growth process. Fig. 2 and 3 show typical HS-FM-AFM images of these different types of steps. These images demonstrate that HS-FM-AFM successfully visualizes atomic-scale structural changes near the step edge during the calcite growth.

The height profiles measured across the four different types of steps (Fig. 2b, d and 3b, d) show that the height difference between the upper and lower terraces corresponds to the single molecular step height (~0.304 nm). Thus, the steps shown in these figures should be monomolecular ones. For the steps with a TR, the height profiles (Fig. 2b and 3b) show a plateau or a shoulder between the upper and lower terraces with a medium height of 0.1–0.2 nm and a width of 1–10 nm. Because of the similarity in these structural features, the observed layer-like structures should correspond to the TRs previously observed during dissolution (*i.e.*, a $Ca(OH)_2$ monolayer formed as an intermediate state). As indicated in

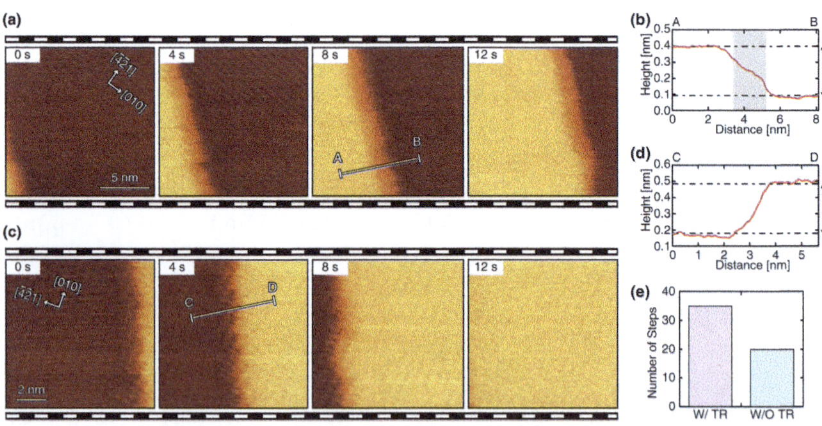

Fig. 2 (a and c) Successive HS-FM-AFM images of the growing calcite step edges with a linear shape obtained on the calcite $(10\bar{1}4)$ surface in the supersaturated solution. (a) With a TR. $\Delta f = 2.67$ kHz. $A = 0.19$ nm. 2 s per frame. 500 × 500 pix^2. (c) Without a TR. $\Delta f = 2.3$ kHz. $A = 0.13$ nm. 2 s per frame. 500 × 500 pix^2. (b and d) Height profiles measured along the lines A–B and C–D. (e) Number of linear steps with and without a TR observed in this study.

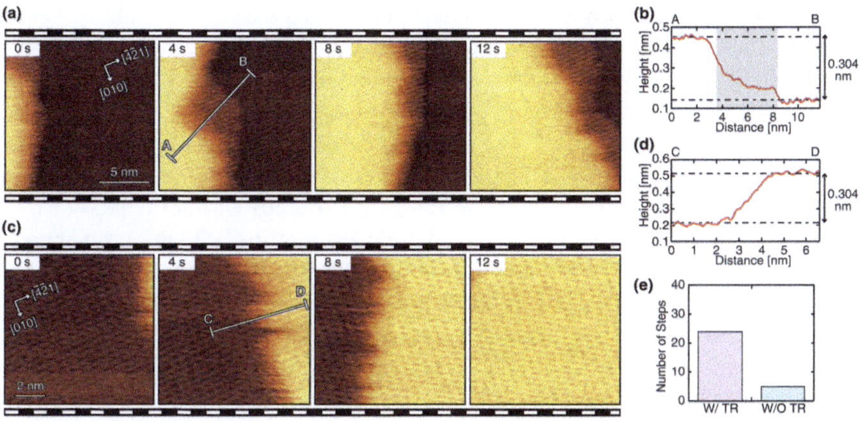

Fig. 3 (a and c) Successive HS-FM-AFM images of the growing calcite step edges with a complex non-linear shape obtained on the calcite (10$\bar{1}$4) surface in the supersaturated solution. (a) With a TR. $\Delta f = 2.2$ kHz. $A = 0.45$ nm. 2 s per frame. 500 × 500 pix^2. (c) Without a TR. $\Delta f = 1.7$ kHz. $A = 0.14$ nm. 2 s per frame. 500 × 500 pix^2. (b and d) Height profiles measured along the lines A–B and C–D. (e) Number of complex steps with and without a TR observed in this study.

Fig. 2e and 3e, the steps with a TR were more often observed than those without a TR, implying that the growth process involving the TR formation should be the primary pathway under these experimental conditions. The same trend was previously confirmed in the dissolution process.[31]

The linear steps typically propagate in one direction with little change in the edge structure (Fig. 2a and c). The TRs formed at these steps also show a linear edge and an almost constant width. This behavior is similar to that observed during dissolution. In contrast, the complex steps and the associated TRs show dynamic changes in their edge structure during their propagation (Fig. 3a and c). In this case, the width of the TRs is not uniform and dynamically changes in the range of 0 to 10 nm. Furthermore, the direction of the step propagation is not necessarily constant but varies during the growth. Such dynamic changes in the step edge structure and the TR width are not observed during dissolution.

In general, the linear step edge structure suggests preferred adsorption of the growth unit to the kinks over the step edges. Thus, there must be a reason to locally weaken such a preference at the location where the complex step edges are formed. One possible reason is the distortion of the crystal lattice caused by defects or dislocations. For example, curved steps are often formed around a screw dislocation to form a spiral step arrangement during the growth process,[10] while linear steps are created around a rectangular pit from the very beginning of the dissolution.[32] Such a difference in the step flow behavior may account for the observed difference in the atomic-scale structures. In addition, local enhancement of the solute concentration may induce adsorption to the step edges. Such variation may occur if the step density is non-uniform.

3.2 Analysis of the TR properties

During calcite growth and dissolution, steps propagate in four different crystallographic orientations (Fig. 1b). Among them, the $[48\bar{1}]_-$ and $[\bar{4}41]_-$ steps are known as acute steps (Fig. 1c), while the $[48\bar{1}]_+$ and $[\bar{4}41]_+$ steps are known as obtuse steps (Fig. 1d). Previous studies have revealed that these two types of steps show different step flow behaviors.[11–13,15,16,20,43–45] To understand the correlation between such a nanoscale step flow behavior and atomic-scale step edge structures, we compare TR properties between obtuse and acute steps and growth and dissolution. Note that the dissolution data was adapted from our previous report.[31]

Fig. 4a shows a comparison of the ratios of the steps without a TR. For the growth process, the acute steps show a slightly lower ratio (29%) than the obtuse steps (32%), although the difference is minor. For the dissolution process, the same relationship is observed but with a much more significant difference. These results suggest that the TR formation does not show a strong preference for the step type in the growth process, while it is preferentially formed at the acute steps in the dissolution.

Fig. 4b shows a comparison of the step velocity. The obtuse steps move slightly faster (1.22 ± 0.736 nm s^{-1}) than the acute steps (0.95 ± 0.43 nm s^{-1}) in the growth process. The same relationship was observed in the dissolution, yet the difference is more evident than that in the growth. As for the dissolution, previous AFM studies have consistently reported a faster velocity for the obtuse steps under various solution conditions.[31,46,47] Meanwhile, this relationship is valid for the growth process only at high supersaturation and is suppressed or even inverted as the supersaturation is reduced.[11,12] While the Ca^{2+} concentration (1.3 mM) in the solution used for this growth experiment is sufficiently high, the CO_3^{2-} concentration (3.6 μM) is relatively low. This low concentration may explain the small difference between the obtuse and acute step velocities in the growth process.

Fig. 4c shows a comparison of the TR width. A significant difference is found not between the acute and obtuse steps but between the growth and dissolution processes: the growing steps present a much broader TR than the dissolving steps. This difference is observed for both linear and complex steps. Fig. 4d and e show the correlations between the velocity and width for the growth and dissolution, respectively. For the growth process, there may be a weak trend that a narrow TR is formed along a fast step while a wide TR is formed along a slow step. Otherwise, no strong correlation is found between the two parameters.

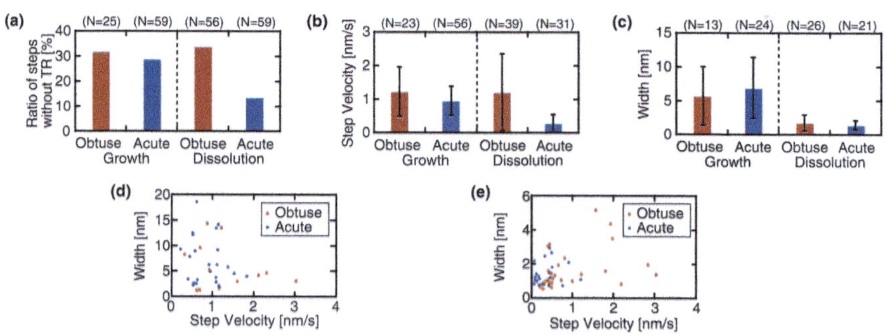

Fig. 4 Comparison of the TR properties between the obtuse and acute steps observed during the calcite growth and dissolution. (a) Ratio of the steps without a TR. (b) Step velocity. (c) TR width. The error bars in (b) and (c) correspond to the standard deviations. (d and e) Relationship between the step velocity and TR width during the (d) growth and (e) dissolution.

Although the mechanism of the TR formation is not fully understood, one possible explanation was suggested by the previous MD simulation.[30,31] According to this simulation, the hydration structure formed on the upper terrace almost seamlessly extends over the TR. Such an extended hydrogen-bonding network may explain the energetic merit to form a TR as an intermediate state during the dissolution or growth. If this model is correct, the wider TR width observed in the growth process may suggest that an extended hydration structure may be more easily formed during the growth process. However, more accurate discussions should require atomic-scale simulations with a model taking into account the TR formation.

3.3 Atomic-scale growth mechanism

This study found that the step edge reaction involving the TR formation is the most favorable pathway in both dissolution and growth. Here, we briefly overview the previously proposed dissolution mechanism and then discuss possible models for the growth mechanism. In the previously proposed model, the dissolution is triggered by the dissociative adsorption of water to the CO_3^{2-} vacancy at the step edge, which produces an ion pair of $CaOH^+$ and HCO_3^-, as described by

$$CaCO_3 + H_2O \rightarrow CaOH^+ + HCO_3^-. \tag{3}$$

The produced ion pairs can form a mobile adsorption layer only when trapped inside a nanoscale pit.[32] Otherwise, the ion pair is soon subject to the following reaction to form a TR consisting of a $Ca(OH)_2$ monolayer.

$$CaOH^+ + HCO_3^- \rightarrow Ca(OH)_2 + CO_2. \tag{4}$$

As for the growth process, an atomistic model was proposed by the recent simulation study.[20] In the proposed model, CO_3^{2-} first adsorbs at the step edge, which induces the subsequent adsorption of Ca^{2+}. Such a model should explain the growth mechanism without involving the TR formation. However, the growth mechanism involving the TR formation has not been investigated by simulation yet, making it difficult to determine the reaction pathway uniquely. Thus, here we only present a few possible reaction pathways that can explain the observed phenomena.

The growth process involves reactions not only at the step edges but also at the TR edges. At the growing edge of a TR, the following reaction should form $Ca(OH)_2$.

$$Ca^{2+} + 2OH^- \rightarrow Ca(OH)_2. \tag{5}$$

This reaction does not progress in a bulk neutral solution. However, the growth of the TR edges observed in this study (Fig. 2a and 3a) suggests that this reaction can progress at the TR edges. This argument is partially supported by our previous MD simulation study,[30] where we confirmed the stable adsorption of a $Ca(OH)_2$ monolayer at the step edge.

As for the reactions at the upper terrace edges, a few different pathways are conceivable. One of them is the inverse reactions of the dissolution process, as described by

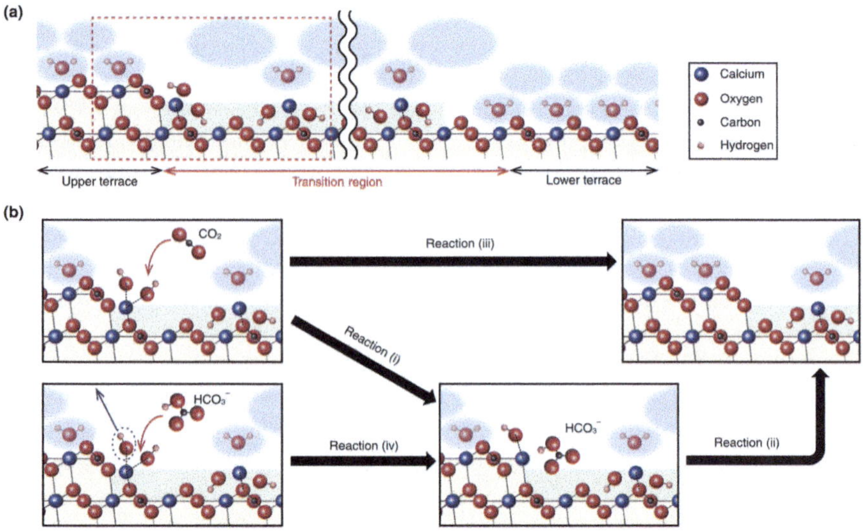

Fig. 5 (a) Atomic-scale model of a step edge with a TR on the calcite (10$\bar{1}$4) surface. (b) Possible chemical reaction pathways for the calcite growth process at the step edge.

$$Ca(OH)_2 + CO_2 \rightarrow CaOH^+ + HCO_3^-, \tag{6}$$

$$CaOH^+ + HCO_3^- \rightarrow CaCO_3 + H_2O. \tag{7}$$

These reactions correspond to reactions (i) and (ii) in Fig. 5b, respectively. Alternatively, these reactions may take place in one step (Fig. 5b(iii)):

$$Ca(OH)_2 + CO_2 \rightarrow CaCO_3 + H_2O. \tag{8}$$

Instead of the adsorption of a CO_2, the growth reaction may be driven by the replacement of a surface OH^- with an adsorbed HCO_3^-, as described by

$$Ca(OH)_2 + HCO_3^- \rightarrow CaOH^+ + HCO_3^- + OH^-. \tag{9}$$

This reaction corresponds to reaction (iv) in Fig. 5b. Afterward, reaction (ii) should progress to produce a $CaCO_3$.

The concentration of HCO_3^- (1.3 mM) is much higher than that of CO_3^{2-} (3.6 μM) in the solution used for this experiment, which suggests that reaction (iv) followed by reaction (ii) may be the major pathway under these experimental conditions. However, a more reliable analysis should require an atomic-level simulation study.

4 Conclusions

In this study, we have investigated the atomic-scale calcite growth mechanism by HS-FM-AFM. We succeeded in the atomic-resolution HS-FM-AFM imaging of structural changes at the growing calcite step edges at 2 s per frame on the calcite

$(10\bar{1}4)$ surface. We compared the results with those previously obtained in the dissolution process and clarified the following points.

Monomolecular steps with and without a TR coexist with a ratio of 7 : 3 in both the dissolution and growth processes. While all the dissolving steps show a linear shape, the growing steps additionally show a complex non-linear shape with many kinks. The TRs formed along the linear steps present a fixed and uniform width, while those along the complex steps present a non-uniform and dynamically varying width. The acute and obtuse steps show similar TR formation probability, TR width, and step velocity in the growth process, while a TR is preferentially formed along an acute step in the dissolution. For both step types, TRs during growth are much wider than those during dissolution. Based on these findings, we have presented possible reaction pathways for the elementary steps in calcite growth. One of them is the inverse of the dissolution reactions and is triggered by adsorption of CO_2 to react with surface $Ca(OH)_2$. In addition, we presented another possible pathway triggered by a replacement of the surface OH^- with an adsorbed HCO_3^-.

So far, the atomic-scale calcite growth mechanism has been mostly investigated with an assumption of the direct adsorption of the growth units to the step edge. However, our HS-FM-AFM observations strongly suggest that the primary growth pathway should involve the formation of a TR as an intermediate state. This finding should improve the accuracy of the initial structure assumed in simulation studies, and lead to a better understanding of the atomistic calcite growth mechanism.

Author contributions

K. M. and T. F. conceptualized the study and designed the experiments. K. M., Y. K., N. M. and T. N. performed the high-speed FM-AFM experiments. T. F. prepared the first draft of the manuscript but all the authors contributed to the data analyses, establishment of the growth model, and the preparation of the final version of the manuscript.

Conflicts of interest

There are no conflicts to declare.

Acknowledgements

This work was supported by the World Premier International Research Center Initiative (WPI), MEXT, Japan, and JSPS KAKENHI grant numbers JP20H00345 and JP20H05212.

Notes and references

1 R. J. Reeder, in *Carbonates, Mineralogy, and Chemistry: Reviews in Mineralogy*, ed. R. J. Reederi, Mineralogical Society of America, 1983.
2 D. M. Sigman and E. A. Boyle, *Nature*, 2000, **407**, 859–869.
3 J. Sarmiento and N. Gruber, *Ocean Biogeochemical Dynamics*, Princeton University Press, 2006.

4 L. Addadi, S. Raz and S. Weiner, *Adv. Mater.*, 2003, **15**, 959–970.
5 S. A. G. Falini, S. Weiner and L. Addadi, *Science*, 1996, **271**, 67–69.
6 M. L. Druckenmiller and M. M. Maroto-Valer, *Fuel Process. Technol.*, 2005, **86**, 1599–1614.
7 E. H. Oelkers, S. R. Gislason and J. Matter, *Elements*, 2008, **4**, 333–337.
8 R. R. Alfano and S. L. Shapiro, *Phys. Rev. Lett.*, 1970, **24**, 592–594.
9 G. Ghosh, *Opt. Commun.*, 1999, **163**, 95–102.
10 A. J. Gratz, P. E. Hillner and P. K. Hansma, *Geochim. Cosmochim. Acta*, 1993, **57**, 491–495.
11 H. H. Teng, P. M. Dove and J. J. D. Yoreo, *Geochim. Cosmochim. Acta*, 2000, **64**, 2255–2266.
12 L. E. Wasylenki, P. M. Dove and J. J. D. Yoreo, *Geochim. Cosmochim. Acta*, 2005, **69**, 4227–4236.
13 G. Nehrke, G. J. Reichart, P. Van Cappellen, C. Meile and J. Bijma, *Geochim. Cosmochim. Acta*, 2007, **71**, 2240–2249.
14 J. J. De Yoreo, L. A. Zepeda-Ruiz, R. W. Friddle, S. R. Qiu, L. E. Wasylenki, A. A. Chernov, G. H. Gilmer and P. M. Dove, *Cryst. Growth Des.*, 2009, **9**, 5135–5144.
15 K. Larsen, K. Bechgaard and S. L. S. Stipp, *Geochim. Cosmochim. Acta*, 2010, **74**, 2099–2109.
16 A. G. Stack and M. C. Grantham, *Cryst. Growth Des.*, 2010, **10**, 1409–1413.
17 E. Ruiz-Agudo, C. V. Putnis, C. Rodriguez-Navarro and A. Putnis, *Geochim. Cosmochim. Acta*, 2011, **75**, 284–296.
18 M. P. Andersson, S. Dobberschutz, K. K. Sand, D. J. Tobler, J. J. De Yoreo and S. L. Stipp, *Angew. Chem., Int. Ed.*, 2016, **55**, 11086–11090.
19 M. De La Pierre, P. Raiteri and J. D. Gale, *Cryst. Growth Des.*, 2016, **16**, 5907.
20 M. De La Pierre, P. Raiteri, A. G. Stack and J. D. Gale, *Angew. Chem., Int. Ed.*, 2017, **56**, 8464–8467.
21 G. Binnig, C. F. Quate and C. Gerber, *Phys. Rev. Lett.*, 1986, **56**, 930.
22 P. E. Hillner, A. J. Gratz, S. Manne and P. K. Hansma, *Geology*, 1992, **20**, 359–362.
23 Y. Liang, D. R. Baer, J. M. McCoy, J. E. Amonette and J. P. Lafemina, *Geochim. Cosmochim. Acta*, 1996, **60**, 4883–4887.
24 R. S. Arvidson, I. E. Ertan, J. E. Amonette and A. Luttge, *Geochim. Cosmochim. Acta*, 2003, **67**, 1623–1634.
25 W. K. Burton, N. Cabrera and F. C. Frank, *Philos. Trans. R. Soc. London, Ser. A*, 1951, **243**, 299–358.
26 I. Kurganskaya and A. Luttge, *J. Phys. Chem. C*, 2016, **120**, 6482–6492.
27 J. M. McCoy and J. P. LaFemina, *Surf. Sci.*, 1997, **373**, 288–299.
28 I. Kurganskaya and S. V. Churakov, *J. Phys. Chem. C*, 2018, **122**, 29285–29297.
29 D. Spagnoli, S. Kerisit and S. C. Parker, *J. Cryst. Growth*, 2006, **294**, 103–110.
30 K. Miyata, J. Tracey, K. Miyazawa, V. Haapasilta, P. Spijker, Y. Kawagoe, A. S. Foster, K. Tsukamoto and T. Fukuma, *Nano Lett.*, 2017, **17**, 4083–4089.
31 K. Miyata, Y. Kawagoe, J. Tracey, K. Miyazawa, A. S. Foster and T. Fukuma, *J. Phys. Chem. C*, 2019, **123**, 19786–19793.
32 K. Miyata, K. Takeuchi, Y. Kawagoe, P. Spijker, J. Tracey, A. S. Foster and T. Fukuma, *J. Phys. Chem. Lett.*, 2021, **12**, 8039–8045.
33 T. Fukuma, M. Kimura, K. Kobayashi, K. Matsushige and H. Yamada, *Rev. Sci. Instrum.*, 2005, **76**, 053704.

34 T. Fukuma and S. P. Jarvis, *Rev. Sci. Instrum.*, 2006, **77**, 043701.
35 T. Fukuma, *Rev. Sci. Instrum.*, 2009, **80**, 023707.
36 T. Fukuma, K. Onishi, N. Kobayashi, A. Matsuki and H. Asakawa, *Nanotechnology*, 2012, **23**, 135706.
37 K. Miyata, S. Usho, S. Yamada, S. Furuya, K. Yoshida, H. Asakawa and T. Fukuma, *Rev. Sci. Instrum.*, 2013, **84**, 043705.
38 S. M. R. Akrami, K. Miyata, H. Asakawa and T. Fukuma, *Rev. Sci. Instrum.*, 2014, **85**, 126106.
39 Y. Mitani, M. Kubo, K. Muramoto and T. Fukuma, *Rev. Sci. Instrum.*, 2009, **80**, 083705.
40 K. Miyata, H. Asakawa and T. Fukuma, *Appl. Phys. Lett.*, 2013, **103**, 203104.
41 K. Miyata and T. Fukuma, *Beilstein J. Nanotechnol.*, 2018, **9**, 1844–1855.
42 S. M. R. Akrami, H. Nakayachi, T. Watanabe-Nakayama, H. Asakawa and T. Fukuma, *Nanotechnology*, 2014, **25**, 455701.
43 M. Xu, X. Hu, K. G. Knauss and S. R. Higgins, *Geochim. Cosmochim. Acta*, 2010, **74**, 4285–4297.
44 M. Xu and S. R. Higgins, *Geochim. Cosmochim. Acta*, 2011, **75**, 719–733.
45 M. E. Smith, K. G. Knauss and S. R. Higgins, *Chem. Geol.*, 2013, **360–361**, 10–21.
46 R. Shiraki, P. A. Rock and W. Casey, *Aquat. Geochem.*, 2000, **6**, 87–108.
47 G. D. Giudici, *Am. Mineral.*, 2002, **87**, 1279–1285.

Faraday Discussions

ROYAL SOCIETY OF CHEMISTRY

DISCUSSIONS

Learning lessons from nature – the future of biomimetics: general discussion

Michael Anderson, Matthew Bennett, Ruel Cedeno, Ian Ford, Takeshi Fukuma, Christian Kuttner, Lucia Maini, Fiona Meldrum, Sten O. Nilsson Lill, Fabio Nudelman, Ivo B. Rietveld, Kevin Roberts, Jan Sefcik, Wenhao Sun, Peter Vekilov and Ran Zhao

DOI: 10.1039/d2fd90024f

Fiona Meldrum opened a general discussion of the paper by Fabio Nudelman: Do the crystals containing encapsulated cells form in bulk solution or only on the walls of the vessel?

Fabio Nudelman responded: It is hard to tell, we didn't check this. The cells will eventually sediment to the bottom of the crystallization vessel, so I would say that both are possible.

Peter Vekilov commented: I think I have an explanation of how one of the crystals overpowers all the other crystals that initially nucleate on the substrate. Owing to their displacements and misorientations with respect to the substrate, the nucleated crystals are strained, but to distinct degrees. The one which is the least strained has the lowest free energy and is exposed to highest degree of supersaturation. It grows the fastest and engulfs all the other crystals.

Fabio Nudelman replied: Yes, I agree.

Ruel Cedeno said: Does the surface charge of the cells play a role in the occlusion? Since the zeta potential is dependent on the pH, is the occlusion process sensitive to changes in pH as well?

Fabio Nudelman replied: The pH of the crystallization solution is basic, as we are mixing $CaCl_2$ with sodium bicarbonate. We didn't try to vary the pH, so it is hard to tell how it will affect the occlusion. Regarding the cells, they normally have a negative surface charge because the coccoliths are coated by acidic, negatively charged polysaccharides. We did zeta-potential measurements that show that as polyallylamine hydrochloride is added, the surface charge of the cells increases, becoming positive. I assume that this is quite important for the occlusion.

Ran Zhao enquired: Are the 'Ehux' algae cells still alive and can they function well, *e.g.* to further produce coccoliths, once they are entrapped? If no, what do you think is the advantage to occluded cells compared to synthetic calcite spheres? Do you think it is possible to obtain other polymorphs or crystal structures if you regulate the growth medium or occlude vaterite/aragonite capsules?

Fabio Nudelman answered: The cells were initially transferred from their growth medium, which is sea water, to a solution containing 10 mM calcium chloride and sodium bicarbonate. They will not survive for long under these conditions. Even if they are still alive when entrapped into calcite, they will eventually die as they cannot exchange gas and nutrients with the environment. Using the cells was a proof-of-principle for the encapsulation of such micron-sized structures into calcite, and using them as carriers. Once we achieved that, we translated the system to a synthetic ones where the cells were replaced by calcitic hollow spheres. In terms of applications, it is better to use the hollow spheres than the cells to avoid, for example, the presence of the biological material, and the need to grow and maintain them.

What we observed regarding other polymorphs was that in the absence of cells, vaterite was obtained, but this was not occluded into calcite.

Christian Kuttner asked: Could you please briefly elaborate on the mechanism of void formation for the hollow calcite spheres? Is there anything special about the final surface chemistry inside or outside of the spheres that is essential for a successful occlusion?

Fabio Nudelman responded: We didn't look in detail at voids that appear in the calcite crystals containing the hollow spheres so I cannot comment on that. The only way we could get the spheres occluded into calcite was by using the positively charged polymer PAH as an additive in the crystallization medium, just like we did for the occlusion of the coccolithophores. There is a sweet spot there in terms of the optimal polymer concentration – too little and nothing happens, too much and it starts interfering with the calcite precipitation by producing those polymer-induced liquid precursor phases. The polymer does change the surface charge of the guest particles (whether the cells or the hollow spheres), making them positively charged. I believe that this is important for the occlusion. The surface chemistry inside the hollow spheres, on the other hand, shouldn't play any role in the occlusion as their inner surface does not interact with the growing calcite crystals.

Lucia Maini queried: In Fig. 2 of the paper (https://doi.org/10.1039/D1FD00095K), the crystal of calcite seems to cover the coccospheres, while in Fig. 8 of the paper, when the crystallization happens on the hollow calcite spheres, it looks like that there is the imprint of the pristine hollow sphere. Does it mean that the synthetic hollow calcite sphere was recrystallized into the main big crystal? Why doesn't this happen with the coccospheres?

Kevin Roberts said: Maybe the crystal is re-nucleating around the edges of the obstruction and then encompassing the "obstacle". If the secondary nucleation process associated with this was epitaxial, then the crystallinity would, in principle, be maintained. This effect was demonstrated by the X-ray tomographic characterisation of agglomeration in hexamine, a high symmetry organic material with a cubic crystal structure.[1]

1 P. Gajjar, T. T. H. Nguyen, J. Sun, I. D. Styliari, H. Bale, S. McDonald, T. L. Burnett, B. Tordoff, E. Lauridson, R. B. Hammond, D. Murnane, P. J. Withers and K. Roberts, Crystallographic tomography and molecular modelling of structured organic polycrystalline powders, *CrystEngComm*, 2021, **23**, 2520–2531.

Fabio Nudelman responded: It is difficult to tell what is really going on. From the images in the paper (https://doi.org/10.1039/d1fd00095k), it is clear that there is Ostwald ripening involved, but we cannot tell the mechanism. Most crystals are clearly dissolving, as we start with the nucleation of several calcite crystals ions on the surface of the coccoliths, and after several hours there is only one left that is growing and encompassing the cells. It is possible that there is secondary nucleation, as you are suggesting, but this needs to be checked.

Ran Zhao asked: If we think of the calcium carbonate protection layers in nature like egg shells or, as shown in the coccospheres, they usually work as intact units but also have porosities at the surfaces which allow for nutrition exchange, can you give an outlook on how we can proceed to achieve this in the field of fabricating biomineral-based delivery vehicles?

Fabio Nudelman replied: Reproducing calcium carbonate structures with morphologies that are similar to those found in coccoliths, or in other biominerals, is still very challenging. To do that, one needs to look at the templating mechanisms that control mineral formation in biological systems, and there are many factors in place. In terms of biomineral-based delivery vesicles, we show that this is possible, not just with coccoliths but also by producing hollow spheres of polycrystalline calcite. The challenge with those hollow spheres is that on their own, they will eventually leak the encapsulated materials. Hence, occluding them inside single crystals of calcite was necessary.

Peter Vekilov opened the discussion of the paper by Takeshi Fukuma: You interpret your results as evidence that dissolution rate is distinct from the growth rate. Asymmetry between growth and dissolution may happen on a macroscopic scale, but not on the molecular scale, where the growth–dissolution symmetry represents a manifestation of the principle of microscopic reversibility, a fundamental tenet of chemical kinetics.

Takeshi Fukuma answered: In the present paper (https://doi.org/10.1039/d1fd00084e), I compared the dissolution and growth from several aspects. As for the step dissolution and growth rates, we reported that they differ and depend on the step types (such as acute and obtuse steps). However, this is not new and has been previously reported.[1,2] I think such a difference in the step flow rates does not contradict the reversibility of the chemical reaction as the solution conditions are significantly different between these two situations. Namely, dissolution experiments are mostly performed in nearly pure water, while the growth experiments are performed in a supersaturated calcium carbonate solution containing various ionic contents.

We also reported that non-linear steps are observed only in the growth process. This difference probably comes from the different step generation mechanisms. In the growth process, steps are generated from screw dislocations, around which steps are not necessarily linear and propagate in various directions. Meanwhile, the dissolution process is triggered by the formation of a rectangular pit, and the surrounding four linear steps linearly propagate in the four specific directions. Again, this does not contradict the reversibility of the chemical reaction at the step edges.

Lastly, we also discussed a few different chemical reaction pathways in the growth process, while we assumed only one reaction pathway for the dissolution. However, this is not due to the asymmetry of the chemical reaction but reflects the variations in the solution conditions. As for the dissolution, we assumed only one solution condition – pure water. For the growth process, we considered different concentrations of CO_3^{2-}, HCO_3^-, and CO_2 in the solution. Therefore, the multiple pathways reflect the multiple solution conditions and do not imply the asymmetry of the chemical reaction processes.

1 R. Shiraki *et al.*, *Aquat. Geochem.*, 2000, **6**, 87.
2 H. H. Teng *et al., Geochim. Cosmochim. Acta*, 2000, **64**, 2255.

Wenhao Sun said: Dear Professor Fukuma, thank you for the wonderful videos and observations. I have a question on your last slide; since you have developed a microkinetic model regarding the transition region in calcite dissolution and growth, it appears that your equations have a dependence on the bicarbonate concentrations. I wonder if you have thought about how pH would change the kinetics of step-edge growth, based on how it influences the width and growth speed of the transition region? I also want to add that if you do an experiment that investigates this pH dependence, it would be very interesting to decouple the thermodynamics aspects (since pH also affects the supersaturation of $CaCO_3$) from the kinetics of the step-edge growth velocity.

Takeshi Fukuma answered: We agree that pH is one of the major factors that may influence the step-growth process. Although we have not performed a systematic study with variable pH or temperature yet, that is one of the studies we would like to perform in the near future.

As for the TR width, we may decouple the contributions from the thermodynamics and kinetics aspects. Our previous MD simulation results showed that an extended hydration layer is formed above the upper terrace and TR. Thus, our current hypothesis is that the TR is formed due to this energetic merit to form the extended hydrogen-bonding network. However, the influence of the upper terrace on the hydration structure should decay with the distance from the upper terrace edge. Therefore, at a certain distance, the energetic merit to form the hydration layer should balance the demerit to constrain the thermal motion of the Ca^{2+} and OH^- ions under the hydration layer. This balance should determine the TR width. If the TR width does not show any dependence on pH, the TR width mostly reflects the thermodynamic balance. In contrast, if it depends on pH, it may also reflect the competition between the reaction speeds at the TR edge and upper terrace edge. These reaction rates also determine the overall step flow rate.

In addition to pH, we should also consider the contributions from the other ionic species, such as CO_3^{2-}, HCO_3^-, and CO_2. As discussed in the paper (https://doi.org/10.1039/d1fd00084e), the ratio of these species could alter the reaction pathway. Therefore, these ionic content ratios could heavily affect both the kinetics and thermodynamics. Due to the large parameter space, it would be reasonable to focus on pH as a first step.

Matthew Bennett asked: Have you tried varying the flow rate of CO_2? Did you see any changes in reaction rates?

Takeshi Fukuma responded: No, we have not. In this experiment, we performed all the experiments with fixed conditions. Although we did not control the CO_2 concentration in the solution or air, we estimated it to be 60 μM from the measured pH value. As the next step, we would like to investigate the dependence of the step edge behavior on the pH or CO_3^{2-}/Ca^{2+} ratio. This is because the strong dependence of the reaction rate on these parameters was previously reported.

Christian Kuttner enquired: What is the difference in the hydration structure between the acute and obtuse steps and how does this mechanistically translate into differences in the transition region (TR) width?

Takeshi Fukuma replied: Strictly speaking, the hydration structures of the acute and obtuse steps are different. According to the previous MD simulation study, the work needed to remove a dissolution unit from the acute step is larger than that from the obtuse step.[1] Such difference should be partially influenced by the difference in the hydration structure. One clear difference between the acute and obtuse steps that we observed in our experiment was in the TR formation probability in the dissolution process. Our previous MD simulation suggests that the energetic merit to form a TR comes from the formation of an extended hydration layer over the upper terrace and TR. Thus, the difference in the hydration structures between the acute and obtuse steps may lead to the difference in the energetic barrier to form an extended hydration layer and, eventually, in the TR formation probability. However, as the TR width does not show a significant dependence on the step type, the stability of the extended hydration layer should not be very different. Of course, these arguments should be verified by MD simulation with a model taking into account the formation of a TR.

1 D. Spagnoli *et al.*, *J. Cryst. Growth*, 2006, **294**, 103.

Michael Anderson commented: The transition region is always the same separation from the main terrace. This means that a $Ca(OH)_2$ layer is growing at precisely the same rate as the carbonate terrace, which seems strange unless they are somehow coordinated. Can you think of a mechanism that would keep these two growth fronts in registry?

Takeshi Fukuma answered: Strictly speaking, the width of the transition region (TR) is not necessarily constant. If we closely look at the images, there are some variations even for the linear steps. For the non-linear steps observed in the growth process, this variation is very evident. These results suggest that the observed structures are not caused by the double tip artifact in the AFM observation.

However, as you pointed out, the TR width is surprisingly uniform for the linear steps observed in the growth and dissolution processes. According to our MD simulation results, an extended hydration layer is formed above the upper terrace and TR. Thus, our current hypothesis is that the TR is formed due to this energetic merit to form the extended hydrogen bonding network. However, the influence of the upper terrace on the hydration structure should decay with the distance from the upper terrace edge. Therefore, at a certain distance, the energetic merit to form the hydration layer should balance with the demerit to constrain the thermal motion of the Ca^{2+} and OH^- ions under the hydration layer. This balance should depend on the local solution condition, temperature, and interaction with the underlying crystal lattice.

The observed constant TR width suggests that these conditions are uniform around the observed step edges. In contrast, at the edge of the non-linear step edges observed in the growth process, the TR width was not uniform and exhibited dynamic changes. This suggests that the aforementioned conditions were not uniform around these step edges. This sounds reasonable, as the growth direction of the non-linear steps was not constant but dynamically changing. This should at least lead to the variation of the interaction with the underlying crystal lattice.

Ivo B. Rietveld commented: The results obtained by AFM are very impressive and the molecular layers of the crystal can clearly been seen. Is it also possible to see the hydration layer on the surface with the method? Moreover, is it possible to see density fluctuations as a result of an increase or decrease in the solute concentration near the surface?

Takeshi Fukuma responded: We recently expanded the capability of our HS-FM-AFM from 2D to 3D and succeeded in observing the 3D hydration structure around a dissolving calcite step edge. The result clearly demonstrates that a hydration layer on the upper terrace is indeed almost seamlessly extended over the TR. Unfortunately, our AFM is not sensitive enough to detect the density fluctuations caused by the inhomogeneity of the solute concentration. However, we sometimes see the contrast fluctuations in the 2D-AFM images, which appear to be related to the ion diffusion on the surface.

Sten O. Nilsson Lill remarked: Would it be possible to use isotopically substituted HCO_3^- to get an understanding of the competing pathways *via* isotope effects on the kinetics and rate-determining steps? Some other technique than AFM could be useful for this.

Takeshi Fukuma replied: Unfortunately, AFM itself is insensitive to isotopic difference. Thus, we plan to vary the ionic contents of HCO_3^-, CO_3^{2-}, and Ca^{2+} to investigate the competitive pathways. However, if the isotopic difference alters the kinetics, we may see such effects through the altered dynamic behavior around the dissolving or growing step edges. Combination with other surface analysis techniques sensitive to the isotopic difference may be useful. However, the precise control of the isotopic labeling may be a great challenge as all the carbon compounds are subject to equilibrium reactions.

Jan Sefcik asked: Would it be useful to think of the transition layer observed at the surface of a growing crystal as non-classical surface nucleation and growth mechanism?

Takeshi Fukuma answered: Yes, I think so. The growth mechanism should involve at least two steps: reactions at the edges of the transition region (TR) and the upper terrace. In addition, the nearly constant width of the TR suggests the interaction between the two steps through the hydration layer above the TR. Such interaction should also be taken into account. Therefore, I think that such a mathematical modeling to describe the calcite growth involving the TR formation is strongly demanded.

Ian Ford asked: What might be the expected scale of fluctuations in the edge position under these conditions? Is it possible that the observations might be consistent with a smeared image of an edge that is fluctuating back and forth?

Takeshi Fukuma replied: Under our experimental conditions, we rarely see switching between the growth and dissolution processes. In the dissolution experiments in pure water, we only see the retreating steps, while in the growth experiments in a supersaturated solution, we only see the growing steps. Only in very limited cases was such switching observed. Even in these cases, the TR edges moving back and forth were clearly visible in the AFM images. Thus, it is very unlikely that the observed TR corresponds to the time-averaged image of the fluctuating step edges. Strictly speaking, at the TR edges, there must be atomic-scale fluctuations corresponding to the adsorption or desorption of the growth or dissolution units, respectively. Although our present AFM is not fast enough to capture such events, the clear atomic-scale step edge structures visualized by our AFM suggest that the scale of such fluctuations should be on the order of the atomic scale.

Conflicts of interest

There are no conflicts to declare.

Faraday Discussions

PAPER

Progress in understanding crystallisation: a personal perspective†

Sarah L. Price *

Received 12th April 2022, Accepted 26th April 2022
DOI: 10.1039/d2fd00077f

After this Discussion meeting, most participants felt that we do not understand crystallisation. However, in the 1980s, I believe that most scientists would have considered that crystallisation was adequately understood. These concluding remarks give a personal impression of the progress that has been made towards appreciating the complexity of crystallisation over the past forty years.

Introduction

This meeting has discussed in depth 23 excellent papers, covering a wide range of systems using diverse techniques. The organisers have brought together many experts on crystallisation from around the world; many have been together in York, but there has also been active participation from those unable to travel because of distance or the Covid epidemic. Nonetheless, a show of hands at the end of the meeting showed, I think unanimously, that we do not understand crystallisation.

This was not a scientific survey, by today's standards, as the people in the room could see each other's responses. However, I think a more independent poll in the 1980s would have shown that the majority of scientists, particularly chemical engineers, felt that we understood crystallisation. It certainly was not a major academic research area. However, such a poll in the 1980s would have been conducted by face-to-face interviews, or possibly by telephone. Computer power was so limited that many scientists dismissed the emerging areas of electronic and atomistic simulations as irrelevant to specific systems. The experimental techniques were also limited in what could be determined at the atomic scale; crystal structures could only be determined from large, high quality crystals, not from powder X-ray data, the range of radiation sources was limited, and atomic force spectroscopy, let alone transmission electron spectroscopy, were a very long

Department of Chemistry, University College London, 20 Gordon St, London, WC1H 0AJ, UK. E-mail: s.l.price@ucl.ac.uk

† This is a personal account from someone who has only done computational simulations of the organic solid state. It may inadvertently include some misinterpretations of the papers. Please check the references carefully.

way from being established techniques that could be applied to crystallisation studies.

Yet today, it would have been difficult for anyone who read the papers, let alone attended this *Faraday Discussion*, to conclude that we did understand crystallisation. The experimental evidence in many papers raised more questions, and a significant number mentioned the further work needed. New techniques were used (https://doi.org/10.1039/d1fd00084e; https://doi.org/10.1039/d1fd00125f; https://doi.org/10.1039/d1fd00090j), but it was clear that many experiments were studying systems that had been chosen or adapted to make the experiments viable. These adaptations, such as using D_2O rather than H_2O to ease the spectroscopy (https://doi.org/10.1039/d1fd00078k), using sugar solution to reduce mobility (https://doi.org/10.1039/d1fd00115a), or humidity control (https://doi.org/10.1039/d1fd00090j), could have affected the crystallisation. Perhaps a more significant development since the 1980s was that most papers were using multiple techniques. I estimate that 11 papers used only experimental techniques, 4 only computer simulations, and there were 8 that combined both experiment and computational modelling. In the 1980s, the emerging electronic structure and atomistic modelling work was done independently, and some cynics felt that the theoretical publications always lagged behind the published experimental results, with more accurate measurements leading to more expensive and, hence, more accurate calculations being performed. Now, there is far more complementary collaboration, with multi-author papers where multi-scale modelling with multi-technique crystallisation characterisation are used to provide a picture of what is going on during crystallisation (https://doi.org/10.1039/d1fd00112d).

We understand the crystallisation of what?

The most popular system for study at this Discussion meeting has been $CaCO_3$, but this has been investigated from many perspectives (https://doi.org/10.1039/d1fd00084e; https://doi.org/10.1039/d1fd00095k; https://doi.org/10.1039/d1fd00082a; https://doi.org/10.1039/d1fd00111f). The other ionic systems with polyatomic ions, $NaBrO_3$ (https://doi.org/10.1039/d1fd00098e) and $SrSO_4$ (https://doi.org/10.1039/d1fd00092f), are not similar in crystallisation behaviour: $NaBrO_3$ has a non-chiral solution but forms enantiomeric crystals (https://doi.org/10.1039/d1fd00098e) and $SrSO_4$ crystallisation can be modelled without considering polymorphism (https://doi.org/10.1039/d1fd00092f). We have discussed simpler systems, such as atomic Ni, but even here the study (https://doi.org/10.1039/d1fd00099c) showed deviations from the classical nucleation theory that held sway in the 1980s. The Ni melt simulations showed that dynamical heterogeneity in the supercooled liquid plays a key role in the mechanism of crystal nucleation. Sodium chloride is also popular as a system to study (https://doi.org/10.1039/d1fd00090j; https://doi.org/10.1039/d1fd00098e; https://doi.org/10.1039/d1fd00089f) because it is a simple system whose crystallisation from water has been utilised by humans from ancient times.

The crystallisation of water in the atmosphere has been discussed (https://doi.org/10.1039/d1fd00115a), as it is vital for understanding environmental science. Water has been heavily studied from all aspects because of its importance to life. This has given rise to a vast literature on its diverse physical properties – which makes it clear that the force-fields for water used in atomistic simulations of aqueous crystallisation are not adequate for the spectroscopy of water clusters, lacking the accurate representation of the intermolecular, particularly many-body, forces.[1,2] However, despite water's position as a uniquely important molecule, its crystallisation behavior and many known phases[3] are unlikely to be uniquely complex, but reflect the extent to which water has been studied. The diversity of the intermolecular interactions and the molecular flexibility of the six organic molecules that have been discussed in this meeting are greater than those of water. Although we now know a lot about the crystallisation of glycine (https://doi.org/10.1039/d1fd00098e; https://doi.org/10.1039/d1fd00101a), diglycine, *p*-aminobenzoic acid (https://doi.org/10.1039/d1fd00112d), isonicotinamide (https://doi.org/10.1039/d1fd00098e), olanzapine and etioporphyrin (https://doi.org/10.1039/d1fd00080b), this hardly spans the types of molecular crystal that are of interest for pharmaceutical development, and *N*,*N*′-bis(*n*-hexyl)naphthalene-1,4,5,8-tetracarboxylicdiimide (NDI-C6) (https://doi.org/10.1039/d1fd00100k) is only one example of the huge range of organic molecules that are of interest as possible functional materials.

The gold nanorod study (https://doi.org/10.1039/d1fd00087j) was informative because it had a simple variation in shape with cylinders of varying aspect ratios, and so clearly showed that anisotropy was a major factor in heterogeneous crystallisation, with nucleation being faster if the rods were more spherical. Although there were significant differences in the nucleation of nanoparticle building units compared to the nucleation of atoms or molecules as building units (https://doi.org/10.1039/d1fd00087j), can we expect the crystallisation of organic molecules, with varying conformations and diverse interactions producing a complex anisotropy, to be simple?

Inorganic crystals can have additional complexities from the variability in composition, particularly if we seek to understand the crystallisation of minerals. The studies of zeolites (https://doi.org/10.1039/d1fd00093d; https://doi.org/10.1039/d1fd00097g; https://doi.org/10.1039/d1fd00096a) provide good examples of the need to consider the chemistry and energetics of the specific system in detail, for example, the morphology of the same zeolite framework being so sensitive to the ions within the channels (https://doi.org/10.1039/d1fd00096a).

This complexity reflects the fact that no two atoms are the same, and nuclear charge is integral. The charge and the number of associated electrons define the electron distribution according to the other atoms present, and cause the differences in the interatomic forces. We are used to recognising substantial changes in the atomic charge distribution with different oxidation or hybridisation states, and with the nature of the bonded atoms. However, the differences extend to the intermolecular interatomic forces, with the specific atomic charge density affecting all the intermolecular forces and, hence, all the properties relevant to crystallisation, such as size and mobility.

This has implications for crystal structures depending on the structural building blocks. In the 1980s, I was using the atomic radii of the cations to separate 172 chalconide spinels, AB_2X_4, into the normal or inverse spinel cation distribution.[4] Only recently have I surveyed the crystal structures of a large group of closely related organic molecules, namely 232 chalcones with small substituents.[5] The isostructural families are small because the most common 1D packing motif is a close packed translation which can have a wide range of substituents sticking out, which pack differently in 3D.[5] Strongly favoured 3D framework structures that can accommodate a wide range of ions or other small species are more prevalent in inorganic than organic systems.

The uniqueness of each atom's charge distribution affects crystallisation more than just determining the final product of the crystal structure. No atom will interact with a given solvent molecule in exactly the same way. We need to carefully distinguish between different ways of modelling solvent effects. Implicit solvation models, where the solvent is represented by a dielectric continuum, are great for modelling the differences between polar and non-polar solvents if the solvent can be assumed to be moving very quickly around the molecule. Explicit solvent modelling is needed if there are long-lived interactions between the solvent and solute, for example water forming a hydrogen bonded network around part of the molecule, which is in competition with hydrogen bonding between the solute molecules. For example, in *p*-aminobenzoic acid, the different hydrogen bonding and $\pi\cdots\pi$ bonding motifs that are found in the polymorphs and in solution (https://doi.org/10.1039/d1fd00112d) are detected by solution NMR.[6] However, the dominant oligomers are remarkably solvent-dependent and also short-lived in molecular dynamics simulations,[6] showing how difficult it can be to link the solution species to the polymorph formed *a priori*. We do tend to focus on hydrogen bonds, but the olanzapine dimer (https://doi.org/10.1039/d1fd00080b) that is found in virtually all its crystal structures (a notable exception is a polymorph that was crystallised within a polymer[7]) is held together with dispersion forces, with the hydrogen bonding between dimers or with other molecules (for the salts and solvates) determining the crystal structure.

Pathways to crystallisation

As pointed out in the opening lecture (https://doi.org/10.1039/d2fd00061j), classical nucleation theory was the theory in the 1980s and it was only in the past couple of decades that the range of mechanisms for crystallisation by particle attachment in synthetic, biogenic and geologic environments became clear. The range of mechanisms in the classic diagram[8] has been illustrated throughout this *Discussion*, but the diversity and complexity has been amplified in the evidence for specific systems. For the simple system of NaCl, simulations suggest that there is a transition between predominantly one-step to two-step nucleation with concentration (https://doi.org/10.1039/d1fd00089f). The occurrence of different nucleation mechanisms for the same system can be explained using a toy model to describe the nucleation pathway, inspired by mesoscopic nucleation theory (https://doi.org/10.1039/d1fd00092f). This Discussion has also produced more detail within each mechanism, specific to variations in the local environment changing the nature and distribution of particles that are attaching. An example is the work on establishing causal relationships between amorphous precursor properties and their effects on the selection of mechanistic pathways of crystallisation and ultimately the properties of the crystalline product (https://doi.org/10.1039/d1fd00096a).

The diversity of crystallisation mechanisms is increased by heterogeneous nucleation with studies using mica, polystyrene sulfonate (PSS) functionalised substrates (https://doi.org/10.1039/d1fd00087j), alkanethiol self-assembled monolayers (https://doi.org/10.1039/d1fd00082a), glass beads (https://doi.org/10.1039/d1fd00101a) and feldspars (https://doi.org/10.1039/d1fd00115a). The relevant surface for nucleation can be within defects (https://doi.org/10.1039/d1fd00115a). The effect may depend on specific intermolecular interactions such as hydrogen bonding (https://doi.org/10.1039/d1fd00101a). Certain "active" defects can affect the diffusion of ions in close proximity to the surface feature (https://doi.org/10.1039/d1fd00082a). In both cases, the molecular flexibility of the growth unit, heterogeneous substrate and how the process of crystallisation reduces the molecular flexibility could have a significant influence. The implication is that the crystallisation vessel surfaces, impurities, and other often unrecorded details, can affect the path of a crystallisation in a complex manner.

"In our opinion, virtually anything is possible" was said in rationalising the change in crystallisation of calcium carbonate on changing solvent from H_2O to D_2O (https://doi.org/10.1039/d1fd00078k), but may be typical of the complexity of the factors influencing crystallisation. "Who cares what you measure?" was said provocatively to emphasise the need for accuracy, reproducibility and careful definition of the experimental system. The papers in the meeting give food for thought – should the ideas emerging, albeit for other systems, be considered for interpreting your results? Examples that struck me as potentially important are:

"The dissolution process does not follow the same structural path as the growth process, or that the timescale for the dissolution process is much shorter than that of the growth process" (https://doi.org/10.1039/d1fd00125f)

"Relative stability is not the primary driving force for transformations between metastable phases" (https://doi.org/10.1039/d1fd00086a).

"… the lack of general rules to correlate the majority solute species and crystal structural blocks that carry the crystal symmetry" (https://doi.org/10.1039/d1fd00080b).

"Consequently, a separation between a specific class of pre-nucleation clusters and the eventual nucleation event becomes less and less evident as the supersaturation value increases. Eventually, the effective energy is so low that all possible pre-critical densities, *i.e.* from monomers and onward, are all likely to be observed, making it harder to differentiate any specific class of precursor clusters" (https://doi.org/10.1039/d1fd00092f).

"This reveals that the highest cooling rates access a much wider range of supersaturations and nucleation cluster sizes when compared to lowest cooling rates" (https://doi.org/10.1039/d1fd00112d).

"…the necessity of generalizing the concept of a critical nucleus to that of a critical ensemble of nuclei" (https://doi.org/10.1039/d1fd00089f).

"Structural pre-ordering in the liquid without a decrease in mobility, therefore, does not promote the nucleation of the crystalline bulk phase" (https://doi.org/10.1039/d1fd00099c).

"Thus, it is possible to tune the pathway of crystallization by controlling the microstructural evolution of amorphous precursors" (https://doi.org/10.1039/d1fd00096a).

High-speed frequency modulation atomic force microscopy has shown that there is sometimes a transition layer of $Ca(OH)_2$ around calcite steps, with the steps sometimes being linear and sometimes complex (https://doi.org/10.1039/

d1fd00084e). I was also intrigued to hear that morphology could be affected by stacking faults (https://doi.org/10.1039/d1fd00091h) or synthesis temperature (https://doi.org/10.1039/d1fd00093d), and it was good to see the progress in modelling morphology (https://doi.org/10.1039/d1fd00097g).

These are just some of the many points raised in the Discussion and conclusion sections of the papers in this *Discussion*, which represent huge progress on the assumptions that were made in the 1980s. We are in a phase of observing the complexities of crystallisation and it will require more applications to different systems for a sense of what concepts are quite widely applicable, and which are very specific to a given experiment. However there are a few more questions that need consideration to define what we are aiming for as an understanding of crystallisation.

What is crystallisation?

Our knowledge of the inapplicability of the atomic level model of classical nucleation theory, and the extent to which we are beginning to replace it, is affected by the use of measures of crystallisation and structural purity that would not have been possible a few decades ago. No student has been mounting hundreds of crystals on a single crystal X-ray diffractometer for these papers. Only electronic structure work has considered infinite perfect static lattices to represent the crystal. Indeed, the molecular dynamics work illustrates the problem of a precise definition of crystallisation. How do you analyse the trajectories of hundreds or thousands of moving molecules (albeit in a periodically repeating simulation box) to have the possibility of identifying a clear transition state for crystallisation (https://doi.org/10.1039/d1fd00089f)?

The experimental definitions have been very careful, with three different probes to determine the time of nucleation (https://doi.org/10.1039/d1fd00092f) or testing to show that the experimental set up has eliminated heterogeneities and memory effects (https://doi.org/10.1039/d1fd00090). The need to control often unreported parameters, such as humidity, has been clearly demonstrated to be important (https://doi.org/10.1039/d1fd00090). A heterogeneous surface needs careful characterisation, given the evidence that certain atomic scale defects can affect the role of the surface in heterogeneous nucleation (https://doi.org/10.1039/d1fd00082a). The key nucleating surface can often be exposed on other surfaces through defects (https://doi.org/10.1039/d1fd00115a).

The role of absorbed species and stacking faults on crystallisation can be very significant, promoting growth in lateral directions and, hence, changing the morphology (https://doi.org/10.1039/d1fd00091h). The effect of defects on morphology has been modelled for zeolites (https://doi.org/10.1039/d1fd00097g). Stacking faults and polymorphic domains are also important in pharmaceuticals; there was considerable controversy over the finding of different polymorphic domains within single crystals of aspirin.[9] The three racemic polymorphs of tazofelone are different stackings of the same sheet,[10] but the different incidence of stacking faults, seen as diffuse streaks in the diffraction, gives rise to variations in melting temperatures and enthalpies between single crystals.

Finding useable definitions for the grey areas

There have been many discussions and findings which illustrate the problems of wanting black and white definitions:

The definition of whether a two component system is a salt or cocrystal can hinge on the location of a proton, but there is a salt–cocrystal continuum.[11] This distinction has been crucial in the emergence of cocrystals as a viable way of changing the crystalline properties of a pharmaceutical in the manufactured form.[12,13] Hence, the refinement of the ΔpK_a rule for predicting whether two compounds will form a salt or cocrystal is important, with the new observation that water in the lattice shifts the range of ambiguity closer to the properties of water (https://doi.org/10.1039/d1fd00081k). For systems in the intermediate region of ΔpK_a, understanding this raises a chicken and egg type question. Does hydration occur because of proton transfer? Or does proton transfer occur because of hydration?

The conclusion that "long-range order structures form through continuous second order transformation from precursors with similar short-range order, but that interconversion between structures with distinct short-range order occurs *via* discontinuous first order transitions" (https://doi.org/10.1039/d1fd00086a) may well apply to systems other than barium calcium carbonates, but needs definitions of "short and long-range order", as well as "similar". How to differentiate between polymorphs and experimental variations of the same structure is not always trivial,[14] as some polymorphs can be very similar.

Confinement is clearly linked to nucleation, but how closely do you have to define the size and shape of the confinement, as well as the nature of the exposed atoms and their charges within the inner surfaces, given that this can clearly determine polymorphic outcome (https://doi.org/10.1039/d1fd00111f)? If a specific surface has a strong influence on nucleation (https://doi.org/10.1039/d1fd00115a) and this is mainly exposed within defects, the convolution of confinement and templating effects will be hard to quantify in the laboratory, let alone model for natural samples. How much will this depend on the system?[15]

The discovery of a new surface polymorph, that is not observed in bulk (https://doi.org/10.1039/d1fd00100k), raises the question of how thin a crystal can be, and whether we can call a structure that diffracts a crystal if it cannot exist except on a specific substrate.

What do you understand by understand?

A pragmatist might say that all that we need is to understand crystallisation processes well enough to be able control industrial crystallisation processes. The observation that secondary nucleation is much more important than primary nucleation in the crystalliser (https://doi.org/10.1039/d1fd00098e) suggests that if this is the only goal, we have been looking in the wrong place.

This meeting has been about fundamental science, using our developing techniques, rather than targeted on improving an industrial crystallisation process. In recent decades, the impetus for scientific understanding, or at least for funding fundamental science, is wanting an ability to design new industrial processes. This meeting has demonstrated potential for novel industrial processes. The role of iodide in controlling the growth of copper, allowing the crystallisation of Cu microplates, has been modelled and realised experimentally (https://doi.org/10.1039/d1fd00091h). We have seen how $CaCO_3$ can be made to crystallise with diverse inclusions of μm size https://doi.org/10.1039/d1fd00095k, which is really impressive when decades ago crystallisation was used for its ability to exclude impurities. This approach for occluding μm hollow calcite spheres within calcite single crystals could be used to create a delivery system for active compounds. In zeolites, the interaction strength of the cation with the pre-nucleation cluster and crystal surface determines crystal morphology and size, a rationalisation of observations that suggests a route for quality control (https://doi.org/10.1039/d1fd00093d). Knowing that alkali infusion significantly reduces the crystallisation times of certain zeolites, and that polymer addition markedly enhances the rates of crystallisation, suggests that these paths can improve the efficiency of hydrothermal syntheses in commercial zeolite production (https://doi.org/10.1039/d1fd00096a).

Increasingly, scientific understanding is defined by having an atomistic model in agreement with experiment. This is going way beyond the expectations in the days before computer simulations could tackle real systems realistically. Indeed, the surprise that a proposed model, despite extensive simplification, adequately describes the kinetics of the observed zeolite crystallisation (https://doi.org/10.1039/d1fd00093d) gives a sense of how much our expectations of scientific modelling have changed. Nonetheless, most of the papers are careful to use terms like "consistent with" or "suggest".

Computer aided design as a measure of understanding

What we aspire to is the quantitative agreement of simulation results with experiments for the observables that experiments can measure, so that we can have enough trust in the underlying atomistic understanding to be able to use it to design new processes. The concept of the digital design of crystals and processes has been emerging over the last decade or so.[16] At a Directed Assembly meeting hosted at Pfizer in 2011, Robert Docherty envisioned that we should aim for the design of a drug product and its manufacturing processes to be like aircraft design – all done computationally before the first piece of steel is cut. This reflects the intimate and inter-dependent relationship among the structure, properties, performance, and processing of a drug.[17] The processing includes crystallisation and the filtering and drying that will depend on the morphology and hydroscopicity of the crystals and the spectra used for monitoring, all properties which will depend on the specific crystal structure. The crystal particles that are within the pharmaceutical product will have the required stability and dissolution rate only if any phase transformations are under control, which will depend on the particle sizes, surfaces and defects.[17]

My own research area is organic crystal structure prediction, and in the opening talk of the 2018 *Faraday Discussion* on this topic, I developed a 2D schematic for progress towards what most experimentalists want from an ideal organic crystal structure prediction (CSP) code.[18] The current form is CSP_0, based on the relative lattice energies of the many computer generated structures containing the same molecule. One axis is the complexity of the system, going from small rigid organic molecules through to more flexible pharmaceuticals with a variety of competing functional groups and range of conformations, to multi-component crystals such as pharmaceutical salt solvates. However, CSP_thd is based on the assumption that the most thermodynamically stable structure will crystallise, and the lattice energy does not include the effects of temperature. It is impossible to crystallise a molecule at 0 K, and many polymorphs are enantiotropically related, *i.e.* the stability order changes below the crystallisation temperature, as seen in NDI-C6 (https://doi.org/10.1039/d1fd00100k) (CSP is mainly useful because some polymorphic phase transitions are not observed experimentally, at least on a time-scale of pharmaceutical development). However, getting the relative thermodynamics right at ambient temperatures is demanding of the calculations as many organic molecules are relatively close to their melting points and may have a large amplitude of motion of parts of the molecule. Pressure and shearing can also affect the bulk relative thermodynamic stability. However, the relative thermodynamic stability of polymorphs can also depend on particle size when the particles are small enough that the surfaces affect the relative thermodynamics (https://doi.org/10.1039/d1fd00111f).[19] The relative thermodynamic stability, particularly of other phases such as hydrates, may also be affected by other variables like the solvent and relative humidity.[20] However, even if we were accurately ranking the thermodynamic stability of the computer-generated structures, there would be a major problem of over-prediction for most molecular systems:[21] far more crystal structures appear to be thermodynamically plausible than there are observed polymorphs.[22] The CSP over-prediction problem is consistent with our discussions, as the suggestion that each pre-nucleation crystal might consist of a different polymorph (https://doi.org/10.1039/d1fd00125f) implies that many low energy structures may be ephemeral phases that are involved in crystallisation. It is clear from this meeting that there is no simple computational model for crystallisation kinetics that could provide an experimental recipe to crystallise a predicted polymorph for the first time. We also do not have sufficient understanding of crystallisation to say that the most thermodynamically stable computer-generated crystal structure could never be found, or at least, is unlikely to affect the crystallisation of a readily crystallised metastable form.[23]

Over-prediction and synthesisability is also a major issue in inorganic crystal structure prediction, though the emphasis is on the composition as well as the structure for targeting novel materials.[24] Organic CSP is also used to help target new functional materials,[24] but there is the additional commercial interest as a complement to pharmaceutical development, where the molecule is chosen for its medicinal effects.

Conclusion

This is a personal perspective on how the understanding of crystallisation has changed during my scientific career. I trained as a theoretical chemist interested in developing models for intermolecular forces from the molecular charge distribution. This was justified by the belief that an accurate model for intermolecular forces between (rigid) molecules would be capable of predicting the behavior of molecules in all phases, gas, liquid and solid.[2] My assertion that in the

1980s most scientists thought that crystallisation was understood is based on having my career path changed by a sudden spectacular increase in interest in polymorphism and CSP. This was mainly due to Abbot Laboratories losing control of the manufacture of ritonavir.[26] The idea that crystallisation was not reliably reproducible was completely anathema to scientists (indeed, "disappearing" polymorphs was the one example from the physical sciences in the book by Rupert Sheldrake, "A New Science of Life: The Hypothesis of Morphic Resonance", whose publication in 1981 lead to a *Nature* editorial entitled "A book for burning?"[27]). Abbott Laboratories quickly reformulated ritonavir to get around the problems of manufacturing the drug product after the appearance of the more stable form II,[28] but it required heroic efforts by their scientists and cost the company dearly. Joel Bernstein, who was one of the very small number of academic experts in polymorphism at the time,[29] did a considerable amount of legal and scientific work on "disappearing" polymorphs.[30] It can require extraordinary precautions, such as work being done in new labs by scientists who have not entered labs where the more stable form has been produced, to crystallise the metastable form again.[30] Our understanding of crystallisation now reflects why polymorphs may appear to "disappear".[31]

There has been a huge amount of work on trying to develop experimental[32] and computational methods to mitigate the risk of the late appearance of a more stable form during drug development, as exemplified by ritonavir.[32,33] This provided a new motivation for developing crystal structure prediction methods, which were initially being developed as a test of the hypothesis that a molecule would crystallise in the most stable structure and as an aid to the discovery of functional organic materials, such as energetics and non-linear optical materials. Arguably, crystal structure prediction is now changing from basic science to an applied technology[34] as it could have prevented the late appearance of a more stable form affecting the treatment of patients suffering from Parkinson's disease.[35] Thus, as understanding crystallisation involves controlling polymorphism, as well as particle size and morphology, there has been considerable progress during my scientific career.

Consider the infamous quote from Donald Rumsfeld in 2011. "There are known knowns, things we know that we know; and there are known unknowns, things that we know we don't know. But there are also unknown unknowns, things we do not know we don't know". There are aspects to crystallisation that very few people knew about in the 1980s that are now well established. The experimental work in this discussion has helped move some "unknowns" to "knowns". The feeling that we currently don't understand crystallisation is a result of how much more we now know, through the application of new techniques and through scientific discussion. We now know a lot more about the complexity of the factors that determine the path of crystallisation. However, if you feel that the more you learn about crystallisation, the more you realize how much you don't know, then you are in good company.

Author contributions

SLP is the sole author, and this is a merely a quickly written version of her concluding remarks at the *Faraday Discussion*.

Conflicts of interest

There are no conflicts to declare.

Acknowledgements

The authors of the papers presented at the *Faraday Discussion* "Understanding Crystallisation" and all the participants are thanked for such a stimulating meeting.

Notes and references

1 A. Albaugh, H. A. Boateng, R. T. Bradshaw, O. N. Demerdash, J. Dziedzic, Y. Z. Mao, D. T. Margul, J. Swails, Q. Zeng, D. A. Case, *et al.*, Advanced Potential Energy Surfaces for Molecular Simulation, *J. Phys. Chem. B*, 2016, **120**(37), 9811–9832, DOI: 10.1021/acs.jpcb.6b06414.

2 A. J. Stone *The Theory of Intermolecular Forces*; Oxford University Press, 2013.

3 C. G. Salzmann, P. G. Radaelli, B. Slater and J. L. Finney, The polymorphism of ice: five unresolved questions, *Phys. Chem. Chem. Phys.*, 2011, **13**(41), 18468–18480.

4 G. D. Price, S. L. Price and J. K. Burdett, The Factors Influencing Cation Site-Preferences in Spinels - a New Mendelyevian Approach, *Phys. Chem. Miner.*, 1982, **8**(2), 69–76.

5 L. S. Price and S. L. Price, Packing Preferences of Chalcones: A Model Conjugated Pharmaceutical Scaffold, *Cryst. Growth Des.*, 2022, **22**(3), 1801–1806, DOI: 10.1021/acs.cgd.1c01381.

6 R. Bobrovs, L. Drunka, A. Auzins, K. Jaudzems and M. Salvalaglio, Polymorph-Selective Role of Hydrogen Bonding and pi–pi Stacking in *p*-aminobenzoic Acid Solutions, *Cryst. Growth Des.*, 2021, **21**(1), 436–448, DOI: 10.1021/acs.cgd.0c01257.

7 S. Askin, J. K. Cockcroft, L. S. Price, A. D. Goncalves, M. Zhao, D. A. Tocher, G. R. Williams, S. Gaisford and D. Q. M. Craig, Olanzapine Form IV: Discovery of a New Polymorphic Form Enabled by Computed Crystal Energy Landscapes, *Cryst. Growth Des.*, 2019, **19**(5), 2751–2757, DOI: 10.1021/acs.cgd.8b01881.

8 J. J. De Yoreo, P. U. P. A. Gilbert, N. A. J. M. Sommerdijk, R. L. Penn, S. Whitelam, D. Joester, H. Z. Zhang, J. D. Rimer, A. Navrotsky, J. F. Banfield, *et al.*, Crystallization by particle attachment in synthetic, biogenic, and geologic environments, *Science*, 2015, **349**(6247), aaa6760, DOI: 10.1126/science.aaa6760.

9 A. D. Bond, R. Boese and G. R. Desiraju, On the polymorphism of aspirin: Crystalline aspirin as intergrowths of two "polymorphic" domains, *Angew. Chem., Int. Ed.*, 2007, **46**(4), 618–622.

10 L. S. Price, J. A. McMahon, S. R. Lingireddy, S. F. Lau, B. A. Diseroad, S. L. Price and S. M. Reutzel-Edens, A molecular picture of the problems in ensuring structural purity of tazofelone, *J. Mol. Struct.*, 2014, **1078**, 26–42, DOI: 10.1016/j.molstruc.2014.01.014.

11 S. L. Childs, G. P. Stahly and A. Park, The salt-cocrystal continuum: The influence of crystal structure on ionization state, *Mol. Pharmaceutics*, 2007, **4**(3), 323–338.

12 S. L. Childs and M. J. Zaworotko, The Reemergence Of Cocrystals: The Crystal Clear Writing Is On The Wall (Introduction to Virtual Special Issue on Pharmaceutical Cocrystals), *Cryst. Growth Des.*, 2009, **9**(10), 4208–4211.

13 C. B. Aakeroy, M. E. Fasulo and J. Desper, Cocrystal or salt: Does it really matter?, *Mol. Pharmaceutics*, 2007, **4**(3), 317–322, DOI: 10.1021/mp060126o.

14 J. van de Streek, Searching the Cambridge Structural Database for the 'best' representative of each unique polymorph, *Acta Crystallogr., Sect. B: Struct. Sci.*, 2006, **62**, 567–579.

15 U. V. Shah, C. Amberg, Y. Diao, Z. Q. Yang and J. Y. Y. Heng, Heterogeneous nucleants for crystallogenesis and bioseparation, *Curr. Opin. Chem. Eng.*, 2015, **8**, 69–75, DOI: 10.1016/j.coche.2015.03.002.

16 M. Bryant, I. Rosbottom, I. Bruno, R. Docherty, C. Edge, R. Hammond, R. Peeling, J. Pickering, K. Roberts and A. Maloney, "Particle Informatics": Advancing Our Understanding of Particle Properties through Digital Design, *Cryst. Growth Des.*, 2019, **19**(9), 5258–5266, DOI: 10.1021/acs.cgd.9b00654.

17 C. C. Sun, Material Science Tetrahedron-A Useful Tool for Pharmaceutical Research and Development, *J. Pharm. Sci.*, 2009, **98**, 1671–1687.

18 S. L. Price, Is zeroth order crystal structure prediction (CSP_0) coming to maturity? What should we aim for in an ideal crystal structure prediction code?, *Faraday Discuss.*, 2018, **211**, 9–30, DOI: 10.1039/c8fd00121a.

19 A. M. Belenguer, G. I. Lampronti, A. J. Cruz-Cabeza, C. A. Hunter and J. K. M. Sanders, Solvation and surface effects on polymorph stabilities at the nanoscale, *Chem. Sci.*, 2016, **7**(11), 6617–6627, DOI: 10.1039/c6sc03457h.

20 D. E. Braun, L. H. Koztecki, J. A. McMahon, S. L. Price and S. M. Reutzel-Edens, Navigating the Waters of Unconventional Crystalline Hydrates, *Mol. Pharmaceutics*, 2015, **12**(8), 3069–3088, DOI: 10.1021/acs.molpharmaceut.5b00357.

21 R. M. Bhardwaj, J. A. McMahon, J. Nyman, L. S. Price, S. Konar, I. D. H. Oswald, C. R. Pulham, S. L. Price and S. M. Reutzel-Edens, A Prolific Solvate Former, Galunisertib, under the Pressure of Crystal Structure Prediction, Produces Ten Diverse Polymorphs, *J. Am. Chem. Soc.*, 2019, **141**(35), 13887–13897, DOI: 10.1021/jacs.9b06634.

22 D. E. Braun, J. A. McMahon, R. M. Bhardwaj, J. Nyman, M. A. Neumann, J. van de Streek and S. M. Reutzel-Edens, Inconvenient Truths about Solid Form Landscapes Revealed in the Polymorphs and Hydrates of Gandotinib, *Cryst. Growth Des.*, 2019, **19**(5), 2947–2962, DOI: 10.1021/acs.cgd.9b00162.

23 D. E. Braun, H. Oberarcher, K. Arnhard, M. Orlova and U. J. Griesser, 4-Aminoquinaldine monohydrate polymorphism: Prediction and impurity aided discovery of a difficult to access stable form, *CrystEngComm*, 2016, **18**, 4053–4067, DOI: 10.1039/C5CE01758K.

24 W. Sun, S. Dacek, S. Ong, G. Hautier, A. Jain, W. Richards, A. Gamst, K. Persson and G. Ceder, The thermodynamic scale of inorganic crystalline metastability, *Sci. Adv.*, 2016, **2**(11), e1600225, DOI: 10.1126/sciadv.1600225.

25 G. M. Day and A. I. Cooper, Energy-Structure-Function Maps: Cartography for Materials Discovery, *Adv. Mater.*, 2018, **30**(37), 1704944, DOI: 10.1002/adma.201704944.

26 S. R. Chemburkar, J. Bauer, K. Deming, H. Spiwek, K. Patel, J. Morris, R. Henry, S. Spanton, W. Dziki, W. Porter, *et al.*, Dealing with the impact of ritonavir

polymorphs on the late stages of bulk drug process development, *Org. Process Res. Dev.*, 2000, **4**(5), 413–417.

27 https://en.wikipedia.org/wiki/Rupert_Sheldrake.

28 J. Bauer, S. Spanton, R. Henry, J. Quick, W. Dziki, W. Porter and J. R. Morris, An extraordinary example of conformational polymorphism, *Pharm. Res.*, 2001, **18**(6), 859–866.

29 J. Bernstein. *Polymorphism in Molecular Crystals*; Clarendon Press, 2020.

30 J. D. Dunitz and J. Bernstein, Disappearing Polymorphs, *Acc. Chem. Res.*, 1995, **28**(4), 193–200.

31 D. K. Bucar, R. W. Lancaster and J. Bernstein, Disappearing Polymorphs Revisited, *Angew. Chem., Int. Ed.*, 2015, **54**(24), 6972–6993, DOI: 10.1002/anie.201410356.

32 S. L. Morissette, S. Soukasene, D. Levinson, M. J. Cima and O. Almarsson, Eludication of crystal form diversity of the HIV protease inhibitor ritonavir by high-throughput crystallisation, *Proc. Natl. Acad. Sci. U. S. A.*, 2003, **100**(5), 2180–2184.

33 C. Wang, I. Rosbottom, T. Turner, S. Laing, A. Maloney, A. Sheikh, R. Docherty, Q. Yin and K. Roberts, Molecular, Solid-State and Surface Structures of the Conformational Polymorphic Forms of Ritonavir in Relation to their Physicochemical Properties, *Pharm. Res.*, 2021, **38**(6), 971–990, DOI: 10.1007/s11095-021-03048-2.

34 J. Nyman and S. M. Reutzel-Edens, Crystal structure prediction is changing from basic science to applied technology, *Faraday Discuss.*, 2018, **211**, 459–476, DOI: 10.1039/c8fd00033f.

35 M. Mortazavi, J. Hoja, L. Aerts, L. Quere, J. van de Streek, M. A. Neumann and A. Tkatchenko, Computational polymorph screening reveals late-appearing and poorly-soluble form of rotigotine, *Commun. Chem.*, 2019, **2**, 70, DOI: 10.1038/s42004-019-0171-y.

Poster titles

Additives manufacturing of porous biominerals, **R. Zhao and E. Amstad,** *École Polytechnique Fédérale de Lausanne (EPFL), Switzerland*

Investigation of anti-freeze and ice nucleating properties of Affimer proteins, **I. Sandei, T. Whale, C. Tiede, T. Gaule and F. Meldrum,** *University of Leeds, UK*

The crystallisation and characterisation of basic magnesium chloride salts studied *in situ* using a laboratory SAXS/WAXS instrument, **K. Pickering, S. Huband, K. Shafran and R. Walton,** *University of Warwick, UK*

Effect of crystallization additives on crystallization of 2,6-dimethoxybenzoic acid polymorphs and their stability, **A. Semjonova and A. Bērziņš,** *University of Latvia, Latvia*

Dual-step reduction and formation mechanism of Cu pseudo-icosahedral microcrystals, **W. Sun and W. Zhou,** *University of St Andrews, UK*

Study of the bulk polymorphism of best performing molecular semiconductors, **P. Pandey, Y. Henri Geerts, E. Modena and L. Maini,** *Università di Bologna, Italy*

Mechanisms for rapid dolomitization at ambient conditions, **J. Kim, B. Puchala, U. Becker and W. Sun,** *University of Michigan, USA*

Understanding polar crystal surfaces in solution, **S. J. Cox,** *University of Cambridge, UK*

Controlling the polymorphism of carbamazepine by droplet-confinement *via* conventional spray drying, **A. Parkes, A. Ziaee, G. Walker and E. O'Reilly,** *University of Limerick, Ireland*

Homogeneous nucleation in the Al(III) system in the presence of organic additives, **M. J. Lukić and D. Gebauer,** *Institute of Inorganic Chemistry, Leibniz University Hannover, Germany*

Controlling urea crystallisation *via* heterogeneous nucleation, **S. Anker, D. McKechnie, P. A. Mulheran, J. Sefcik and K. Johnston,** *University of Strathclyde, UK*

Nucleation studies of a THF-hydrate phase-change material, **H. T. Logan, D. E. Oliver and C. R. Pulham,** *University of Edinburgh, UK*

Composition of polymer stabilized amorphous calcium carbonates, **M. Gindele and D. Gebauer,** *Institute of Inorganic Chemistry, Leibniz University Hannover, Germany*

Exploiting the surface properties of graphene for polymorph selectivity, **M. Boyes, A. Alieva, J. Tong, V. Nagyte, M. Melle-Franco, T. Vetter and C. Casiraghi**, *University of Manchester, UK*

Nucleation induced by sheared flow in two geometrically well-defined systems, **C. Devos, J. Yin, T. van Gerven and S. Kuhn**, *Katholieke Universiteit Leuven, Belgium*

Study of bulk polymorphism of NDI derivatives, **I. de Oliveira Martins, E. Modena and L. Maini**, *Università di Bologna, Italy*

One dimensional crystal nucleation rates, **K. E. Blow, D. Quigley and G. C. Sosso**, *University of Warwick, UK*

In-situ monitoring crystallization from solution by electrical readout, **J. Tong, A. Doumbia, R. U. Khan, A. Rahmanudin, M. Turner and C. Casiraghi**, *University of Manchester, UK*

Metal organic frameworks: understanding crystal growth at the nanoscale, **R. J. S. Pooley, M. W. Anderson, M. Attfield, S. O'Shea and K. Banas**, *University of Manchester, UK*

In-situ optical spectroscopy of crystallization: one crystal nucleation at a time, **O. Urquidi, J. Brazard, N. LeMessurier, L. Simine and T. Adachi**, *University of Geneva, Switzerland*

Crystal structure prediction – guided crystallization of elusive polymorphs of meloxicam, **M. K. Dudek, P. Paluch, J. Zając and A. Jeziorna**, *Center of Molecular and Macromolecular Studies, Polish Academy of Sciences, Poland*

Stabilisation of ice clusters by hydrogen-disordering accounts for anomalous enhancement of heterogeneous ice nucleation, **T. Whale**, *University of Warwick, UK*

Nucleation and growth of inorganic crystals at electrified interfaces: calcium carbonate synthesis from electrooxidation of calcium formate, **I. J. McPherson, P. D. Morris and P. R. Unwin**, *University of Warwick, UK*

Single crystal nucleation spectroscopy of amino acids with hydrophobic side chains: A step toward the general nucleation model of amino acids, **O. Urquidi, N. Barbosa, J. Brazard and T. Adachi**, *University of Geneva, Switzerland*

Two-dimensional colloidal crystals on non-uniformly curved surfaces, **M. A. Miller, J. O. Law, J. M. Dean, A. F. Gerrand and H. Kusumaatmaja**, *Durham University, UK*

Homogeneous ice nucleation in an *ab initio* machine learning model of water, **P. M. Piaggi, J. Weis, A. Z. Panagiotopoulos, P. G. Debenedetti and R. Car**, *Princeton University, USA*

The use of nanopipettes for electrochemically controlled precipitation, **P. D. Morris, B. Keating, I. J. McPherson, P. R. Unwin, M. A. Edwards and R. I. Walton**, *University of Warwick, UK*

Nanoscale X-ray chemical imaging, diffraction and spectroscopy of crystallisation using an *in situ* liquid cell, **J. M. Walker, M. A. Gomez-Gonzalez, P. Quinn and J. E. Parker,** *Diamond Light Source, UK*

Following crystallisation of inorganic systems in the confined space of miniemulsion droplets, **F. Tajoli, P. Dolcet and S. Gross,** *University of Padova, Italy*

Reversing the polymorphic outcome of crystallization and the apparent relative stability of nitrofurantoin using additives, **A. Bērziņš, A. Trimdale and J. H. ter Horst,** *University of Latvia, Latvia*

SECCM-IRM: A new tool for quantitative *in situ* studies of crystal dissolution and growth, **B. Keating, I. J. McPherson, D. Valavanis, P. Ciocci, G. N. Meloni, A.-J. Agyei, P. Morris, J.-F. Lemineur, F. Kanoufi and P. R. Unwin,** *University of Warwick, UK*

Exploring the role of crystal habit in the Ostwald rule of stages, **E. Hadjittofis, D. Aytkhozhina, S. Vargas and K. L. Sedransk Campbell,** *UCB Pharma, Belgium*

Alkali metal nitrates: a new model for crystallisation, **V. Fantauzzo, S. R. Yeandel, C. L. Freeman and J. H. Harding,** *University of Sheffield, UK*

First principles calculation of the melting curves of ice polymorphs in the vicinity of the liquid-liquid critical point, **T. E. Gartner III, P. M. Piaggi, R. Car and P. G. Debenedetti,** *Princeton University, USA*

Single particle reconstruction of iron mineral formation within Apoferritin from cryo-TEM, **J. Galloway, O. Nahi, D. P. Maskell, R. F. Thompson, E. L. Hesketh and F. Meldrum,** *University of Leeds, UK*

In-situ observation of crystallisation with terahertz spectroscopy, **J. Kölbel, Q. Li, R. Zhang, A. Bond, M. Davis, T. Korter, T. Threlfall and J. A. Zeitler,** *University of Cambridge, UK*

Directional crystallization from the melt of semiconductor blend, **G. Liu, J. Liu, A. Dunn, P. Nadazdy, P. Siffalovic, R. Resel, M. Abbas, G. Wantz and Y. H. Geerts,** *Laboratoire de Chimie des Polymères, Belgium*

Interfacial free energies from MD simulations: application to $CaSO_4 \cdot xH_2O$, **S. R. Yeandel, C. L. Freeman and J. H. Harding,** *University of Sheffield, UK*

Polymorph selection of zinc zeolitic imidazolate frameworks *via* kinetic and thermodynamic control, **E. Balog, G. Varga, Á. Kukovecz, Á. Tóth, D. Horváth, I. Lagzi and G. Schuszter,** *University of Szeged, Hungary*

Spatial separation of precipitate in reactive media, **P. Papp, Á. Tóth and D. Horváth,** *University of Szeged, Hungary*

First-principles modelling of organic-inorganic interfaces in urinary calculi, **R. Morris, H. Chappell, A. Scott, A. Borissova and J. Smith,** *University of Leeds, UK*

Understanding how green additives control crystallization at the nanoscale, **A. Baken, A. van Driessche, A. Fernandez-Martinez, M. Di Michiel and M. Kellermeier,** *European Synchrotron Radiation Facility (ESRF), Université Grenoble*

Alpes, Université Savoie Mont Blanc, Centre national de la recherche scientifique (CNRS), Institut de Recherche pour le Développement (IRD), Institut français des sciences et technologies des transports, de l'aménagement et des réseaux (IFSTTAR), Institut des sciences de la Terre (ISTerre), France

Kinetic control of competing nucleation in a system of interacting dimers on square lattice, **D. Mandal and D. Quigley,** *University of Warwick, UK*

Observing calcium sulfate crystallisation and phase transformation using *in situ* transmission electron microscopy techniques, **M. Ilett, H. M. Freeman, Z. Aslam, J. M. Galloway, D. P. Klebl, S. P. Muench, I. J. McPherson, O. Cespedes, Y.-Y.Kim, F. Meldrum, S. R. Yeandel, C. L. Freeman, J. H. Harding and, R. M. Drummond-Brydson,** *University of Leeds, UK*

Calcite growth under AFM is transport limited, **R. Darkins, D. Duffy, I. Ford, I. McPherson, P. Unwin,** *University College London, UK*

Kinetic characterization of flow-driven precipitation, **M. Emmanuel, P. Papp, G. Schuszter, Á. Tóth and D. Horváth,** *University of Szeged, Hungary*

Observation and simulation of the phase transition of a plastic crystal of nonadecane, **T. Tanno,** *Akita University, Japan*

The effect of water on polymorph selection in amorphous calcium carbonate systems, **E. Armstrong, C. Freeman and J. Harding**, *University of Sheffield, UK*

Impact of amorphous mediated cocrystallization on dissolution of Telmisartan-alpha ketoglutaric acid solid form, **J. Haneef,** *Jamia Hamdard, India*

Control of crystallization pathways and kinetics of mullite-type $Bi_2M_4O_9$ (M = Fe^{3+}, Al^{3+}, Ga^{3+}) and $Bi_2Mn_4O_{10-}$, A. Kirsch, **N. Lefeld, M. Gogolin, S. Banerjee and K. M. Ø. Jensen,** *University of Copenhagen, Denmark*

From interaction to inhibition: simulations reveal the growth-altering effects of aspartate on calcite, ***A. Broad, R. Darkins, I. J. Ford and D. M. Duffy,*** *University College London, UK*

Evolution trailing of embryonic zeolites into hierarchical ZSM-5, **M. Alonso-Doncel, A. Peral, C. Ochoa-Hernández, R. Sanz and D. P. Serrano,** *IMDEA Energy Institute, Spain*

Effect of agitation on primary nucleation indicates that colloidal scale clusters play role in nucleation mechanism, **C. Forsyth, G. Zimbitas, M. J. Vesga, D. McKechnie, M. Haw, P. Mulheran, K. Johnston and J. Sefcik,** *University of Strathclyde, UK*

Engineered growth of polycrystalline amino acid films for eco-friendly piezoelectric sensing, **S. Guerin,** *University of Limerick, Ireland*

Design of experiment studies of $MgSO_4$ crystallisation, **J. Kölbel, S. Coles, A.-L. Davesne, D. Robinson and T. Threlfall,** *University of Southampton and University of Cambridge, UK*

Controlled crystallisation of APIs by reverse anti-solvent crystallization using micromixing technology, **M. J. Bennett, E. Beveniou and A. Kerr,** *Micropore Technologies, UK*

Mapping thermodynamics to control polymorphism, **I. B. Rietveld, K. Li, G. Gbabode, M. Barrio and J.-L. Tamarit,** *Université de Rouen and Université Paris Cité, France*

Facet specific crystal growth rates of α-glycine crystals in the presence of impurities, **C. Offiler, A. J. Cruz-Cabeza, R. J. Daveya and T. Vetter,** *University of Manchester, UK*

From molecules to materials: structural landscape of zinc terephthalate from solution, **A. Ramanan, M. Jadon, M. Srivastava and P. K. Roy,** *Indian Institute of Technology Delhi, India*

Rational design and control of crystallisation processes using X-ray PDF analysis, **A. R. Pallipurath, B. Evans, A. Pugejs, L. H. Al-Madhagi and S. L. M. Schroeder,** *University of Leeds, UK*

Early stages of the nucleation of organic aromatic carboxylic acids, **F. Lücke and D. Gebauer,** *Institute of Inorganic Chemistry, Leibniz University Hannover, Germany*

Kinetic effects in calcium phosphate nucleation, **D. McDonogh and D. Gebauer,** *Institute of Inorganic Chemistry, Leibniz University Hannover, Germany*

Enzyme-induced crystallization of guanine from aqueous solution, **J. Hasselt, Y.-Y. Kim and F. Meldrum,** *University of Leeds, UK*

The affiliation of the presenting author is listed above.

The Faraday Division and CrystEngComm Poster Prize for the best poster was jointly awarded to Alice Parkes of University of Limerick, Ireland for their poster on "Controlling the polymorphism of carbamazepine by droplet-confinement *via* conventional spray drying" and Alex Broad of University College London, UK for their poster on "From interaction to inhibition: simulations reveal the growth-altering effects of aspartate on calcite".

List of participants

Dr Nathan Abraham, *AbbVie, USA*
Miss Katie Ackermann, *Royal Society of Chemistry, United Kingdom*
Dr Takuji Adachi, *University of Geneva, Switzerland*
Ms Dilafruz Allanazarova, *Leibniz University Hannover, Institute of Inorganic Chemistry, Germany*
Dr María del Mar Alonso Doncel, *IMDEA Energy Institute, Spain*
Professor Michael Anderson, *University of Manchester, United Kingdom*
Ms Samira Anker, *University of Strathclyde, United Kingdom*
Miss Emma Armstrong, *University of Sheffield, United Kingdom*
Ms Isha Bade, *Imperial College London, United Kingdom*
Ms Annet Baken, *European Synchrotron Radiation Facility (ESRF), Université Grenoble Alpes, Université Savoie Mont Blanc, Centre national de la recherche scientifique (CNRS), Institut de Recherche pour le Développement (IRD), Institut français des sciences et technologies des transports, de l'aménagement et des réseaux(IFSTTAR), Institut des sciences de la Terre (ISTerre), France*
Dr Matthew Bennett, *Micropore Technologies, United Kingdom*
Ms Elina Beveniou, *Micropore Technologies, United Kingdom*
Mr Harmeet Bhoday, *Missouri University of Science and Technology, USA*
Ms Katarina Blow, *University of Warwick, United Kingdom*
Dr Matthew Boyes, *University of Manchester, United Kingdom*
Dr Robin Brabham, *Royal Society of Chemistry, United Kingdom*
Mr Alexander Broad, *University College London, United Kingdom*
Dr Agris Bērziņš, *University of Latvia, Latvia*
Mr Ziga Casar, *École polytechnique fédérale de Lausanne (EPFL), Switzerland*
Dr Ruel Cedeno, *Centre Interdisciplinaire de Nanoscience de Marseille, Centre national de la recherche scientifique (CINaM-CNRS), France*
Professor Basab Chattopadhyay, *Norwegian University of Science and Technology, Norway*
Ms Rachel Cole, *Royal Society of Chemistry, United Kingdom*
Professor Eric Collet, *Université de Rennes 1, France*
Mr Dave Collins, *University of Leeds, United Kingdom*
Mr Joseph Counte, *Sensor Coating Systems, United Kingdom*
Dr Stephen Cox, *University of Cambridge, United Kingdom*
Dr Aurora Cruz-Cabeza, *University of Manchester, United Kingdom*
Professor Dr Helmut Cölfen, *University of Konstanz, Germany*
Dr Sourav De, *Research Centre for Natural Sciences, Hungary*
Ms Inês de Oliveira Martins, *Università di Bologna, Italy*
Dr Jim De Yoreo, *Pacific Northwest National Laboratory, USA*
Mr Cedric Devos, *Katholieke Universiteit Leuven, Belgium*
Dr Maria Diaz-Lopez, *Diamond Light Source, United Kingdom*
Mr Florian Dietrich, *University College London, United Kingdom*
Dr Huan Doan, *University of Bristol, United Kingdom*
Professor Rik Drummond-Brydson, *University of Leeds, United Kingdom*

Dr Marta Dudek, *Center of Molecular and Macromolecular Studies, Polish Academy of Sciences, Poland*
Dr Katharina Edkins, *The University of Manchester, United Kingdom*
Miss Vittoria Fantauzzo, *University of Sheffeld, United Kingdom*
Dr Daniela Farmer, *SENFI UK, United Kingdom*
Ms Marta Ferran Marques, *Sensor Coating Systems, United Kingdom*
Professor Kristen Fichthorn, *Pennsylvania State University, USA*
Dr Aaron Finney, *University College London, United Kingdom*
Professor Ian Ford, *University College London, United Kingdom*
Dr Colin Freeman, *University of Sheffield, United Kingdom*
Professor Takeshi Fukuma, *Kanazawa University, Japan*
Dr Johanna Galloway, *University of Leeds, United Kingdom*
Professor Dr Denis Gebauer, *Leibniz University of Hannover, Germany*
Dr Sudeshna Ghosh, *Indian Institute of Technology Kanpur, India*
Mr Maxim Gindele, *Leibniz University Hannover, Institute of Inorganic Chemistry, Germany*
Mr Stuart Govan, *Royal Society of Chemistry, United Kingdom*
Dr Chandler Greenwell, *XtalPi, USA*
Professor Dr Silvia Gross, *Università di Padova, Italy*
Dr Romain Grossier, *Centre Interdisciplinaire de Nanoscience de Marseille, Centre national de la recherche scientifique (CINaM-CNRS), Aix-Marseille Université, France*
Dr Sarah Guerin, *University of Limerick, Ireland*
Ms Mingxia Guo, *Imperial College London, United Kingdom*
Mr Lennart Günzel, *Leipzig University, Germany*
Dr Eftychios(Efty) Hadjittofis, *UCB Pharma, Belgium*
Ms Ingvild Uri Hageberg, *Norwegian University of Science and Technology, Norway*
Dr Jamshed Haneef, *Jamia Hamdard, India*
Professor John Harding, *University of Sheffield, United Kingdom*
Dr Alan Hare, *without affiliation (retired), United Kingdom*
Ms Maya Harris, *École polytechnique fédérale de Lausanne (EPFL), Switzerland*
Mr Jorin Hasselt, *University of Leeds, United Kingdom*
Dr Jim Hickey, *Sensor Coating Systems, United Kingdom*
Dr Adam Hill, *University of Manchester, United Kingdom*
Professor Dezső Horváth, *University of Szeged, Hungary*
Dr Liam Hunter, *University of Leeds, United Kingdom*
Miss Sarah Huynen, *University of Glasgow, United Kingdom*
Dr Martha Ilett, *University of Leeds, United Kingdom*
Dr Adam Keates, *Syngenta, United Kingdom*
Mr Blane Keating, *University of Warwick, United Kingdom*
Mr Joonsoo Kim, *University of Michigan, USA*
Dr Yi-Yeoun Kim, *University of Leeds, United Kingdom*
Professor Dr Yuki Kimura, *Institute of Low Temperature Science, Hokkaido University, Japan*
Dr Andrea Kirsch, *University of Copenhagen, Denmark*
Professor Christine Kirschhock, *Katholieke Universiteit Leuven, Belgium*
Dr Alexei Kiselev, *Karlsruhe Institute of Technology, Institute of Meteorology and Climate Research, Germany*
Miss Weronika Kras, *University of Manchester, United Kingdom*
Dr Christian Kuttner, *Nature Communications / Springer Nature, Germany*
Miss Johanna Kölbel, *University of Cambridge, United Kingdom*
Miss Sarah Latham, *Royal Society of Chemistry, United Kingdom*
Dr Alfred Lee, *Merck Sharp & Dohme, USA*
Miss Qi Li, *University of Cambridge, United Kingdom*

Mr Enshu Liang, *Imperial College London, United Kingdom*
Dr Zhiyu Liao, *University of Glasgow, United Kingdom*
Dr Maya Lipert, *AbbVie, USA*
Dr Guangfeng Liu, *Laboratoire de Chimie des Polymères, Belgium*
Miss Hannah Logan, *University of Edinburgh, United Kingdom*
Professor Dr Olga Lopez Acevedo, *Universidad de Antioquia, Colombia*
Dr Miodrag J. Lukic, *Institute of Inorganic Chemistry, Leibniz University Hanover, Germany*
Mr Felix Lücke, *Institute of Inorganic Chemistry, Leibniz University Hanover, Germany*
Mr Joshua Magiera, *University of Warwick, United Kingdom*
Professor Dr Lucia Maini, *Università di Bologna, Italy*
Dr Dipanjan Mandal, *University of Warwick, United Kingdom*
Dr Ivan Marziano, *Pfizer UK, United Kingdom*
Dr Adriana Matamoros Veloza, *University of Leeds, United Kingdom*
Mr David McDonogh, *Leibniz University Hannover, Germany*
Dr Ian McPherson, *University of Warwick, United Kingdom*
Professor Fiona Meldrum, *University of Leeds, United Kingdom*
Emer Prof Gerald Miller, *University of Maryland, USA*
Dr Mark Miller, *Durham University, United Kingdom*
Mr Hamish Mitchell, *Imperial College London, United Kingdom*
Mr Peter Morris, *University of Warwick, United Kingdom*
Miss Rhiannon Morris, *University of Leeds, United Kingdom*
Miss Fiona Nalden, *Royal Society of Chemistry, United Kingdom*
Mr Petros Neoptolemou, *University of Manchester, United Kingdom*
Dr Sten Nilsson Lill, *AstraZeneca, Sweden*
Dr Fabio Nudelman, *University of Edinburgh, United Kingdom*
Miss Samantha O'Connell, *Atomic Weapons Establishment (AWE), United Kingdom*
Miss Caroline Offiler, *University of Manchester, United Kingdom*
Dr Anuradha Pallipurath, *University of Leeds, United Kingdom*
Miss Priya Pandey, *Università di Bologna, Italy*
Mr Paszkál Papp, *University of Szeged, Hungary*
Dr Julia Parker, *Diamond Light Source, United Kingdom*
Ms Alice Parkes, *University of Limerick, Ireland*
Mr Cristian Alirio Parra Jimenez, *Universidad de Antioquia, Colombia*
Mr Nick Pellens, *Katholieke Universiteit Leuven, Belgium*
Dr Pablo Piaggi, *Princeton University, USA*
Miss Katie Pickering, *University of Warwick, United Kingdom*
Miss Samantha Piper, *Monash University, Australia*
Professor Boaz Pokroy, *Technion Israel Institute of Technology, Israel*
Miss Rachel Pooley, *University of Manchester, United Kingdom*
Professor Sarah Price, *University College London, United Kingdom*
Dr Maria Cinta Pujol Baiges, *Universitat Rovira i Virgili, Spain*
Professor David Quigley, *University of Warwick, United Kingdom*
Professor Dr Arunachalam Ramanan, *Indian Institute of Technology Delhi, India*
Dr Ivo Rietveld, *University of Rouen, France*
Professor Jeffrey Rimer, *University of Houston, USA*
Dr Roberto Risi, *Abbvie, USA*
Professor Kevin Roberts, *University of Leeds, United Kingdom*
Dr Karen Robertson, *University of Nottingham, United Kingdom*
Dr Jutta Rogal, *New York University, USA*
Mr Pietro Sacchi, *University of Manchester, United Kingdom*
Dr Ghazala Sadiq, *Cambridge Crystallographic Data Centre, United Kingdom*

Dr Matteo Salvalaglio, *University College London*, *United Kingdom*
Miss Ilaria Sandei, *University of Leeds*, *United Kingdom*
Mr Kristaps Saršūns, *University of Latvia*, *Latvia*
Dr Gábor Schuszter, *University of Szeged*, *Hungary*
Mrs Rosalind Searle, *Royal Society of Chemistry*, *United Kingdom*
Dr Kyra Sedransk Campbell, *University of Sheffield*, *United Kingdom*
Professor Jan Sefcik, *University of Strathclyde*, *United Kingdom*
Mrs Aina Semjonova, *University of Latvia*, *Latvia*
Dr Andrew Shore, *Royal Society of Chemistry*, *United Kingdom*
Dr Paul Smith, *Cargill*, *Belgium*
Dr Michael Spencelayh, *Royal Society of Chemistry*, *United Kingdom*
Professor Wenhao Sun, *University of Michigan, Ann Arbor*, *USA*
Mr Weihao Sun, *University of St Andrews*, *United Kingdom*
Mrs Francesca Tajoli, *University of Padova*, *Italy*
Professor Dr Takenori Tanno, *Akita University*, *Japan*
Professor Dr Joop ter Horst, *University of Rouen*, *France*
Mr Shailabh Tewari, *Indian Institute of Technology Delhi*, *India*
Mrs Sarah Thirkell, *Royal Society of Chemistry*, *United Kingdom*
Professor Damien Thompson, *University of Limerick*, *Ireland*
Dr Jincheng Tong, *University of Manchester*, *United Kingdom*
Mrs Aija Trimdale-Deksne, *University of Latvia*, *Latvia*
Miss Mollie Trueman, *University of Manchester*, *United Kingdom*
Dr Marko Ukrainczyk, *APC*, *Ireland*
Professor Patrick Unwin, *University of Warwick*, *United Kingdom*
Mr Oscar Urquidi, *University of Geneva*, *Switzerland*
Mr Alexandre Van Bronkhorst Ferreira, *University College London*, *United Kingdom*
Dr Alexander E.S. van Driessche, *Instituto Andaluz de Ciencias de la Tierra (IACT), Consejo Superior de Investigaciones Científicas (CSIC), University of Granada*, *Spain*
Dr Stéphane Veesler, *Centre Interdisciplinaire de Nanoscience de Marseille, Centre national de la recherche scientifique (CINaM-CNRS), Aix-Marseille Université*, *France*
Professor Peter Vekilov, *University of Houston*, *USA*
Dr Vivek Verma, *Imperial College London*, *United Kingdom*
Dr Jessica Walker, *Diamond Light Source*, *United Kingdom*
Professor Michael Ward, *New York University*, *USA*
Dr Monika Warzecha, *University of Strathclyde*, *United Kingdom*
Dr Susan Weatherby, *Royal Society of Chemistry*, *United Kingdom*
Professor Andrew Weller, *University of York*, *United Kingdom*
Dr Thomas Whale, *University of Warwick*, *United Kingdom*
Dr Helen Wheatcroft, *AstraZeneca*, *United Kingdom*
Dr Michael Whittaker, *Lawrence Berkeley National Laboratory*, *USA*
Dr Stephen Yeandel, *University of Sheffield*, *United Kingdom*
Mr Daniyal Younas, *Norwegian University of Science and Technology*, *Norway*
Dr Jacek Zeglinski, *APC*, *Ireland*
Miss Ruohan Zhang, *University of Cambridge*, *United Kingdom*
Ms Ran Zhao, *École polytechnique fédérale de Lausanne (EPFL)*, *Switzerland*